MEASURING, MONITORING AND MODELING CONCRETE PROPERTIES

Measuring, Monitoring and Modeling Concrete Properties

An International Symposium dedicated to Professor Surendra P. Shah, Northwestern University, U.S.A.

Edited by

MARIA S. KONSTA-GDOUTOS
Democritus University of Thrace, Xanthi, Greece

Springer

A C.I.P. Catalogue record for this book is available from the Library of Congress.

ISBN-10 1-4020-5103-4 (HB)
ISBN-13 978-1-4020-5103-6 (HB)
ISBN-10 1-4020-5104-2 (e-book)
ISBN-13 978-1-4020-5104-3 (e-book)

Published by Springer,
P.O. Box 17, 3300 AA Dordrecht, The Netherlands.

www.springer.com

Cover picture

10 m x 10 m image of C-S-H Gel from triboindentor.
Courtesy of Ms. Paramita Mondal, Ph. D. student, Center for ACBM, Northwestern University

Printed on acid-free paper

Printed in the Netherlands.

Contents

Fracture and Deformation of Cement Based Composites

Quantifying Damage for Early Age Concrete

Designing Concrete for Unconventional Properties

Quantitative Image Analysis for Microstructural Characterization of Concrete

Concrete Deterioration, Repair and Rehabilitation

FRPs and Textiles in Cement Composites

Preface of ECF16 Chairman Emmanuel E. Gdoutos

The "16th European Conference of Fracture," (ECF16), was held in the beautiful town of Alexandroupolis, Greece, site of the Democritus University of Thrace, July 3-7, 2006. Within the context of ECF16 forty six special symposia and sessions were organized by renowned experts from around the world. The present volume is devoted to the symposium on "Measuring, Monitoring and Modeling Concrete Properties" (MMMCP) organized by my wife Dr. Maria Konsta-Gdoutos in honor of our good friend Surendra P. Shah of Northwestern University, USA. I am greatly indebted to Maria for undertaking the difficult task to organize this symposium with great success and edit the symposium volume.

Started in 1976, the European Conference of Fracture (ECF) takes place every two years in a European country. Its scope is to promote world-wide cooperation among scientists and engineers concerned with fracture and fatigue of solids. ECF16 was under the auspices of the European Structural Integrity Society (ESIS) and was sponsored by the American Society of Testing and Materials, the British Society for Stain Measurement, the Society of Experimental Mechanics, the Italian Society for Experimental Mechanics and the Japanese Society of Mechanical Engineers. ECF16 focused in all aspects of structural integrity with the objective of improving the safety and performance of engineering structures, components, systems and their associated materials. Emphasis was given to the failure of nanostructured materials and nanostructures and micro- and nanoelectromechanical systems (MEMS and NEMS). The technical program of ECF16 was the product of hard work and dedication of the members of the Scientific Advisory Board, the pillars of ECF16, to whom I am greatly indebted. As chairman of ECF16 I am honored to have them on the Board and work closely with them for the success of ECF16.

ECF16 has been attended by more than nine hundred participants, while more than eight hundred papers have been presented, far more than any other previous ECF over a thirty year period. I am happy and proud to have welcomed in Alexandroupolis well-known experts, colleague, friends, old and new acquaintances who came from around the world to discuss problems related to the analysis and prevention of failure in structures. The tranquility and peacefulness of the small town of Alexandroupolis provided an ideal environment for a group of scientists and engineers to gather and interact on a personal basis.

I wish to thank very sincerely the editor Dr. Maria Konsta-Gdoutos for the excellent appearance of this volume and the authors for their valuable contributions. Finally, a special word of thanks goes to Mrs. Nathalie Jacobs of Springer who accepted my proposal to publish this special volume and her kind and continuous collaboration and support.

January 2006
Xanthi, Greece

Emmanuel E. Gdoutos
ECF16 Chairman

Editor's Preface

This volume contains 94 papers presented at the symposium on "Measuring, Monitoring and Modeling Concrete Properties," (MMCP), which was organized in honor of Surendra P. Shah, Walter P. Murphy Professor of Northwestern University. The symposium took place under the umbrella of the 16th European Conference of Fracture in Alexandroupolis, Greece, on July 3-7, 2006. The book is dedicated to Surendra P. Shah, a researcher, teacher, and advocate for the cement and concrete sciences, in recognition of his continuous, original, diversified and outstanding contributions for half a century.

The book consists of invited papers written by leading experts in the field. It contains original contributions concerning the latest trends and developments in measuring, modeling, and monitoring concrete properties. Fourteen keynote papers were contributed by B.L. Karihaloo, H. Stang, M. Corradi, G.W. Scherer, K. Pilakoutas, M.A. Tasdemir, J. van Mier, F. Dehn, D. Cusson, T. Sui, B.H. Oh, B. Mobasher, N.R. Buenfeld and C. Andrade.

The papers cover a wide range of subjects including fracture and mechanisms of deterioration of cementitious composites, engineering performance and modeling for early-age, high-performance fiber-reinforced cementitious composites and development of innovative cementitious materials. They are arranged in the following 8 sections:

The first section on *engineering performance and modeling for high-performance cementitious composites (HPFRCs),* contains nineteen papers dealing with computational and experimental micro-mechanical modeling of the mechanical behaviour of HPFRCCs and the micro-structural modeling of their durability characteristics.

The second section on *fracture and deformation of cement based composites* contains eleven papers dealing with a mechanical behavior and fracture of conventional and new cement based composites, advanced methods for crack detection in concrete, obtaining three-dimensional information of fracture processes, and new developments on the fracture toughness of concrete including high strength concrete and size effects.

The third section on *quantifying damage for early age concrete* contains nine papers dealing with realistic assessment and modeling for mechanical behavior of early age concrete, autogenous shrinkage and microcracking in heterogeneous systems.

The fourth section on *development of innovative cementitious materials* contains twelve papers dealing with performance criteria on the incorporation of industrial byproducts in concrete and cementious composites.

The fifth section on *designing concrete for unconventional properties* contains eight papers dealing with the use of chemical admixtures such as superplasticizers, shrinkage reducing admixtures and viscosity modifying agents for designing the composition of concretes with special rheological properties, such as self compacting concrete.

The sixth section on *quantitative image analysis for microstructural characterization of concrete* contains four papers dealing with the use of imaging technology as a basis to quantify and model the variability in the nano-, micro-, and meso-structure of concrete.

The seventh section on *concrete deterioration, repair and rehabilitation* contains twenty two papers dealing with new concepts for the processing of repair and rehabilitation measures of concrete structures, the development of predicting models characterizing the corrosion process of steel in concrete and a special session on the deterioration of historic building materials and the development of crystallization pressure and the initiation and propagation of cracks from growth of salt crystals.

Finally, the eighth section on *FRPs and textiles in cement composites* contains nine papers dealing with the bond characteristics and structural behaviour of new reinforcing materials such as fiber reinforced plastics and hybrid fabrics, and the modeling of textile reinforced concrete.

I consider it an honor and privilege I have had the opportunity to edit this volume. I wish to thank very sincerely the authors who have contributed to this volume and all those who participated in the symposium on "Measuring, Monitoring and Modeling Concrete Properties" to honor Surendra P. Shah, a great friend and a colleague, whose work will be indelibly imprinted on the pages of cement and concrete science history.

All those involved in the work of this symposium are gratefully acknowledged, in particular Professor E.E. Gdoutos for organizing ECF16 from start to finish and the members of the Scientific Committee for soliciting and reviewing of papers. Finally, a special word of thanks goes to Ms Nathalie Jacobs of Springer for her interest in publishing this work and her kind collaboration and support.

March, 2006
Xanthi, Greece

Maria S. Konsta-Gdoutos
Editor

Surendra P. Shah

Walter P. Murphy Professor
Northwestern University, Evanston, Illinois, USA

Professor Surendra P. Shah's career began as most of ours begin. Shah, who grew up in India, began his academic career there where he received his undergraduate degree from B.V.M. College. His graduate work began at Lehigh University where he completed his Masters of Science. He then took two years to work as a design engineer at Modjeski and Masters. During this time he met and married Dorothie Crispell. Suru then attended Cornell University, where he received his Ph.D. in civil engineering under the advisement of Professors George Winter, Richard White, and Floyd Slater. After receiving his Ph.D. he was ready to begin the journey that would establish him as a leading figure in the research and teaching in the field of cement and concrete sciences.

In 1965, Suru joined the faculty of the Materials Engineering Department at the University of Illinois at Chicago (UIC). There he taught courses in civil engineering and materials science while developing a state-of-the-art research laboratory. In 1981, he joined the faculty of Northwestern University where he is now a Walter P. Murphy Professor. His research continues to focus on synthesizing materials science, mechanics and structural engineering by combining our knowledge of different scales. He pioneered research to better understand and develop new materials. He has written over 400 journal articles, co-edited 20 symposium proceedings, co-authored two textbooks, and served as the editor-in-chief for the journal, Materials and Structures. His foresight in research led to the establishment of the National Science Foundation Center for Science and Technology of Advanced Cement-Based Materials (ACBM). His leadership at ACBM has provided the opportunities for growth of research in the field, as well as the growth of cement-based curriculum in undergraduate programs around the world. As he built a strong research base, he was developing the next wave of researchers and educators.

Suru Shah has been a strong advocate in the training of future scientists in the cement and concrete field. Through his work at ACBM, he established a network of faculty whose goal is to increase the amount of time spent teaching cement sciences as well as improving the tools for teaching. Through the establishment of the Undergraduate Faculty Workshop, ACBM has reached over 100 faculty and in turn, influenced over 10,000 students. This is not to say that Suru has not influenced his fair share of students on his own. During his tenures at UIC and Northwestern, he as advised over 160 graduate students and 80 post-doctoral fellows. His influence in these student's academic careers has been recognized at both UIC and Northwestern as the recipient of Excellence in Teaching awards. Suru's strong commitment to teaching is evident by the number of his students who have joined the ranks of academia. As a result of the enduring partnership of ACBM with industry, many of his students are enjoying successful careers in the industrial sector.

Surendra Shah's dedication to the field is evidenced by the innovations he has brought about through his research and teaching. Acknowledgement of this work is further established by his recent election to the National Academy of Engineering, the most prestigious award given in the engineering field. Over his career, he has received many other awards including the ACI Phiello Award, the Swedish Concrete Award, the RILEM Gold Medal, the ASCE-CERF Charles Pankow Award, and the ASTM Thompson Award. ACI, RILEM and the University of Dundee have organized symposia in his honor. In addition to his work in cement and concrete sciences, Suru is a strong proponent of the arts. He is an avid stage fan, including opera, theatre, and the symphony. He is a movie buff and, it is no secret that, he is a wine and food aficionado. Surendra P. Shah epitomizes the phrase "Renaissance Man."

Suru Shah is a man who has pushed the envelope in all areas of cement and concrete technology and in doing so, has inspired professionals, researchers, and, students to do the same. He has no immediate plans for retirement. He will continue his mission as a teacher, researcher, and advocate for the cement and concrete sciences. It is with a deep sense of honor and respect that this book and symposium are dedicated to the work and legacy of Surendra P. Shah.

March, 2006
Xanthi, Greece

Maria S. Konsta-Gdoutos
Editor

SCIENTIFIC ADVISORY BOARD

Engineering Performance and Modeling for High-Performance Cementitious Composites

EFFECT OF FIBRE DISTRIBUTION ON THE FATIGUE PERFORMANCE AND AUTOGENOUS SHRINKAGE OF CARDIFRC®

D. Nicolaides, A. Kanellopoulos and B.L. Karihaloo
School of Engineering, Cardiff University, Queen's Buildings, P.O. Box 925, Cardiff CF24 0YF, UK

Abstract: This paper describes a recent fatigue performance check on the high performance fibre-reinforced cementitious composite CARDIFRC®. It is shown that an even distribution of fibres throughout the bulk of the material is crucial to its excellent fatigue performance. Moreover, the even distribution of fibres is also a key factor in the reduction of the autogenous shrinkage strains in this material. The aim of this investigation is to reveal the reason behind the low fatigue life and large scatter in the autogenous shrinkage strains of CARDIFRC, when large test specimens were used. It is confirmed to be due to poor distribution of fibres in the large specimens.

Key Words: Fatigue, shrinkage, fibre distribution

1. INTRODUCTION

The mechanical performance of any HPFRCC depends to a high degree on the even distribution of fibres in the bulk of the material. Any regions with a low concentration of fibres or with no fibres are potential sites of weakness. The distribution of fibres in the mix depends on several factors, e.g. on how the fibres were introduced into the mix, on the vibration frequency during compaction, and on the size and shape of the object cast from CARDIFRC. The distribution of fibres within the matrix of the beam is a critical parameter affecting the fatigue performance of steel fibre reinforced concrete. It is, however, extremely difficult to achieve an even distribution of fibres in large specimens. Failure to attain this goal may result in an extremely low fatigue life, whereas a proper and even fibre distribution can guarantee an extremely long fatigue life. This is more evident in the case of HPFRCCs, where the interfacial bond between the fibres and the matrix is particularly strong, due to the dense structure of the material. The difficulty in achieving an even distribution of fibres is more pronounced in thicker specimens (e.g. 100 mm),

M.S. Konsta-Gdoutos, (ed.), Measuring, Monitoring and Modeling Concrete Properties, 3–16.

whereas the even distribution can be achieved without difficulty in specimens with a relatively small thickness (e.g. 35 mm). The uneven fibre distribution in the matrix also affects measurement of autogenous shrinkage in large specimens of CARDIFRC®.

2. FATIGUE EXPERIMENTAL PROCEDURE

Specimens of Dimensions 100x100x500 mm

Fatigue tests were carried out in three-point bending in a stiff self-straining testing frame fitted with a DARTEC 2500 kN dynamic-static actuator on 100x100x500 mm beams made of CARDIFRC® mix I. The beams were simply supported over a span of 400 mm. Four short cylindrical clamps were set on the supports to prevent the beam from moving, during the cyclic load application. These clamps did not actually come in contact with the specimen, unless it started moving from its original position. The machine was powered by a 23 lit/min DARTEC hydraulic pump and was connected with a DARTEC 9600 Digital Feedback Controller. Four types of measurement were recorded for each beam: (1) the load from the load cell of the testing machine; (2) the vertical deflection at the mid-span; (3) the time from the start of load application; and (4) the number of cycles to failure.

The vertical mid-span deflection was measured by a single LVDT. This LVDT was calibrated for a very narrow range of deformation (±2.5 mm), because the deformation of the beam during its fatigue life was expected to be very small. In this way, it was also anticipated to minimise the noise in this particular measurement. A mechanical stop was installed 10 mm below the centre of the beam in order to prevent damage to the LVDT if the beam suddenly failed. All the data were acquired using DT800 dataTaker logger and stored temporarily in its memory before they were downloaded to the connected computer. The DT800 dataTaker is a data acquisition and logging instrument, which has the ability to operate in burst mode. In this mode, DT800 can sample at very high rates, but only for short periods of time. This mode of operation is appropriate for fatigue data acquisition and logging. The logging procedure was controlled by appropriate software, called DeLogger Plus. This software package contains a powerful set of tools for working with DT800, like the programme builder option. This option allows the user to define the number and types of the scan channels, the time interval to trigger each scan, the quantity to be read (e.g. voltage) and how to convert the reading into appropriate engineering units.

Fatigue tests are generally carried out for a given constant minimum stress or for a constant ratio between the minimum and maximum stress levels. In this study, a constant minimum stress level of 10% of the static flexural strength was maintained. The cyclic tests were carried out at maximum stress levels ranging from 70% to 85% of the monotonic strength. The fatigue tests were performed in load control. The ultimate static load in three-point bending equalled P_u = 57 kN. This value resulted if the standard deviation of the peak load measured in static three point bend tests was subtracted from the mean value. Each specimen was first subjected to three slow cycles between 2 kN and 28 kN, which corresponded to a stress level of 50% of the monotonic strength. It should be noted that specimens tested in this study were not pre-cracked before cyclic loading, so the selected stress level of 50% lies within the elastic range of the material response. This is an essential characteristic of this study compared with previous studies on fatigue of fibre reinforced concrete. The specimens were preloaded for stabilisation purposes. The test stopped automatically after the specimen failure or after one million cycles, whichever occurred first.

The specimens were subjected to a sinusoidal cyclic loading with a frequency of 6 Hz. The choice of the frequency was dictated by the time required for the test beam to reach 1000000 cycles and by the need to avoid undesirable side effects because of high frequencies. Despite the fact that several researchers have conducted fatigue tests with frequencies up to 20 Hz, these values were considered too high to maintain an accurate load range and to minimise the effects of inertia. On the other hand, frequencies of 1-2 Hz were considered too low to complete a full test (up to 1000000 cycles) within a reasonable time period. Therefore, a frequency of 6 Hz was selected and a full test was completed in less than 48 hours.

3. RESULTS AND DISCUSSION

Tests have been performed in the load ranges between 10%-85%, 10%-80% and 10%-70% of the ultimate flexural strength. Table 1 and Figure 1 show the fatigue life of specimens 100x100x500 mm (i.e. number of cycles to failure, N). It is clear that there is a large scatter in the experimental fatigue life, which is a characteristic of the fatigue tests. This is attributed to the nature of the material and also to errors in test variables, which are repeated in a large number of cycles. The large scatter in the experimental results did not give an opportunity to discern a trend in the fatigue life of the material , not even after a large number of tests were performed.

The large scatter in the fatigue life of the tested beams is mainly attributed to the inhomogeneous nature of the material. The addition of fibres to the concrete matrix can dramatically improve the fatigue performance of the composite and also impart it with additional strength in tension, shear and flexure. It is, however, extremely difficult to achieve an even distribution of fibres within the mix, because of the large quantity of fibres used. This has the effect that some internal vertical planes of the specimen under flexure will be devoid of fibres. In the worst case that the planes devoid of fibres are located in the region of the maximum bending moment, they become the failure planes after only very few cycles. Another important consideration is the orientation of the fibres in the region of the maximum bending moment.

The failure surfaces of the tested beams were thoroughly examined visually. It was observed that beams that failed after a very small number of cycles, had large areas in these planes almost devoid of fibres, especially in the zones of tensile stress. Others had a significant number of fibres in these planes, but these were oriented vertically, i.e. they did not exert any closure pressure across the plane of failure. On the other hand, specimens that withstood a large number of cycles before failure, had an even distribution of fibres crossing the plane of failure, and most of them applied substantial closure pressure, thus extending the fatigue life of the beams. In the case of the beams that withstood 1000000 cycles without failure, the planes of failure (revealed after the specimens were broken) were full of fibres (most of them long), all of them exerting closure pressure. It can be concluded that the distribution of fibres within the matrix of the beam is a very important parameter affecting the fatigue life of the tested specimens.

Table 1. Number of cycles sustained by cardifrc - mix i beams (100x100x500 mm)

Fatigue Tests - CARDIFRC® - Mix I Beams (100x100x500 mm)			
Beam Number	10% - 85% Pu	10% - 80% Pu	10% - 70% Pu
1	706	9918	5910
2	181738	2144	18
3	18	14	437
4	2510	4036	1000000
5	41539	28733	279
6	169279	911453	4
7	1000000	195703	1000000
8	6174	9337	3
9	4918	131237	527988
10	161839	78940	110999
11	493	-	-
12	54977	-	-

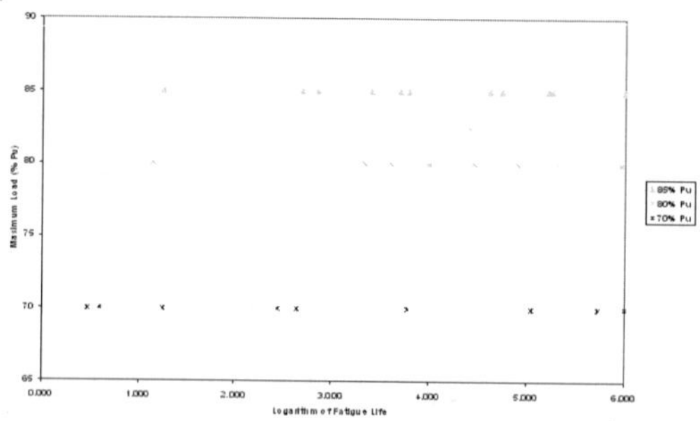

Figure 1. Fatigue life of CARDIFRC - Mix I beams (100x100x500 mm)

It is, however, extremely difficult to achieve an even distribution of fibres in the mix. For the improvement of fibre distribution, the fibres are added to the mix through vibrating sieves; for 6 mm long fibres, a 5 mm sieve is used and for 13 mm long fibres a 12 mm sieve is used. Many interesting observations regarding the fibre distribution in CARDIFRC were made by the image analysis of specimens, which will be discussed later.

There was a suspicion that the poor distribution of the fibres within the specimen was also a result of the high frequency of compaction 100 Hz (on an electric vibrating table) used during the casting of the specimens as a result of which the fibres were forced to the sides of the beams. In order to confirm or dispel this suspicion, some of the remaining 100x100x500 mm specimens were cut into six smaller specimens of dimensions 33x100x250 mm, as shown in Figure 2.

The cut specimens were tested statically in three-point bending under displacement control, and the peak loads were noted. The average peak loads for top, middle and bottom specimens were:

Top Specimens: $P_{u,top} = 4.30$ kN
Middle Specimens: $P_{u,mid} = 6.09$ kN
Bottom Specimens: $P_{u,bot} = 11.31$ kN

It can be clearly observed that the ultimate load from the flexural static tests is much lower at the top, and is significantly increased when moving to middle and finally bottom specimens. The bottom specimens give the highest value of peak load, which is a clear indication that these specimens had more fibres, in comparison with the top and middle specimens. These experimental results undoubtedly confirm the suspicion of the poor distribution of fibres within the original specimens, as a result of the high frequency of almost 100 Hz used during the compaction of the cast specimens.

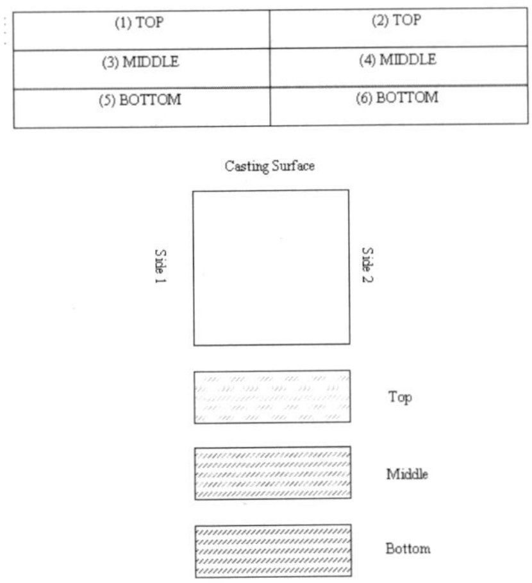

Figure 2. Schematic presentation of the cut specimens and the fibre density in the top, middle and bottom specimens

Specimens of Dimensions 35x90x360 mm

After the completion of the static tests on the cut specimens, where it was clearly revealed that the high vibration frequency had produced poor distribution of the fibres, it was decided to cast smaller CARDIFRC®, Mix I beams, 360x90x35 mm, and use a frequency during compaction not exceeding 50 Hz. The selected dimensions were regarded to be more realistic, in the sense that CARDIFRC® is a material that is mainly intended for use for repairing and strengthening in thin strips of about 20 mm thickness.

For the fatigue tests, it was decided to use a value of ultimate load equal to $P_u = 10$ kN. This value is two standard deviations less than the mean value of the peak load obtained earlier from static three-point bend experiments, i.e. $P_u = P_{avg} - 2S.D.$ The subtraction of two standard deviations from the average peak load increases the probability that the ultimate monotonic load of all specimens tested in fatigue would be lower than the applied P_u, to about 95.5%. This choice was aimed at minimising the factors that caused the huge scatter in fatigue life of the 100x100x500 mm beams, and eventually at obtaining more consistent results.

The fatigue tests were carried out in three-point bending, in the same way as was described earlier for the 100x100x500 mm beams. The tests were performed under load control between two limits (with a sinusoidal force variation in time). The minimum stress level, S_{min}, was 10% of the monotonic strength and the maximum stress level, S_{max}, ranged from 80% to 90% of the monotonic strength. Before the cyclic loading was applied, the beams were preloaded with 3 static loading/unloading cycles between 1 kN and 5 kN, the higher load corresponding to 50% of the monotonic flexural strength of the material. The specimens were preloaded for stabilisation purposes. The frequency of loading used was 6 Hz. The tests stopped after specimen failure or after one million cycles, whichever occurred first. In the special case of two specimens, it was decided to test them up to a larger number of cycles.

Tests have been performed in the load ranges between 10%-80%, 10%-85% and 10%-90% of the ultimate flexural strength. Table 2 gives the fatigue life of specimens (i.e. number of cycles to failure, *N*). It is immediately noticeable that there is an excellent consistency in the fatigue life for the load ranges between 10%-80% and 10%-85%, since all of the eight specimens sustained 1000000 or more cycles without failure. This is attributed to the smaller thickness of these specimens (35 mm), the lower frequency used during the vibration of these specimens, which ensures a more even distribution of the fibres within the specimens, and also to the choice of P_u as equal to P_{avg}-2*SD*, as explained above.

Table 2. Flexural fatigue tests experimental results (CARDIFRC® - MIX I, 360x90x35 mm)

Load Amplitude (% P_u)	Fatigue Life (*N*)				
	N_1	N_2	N_3	N_4	Average Fatigue Life (*N*)
10-90%	1000000	21564	9315	1000000	-
10-85%	1000000	1000000	1000000	20000000	1000000(*1)
10-80%	1000000	1000000	1000000	2000000	1000000(*1)

*1 The average fatigue life considers only the first 1000000 cycles sustained by specimen No.4.

An important observation from the tests performed at maximum load levels of 80% and 85% P_u, is the fact that none of the eight tested specimens developed any visible cracks during the 1000000 cycles, attesting once again to the improved distribution of the fibres within these specimens. It also implies that probably no specimen will fail at lower maximum stress levels. From the above experimental results, it can be concluded that the endurance limit of the material is approximately at 85% of its flexural strength. Below this limit none of the tested specimens failed, not even after a very large number of cycles (e.g. 20000000 cycles). Slightly above this load limit, some specimens did not fail after 1000000 cycles, whereas some others failed after a relatively small number of cycles. Another remark about the observed fatigue limit is that it is very high, not very often observed in the relevant literature. This is an indication that CARDIFRC® has an extensive elastic zone. This is confirmed by direct tension tests performed by Benson (2003).

In order to check whether internal cracks had developed in the specimens that sustained 1000000 or more cycles without failure, they were tested afterwards in static three-point bending. The purpose of this static testing was to compare the post-fatigue flexural strengths and static envelope curves with the pre-fatigue test results. None of the specimens had any visible external cracks at the end of fatigue testing. The specimens tested at 80%, 85% and 90% P_u in fatigue, showed a small increase in their flexural strength (Table 3). This increase seemed to be higher than could be attributed to the increase due to age alone. It is believed to depend on the maximum flexural fatigue stress (S_{max}) to which the specimens were subjected earlier. From the available evidence, it was anticipated that with lower S_{max} values the increase in flexural strength would be higher. This was however not confirmed experimentally, since the increase of the flexural strength for specimens tested earlier up to 80%, 85% and 90% P_u was of the same order of magnitude. A remarkable observation was that the specimens which did not fail after 1000000 cycles up to 90% P_u, showed the highest increase (7%), whereas the lowest increase (5%) was observed for specimens tested earlier up to 80% P_u. The increase observed in specimens tested earlier up to 85% P_u was about 6%.

Although it seems that there exists a linear correlation between the fatigue loading stress and the corresponding post-fatigue static flexural strength, no definite conclusions can be drawn. The difference is believed to be very small; therefore the increase in the flexural strength is assumed to be of the same order of magnitude for the three groups. Moreover, the result of the average strength for specimens tested up to 90% P_u, was based only on two specimen results, which did not fail after fatigue. It is believed that the increase in flexural strength is approximately constant for specimens subjected to fatigue stress close to the fatigue limit of the material. This result confirms previously noted results in the literature, that prior cycling may lead to an improvement in strength (Naaman and Hammoud, 1998; Ramakrishnan et al., 1996; Naaman and Harajli, 1990). It has been suggested that this increase in strength is due to densification of the material, caused by stress cycling. It is also known that most FRCs are linearly elastic up to about 80% or more of the matrix tensile strength and that the microcracking process starts beyond this point. This leads to the conclusion that beam specimens subjected to cyclic flexural stress below this level are not likely to have a reduced first crack flexural strength (Ramakrishnan and Lokvik, 1992). However, the mechanisms behind the increase of the post-fatigue flexural strength in HPFRCCs are still unclear.

Table 3. Experimental peak loads and tensile/flexural strengths of CARDIFRC® - MIX I beams (360x90x35 mm), tested in 3 – point bending, after they have been subjected first to fatigue loading

Beam Number	**10%-90% Pu**		**10%-85% Pu**		**10%-80% Pu**	
	P_u (kN)	f_t (MPa)	P_u (kN)	f_t (MPa)	P_u (kN)	f_t (MPa)
1	14.65	55.81	12.65	48.19	14.30	54.48
2	-	-	14.64	55.77	13.70	52.19
3	-	-	14.30	54.48	12.40	47.24
4	12.88	49.07	13.33	50.78	13.60	51.81
Average Strength		52.44		52.30		51.43

4. IMAGE ANALYSIS OF FATIGUE SPECIMENS

Due to the large scatter observed in the fatigue life of large CARDIFRC® - Mix I beams (100x100x500 mm), it was decided to investigate the fibre distribution in the planes of failure of six specimens tested under fatigue loading (between 10-70% P_u). The fatigue life of the selected specimens varied between a very small number of cycles (less than a decade) and a very high number of cycles (more than 500000 cycles). In addition, the distribution of fibres in a section of a specimen that did not fail after 1000000 cycles was also investigated. This particular specimen was cut along a predetermined section, located in the centre of the beam, using a diamond saw. In the case of specimens that failed under fatigue loading, the planes of failure had to be flattened first before examination under the microscope. This was also done using a diamond saw, as close as possible to the actual plane of failure. The exact number of cycles sustained by each of the selected specimens and the average number of fibres (/cm^2) resulting from the image analysis of their planes of failure are shown in Table 4. Other statistical measures are given in Table 5.

Table 4. Average number of fibres resulting from image analysis of specimens tested under fatigue loading

No. of Cycles	Average No. of Fibres (/cm^2)	Max. No. of Fibres (/cm^2)	Notes
3	120	Theoretical result based on the solution of "*the Buffon needle problem*". For a 10x10x10 mm cube, the total number of fibres in a cut section, equals 215/cm^2.	All specimens (100x100x500 mm) made of CARDI-FRC® - Mix I and tested in fatigue between 10-70% P_u.
4	116		
437	125		
5910	131		
110999	159		
527988	174		
>1000000	194		

The conclusion from this investigation is that a high average number of fibres in the plane of failure is a guarantee for an extended fatigue life of the specimen. The case of the specimen that sustained just 4 cycles is an exception, since the average number of fibres is lower than the specimen that sustained 3 cycles. It is, however, important to note that both specimens have an average number of fibres lower than that in the specimens that sustained higher number of cycles. All planes of failure examined have at least one grid where the maximum counted number of fibres is equal or very close to the theoretical maximum resulting from the solution of the ***"Buffon needle problem"*** (Nicolaides, 2004). On the other hand, the minimum fibre count varies significantly between the selected specimens. This discrepancy is reflected clearly in the values of standard deviation and coefficients of variation. It is apparent that the decreased standard deviation, which means a more even distribution of fibres within the matrix, leads to a higher number of cycles to failure. The same conclusion is also supported by the coefficients of variation, which also decrease as the fatigue life increases.

Table 5. Statistical analysis of the results of the image analysis of specimens tested under fatigue loading

No. of Cycles	Average No. of Fibres (/cm^2)	Max. Count	Min. Count	Standard Deviation (SD)	Coefficient of Variation (COV) (%)
3	120	215	5	70.30	61
4	116	215	5	66.02	55
437	125	215	28	58.89	47
5910	131	202	27	37.73	29
110999	159	212	72	28.32	18
527988	174	213	127	20.82	12
>1000000	194	215	138	14.54	7

Specimens that sustained a very small number of cycles (3 and 4 cycles) have a very heterogeneous distribution of fibres in their planes of failure. It is clear from Figures 3-5 that large areas of these sections have a significantly lower density of fibres. The areas with lower density of fibres are located in both cases in the bottom of the specimens. This is very important, since these parts of the beams were subjected to tension, and a lower number of fibres made it easy for a crack to initiate, resulting in extremely low fatigue life. Although the upper parts of these sections have a higher number of fibres, this was not sufficient to prevent their fast fracture. It must be mentioned that at some locations in the examined sections the number of counted fibres was extremely small, less than ten (Table 5). The average number of fibres in these two sections was 118 (/cm^2).

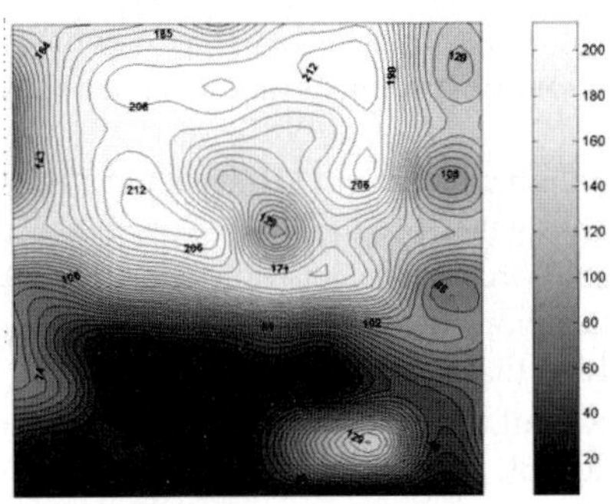

Figure 3. Contour plot showing the fibre distribution for the 100x100 mm beam plane of failure, after fatigue testing between 10-70% P_u (failure after 4 cycles)

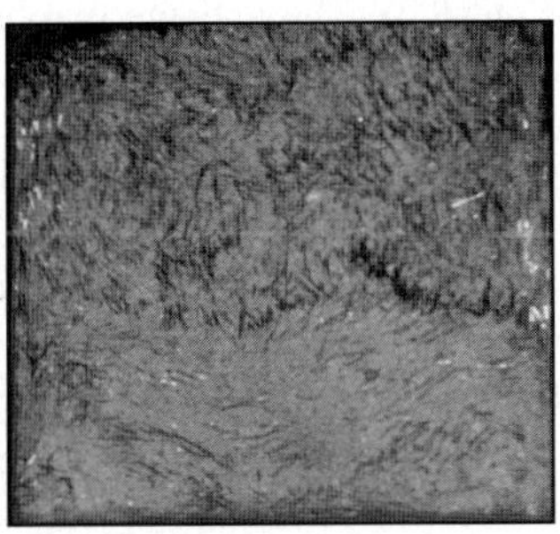

Figure 4. Plane of failure of the 100x100 mm beam after fatigue testing between 10-70% P_u (failure after 4 cycles)

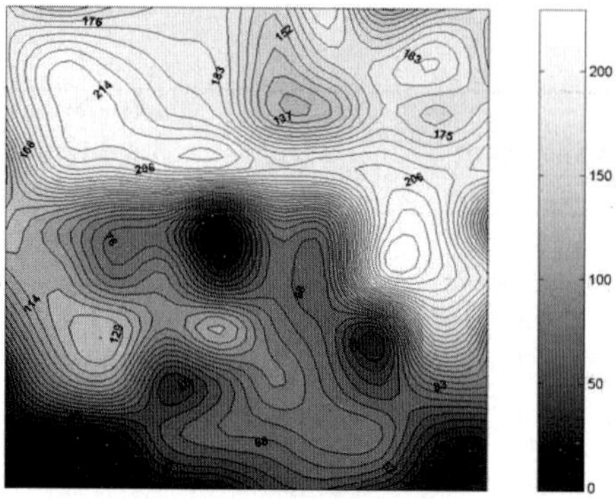

Figure 5. Contour plot showing the fibre distribution for the 100x100 mm beam plane of failure, after fatigue testing between 10-70% P_u (failure after 3 cycles)

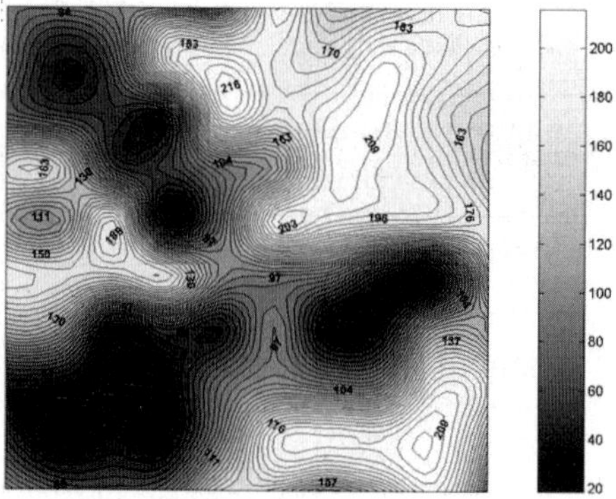

Figure 6. Contour plot showing the fibre distribution for the 100x100 mm beam plane of failure, after fatigue testing between 10-70% P_u (failure after 437 cycles)

The third specimen under examination that sustained a slightly higher number of cycles (437 cycles) also had a heterogeneous distribution of fibres in the plane of failure. It is very apparent that three main regions, covering a large fraction of the total section, have a considerably lower density of fibres, as shown in Figure 6. The average number of fibres in this section is 125 (/cm^2), which is higher than the number of fibres counted in the sections of specimens that sustained lower number of cycles.

The fourth specimen under investigation, which failed after 5910 cycles, had a generally uniform distribution of fibres in the plane of failure, apart from an area at the bottom of the beam and extending up to the centre, where the fibre concentration was noticeably lower (Figure 7). The average number of fibres in this section is 131 (/cm^2), which is higher than the number of fibres counted in the sections of specimens that sustained lower number of cycles.

Specimens that sustained significantly higher number of cycles (110999 and 527988), have a considerably higher number of fibres in their planes of failure (159/cm^2 and 174/cm^2, respectively) (Figures 8, 9). The distribution of fibres is also generally even, with the exception of some areas with lower fibre concentration. It is believed that these areas facilitated the crack initiation and propagation within these sections.

Finally, in the case of the specimen that did not fail after the application of 1000000 cycles, the even distribution of fibres is very noticeable, which in combination with the significantly higher average number of fibres (194/cm^2), satisfactorily explains why the specimen did not fail during testing (Figure 10).

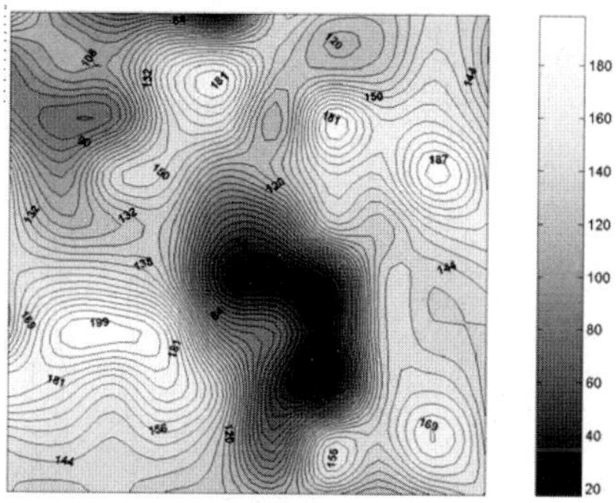

Figure 7. Contour plot showing the fibre distribution for the 100x100 mm beam plane of failure, after fatigue testing between 10-70% P_u (failure after 5910 cycles)

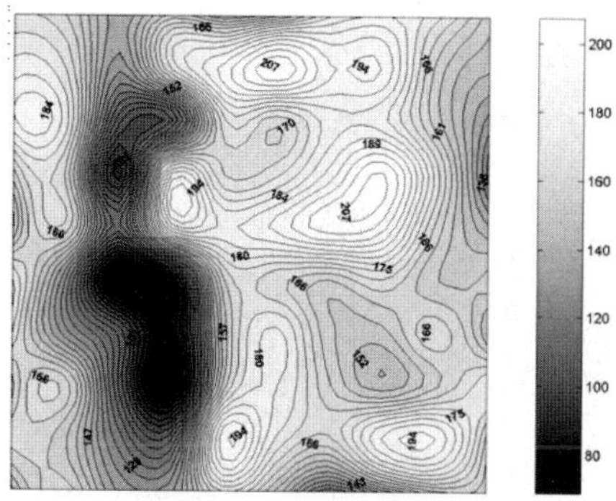

Figure 8. Contour plot showing the fibre distribution for the 100x100mm beam plane of failure, after fatigue testing between 10-70% P_u (failure after 110999 cycles)

Considering that the theoretical maximum number of fibres per cm^2 resulting from the solution of the ***"Buffon needle problem"*** (Nicolaides, 2004) is 215, it can be concluded that the closer the number of fibres in the plane of failure is to the theoretical maximum, the higher the number of cycles it will sustain (Figure 11). This result is very important in the light of an existing correlation between the image analysis and CT-scanning analysis, as shown by Nicolaides (2004). By applying the non-destructive CT imaging method, the maps of the X-ray absorption density of the specimen can be produced. The magnitude of the X-ray absorption in several areas of the section can lead to an estimate of the corresponding numbers of fibres in those areas (based on the correlation between X-ray absorption density and corresponding number of fibres), and therefore to an estimation of the expected fatigue life of the specimen.

5. AUTOGENOUS SHRINKAGE

Autogenous shrinkage experiments conducted on large (100x100x500 mm) CARDIFRC® members revealed a large scatter in the measured strains as a consequence of the uneven distribution of fibres into the matrix, as shown in the Figures 3-10. This large scatter is clearly seen from the entries in Table 6 and Figure 12. The spread of the results is very large, since even at the end of the first 10 hours the COV is relatively high for such set of data. These tests were therefore discontinued after 90 hours. Smaller specimens with the same dimensions as those used for fatigue testing were used for measuring autogenous shrinkage strains. These specimens gave very consistent results with very low scatter.

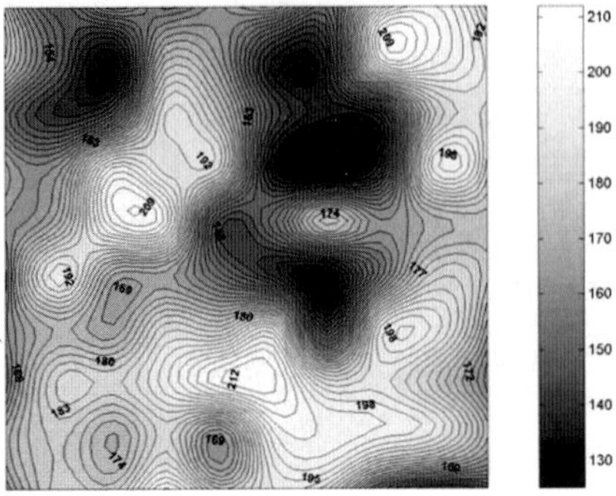

Figure 9. Contour plot showing the fibre distribution for the 100x100 mm beam plane of failure, after fatigue testing between 10-70% P_u (failure after 527988 cycles)

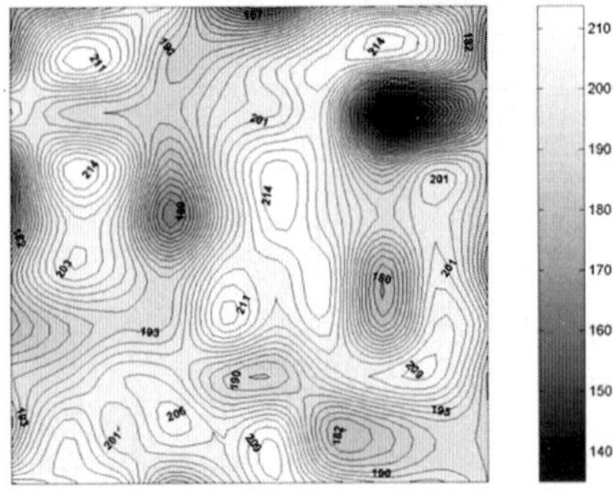

Figure 10. **C**ontour plot showing the fibre distribution for the 100x100 mm beam plane of failure, after fatigue testing between 10-70% P_u (no failure after 1000000 cycles)

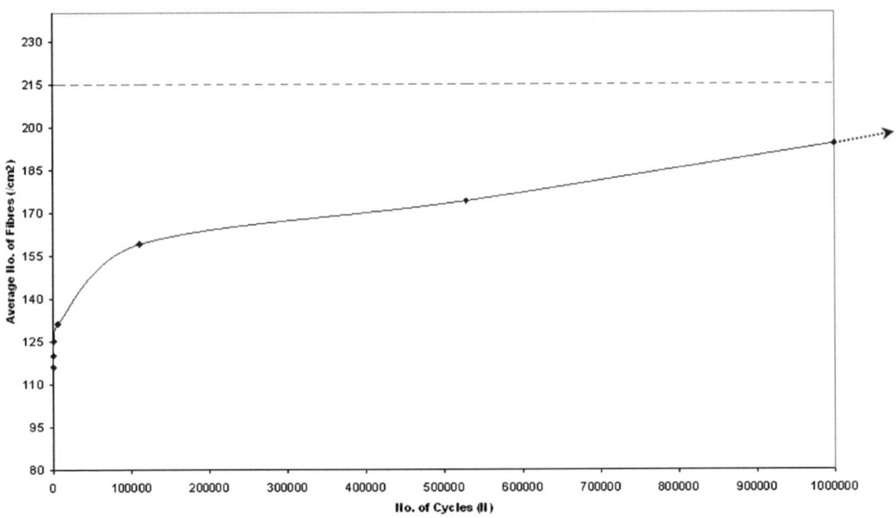

Figure 11. Relation between average number of fibres and number of cycles to failure

Table 6. Summary of results for large beams with fibres at the end of various time intervals

	Time Intervals (hours)								
	10	20	30	40	50	60	70	80	90
Beam 1	70.59	112.33	144.07	172.86	200.73	226.62	251.49	272.65	293.75
Beam 2	139.12	195.04	236.09	270.35	302.88	333.53	361.07	384.93	408.03
Beam 3	77.95	132.37	172.58	208.45	242.78	274.17	303.30	327.52	351.17
Beam 4	63.75	101.41	130.03	155.97	181.07	204.36	226.73	245.75	264.71
Beam 5	84.28	143.18	186.73	225.62	262.84	296.91	328.53	354.84	380.55
Beam 6	153.53	215.35	260.78	298.72	334.77	368.74	399.30	425.79	451.44
Mean	98.20	149.95	188.38	221.99	254.18	284.05	311.74	335.25	358.28
COV	35.49	27.81	24.84	22.67	21.11	20.08	19.12	18.48	17.88

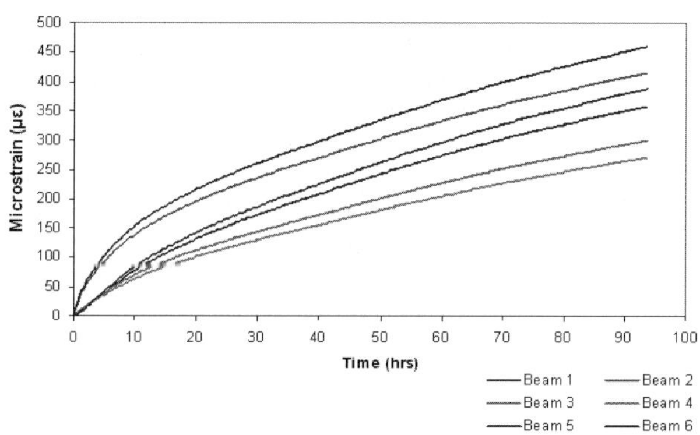

Figure 12. Autogenous shrinkage strain variation with time for CARDIFRC® Mix I

6. CONCLUSIONS

It can be conclusively said that the fatigue performance and autogenous shrinkage strain level of steel fibre reinforced concrete is strongly related to the fibre distribution within the specimens. A proper and even fibre distribution can lead to an extremely long fatigue life and a consistent measurement of autogenous shrinkage strains, especially in the case of HPFRCCs, where the interfacial bond between the fibres and the matrix is particularly strong, due to the dense structure of the material. On the other hand, specimens with poor fibre distribution in the potential planes of failure may exhaust their fatigue life after just a few cycles. Especially for CARDIFRC®, the better fibre distribution achieved in specimens of smaller dimensions 360x90x35 mm revealed that it can have an excellent fatigue life, extended up to an extremely high number of cycles, without significant internal damage. Regarding the autogenous shrinkage of CARDIFRC®, it has been observed that it is also strongly affected by the fibre distribution in the matrix. The autogenous shrinkage strains measured on large CARDIFRC® members revealed a relatively large scatter as a result of the uneven distribution of fibres, just as in fatigue.

7. REFERENCES

Benson S.D.P. *CARDIFRC - Development and constitutive behaviour.* PhD Thesis, Cardiff University, UK, 2003.

CARDIFRC® patent number GB 2391010, Karihaloo B.L., Benson S.D.P. and Alaee F.J., 2001.

Kanellopoulos A. *Autogenous shrinkage of CARDIFRC®*. PhD Thesis, Cardiff University, UK, 2004.

Naaman A.E. and Hammoud H. *Fatigue characteristics of high performance fiber-reinforced concrete.* Cement and Concrete Composites, 20, 1998, pp. 353-363.

Naaman A.E. and Harajli M.H. *Mechanical properties of high performance fiber concretes: a state-of-the-art report (SHRP-C/WP-90-004).* SHRP National Research Council, Washington DC, 1990.

Nicolaides D. *Fracture and fatigue of CARDIFRC®*. PhD Thesis, Cardiff University, UK, 2004.

Ramakrishnan V. and Lokvik B.J., *Flexural fatigue strength of fiber reinforced concretes.* High performance fiber reinforced cement composites (eds. Reinhardt H.W. and Naaman A.E.), RILEM Proceedings 15, Chapman and Hall, London, 1992.

Ramakrishnan V., Meyer C., Naaman A.E., Zhao G., Fang L. *Cyclic behaviour, fatigue strength, endurance limit and models for fatigue behaviour of FRC.* High performance fiber reinforced cement composites (eds. Naaman A.E. and Reinhardt H.W.), E & FN Spon, London, ISBN 0 419 21180 2, UK, 1996.

STRUCTURAL APPLICATIONS OF HPFRCC IN JAPAN

K. Rokugo[1], M. Kunieda[2] and S. Miyazato[3]
[1]*Gifu University, 1-1 Yanagido, Gifu, 501-1193, Japan;* [2]*Nagoya University, Furo-cho, Chikusa-ku, Nagoya 464-8603, Japan;* [3]*Kanazawa Institute of Technology, 7-1 Ogigaoka, Nonoichi-cho, Ishikawa-gun, Ishikawa 921-8501, Japan*

Abstract: High Performance Fiber Reinforced Cement Composites (HPFRCC) show multiple cracking and strain-hardening behaviors in tension. Current applications in Japan include bridge decks, building dampers, retaining wall, irrigation channels and so forth. In addition to tensile load bearing capacity of HPFRCC, protection against penetration of substance through fine cracks is one of the advantages in these applications. This paper introduces recent applications using HPFRCC in Japan.

Key words: HPFRCC; strain hardening; multiple fine cracks; required performance; applications.

1. INTRODUCTION

High Performance Fiber Reinforced Cement Composites (HPFRCC), which contain synthetic and/or metallic short fiber, show multiple cracking and strain-hardening behaviors in tension. One type of HPFRCC that has novel mechanical properties in tension is Engineered Cementitious Composites (ECC), which was developed by Li[1]. Figure 1 shows an image of the material response of HPFRCC under uniaxial tension. A lot of research for the evaluation and improvement of material properties, and the development of process technology, design concepts and ideas for applications has been performed [2,3]. The Research Subcommittee on Fiber Reinforced Mortar with Multiple Cracks (JSCE) has investigated the evaluation and utilization methods of HPFRCC, and summarized the committee report [2]. This paper introduces recent applications using HPFRCC in Japan, referring to the committee report [2].

M.S. Konsta-Gdoutos, (ed.), Measuring, Monitoring and Modeling Concrete Properties, 17–23.

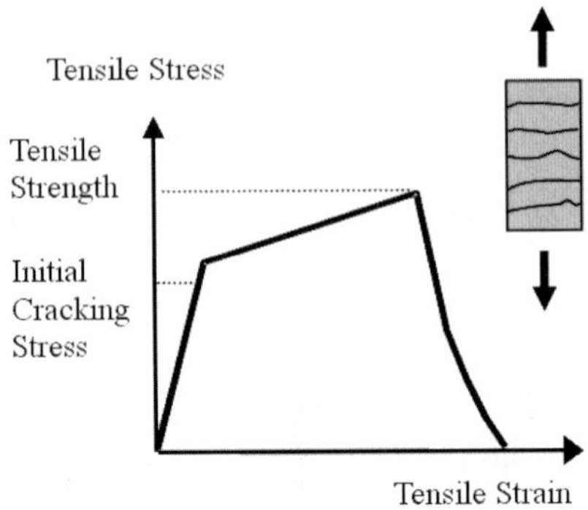

Figure 1. Image of material response of HPFRCC in tension

2. RECENT APPLICATIONS IN JAPAN

Recent applications of HPFRCC in Japan take advantage of the superior mechanical properties and fine cracking mode of such composites. Applications in Japan include the following:

- Bridge decks to improve fatigue resistance through tensile force bearing capacity of HPFRCC
- Dampers in reinforced concrete buildings to increases energy absorption and suppress vibration during earthquakes, in addition to minimizing repair after severe loading
- Surface repair of dams and irrigation channels to improve shielding properties of deteriorated concrete surfaces.
- Surface repair of retaining walls deteriorated by alkali silica reaction to improve aesthetic appearance
- Surface repair of viaducts for carbonation retardation

2.1 Bridge decks

Since HPFRCC can bear tensile forces, members made of HPFRCC in combination with steel plates provide higher flexural resistance with a thinner cross section than normal steel-concrete members. Figure 2 shows construction site of Mihara Ohashi[4] in Hokkaido with a bridge length of 972 m and central span of 340 m. In 2004, half the depth of the asphalt overlay on the steel decking of this bridge was replaced with 40 mm thick HPFRCC to increase the load bearing capacity and stiffness of the decks while reducing the stress generated, thereby improving the fatigue resistance of the stiffener for the steel deck. This became necessary because the requirements for fatigue resistance in the standard specifications were revised while the bridge was under construction.

Plate-type dowels as shown in Fig. 2(a) were adopted to ensure the bond between the HPFRCC and the steel deck[5]. HPFRCC was mixed at ready-mixed concrete plants, hauled to the construction site by large agitating trucks, and subjected to secondary mixing at the construction site. Approximately 30 m^3 of HPFRCC were placed each day, with the amount of mixed HPFRCC totaling approximately 800 m^3. The placing process is shown in Figure 2b.

(a) Adopted plate-type dowels

(b) Placement of HPFRCC

Figure 2. Construction site for bridge decks

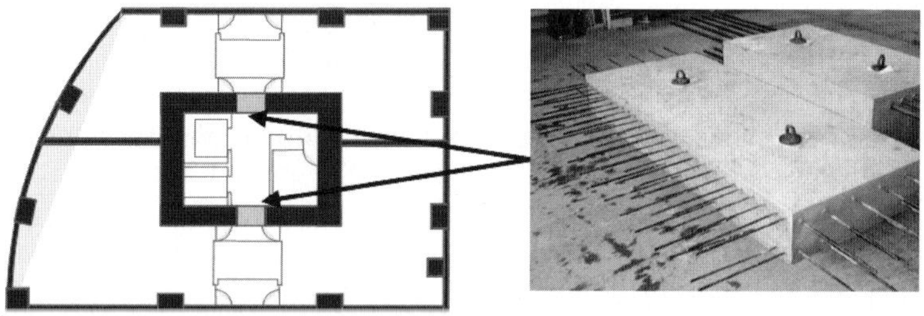

Figure 3. Pre-cast segment of damper for buildings

2.2 Dampers for buildings

An HPFRCC member reinforced with steel bars is capable of absorbing a large amount of energy under alternate loading. As shown in Fig. 3, HPFRCC members were incorporated as dampers in reinforced concrete buildings in Tokyo and Yokohama in 2004 and 2005, respectively[6]. The HPFRCC members were connection elements with main (core) frames of high-rise buildings, and the elements involved both higher energy consumption and minimized repair work after earthquake. To design the elements considering structural response, shear tests using 1/2.5 scale specimens were conducted. The test results show that the elements have novel structural performance and narrow crack width (smaller than 0.3 mm) after cyclic loading.

2.3 Surface repair for dams

The shielding performance of HPFRCC is excellent owing to its small crack widths, which minimize water permeation.

The dam height of Mitaka Dam[7], a gravity concrete dam in Hiroshima Prefecture, was increased from approximately 33 m to 44 m by placing new concrete onto the existing dam body on the downstream side (Fig. 4). In 2003, HPFRCC of 30 m^3 was sprayed on the upstream dam surface (area: 500 m^2) with a thickness of 30 mm, to improve the shielding performance of the deteriorated existing concrete surface. Anchors were driven at 1.5 m^2 intervals to ensure a strong bond between the substrate and HPFRCC.

Figure 4. Spraying on dam surface

2.4 Surface repair of irrigation channels

Many irrigation channels suffer deterioration due to abrasion, having been in service for several decades. The Central Main Channel[2] (side wall height: 1.1 m, bottom slab width: 1.5 m) in Shiga Prefecture, shown in Fig. 5, was so deteriorated that coarse aggregate was exposed on the surfaces and the edges were partially lost, with cracks approximately 1 mm wide and approximately 1m long. At Seridanno Channel in Toyama Pref. (side wall height: 1.1 m, masonry side wall height: approx. 2.4 m, bottom slab width: approx. 2.05 m), coarse aggregate was exposed on the surfaces and the bottom slabs were partially spalled off. Part of the filling mortar at the bottom of the masonry sidewall was also lost.

A water jet was used for substrate treatment to remove deteriorated mortar. Each masonry joint in the sidewalls of the Seridanno Channel was filled with mortar. HPFRCC was troweled and sprayed at the Central Main Channel and Seridanno Channel, respectively, in 2005.(thicknesses of 6 and 10 mm for sidewalls and bottom slabs, respectively.)

Conventional repair mortar and ultrahigh strength polymer cement mortar were also used at the Central Main Channel for comparison with HPFRCC. Cracks were observed in conventional mortar and ultrahigh strength polymer cement mortar one month after application, whereas no cracking was found in HPFRCC.

2.5 Surface repair of retaining walls

Because of its small cracking widths, HPFRCC is suitable for application on the surface of cracked concrete structures in terms of aesthetics. A gravity concrete retaining wall in Gifu Prefecture measuring approximately 18 m in width and 5 m in height was constructed in the 1970s.

As cracks due to alkali-aggregate reaction developed in the wall, these cracks were injected with epoxy resin and the wall surface was coated with an organic coating material in 1994. When the surface repair material also cracked, the wall was subjected to surface repair using sprayed HPFRCC and other materials[8] in 2003, as shown in Fig. 6.

(a) before repair (b) after repair

Figure 5. Surface repair of irrigation channels

(a) before repair (b) after repair

Figure 6. Surface repair of retaining walls

The wall was divided into nine sections on which different types of repair were carried out. These were combinations of three repair materials (two HPFRCCs and one repair mortar), three reinforcement levels (welded wire mesh, expanded metal, and no reinforcement), and two crack treatment levels (with and without sealing to increase the bondless areas). A shotcreting thickness of 50 to 70 mm was adopted to accommodate the reinforcement.

No cracking was observed until seven months after repair by HPFRCC. The crack widths then developed to not more than 0.05 and 0.12 mm at 10 and 24 months after repair, respectively. Meshes of fine cracks were similarly observed. Cracking was harder to observe 24 months after repair compared to 12 months after, being obscured by dirt accumulated on the surface. On the other hand, cracking was visually observed on repair with normal repair mortar just one month after repair, with crack widths developing to 0.03, 0.2, and 0.3 mm at 3, 10, and 24 months after repair, respectively.

2.6 Surface repair of viaducts

Owing to its low air- and water-permeability due to small crack widths, HPFRCC surface repair is expected to retard carbonation of concrete structures. Surface protection with an organic paint type lining may be applied to railway viaducts, which require durability, for maintenance against carbonation. However, such lining is prone to early cracking due to the movement of cracks (opening and closing action) under the loads of railway traffic. Sprayed HPFRCC was applied to a thickness of 10 mm to viaduct girders having bending cracks for surface protection in 2005 (Fig. 7)[9,10]. Anchors were used to ensure bonding between the substrate concrete and HPFRCC.

Prior to the trial tests, alternate loading tests were conducted with small stress amplitudes to assume railway traffic loading on reinforced concrete beams having bending cracks, with HPFRCC sprayed onto the bottom surfaces. After alternate loading for 17 million cycles, the crack width of reinforced concrete beams with sprayed HPFRCC was 0.13 mm, half that of beams without HPFRCC (0.25 mm), clearly demonstrating the effect of HPFRCC. The carbonation-suppressing effect of HPFRCC sprayed onto existing concrete surfaces was also confirmed by accelerated carbonation testing.

Figure 7. Surface repair of viaducts

3. CONCLUDING REMARKS

This paper introduces recent applications using HPFRCC in Japan. Various tasks including those listed below remain to be done to facilitate the application of HPFRCC to existing structures.

(1) Required performance and its criteria through structural levels

For instance, it is well know that the ductility of HPFRCC helps to reduce damage of structures. HPFRCC also provides higher spalling resistance. However, how much ductility is required for structures? Quantitative requirements such as the deformation capacity of HPFRCC should be determined, as a structural response level rather than a material and/or elemental response level. Trial tests as described in the previous sections may be useful for a discussion of the required performance of structures using HPFRCC and its criteria.

(2) Supply of large amounts of HPFRCC in situ

Only a few companies can supply HPFRCC in situ, using ready mixed concrete plants. Most concrete-related institutes should define HPFRCC production techniques in addition to establishing quality control methods.

(3) Prediction tools for performance of structures with HPFRCC

Predicting the performance of structures that use HPFRCC through numerical analysis is important. Not only mechanical performance (structural response) but also time dependent performance (durability) should be accurately simulated to determine the advantages of HPFRCC.

(4) Cost reduction

The material cost of HPFRCC is more expensive than that of ordinary concrete at this time. In addition to direct material cost reductions, maintenance taking into consideration life cycle cost should be promoted.

(5) Reduction of environmental impact

Ordinary construction materials has been tried to contribute to the reduction of environmental impact by using waste, to propose recycle procedure of the materials. The attainment of long life structures through the use of HPFRCC contributes to sustainable development, and this concept should be simulated in existing structures. Strategies to reduce environmental impact such as recycling, use of waste, and rebuilding should also be prepared for HPFRCC.

Acknowledgements

The authors would like to thank the members of the JSCE 334 Committee on Fiber Reinforced Mortar with Multiple Cracks for their helpful advice. The Committee's report on HPFRCC in particular provided valuable suggestions for summarizing this paper.

4. REFERENCES

1. V.C. Li, From Micromechanics to Structural Engineering -The Design of Cementitious Composites for Civil Engineering Applications, J. Struct. Mech. Earthquake Eng., JSCE, 10(2): pp.37-48 (1993).
2. JSCE, 2005, Evaluation and Application of Fiber Reinforced Mortar with Multiple Fine Cracks, Japan Society of Civil Engineers, Concrete Engineering Series 64. (in Japanese)
3. Japan Concrete Institute, 2002, Proc. of the JCI International Workshop on Ductile Fiber Reinforced Cementitious Composites, JCI.
4. Mitamura, H., Sakata, N., Shakushiro, K., Suda, K. and Hiraishi, T., Application of Overlay Reinforcement Method on Steel Deck Utilizing Engineered Cementitious Composites – Mihara Bridge-, Bridge and Foundation Engineering, **39**(8): pp.88-91 (2005).
5. Fukuda, I., Mitamura, H., Imano, H. and Matsui, S., 2004, Effect of ECC Overlay Reinforcement Method on Steel Plate Deck Attached with FRP Dowels, Proc. of the Japan Concrete Institute, **26**(2): pp.1693-1698. (in Japanese)
6. Maruta, M., Kanda, T., Nagai, S. and Yamamoto, Y., 2005, New High-Rise RC Structure Using Pre-Cast ECC Coupling Beam, Concrete Journal, **43**(11): pp.18-26.(in Japanese)
7. Kojima, S., Sakata, N., Kanda, T. and Hiraishi, T., Application of Direct Sprayed ECC for Retrofitting Dam Structure Surface -Application for Mitaka-Dam -, Concrete Journal, **42**(5): pp.135-139 (2004). (in Japanese)
8. Rokugo, K., Kunieda, M. and Lim, S.C., 2005, Patching Repair with ECC on Cracked Concrete Surface, Proc. of ConMat05, Vancouver, Canada [CD-ROM].
9. Suda, K. and Rokugo, K., Anti-Carbonization Process Utilizing Direct Sprayed ECC Applying to Railway Viaduct Involving Flexural Fatigue Cracks, Concrete Journal, 43(5), pp.162-167 (2005). (in Japanese)
10. Inaguma, H., Seki, M., Suda, K. and Rokugo, K., 2005, Experimental Study on Crack-Bridging Ability of ECC for Repair under Train Loading, Proc. of International Workshop on High Performance Fiber Reinforced Cementitious Composites in Structural Applications, Honolulu, Hawaii.

SIMULATION OF THE TENSILE STRESS-STRAIN BEHAVIOR OF STRAIN HARDENING CEMENTITIOUS COMPOSITES

J. Yang and G. Fischer
Department of Civil and Environmental Engineering, University of Hawaii, USA; Department of Civil Engineering, Technical University of Denmark, Denmark

Abstract: In this paper, a method to simulate the tensile stress-strain behavior of strain hardening cementitious composites (SHCC) is described. Experimentally obtained information on the fiber bridging stress-crack opening relationship of individual cracks is used to simulate the formation of multiple cracking in the strain hardening stage of SHCC and the development of crack spacing and widths are captured quantitatively. The simulation results compare well with experimentally obtained results of the composite response in the uniaxial tension test.

Keywords: SHCC; stress-crack opening curve; stress-strain behavior; simulation model.

1. INTRODUCTION

Fiber reinforced cementitious composites are categorized into conventional Fiber Reinforced Concrete (FRC) with tension softening behavior and crack localization and Strain Hardening Cementitious Composites (SHCC) with increasing tensile loading capacity accompanied by multiple crack formation. Compared to plain concrete and conventional FRC, SHCC shows a significant improvement in ductility characterized by reaching its ultimate strength in the post cracking deformation regime and a relatively large inelastic deformation capacity. SHCC characteristics have been obtained at moderate fiber content in Engineered Cementitious Composites (ECC)[1], such that industrial application of this composite material has become economically feasible. In the last decade, research efforts have focused on the micro mechanics of SHCC material and models have been established to characterize the fiber bridging behavior leading to such composite properties. An analytical model that accounts for the multiple cracking of short randomly distributed fiber composites was introduced[2] under the assumption that fiber rupture does not occur in the fiber bridging-crack opening process. Later, the model was extended[3] to account for the possibility of fiber rupture, which is referred to as fiber pull-

M.S. Konsta-Gdoutos, (ed.), Measuring, Monitoring and Modeling Concrete Properties, 25–31.

out and rupture model (FPRM). Furthermore, a model was proposed[4] to predict the bridging stress-crack opening relationship (hereafter abbreviated as σ_B -δ curve) – a fundamental material property used in the design of Engineered Cementitious Composites (ECC). The model was derived for fibers having a statistical distribution of tensile strength. For SHCC to develop multiple cracking in the strain hardening stage, the flaw size distribution of the matrix and with that the distribution of cracking strength in the cementitious matrix plays an important role. Wu & Li[5] predicted the multiple cracking process of short random fiber reinforced brittle matrix composites through Monte Carlo simulation of the flaw size distribution. Kanda et al.[6], by employing a probabilistic description of initial flaw size distribution, proposed a micro-mechanics based theoretical approach for predicting the tensile stress-strain relationship of random short fiber reinforced cementitious composites showing strain hardening behavior.

The above-mentioned models account for the influences of the variation of fiber tensile strength and matrix flaw size and are combining the micro mechanics of fiber bridging with random properties of fiber and matrix. However, due to the difficulty in realistically estimating the variability of each parameter and its influence, the resulting predictions of the tensile stress-strain behavior may differ from experimentally obtained results.

In this paper, the experimentally obtained σ_B -δ curve is directly examined and the obtained information is utilized to simulate the multiple cracking and strain hardening behavior of SHCC in uniaxial tension. In this procedure, parameters that are difficult to realistically predict, such as variability of fiber tensile strength, orientation, dispersion, matrix flaw size distribution, fiber matrix interfacial bonding are inherently included in the input parameters for the simulation. Based on the experimentally obtained data of the fiber bridging stress-crack opening relationship, a simulation method is proposed and some preliminary results are shown.

2. σ_B -δ CURVE OF PVA ECC

The particular SHCC investigated in this study was an Engineered Cementitious Composite (ECC) reinforced with PVA (Polyvinyl Alcohol) fibers (2.0 Vol-%). The PVA fibers have a nominal tensile strength of 1600 MPa and are 8 mm long with a diameter of 39 μm. The ECC mix proportions are listed in Table 1, with proportions of ingredients given by ratio of weight, except for the fibers, which are quantified by volume fraction.

Table 1. Mix proportions of PVA FRCC

Cement	Fly Ash	Sand	Water	SP	PVA Fiber(Vol. %)
1	2	1.4	1	0.023	2

The fiber reinforcement was added to the cementitious matrix during the mixing procedure and the composite was cast into blocks (3in x 4in x 16 in) from which the test specimens were cut into the shape shown in Figure 1. The specimens are notched on all sides to facilitate the formation of a single crack. Ideally, a single crack should be generated at the notch when a tensile load is applied and the crack opening is measured using a clip gauge. The stress is calculated by normalizing the tensile load by the cross-sectional area at the notch and is plotted against the measured crack opening to obtain the σ_B -δ curve. To check whether there is indeed only a single crack at the predefined crack location, the test is terminated immediately after the load has reached the peak level, and the specimen is cut vertically along the center line of its thickness to visually inspect the crack formation. Figure 2 shows typical observations of cut specimens with a single crack and the corresponding σ_B -δ curve.

The tensile tests were carried out on specimens that were cured in air for 28 days. Figure 3 shows the σ_B -δ curves obtained. From these curves it can be observed that while the ultimate fiber bridging stress has relatively large variability, the crack opening at the peak load remains relatively constant.

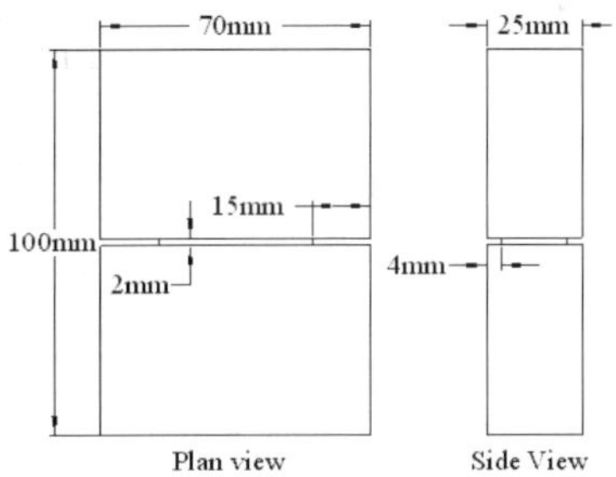

Figure 1. Specimen dimensions

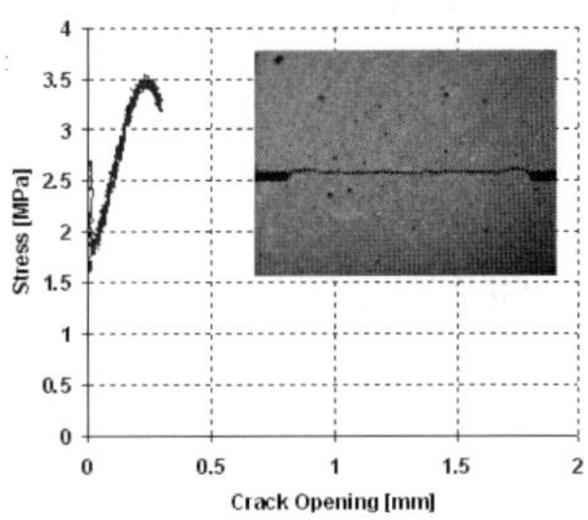

Figure 2. σ_B -δ curve of a single crack

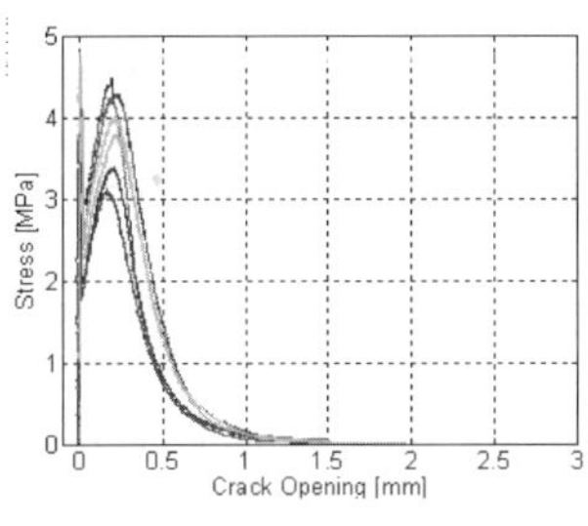

Figure 3. σ_B -δ curves of specimens

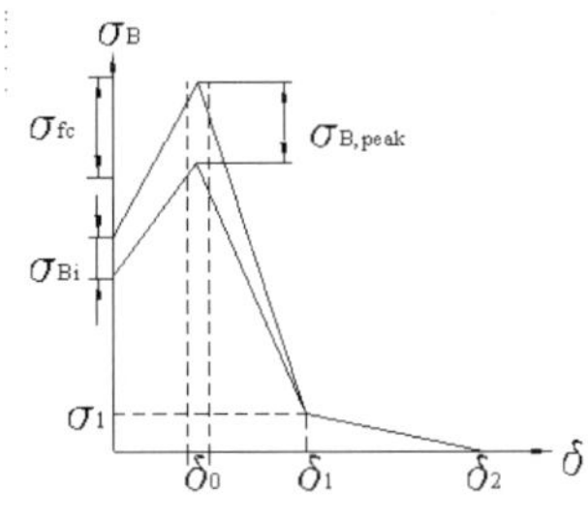

Figure 4. Envelope of the simplified σ_B -δ curves

3. SIMULATION CONCEPT

Based on the range of fiber bridging stress-crack opening responses obtained from the experimental investigation, a tri-linear simplification as shown in Figure 4 is chosen to represent the experimentally obtained responses. To account for the variability in the σ_B -δ curve, a range in characteristic composite parameters is assumed in the model based on the experimentally obtained data. The parameters that affect the behavior of the composite are the first cracking strength σ_{fc}, the peak bridging stress $\sigma_{B,peak}$, the crack opening δ_0 at peak bridging stress, and the fiber bridging stiffness K_f. These parameters may vary randomly within the experimentally defined range and will inherently account for the variability in fiber tensile strength, orientation, interfacial bond characteristics, and matrix flaw size distribution. The simulation model for the composite stress-strain behavior developed in this study will include potential multiple cracking and strain hardening features of the ECC in direct tensile loading. For general applicability of the model to SHCC materials as well as tension softening FRCC, model parameters can be adjusted to experimental data obtained for the fiber bridging stress-crack opening relationship of other FRCC materials.

During the displacement controlled deformation process of the composite, the tensile specimen initially behaves like a spring with an effective stiffness K_m of the un-cracked composite. At increasing deformations, the tensile load increases until the first crack forms at the largest flaw and lowest first cracking strength in the specimen. After the first tensile crack is formed, the cracked composite is represented by inserting a second spring element with a stiffness corresponding to the fiber bridging stiffness K_{f1}, as shown in Figure 5. The load will subsequently drop to a value where force equilibrium between the fiber bridging section and adjacent uncracked composite section is achieved. When the induced tensile deformations in the composite are further increased, the load will again increase with a modified total stiffness K until the tensile stresses are sufficient to cause the formation of a second tensile crack in the composite. This process is continued and multiple cracking can initiate until the tensile stress reaches the lowest peak bridging strength as defined by the parameter envelope (Figure 4). At this point, localization of cracking occurs and the composite fails in a tension softening manner.

The maximum number of potential cracks in a given composite is governed by the specimen length and the minimum crack spacing x_d, which has been theoretically derived[5] for the case of random short fiber reinforced composites. A crack spacing between x_d and $2x_d$ is expected at crack saturation. However, due to the variation of matrix properties, fiber matrix interfacial bonding, and fiber distribution, the observed minimum crack spacing often exceeds twice the derived minimum crack spacing. For PVA-ECC with preexisting flaws, a minimum crack spacing of 1.7 mm to 2.5 mm at crack saturation was reported[7].

Experimental results of the stress-strain response of PVA-ECC show a minimum crack spacing of 2 mm, which is adopted as a parameter for the simulation model presented in this paper. At a given specimen length and minimum crack spacing, the maximum number of potential cracks and their locations are identified along the specimen. Then, a randomly selected σ_B - δ curve from the envelope (Figure 4) characterized by its associated parameters (see e.g. Table 2) is assigned to each potential crack location and the simulation of the uniaxial tension test can be run using the procedure described above.

4. PRELIMINARY SIMULATION RESULTS AND EXPERIMENTAL VERIFICATION

The proposed simulation model is applied to predict the stress-strain behavior of PVA-ECC specimens tested in uniaxial tension. The specimens have a dogbone shape (Figure 6) and are 305 mm long, 25 mm thick, and have a gauge length of 100 mm. Specimens of the shape shown in Figure 1 were made along with the dogbone specimens from the same mix to identify the σ-δ curves of this composite.

Both the notched specimens and the dogbone specimens were tested under uniaxial load at the age of 28 days. The experimentally obtained σ_B -δ curves of the notched specimens are shown in Figure 3. The parameters that describe the envelop of the σ_B -δ curves are summarized in Table 2. This information is taken as input parameters of the simulation model and the predicted stress-strain curves for the dogbone specimens are compared with experimentally obtained stress-strain curves in the uniaxial tension test in Figure 7(a) and (b).

Since a random σ_B -δ curve from within the experimentally obtained envelope is assigned to each potential crack location in each simulation run, the obtained results for the composite stress-strain curve differ as well in each simulation. Figure 7(a) shows three simulated results plotted in the same figure. The comparison shows that the simulation results agree well with experimental data. In addition, the plotted results demonstrate that the simulation program developed using aforementioned concept is capable of capturing the multiple cracking phenomenon and the strain hardening behavior of ECC under direct tensile loading.

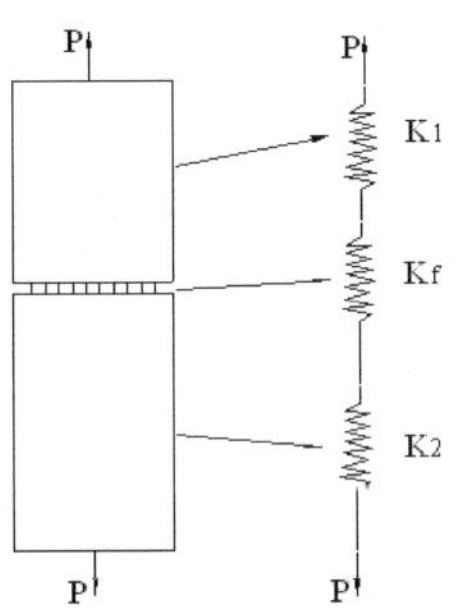

Figure 5. Equivalent spring system

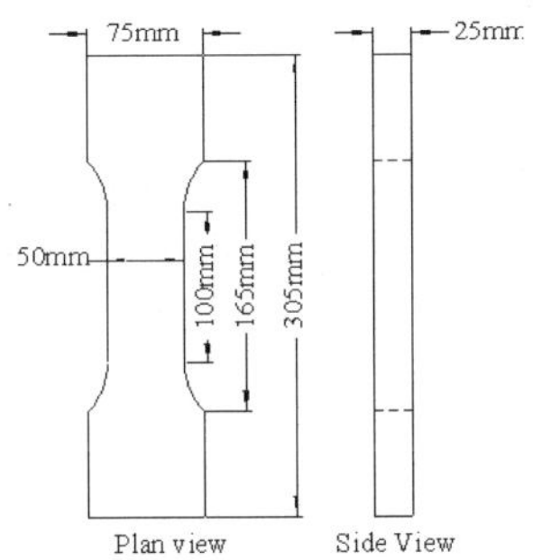

Figure 6. Dimensions of tensile specimen

Table 2. Parameters of σ_B -δ envelope

Parameter Name	Value
σ_{fc}	3.2 – 4.8 MPa
σ_i	1.82 – 2.97 MPa
$\sigma_{B,peak}$	2.9 – 4.6 MPa
δ_0	0.17 - 0.22 mm
σ_1	1.0 MPa
δ_1	0.60 mm
δ_2	2.0 mm

This simulation procedure can offer other related results such as evolution of crack spacing and a quantitative assessment of crack widths. A typical simulated stress-strain curve and crack distributions for the dogbone specimens are shown in Figure 8(a) and (b). The large crack in Figure 8(b) indicates the position of crack localization prior to failure of the specimen. As multiple cracking occurs in the specimen, the evolutions of average crack spacing, average crack width as function of strain are plotted in Figure 9(a) and (b), and the evolution of the maximum and minimum crack width are plotted in Figure 10.

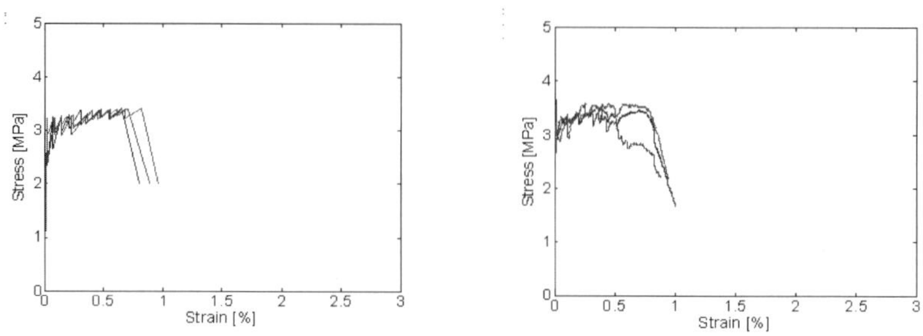

Figure 7. (a) Simulated σ–ε curves, (b) Experimentally obtained σ–ε curves

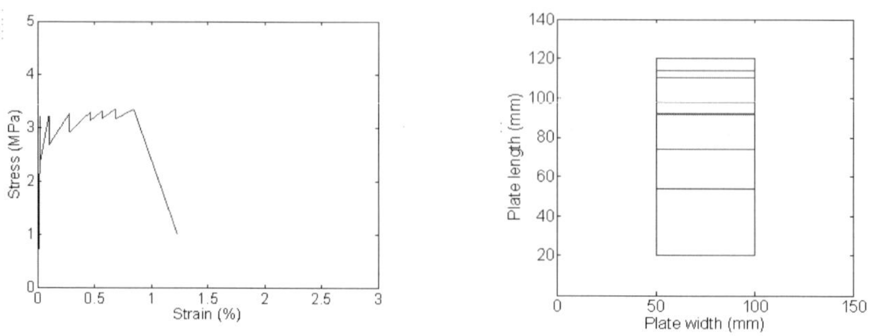

Figure 8. (a) Typical simulated σ–ε curve of PVA-ECC, (b) Simulated crack positions

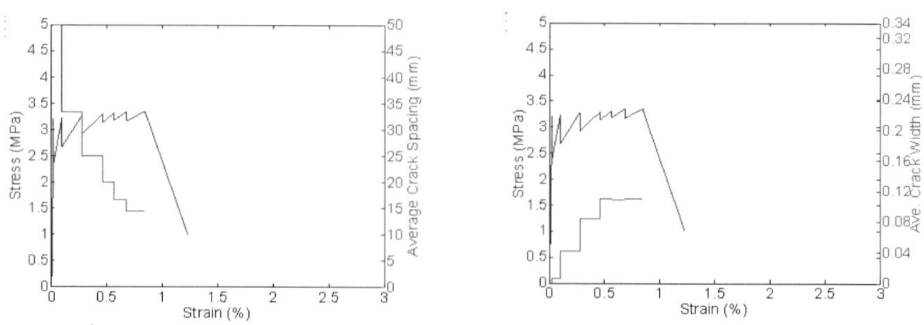

Figure 9. (a) Evolution of average crack spacing, (b) Evolution of average crack width

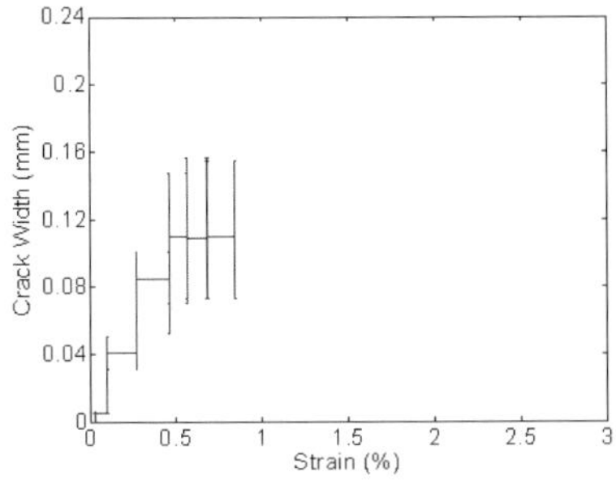

Figure 10. Evolution of maximum and minimum crack width

5. CONCLUSION

A method to simulate and predict the multiple cracking and strain hardening behavior of Strain Hardening Cementitious Composites (SHCC) under uniaxial tension was developed based on experimental information of the fiber bridging stress-crack opening relationship of the composite. Utilizing this information as input parameters of the simulation of the composite tensile stress-strain behavior, the variability of composite material properties, such as matrix flaw size, fiber tensile strength, fiber matrix interface characteristics, and fiber orientation, can be realistically incorporated. The multiple cracking and strain-hardening behavior can be captured by the suggested simulation model. In addition, the evolution of crack width and spacing can be quantified. The simulated response and experimentally obtained stress-strain behavior of SHCC are in agreement. The proposed method can serve as a tool in estimating the stress-strain behavior of SHCC based on σ_B -δ information, which is essential in the design of structural applications using SHCC. The simulation model can also be used to design SHCC materials with a target tensile stress-strain response and crack width limit by identifying the optimal range of matrix first cracking strength, peak fiber bridging strength and fiber bridging stiffness.

6. REFERENCES

1. Li, V.C., Kanda, T., Engineered Cementitious Composites for Structural Applications, *Innovations Forum in ASCE J. Materials in Civil Engineering*, 10(2), 66-69 (1998).
2. Li, V.C., and Leung, C.K.Y., Steady state and multiple cracking of short random fiber composites, *ASCE J. of Engineering Mechanics*, 188 (11), 2246-2264 (1992).
3. Maalej, M., Li, V.C. and Hashida, T., Effect of fiber rupture on tensile properties of short fiber composites, *ASCE J. of Engineering Mechanics*, 121 (8), 903-913 (1995).
4. Maalej, M., Tensile properties of short fiber composites with fiber strength distribution, *J. of Material Science*, 36, 2203 – 2212 (2001).
5. Wu, H.C. and Li, V.C., Stochastic process of multiple cracking in discontinuous random fiber reinforced brittle matrix composite, *Int'l J. of Damage Mechanics* 4(1), 83-102 (1995).
6. Kanda, T., Lin, Z., and Li, V.C., Tensile stress-strain modeling of pseudostrain hardening of cementitious composites, *Journal of Materials in Civil Engineering*, 147-156 (May, 2000).
7. Wang, S. and Li, V.C., Tailoring of pre-existing flaws in ECC matrix for saturated strain hardening, *Proceedings of FRAMCOS-5*, Vail, Colorado, USA, April 2004, 1005-1012.

EFFECT OF THE TEST SET-UP AND CURING CONDITIONS ON FRACTURE BEHAVIOR OF STRAIN HARDENING CEMENT-BASED COMPOSITES (SHCC)

V. Mechtcherine and J. Schulze

Institute for Building Materials, Technical University of Kaiserslautern, Postfach 3049, 67653 Kaiserslautern, Germany

Abstract: This paper addresses the group of Strain Hardening Cement-based Composites (SHCC), which possesses an ultimate strain of approx. 5 % due to the bridging of fine multiple cracks by short, well distributed fibres. An extensive mechanical testing program on an ultra-ductile concrete is presented and conclusions with regard to an appropriate mechanical testing procedure for such cement-based composites are eventually drawn.

Key words: SHCC; ECC; strain hardening; tension tests; bend tests; test set-up; unnotched and notched specimens; load boundaries; curing conditions; standard test

1. INTRODUCTION

SHCC is a ductile cement-based composite, which exhibits a pronounced strain hardening behaviour due to the bridging of fine multiple cracks by short, well distributed fibres. Such materials have considerable potential practical applications (Li, 2003; Mechtcherine, 2005). Due to their advantageous and quite easily describable stress-strain behaviour, which is generally similar to that of steel, the usage of SHCC might soon revolutionize the design of concrete structures and their repair in particular cases.

In this paper, the challenge of an adequate mechanical testing of SHCC is addressed and the results of an extensive experimental program are presented and discussed. Eventually, an appropriate procedure for the testing of such ductile cement-based composites in tension is proposed based on the evaluated test results.

M.S. Konsta-Gdoutos, (ed.), Measuring, Monitoring and Modeling Concrete Properties, 33–39.

2. COMPOSITION OF SHCC USED IN THIS STUDY

In this investigation, an ultra-ductile SHCC with polymeric fibres was used (cf. Table 1). Such composites, which material design is based on micromechanical modelling, are often referred to as Engineered Cementitious Composites (ECC) (Li, 2003).

A mix of a Portland cement 42.5 R (30% by mass) and fly ash (70% by mass) was used as a binder. The fine aggregate was quartz sand. Furthermore, 2.25% by volume of PVA fibres with a length of 12 mm were used. A superplasticizer (SP) and a viscosity agent (VA) were added to the mix in order to adjust its rheological properties in the fresh state. The average compressive strength of this concrete was 30 MPa and the Young modulus was 15.8 GPa.

Table 1. Composition of SHCC used for the mechanical experiments

Cement [kg/m³]	Fly ash [kg/m³]	Quartzsand [kg/m³]	Water [kg/m³]	SP [kg/m³]	VA [kg/m³]	PVA fibers [kg/m³]
320	750	535	335	16.1	3.2	29.3

3. OVERVIEW OF THE TEST PROGRAM

Figure 1 gives an overview of the experimental program and a schematic view of the specimen geometries. Unnotched and notched specimens were used for the uniaxial tension tests. The unnotched specimens were dog-bone shaped prisms with a cross-section of the narrow section (the corresponding gross cross-section is given in parentheses) of 24 (40) mm by 40 mm and 60 (100) mm by 100 mm, respectively. The gauge length was 100 mm in the tests on the unnotched specimens with a thickness of 40 mm (further referred to as "small" prisms), and 250 mm for the prisms with a thickness of 100 mm ("large" specimens). Notched prisms possessed the same effective cross-sections as the corresponding dog-bone shaped prisms. Here, a gauge length of 20 mm was used for small specimens and a gauge length of 50 mm for the larger ones.

A series of both three-point and four-point bend tests were performed on unnotched beams with a cross-section of 40 mm by 40 mm (span of 120 mm) and 100 mm by 100 mm (span of 300 mm), respectively. In the four-point bend tests, the beams were loaded at the one-third points of the span. Additionally, three-point tests were carried out on notched beams with an effective cross-section of 24 mm by 40 mm (depth of the sawn notch = 16 mm) and 60 mm by 100 mm (notch depth of 40 mm), respectively.

All specimens were cast horizontally in metal moulds. After demoulding, the specimens were stored at different curing conditions in order to investigate their effect on the mechanical behaviour of SHCC. Curing conditions were subsequently implemented, which included either sealing (wrapped in a plastic foil), storing at the ambient atmosphere (unsealed) or storing in water. All specimens were tested at a material age of 28 to 35 days. For each investigated parameter, at least three specimens were tested. The uniaxial tension tests were performed both with rotational and non-rotational boundaries, respectively. The displacement rate of the machine was 0.01 mm/s.

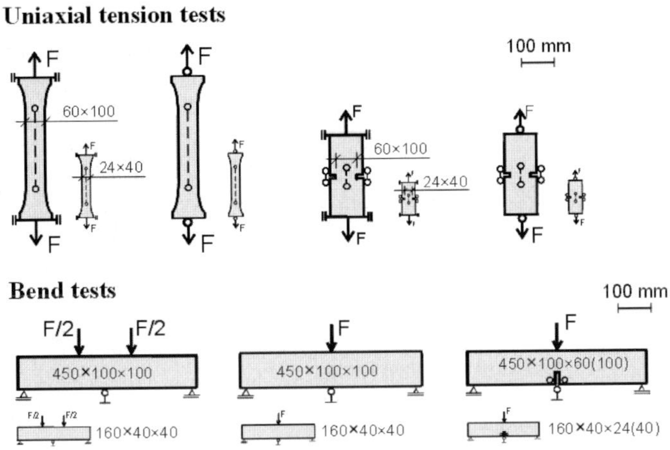

Figure 1. Overview of the used specimen geometries and test set-ups

4. RESULTS FROM THE DIRECT TENSION TESTS

4.1 Effect of the notches

Figure 2 shows typical stress-deformation relations obtained from the uniaxial tension tests with non-rotational loading platens, which were performed on unnotched and notched prisms, respectively, indicating the characteristic values derived from the curves. These and some additional data from the tests on small prisms are given in Table 2 for chosen combinations of test parameters. Further data may be found in Mechtcherine and Schulze (2005a).

Typically for ductile materials (e.g. steel), when considering the investigated SHCC, not only was there no reduction in the tensile strength f_t, but a slight increase was rather observed in the direct tension tests on notched specimens in comparison to the test on unnotched prisms (Figure 2 and Table 2).

As a result of the concentration of cracks around the notched cross-section, much smaller deformations δ_u at the maximum load could be measured in the tests on notched specimens. Consequently, the values of the specific work of fracture $W_{F,ft}$ (the specific energy consumed until the tensile strength f_t was reached) as well as W_F (calculated from the area under the complete stress-deformation curve) were considerably smaller than the corresponding values obtained from the tests on unnotched prisms, see Table 2.

It is worthy to note that the ultimate strain ε_u (equal to ultimate deformation δ_u divided by the gauge length) was higher for the notched specimens, indicating denser cracking – a phenomenon, which could be clearly confirmed by the observation of the crack patterns on the surfaces of the specimens.

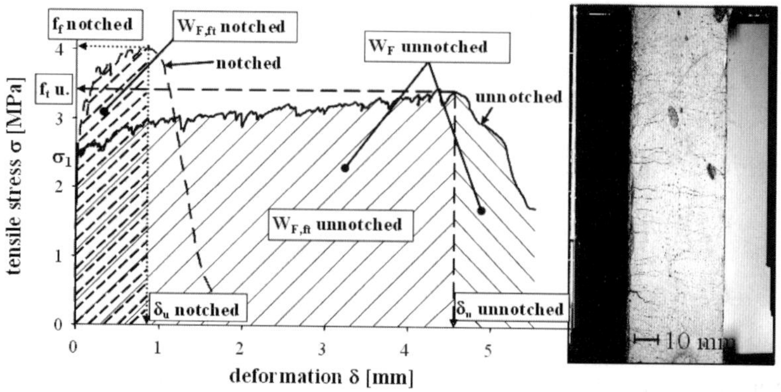

Figure 2. Tension tests: typical stress-deformation relations obtained from small specimens (left), and crack pattern for unnotched specimens (right)

4.2 Effect of the load boundaries

The specimens were wedged between the load cell and the base plate of the machine during the tension tests with hindrance of the specimen rotation. Calottes were installed between the load carrying portion of the machine and the metal adapters glued to the ends of the specimens for the tension tests without hindrance of the specimens' rotation.

The measurements of the deformations on the opposite sides of the prism show that the deformations increase more significantly on the side of the prism where the first crack appeared for the test with rotational load boundaries. A multiple crack formation appears on both sides of the specimen during the tests, whereby the measured deformation discrepancy continuously increases. This implies that the crack formation proceeds faster on one side, whereby the specimen loading becomes increasingly eccentric.

The crack formation irregularity of the case just described does not occur during the test with a specimen rotation hindrance. Both sides of the specimen display almost identical curves until ultimate strain occurs. A crack formation irregularity along the specimen cross-section is only then possible to determine.

When considering the average curves of both sides of the specimen, which are usually considered when characterising the material behaviour, then it is determined that the measured σ-ε relationships, obtained from the test with rotational loading plates, are considerably lower than the σ-ε curves, obtained from the test where rotational hindrance played a factor. Smaller values of the first crack stress σ_1 and the tensile strength f_t are determined accordingly. Furthermore, a smaller ultimate strain ε_u at failure is reached in the test with rotational boundaries. Further details may be found in Mechtcherine and Schulze (2005b). All further tensile tests results displayed were attained by implementing the tests with rotational hindrance of the loading plates.

4.3 Effect of the specimen size

With regard to the tensile strength or the stress at the first crack, no size effect could be observed in the tensile tests. While a decrease of the ultimate strain was observed with increasing specimen size, a significant increase of the specific work of fracture W_F was measured for larger specimens. The main reason for this increase was the fact that larger specimens were 2.5 times longer and correspondently had more cracks then the smaller prisms. The formation and opening of a greater number of cracks result in a higher energy consumption per unit area of the specimen cross-section. Further details may be found in Mechtcherine (2005).

4.4 Effect of curing conditions

Curing conditions had a considerable effect on the results of the tension tests. From the tests on unsealed specimens, slightly higher values for the stress at the first crack were obtained in comparison to the results from the tests on sealed specimens, see Table 2. The values of the tensile strength f_t and the ultimate strain ε_u, and as a result, the specific work of fracture $W_{F,ft}$ and W_F, were generally slightly lower in the case of unsealed specimens, both for unnotched and notched prisms. Figure 3 shows typical curves obtained from the tests on both types of specimens under different curing conditions.

The storage of the specimens in water resulted in a significant reduction of the tensile strength in comparison with the sealed or unsealed specimens. At the same time, in the experiments on notched specimens, no considerable effect of the curing in water was observed with regard to the specific work of fracture. The corresponding stress-deformation curves clearly show a less steep descending branch in the case of this type of curing (Figure 3, right). This phenomenon definitely results from the specific properties of the matrix and especially of the interface between the matrix and fibres, which are a result of different curing conditions. However, the particular mechanisms must still be clarified.

5. MAIN RESULTS FROM THE BEND TESTS

In the bend tests, multiple cracking was observed for every investigated parameter; it was however most pronounced in the four point-bend tests on unnotched specimens. Some results from the bend tests are given in Table 3 (an extended presentation of the results may be found in Mechtcherine and Schulze (2005a, 2005b). Unlike the tensile strength f_t, the flexural strength $f_{t,fl}$ decreased with increasing specimen size. This effect was however only slightly pronounced in the tests on notched specimens. The notches generally showed an effect which is similar to the effect observed in the uniaxial tension tests.

As expected, the deflection hardening was very pronounced. The values of the flexural strength $f_{t,fl}$ were manifold of the tensile strength values f_t obtained from uniaxial tension tests. This results from the crack zone making a considerable contribution to the transmission of tensile stress while the neutral axis shifted. With increasing specimen size, a significant increase of the specific work of fracture was observed in the bend tests, especially with the three-point arrangement. The four-point bend tests gave significantly higher values $W_{F,ft,fl}$ and W_F than the three-point bend tests. This is a result of a wider region, where the cracking occurred. Characteristically, in a few of the four-point bend tests, larger shear cracks developed leading to a shear failure of the specimen in some cases.

For the effect of the curing conditions, similar tendencies were observed as in the uniaxial tension tests, cf. Table 3 and Table 2.

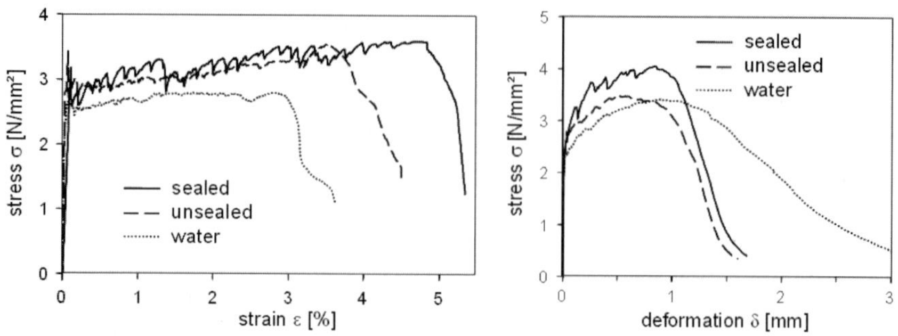

Figure 3. Effect of the curing on the results of tension tests performed on small unnotched (left) and notched (right) prisms

Table 2. Effect of the notches and curing conditions on the results from the uniaxial tension tests on small prisms, average values

Geometry/curing conditions	Stress at first crack σ_1 [MPa]	Tensile strength f_t [MPa]	Strain at f_t ε_u [%]	Deformation at f_t δ_u [mm]	Specific work of fracture until f_t $W_{F,ft}$ [N/m]	Specific work of fracture W_F [N/m]
unnotched/sealed	2.41	3.45	4.75	4.75	15320	16650
unnotched/unseal.	2.50	3.20	3.98	3.98	11240	13170
unnotched/water	2.30	2.57	2.57	2.57	6210	7590
notched/sealed	2.44	3.83	5.65	1.13	3810	6270
notched/unsealed	2.44	3.68	4.49	0.90	3050	4720
notched/water	1.33	2.97	6.86	1.37	3790	6480

Table 3. Effect of the notches and curing conditions on the results from the bend tests on large beams, average values

Test set-up	Bending stress at first crack $\sigma_{1,fl}$ [MPa]	Flexural strength $f_{t,fl}$ [MPa]	Deflection at $f_{t,fl}$ δ_u [mm]	Specific work of fracture until $f_{t,fl}$ $W_{F,ft,fl}$ [N/m]	Specific work of fracture W_F [N/m]
4-pt, unotched, sealed	5.0	11.3	5.68	13420	19210
3-pt, unotched, sealed	5.8	12.7	4.50	8820	15720
3-pt, notched, sealed	5.8	15.3	2.40	3190	10130
3-pt, notched, unsealed	6.1	14.6	2.24	2750	8280
3-pt, notched, water	4.6	13.9	2.55	3030	10540

6. CONCLUSIONS

Testing the influence of diverse parameters on the measured results was performed on an ultra-ductile SHCC with the goal of mapping out a suggestion for suitable test methodology, capable of being adapted as standard testing.

The most important conclusions with regard to the individual test modalities are listed in the following.

Specimen notching

Only the test on the unnotched specimen reproduces the characteristic material behaviour of SHCC under tensile stress. The stress-strain relationships obtained from these tests can be considered as material-specific. An increase in the measured strength due to the notching of specimens can be traced back to the multi-axial stress exerted on the notches (similar to steel). These complex stress conditions, which depend on the notch geometry, as well as irregular cracking in the area limited by the gauges render the interpretation of the results obtained from tests on notched specimens extremely difficult. Such tests are not suitable for an unerring derivation of the characteristic material behaviour.

Load boundaries

The usage of non-rotational loading plates has been proven to be beneficial for the implementation of the tensile tests. A uniform strain distribution is attained throughout the cross-section when implementing such a test set-up in contrast to tests with rotational loading plates.

Sealing

The type of curing resp. storage influences the performance of SHCC. Choosing the curing type for the specimens is to be oriented on the prospective conditions when manufacturing the concrete as well as how the building is to be used. Otherwise, the most neutral curing method best suited for a standard test is sealing (no water absorption, no water loss).

Indirect test methods

Four-point bend tests on unnotched beams can – after previous verification by means of uniaxial tensile tests – be used as quality control when manufacturing SHCC. A derivative of the characteristic stress-strain relationships from bend tests by means of a reliable algorithm has not yet been assured.

7. REFERENCES

Li, V.C., 2003, On Engineered Cementitious Composites (ECC): A Review of the Material and its Applications. *J. Advanced Concrete Technology*, **1**(3):215-230.

Mechtcherine, V., ed., 2005, *Ultra-ductile concrete with short fibres – Development, Testing, Applications*. ibidem-Verlag, Stuttgart.

Mechtcherine, V., and Schulze, J., 2005a, Testing the behaviour of strain hardening cementitious composites in tension. Proceedings of the International Workshop on High Performance Fiber Reinforced Cementitious Composites in Structural Applications, Honolulu, Hawaii, USA, May 23-26.

Mechtcherine, V., and Schulze, J., 2005b, Ultra-ductile concrete – material design concept and testing. *CPI Concrete Plant International*, **5**.

CONDITION FOR STRAIN-HARDENING IN ECC UNIAXIAL TEST SPECIMEN

L. Dick-Nielsen, H. Stang and P.N. Poulsen

Department of Civil Engineering, Technical University of Denmark, Lyngby, Denmark.

Abstract: This paper discusses the adequateness of the steady state flat crack criterion for crack propagation in Engineered Cementitious Composites. The requirement of a minimum complementary energy for the fiber bridging curve is found not to be relevant in the aim of controlling the crack opening under crack propagation.

Key words: ECC; strain-hardenings criteria; flat crack propagation; the fictitious crack model; specimen geometry.

1. INTRODUCTION

This paper discusses the fundamental principles behind the strain-hardening process of Engineered Cementitious Composite (ECC), which is a high performance fiber reinforced cementitious composite. For strain-hardening to occur in an ECC specimen it is required, that the criteria for multiple cracking are satisfied. These criteria insure that (1) the maximum fiber bridging stress is higher than the stress at which cracking is initiated and that (2) the cracks propagate in a steady state manner in an infinitely large specimen. The criteria are simplified and do not take into account geometry of the specimen, boundary conditions and interaction between cracks and initial defects (localization). A series of numerical simulations of uniaxial tensile tests are carried out on specimens containing one predefined flaw. The simulations are performed to get a better understanding of the governing mechanisms behind crack propagation in ECC. The applied cohesive law takes into account mortar as well as fiber properties. A parameter study is performed to investigate the influence of mortar and fiber properties, specimen size, boundary conditions and the position and size of initial flaw. The simulations will shed light on the adequateness of the proposed criterion for steady state flat crack propagation.

M.S. Konsta-Gdoutos, (ed.), Measuring, Monitoring and Modeling Concrete Properties, 41–47.

2. FRACTURE MECHANICS BASIS

For strain-hardening to occur in an ECC specimen, the maximum crack opening has to be small during crack propagation compared to deformations at peak bridging stress. In contrast if the maximal crack opening is not controlled during crack propagation it could lead to fiber rupture or pull-out. In order to avoid the maximal crack opening to increase during crack propagation it is convenient to have the crack propagating in a flat crack mode. The flat crack propagation was first analyzed by Marshall and Cox[1] applying the J-integral approach. According to Marshall and Cox the complementary energy J'_b of the fiber bridging curve has to be larger than the matrix toughness J_{tip}. This criterion as well as the Griffith theory is based on LEFM. By use of the *fictitious crack model* (FCM), attributed to Hillerborg[2], it has been shown[3] that even though mortar commonly is regarded as brittle LEFM can not be used to calculate the crack propagation. Constructing the cohesive laws for the materials is a fundamental issue when dealing with the FCM. The cohesive law for the mortar is found through a wedge splitting test (WST) and an inverse analysis[4]. For the fibers the cohesive law is found by use of a closed form solution[5]. The cohesive law for ECC is then found through superposition of the cohesive laws for the mortar and fibers; this approach has earlier been suggested[6] and proved[7].

3. GEOMETRY AND MATERIAL PARAMETERS

At the present time no global standard specimen geometry or test set-up exists for testing the strain-hardening capacity in ECC. However a common tendency is that one dimension of the test specimen is significant smaller than the two others. In order to arrive at some general conclusions the majority of simulations in the present paper are performed for an infinite sheet. Simulations for selected finite geometries are performed to relate these general conclusions to finite geometry. The simulations will all be for sheets loaded in uniaxial tension containing one initial stress free slit like flaw.

At the present time no global standard specimen geometry or test set-up exists for testing the strain-hardening capacity in ECC. However a common tendency is that one dimension of the test specimen is significant smaller than the two others. In order to arrive at some general conclusions the majority of simulations in the present paper are performed for an infinite sheet. Simulations for selected finite geometries are performed to relate these general conclusions to finite geometry. The simulations will all be for sheets loaded in uniaxial tension containing one initial stress free slit like flaw.

In this paper the cohesive law for the mortar is simplified by a bilinear cohesive law (see Figure 4 A)), where the area under the curve can be interpreted as the mortar toughness J_{tip} (or fracture energy G_f). The material data for the mortar has been obtained from an inverse analysis of a wedge splitting test[8,4]. The following material data were found for the mortar: the tensile strength $f_t = 2.83$ MPa, the stress-separations constants $a_1 = 156$ mm^{-1}, $a_2 = 9.7$ mm^{-1} and $b_2 = 0.24$, the mortar toughness $J_{tip} = 14.05$ N/m and the elastic modulus E = 31 GPa.

To calculate the fiber bridging curve a closed-form solution[5] is used. For the fibers following material constants are used: the slip-hardening coefficient $\beta = 2.21$, the fiber volume fraction $v_f = 2.21$, the snubbing coefficient f = 0.3, the fiber strength reduction coefficient f' = 0.3, the fiber length L_f = 12 mm, the frictional stress τ_0 = 0.3, the modulus of the fiber E_f = 42.8 GPa, Young modulus of the matrix E_f = 31 GPa, the chemical bond strength G_d = 4.71 N/m, the fiber diameter d_f = 39.6 μm and the in-situ fiber strength σ_{fu}=1400MPa. The cohesive laws for the mortar, fibers and ECC are shown in Figure 3 C).

4. RESULTS AND DISCUSSIONS

4.1 Influence of specimen geometry

In order to investigate the influence of infinite and finite specimen geometry eight Finite Element Method (FEM) models is employed. The FEM models contain an interface in which the crack can propagate. The FEM model for the infinite sheet has a height of 1000 mm and a width of 1200 mm and contains an initial slit like center flaw with the length, $2a_0$ of 4 mm. When the dimensions of the sheet are large compared with the length of the initial flaw, $2a_0$, the sheet can be regarded as infinite[9]. The model consists of 30x206 (height x width) quadrilateral, 8 nodes plane stress elements. The element size increases with a factor 1.03 from the flaw tip towards the edge along the width, and with a factor 1.75 from the flaw towards the ends along the height. In all the FEM simulations the boundary conditions along the two loaded edges results in a uniform displacement in the load direction along the width. The results obtained with the FEM model for an infinite sheet containing a center flaw, show good agreement with the corresponding results obtained with a semi-analytical approach[3]. In Figure 1A) and B) results for different geometries are shown for a center crack and an edge crack respectively. The center flaws all have a total length, $2a_0$ of 4 mm, while the edge flaws have a total length, a_0 of 2 mm. In Figure 1C) the opening at crack middle is shown for different crack lengths, for a finite geometry as well as for an infinite geometry.

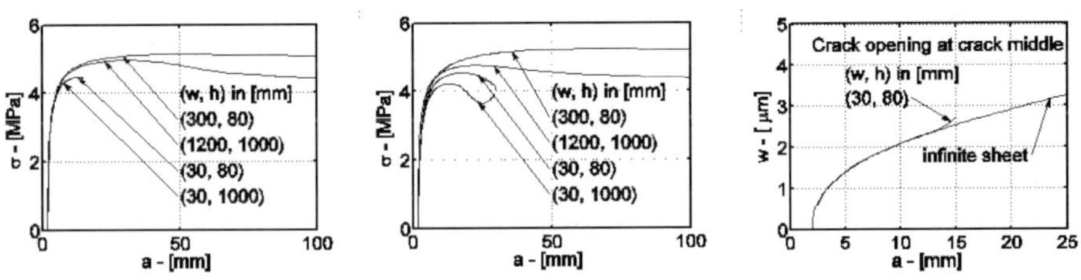

Figure 1. Influence of geometry. **A)** Center crack - Half crack length, a and matching far-field stress, σ. **B)** Edge crack - Total crack length, a and matching far-field stress, σ. **C)** Center crack - Opening at crack middle, w and matching half crack length, a

In order to achieve some general results independent of specimen geometry the remaining results dealing with center cracks are simulated using the semi-analytical approach, while the FEM-model with the dimensions 1000 mm x 1200 mm (height x width) is used for the edge cracks. In this way the results are for a single crack propagating in an infinite or semi-infinite sheet.

When investigating the influence of the specimen geometry, the sheet with finite dimensions 30 mm x 80 mm (width x height) are taken as a point of reference. These dimensions are chosen so that they match with the dimensions of the gauged part of test specimen in the JSCE - Tentative Guideline for Design and Constructions of Engineering Cementitious Composites - ECC[10]. For center cracks as well as edge cracks the force needed to drive the crack in this sheet is smaller compared to the corresponding for an infinite sheet; see Figure 1 A) and B). This is due to the smaller stiffness, caused by the small width in the finite sheet. When keeping the width fixed at 30 mm and increasing the height to 1000 mm, the results don't change for the center crack. For the edge crack on the other hand increasing the height causes the first crack stress, σ_{fc} (peak stress) to decrease 7 %. When increasing the height the model becomes less stiff, caused by the increased distance to the stiff boundaries. Due to symmetry in the center crack model it is not that sensitive to the change of stiffness. When keeping the height fixed at 80 mm and increasing the width to 300 mm, the first crack strength in both models increases, compared to the one obtained for a width of 30 mm. This is due to the fact that the stiffness is increased and for these dimensions the center crack model is affected as well.

In Figure 1C) the opening at crack middle during crack propagation is shown for the infinite sheet containing a center crack (semi-analytical model) and for a sheet with the dimensions 30 mm x 80 mm. As shown in the figure the crack opening at crack middle in the finite sheet is identical to the opening in the infinite sheet up till a crack length, a of 13 mm. After this crack length the opening in the finite sheet becomes larger than the one in the infinite sheet. This was expected due to the stiff boundary conditions in the small specimen. At a crack length, a of 15 mm in the finite sheet the crack has run through the sheet.

4.2 Influence of initial flaw size

Simulations have been performed for sheets containing initials flaws with different realistic lengths. The results are shown in Figure 2 and Figure 3 for center cracks and edge cracks respectively. The entire length of the center flaw is denoted $2a_0$ while the entire length of the edge flaw is denoted a_0. Due to this definition flaw lengths in the two situations can be directly compared.

As stated in[3] increasing length of the initial stress free flaw results in a decrease of the first crack strength (see Figure 2 C)). For center cracks the decrease in first crack strength is weak while it for edge cracks is more pronounced.

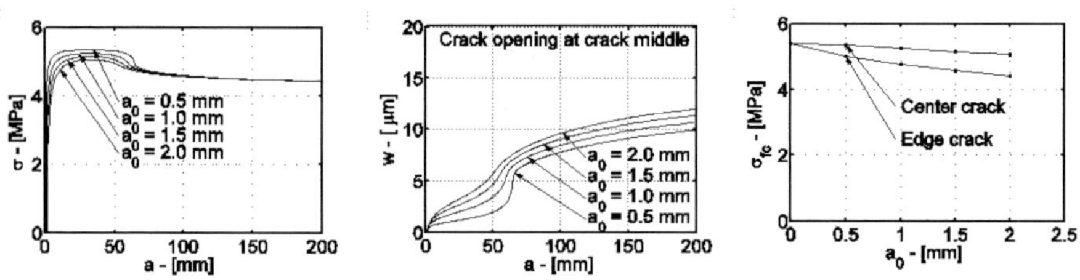

Figure 2. Influence of initial crack length, a_0 – center crack. **A)** Half crack length, a and matching far-field stress, σ. B) Opening at crack middle, w and matching half crack length, a. **C)** Relation between first crack strength, σ_{fc} and initial crack length, a_0

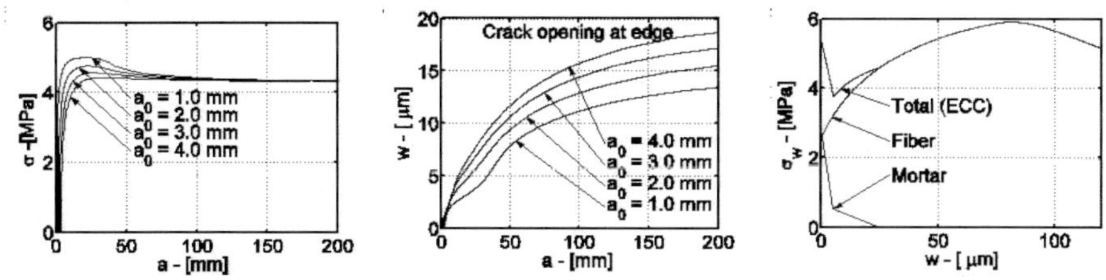

Figure 3. Influence of initial crack length, a_0 – edge crack. **A)** Total crack length, a and matching far-field stress, σ. B) Opening at crack middle, w and matching total crack length, a. **C)** Cohesive laws - cohesive stresses, σ_w and matching crack openings, w

According to the criteria for multiple cracking the complementary energy of the fibers has to be larger than the toughness of the mortar in order for the crack to propagate in a flat crack mode. The total complementary energy, J_{tip} for the fibers is 81.2 N/m and the mortar toughness, J_{tip} is 14.05 N/m. In Figure 2 A) the peak for the specimen containing an initial flaw with a length, a_0 of 2 mm, occur at a load of 5.06 MPa and a crack length, a of 33.5 mm. At this point the crack opening at the middle of the crack is 3.87 μm as shown in Figure 2 B) and 3.30 μm at x = 2 mm (the beginning of the cohesive zone). For crack lengths, a smaller than 200 mm the maximal crack opening is less than 20 μm independent of crack position and initial flaw size. At no time during these crack propagations is the process zones fully evolved.

4.3 Influence of the stress-separation constant a_1

The influence of the stress-separation constant, a_1 for the mortar is investigated (see Figure 4 A)); this is done by altering the shape of the cohesive law for the mortar. The opening, w in the second point in the cohesive law is varied while keeping the stress at this point fixed. The tensile strength, f_t and the opening at the end of the cohesive law are kept fixed as well (shown in Figure 4 B)). The new cohesive laws for the ECC material are shown in Figure 4 C). The results from the simulations are shown in Figure 5 and Figure 6 for center cracks and edge cracks respectively.

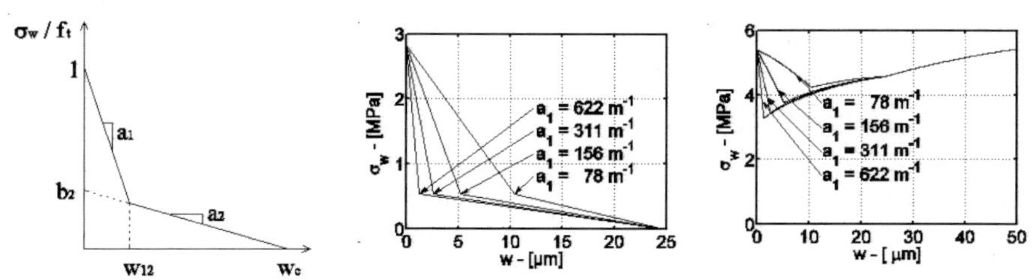

Figure 4. Cohesive laws for different a_1 - cohesive stresses, σ_w and matching crack openings, w. **A)** Stress-separation constants **B)** Mortar. **C)** Total (ECC)

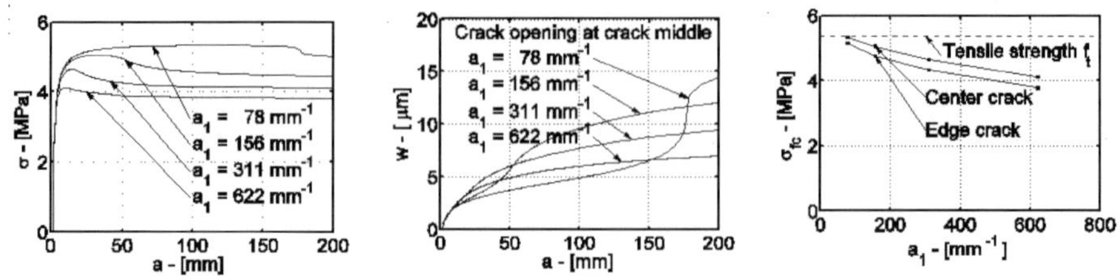

Figure 5. Influence of stress-separation constant, a_1 – center crack. **A)** Half crack length, a and matching far-field stress, σ. B) Opening at crack middle, w and matching half crack length, a. **C)** Relation between first crack strength, σ_{fc} and stress-separation constant, a_1

As shown in Figure 5 C) the first crack strength decreases when the stress-separation constant, a_1 increases. The decrease in the first crack strength is independent of the position of the initial flaw. In Figure 5 B) and Figure 6 B) the opening at the middle of the crack during crack propagation is shown. When interpreting these results one should keep in mind that the hardening branch in the cohesive law begins at following openings 1.3 μm, 2.6 μm, 5.2 μm and 10.4 μm for the respectively cohesive laws.

When altering the stress-separation constant, a_1, in the manner as described, the toughness of the mortar is altered as well. An increase in a_1 results in a decrease of the mortar toughness. According to the criteria for flat crack propagation a decrease in mortar toughness should increase the chances for the crack to propagate in a steady-state flat crack mode. As shown in Figure 5 B) and Figure 6 B) an increase in a_1 results in smaller crack openings, but only for relative large crack lengths (200 μm). All crack openings in these simulations are smaller than 20 μm.

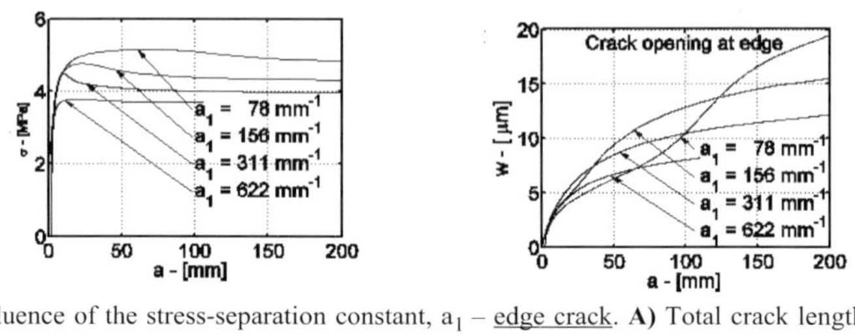

Figure 6. Influence of the stress-separation constant, a_1 – edge crack. **A)** Total crack length, a and matching far-field stress, *σ*. B) Opening at crack middle, w and matching total crack length, a

5. CONCLUSIONS

In the present investigation it has been examined whether the criterion for steady-state crack propagation is adequate. A parameter study of the influence of specimen size, initial flaw sizes and the stress-separation constant, a_1 for the mortar is performed. The study shows that the first crack strength and the evolution of the opening at crack middle are influenced by these parameters. The maximum crack opening observed during crack propagation in various simulations is small, 20 μm and also small compared to typical deformations at peak bridging stress. The complementary energy criterion is found not to be relevant, in the aim of achieving steady-state flat crack growth.

6. REFERENCES

1. D.B. Marshall, and B.N. Cox. A J-integral method for calculating steady-state matrix cracking stresses in composites. Mech. Mat., 8, 127-133, (1988).
2. A. Hillerborg, M. Modeer and P.E. Petersson. Analysis of Crack Formation and Crack Growth in Concrete by Means of Fracture Mechanics and Finite Elements. Cem. Concr. Res.,6, 773-782, (1976).
3. L. Dick-Nielsen, P.N. Poulsen H. Stang and J.F. Olesen. Semi-analytical cohesive crack model for the analysis of first crack strength of mortar. Proc. of the 17th Nordic Seminar on Computational Mechanics, 183-186, (2004).
4. L. Østergaard, J. F. Olesen, H. Stang and D. A. Lange. A simple and fast method for interpretation and inverse analysis of the wedge splitting test. Submitted for publication.
5. Z. Lin, T. Kanda and V.C. Li. On interface property characterization and performance of fiber-reinforced cementitious composites Concrete Science and Engineering,1, 173-184, (1999).
6. V.C. Li, H. Stang, and H. Krenchel, Micromechanics of Crack Bridging in Fiber Reinforced Concrete, Materials and Structures, **26**, 486-494, (1993).
7. L. Dick-Nielsen, H. Stang and P.N. Poulsen. Micro-mechanical Analysis of Fiber Reinforced Cementitious Composites using Cohesive Crack Modeling. Proceedings of the Knud Højgaard conference, (2005).
8. S. Wang, Private communication. DTU, 2004
9. H. Stang, J. F. Olesen, P. N. Poulsen and L. Dick-Nielsen. On the Application of Cohesive Crack Modeling in Cementitious Materials. Proceedings of the Knud Højgaard conference, (2005).
10. SCE. Tentative Guideline for Design and Constructions of Engineering Cementitious Composites-ECC. (Draft) JSCE TC ,334, (2005).

EXPERIMENTAL AND NUMERICAL ANALYSIS OF UHPFRC PLATES AND SHELLS

E.M.R. Fairbairn[1], R.D. Toledo Filho[1], R.C. Battista[1], J.H. Brandão[2], J.I. Rosa[1] and S. Formagini[1]

[1]COPPE/Universidade Federal do Rio de Janeiro; [2]Universidade Federal de Mato Grosso,Rio de Janeiro-RJ-Brasil

Abstract: In this paper the experimental and numerical analysis of Ultra High Performance Fiber Reinforced Concrete UHPFRC plates and shells are presented. The 900x900x15mm mm plate and the 3000x3000x10mm pyramidal shell were constructed without reinforcing bars and were tested to failure with a perpendicular concentrated force applied in its center. Two types of UHPFRC were used: CONAD, a home-made material, and Lafarge's DUCTAL®. The tensile constitutive relations of the materials were determined by inverse analysis, based on the results obtained from four points bending tests performed on prisms and plates. For the FE analysis it was used the TNO-Delft DIANA commercial code with a smeared cracking model. The good correlation between numerical and experimental results (load-deflection and crack pattern) indicated that the used methodology seems to be consistent to determine the behavior of the UHPFRC plates and shells up to failure.

1. INTRODUCTION

Many thin concrete plates and shells have been built around the world from the 50's to the 60's, but their use has gradually declined over the past few decades. This decline has been mainly due to the high cost of construction, difficulties in dealing with reinforcement and formwork and the complexity of the analysis.

However, advances in computer technology, concrete modeling and concrete technology have created a new paradigm for the design and construction of concrete plates and shells. This is the case of the advances in concrete modeling using the Finite Elements Method (FEM). For the last decades the crushing and cracking models, developed mainly within the framework of academic research, became operational, and have been implemented in commercial codes. Also the advances in computer technology have combined performance of meshing, modeling and visualization allowing the effective use of complex FE models for the most challenging analyses. In what concerns the evolution of the

M.S. Konsta-Gdoutos, (ed.), Measuring, Monitoring and Modeling Concrete Properties, 49–58.

cementitious materials, the High Performance Concretes (HPC) and the Ultra High Performance Fiber Reinforced Concretes(UHPFRC) left the academic world to become industrial products.

In this paper the analysis of thin concrete plates and shells is revisited from the point of view of these new paradigms. We used two types of UHPFRC: CONAD, a home-made concrete that was used to build a 900x900x15mm mm plate, and Lafarge's DUCTAL® that was used to build a 3000x3000x10mm pyramidal shell. The plate and the shell were built in the laboratory and an experimental program was performed encompassing testing until failure of the structures.

Even though there are mechanical and numerical models implemented in software that are dedicated to fiber reinforced concrete (see, for instance [1-2]) we decided to use a commercial code (DIANA 8.1) for the sake of simplicity, and aiming to use a software that is commercially available for the design engineer.

The constitutive relation in tension was determined by an inverse analysis based on the experimental results of bending and direct tension tests. Since the development of compressive stresses was not determinant for the plates we used a Drucker-Pragger plasticity model for which the parameters were determined from compressive tests on cylinders.

The experimental and numerical results presented good correlation indicating that the FEM analysis can be used by the engineer as an operational and simple tool for the design of UHPFRC plates and shells.

2. MATERIALS

For the mix design of the UHPFRC named CONAD (abbreviation of the Portuguese phrase *Concreto de Altíssimo Desempenho*) we used the compressive packing model (CPM) developed by de Larrard and collaborators [3-4]. Therefore, as required by the model, the following experimental properties have been determined for the constituents: the virtual compactness of the individual classes; the size grading distributions; the specific gravity; the cement contribution to compressive strength; and the saturation dosage of the chemical additive. A type III slag Portland cement (PC) was used in the composite production. Two classes of quartzite sand (S1 and S2) with nominal size of 150-300 μm and 425-600 μm, respectively, and quartz silica flour (QG) of 18 μm size were used as aggregate. The silica fume (SF) content was 5.8% of the cement weight and the water/binder ratio was 0.17. The superplasticizer (SP) was a polycarboxylate with a solid content of 32.5%. A fiber volume fraction of 2% of steel fiber (12 mm long with diameter of 0.18 mm) and 2.6% of wollastonite micro-fiber (WO) were used as reinforcement. The grain size distribution of the powder materials are shown in Figure 1 and the UHPFRC mix composition is presented in Table 1. Detailed information about the mix design and production of CONAD is provided by Formagini [5].

The CONAD presented, at 28 days, compressive strength and Young's modulus of 162 MPa and 47.7 GPa respectively. Its rheological characteristics corresponds to a self-compacting concrete.

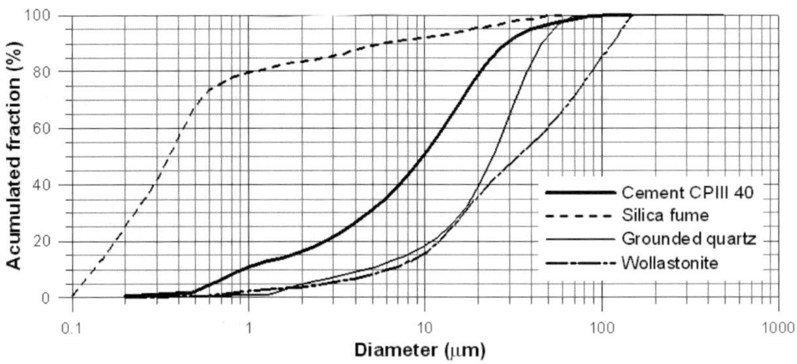

Figure 1. Grain size distribution of the powder materials of CONAD

Table 1. Composition of CONAD

Consumption	PC	SF	QG	S1	S2	WO	Steel fibre	SP	W	W/B
(kg/m3)	1011	58	79	60	823	76	158	50	162	0.17

A typical bending load-deflection curve obtained for the specimen tested after 28 days of cure is presented in Figure 2. This figure reveals the ductile nature of the designed UHPFRC. A maximum post-cracking stress of about 35.0 MPa was achieved at a deflection of about 7.9mm due to the fine multiple cracking in the area subjected to the higher moment (see Figure 3). After the peak load the main crack's localization occurred and a strain softening behaviour is observed.

A typical tensile stress-elongation curve obtained for the specimen submitted to a direct tensile test after 28 days of cure is presented in Figure 4. CONAD exhibited an elastic behavior from point A to B when the first crack appeared. An average first-crack stress of 10.2 MPa was reached at a deflection of about 0.026 mm while the maximum post-cracking stress (point C in Figure 4) of about 11.1 MPa was achieved at a elongation of about 0.213mm. Note that the elongation at maximum stress is 8 times higher than that observed at first crack. From point C, due to the main crack localization, a tension softening behavior is observed. It is worth to mention that even at an elongation as high as 0.4-0.5mm (15-20 times the first crack elongation) CONAD still resists an effort equivalent to the first crack stress.

The other UHPFRC used in this research is DUCTAL®, a commercial cementitious material produced by Lafarge, developed in a joint venture with Rhodia and Bouygues. It is a new fiber reinforced cementitious material designed on a multi-scale reinforcement concept. We used DUCTAL® FM (FM standing for metallic fibers). Its main mechanical and rheological characteristics are listed in Table 2. The material has been designed aiming to develop excellent characteristics of durability, which corresponds to the low porosity of the matrix obtained by means of a mix design based on maximum compacity of the granular classes. The main characteristics of the durability of DUCTAL® are listed in Table 3 [6-7].

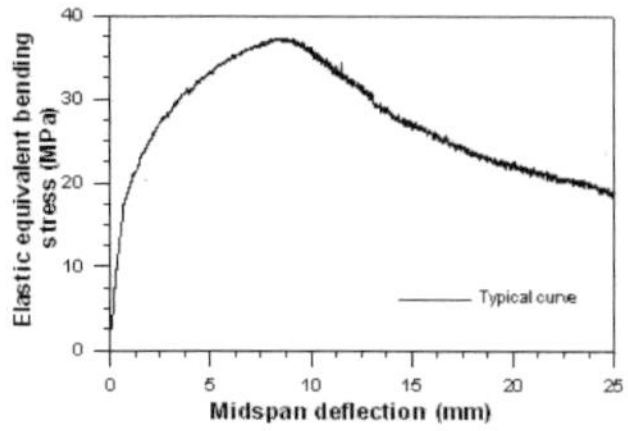

Figure 2. Load-deflection curve of CONAD

Figure 3. Typical multiple cracking at bending of CONAD

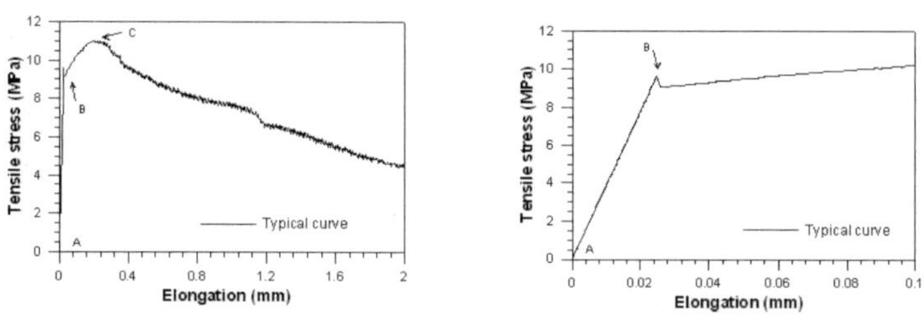

a) Scale 0-2mm b) Zoom to scale 0-0.1mm

Figure 4. Typical tensile stress-elongation curve of the UHPFRC

The highly ductile behavior of DUCTAL®, both in tension and flexure is close to the behavior of elastic-plastic materials. This deformation performance level results from an improvement in the micro-structural properties of the mineral matrix (micro-scale reinforcement) and the control of the link between matrix and fibers (macro-scale reinforcement) [6]. Natural inorganic micro-fibers might be employed as micro-scale reinforcement. In this way, mica flakes, or wollastonite micro-fibers can be incorporated by partial sand substitution. Macro-fibers, made of high-grade steel, or organic materials are between 3 and 50 times longest than the largest particle, and have a small cross-section to ensure an adequate bond. Besides its ability to be virtually self-placing or dry-cast, the improved physical characteristics of DUCTAL® eliminates the need for reinforcing steel bars. Also, it can be produced with customary industrial tools by casting, injection, or extrusion. To illustrate the ductile behavior of DUCTAL®, a typical load-displacement curve for a beam submitted to three-point bending test is displayed in Figure 5 and a photograph of a typical behavior of multiple cracking in bending is shown in Figure 6 [9].

3. STRUCTURES

The structure built with CONAD was a simply supported plate without reinforcement bars, having the dimensions 900x900x15 mm with a 300x300x15mm capital in its center. Loading was applied until failure under displacement control by an actuator acting on the center of the plate. The main geometric characteristics and the test rig are shown in Figure 7.

Table 2. Main mechanical and rheological characteristics of DUCTAL® FM

Compressive strength	150 – 180 MPa
Flexural strength	32 MPa
Tensile strength	8 MPa
Young's Modulus	50 GPa
Fracture Energy Gf	20.000 – 30.000 J/m²
Cracking toughness Gc	20 – 30 J/m².
Density	2500 kg/m3
Creep coefficient	0.15 – 0.30
Abrahms cone	500-700 mm
ASTM shock table	250 mm

Table 3. Main durability characteristics of DUCTAL®

Choride ion diffusion	$0.02 \cdot 10^{-12}$ m²/s
Carbonation (depth of penetration)	<0.5 mm
Abrasion (relative volume loss)	1.2 index 1
Freeze-thaw (resisual E-mod after 300 cycles)	100%
Capillary porosity (>10μm)	<1%
Total porosity	2-6%
Choride ion diffusion	$0.02 \cdot 10^{-12}$ m²/s
Carbonation (depth of penetration)	<0.5 mm
Abrasion (relative volume loss)	1.2 index 1
Freeze-thaw (resisual E-mod after 300 cycles)	100%

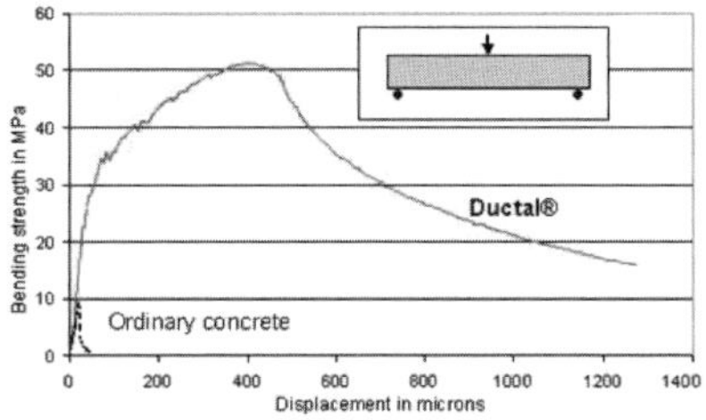

Figure 5. Load-deflection curve of DUCTAL®

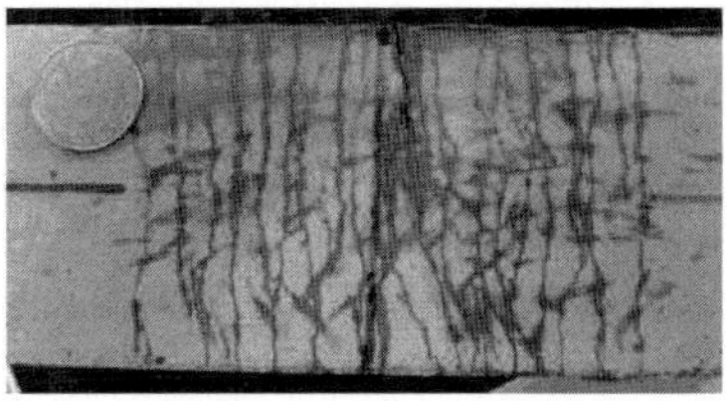

Figure 6. Typical multiple cracking at bending of DUCTAL®

The shell built with DUCTAL® is a 3000x3000x10mm pyramidal shell without reinforcement bars. This shell may be used as a module within a modular roofing system, allowing elegant architectural solutions. Figure 8 shows a picture of an architectural application of a normal concrete pyramidal shell for roofing. Figure 9 displays a sketch of the geometry of the shell and Figure 10 the test rig, indicating that the shell is submitted to a load vertically applied in its center, being simply supported at its perimeter. This loading is the "inverse" of the loading acting on the shell in real structures such as the one shown in Figure 8. The formwork of the shell was constructed under strictly controlled technical conditions, and had the upper surface in acrylic (Figure 11) in such a way that we could visually observe the placing of DUCTAL®. After curing in a wet environment, the shell was placed on the reinforced concrete frame (Figure 12), instrumented, and loaded with a displacement controlled actuator until failure.

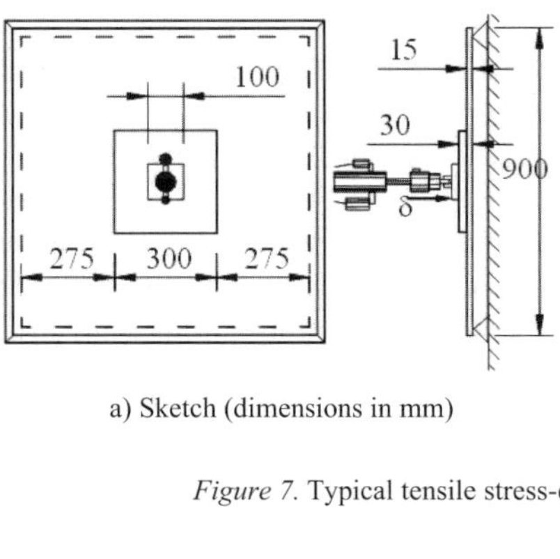

a) Sketch (dimensions in mm)

b) Photograph

Figure 7. Typical tensile stress-elongation curve of the UHPFRC

Figure 8. Pyramidal shell roofing (normal concrete) at the research center of PETROBRAS-Rio de Janeiro, Brasil

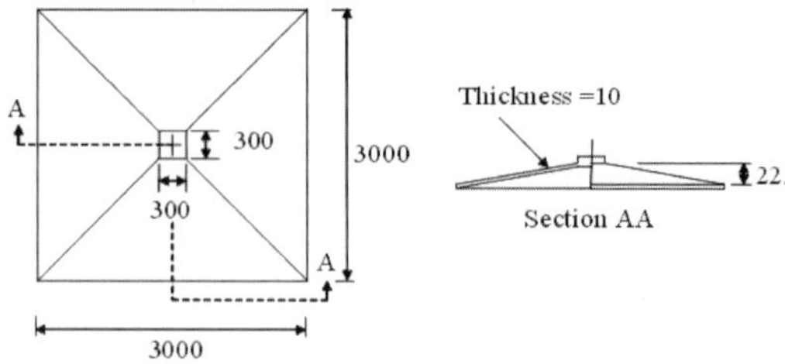

Figure 9. Sketch of the shell built with DUCTAL®

4. FINITE ELEMENT ANALYSIS

FE modeling was performed with commercial code DIANA 8.1 from TNO-Delft-The Netherlands [10]. This software has several modules and elements dedicated to concrete, including a set of constitutive equations in tension and compression, with various models for cracking and crushing initiation, smeared and discrete crack modeling, interface cohesive cracking elements, reinforcement elements, prestressing, early-age effects, etc. We used smeared multi-directional fixed cracking with strain decomposition and linear tension cut-off initiation criterion for representing the cracked composite. A crack band length *h* that assure the objectivity of the mesh is taken as the square root of the area of the finite element.

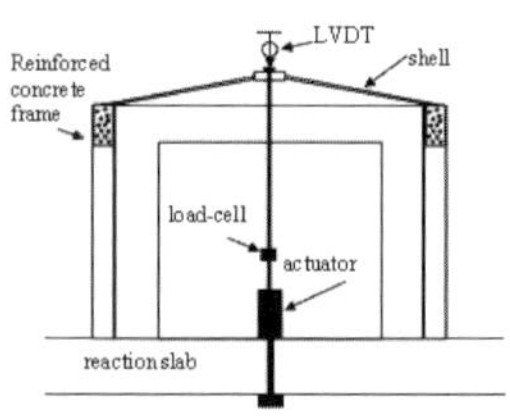

Figure 10. Test rig

Figure 11. Formwork of the shell

Figure 12. View of the shell being placed on the concrete frame

The characteristics of the tension behavior were deduced from inverse analysis, which used as starting point some results obtained from the experimental σ-CMOD diagram of the direct tension test [11-12]. Therefore, the goal of the inverse analysis was to adjust the constitutive relation of the cracked material until the load-deflection curve of the 4 point bending test were fitted by the numerical results.

Since the structural behavior of the plate and the shell here analyzed is not governed by crushing in regions subjected to compressive stresses, we used perfect associated plasticity with Druker-Prager criterion. Therefore, the formula relating the friction angle, the compressive strength, and the cohesion may be written as follows:

$$c = f_c \frac{1 - \sin\phi}{2\cos\phi} \tag{1}$$

In what concerns shearing after cracking, we used shear retention according to equation (2) for considering the effect of aggregate interlock and dowel action introduced by the fibers after cracking.

$$G_{cr} = \beta G \quad ; \quad \beta = \frac{1}{1 + 4447\, \varepsilon_{nn}^{cr}} \tag{2}$$

5. NUMERICAL ANALYSIS OF THE CONAD PLATE

The constitutive equation found by inverse analysis is shown in Figure 13, and the good agreement between the experimental and numerical load-deflection curves is shown in Figure 14.

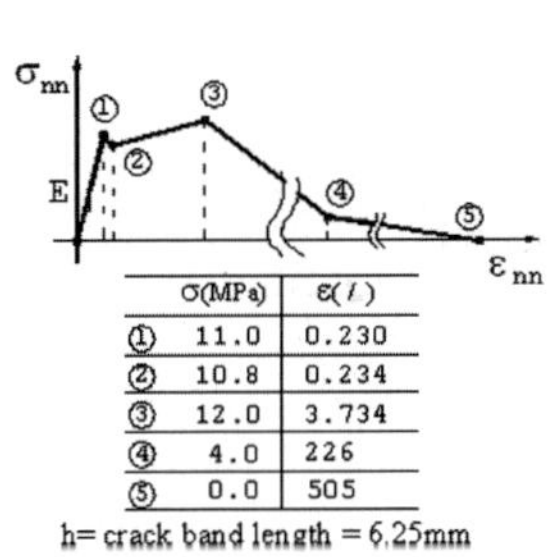

Figure 13. Constitutive equation of CONAD found by inverse analysis

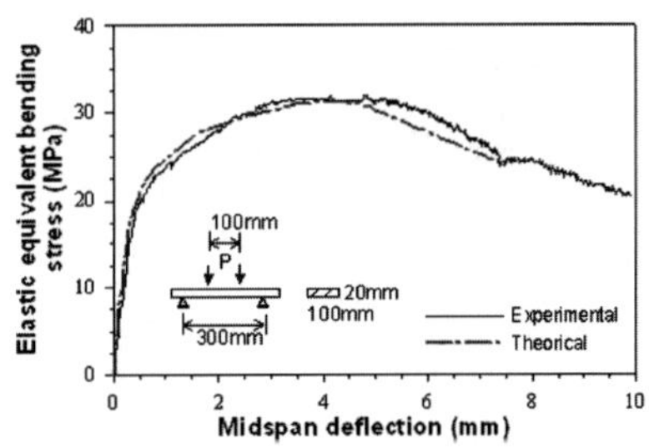

Figure 14. Comparison between experimental and numerical results for CONAD's 4-points bending test inverse analysis

The main characteristics of the analysis of the CONAD square plate are shown in Figure 15. We used eight-node isoparametric quadratic element (CQ40S). However, since the crack band length h for the plates is equal to 23.6mm, the values of the strains in the constitutive equation given in Figure 13 has been changed to assure the constancy of Gf (e.g., ε_4=61‰; ε_5=134‰). It should be pointed out that, in the experiment, the supports are free to displace from down to bottom. For this reason we used an interface element to simulate the fact that the plate can loose contact with the supports.

The experimental and numerical load-displacement curves at mid-span are shown in Figure 16. These results indicate good correlation, showing the accuracy of the analysis. It should be noted the ductile behavior of the plate, which presented hardening behavior after cracking, corresponding to multiple cracking of the structure before failure.

Comparisons between cracking patterns obtained from numerical and experimental analysis are shown in Figure 8 and Figure 9. Once more a good agreement between experiment and simulation was found indicating that the numerical modeling was able to reproduce the multiple cracking behavior of the UHPFRCC plate.

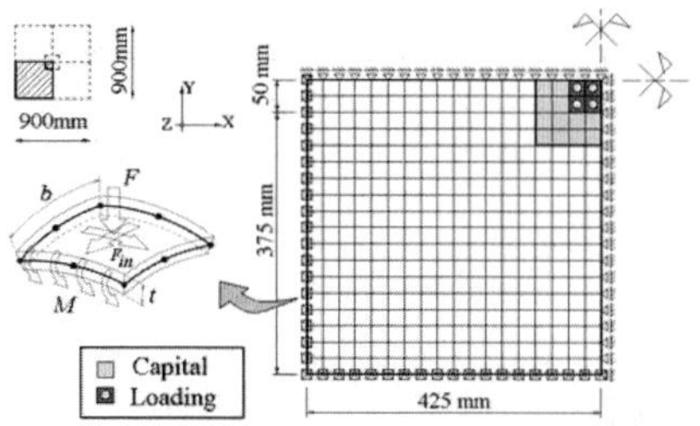

Figure 15. Main characteristics of the FEM analysis of the CONAD plate

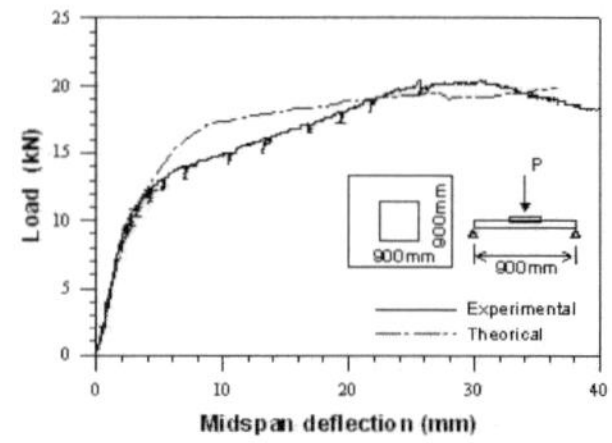

Figure 16. CONAD plate experimental/numerical correlation

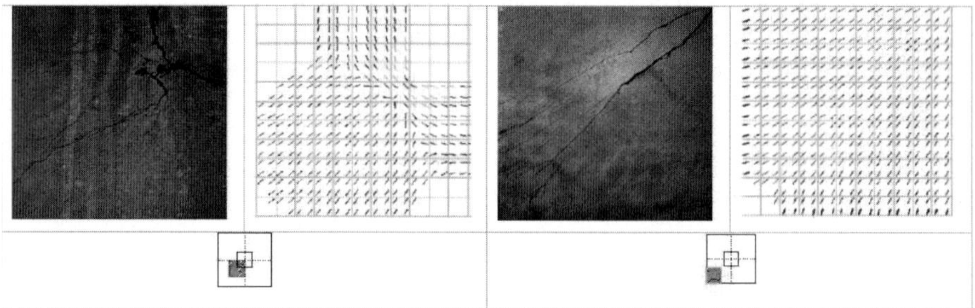

Figure 17. Cracking at failure: experimental and numerical

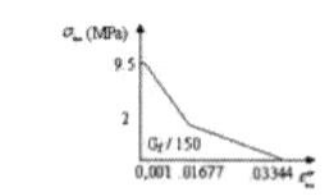

Figure 18. Constitutive relation.

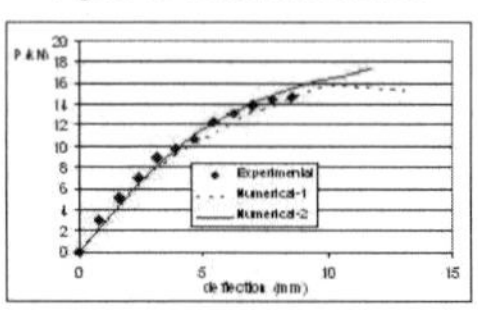

Figure 19. Load-deflection curve at the center of the shell

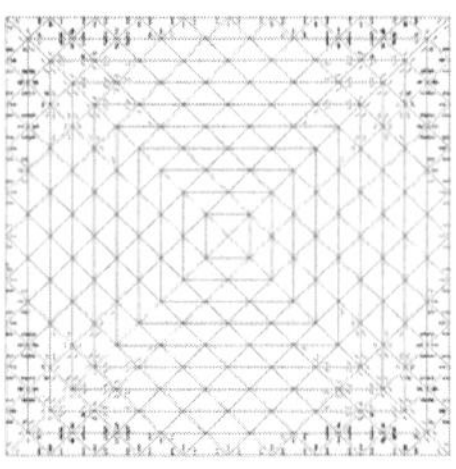

Figure 20. FEM mesh and cracking pattern at failure

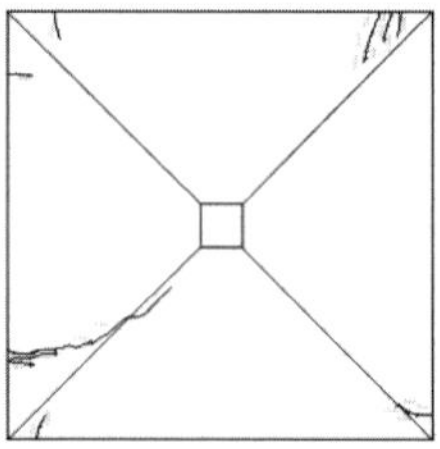

Figure 21. Experimental cracking pattern at failure

6. NUMERICAL ANALYSIS OF THE DUCTAL® SHELL

An inverse analysis similar to the one performed to the CONAD plate was carried out to the DUCTAL® shell, indicating the constitutive relation in tension shown in Figure 18. Similarly to the CONAD plate, the supports of the shell are free to displace from down to bottom. For this reason interface elements were also used to simulate the contact of the shell with the supports. The numerical and experimental load-deflection curves for the point located at the center of the shell are displayed in Figure 19. The curve labeled *Numerical-1* indicates the analysis corresponding to the information contained in this paper. The curve labeled *Numerical-2* indicates a more accurate – as built - analysis for which the imperfections of the shell have been introduced in the FEM model (see details in reference 13). The FEM mesh together with the numerical cracking pattern at failure is shown in Figure 20, whereas the experimental cracking pattern at failure is shown in Figure 21. The results shown in Figure 19 to Figure 21 display the accuracy of the analysis.

7. CONCLUDING REMARKS

This paper presented the FEM analysis of a plates and shells made of Ultra High-Performance Fiber Reinforced Concretes, which were tested in the laboratory and analyzed by a FE procedure. The results indicated the accuracy of the procedures here presented, displaying the high-performance of the materials, which are ready to be used for challenging applications in civil engineering.

The software used for performing the analysis presented in this paper is a FEM code commercially available for the design engineers. In this way, the present research confirmed other studies reported in the technical literature, which have shown that the FEM can be a useful tool for the design of UHPFRC structures if accurate cracking models and constitutive relations are provided.

Acknowledgments

The authors gratefully acknowledge the support of part of this work by the Brazilian Research Council CNPq and by the Brazilian National Agency for Electrical Energy (ANEEL). We also acknowledge Paul Acker from Lafarge for the supply of DUCTAL®

8. REFERENCES

1. Chuang, E. Y., Ulm, F.-J., "Two-phase composite model for high performance cementitious composites", ASCE J. Engng. Mech. **128**, 1314–1323 (2002).
2. Rossi, P., Une modélisation numérique de la fissuration des structures en béton fibré, Bull. Lab. Ponts et Chaussées, **216**, 41-48 (1998).
3. de Larrard, F., Concrete mixture proportioning: a scientific approach, E&FN SPON, London (1999).
4. Sedran, T., Rhéologie et rhéométrie des bétons. application aux bétons autonivelants, PhD Thesis of Ecole Nationale des Ponts et Chaussées (1999).
5. Formagini, S., Scientific Mix design and mechanical characterization of ultra high performance fiber reinforced concrete, PhD Thesis, COPPE/UFRJ, Rio de Janeiro, Brazil (2005), in Portuguese.
6. Acker, P., Ultra-high performance concretes – properties and applications, New concrete products, Lafarge Group, Paris. (1999).
7. Behloul, M., Durukal, A., Batoz, J.F., Chanvillard, Gilles., "Ductal®: Ultra High-Performance Concrete Technology Hith Ductility", In: Proceedings of the Sixth Int. RILEM Symp. on FRC - BEFIB 2004, v. 2, pp 1281-1290, Varenna, (2004).
8. Orange, G., Dugat, J., Acker, P., "Ductal® New Ultra-High Performance Concretes – Damage Resistance and Micromechanical Analysis", In: Proceedings of the Fifth

International RILEM Symposium on Fibre-Reinforced Concretes (FRC)-BEFIB 2000, pp. 781–800, Lyon, France (2000).
9. Lafarge, http://www.ductal-lafarge.com/ (2005).
10. Diana, 'User's Manual', TNO Build and Construction Research, Netherlands, Lakerveld b.v., (2003).
11. Fairbairn, E.M. R., Toledo-Filho, R.D., Formagini, S., Rosa, J.I., Battista, R.C., "Experimental analysis and modeling of ultra high performance fiber reinforced concrete plates", in Int. RILEM Workshop on HPFRCC in Struct. Appl., Honolulu, (2005).
12. Formagini, S., Toledo-Filho, R.D., Fairbairn, E.M. R., 'Mix design and mechanical caracterization of an ultra high performance fiber reinforced cement composites (UHPFRCC)', in Int. RILEM Workshop on HPFRCC in Struct. Appl., Honolulu, (2005).
13. Rosa, J.I, Numerical and experimental modeling of prisms, plates and shells made of high-strength fiber composites, M.Sc. Thesis, COPPE/UFRJ, Rio de Janeiro, Brazil (2005), in Portuguese.

FRC AND HPFRC COMPOSITES: FROM CONSTITUTIVE BEHAVIOUR TO STRUCTURAL APPLICATIONS

M. di Prisco and M. Colombo
Department of Structural Engineering, Politecnico di Milano, P.za Leonardo da Vinci 32, 20133 Milano, Italy

Abstract: Four applications are considered in the paper: SFRC roof elements, sheltering structures, ground slabs and GFRC light faade panels. The common point for all these structures is the use of structural redundancy in order to guarantee ductility in the structural behaviour, although the constitutive behaviour owing to the reduced amount of fibers is always characterized by softening in uniaxial tension. Three different sources of redundancy are discussed: the heterogeneity of the mechanical characteristics, the volume size and the ability of the structure to redistribute stresses on the basis of its shape and boundary conditions. The main purpose is to debate the use of the characteristic material strength values to compute the bearing capacity of structures characterized by a significant statical redundancy.

Keywords: Steel fibres, glass fibres, uniaxial tension, bending, strength scattering, characteristic strength, structural ductility

1. INTRODUCTION

Fibre reinforced concrete is characterized by a residual tensile strength after the cracking of the matrix, when it is subjected to uniaxial tension. The residual strength depends on the fibre amount, the fibre geometry, the fibre orientation, the fibre dispersion, the cementitious matrix mix design and the concrete placement. The complex interaction between all these parameters suggests the designer to regard the composite material as a unique material whose mechanical features have to be identified by means of suitable tensile tests. Such tests are usually direct tension or bending tests and are carried out on unnotched or notched specimens that have to reproduce the real conditions of the analysed structure.

M.S. Konsta-Gdoutos, (ed.), Measuring, Monitoring and Modeling Concrete Properties, 59–68.

In the concrete structure market the fibre amount, usually regarded as economically acceptable, prevents the reaching of a hardening behaviour in uniaxial tension and therefore the residual strength is always a reduced percentage of the peak strength (in the range of 25-50%) that correspond to the onset of the unstable propagation in the matrix. This mechanical feature causes localization in uniaxial tension and a brittle response when the unstable crack propagates in the cementitious matrix. The residual strength is mainly associated to the fibre pull-out and this mechanism represents the main source of energy dissipation in the progressive crack opening.

The residual strength value is strongly affected by both the topological and the time-dependence variability and the large scattering measured in the softening branch causes a strongly reduced characteristic value, when the computation is carried out according to the usual relation:

$$f_k = f_m - ks \quad (1)$$

where k is a factor depending on the specimen number and on the interpolation curve adopted in the statistical approach and s is the standard deviation.

Recent results carried out on square ground slabs made of steel fibre reinforced concrete centrally loaded (Fig.1) have clearly shown as a significant scattering on the constitutive laws, experimentally observed in the bending tests (Fig.1a), does not correspond to a similar scattering in the structural response (Fig.1b). It is important to underline that the square plates were reinforced only by steel fibers, but the variability of the material response is strongly smoothed by the statical redundancy degree of the structure (di Prisco[1]). The main cause of this surprising result has to be searched in essentially two reasons.

The first reason is related to the choice of the ultimate displacement in the structure. In fact a vertical deflection of a square slab with a 3m long side causes a maximum crack opening displacement in the radial cracks significantly smaller than that set as the ultimate limit reference in bending tests. The ratio between the two values is controlled by the crack number that is in its turn affected by the ratio between the Limit Of Proportionality (LOP) and the ultimate loads.

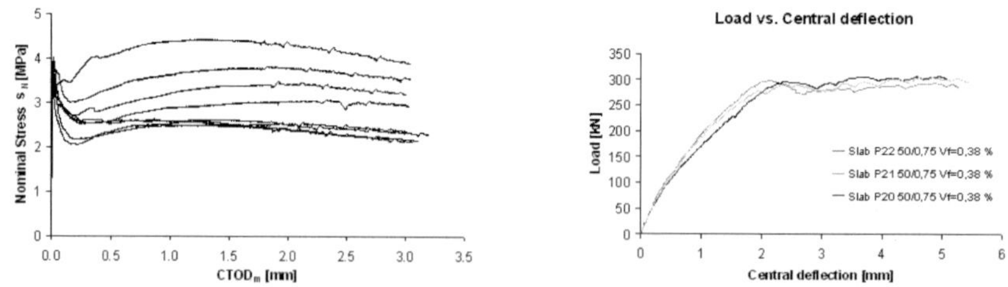

(a) (b)

Figure 1. SFRC square ground slabs centrally loaded: four point bending tests on notched beam (a) and load vs. deflection measured in the structural test (b) (di Prisco et al[1])

Further on, the crack opening distribution in the structure regions interested by crack propagation is not the same of the bent beam and therefore the integral of the residual stresses cannot be the same measured in the specimen used for the material characterization. The second reason is mainly related to the topological fibre distribution in the structure[2]. When the peak load significantly exceeds the load threshold that causes the onset of the cracking process, each representative volume of the structure[3] gives its contribution and the total load is unavoidably redistributed between each volume as occurs when several springs work in a parallel system. Of course this effect depends on the volume of the structure regions involved in the failure resistant mechanism, but the final result is to induce a total bearing capacity that is controlled by the average strength rather than the characteristic one. With reference to the latter reason, some national provisions introduced particular tests which are characterized by a high statical indetermination degree and are strictly related to the specific application: the square plate used in the shotcrete technology and the centrally – loaded round panel introduced in the ASTM C1550-2003[4] as a standard test method to measure the flexural toughness of fiber-reinforced concrete are an example[4,5].

In this framework, the present contribution would like to highlight some basic aspects with reference to uniaxial tension and bending tests, which are sometimes hidden in the identification procedure that is often simplified as much as possible in order to generate a robust design procedure.

2. STRUCTURAL APPLICATIONS

The new code provisions presented in Italy[6], allow the designer to conceive a FRC structure that is not forced to respect the specific ductility requirement related to the rectangular cross section, although a ductile behaviour has to be guaranteed in terms of the total structural response. This constraint is imposed in terms of a minimal ratio between the ultimate and the cracking load, or in terms of a minimum ratio between the maximum deflection at the ultimate state and the deflection associated to the onset of cracking.

A first example that can be used to highlight the effect of statical indetermination on the total response of a simple structure is shown in Fig. 2. In this case although the uniaxial tension tests reveal a marked softening behaviour (Fig. 2a), the bending response of the notched beams (Fig. 2b) progressively increases its ductility with the growth of the fiber content, even if the notch prevents a multilocalization process also for the hardening behaviour shown by the largest fiber content. The same material was used also to carry out some small slabs tested both in an external perimeter simply supported set up (Fig. 2c) and rested on two different continuous supports made of rubber or sand (Fig. 2d). Also in this case, the comparison between the plain concrete specimens and the fibre reinforced ones emphasize the fibre contribution in terms of both peak load and toughness.

It is obvious that every approach that computes the total bearing capacity of the structure on the basis of a limit analysis approach could largely underestimate the real bearing capacity of the structure. In fact the bearing capacity is computed by multiplying a coefficient calibrated on equilibrium by a specific characteristic strength extracted by uniaxial tension tests. Thus operating, the local dispersion associated to the strength at a selected crack opening value, assumed as a reference for the ultimate limit state of the material, could directly affect the resistant load capacity of the structure. Similar considerations can be extended to the particular sheltering structure designed to stabilize a morenic slope shown in Fig. 3[8]. Also in this case, the redundancy degree offered by the 4 ground anchors interacting with the framed structure, post tensioned in both the mean plane directions, increases not only the bearing capacity of the FRC frame, but also guarantees an adequate safety to the construction that makes use of permanent anchors.

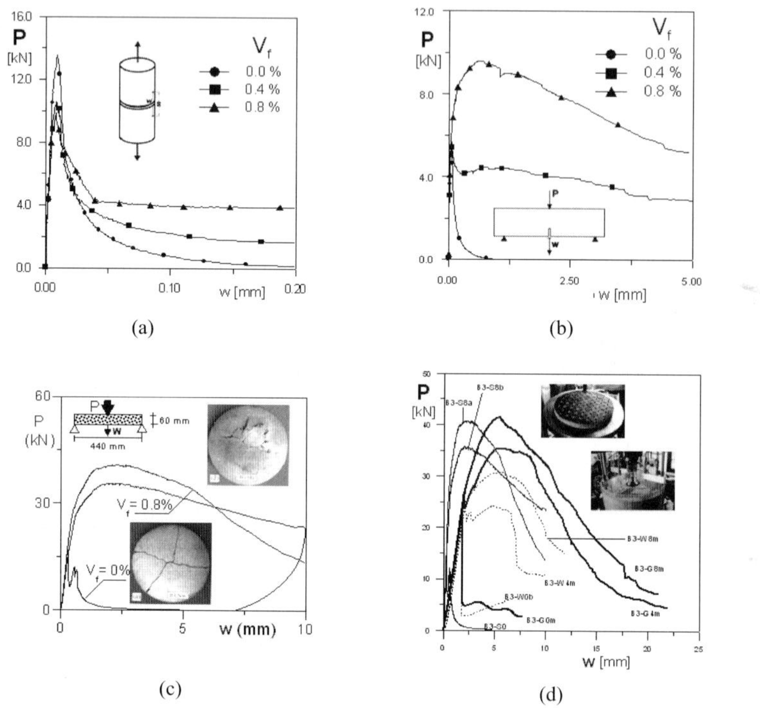

Figure 2. Influence of statical indetermination on global ductility by using SFRC material (hooked-end; l_f=30mm; d_f=0.6): (a) uniaxial tension tests (b) bending tests; (c) slabs simply supported on the external perimeter (d) slabs resting on continuous support made of rubber or sand [W – Winkler support/rubber; G – ground support/sand; S – simply supported on the external perimeter]. (di Prisco et al.[7])

Passing from fiber reinforced concrete to high performance fibre reinforced cementitious composites, an interesting application regards the covering roof elements used in industrial buildings. These structures are usually designed by assuming the longitudinal bending assured by the prestressed reinforcement as the critical resistant mechanism. A wide experimental investigation supported by an Italian Company aimed at the substitution of the traditional transversal reinforcement, consisting in a welded steel fabric, with hooked-end steel fibres was recently carried out.

The conventional reinforcement in the ends was maintained to prevent any collapse in the D-regions. The main problem in this application is to ascertain that the transversal bending does not anticipate the longitudinal bending collapse, because the reduced thickness of the wings in the current cross section can introduce a significant interaction between the longitudinal and the transversal bending, that causing a brittle failure of the structure (Figs. 4a,b).

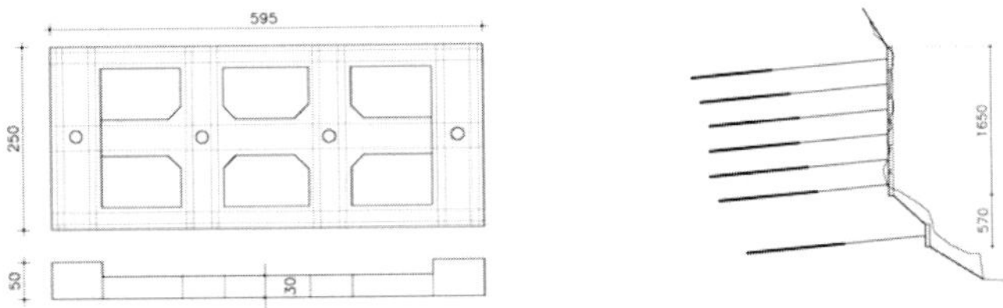

Figure 3. Sheltering structure designed in SFRC post-tensioned precast elements

Figure 4. SFRC roof elements: (a,b) failure in full-scale tests (c) View of ribs added along the wings

The introduction of suitable transversal ribs along the wings in the zone where the large longitudinal curvature causes a significant increase of the transversal bending moment, can prevent the anticipated failure of the wing, guaranteeing the same bearing capacity of the conventional reinforced structure (Fig. 4c). It is worth noting that in this application, a realistic estimation of the transversal bearing capacity, able to take into account the statical indetermination introduced by the ribs, can allow the designer an optimized thickness of the cross section profile.

3. FROM UNIAXIAL TENSION TO BENDING

The scientific community [2] has accepted both the uniaxial tension and the four or three point bending tests to characterize the tensile behaviour of FRC materials. Previous experimental investigations have shown as uniaxial tension tests are always more conservative than bending tests [9]. In order to compare the two approaches, three notched beam specimens made of SFRC were tested according to UNI specifications [10] (Fig. 5a). The nominal strength vs. the CTOD is shown in Fig. 5d. From the same beam specimens, three cylindrical specimens were cored (Fig. 5a). The cylindrical specimens were also notched with a notch/diameter ratio equal to 0.2. The surface of the critical section in the bending test was about 3.5 times that of the uniaxial tension test. The uniaxial tension test was fixed-end platens and the minimum and maximum response curves are presented in terms of nominal strength vs. crack opening displacement measured on a gauge length of 50 mm (Fig. 5b). By adopting the experimental curves obtained in uniaxial tension as constitutive relationships in direct tension, it is possible to reproduce the bending test of the notched test by using the Plane Section model.[11,12] The computed peak load is higher than the experimental value because the notch effect are neglected by the PS model and the fixed-end platens test usually overestimates the uniaxial tensile peak strength. With reference to the scattering between the maximum and the minimum structural response, it is possible to observe that the absolute values are similar but correspond to a reduced percentage of the reference strength, because the nominal bending strength is higher. This result shows how the redistribution of the stresses guaranteed by the planarity of the cross section favours a reduction of the variation coefficient (standard deviation/mean value). Further on the reproduced curves fit reasonably well the experimental ones even though for large crack opening displacements they underestimate the experimental strengths.

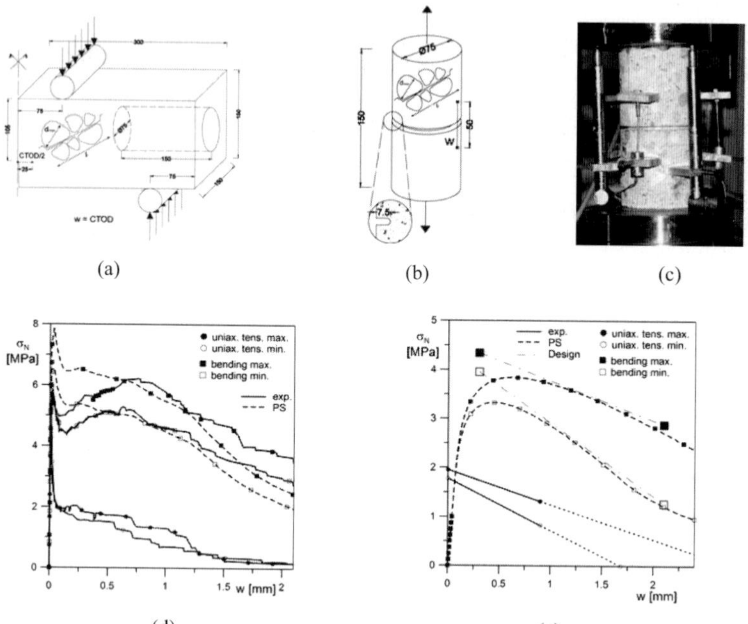

Figure 5. SFRC: (a) four point bending tests set-up; (b,c) uniaxial tensile tests set-up; (d) uniaxial test and 4PB test experimental results and PS simulations; (e) uniaxial tensile constitutive law for design, numerical simulations and design predictions

If uniaxial tension tests are used to identify a linear model relationship stress-crack opening displacement for both the maximum and the minimum tension test response, it is possible to compute the difference |max-min| related to the linear constitutive laws identified in uniaxial tension and the related difference between the corresponding nominal strengths obtained in bending by using the PS approach. The identification procedure adopted to obtain the linear constitutive relationships in uniaxial tension takes as a reference for the COD the serviceability threshold w = 0.15 mm and the ultimate threshold w= 0.9 mm that correspond to the half values chosen in UNI Provisions for bending, and consider an integration range of ± 20% with respect to the specified values.

The particular choice of the ultimate range chosen for the identification produces the effect to diverge the linear softening curves describing the identified maximum and minimum constitutive laws. This occurrence is amplified by bending curves obtained by both the PS approach and the Design approach based on simplified equilibrium assumptions.

The latter approach considers an elastoplastic behaviour in uniaxial tension with an elastic behaviour in compression at the serviceability conditions (w = 0.3 mm) and a rigid linear softening behaviour in uniaxial tension at the Ultimate Limit State, thus implying a compression force concentrated at the extrados of the rectangular cross section. The example shows how it is possible to significantly enlarge the standard deviation of the identified linear relationships if the ultimate range of integration is too reduced or even the single value related to a fixed crack opening is considered. Similar comments can be formulated if a mortar specimen reinforced with glass fibres is analysed (Fig. 6). This kind of material is used for façade panels which are mainly loaded by the wind. Also in this case by assuming uniaxial tension tests as constitutive laws, it is possible to reproduce bending behaviour. The material is generally used for thin layers. It is worth noting that by assuming the characteristic length equal to the cross section depth, the ultimate value of the crack opening can significantly change: if the thickness ranges between 8 and 50 mm, the associated ultimate crack opening $w_d = 0.02\ l_{cs}$ oscillates between 0.16 and 1 mm in bending and between 0.08 and 0.5 mm in uniaxial tension ($w_d = 0.01\ l_{cs}$). A strong variation of the thickness can produce huge approximations when the Design approach is used for small crack opening values (Fig. 6b), even if the scattering (max-min) for Design and PS approach remains of the same order.

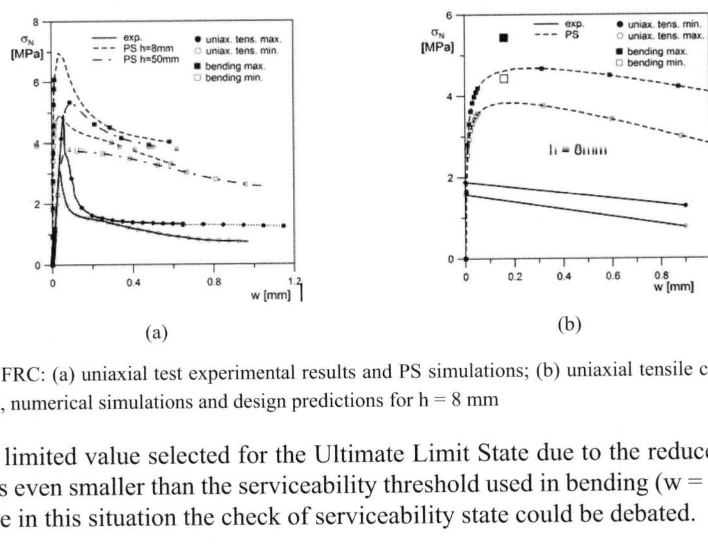

Figure 6. FRC: (a) uniaxial test experimental results and PS simulations; (b) uniaxial tensile constitutive law for design, numerical simulations and design predictions for h = 8 mm

The limited value selected for the Ultimate Limit State due to the reduced thickness becomes even smaller than the serviceability threshold used in bending (w = 0.3mm) and therefore in this situation the check of serviceability state could be debated.

4. FROM BENDING TO UNIAXIAL TENSION

According to CNR–DT 204 [6], it is possible to identify the characteristic uniaxial tension constitutive law from bending tests carried out on notched specimens (UNI 11039[10]), by using a coefficient k set equal to 1.48. The linear relation uses the equivalent nominal strengths $\sigma_{0\text{-}0.6}$ and $\sigma_{0.6\text{-}3.0}$ computed on the indicated crack opening ranges. With reference to the sheltering structure material, the uniaxial tension curves, so identified, are used to reproduce the same bending tests: the result of this procedure shows a reliable fitting (Fig. 7b). It is interesting to underline as the use of the characteristic ductility indexes ($D_0=f_{IF}/\sigma_{0-0.6}$ and $D_1=\sigma_{0.6-3.0}/\sigma_{0-0.6}$) computed on the basis of each single curve rises to characteristic bending equivalent strengths much smaller that those computed by using directly the equivalent strengths (Fig. 7a). Once again the variation coefficient related to the uniaxial tensile strengths computed on the identified linear relationships results larger than that related to nominal strengths computed in bending. By analysing a reinforced beam made of the same material (Fig. 8a), the increase of the statical redundancy in the cross section, due to the reinforcement, implies a more reduced distance between the maximum and the minimum curves as clearly shown in Fig. 8b. In this case, the structural characteristic length becomes the crack distance and is computed according to the relation introduced in the RILEM Recommendations[13], taking into account the reduction due to fibre addition, because the aspect ratio chosen ($l_f/d_f=75$) is lower than 50.

Similar comments can be drawn by substituting the bonded steel reinforcement with post-tensioned reinforcement characterized by the same mechanical ratio, as it occurs in the designed sheltering structure. The lack of a continuous bonded reinforcement gives rise to the problem to determinate the characteristic structural length l_{cs}. It could be expected that the crack distance could decrease with the increase of the axial force introduced by the post-tension. In Fig. 8b two bound values for l_{cs} are investigated: the cross section depth and the fibre length. The large gap between the curves highlights the need to investigate this case by changing the post-tensioning force.

5. CONCLUDING REMARKS

In this research contribution it has been shown that a high standard deviation of SFRC tensile strengths measured in the post-cracking uniaxial tension curves does not correspond to a high standard deviation in the bearing capacity of statically not determined structures. The main reasons must be searched in the strong confinement that induces many cracks relatively close rather than few cracks very opened and in the activation of a sort of parallel system of each representative volume that keeps the full size structural behaviour close to the mean strength value rather than to the characteristic one.

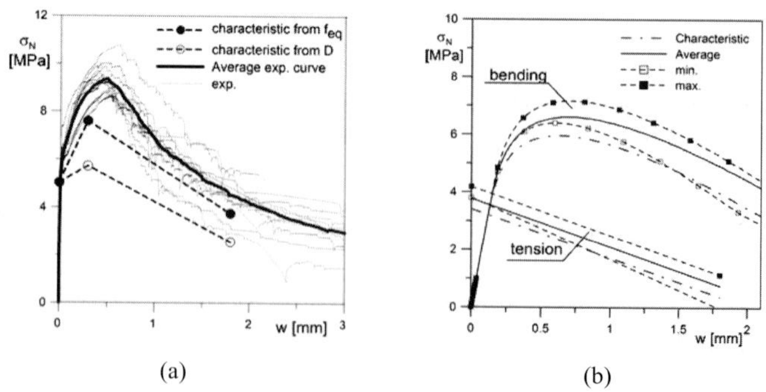

Figure 7. SFRC: (a) four point bending experimental results. (b) Uniaxial tension constitutive law and PS simulations

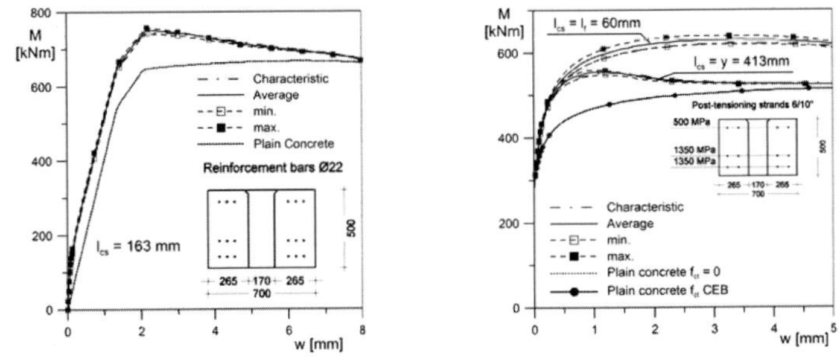

(a) (b)

Figure 8. SFRC: (a) R/C cross section and PS simulations; (b) Post-tensioned cross section and PS simulations

When the constitutive law in tension is identified from uniaxial tension test, the scattering usually is larger than that directly obtainable from bending tests (notched or unnotched). The redistribution that takes place over the cross section is able to guarantee a more limited scattering in relation to uniaxial tension tests. Moreover, by choosing single values corresponding to certain thresholds of the crack opening displacement, it is possible to underestimate the real mechanical behaviour: an integration procedure performed over fixed intervals of the crack opening displacement can assure a more reliable fitting of the experimental behaviour.

In order to guarantee a good ductility in a structure that fails in bending, the ultimate crack opening has to be related to an acceptable limit for the relative rotation in the hinge. When no particular prescriptions are required, a value of 2% can be proposed. In this case, when very thin FRC mortar layers are considered, the ultimate crack opening becomes very small due to the reduced element depth, and the use of simplified equilibrium conditions, which introduce a rigid-linear constitutive model in tension and a compressive force concentrated in the extrados fibre, can introduce large approximations in the computation of the ultimate bending moment.

The computation of the characteristic curves in the post-cracking behaviour has to be performed by using directly the strengths associated to the selected crack opening ranges and the passage through the toughness indexes (UNI 11039[10]) must be discouraged, because it keeps the characteristic curve on the safe side, reducing the toughness contribution of fibres too much. The increase of the redundancy degree obtained in the bending of a r/c beam by the introduction of conventional reinforcement reduces the standard deviation of the cross section response curves which result less affected by the standard deviation measured in the uniaxial tension laws experimentally identified. The same effect should be shown by the introduction of post-tensioned reinforcement when it is unbounded: the right evaluation of the crack distance in this case can significantly change the total response of the member cross section in bending.

Acknowledgments

The research has been carried out with the financial support "PRIN 2004" given by the Italian Ministry of Instruction, University and Research to the project "Fibre Reinforced Concrete for Durable and Economical Structures and Infrastructures."

6. REFERENCES

1. di Prisco, M., Failla, C., Plizzari, G.A., Toniolo, G. Italian guidelines on sfrc, *Fiber-reinforced concrete: from theory to practice*, Ahmad, S., di Prisco, M., Meyer, C., Plizzari, G.A., Shah, S., ed., International Workshop on Advances in Fiber Reinforced Concrete, Bergamo, Italy, September 24-25, 39-72 (2004).
2. Ferrara, L., Meda, A., Lamperti, T., Pasini, F., Connecting fibre distribution, workability and mechanical properties of SFRC: an industrial application to precast elements, *Fibre reinforced concretes*, M. di Prisco, R. Felicetti, G.A. Plizzari, ed., BAGNEUX Rilem Publications S.a.r.l. (FRANCE),Proceeding of the International Conference, 101-110, (2004).
3. Ahmad, S., di Prisco, M., Meyer, C., Plizzari, G.A., Shah, S. (Eds.), Fiber-reinforced concrete: from theory to practice, *International Workshop on Advances in Fiber Reinforced Concrete*, (Bergamo, Italy, September 24-25, 2004).
4. ASTM C1550, Standard test methods for flexural toughness of fiber-reinforced concrete (using centrally – loaded round panel), (2003).
5. Dupont, D., Vandewalle, L., Comparison between the round plate test and the RILEM 3-point bending test, *Fibre reinforced concretes*, M. di Prisco, R. Felicetti, G.A. Plizzari, ed., BAGNEUX Rilem Publications S.a.r.l. (FRANCE), Proceeding of the International Conference, 101-110, (2004).
6. CNR-DT 204, Instruction for design, execution and control of fibre reinforced concrete structures (in Italian), (2006).
7. di Prisco, M., Felicetti, R., Iorio, F., Bending behaviour of HPC thin-web elements, (in Italian), in *The fracture mechanics in HP concrete*, M. di Prisco and G. Plizzari, ed., Starrylink, Brescia, 157-182 (2003).
8. di Prisco, C., di Prisco, M., Mauri M., Scola, M., A New Design for Stabilizing Ground Slopes, *FIB conference*, Naple, (in printing), (2006).
9. di Prisco, M., Felicetti, R., Lamperti M. & Menotti, G, On size effect in tension of SFRC thin plates, *Fracture Mechanics of Concrete Structures*, V.C. Li C.K.Y. Leung, K.J. Willam, S.L. Billington, ed., vol.2, B.L.Schmick and A.D.Pollington (USA), 1075-1082, (2004).
10. UNI 11039, Concrete reinforced with steel fibres. Part II: test method for the determination of first cracking strength and ductility indexes (In Italian), (2004).
11. Hordijk, D. Local approach to fatigue of concrete, *Ph.D.Thesis*, Delft University of Technology, 1-207, (1991).
12. Kooiman A., Modelling steel fibre reinforced concrete for structural design, *Ph..D. Thesis*, Technical University Delft, Optima Grafische Communicatie, Rotterdam, (2000).
13. RILEM TC 162-TDF, Test and design methods for steel fibre reinforced concrete. Design of steel fibre reinforced concrete using the σ-w method: principles and applications, *Materials and Structures*, 35, June, 262-278, (2002).

TAILORED COMPOSITE UHPFRC-CONCRETE STRUCTURES

E. Denarié and E. Brühwiler
Division of Maintenance and Safety (MCS), Ecole Polytechnique Fédérale de Lausanne (EPFL), CH-1015 Lausanne

Abstract: The extremely low permeability of Ultra High Performance Fiber Reinforced Concretes (UHPFRC) associated to their outstanding mechanical properties make them especially suitable to locally "harden" reinforced concrete structures in critical zones subjected to an aggressive environment and to significant mechanical stresses. UHPFRC materials can be applied on new structures, or on existing ones for rehabilitation, as thin watertight overlays in replacement of waterproofing membranes, as reinforcement layers combined with reinforcement bars, or as prefabricated elements such as curbs. Tailored composite UHPFRC-concrete structures promise a long-term durability which helps to avoid multiple interventions during the service life of structures. With this aim in view, an extensive research and development program was conducted to demonstrate the applicability of UHPFRC for the improvement of structures, including laboratory tests as well as full-scale pilot tests of application on site. This paper reports on two main aspects of the project: (1) the experiences gathered during first full scale tests of applications on a bridge, in Switzerland, (2) the underlying more fundamental aspects related to the potential of UHPFRC in the tensile hardening domain and the sound characterization of this important mechanical property.

Key words: UHPFRC, cracking, restraint, eigenstresses, composite construction, tensile hardening.

1. INTRODUCTION

The premature deterioration of reinforced concrete structures is a heavy burden for society. In order to manage structures effectively and to reduce this burden to the minimum, the number and extent of interventions have to be kept to the lowest possible level. The outstanding mechanical and protective properties of Ultra High Performance Fiber Reinforced Concretes (UHPFRC) make them especially suitable to locally "harden" reinforced concrete structures in critical zones subjected to an aggressive environment and to

M.S. Konsta-Gdoutos, (ed.), Measuring, Monitoring and Modeling Concrete Properties, 69–75.

significant mechanical stresses, Brühwiler et al. (2004, 2005), Denarié (2005), Habel (2004). The unique properties of these materials applied in tailored composite structures enable one to (1) decrease the time spent for the rehabilitation works, (2) increase the durability to an extent that will make the rehabilitated structure fulfill all requirements of functionality, serviceability and resistance, for the planned service life, (3) increase the load carrying capacity with compact and efficient geometries.

This paper gives an overview of the concepts and describes a first application of UHPFRC for the rehabilitation of a reinforced concrete structure, with a special emphasis on the characterization of the tensile strain hardening behavior.

2. UHPFRC MATERIALS

UHPFRC are characterized by an ultra-compact matrix with an extremely low permeability, Roux et al. (1995) and by a high tensile strength (above 10 MPa). They are reinforced by a multilevel cocktail of steel fibres. With an optimized fibrous reinforcement, the tensile strain-hardening of these materials can reach the range of the yield strain of construction steel (up to 0.2 %). The very low water/binder ratio of UHPFRC (0.130 to 0.160) prevents the complete hydration of a major part of the cement (around 70 %) and gives the material a significant hydrophilic behaviour, and a self healing capacity for microcracks, Charron et al, (2004), Parant (2003). In the fresh state, despite their very low water/binder ratio, UHPFRC can be tailored to be self-compacting and at the same time tolerate slopes up to 3 %. In the context of these works, the tensile strain hardening UHPFRC family CEMTEC$_{multiscale}$®, developed at LCPC, Rossi et al. (2002), Boulay et al. (2003) was used and optimized for rehabilitation applications at MCS-EPFL, Habel (2004), Denarié (2005), SAMARIS D25b (2006).

3. COMPOSITE UHPFRC-CONCRETE STRUCTURAL MEMBERS

The concept of application of UHPFRC for the rehabilitation of structural members is schematically illustrated in Figure 1, Brühwiler et al. (2004), (2005a, 2005b). An "everlasting winter coat" is applied on the bridge superstructure only were it is needed, in zones of severe environmental and mechanical loads. Critical steps of the construction process such as application of waterproofing membranes or compaction by vibration can be prevented, and the associated sources of errors avoided. The construction process becomes then simpler, quicker, and more robust, with an optimal use of composite construction.

A comprehensive series of tests in the laboratory on composite UHPFRC-concrete structural members have successfully validated this concept for various geometries, and boundary conditions, with various degrees of restraint, with or without reinforcement bars in the UHPFRC layer, Habel (2004), SAMARIS D18a (2005), SAMARIS D18b (2005).

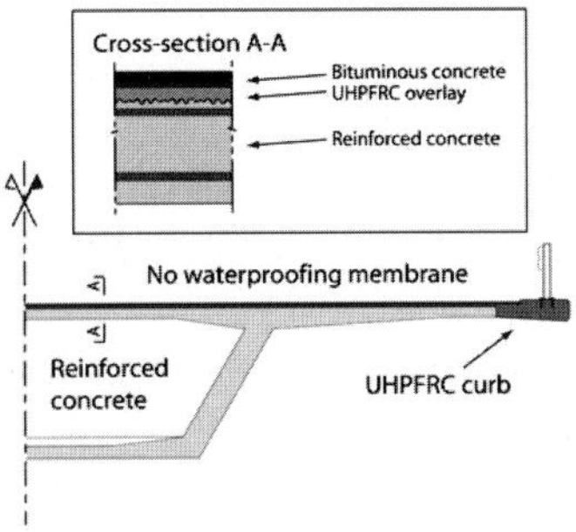

Figure 1. Concept of application of the local "hardening" of *bridge s*uperstructures with UHPFRC

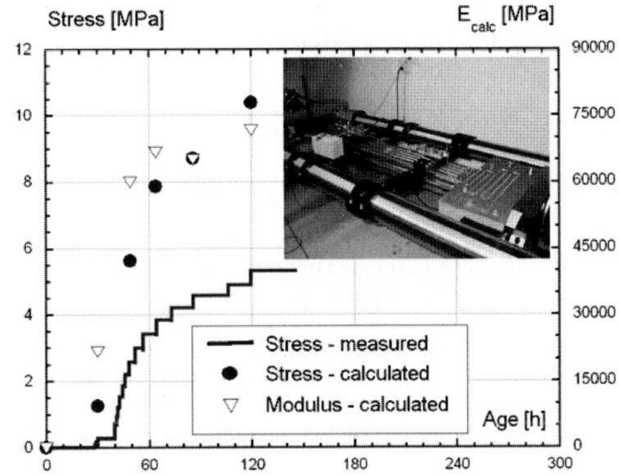

Figure 2. Stresses under incremental full restraint and TSTM test device, after Kamen et al. (2005)

UHPFRC materials provide at the same time a very low intrinsic permeability, Charron et al. (2004), and an overall deformation capacity sufficient to avoid cracking under restrained shrinkage deformations, Kamen et al. (2005), Habel (2004).

The dominant source of deformations in UHPFRC is autogenous shrinkage, with maximum values of 600 μm/m at long term, Habel (2004), Kamen et al. (2006), which is comparable to the values obtained for usual concretes subjected to drying. Incremental restrained shrinkage tests on UHPFRC specimens at early age show that the development of eigenstresses under full restrained shrinkage, remain moderate (45 % of the tensile first crack strength) with respect to the uniaxial tensile characteristics of the UHPFRC tested, Kamen et al. (2005), as shown on Figure 2.

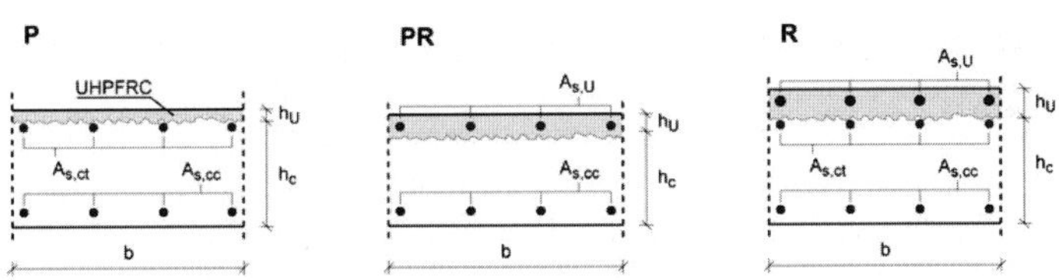

Figure 3. Three types of cross sections for improved composite bridge deck slab, Habel (2004), Brühwiler et al. (2005)

This new construction technique is specially well-suited for bridges but can also be implemented for galleries, tunnels, retaining walls, following the same concept. When it is required, the combination of the protective properties and deformation capability of UHPFRC with the mechanical performance of reinforcement bars (normal or high grade) provides a simple and efficient way of increasing the stiffness and load-carrying capacity with compact cross sections, Habel (2004), Brühwiler et al. (2004). A summary of possible geometries for the protection (configurations P and PR) or reinforcement (R) of existing structures is given on Figure 3.

Strain hardening UHPFRC appear to be an excellent compromise of density, high tensile strength, and significant deformation capability, perfectly suited for combination with normal concretes, in existing or new structures.

4. FIRST APPLICATION

With the support of the road administration of the Swiss canton Wallis, and under the guidance of MCS-EPFL, the bridge over the river la Morge, nearby Sion, has been rehabilitated and widened in an unusual way by using Ultra High Performance Fiber Reinforced Concretes (UHPFRC). It was indeed the very first time that UHPFRC of the $CEMTEC_{multiscale}$® family, were cast in-situ, for the rehabilitation of a bridge. The entire surface of the bridge with a span of 10 m was improved in three steps during the autumn 2004, Figure 4.

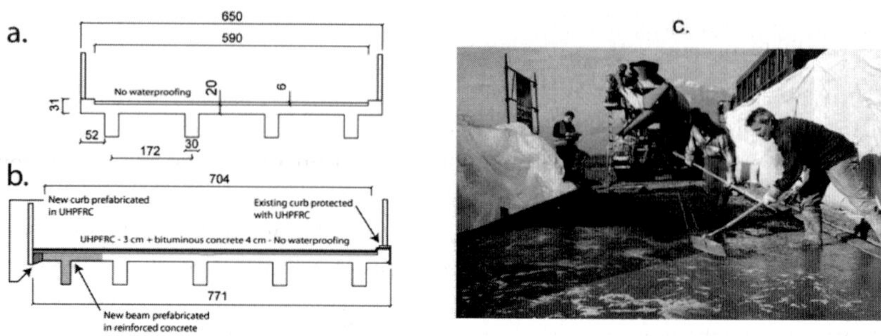

Figure 4. Cross section of the bridge, a, before, and b., after, the rehabilitation (dimensions in cm), and c. pouring of the UHPFRC (Photo A. Herzog)

Two different recipes of $CEMTEC_{multiscale}$® were used, with similar components (Cement CEM I 52.5, Microsilica, fine sand D_{max}=0.5 mm), with a Microsilica/Cement ratio of 0.26. The reinforcement of the ultra compact matrices was provided by a mix of micro (steel wool) and macrofibers (l_f=10 mm, aspect ratio: 50) with a total dosage of 706 kg/m^3 (9% vol.). Recipe CM22, more liquid, had 1410 kg/m^3 cement, and a Water/Binder ratio of 0.131. Recipe CM23, was designed with 1434 kg/m^3 cement, and a lower Water/Binder ratio of 0.125 to guarantee a tolerance to a slope of the substrate of 2.5 %.

Firstly, the downstream curb was replaced by a new prefabricated UHPFRC CM23 curb on a new reinforced concrete beam. Secondly, the chloride contaminated concrete of the upper surface of the bridge deck was replaced by 3 cm of UHPFRC CM23, on October 22 for the first lane and November 5, for the second lane. Finally, the concrete surface of the upstream curb was replaced with 3 cm of UHPFRC CM22 on November 9. Air permeability tests realized on site, on the rehabilitated deck of the bridge, delivered, as expected, outstanding protective properties: air permeability kT=0.004*10^{-16} m^2 on average for UHPFRC, compared to 0.010 – 0.10x10^{-16} m^2 for good concretes and > 1x10^{-16} m^2 for bad concretes.

Thanks to the extremely low permeability of the UHPFRC, no water-proofing membrane is needed, and the thickness of the bituminous concrete can be reduced to a minimum. This full scale realization in realistic site conditions clearly demonstrates that the technology of UHPFRC is now mature for cast in-situ applications of rehabilitation, using standard equipments.

5. TENSILE BEHAVIOUR OF UHPFRC

The strain hardening behavior is one of the most appealing features of UHPFRC such as CEMTEC$_{\text{multiscale}}$®. This mechanical property has however to be characterized in a sound way and a clear link has to be found between the response of laboratory specimens and structural applications.

With these aims in view, an instrumented uniaxial tensile test was developed to determine in perfectly rigid conditions the stress-strain response and the displacement field over a prism of constant cross section of 350 mm length, in the central part of an unnotched dog bone shaped specimen (l= 700 mm, minimum cross section: 50 x 100 mm). The test procedure was applied at 28 days to UHPFRC specimens cast-on site during the first application of UHPFRC with the material CM23. The results delivered, as expected, remarkable average properties as illustrated on Figure 5a: maximum tensile strength of 13.7 MPa and a maximum strain in the strain-hardening domain of 1.5 ‰ .

The surface of the specimens was coated with lime to highlight fine cracks. Multiple cracking was visually observed on the specimens, in the tensile hardening domain, with a spacing of 5 to 7 cm. Further, the readings of the Omega gauges distributed over the length of the specimen confirmed the distributed character of cracking in the strain hardening domain. Localization of fracture in one single band could be observed only after the maximum force (after point 2, on Figure 5a), as shown on Figure 5b.

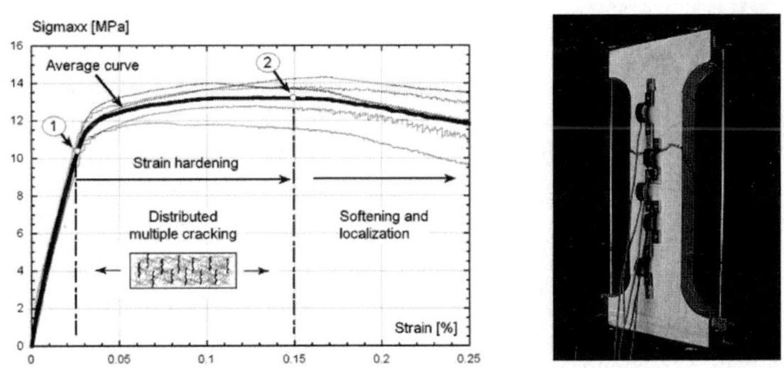

a. b.

Figure 5. Uniaxial tensile tests at 28 days, material CM23, cast on site; a. 5 individual tests results and average curve, b. specimen and instrumentation, after end of test

6. CONCLUSIONS

- A new concept of structural rehabilitations with Ultra High Performance Fiber Reinforced Concretes is proposed to increase the durability of structures and their mechanical performance, and to decrease the number of interventions during their service life.
- This concept has been validated by numerous laboratory tests with configurations corresponding to various practical applications.
- A first application of this concept has been successfully realized and the required properties of the UHPFRC were achieved with standard equipments, and verified in-situ.
- This first application opens the way to tailored composite UHPFRC-concrete structures with improved durability and resistance.
- The uniaxial tensile behavior of UHPFRC specimens cast on site has been characterized in the laboratory. A significant strain hardening response (average of 0.15 %) could be observed on all 5 specimens, with distributed damage and multiple cracks in the hardening domain.

Acknowledgements

This project was financially supported by the Swiss Secretary of State for Education and Research (SER) in the context of the European project "Sustainable and Advanced Materials for Road Infrastructures" (SAMARIS), and by the Swiss National Science Foundation. The pilot application has been made possible thanks to the support of the road administration of the Swiss Canton Valais.

7. REFERENCES

Brühwiler, E., Denarié, E., Habel, K., 2005, Ultra-High Performance Fiber Reinforced Concrete for advanced rehabilitation of bridges, *Proceedings Fib symposium "Keep Concrete Attractive"*, Budapest, Hungary, 23 to 25 may 2005, Eds G.L. Balasz & A. Borosnyoi, pp. 951-956.

Brühwiler, E., Habel, K., Denarié, E., 2004, Advanced reinforced concrete for the improvement of bridges, *Proceedings Second International Conference on Bridge Maintenance, Safety and Management (IABMAS'04)*, Kyoto, Japan.

Charron J.-P., Denarié E., Brühwiler E., 2004, Permeability of UHPFRC under high stresses, *Proceedings RILEM Symposium, Advances in Concrete Through Science and Engineering*, March 22-24, CD-ROM, Chicago, USA.

Denarié E., 2005, Structural rehabilitations with Ultra-High Performance Fiber Reinforced Concretes (UHPFRC), Keynote lecture, *Proceedings International Conference on Concrete Repair, Rehabilitation and Retrofitting – ICCRRR 2005*, 21-23, November 2005, Cape Town, South Africa,

Habel K., 2004, Structural Behaviour of Elements Combining Ultra-High Performance Fiber Reinforced Concretes (UHPFRC) and Reinforced Concrete, *Doctoral thesis n° 3036, Swiss Federal Institute of Technology, Lausanne*, Switzerland.

Kamen A., Denarié E., Brühwiler E., 2005, Mechanical Behaviour of Ultra High Performance Fiber Reinforced Concretes (UHPFRC) at early age, and under restraint, *Proceedings*

CONCREEP 7, September 12-14, 2005 – Nantes, France, Eds. G. Pijaudier-Cabot, B. Gérard, P. Acker, Hermès Publishing, pp. 591-596.

Parant, E., 2003, Mécanismes d'endommagement et comportements mécaniques d'un composite cimentaire fibré multi-échelles sous sollicitations sévères: fatigue, choc, corrosion. *Ph.D. thesis of Ecole nationale des Ponts et Chaussées*, Paris (in French).

Rossi P., 2002, Development of new cement composite material for construction, Innovations and Developments in Concrete Materials and Construction, *Proceedings International Conference, University of Dundee, Edited by R. K. Dhir, P. C. Hewlett and L. J. Csetenyi, Dundee,* Scotland, pp 17-29.

Roux, N., Andrade, C., Sanjuan, M.A., 1995, Etude Expérimentale sur la durabilité des Bétons de Poudres Réactives, *Annales de l'Institut Technique du Bâtiment et des Travaux Publics (ITBTP), Les Bétons de Poudres Réactives (BPR) à Ultra Haute Résistance (200 à 800 MPa)*, 532, Série Béton 320, pp. 133-141 (in French).

SAMARIS D18a, 2005, Report on laboratory testing of UHPFRC, part a, *European project 5th FWP / SAMARIS – Sustainable and Advanced MAterials for Road Infrastructures – WP 14: HPFRCC*, http://samaris.zag.si/.

SAMARIS D18b, 2006, Report on laboratory testing of UHPFRC, part b., *European project 5th FWP / SAMA-RIS – Sustainable and Advanced MAterials for Road Infrastructures – WP 14: HPFRCC*, http://samaris.zag.si/.

SAMARIS D22, 2005, Full scale application of UHPFRC for the rehabilitation of bridges – from the lab to the field, *European project 5th FWP / SAMARIS – Sustainable and Advanced MAterials for Road Infrastructures – WP 14: HPFRCC*, http://samaris.zag.si/.

SAMARIS D25b, 2006, Guidelines for the use of UHPFRC for rehabilitation of concrete highway structures, *European project 5th FWP / SAMARIS – Sustainable and Advanced MAterials for Road Infrastructures – WP 14: HPFRCC*, http://samaris.zag.si/.

HYBRID FIBER REINFORCED CONCRETE

L. Vandewalle
K.U.Leuven, Department of Civil Engineering, Kasteelpark Arenberg 40, 3001 Heverlee, Belgium

Abstract: This paper describes the mechanical performance of hybrid fiber concrete with a maximum fiber dosage of 0.75 Vol.%. Two types of fibers were used: short straight steel fibers and long hooked-end steel fibers. The concrete compressive strength ($f_{c,cub150}$) varied between 55 and 65 MPa, dependent on the fiber mixture. Several dosages of the two types of fibers were examined. Three-point bending tests[1] using notched prisms were performed. The obtained results confirm that the flexural strength at small crack widths and the ductility of the hybrid fiber concrete with rather low fiber content is remarkably improved in comparison with plain concrete. Moreover, the postcracking behaviour of the hybrid fiber concrete is not only dependent on the total fiber content but also on the fiber mixture.

Key words: hybrid mixture; steel fibers; crack width; ductility; bending test.

INTRODUCTION

Durability and ductility of concrete structures are very important issues; especially because concrete is a brittle material with a low tensile strength and strain capacity. Both "D's" can be achieved by using steel fibers[2-5].

However, to have a durable construction, small crack widths (0.2 to 0.3 mm) are required in the serviceability limit state. Short straight steel fibers can serve as bridging mechanisms during cracking. As a consequence, the use of short steel fibers, especially those with a high aspect ratio, is beneficial for that. Ductility, on the other hand, refers to large deformations, i.e. a good bridging effect of the fibers at large crack widths is necessary. To perform this task, long deformed fibers are more obvious.

So, to achieve both durability and ductility, the application of multiple fiber types, i.e. a mixture of short and long fibers, is logical. The application of different types of fibers together in a concrete mixture, was proposed for the first time by Rossi[6], as the so-called multi-modal fiber reinforced concrete.

This paper describes the mechanical performance of hybrid fiber concrete with a maximum fiber dosage of 0.75 Vol.%. Two types of fibers were used: short straight steel fibers and long hooked-end steel fibers. The concrete compressive strength ($f_{c,cub150}$)

M.S. Konsta-Gdoutos, (ed.), Measuring, Monitoring and Modeling Concrete Properties, 77–82.

varied between 55 and 65 MPa, dependent on the fiber mixture. Several dosages of the two types of fibers were examined. Three-point bending tests[1] using notched prisms were performed. The obtained results confirm that the flexural strength at small crack widths and the ductility of the hybrid fiber concrete with rather low fiber content is remarkably improved in comparison with plain concrete. Moreover, the postcracking behaviour of the hybrid fiber concrete is not only dependent on the total fiber content but also on the fiber mixture.

2. RESEARCH PROGRAM

At the department of Civil Engineering of the K.U.Leuven, a test program was executed that involved RILEM 3-point bending tests[1] to measure the postcracking behaviour. The test set-up is shown in Figure 1.

The bending test is executed by means of CMOD (crack mouth opening displacement) control, i.e. the machine shall be operated in such a manner that the CMOD increases at a constant rate of 50 μm/min for CMOD from 0 to 0.1 mm, until the end of the test at a constant rate of 0.2 mm/min.

The compressive strength is measured on cubes with side = 150 mm. Two types of fibers are applied, i.e. one short straight steel fiber (S-fiber) with a length of 13 mm, a diameter of 0.16 mm respectively and one long hooked-end steel fiber (L-fiber) with a length of 35 mm, diameter of 0.55 mm respectively. The total fiber content ranges from 0 (reference mix) to 60 kg/m^3. Eight mixtures in total are tested as shown in Table 1.

The concrete composition is identical for all mixtures (see Table 2). Only the dosage of superplasticizer changed since the application of fibers has an impact on the workability.

3. TEST RESULTS

3.1 Compressive strength

The mean value of the cube compressive strength is for the different series given in Table 1. The addition of steel fibers results in a higher compressive strength. Moreover, from the results, it can be concluded that short steel fibers provide a higher improvement of $f_{cm,cube}$ in comparison with the long hooked-end fibers.

3.2 Postcracking behaviour

3.2.1 Scatter

Each series contains 6 specimens. The load-CMOD-diagrams of mix L00K60 are shown in Figure 2, of mix L60K00 and L30K30 in Figures 3 and 4 respectively. The detailed results of the other series are given elsewhere[7].

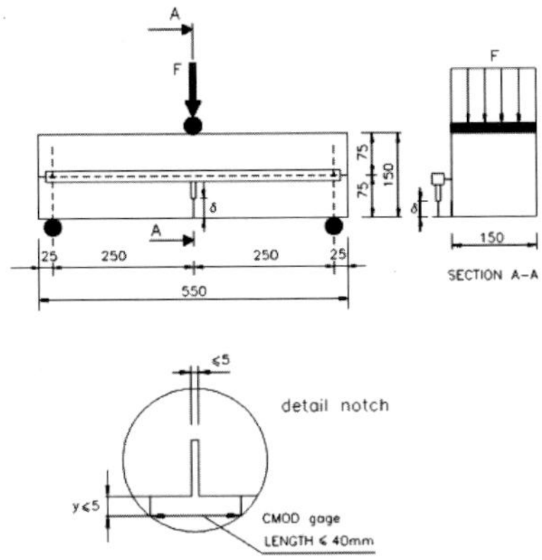

Figure 1. Test set-up for 3-point bending test[1]

Table 1. Overview of research program.

Series	S-fiber kg/m³	L-fiber kg/m³	$F_{cm,cube}$ MPa
L00K00	-	-	54.5
L00K30	30	-	62.8
L00K60	60	-	66.9
L30K00		30	55.9
L60K00	-	60	57.2
L30K30	30	30	65.2
L20K40	40	20	67.2
L40K20	20	40	61.6

Table 2. Concrete composition

	kg/m³
Gravel 4/16	1012
Sand 0/5	865
Cement CEM I 52.5 N	350
Water	175
W/C	0.5

From this limited number of series it can be concluded that the application of short fibers provides a much lower scatter of the results in comparison with the use of long fibers. This is logical since the number of fibers in one kg is much higher for short fibers than for long fibers. This is particularly important for test specimens with a relatively small cross section. Fiber counts[8] have shown that the postcracking behaviour was directly related to the number of fibers intersecting the fracture surface. A small variation or difference in number of fibers has a direct and relatively large influence on the toughness of the materials tested. This phenomenon would be more pronounced in specimens which contain a lower absolute number of fibers.

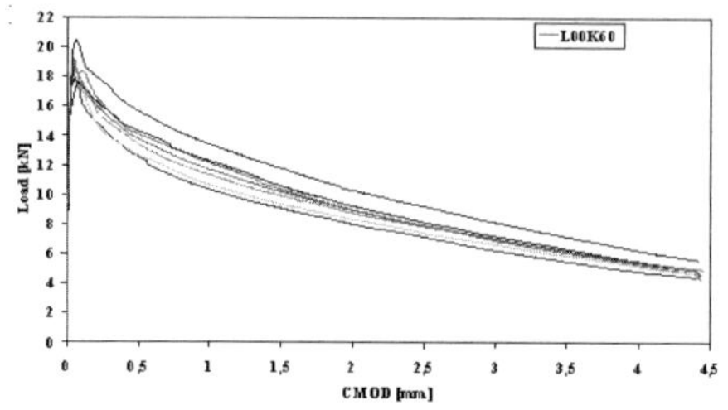

Figure 2. Load-CMOD-diagrams for L00K60

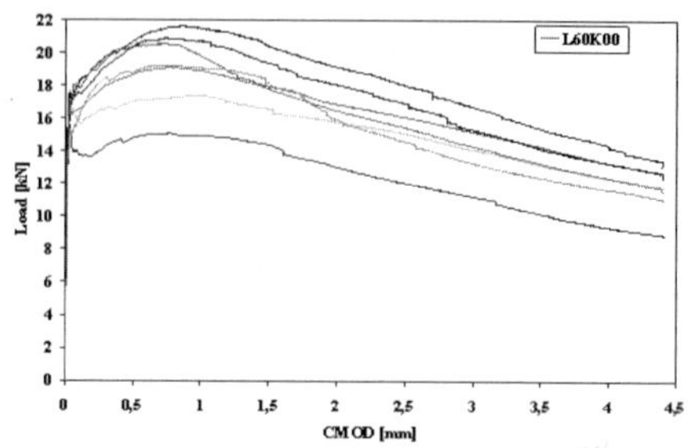

Figure 3. Load-CMOD-diagrams for L60K00

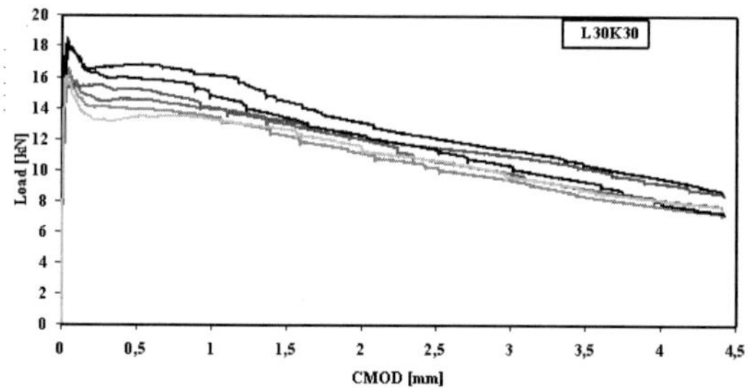

Figure 4. Load-CMOD-diagrams for L30K30

3.2.2 Tensile strength –ductility

The mean load-CMOD-diagram for series L30K00 and L00K30 is shown in Figure 5 and for series L60K00 and L00K60 in Figure 6 respectively. From both figures, it follows that the tensile strength of mixes L30K00, L00K30 and L60K00 is almost equal to each other. However, the tensile strength of L00K60 is some 10 % higher. The series with the short fibers show a slightly better postcracking behaviour than the mixes with the long fibers up to a CMOD-value of 0.15 mm. Short fibers can bridge microcracks more efficiently because they are very thin and their number in concrete is much higher than that of the long thick fibers for the same fiber volume quantity.

Taking into account that microcrack formation and crack bridging by short fibers occur in the first part of tensile loading, the short fibers can have an influence on the post-cracking behaviour in the region of small crack widths. However, for larger crack widths the ductility of the mixes with the long fibers is much better than that of the corresponding mixes with the short fibers. As the microcracks grow and join into larger macrocracks, the long hooked-end fibers become more and more active in crack bridging.

The origin of the higher residual forces for long hooked-end steel fibers at larger CMOD-values is twofold:

1. The presence of a hooked-end
2. Long embedded length.

Both aspects provide a higher pull-out force for long hooked-end fibers in comparison with short fibers, particularly at larger crack widths. Long fibers can therefore provide a stable post-peak response. Short straight fibers will be less active because they are being pulled out more and more as the crack width increases.

The mean load-CMOD-diagram of all hybrid mixes is shown in Figure 7. The post-cracking behaviour of the mixes in the region of the larger crack widths is consistent with the previous statements : the higher the volume percentage of long hooked-end steel fibers the better is the postcracking performance. The same is not fully true for the region of microcracks (CMOD < 0.05 mm) since the postcracking behaviour of L40K20 in that CMOD-region is better than that of L20K40 although the contrary should be expected: the higher the amount of short fibers in the mix the better the performance of that mix at microcracking.

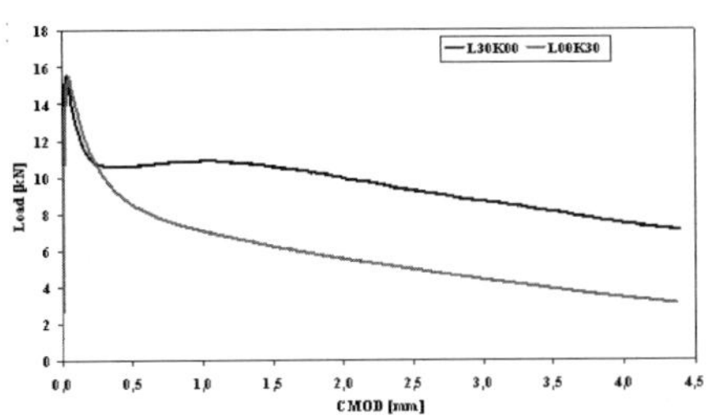

Figure 5. Mean load-CMOD-diagrams for a total fiber content of 30 kg/m^3

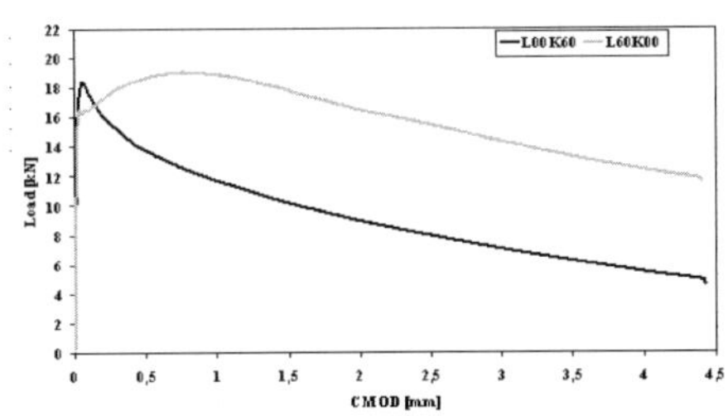

Figure 6. Mean load-CMOD-diagrams for a total fiber content of 60 kg/m^3

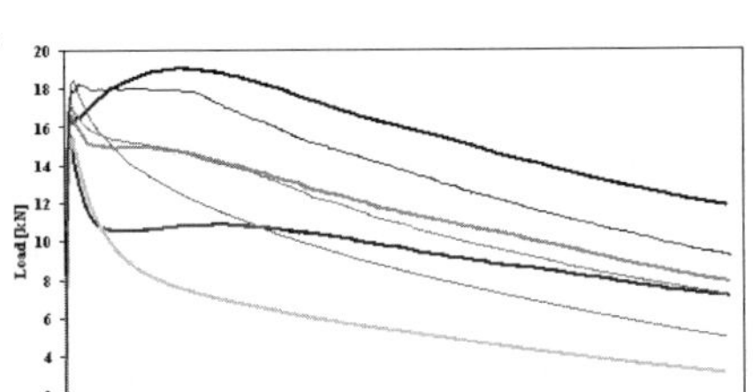

Figure 7. Mean load-CMOD-diagrams of hybrid mixes

4. CONCLUSIONS

From the research program on hybrid fiber concrete executed at the department of Civil Engineering of the K.U. Leuven, it can be concluded that:

1. For the same applied fiber volume percentage much more short fibers than long fibers are present in the cross section of the specimen. This explains the lower scatter on the test

results of hybrid mixes with short fibers and also the better ability of short fibers to bridge the microcracks.

2. Long hooked-end fibers, on the contrary, perform better at larger crack widths due to the presence of the hook and the longer embedded length.

5. REFERENCES

1. Vandewalle L. et al., Recommendation of RILEM TC162-TDF : Test and design methods for steel fiber reinforced concrete : final recommendation for bending test, *Materials and Structures,* Vol.35, pp.579-582.
2. Shah, S.P. and Kuder, K.G., Hybrid and high- performance fiber-reinforced cementitious composites, *Fiber reinforced concrete. From theory to practice,* edited by S.Ahmad, M.di Prisco, C.Meyer, G.A.Plizzari and S.Shah (Starrylink Editrice 2004), pp.83-92.
3. Markovic, I., van Mier, J., Walraven, J.C., Development of high performance hybrid fibre concrete, *Proceedings of the 4th International Workshop on HPFRCC,* edited by Naaman and Reinhardt, Ann Arbor 2003, pp.277-300.
4. Markovic, I., van Mier, J., Walraven, J.C., Tensile behaviour of high performance hybrid fibre concrete, *Proceedings of the 5th International Symposium on Fracture Mechanics of Concrete and Concrete Structures,* Vail Colorado 2004, Vol.2, pp.1113-1121.
5. Vandewalle, L., Dupont, D., Crack formation in SFRC beams containing longitudinal reinforcement, *Proceedings of the 6th CANMET/ACI International Conference on "Durability of Concrete",* Thessaloniki 2003, pp.367-384.
6. Barr, B.I.G., Lee, M.K., de Place Hansen, E.J., Dupont, D., Erdem, E., Schaerlaekens, S., Schnütgen, B., Stang, H. and Vandewalle, L., Round-robin analysis of the RILEM TC162-TDF bending test – Part 3 - Fibre distribution, *Materials and Structures*, Vol.36, pp.631-635.
7. De Smedt, K. and Rolies, K., Onderzoek naar de fysische en mechanische eigenschappen van hybride staalvezelbeton, *Master thesis K.I.H.De Nayer Belgium,* 2005 (in Dutch).
8. Rossi, P., Acker, P., Malier, Y., Effect of steel fibers on two stages: the material and the structure, *Materials and Structures,* Vol.20, pp.436-439.

PREVENTING AUTOGENOUS SHRINKAGE OF HIGH-PERFORMANCE CONCRETE STRUCTURES BY INTERNAL CURING

D. Cusson and T. Hoogeveen
National Research Council Canada, Ottawa, Canada, K1A 0R6

Abstract: The effect of internal curing on the structural behaviour of large high-performance concrete specimens having different amounts of pre-soaked porous lightweight aggregate was investigated. The results show that the use of lightweight aggregate in high-performance concrete can effectively reduce autogenous shrinkage through improved hydration of cement.

Key words: autogenous shrinkage; high-performance concrete; internal curing; restrained shrinkage; self-desiccation; tensile creep.

1. INTRODUCTION

High-performance concrete (HPC) structures made with low water-cement ratio often exhibits early cracking due to self-desiccation, which may lead to a reduction of their service lives. This paper presents the test results on large prismatic HPC specimens under restrained autogenous shrinkage. Pre-soaked lightweight aggregate (LWA), made of porous expanded shale, was used to provide effective internal curing in order to reduce self-dessication and autogenous shrinkage cracking in the HPC specimens.

Testing systems and methods were developed at NRC (Cusson et al., 2005) for studying restrained shrinkage and creep of large HPC specimens, which present some new features over existing approaches (Kovler, 1994; Bjontegaard et al., 1999). Large size specimens enable the study of the behaviour of concrete made with large coarse aggregate and reinforcing bars. The system can impose a partial (or full) degree of restraint through embedded reinforcement, which is representative of the field conditions. An advantage of using a partial degree of restraint is that restrained shrinkage testing can be conducted without failing high-shrinkage concrete specimens.

M.S. Konsta-Gdoutos, (ed.), Measuring, Monitoring and Modeling Concrete Properties, 83–89.

2. EXPERIMENTAL PROGRAM

The control concrete used ASTM Type 1 cement, had a water-cement ratio of 0.34, and a cement-sand-stone ratio of 1:2:2. Normal density aggregates were used. Three variations of this mix design included 6%, 12% and 20% replacements of normal-weight sand by pre-soaked lightweight aggregate (LWA). For each mix design, all large concrete specimens and small concrete samples were prepared from the same batch and sealed with plastic sheets to prevent external drying during testing. After placing the concrete, the forms were not removed during testing to protect the sensors and prevent thermal shock.

Figure 1 presents the main setup for testing free and restrained shrinkage of large prismatic concrete specimens (200 x 200 x 1000 mm). For the restrained specimen, the axial strain was measured with electrical strain gauges (SG) centered on the four 10-mm reinforcing bars. The test apparatus included a closed loop servo-hydraulic system to control the actuator, using the rebar-mounted strain gauges as the feedback signal. The force, measured by a load cell, was transmitted to the concrete by the steel bars, which have their ends welded to the stiff end plates connected to the rigid test frame.

An unrestrained companion specimen was prepared with no reinforcement. Free shrinkage was measured with LVDTs placed at both ends of the specimen. Two relative humidity (RH) sensors were placed in the specimen to measure the extent of self-desiccation. Thermocouples (TC) were embedded in concrete at the locations shown in Fig. 1.

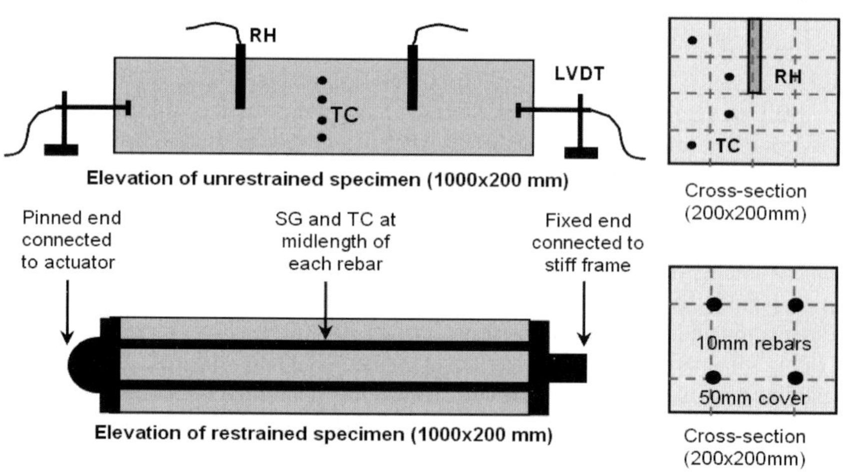

Figure 1. Test setup and dimensions of specimens

Additional tests were also conducted in parallel on smaller concrete samples at different times, including the determination of the thermal expansion coefficient, compressive strength, splitting tensile strength, compressive modulus of elasticity and Poisson ratio. The temperature history was measured for each concrete specimen and sample, and maturity was calculated for each size of specimen in order to use consistent sets of results in the calculations. The theoretical background, as well as the strain and stress calculation procedures are presented elsewhere (Cusson et al., 2005).

3. RESULTS AND ANALYSES

All specimens were tested under realistic temperature regimes. The concrete temperature reached 45°C at approximately 12 hours after casting and returned to 25°C two days after casting, regardless of the amount of LWA used in the concrete. For the control specimen, the RH reduced from 100% at set time to 94% after 2 days and 91% after 7 days. Due to the added internal curing, the 20%-LWA specimen had reduced drying with the RH reducing to 98% after 2 days and 96% after 7 days. The improved hydration also resulted in an increased 7-day compressive strength going from 50 MPa for the control concrete to 57 MPa for the 20%-LWA concrete.

3.1 Free shrinkage

Figure 2 presents the evolution of autogenous shrinkage after removal of the thermal strain from the total strain measured in the unrestrained specimens. Note that the thermal strain curves of the different specimens were very similar. With increased amounts of LWA used in the concrete specimens, autogenous shrinkage reduced considerably, especially for the 12%- and 20%-LWA concrete specimens. The results also show that most of the autogenous shrinkage caused by internal drying developed within one day of setting. This observation suggests that the prevention of excessive self-dessication and autogenous shrinkage cracking in HPC structures should involve techniques that are effective shortly after the setting of concrete.

For each concrete specimen, Figure 2 also indicates the critical autogenous shrinkage strain measured at early age (i.e. between 1 and 2 days) when the rate of shrinkage had reduced considerably. As suggested in the figure, the part of autogenous shrinkage strain that will result in tensile stress should be measured from the peak strain corresponding to the maximum expansion due to swelling, if any. In fact, significant swelling was observed for the 20%-LWA concrete specimen, and was most likely due to the large amount of internal curing water supplied to the cement by the LWA for continuous hydration.

3.2 Restrained shrinkage

In order to avoid failing the control and the 6%-LWA specimens during testing, a degree of restraint close to 0.9 was required. For the 20%-LWA specimen, a degree of restraint close to 1.1 was experimented (i.e. the loading system was actually pulling slightly on the specimen). Figure 3 illustrates the tensile stress/strength ratio for each restrained concrete specimen. The ratio is the actual tensile stress measured in concrete (resulting from thermal, autogenous and creep strains) divided by the actual splitting tensile strength measured on 100x200mm concrete cylinders tested at different times. Each stress/strength curve was also normalised by the actual degree of restraint used in the experiment. The results presented in Fig. 3 can be considered equivalent to those obtained under a constant 100% degree of restraint. This normalisation is required in order to allow a comparison between restrained concrete specimens tested under different degrees of restraint.

The control and 6%-LWA specimens performed similarly with a predicted failure at about 1.5 days had they been tested under a full restraint. It is clear that adding only 6% LWA in the concrete was not enough to prevent failure. The 12%-LWA concrete specimen performed well and reached a maximum stress/strength ratio of 90% after nearly 3 days.

The 20%-LWA specimen experienced compressive stress during the first day (due to thermal expansion and swelling), followed by low tensile stress developing after the cooling period. With a maximum tensile stress/strength ratio of 50% obtained after nearly 3 days, the use of 20% LWA in the concrete proved very effective in reducing the risk of cracking.

Figure 4 compares the tensile modulus of elasticity obtained for each specimen. The secant modulus is a best-fit calculation from stress/strain data measured during partial unloading/reloading conducted periodically on the restrained concrete specimens.

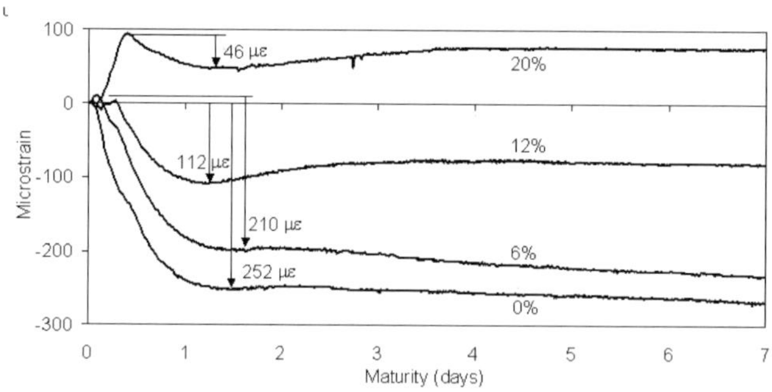

Figure 2. Autogenous shrinkage strain measured in unrestrained specimens

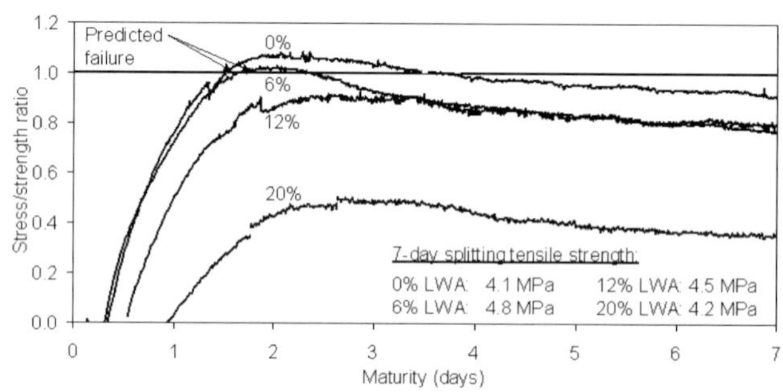

Figure 3. Normalised concrete stress/strength ratio measured in restrained specimens

The effective modulus was obtained from the secant modulus and the time-dependent creep coefficient measured for each specimen (which varied from 1.3 to 2.0 at 7 days). It can be seen that the addition of LWA sand in the concrete had a marginal effect on the effective modulus of elasticity. Some variation in the secant modulus is seen between the different specimens; however, 7-day tests on 100x200mm concrete cylinders for the compressive elastic modulus yielded very similar values such as 31600, 31200, 32000 and 31400 MPa obtained for the 0%-, 6%-, 12%- and 20%-LWA concrete specimens, respectively.

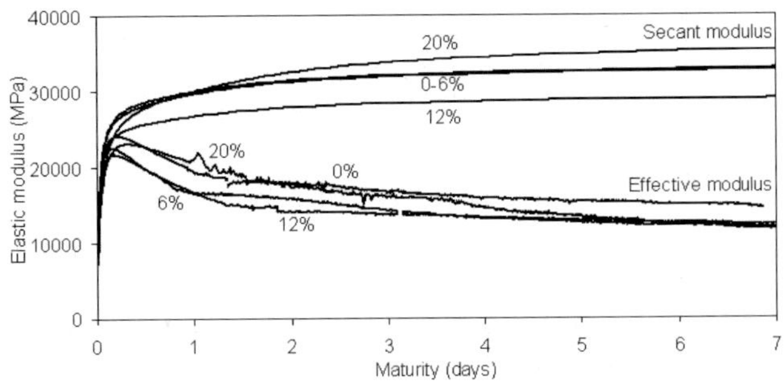

Figure 4. Tensile elastic modulus of concrete measured in restrained specimens

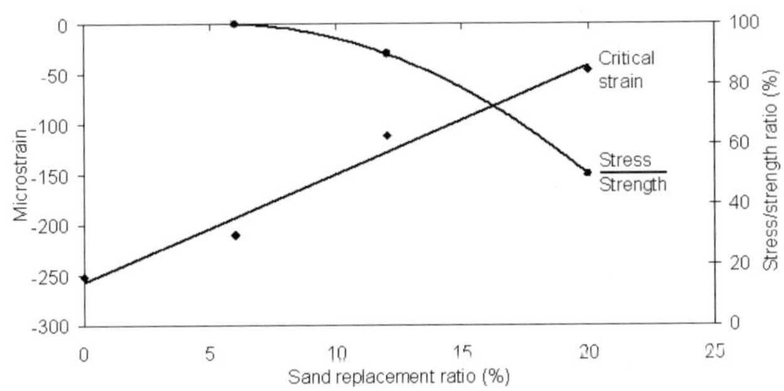

Figure 5. Effect of LWA sand replacement ratio on strain and stress reductions

3.3 Effectiveness of internal curing

Figure 5 illustrates the effectiveness of internal curing on the reduction of both shrinkage and tensile stress in the HPC specimens. As shown in the figure, the critical autogenous shrinkage strain (as defined in Fig. 2) and the stress/strength ratio were reduced considerably with an increased amount of LWA, especially for the 20%-LWA specimen. Both curves indicate that an amount of 25% in this concrete formulation would be required to completely eliminate autogenous shrinkage and tensile stress. An amount higher than 25% is not recommended since excessive swelling might occur.

Note that the water used to pre-soak the LWA was accounted for in the calculation of the water-cement ratio, which was maintained constant for the different concretes. This particular requirement made the evaluation of the internal curing effectiveness more severe than if additional water had been used to pre-soak the LWA.

4. INTERNAL CURING WATER REQUIREMENT

4.1 Theoretical estimation

The amount of internal curing water required to eliminate self-desiccation can be calculated from chemical shrinkage. Knowing the degree of saturation and absorption capacity of the aggregate, one can estimate the required mass of LWA to introduce in concrete in order to provide adequate internal curing. Researchers, such as Zhutovsky et al. (2004) and Bentz et al. (2005), suggested a similar form of the following equation:

$$M_{LWA} = \frac{M_c \cdot CS \cdot \alpha_{\max}}{S_{LWA} \cdot \phi_{LWA} \cdot \eta_{LWA}} \qquad \alpha_{\max} = \frac{w/c}{0.36} \leq 1 \tag{1}$$

where M_{LWA} is the mass of LWA per unit volume of concrete (kg/m^3); Mc is the mass of cement in concrete (kg/m^3); CS is the chemical shrinkage (kg of water per kg of hydrated cement); α_{max} is the maximum expected degree of hydration; S_{LWA} is the saturation degree of LWA; ϕ_{LWA} is the absorption capacity of LWA (kg of water per kg of dry LWA); η_{LWA} is an efficiency factor for the LWA (to be determined in the next section); and w/c is the water-cement ratio of concrete. Note that the equation for α_{max} suggests that the lowest w/c that can achieve complete hydration of the cement is 0.36.

4.2 Experimental validation

In practice, elimination of self-dessication and autogenous shrinkage in HPC structures may require more LWA than calculated with Eq. 1 since the water supplied by the LWA for internal curing may not be completely available at an early age. Factors such as the size of pores in the LWA and the porosity of the cement paste around the aggregate particles can significantly influence the effectiveness of internal curing.

The test results of this study can be used to evaluate the prediction accuracy of Eq. 1, or more specifically the efficiency factor of LWA. For instance, Fig. 5 suggests that the use of 25% LWA of total sand content would enable complete elimination of autogenous shrinkage in the HPC specimens of this study. Equation 1 predicts a sand replacement ratio of 20% based on the following data: cement content of 445 kg/m^3; chemical shrinkage of 0.07 (measured by XRD analysis); max. hydration degree of 0.94 (Eq. 1); saturation degree of 1.0; absorption capacity of 0.17 at 24 hours (measured by the LWA manufacturer), and total sand content of 890 kg/m^3.

For the particular LWA used in the study, the efficiency factor is:

$$\eta_{LWA} = 0.8 \tag{2}$$

Zhutovsky et al. (2004) reported that efficiency factors as high as 80% had previously been observed. It is therefore suggested to use an efficiency factor no greater than 80% unless new experimental data can demonstrate higher efficiency factors from lightweight aggregates.

It should be noted that other types of LWA included in concretes that are different from those used in this study may require different amounts of LWA for effective internal curing. The effects of supplementary cementing materials (such as silica fume, fly ash and slag) on the effectiveness of internal curing are still unknown and are not accounted for in Eq. 1.

5. SUMMARY AND CONCLUSIONS

Restrained shrinkage testing of large high-performance concrete (HPC) specimens was conducted to study the effect of internal curing on the structural performance of concrete specimens made with lightweight aggregate (LWA) sand. The following conclusions are drawn:

- Autogenous shrinkage, if not controlled, can reach very high values within 24 hours, leading to rapid failure under restrained shrinkage.
- Tensile stress and cracking due to self-desiccation in HPC specimens can be completely eliminated through proper internal curing at early age.
- Wet LWA, in the amount of 20% of total sand, provided enough internal curing water to the hydrating cement in order to eliminate autogenous shrinkage and maintain the tensile stress/strength ratio under 50%.
- An optimum amount of 25% LWA in the HPC tested in the study was predicted to completely eliminate tensile stresses due to the simultaneous effects of autogenous, thermal and creep strains.
- The amounts of LWA used in the concrete specimens (up to 20%) did not adversely affect the strength or elastic modulus of concrete.

Acknowledgments

The authors would like to thank Mr. John Roberts of Northeast Solite Corporation for providing the lightweight aggregate, as well as Dr. Lyndon Mitchell and Mr. Glendon Pye of NRC for their technical assistance.

6. REFERENCES

Bentz, D.P., Lura, P., Roberts, J.W., 2005, Mixture proportioning for internal curing, Concrete International, February, 1-6.

Bjontegaard, O., Kanstad, T., Sellevold, E.J., Hammer, T.A., 1999, Stress inducing deformations and mechanical properties of high-performance concrete at very-early-ages, 5th Int. Symposium on Utilization of High-Strength/High-Performance Concrete, Sandefjord, Norway, June 20-24, 1027-1040.

Cusson, D., Hoogeveen, T.J., Mitchell, L.D., 2005, Restrained shrinkage testing of high-performance concrete modified with structural lightweight aggregate, 7th Int. Symposium on Utilization of High-Strength/High Performance Concrete, Washington D.C. USA, June 20-24, ACI SP 228-87, **1**:1335-1372.

Kovler, K., 1994, Testing system for determining the mechanical behaviour of early-age concrete under restrained & free uniaxial shrinkage, Materials & Structures, **27**:324-330.

Zhutovsky, S., Kovler, K., Bentur, A., 2004, Influence of cement paste matrix properties on autogenous curing of high-performance concrete, Cement & Concrete Composites, **26**:499-507.

THERMO-MECHANICAL ANALYSIS OF YOUNG CONCRETE

Application to a Restrained Slab

M. Azenha, R. Faria and J.A. Figueiras

Faculty of Engineering of the University of Porto, Laboratory for the Concrete Technology and Structural Behaviour, Rua Dr. Roberto Frias, s/n, 4200-465 Porto, PORTUGAL

Abstract: For freshly cast concrete structures in presence of strong restraints, the thermal stresses associated with the heat of hydration of cement can lead to cracking, with detrimental consequences from both aesthetical and serviceability points of view. In this paper, a methodology for the thermo-mechanical analysis of concrete structures is presented, along with an application to a highly restrained slab.

Key words: cement hydration; thermo-mechanical analysis; early-age; concrete creep; thermal stresses

1. INTRODUCTION

Thermal cracking of concrete at early ages is a consequence of the deformations that occur due to the cement hydration induced temperature changes. If such deformations are internally or externally restrained, tensile stresses may develop and reach the still developing tensile strength of concrete, culminating in the occurrence of thermal cracking.

It is often mentioned that thermal cracking is more likely to occur when high performance concretes are used or when structures are massive. Yet, it is known that the reasons for cracking in concrete structures due to thermally induced stresses are essentially related to the heat generation capacity of the mix, the volume/shape of concrete involved and the presence of internal or external restraints. So, there may exist restraint conditions under which the risk of thermal cracking is elevated, both for high performance or ordinary concrete, and irrespective to the structure being massive or not.

For numerical analysis of thermal cracking, prediction of the thermal field within concrete is necessary, as well as the correspondent stresses, thus calling for a thermo-mechanical analysis framework. In this paper a methodology for the execution of such task is presented, with the simplifying assumption that the thermal field can affect the mechanical one, but the opposite influence is negligible.[1] Therefore, in the methodology

M.S. Konsta-Gdoutos, (ed.), Measuring, Monitoring and Modeling Concrete Properties, 91–97.

to be presented the thermal field is computed first, and the obtained results are used as an input for the mechanical analysis to be performed subsequently. Allowance is taken to the continuous evolution of the mechanical properties of concrete during cement hydration, as well as to creep.

The application that is presented at the end of this paper concerns to a slab supported on a regular mesh of piles. Because of the strong restraints induced by the piles, this problem is quite suitable for validation and exemplification purposes.

2. THERMAL ANALYSIS

The thermodynamic equilibrium of a domain under thermal transient conditions is expressed by the Fourier's Law

$$k\nabla\cdot(\nabla T) + \dot{Q} = \rho c\,\dot{T} \tag{1}$$

where k is the thermal conductivity of concrete, T is the temperature, $\dot{Q}$ is the rate of internal heat generated by the cement hydration and is the volumetric specific heat. Switching to an adequate format for finite element (FE) implementation, assuming the usual interpolation strategy T=***NTe***, where ***N*** denotes the interpolation matrix and $\boldsymbol{T}^e$ designates the nodal temperatures for a given FE "*e*" with volume Ω_e and boundary Γ_{qe}, and adopting a temporal discretization where $\dot{T}_{n+1} = (T_{n+1} - T_n)/\Delta t$ (n and n+1 relate to two consecutive time steps separated by Δt), the Fourier's Law becomes 2:

$$\frac{1}{\Delta t} C^e \left(T^e_{n+1} - T^e_n\right) + K^e T^e_{n+1} = F^e_T + F^e_Q \tag{2}$$

with

$$C^e = \int_{\Omega_e} N^T \rho c\, N\, d\Omega \qquad K^e = \int_{\Omega_e} \nabla N^T k \nabla N\, d\Omega + \int_{\Gamma_{q_e}} N^T h\, N\, d\Gamma_q \tag{3}$$

$$F^e_T = \int_{\Gamma_{q_e}} N^T h\, T_{env}\, d\Gamma_q \qquad F^e_Q = \int_{\Omega_e} N^T \dot{Q}_{n+1}\, d\Omega \tag{4}$$

Here h is the convection/radiation coefficient in the boundary and T_{env} stands for the environmental temperature. Before any further description of the parameters involved in equation (1), an important concept for the modelling of concrete behaviour at early ages should be discussed: the degree of hydration , which can be defined as the ratio between the quantity of cement that has reacted up to a given instant and the initial amount of cement added to the concrete mix. This scalar quantifies the extent of the hydration reactions of cement, and so it will be used as an indication about the state of formation of cement micro-structure, which in turn influences the evolution of both the thermal and mechanical concrete properties. The direct determination of the degree of hydration calls for sophisticated experimental methods, such as the X-Ray diffraction techniques. Nevertheless, for practical applications the degree of hydration may be obtained indirectly by the ratio between the heat released up to a certain instant and the total heat expected at completion of the cement hydration reaction.[3]

The thermal conductivity k of concrete is known to be strongly dependent on the kind of aggregates used in the mixture, and it usually ranges between 1.2 and 2.5 W/m°C.[4] For practical purposes it is plausible to assume that during the cement hydration this property is constant and equal to its final value (even though it exhibits slightly higher values in the beginning of the process)[5]. In what concerns the specific heat of concrete, ρc, it has a relatively constant value along the cement hydration reaction, exhibiting variations bellow 5% of its final value[6]; therefore, it is feasible to adopt a constant value for this property, matching the hardened concrete specific heat, which typically ranges between 800 and 1170 J/kg°C.

In what regards the internal generation rate of heat of hydration $\dot{Q}$, an Arrhenius formulation is adopted [7]:

$$\dot{Q} = a\, f(\alpha)\, e^{-\frac{E_a}{R\,T}} \tag{5}$$

where E_a is the activation energy (J/mol), R is the universal gas constant (8.314 J/mol K^{-1}), a is the maximum value of the heat production rate (J/s) and $f(\alpha)$ describes the evolution of the normalized heat production rate. Equation (5) can be directly calibrated from an adiabatic experimental test of the concrete mixture, or indirectly through other calorimetric techniques, like semi-adiabatic or isothermal tests.[8,9] Numerical micro-structural models have also been developed with the capability to predict the heat release of hydrating mixes. [3, 10, 11]

3. MECHANICAL ANALYSIS

Due to the strong micro-structural transformations that occur in cement during hydration, the mechanical characteristics of concrete exhibit a significant evolution, which may be expressed as a function of the degree of hydration using equations like[12]:

$$X_i(\alpha) = X_{i1}\left[(\alpha - \alpha_0)/(1 - \alpha_0)\right]^{\eta_i} \tag{6}$$

where X_i is a relevant mechanical property (such as the tensile strength f_{ct} ,the compressive strength f_c or the elasticity modulus E_c), X_{i1} is the hypothetical value of X_i for α=1, α_0 is a threshold degree of hydration under which mechanical properties of concrete are negligible and η_i is a coefficient for property X_i (1 for f_{ct} , 2/3 for f_c, 1/2 for E_c). The thermal dilation and the Poisson's coefficient were considered to be constant during hydration, with the typical values usually considered for hardened concrete, that is, α_T=10×10^6 K^{-1} and υ = 0.2, respectively.

Creep is widely recognized to play an important role in the dissipation of early-age tensile stresses. One of the most widely used functions for the early-age creep is the Double Power Law (DPL) [13]

$$J(t,t') = 1/E_0 + \phi_1/E_0\,(t')^{-m}\,(t - t')^{n} \tag{7}$$

where J is the creep compliance function, E_0 is the asymptotic elastic modulus (corresponding to short term loads), t is time, t' is the instant at loading and ϕ_1, m and n are material parameters. Because of the large stress fluctuations that occur in concrete during early ages, the compliance function is approximated by using a Taylor series expansion, rather than a Dirichlet expansion (see de Borst and van den Boogaard[1] for further details).

Regarding shrinkage, for the structure that will be discussed in this article its overall effect may be considered negligible.

4. Application: restrained slab

The described methodology will be used for analyzing the early-age thermal stresses of a 0.35m thick concrete slab, supported on a 3m×4m rectangular mesh of piles. Concrete with a water-to-cement ratio of 47% and a cement dosage of 285 kg/m^3 was used. The slab covers the ground floor of a 140×41m^2 warehouse, and it was cast without contraction joints (see Figure 1). The construction of the slab was made by casting in an alternate manner 6m wide slab strips. The slab is not really much thick and the cement dosage is relatively small, so a fairly small temperature rise and subsequent drop is to be expected. Yet, because of the large plan dimensions, as well as the existence of the supporting piles, a strong in-plan mechanical restriction takes place, leading to significant thermal stresses, even though the temperature changes might not be very much relevant. Temperature evolution inside concrete was monitored in selected spots 5cm inside the slab, from both the top and bottom surfaces. Measurements of the environmental temperature were also performed. As far as the thermal field within the slab is concerned, the heat flow occurs mainly along the vertical direction (perpendicular to the slab middle plane), and therefore a 1D model could be used in the numerical simulation. Such analysis was carried out until the age of 9.2 days, with time steps of 1h, and adopting the following material properties and relevant data: $k = 2.6$ W/mK; $\rho c = 2400$ J/m^{3}K; h_{top}=4.5 W/m^{2}K; h_{soil} =7.5 W/m^{2}K; $T_{soil,fic}$=17°C; E_α=17 kJ/mol, adiabatic temperature rise $T_{ad} = 14+47(1-e^{-1.5tdays})$. It is worth mentioning that the bottom boundary condition corresponds to a simplified idealization of the soil, which actually has thermal inertia and variable temperature along time. Also, the adiabatic temperature rise of the mix was obtained through the use of the micro-structural hydration model by Maekawa et al.[11]

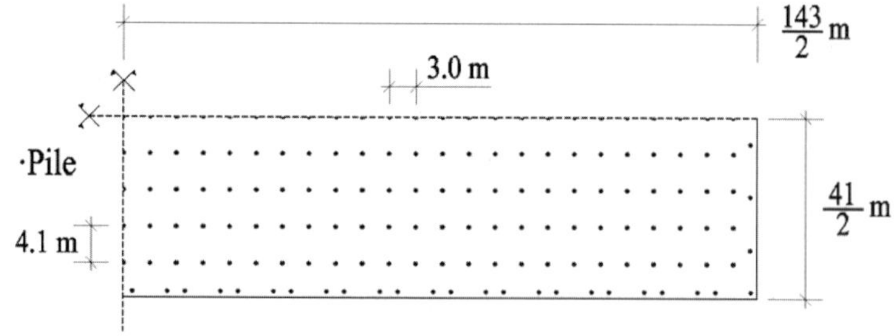

Figure 1. Structural plan

The coherence between the numerical simulation and the field temperature measurements is quite satisfactory, as depicted in Figure 2. From the obtained results it is also remarked that the maximum temperature gradient within the slab is of about 4°C, and consequently the degree of hydration across the slab thickness is quite homogenous.

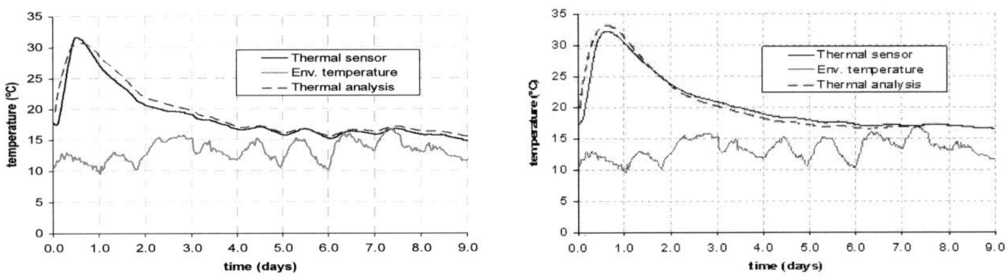

Figure 2. Temperatures: 5cm from the top of the slab (left); 5cm from the bottom (right)

This 1D thermal analysis was complemented with a 2D mechanical one, to investigate the implications of the adopted sequence of construction of the slab strips. The 2D model simulates the central plane of the slab, where the evolutions of the temperature and hydration degree at each point (idealized as representative of all points along the corresponding normal) were assumed to coincide with the ones obtained at the middle plane in the 1D model. The geometry of this 2D idealization is depicted in Figure 3 (axes P1-P3, P3-P8 and P6-P8 are symmetries), and it aims to reproduce a 30m long and 6m wide hardening concrete strip (assumed to be representative of the real 137m long strips), completely surrounded by hardened concrete of previously cast strips (reproduced by the broken line P1-P2-P5-P4-P8-P6). Friction between soil and concrete was considered to be negligible, and the piles were assumed to prevent all the displacements on the points of connection to the slab. On the interface between the hardening and the hardened concrete perfect bond was assumed. The mechanical properties adopted for this simulation were the following: $E = 29 \times 10^9 \alpha^{0.5}$ Pa, $\varphi_1 = 2.26$, m=0.35, and n=0.30.

The mechanical analysis indicates that the maximum tensile stress takes place along the X direction on a considerable length of the hardening slab, as it can be observed in Figure 4, reaching about 2.5MPa. Bearing in mind that the used concrete class was C30/37, the average tensile strength should be 2.9MPa at the age of 28 days, which according to the above reported normal stress computed for the slab leads to the conclusion that transverse cracking should be expected. The actual slab really cracked according to the predictions of the thermo-mechanical analysis (see Azenha[14]), confirming the plausibility of the numerical results. After the above described combination of simplified models, a full 3D thermo-mechanical analysis was also performed with the FE mesh reproduced in Figure 5a. The thermal and mechanical properties, as well as boundary conditions, were adopted as to be consistent with the ones used for the previous analyses. The results obtained from the 3D model in terms of the thermal field were quite similar to the ones obtained in the 1D thermal simulation, except for the edge areas of the slab (as expected). The normal stresses at the middle plane of the slab predicted by the 3D model matched quite well the previously mentioned ones for the 2D idealization, as it can be noted in Figure 5b, where the stress evolution for point P3 is depicted for both models.

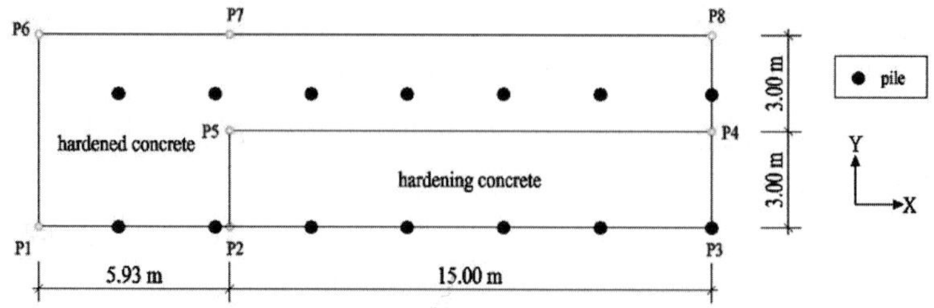

Figure 3. Schematic representation of the 2D model

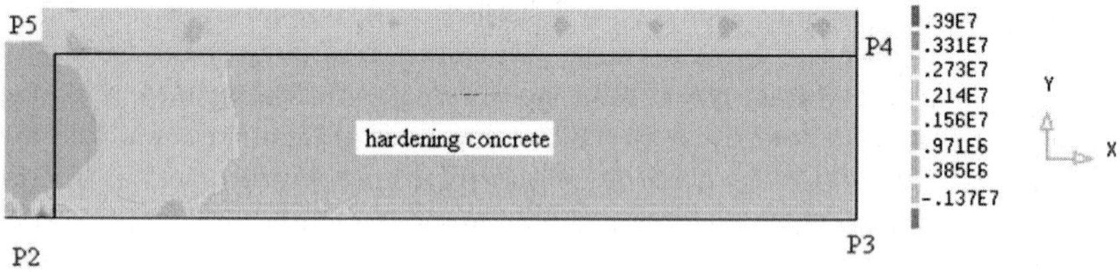

Figure 4. Normal stresses at t=9.2 days (Pa)

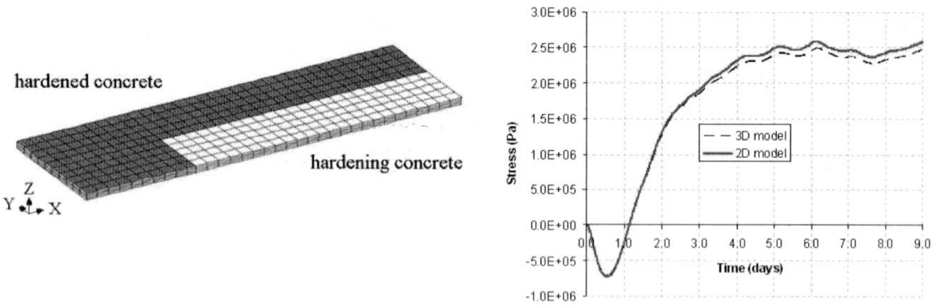

Figure 5. a) 3D FE mesh; b) Evolution of normal stresses at P3 (2D v.s.3D)

5. CONCLUSIONS

A thermo-mechanical methodology for predicting thermal stresses in concrete at early ages was presented in this paper. The exothermal tendency of the cement hydration reactions was accounted for, and the finite element method was used for computing both the thermal and the stress fields. In the mechanical submodel evolution of the concrete properties during hydration, as well as the creep phenomenon that plays a decisive role in the stress distribution, were taken into consideration.

An application concerning a strongly restrained slab was presented, where good coherence was obtained between the measured and calculated temperatures. Also, the orientation of the cracking pattern observed in-situ matched the one predicted by the performed numerical analyses. The validity of combining a 1D through-thickness thermal analysis with a 2D mechanical analysis was confirmed, in view of both the experimental results and a more complex 3D analysis.

The overall results of this research point to a good accuracy of the described numerical methodology in what regards to the prediction of thermal stresses, and therefore allowing proper decisions to be made before construction in order to avoid thermal cracking.

Acknowledgements

Financial support from the Portuguese Foundation for Science and Technology, through a PhD grant provided to the first author (SFRH/BD/13137/2003) and a research project (POCTI/ECM/56458/2004), is gratefully acknowledged.

6. REFERENCES

1. de Borst, R. and van den Boogaard, A. Finite-element modeling of deformation and cracking in early-age concrete. J. Eng Mech Div., ASCE, 1994;120(12): 2519-2534.
2. Cervera, M., Faria, F., Oliver, J. and Prato, T. Numerical modelling of concrete curing, regarding hydration and temperature phenomena. Computers and Structures, 2002;80(18-19): 1511-1521.
3. Breugel, K. Simulation of hydration and formation of structure in hardening cement-based materials. Doctoral Thesis, Delft, 1991.
4. Breugel, K. Prediction of temperature development in hardening concrete. In: Prevention of thermal cracking in concrete at early ages. Report 15, R. Springenschmid, E&FN SPON, 1998.
5. Ruiz, J., Schindler, A., Rasmussen, R., Kim, P. and Chang, G. Concrete temperature modeling and strength prediction using maturity concepts in the FHWA HIPERPAV software, 7th International Conference on Concrete Pavements, Orlando, USA, 2001.
6. De Schutter, G. Thermal properties. In: Early age cracking in cementitious systems. Report 25, A. Bentur, RILEM Publications s.a.r.l., 2001.
7. Reinhardt, H., Blaauwendraad, J. and Jongedijk, J. Temperature development in concrete structures taking account of state dependent properties, Int. Conf. Concrete at Early Ages, Paris, France, 1982.
8. Morabito, P. Methods to determine the heat of hydration of concrete. In: Prevention of thermal cracking in concrete at early ages. Report 15, R. Springenschmid, E&FN SPON, 1998.
9. Wadsö, L. An experimental comparison between isothermal calorimetry, semi-adiabatic calorimetry and solution calorimetry for the study of cement hydration., Nordtest report TR 522, 2003.
10. Bentz, D. Three-dimensional computer simulation of Portland cement hydration and microstructure development. J. Am. Ceram. Soc., 1997;80(1).
11. Maekawa, K., Chaube, R. and Kishi, T. Modelling of concrete performance, E&FN SPON, 1999, 308.
12. Rostásy, F., Gutsch, A. and Krauß, M. Computation of stresses and cracking criteria for early age concrete - Methods of iBMB. IPACS, Task 3, 2001.
13. Bazant, Z. and Osman, E. Double power law for basic creep of concrete. Materials and Structures, Research and Testing, 1976;9(49): 3-11.
14. Azenha, M. Behaviour of concrete at early ages. Phenomenology and thermo-mechanical analysis. (in Portuguese). MSc Thesis, Faculty of Engineering of the University of Porto, 2004.

MODELING HIGH STRENGTH CONCRETE USING FINITE ELEMENT WITH EMBEDDED COHESIVE CRACK

A.M. Fathy[1], J. Planas[2], J.M. Sancho[2], D.A. Cendón[2] and J.C. Gálvez [2]

[1]*Faculty of Engineering, Ain Shamas University, Cairo, Egypt;* [2]*Universidad Politécnica de Madrid, ETS de Ingenieros de Caminos, Profesor Aranguren s/n, 28040 Madrid, Spain.*

Abstract: A proposal for a standard test to determine fracture parameters of concrete was recently submitted to ACI committee 446 and to RILEM TC 187-SOC. This paper first summarizes the test procedure and over lights the most important aspects of it. To analyze the reliability of that test, an experimental study was carried out applying this test to four concrete mixes, two for normal strength concrete (NSC) and two for high strength concrete (HSC). The calculations were carried out with special finite elements recently developed by the authors that incorporate the crack behavior as an embedded cohesive crack with limited adaptability. The results show that both the test and the computational method are suitable for HSC as well as for NSC.

Key words: high strength concrete; finite element; cohesive crack; softening curve; fracture energy

1. INTRODUCTION

The cohesive crack model introduced for concrete by Hillerborg et al. thirty years ago[1] is able to describe concrete fracture in tension with a good accuracy/complexity ratio. The main input of the model is the softening curve which is known to be strongly nonlinear[2,3,4]. One simplified model used since the 80 is that of a bilinear curve (see [2] for a review of proposed bilinear curves). In the present paper, an updated implementation is presented of the experimental method proposed in[5] and later subjected to several improvements[2,6,7]. Since a very limited amount of data exists in the literature for fracture properties of high strength concrete HSC and some difficulties may be expected in the experiments for such materials due to increased strength and potential brittleness, the method is applied to high strength concrete as well as normal strength concrete NSC. A recently developed method of numerical analysis [8] is also applied to predict the behavior

M.S. Konsta-Gdoutos, (ed.), Measuring, Monitoring and Modeling Concrete Properties, 99–106.

of the beams based on the resulting stress-crack opening curve, in order to see to what an extent the numerical method that proved to be suitable for NSC, is also suitable for HSC. The essential conclusion of this combined experimental-numerical research is that both the experimental and the numerical methods investigated are suitable for HSC as well as for NSC.

2. PROPOSED TEST METHOD

The test is based on combining the results of splitting tensile strength tests and of three-point-bending tests on notched beams [2,5-7]. The splitting tensile test is carried out on standard cylinders according to ASTM C 496 with slight improvements (narrower loading strips, limited rate) [9,10]. In the bending test, a beam of rectangular cross section with a central notch is subjected to stable three-point-bending with a loading span of three times its depth. To ensure stability, the test is carried out under crack mouth opening (CMOD) control. Weight compensation is used to obtain stable test up to complete failure. The values of the CMOD, the load and the deflection at mid span are measured. To avoid measuring the inelastic deformation originated at the supports, the displacement is determined relative to the points directly above and below the loading points by means of a reference frame. In our implementation of the test (Figure 1), beams of 500x100x100 mm are with a notch depth of 33 mm. Weight compensation is achieved by means of steel deadweights clamped at the end of the beam. Two inductive displacement transducers are used to measure the deflection, and knifes for the CMOD are fixed on two steel plates which are connected to the specimen with screws (Figure 1, right). One of the steel plates is wide enough to be used as a reference point for the displacement transducers. Roller supports over hard steel plates have to be used and one of them must be free to rotate about an axis normal to the loading plane to minimize torsion.

Figure 1. Overview of all apparatus of the three point bending test

After machining and installing the measured devices, the specimen is preloaded with 5 to 10% of the estimated maximum load and then the machine control is transferred to CMOD and the test is run at an initial CMOD rate selected to reach the maximum load within about 3 to 5 min. After the load goes through the peak and decreases to about 90% of the maximum, the CMOD rate is increased progressively to keep an approximately constant rate of load decay until the end of the test, which is conventionally reached when CMOD = 4D/300 (= 1.333 mm in our tests).

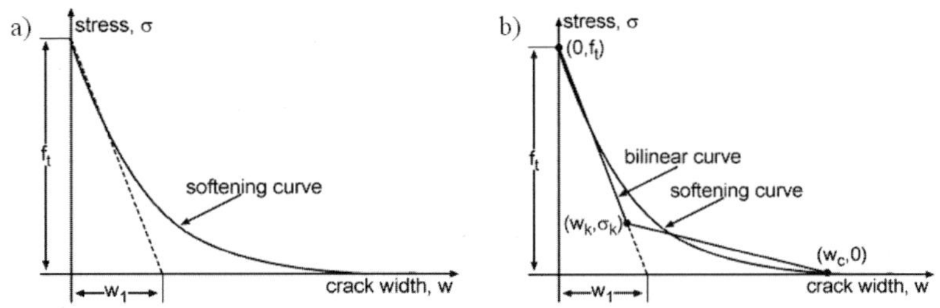

Figure 2. Initial linear approximation (a) and full bilinear approximation (b)

The interpretation of the test results is based on the following: The general softening curve of the concrete can be approximated in the first part with a straight line as shown in Figure 2a. This line is defined by the tensile strength f_t and the horizontal intercept w_1. The work done by Rocco[9,10] shows that the tensile strength f_t can be taken as the indirect tensile strength of the Brazilian test f_{ts}, provided that the load bearing wooden strip is not wider than 8% of the cylinder diameter and that the stress rate all up to the peak is less than about 1 MPa/min. On the other hand, the maximum load in bending depends only on the initial linear part of the softening curve [5]. This leads to the definition of a brittleness length ℓ_1 which can be calculated as[6]:

$$\ell_1 = (1-\alpha_0)^m D\left[c_1/\left(x^2-1\right)^2 + c_2/x^2\right] = Ew_1/2f_t \tag{1}$$

where $\alpha_0 = a_0/D$, m, c_1 and c_2 are function of the S/D ratio and take the value of 1.7, 11.2 and 2.365 respectively for a span-to-depth ratio of 3. (NOTE: in[6] values of the constants were determined for a span-to-depth ratio of 4; the present values were recomputed using the same procedure for the span-to-depth ratio of 3, which is standard in testing concrete prisms in bending in ASTM test methods). The value x is equal to the ratio f_t/f_p, where the plastic resistance f_p is calculated as:

$$f_p = P_u S/2B\left(D-a_0\right)^2 \tag{2}$$

Therefore after calculating the value of ℓ_1 we can get the value of w_1, provided the modulus of elasticity E is known. E can be easily computed from the initial part of the load vs. CMOD curve as shown later.

To determine the remaining part of the bilinear softening curve, the fracture energy is determined from the work of fracture in a classical way[11,12], and the center of gravity of the area enclosed by the softening curve and the axes is obtained from the behavior of the tail of the load-CMOD curve[5,2]. Indeed, the load in the far end of the test can be expressed as a function of the deflection or the CMOD w_M assuming that the movement of the specimen can be approximately described as two rigid halves linked by a hinge in the middle. Then $\delta = w_M S/4D$ and P can be expressed as:

$$P = A/\delta^2 = A\left(4D/S\right)^2/w_M^2 \tag{3}$$

where A is a constant that was shown in [5] to be related to the fracture energy G_F and the centre of gravity of the area enclosed by the softening curve w_G as $A = BSG_F w_G/4$.

Based on these ideas, the detailed calculation procedure is as follows:

a) Using the equation $f_t = f_{ts} = 2P_u/(\pi DL)$ from the Brazilian test, the value f_t is determined, where P_u, D and L is the ultimate load, diameter and length of the specimen respectively.

b)The average modulus of elasticity E_m can be determined as follows:

- The test points with measured load P' between 15 and 55% of the maximum load on the raising branch of the P'-CMOD curve are selected.
- Straight line is fitted to these points by linear regression of CMOD vs. load to calculate the initial compliance of the specimen $C_i = \Delta w_m / \Delta P'$
- The elastic modulus E is calculated as $E = 6Sa_0V_1(\alpha'_0)/(C_iBD^2)$, where $\alpha'_0 = (a_0+h)/(D+h)$, h is the thickness of the steel knife used to clamp the CMOD gauge and

$$V_1(\alpha) = 0.8 - 1.7\alpha + 2.4\alpha^2 + 0.66/(1-\alpha)^2 + 4D(-0.04 - 0.58\alpha + 1.47\alpha^2 - 2.04\alpha^3)/S \quad (4)$$

- Then, the average modulus of elasticity E_m is calculated.

c) The constant A is next determined as follows:

- The points with CMOD > $4D/300$ are eliminated from the record.
- The values of the CMOD and the load in the last point of the curve are recorded (w_{MR} and P'_R).
- A corrected load P_1 is recalculated as $P_1 = P' - P'_R$
- The curve P_1 -CMOD is drawn and the CMOD value for zero load w_{MA} is determined.
- The values X for all points of the tail such that $P_1 < 0.05\,P_{1u}$, where P_{1u} is the maximum corrected load, are calculated as

$$X = (4D/S)^2\left[1/(w_M - w_{MA})^2 - 1/(w_{MR} - w_{MA})^2\right] \quad (5)$$

- The values of P_1 are plotted against X and a curve fit to the equation $P_1 = X(A+kX)$ is performed.
- The value of A is determined to 3 significant digits.

d) The net plastic flexural resistance f_{pm} is determined as follows

- The effective maximum load P_u is calculated as $P_u = P_{1u} + A/(w_{MR} - w_{MA})^2$
- For every specimen; f_p is calculated using the Eq. (2) and then the average f_{pm} is calculated.

e) The average fracture energy G_{Fm} is determined as follows:

- The corrected load P_1 is plotted against the deflection δ and the deflections δ_A and δ_R (in the loading and unloading parts) correspond to zero load are determined.
- For every specimen, the measured fracture work W_{Fm} is calculated as the area enclosed by the curve P_1 - δ and the δ axis.

- Then, the total fracture work is calculated as $W_F = W_{Fm} + 2A/\delta_R - \delta_A$
- The fracture of energy G_F for every specimen is computed using the equation $G_F = W_F/[B(D\text{-}a_0)]$
- The average fracture energy G_{Fm} is, then calculated.

f) The average value for the abscissa of the centre of gravity of the softening curve w_{Gm} can be determined after computing the value w_G for every specimen using the equation $w_G = 4A/(BSG_F)$

g) The characteristic points of the bilinear approximation of the softening curve can be determined as follows:

- The brittleness length ℓ_1 for each specimen is determined using the first part of Eq. (1) and then the average value ℓ_{1m} is calculated.
- w_1 is computed for the concrete as $w_1 = 2f_t\ell_{1m}/E_m$
- The characteristic crack opening w_{ch} is computed as $w_{ch} = G_{Fm}/f_t$
- The critical length of the bilinear approximation is calculated using the following equation:

$$w_c = w_{ch}\frac{3w_{Gm} - w_1}{2w_{ch} - w_1}\left[1 + \sqrt{1 - \frac{2w_1(3w_{Gm} - 2w_{ch})(2w_{ch} - w_1)}{w_{ch}(3w_{Gm} - w_1)^2}}\right] \tag{6}$$

- The coordinate (σ_k, w_k) of the angular point of the bilinear curve (as shown in Figure 2b) is then given by:

$$\sigma_k = f_t(2w_{ch} - w_1)/(w_c - w_1) \quad \text{and} \quad w_k = w_1(w_c - 2w_{ch})/(w_c - w_1) \tag{7}$$

3. EXPERIMENTAL WORK

Four types of concrete have been produced, two with design strength 25 and 40 MPa (NSC) and two with design strength 80 and 100 MPa (HSC). Normal cement was used with graduated sand and gravel to produce the NSC. For the production of HSC high strength cement was used in addition to silica fume and super plasticizers. The ACI method for mix design of concrete was used to evaluate the mix proportions of every material for the NSC. As the aim of the investigation was not looking for the best mix design, proportions used in another research were adopted to produce the HSC. Table 1 shows the specimen dimension and number as well as the applied tests. The test results are summarized in Table 2.

The compressive strengths achieved were 28.7 and 39.1 (MPa) for (NSC) and 85.2 and 87.7 (MPa) for (HSC). Test results show that f_t, E_m and G_{Fm} and ℓ_1 increase with the increase of the compressive strength, while ℓ_{ch} decreases. The ratio σ_k/f_t is low for NSC (published ranges are between 0.15 and 0.25) and very low for HSC.

Table 1. Used specimens for different tests

specimen			Test	
No.	Type	Dimensions (mm)	Type	Standard
3	Cylinder	(diameter x height) 150x300	Compressive	ASTM C39
3			Brazilian	ASTM C496
6	Notched Beam	(length x width x depth) 500x100x100	3 point bending test	Proposal to ACI committee 446

Table 2. Experimental results

Mix n°.		f_c (MPa)	f_t (MPa)	E_m (GPa)	w_c (mm)	G_{Fm} (N/mm)	σ_k (MPa)	w_k (m)	ℓ_1 (mm)	ℓ_{ch} (mm)
NSC	1	28.8	2.24	31.4	0.379	77.2	0.350	9.76	81.5	484
	2	39.1	2.83	32.3	0.393	94.4	0.387	13.0	88.6	379
HSC	3	85.2	5.39	45.0	0.466	134.4	0.313	22.8	103	208
	4	87.7	5.80	48.3	0.461	138.9	0.305	23.6	107	199

f_c: compressive strength, f_t: tensile strength, E_m: modulus of elasticity, w_c: critical crack opening, G_{Fm}: The fracture energy, σ_k and w_k: coordinates of the corner-point of the bilinear softening curve as shown in Figure2b, ℓ_1: as defined in Eq. (1) and ℓ_{ch}: Hillerborg's characteristic length $\ell_{ch} = E_m G_{Fm} / f_t^2$

The tests were stable until the end of the test and therefore the proposed standard test can be used with high strength concrete as well as ordinary one. Note that ℓ_{ch} is a measure of ductility, its inverse of brittleness, but assumes that all the fracture energy can be spent in the process, thus, HSC is more brittle, in this sense than NSC. However, for ordinary sizes, the structural elements fail (reach the peak load) much before the fracture energy is fully spent and thus a better measure of ductility or brittleness (which we call *operational*) is the brittleness length ℓ_1, which increases with strength, and thus HSC appears to be less brittle than NSC.

4. NUMERIC SIMULATION OF TESTS BY FE

As it was mentioned before, the tests were numerically simulated using a special element recently presented by the authors [8]. The finite element kinematics is based on the Strong Discontinuity Analysis. To avoid the necessity of tracking algorithms, the element uses a novel approach based on a combination of damage-like cohesive embedded crack with a central traction-separation law, and limited crack adaptability. That allows to describe the cohesive crack growth with adequate accuracy while keeping the formulation strictly local. Although the formulation may look in some aspects similar to a traditional smeared crack approach, further analysis show that this is not so, and that the strong discontinuity kinematics is an essential ingredient to get good results. For the sake of completeness, simulations were also performed with another technique based on the boundary integral method implemented in the program Splitting Lab developed by the second author [2]. The results obtained with both techniques and the experimental results are compared in Figure 3.

The results show that for the range of material properties envisaged in the study, the adaptable embedded crack FEM performs as well as the boundary integral method and that the overall fit to the experimental results is excellent for HSC and only acceptable for NSC, which show a too fast initial decrease of load followed by a too strong response in the central part, which is probably due to a very rounded softening curve that is hard to approximate by a bilinear curve.

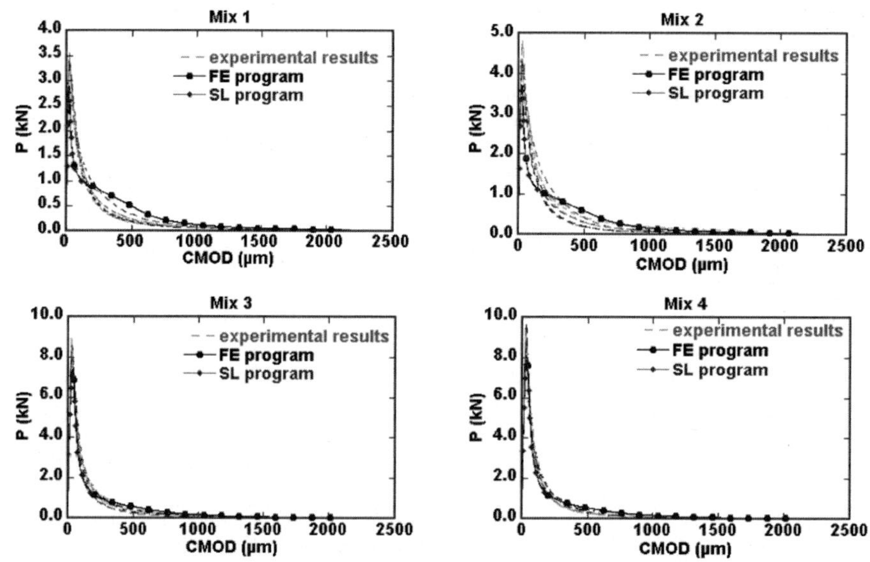

Figure 3. The load (kN) vs CMOD (m) for the four mixes. (FE program: the new finite element program developed by the authors, SL program: Splitting Lab program.)

5. FINAL REMARKS

The main conclusion to be drawn from the foregoing results is that both the proposed experimental method and the computational procedure are robust enough to be applied to NSC and to HSC up to 90 MPa.

On the other hand, the strength-like properties are, as expected, larger for HSC, and the traditional, overall, brittleness (inverse of characteristic length ℓ_{ch}) is also larger for HSC. However, the operational brittleness (inverse of brittleness length ℓ_1) is less for HSC than for NSC.

Acknowledgments

The authors gratefully acknowledge partial financial support for this work from the Spanish Ministerio de Educación y Ciencia under grants BIA2005-09250-C03-01 and BIA2005-09250-C03-02.

6. REFERENCES

1. Hillerborg A., Modéer M. and Petersson P.E. "Analysis of crack formation and crack growth in concrete by means of fracture mechanics and fracture elements", *Cement and Concrete Research*, 6,773-782 (1976).Bazant, Z.P. and Planas, J, *Fracture and Size Effect in Concrete and Other Quasibrittle Materials,* CRC Press, Boca Raton, FL (1998).
2. Elices, M., Guinea, G.V., Gómez, J. and Planas, J. "The cohesive zone model: advantages, limitations and challenges", *Engineering Fracture Mechanics*, **69**, 137-163 (2002).
3. Planas, J., Elices, M., Guinea, G.V., Gómez, F.J., Cendón, D.A. and Arbilla, I. "Generalizations and specializations of cohesive crack models", *Engineering Fracture Mechanics*, **70**(14),1759–1776(2003)
4. Guinea, G.V., Planas, J. and Elices, M., "A general bilinear fitting for the softening curve of concrete", *Materials and Structures,* **27**, 99-105, (1994).
5. Planas, J. , Guinea, G.V. and Elices, M. "Size Effect and Inverse Analysis in Concrete Fracture", *International Journal of Fracture,* **95**, 367-378 (1999).
6. Planas, J. , Guinea, G.V. and Elices, M. "Standard Test Method for Bilinear Stress-Crack Opening Curve of Concrete," Proposal submitted to ACI Committee 446, (2002), revised (2005).
7. Sancho, J.M., Planas, J., Cendón, D.A., Reyes, E. and Gálvez J.C. "An embedded cohesive crack model for finite element analysis of concrete fracture", *Engineering Fracture Mechanics,* Accepted for publication (2005). Also in *Fracture Mechanics of Concrete Structures, Li et al (eds),* Ia-FraMCos, ISBN 0-87031-135-2, pp. 107-114 (2004).
8. Rocco, C., Guinea, G.V., Planas, J. and Elices, M. "Mechanisms of rupture in the splitting test," *ACI Materials Journal,* **96**(1), 52-60 (1999).
9. Rocco, C., Guinea, G. V., Planas, J. and Elices, M. "Size effect and boundary conditions in the Brazilian test: experimental verification," *Materials and Structures,* **32**, 10-217 (1999).
10. RILEM (1985) "Determination of the fracture energy of mortar and concrete by means of three-point bend tests on notched beams", *Materials and Structures,* **18**, 285-290 (1985). (RILEM Draft Recommendation, TC 50-FMC Fracture Mechanics of Concrete.)
11. Elices, M., Guinea, G. V. and Planas, J. "On the measurement of concrete fracture energy using three point bend tests," *Materials and Structures,* **30**, 375-376 (1997).

SIZE EFFECT OF CONCRETE: UNIAXIAL AND FLEXURAL COMPRESSION

A.L. Gamino, J.U. A. Borges and T.N. Bittencourt
University of São Paulo, Av. Prof. Almeida Prado, 271, CEP05508-900, São Paulo-SP, Brazil

Abstract: This paper presents an analytical and a numerical approach to evaluate the size (slenderness) effect on the post-peak behavior of concrete in compression. The analytical approach takes into account the specimen height in the calculation of the ductility of plain concrete under uniaxial compression and the uniform moment zone length in the calculation of the ductility of reinforced concrete (RC) beams. A numerical modeling using the computer program DIANA was carried out to evaluate the size dependence for concrete in compression and to confirm the predictions of the proposed analytical model. In addition, recent experimental results of beams with different sizes have been found to correlate reasonably well with the ones predicted by both the analytical approach and the numerical modelling.

Keywords: size effect; DIANA; reinforced concrete beams; concrete in compression; post-peak behavior; FEM analysis.

1. INTRODUCTION

The flexural response of reinforced concrete beams depends upon several factors such as concrete strength, location and quantity of the reinforcement, and the section geometry. In addition, the stress-strain response of concrete in compression (particularly the post-peak) has a strong influence on the behavior of reinforced concrete elements.To accurately predict the response of reinforced concrete beams, sufficient knowledge of the stress-strain curve of the concrete, including the post-peak behavior (i.e., descending branch of the stress-strain curve) is required.

Unfortunately, to date the compressive response of concrete and its contribution to the post-peak response of a reinforced concrete beam has still not been clearly understood. As in the case of tensile fracture, compressive failure involves post-peak strain-softening. However, unlike tensile failure in which a thin crack forms, compressive failure results in the development of a localized damage zone (Jansen and Shah 1997). This damage zone has a finite length, which may occupy a portion of a large specimen, or the entire speci-

M.S. Konsta-Gdoutos, (ed.), Measuring, Monitoring and Modeling Concrete Properties, 107–113.

men for small-sized specimens. As a result, it is generally accepted that the post-peak behavior of concrete (following localization of damage) subjected to uniaxial compression is influenced by specimen size and boundary conditions (Vonk 1992; van Vliet and van Mier 1996; Jansen and Shah 1997). This implies that the stress-strain response of concrete after localization is dependent on specimen size, larger specimens exhibiting a more brittle post-peak response.

2. CONCRETE UNDER UNIAXIAL COMPRESSION

The analytical approach adopted herein is similar in concept to the one proposed previously for damage localization in concrete under tensile loading (Hillerborg *et al.* 1976; Bažant 1989). In this idealization, the response of the specimen is homogeneous up to the peak load and the damage is assumed to localize into a band of finite height along the length of the specimen at the peak load. After localization, the responses of the damaged and bulk concrete (i.e., unloading zone) are considered separately, and each of these behaviors are treated as material properties. After peak load, the damage zone continues to accrue damage and exhibit an increase in deformation while the remainder of the specimen unloads elastically. After the peak load, the total displacement of the specimen is obtained by adding the displacements inside and outside the damage zone. A linear unloading path is assumed for the bulk concrete. This paper uses this approach to predict the flexural behavior of reinforced members with different constant moment zone lengths. Specific details can be found elsewhere (Borges *et al.* 2002; Borges *et al.* 2004).

The length of the damage zone has to be estimated accurately for proper implementation of this type of model. Rokugo and Koyanagi (1992), Markeset (1994) and Jansen and Shah (1997) estimated this value to be 2 to 3 times the width of the specimen at the end of the tests.

Figure 1 shows the localization approach for concrete in uniaxial compression.

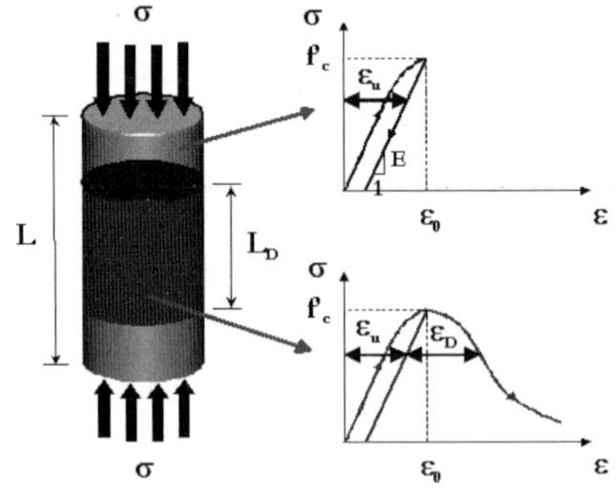

Figure 1. Localization approach for concrete in uniaxial compression

The length of the damage zone in uniaxial compression is taken equal to 1.5 times the specimen diameter throughout the loading process (Borges *et al.* 2002). In flexure, this length is assumed to be equal to 4 times the depth of the neutral axis at peak moment.

For any specimen containing a localized damage zone, the post-peak displacement of the overall specimen is given by

$$\delta = \varepsilon L = \varepsilon_u L + \varepsilon_D L_D \tag{1}$$

where ε is the overall strain measured along the entire length of the specimen, ε_u is the strain in the bulk concrete outside the damage zone that unloads elastically, L is the length of the specimen, L_D is the length of the damage zone, and ε_D is the additional inelastic strain within the damage zone. It is assumed that ε_D is a characteristic material parameter depending only on the type of concrete. According to the linear bulk unloading path, i.e., the path along which the bulk concrete unloads, the unloading strain is given as

$$\varepsilon_u = \varepsilon_0 - \frac{f_c' - \sigma}{E} \tag{2}$$

where ε_u is the strain corresponding to peak stress, f_c' is the concrete compressive strength, E is the elastic modulus of the concrete and σ is the stress at a given point on the curve.

Combining equations 1 and 2, the overall post-peak strain for a specimen containing a localized damage zone can be written as

$$\varepsilon = \varepsilon_0 - \frac{f_c' - \sigma}{E} + \frac{\varepsilon_D L_D}{L} \tag{3}$$

The parameter D for a given material can be determined from uniaxial compression test results of specimens of different lengths. According to Eq. (3), the post-peak inelastic strain within the damage zone can be expressed

$$\varepsilon_D = (\varepsilon - \varepsilon_0 + \frac{f_c' - \sigma}{E})\frac{L}{L_D} \tag{4}$$

3. RC BEAMS SUBJECTED TO PURE BENDING

In flexure, the length L_D of the damage zone is assumed to be proportional to the neutral axis depth. This was initially suggested by Hillerborg (1988) and Markeset (1993). Weiss *et al.* (1999) carried out an experimental investigation with beams of various sizes subjected to four-point bending and found a value equal to four times the neutral axis depth at peak load. This has been confirmed by the test results of Borges (2002). Using any hypotheses, the post-peak response within the damage zone is determined inserting L = L_D in Eq. (3). Together with a linear softening curve, this yields the following compressive stress-strain relation within the damage zone

$$\sigma = \frac{\varepsilon_{DC,f} - \varepsilon_D + \varepsilon_0 - \dfrac{f_c'}{E}}{\dfrac{\varepsilon_{DC,f}}{f_c'} - \dfrac{1}{E}} \tag{5}$$

where $\varepsilon_{DC,f}$ is the critical damage strain in flexure and ε_D is the post-peak inelastic strain within the damage zone. To take the confinement effect in bending, the value of the critical damage strain in flexure is assumed to be two times that in uniaxial compression, i.e., $\varepsilon_{DC,f} = 2\ \varepsilon_{DC}$. For a given value of bending moment, the overall top fiber strain is determined by

$$\varepsilon = \frac{\varepsilon_u\,(L_M - L_D) + \varepsilon_D\, L_D}{L_M} \tag{6}$$

4. EXPERIMENTAL RESULTS

An experimental investigation was undertaken to verify the validity of the proposed approach to predict the response of reinforced concrete beams with different lengths under pure bending.

Twelve simply supported reinforced beams were cast and tested under four-point bending. The test variables were the beam size (three different uniform moment zone lengths) and the reinforcement ratio (low and high reinforcement ratio). The concrete compressive strength and the steel yield strength for all beams were 110 MPa and 560 MPa, respectively. The age at testing was 120 days for both the beams and the control specimens. The beams were designed with a reinforcement ratio close to that for a balanced failure, i.e., $\rho/\rho_b = 0{,}96$. Stirrups were provided along the shear span to ensure a flexural mode of failure. Three values for the uniform moment zone length were adopted: 300, 500 and 700 mm. The value of the shear span was kept constant for all beams. Figure 2 illustrates the beams tested.

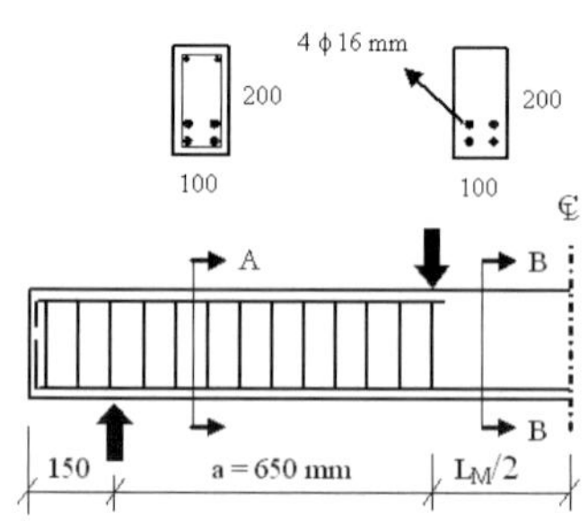

Figure 2. Detail of the test specimens

Experimental and calculated moment-strain curves has been shows in Figure 5. It can be seen that the analytical results are in good agreement with the experimental test response, although the maximum moment was slightly overestimated. The reduction in post-peak ductility with increasing beam length was correctly captured by this simple modeling approach.

5. NUMERICAL ANALYSIS

The objective is to evaluate the size effect by means of smeared crack models for the concrete and confirm the observed behavior and the predictions of the analytical model. In the numerical simulations the finite element-based code DIANA is used. The non-linear hardening model of *Thorenfeldt* and the non-linear softening model of *Hordijk* was adopted. The non-linear hardening model of *Thorenfeldt* presented in Figure 3 presents a equation between the tensions of compression and the deformations based in the adoption of diverse parameters in agreement the Eq. (7).

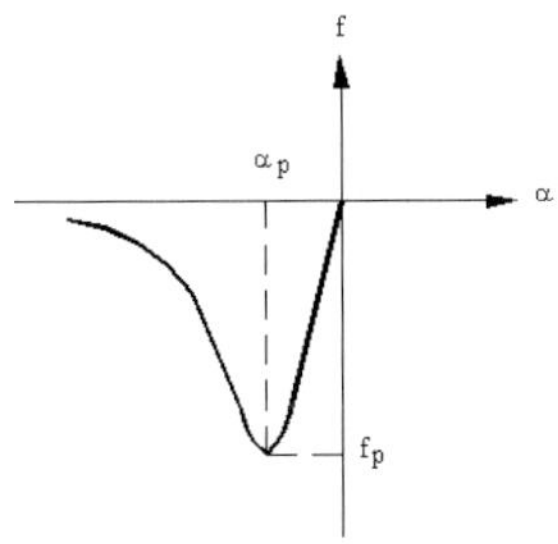

Figure 3. Thorenfeldt hardening curve

$$f = -f_p \frac{\alpha_j}{\alpha_p}\left(\frac{n}{n-1+\left(\frac{\alpha_j}{\alpha_p}\right)^{nk}}\right) \quad n = 0{,}80 + \frac{f_c}{17}\ ; \quad k = \begin{cases} 1 & \text{if } 0 > \alpha > \alpha_p \\ 0{,}67 + f_c/62 & \text{if } \alpha \le \alpha_p \end{cases} \tag{7}$$

The non-linear softening model of *Hordijk* presented in Figure 4 and Eq. (8) uses an exponential relation between the normal stress of traction and the deformations, with "$c_1 = 3$" and "$c_2 = 6{,}93$".

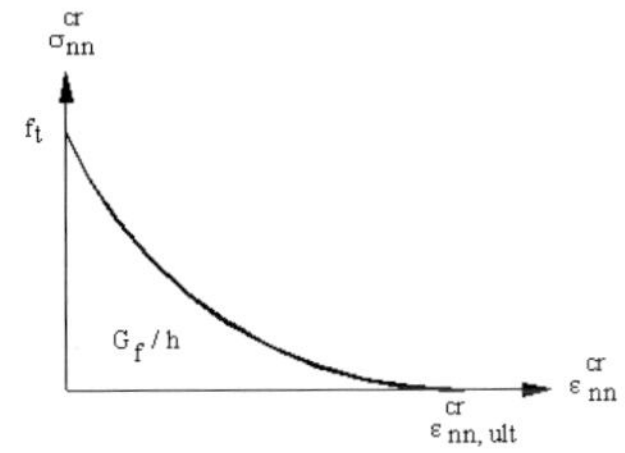

Figure 4. Hordijk non-linear softening curve

$$\frac{\sigma_{nn}^{cr}\left(\varepsilon_{nn}^{cr}\right)}{f_t} = \begin{cases} \left(1+\left(c_1\frac{\varepsilon_{nn}^{cr}}{\varepsilon_{nn,ult}^{cr}}\right)^3\right)\exp\left(-c_2\frac{\varepsilon_{nn}^{cr}}{\varepsilon_{nn,ult}^{cr}}\right) - \frac{\varepsilon_{nn}^{cr}}{\varepsilon_{nn,ult}^{cr}}\left(1+c_1^{\,3}\right)\exp\left(-c_2\right) \xrightarrow{IF} 0 < \varepsilon_{nn}^{cr} < \varepsilon_{nn,ult}^{cr} \\ 0 \xrightarrow{IF} \varepsilon_{nn,ult}^{cr} < \varepsilon_{nn}^{cr} < \infty \end{cases} \tag{8}$$

The following parameters has been used for the concrete constitutive model: β=0,2; G_f=0,09N/mm (in agreement with Hordijk, 1991); f_c=107MPa; f_t=7,85MPa; E=45021MPa.

An elastoplastic *Von Mises* model has been utilized for the reinforcement simulation; for the cracked concrete, smeared rotating model has been utilized using Q8 quadratic plane elements. The TRUSS rebars elements have been embedded in quadratic Q8 concrete elements. Figure 5 shows the numerical, experimental and calculated moment-strain curves.

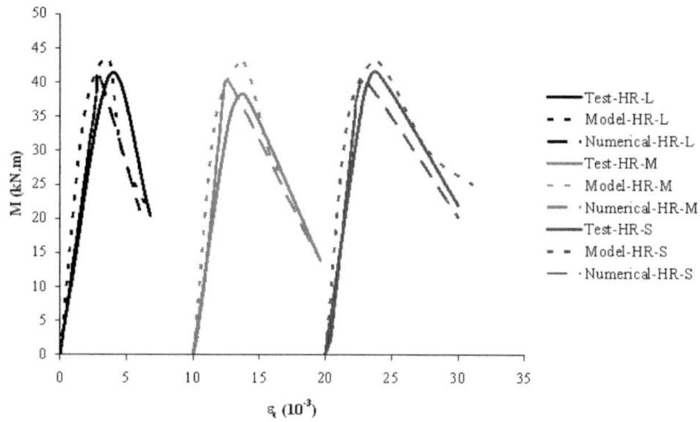

Figure 5. Numerical moment-strain curves

The reduction in post-peak ductility with increasing beam length has been correctly captured in numerical simulation in agreement to experimental and analytical approach. Figure 6 shows the compressive stress field in the beams. The increase in compressive stress has been proportional to the increase of the beam length.

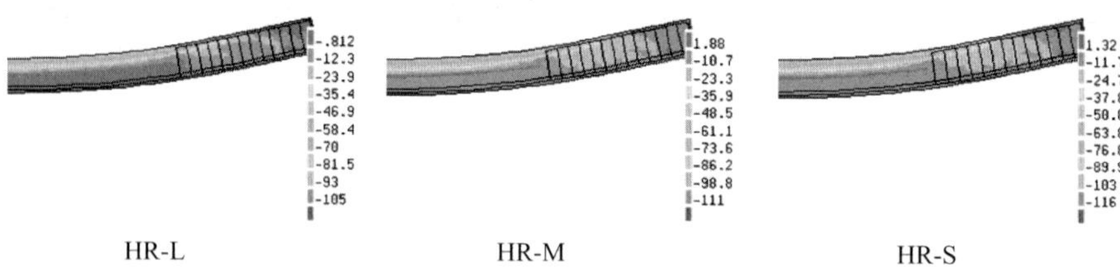

Figure 6. Numerical compressive stress in concrete (in MPa)

6. CONCLUSIONS

For the studied RC beams the analytical results are in good agreement with the experimental test responses, although the maximum moment was slightly overestimated. The reduction in post-peak ductility with increasing beam length was correctly captured by this simple modeling approach. The same has been observed and confirmed by the numerical results. The method of separating the post-peak behaviors for cross sections within and outside the damage zone can provide a way to perform simple incremental cross-sectional analyses for the beams to simulate the length effect on the post-peak ductility. Due to strain localization, there is a remarkable length effect on the post-peak behavior of specimens under uniaxial compression as well as reinforced beams under bending. This effect is primarily manifested in ductility behavior as opposed to changes in load carrying capacity.

7. REFERENCES

Bazant, Z.P. "Identification of strain-softening constitutive relation from uniaxial tests by series coupling model for localization," *Cement and Concrete Research.*, v.19, 1989.

Borges, J.U.A., Subramanian, K., Weiss, K.J., Shah, S., Bittencourt, T.N., "Slenderness Effect on the Ductility of Concrete in Uniaxial and Flexural Compression", *ACI Structural Journal*, 2004. (accepted)

Borges, J.U.A., Bittencourt, T.N., "Analytical Model for Prediction of Size-Dependent Stress-Strain Curves of High-Strength Concrete Cylinders in Uniaxial Compression", *6th Int. Symp. on Util. of High Strength Concrete*, Leipzig, Germany, Vol.1, pp. 165-176, 2002.

Hillerborg, A.; Modeer, M.; Petersson, P.E. "Analysis of crack formation and crack growth in concrete by means of fracture mechanics and finite elements," *Cement and Concrete Research*, v.6, n.6, 1976, p.773-781.

Hillerborg, A. "Rotational capacity of reinforced concrete beams," *Nordic Concrete Research*, 7, 121-134, 1988.

Hordijk, D.A., "Local approach to fatigue of concrete," PhD thesis, Delft Un. of Tech., 1991.

Jansen, D.C., Shah, S.P. "Effect of length on compressive strain softening of concrete," *J. Engrg. Mech.*, ASCE, 123(1), 25-35, 1997.

Markeset, G. "Failure of concrete under compressive strain gradients," PhD thesis, Norwegian Inst. of Technol., Trondheim, Norway, 1993.

Markeset, G. "Comments on size dependence and brittleness of high strength concrete," *SINTEF Report STF70 A95029*, 1994.

Rokugo, K., Koyanagi, W. "Role of compressive fracture energy of concrete on the failure behavior of reinforced concrete beams," *Applications of fracture mechanics to reinforced concrete*, Carpinteri, A., ed., Elsevier, 1992.

van Vliet, M.R.A., van Mier, J.G.M. "Softening behaviour of concrete under uniaxial compression," *Proc., 2nd Int. Conf. on Fracture Mechanics of Concrete and Concrete Structures (FRAMCOS)*, Zurich, Switzerland, 383-396, 1996.

Vonk, R. "Softening of concrete loaded in compression," PhD Thesis, Eidhoven Univ. of Technol., Eindhoven, The Netherlands, 1992.

Weiss, W.J.; Güler, K.; Shah, S.P. "An experimental investigation to determine the influence of size on the flexural behavior of high strength rc beams," *Proc., 5th Int. Symp. on Ut. of High-Strength/High-Performance Concrete*, Sandefjord, Norway, 709-718, 1999.

EMBEDDED CRACK ELEMENTS WITH NON-UNIFORM DISCONTINUITY MODES

O.L. Manzoli and P.B. Shing

Department of Civil Engineering, São Paulo State University, Av. Luiz E. C. Coube, S/N, 17033-360, Bauru-SP, Brazil

Department of Structural Engineering, University of California at San Diego, La Jolla, CA 92093, USA

Abstract: The consequences of the use of embedded crack finite elements with uniform discontinuity modes (opening and sliding) to simulate crack propagation in concrete are investigated. It is shown the circumstances in which the consideration of uniform discontinuity modes is not suitable to accurately model the kinematics induced by the crack and must be avoided. It is also proposed a technique to embed cracks with non-uniform discontinuity modes into standard displacement-based finite elements to overcome the shortcomings of the uniform discontinuity modes approach.

Key words: Finite elements; embedded crack element; concrete fracture; material failure, non-uniform discontinuity modes; stress locking.

1. INTRODUCTION

In recent years, techniques to embed crack into finite elements have been shown to be efficient for modeling arbitrary crack propagation in solids using a fixed finite element mesh (e.g., see references[1,2]). These techniques allow crack occurrence to be introduced into standard finite elements independent of the location and orientation of the element boundaries when a given crack initiation criterion is reached during the analysis. Therefore, the need for interface elements and mesh reconstruction techniques[3] are no longer required to model crack propagation. This approach has been used to model fracture in reinforced concrete members [4].

Although many formulations have been proposed to embed cracks into different underlining elements[1], most of them are limited to modeling cracks with uniform discontinuity modes (opening and sliding) in an element. This simplification can introduce severe stress locking for some problems. In this paper, the deficiency of using embedded crack elements with uniform discontinuity modes is demonstrated, and a method to intro-

M.S. Konsta-Gdoutos, (ed.), Measuring, Monitoring and Modeling Concrete Properties, 115–123.

duce non-uniform discontinuity modes to circumvent the aforementioned problem is presented.

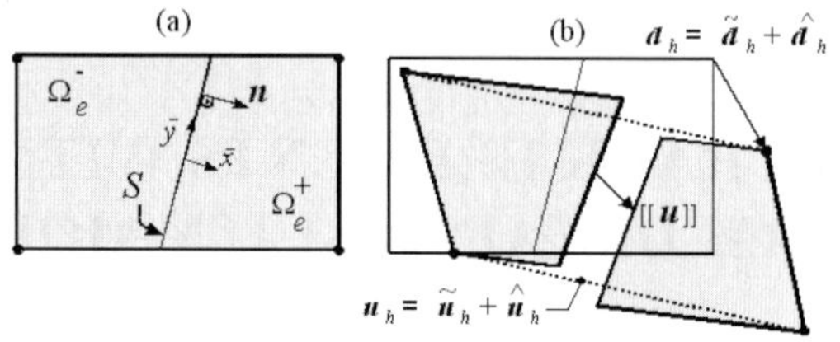

Figure 1. Decomposition of the displacement field

2. EMBEDDED CRACK ELEMENT

The displacement field $\mathbf{u}$ inside an element can be decomposed into a component $\tilde{\mathbf{u}}$ associated with the deformation of the continuum portion and a component $\hat{\mathbf{u}}$ related to the rigid-body relative motion between the two parts of the element (see Figure 1.b):

$$\mathbf{u} = \tilde{\mathbf{u}} + \hat{\mathbf{u}} \tag{1}$$

Similar to the displacement field, the vector of nodal displacements $\mathbf{d}$ can be decomposed into a part due to the deformation of the continuum portion and a component $\hat{\mathbf{d}}$ due to the relative rigid-body motion. If the relative interface motion is assumed uniform in the element, $\hat{\mathbf{d}}$ is given by

$$\hat{\mathbf{d}} = \mathbf{P}[\![\mathbf{u}]\!] \qquad \mathbf{P} = \begin{bmatrix} \mathcal{H}(\bar{x}_1) & 0 \\ 0 & \mathcal{H}(\bar{x}_1) \\ \vdots & \vdots \\ \mathcal{H}(\bar{x}_{n_{ed}}) & \\ & \mathcal{H}(\bar{x}_{n_{ed}}) \end{bmatrix} \tag{2}$$

where n_{ed} is the number of element nodes, $\bar{x}_i$ (i =1,2, n_{ed}) are the nodal coordinates in the local coordinate system $\bar{x}$ - $\bar{y}$ aligned with the crack (see figure 1), $\mathcal{H}(\bar{x})$ is the Heaviside function (i.e., $\mathcal{H}(\bar{x}) = 1$ if $\bar{x} > 0$ and $\mathcal{H}(\bar{x}) = 0$ otherwise), and $[\![\mathbf{u}]\!]$ is the vector containing the components of the displacement jump.

The strain field of the continuum portion can be approximated by

$$\tilde{\boldsymbol{\varepsilon}}_h = \mathbf{B}(\mathbf{d} - \mathbf{P}[\![\mathbf{u}]\!]) \tag{3}$$

where $\mathbf{B}$ is the standard strain-displacement matrix.

The finite element equilibrium equations can be written as

The strain field of the continuum portion can be approximated by

$$\tilde{\boldsymbol{\varepsilon}}_h = \mathbf{B}(\mathbf{d} - \mathbf{P}[\![\mathbf{u}]\!]) \tag{3}$$

where **B** is the standard strain-displacement matrix.

The finite element equilibrium equations can be written as

$$\mathop{\mathrm{A}}_{e=1}^{n_{el}} \mathbf{f}_{\mathrm{int}_e} - \mathop{\mathrm{A}}_{e=1}^{n_{el}} \mathbf{f}_{ext_e} = \mathbf{0} \tag{4}$$

where **A** represents the finite element assembly operator, and for a linearly elastic continuum, $\mathbf{f}_{\mathrm{int}_e}$ and $\mathbf{f}_{ext_e}$ are the vectors of internal and external nodal forces given by

$$\mathbf{f}_{\mathrm{int}_e} = \tilde{\mathbf{K}}_e(\mathbf{d} - \mathbf{P}[\![\mathbf{u}]\!]) \tag{5}$$

$$\mathbf{f}_{ext_e} = \int_{\Omega_e} \mathbf{N}^T \bar{\mathbf{b}}\, d\Omega + \int_{\Gamma_{e_t}} \mathbf{N}^T \bar{\mathbf{t}}\, d\Gamma \tag{6}$$

where Ω_e is the domain of the element, Γ_{e_t} is the part of the boundary subjected to prescribed traction $\bar{\mathbf{t}}$, $\bar{\mathbf{b}}$ is the prescribed body forces, $\mathbf{N}$ is the standard finite element conforming shape function matrix, and $\tilde{\mathbf{K}}_e$ is the elastic stiffness matrix.

In the regularized strong discontinuity approach[5, 6], the behavior of the interface is described by a continuum (stress vs. strain) constitutive law. In this case, the traction vector in the interface can be given by:

$$\mathbf{t}_S = \mathcal{N}_{\mathbf{n}}\, \Sigma^c(\boldsymbol{\varepsilon}_S) \text{ in } S \tag{7}$$

where $\Sigma^c(\bullet)$ returns the stress from a given strain and its history. For two-dimensional problems, $\mathcal{N}_{\mathbf{n}}$ is expressed as:

$$\mathcal{N}_{\mathbf{n}} = \begin{bmatrix} n_x & 0 & n_y \\ 0 & n_y & n_x \end{bmatrix} \tag{8}$$

where n_x and n_y are the components of the unity vector $\mathbf{n}$ normal to the crack.

By considering the fracture process zone as a very thin band of width *k*, the strains in the crack can be approximated by[2]

$$\boldsymbol{\varepsilon}_S = \tilde{\boldsymbol{\varepsilon}}_h + \frac{1}{k}\mathcal{N}_{\mathbf{n}}^T[\![\mathbf{u}]\!] = \mathbf{B}(\mathbf{d} - \mathbf{P}[\![\mathbf{u}]\!]) + \frac{1}{k}\mathcal{N}_{\mathbf{n}}^T[\![\mathbf{u}]\!] \tag{9}$$

The continuum and the interface can be coupled by imposing of the following condition:

$$\mathbf{t}_S - \mathcal{N}_{\mathbf{n}}\mathbf{E}\tilde{\boldsymbol{\varepsilon}}_h = \mathbf{0} \quad \text{in } Q \in S \tag{10}$$

in which **E** is the elastic material matrix. Equation (10) enforces the traction continuity locally at a collocation point *Q* of the crack.

With a nonlinear interface, Eq. (4) has to be solved in an iterative manner. In each iteration, the internal forces must be evaluated for a trial nodal displacement vector. To do this, for a given nodal displacement vector **d**, the traction continuity equations, Eq. (10), has to be solved for the jumps and the internal forces can be evaluated by Eq. (5).

To describe the non-linear behavior of a crack, we use a standard elastoplastic constitutive model that can be described by the following set of incremental equations[7]:

With a nonlinear interface, Eq. (4) has to be solved in an iterative manner. In each iteration, the internal forces must be evaluated for a trial nodal displacement vector. To do this, for a given nodal displacement vector **d**, the traction continuity equations, Eq. (10), has to be solved for the jumps and the internal forces can be evaluated by Eq. (5).

To describe the non-linear behavior of a crack, we use a standard elastoplastic constitutive model that can be described by the following set of incremental equations[7]:

$$\dot{\boldsymbol{\sigma}}_h = \mathbf{E}(\dot{\boldsymbol{\varepsilon}} - \dot{\boldsymbol{\varepsilon}}^p)\,;\ \dot{\boldsymbol{\varepsilon}}^p = \dot{\lambda}\frac{\partial \phi}{\partial \boldsymbol{\sigma}}\,;\ \dot{q} = H(\lambda)\,\dot{\lambda} \tag{11}$$

where $\boldsymbol{\varepsilon}^p$ is the plastic strain vector, $\dot{\lambda}$ is the plastic multiplier, q is the hardening/softening internal variable, and H *is the* hardening/softening modulus. The loading and unloading situations are distinguished by the Kuhn-Tucker conditions:

$$\phi(\boldsymbol{\sigma}, q) \le 0, \quad \dot{\lambda} \ge 0, \quad \dot{\lambda}\,\phi(\boldsymbol{\sigma}, q) = 0 \tag{12}$$

where ϕ is the yield surface with $\phi(\boldsymbol{\sigma}, q) \le 0$ defining the elastic domain.

The following expressions for the yield surface and softening law are adopted:

$$\phi(\boldsymbol{\sigma}, q) = \sqrt{\frac{2}{3}}\,\|\mathbf{S}\| + p - q\,;\ H(\lambda) = \frac{\partial q(\lambda)}{\partial \lambda} = k\left(-0.95\frac{f_t^2}{G_F}e^{-\frac{f_t}{G_F}k\lambda}\right) \tag{13}$$

where $p = Tr(\boldsymbol{\sigma})/3$ is the mean stress, $\mathbf{S} = \boldsymbol{\sigma} - p\mathbf{I}$ is the deviatoric stress, f_t is the tensile strength, and G_f is the fracture energy.

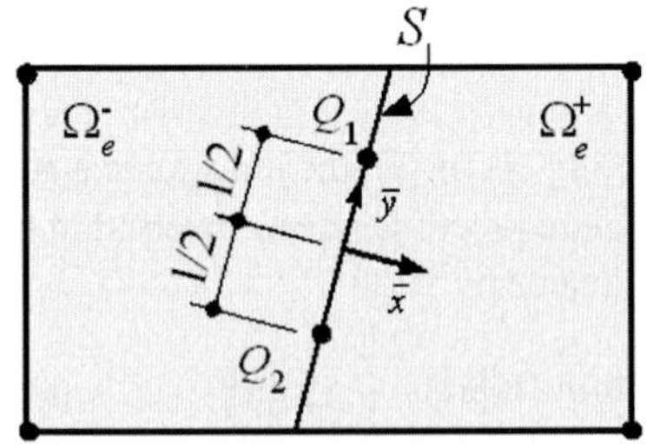

Figure 2. Collocation points on an interface with non-uniform discontinuity modes

3. Non-uniform discontinuity modes

Consider an interface discontinuity with two collocation points, Q_1 and Q_2, as shown in Figure 3. Each collocation point introduces a discontinuous jump, $[\![\mathbf{u}]\!]_i$ (where i =1, 2). It is possible to define $\mathbf{P}_1$ and $\overline{\mathbf{x}}_i$ matrices related to the collocation points such that:

$$\hat{\mathbf{d}} = [\mathbf{P}_1 \quad \mathbf{P}_2]\{[\![\mathbf{u}]\!]_1 \quad [\![\mathbf{u}]\!]_2\}^T \text{ with } \mathbf{P}_k = \begin{bmatrix} M_k(\overline{\mathbf{x}}_1) & 0 \\ 0 & M_k(\overline{\mathbf{x}}_1) \\ \vdots & \vdots \\ M_k(\overline{\mathbf{x}}_{n_{en}}) & 0 \\ 0 & M_k(\overline{\mathbf{x}}_{n_{en}}) \end{bmatrix} (k = 1,2) \tag{14}$$

where $\overline{\mathbf{x}}_i$ ($i = 1, 2,..., n_{en}$) are the nodal coordinates in the local coordinate system $\overline{x}$ - $\overline{y}$ aligned with the crack (see Figure 2). The functions $M_k(\overline{\mathbf{x}})$ can be constructed from the linear interpolation functions on S:

$$M_1(\overline{\mathbf{x}}) = \mathcal{H}(\overline{x})\frac{l/2+\overline{y}}{l}; \quad M_2(\overline{\mathbf{x}}) = \mathcal{H}(\overline{x})\frac{l/2-\overline{y}}{l} \tag{15}$$

4. NUMERICAL TESTS

4.1 Bending Test

This numerical study is performed using elements with embedded discontinuity based on the following displacement-based finite element formulations: triangular three-node element (T3) and the bilinear quadrilateral four-node element (Q4)[8]. The test is performed on a square plane stress element, whose geometry and boundary conditions are shown in Figure 3. The material parameters are: E=30 000 MPa; f_t = 3.0 MPa; G_F = 0.1 N/mm; ν=0.2. The numerical analysis is carried out with the following embedded crack finite element approximations: (a) a single quadrilateral element with uniform discontinuity modes (Q4-U), (b) a single quadrilateral element with non-uniform discontinuity modes (Q4-NU), (c) two quadrilateral elements with uniform discontinuity modes (2XQ4-U), and (d) two triangular elements with uniform discontinuity modes (2XT3-U). Figure 4 shows the different finite element discretizations used for the test.

The initially uncracked specimen is subjected to two consecutive loading paths. First, a monotonically increasing uniaxial tensile stress is applied by increasing the horizontal displacement of the right nodes (see Figure 3.a). This loading path ceases when the stress reaches the tensile strength of the material and the first crack line, S_1, forms perpendicular to the first principal stress (see Figure 3.a). The second loading path consists of a translation accompanied by a rotation of the right edge of the specimen with respect to the left one, as shown in Figure 3.b.

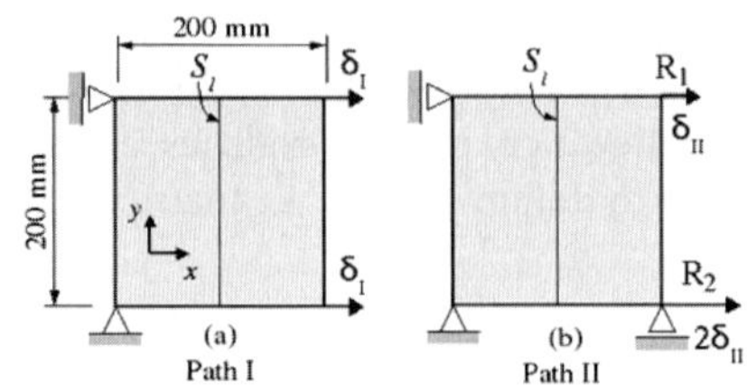

Figure 3. Bending test.

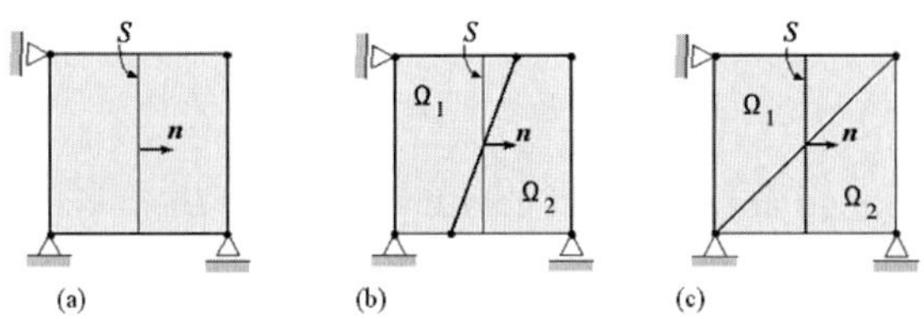

Figure 4. Different mesh discretizations for the bending test: (a) Q4-U or Q4-NU, (b) 2xQ4U, and (c) 2xT3-U

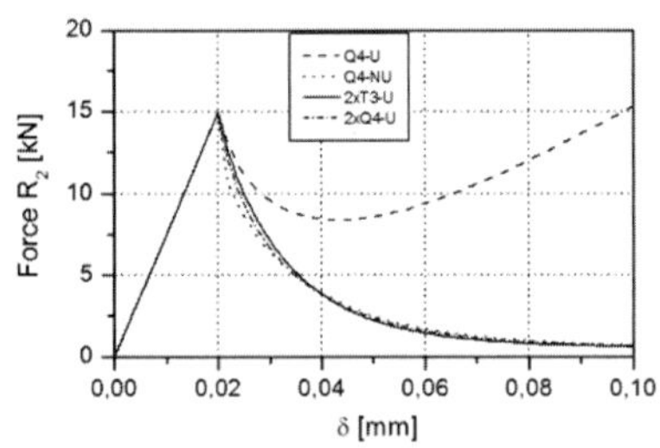

Figure 5. Resisting forces vs. displacement for the bending tests

Figure 5 shows the evolution of the resisting force $R2$ at the two right nodes (see Figure 3.b) for the different finite element approximations. The approximation using a single quadrilateral element with uniform discontinuity modes (Q4-U) exhibits a strong stress locking, which prevents the relaxation of the resisting forces. The approach with uniform discontinuity modes is not able to describe a relative rigid-body rotation between the two fragments of the element. As a consequence, the imposed rotation of the right edge mobilizes the strain of the elastic solid portion, inducing stress locking. The resisting forces obtained with the elements Q4-NU show that the stress locking is completely removed by the consideration of a non-uniform discontinuity mode, which is able to accommodate relative rigid-body rotation.

The approximation using two elements with uniform discontinuity mode (2xT3 and 2xQ4) does not present any stress-locking effect. As shown in the deformed configurations of Figure 6, the imposed bending mode can be described by means of the rigid-body rotation of the fragmented components of an element with uniform crack opening mode. Note that a mixed-mode discontinuity was obtained even for a bending situation in which a mode-I behavior was expected.

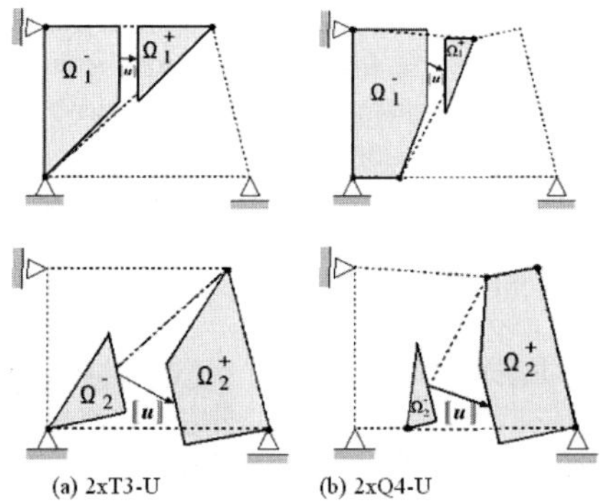

Figure 6. Deformed configuration for the bending test

Then, stress locking would be expected if an interface constitutive model that is not able to describe mixed-mode opening had been used or if the crack line had crossed the two opposite sides of a single quadrilateral element.

These tests elucidate why no stress locking due to bending-type deformation has been reported in the literature for numerical analysis using embedded triangular elements with uniform discontinuity modes. The same can be said for numerical analysis with quadrilateral elements in which the crack line cross most of the elements through two adjacent sides. In the latter case, stress locking occurs only in the elements with a crack passing through two opposite sides, generating a small effect in the structural response. Only a deeper look in the stress field of these elements would show the stress-locking effect, as it is clear in the next test.

4.2 Double-notched Specimen

In this test performed by Nooru-Mohamed[9], a concrete square with deep notches on both sides was subjected first to a shear (horizontal) load PS=5 kN and then to a normal (vertical) tensile load PN under displacement control, while keeping the shear load constant. The geometry and boundary conditions of the test are depicted in Figure 7.a. The assumed parameters are: E= 32 GPa, ν=0.18, ft =2.5 MPa, GF =0.1 N/mm, and k=1 mm. Cracks are allowed to initiate from the two elements located near the notch tips and an algorithm to track the crack that propagates during the loading is used, maintaining the continuity of the crack across the element boundaries.

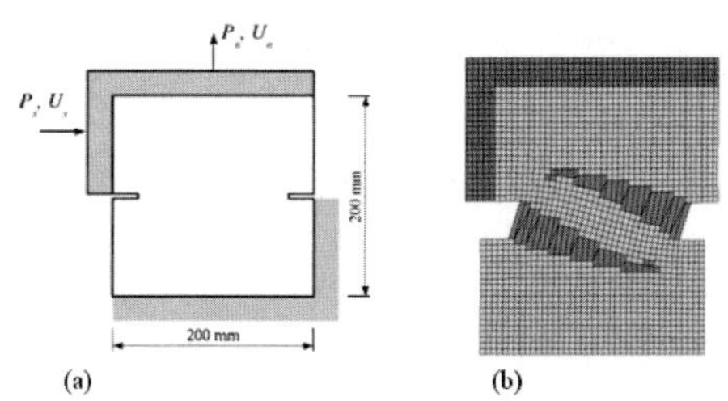

Figure 7. Double notched specimen: (a) problem definition, (b) deformed mesh

The numerical analyses were performed using triangular quadrilateral (Q4-U and Q4-NU) finite elements. Figures 7.b shows the deformed mesh exaggerated by 500 times at a normal displacement of 0.06 mm. The shaded elements show the crack paths. Since the crack lines propagating from the notch tips cross a large number of elements through two adjacent sides, the structural response obtained with the Q4-U element does not show any pronounced stress locking. However, the contours for maximum principal stress shown in Figure 8.a reveals the spurious oscillations of the stress field obtained with this element, particularly in the vicinity of the elements that are cracked.

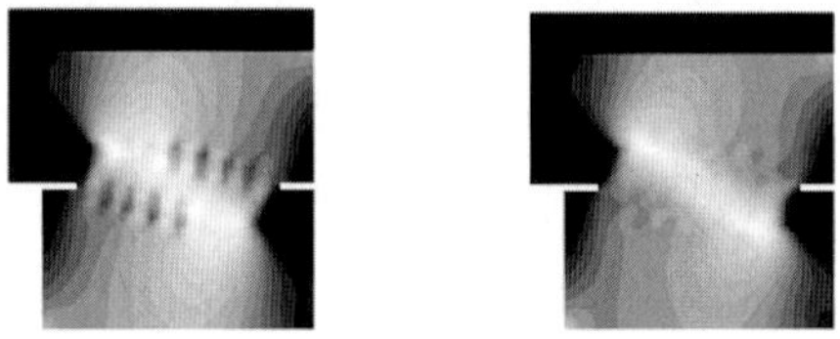

Figure 8. Contours of maximum principal stress for the double-notched specimen: (a) Q4-U, (b) Q4-NU elements

5. CONCLUSIONS

A general technique to embed cracks into standard finite elements, which allows for both uniform and non-uniform discontinuity modes to be considered to represent the kinematics of a crack, has been presented. It has been demonstrated that a non-uniform crack opening mode is crucial to avoid stress locking in quadrilateral elements subjected to bending-type deformation. Stress locking will occur when an interface line crosses an element through two opposite sides. It has also been elucidated why non-uniform modes are not required for triangular elements or for quadrilateral elements crossed by an interface passing through two adjacent sides. The price of this is that a mixed-mode discontinuity develops even for bending-type deformation, for which only mode-I discontinuity should be expected. Therefore, for this case, the use of an interface constitutive model that allows for mixed-mode fracture is essential to avoid stress locking.

Acknowledgments

The first author acknowledges the financial support from the State of São Paulo Research Foundation (FAPESP).

6. REFERENCES

1. Jirásek, M., "Comparative Study on Finite Elements with Embedded Discontinuities", *Comp. Methods Appl. Mech. Engng.*, ***188***, 307-330, (2000).
2. Oliver, J., "Modeling Strong Discontinuities in Solid Mechanics via Strain Softening Constitutive Equations. part 2: Numerical simulation ", *Int. J. Num. Meth. Eng.*, **21**(39), 3575-3600 (1996).

3. Arrea M. and Ingraffea A.R. "Mixed-mode crack propagation in mortar and concrete".Technical Report 81-13, Dept. of Struct. Engng., Cornell University (1982).
4. Spencer, B.W., "Finite Elements with Embedded discontinuities for Modeling Reinforced Concrete Members", Ph.D. Thesis, Department of Civil, Environmental and Architectural Engineering, University of Colorado (2002).
5. Oliver, J, "Modeling Strong Discontinuities in Solid Mechanics via Strain Softening Constitutive Equations. part 1: Fundamentals", *Int. J. Num. Meth. Eng.*, **21**(39), 3575-3600 (1996).
6. Oliver.J. Cervera, M. and Manzoli, O.L., "Strong Discontinuities and Continuum Plasticity Models: the Strong Discontinuity Approach", *Int. J. of Plasticity*, **3**(15), 319-351(1999).
7. Simo, J.C. and Hughes, T.J. R., "*Computational Inelasticity*", Springer-Verlag (1998).
8. Hughes, T.J.R., "*The Finite Element Method; linear static and dynamic finite element analysis*", Prentice-Hall, Englewod Cliffs, N.J. (1987).
9. Nooru-Mohamed, M.B., "Mixed-mode fracture of Concrete: an experimental approach". Ph.D. Thesis, Delft University of Technology, Delft (1992).

EFFICIENT STRENGTHENING TECHNIQUE FOR REINFORCED CONCRETE SLABS

SFRC and CFRP laminate strips

E. Bonaldo[1], J.A.O. de Barros[2] and P.B. Lourenço[2]

[1]*PhD Student;* [2]*Associate Professor University of Minho, Department of Civil Engineering, P-4800-058 Guimarães, PORTUGAL*

Abstract: A promising strengthening strategy, using carbon fiber reinforced polymer (CFRP) materials, consists in applying CFRP laminate strips into pre-cut slits opened in the concrete cover of the elements to strengthen. This strengthening technique is designated by Near Surface Mounted (NSM) and has been successfully used to increase the flexural and the shear resistance of concrete and masonry structures. The present work describes an efficient strategy, using steel fiber reinforced concrete (SFRC) and NSM CFRP laminates, for the strengthening of existing reinforced concrete (RC) slabs. The use of a SFRC compression overlay can provide the necessary ductility for attaining high level of tensile stress in the CFRP strengthening system and therefore preventing the concrete crushing failure mode. In the present work, the effectiveness of this technique to increase the service and ultimate load carrying capacity of RC slabs is assessed by an experimental program. A numerical strategy was developed to predict the load-deflection relationship of this type of elements. The results are presented and analyzed, and the performance of the numerical model is appraised.

Key words: flexural strengthening; reinforced concrete slabs; CFRP laminate; thin bonded overlay; steel fibre reinforced concrete; epoxy adhesive.

1. INTRODUCTION

The Near-Surface Mounted (NSM) strengthening technique has been used in the recent years, with remarkable efficiency, to increase the flexural strength[1-4] and the shear resistance[1,5] of reinforced concrete elements. The NSM technique involves the embedment of CFRP bars - of circular, square or rectangular cross-section - into grooves opened on the concrete surface. When compared to the Externally Bonded Reinforcing (EBR) technique, the NSM technique assures a higher anchoring capacity to the FRP. As a consequence, a high tensile stress can be mobilized in the CFRP, as long as the member load

M.S. Konsta-Gdoutos, (ed.), Measuring, Monitoring and Modeling Concrete Properties, 125–131.

carrying capacity is not limited by a premature failure mode. For RC slabs of low or medium concrete strength, the increment of the flexural resistance that NSM can provide might be limited by the maximum allowable compression strain in the most compressed concrete fibre. This drawback can be overcome by adding a concrete layer in the compression zone of the existent slab[6]. To attain the desired structural performance (e.g. full composite action), the new concrete overlay and the existent concrete slab should behave monolithically. A sound bond between the new layer and the existing concrete slab can be guaranteed if a proper epoxy compound is used[7,8].

2. EXPERIMENTAL WORK

2.1 Slab specimens, test set-up and materials

To assess the efficiency of the hybrid strengthening technique for the increase of flexural load carrying capacity of RC slabs, the slab strip specimens represented in Figure 1 were used. The cross section dimensions and the test set up of the tested slab strip specimens are also illustrated in Figure 1. Two unstrengthened RC slabs formed a control set (SL01 and SL06), three slabs were strengthened with CFRP laminates according to NSM technique (SL03S, SL04S and SL08S), and three were strengthened with NSM laminates and a compression SFRC overlay (SL02SO1, SL05SO1 and SL07SO2). The number of CFRP laminate strips applied in each RC slab was evaluated in order to obtain an increase of 50% in the service load, which was assumed as the load producing a mid-span displacement of $\ell/250 = 1800\ \text{mm}/250 = 7.2$ mm. Each slab specimen was tested in simply supported conditions, with a clear span of 1.8 m, and under line loads at 0.6 m from the supports, see Figure 1(b). The monotonic loading was controlled by the LVDT placed at slab mid span (see Figure 1(b)), using a displacement velocity of 20 μm/s up to failure of the slab. Figure 2 outlines the arrangement of the strain gauges (SG) applied to measure the strains in the CFRP laminates, steel bars and concrete. Tables 1 and 2 include the main mechanical properties of the materials used in the present work. In Table 1 f_{cm} is the compressive strength and E_c the elastic modulus, $f_{ctm,fl}$ and $f_{ctm,ax}$ the flexural and axial tensile strength, respectively, of plain concrete. In Table 2, E is the elastic modulus, σ_u and ε_u the ultimate strength and strain, respectively; σ_{sy} and ε_{sy} the steel yield stress and strain in tension; f_{cm} the compressive strength, $f_{eqm,2}$ and $f_{eqm,3}$ the equivalent flexural tensile strength parameters of SFRC overlays; t_f and W_f the thickness and width of the CFRP.

2.2 Results and comments

Table 3 includes the maximum load, the maximum concrete compression strain, the maximum strain in the CFRP laminates and the failure modes of the tested slab strips. Due to problems with the data acquisition system, the strains in the SL04S were not measured. The maximum load of the strengthened slabs was about five times higher than the maximum load of the corresponding unstrengthened slabs. The maximum concrete compression strain has exceeded the strain corresponding to the concrete strength. The maximum strains recorded in the CFRP laminates are about 80% of its ultimate strain, but these values do not correspond to the maximum load since the strain gage data acquisition was interrupted for the load values included in brackets. Due to the significant increase of the slab load carrying capacity, provided by the NSM strengthening technique and the hybrid system, some slabs have failed in a flexure/shear combined mode, but for a deflection that was several times the deflection at the yielding of the reinforcement.

Using the strains recorded in the strain gauges installed on the laminates, the average laminate-concrete bond stresses (τ_{bm}^{RL}) developed along the CFRP laminate strips was evaluated (refer to Figure 3(a)). A typical bond stress variation in the CFRP laminate strips is shown in Figure 3(b).

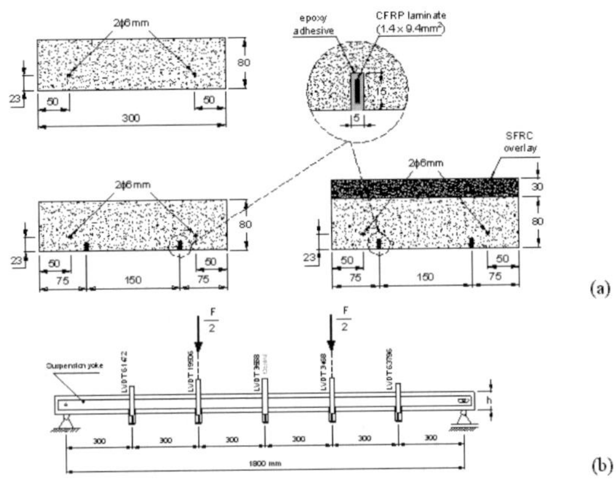

Figure 1. (a) Slab cross-section dimension and disposition of the steel bars and CFRP laminates and (b) load configuration and arrangement of the LVDTs (dimensions in mm)

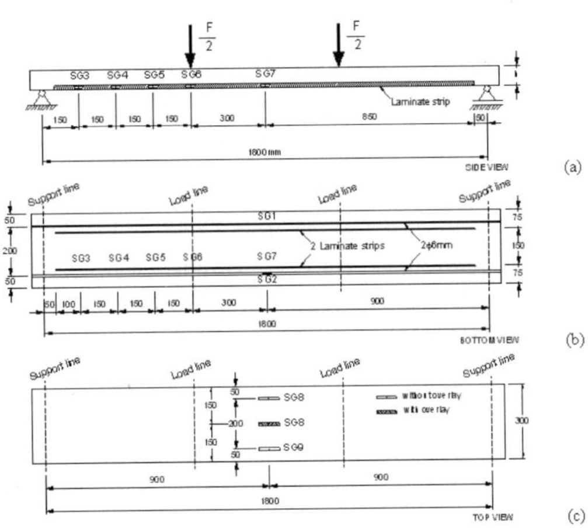

Figure 2. Disposition of the strain gauges: (a) side, (b) bottom and (c) top views

Table 1. Characteristics of the plain concrete (obtained experimentally[9])

Property	Control Slabs		NSM Strengthened Slabs			NSM & SFRC Strengthened Slabs		
	SL01	SL06	SL03S	SL04S	SL08S	SL02SO1	SL05SO1	SL07SO2
f_{cm} (MPa)	45.65	49.39	43.13	32.41	49.35	47.76	49.56	47.80
$f_{ctm,fl}$ (MPa)	4.98	6.10	5.92	4.73	5.80	5.86	5.59	6.04
$f_{ctm,ax}$ [a] (MPa)	2.75	3.37	3.27	2.61	3.21	3.24	3.10	3.32
E_c [a] (GPa)	35.67	36.61	35.00	31.82	36.60	36.21	36.66	36.22

[a] Derived from CEB FIP model code 1990

Table 2. Summary of the characteristics of the steel reinforcement, CFRP and its adhesive, and SFRC and its adhesive (obtained experimentally[9])

Steel reinforcement	CFRP laminate	Laminate adhesive	SFRC overlay[a]		Overlay adhesive
$\phi_s = 6mm$ $E_s = 217.3GPa$ $\sigma_{sy} = 548.8MPa$ $\varepsilon_{sy} = 2.70‰$	$t_f = 1.41mm$ $W_f = 9.37mm$ $E_f = 156.1GPa$ $\sigma_{fu} = 2879.1MPa$ $\varepsilon_{fu} = 18.45‰$	$E_a = 7.47GPa$ $\sigma_{au} = 33.0MPa$ $\varepsilon_{au} = 4.83‰$	O1	$f_{cm} = 38.93MPa$ $f_{eqm,2} = 5.00MPa$ $f_{eqm,3} = 4.12MPa$	$E_e = 3.62GPa$ $\sigma_{eu} = 26.56MPa$ $\varepsilon_{eu} = 10.74‰$
			O2	$f_{cm} = 53.10MPa$ $f_{eqm,2} = 4.83MPa$ $f_{eqm,3} = 3.86MPa$	

[a] Evaluated according to RILEM TC 162 TDF recommendations as reported in Barros et al.[10]

Table 3. Summary of the slab test results

Strengthening	Slab I.D.	Average ultimate load (kN)	Strength increasing ratio (%)	Concrete compression strain[c] (‰)	CFRP laminate strain[d] (‰)	Type of Failure
Reference	SL01	5.03	NA	2.26[5.35]	NA	Flexure
	SL06	*(9.00%)*		1.96[4.71]		Flexure
CFRP laminate strengthening	SL03S	24.48	386.68	3.40[24.24]	14.10[23.13]	Flexo-shear
	SL04S	*(1.59%)*		NE	NE	Flexure
	SL08S			2.90[24.00]	12.70[18.70]	Flexure
CFRP + SFRC strengthening	SL02SO1	33.79	571.77[a]	2.66[35.42]	12.95[34.42]	Flexo-shear
	SL05SO1	*(5.56%)*	38.03[b]	2.53[31.66]	13.50[31.50]	Flexo-shear
	SL07SO2			NE	12.58[26.84]	Flexure

(*value*) Coefficient of Variation (COV) = (Standard deviation/Average) x 100
[a] With respect to the reference; [b] With respect to the CFRP laminate strengthening; [c] Value in square brackets is the maximum load; [d] Maximum value recorded in SG7 and corresponding load in square brackets; NA: not applicable; NE: not evaluated

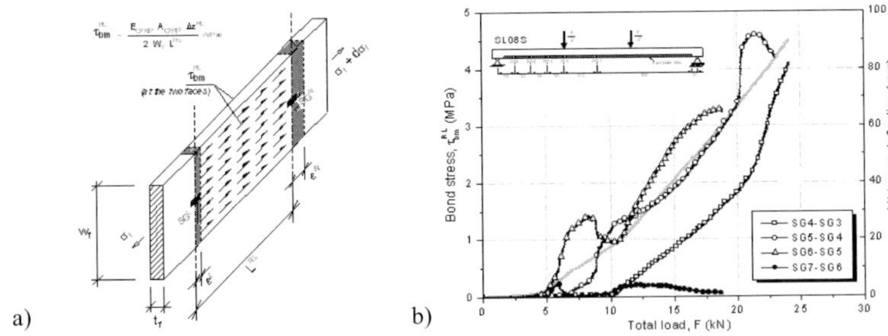

Figure 3. (a) average bond stress calculation and (b) typical bond stress variation in monitored CFRP laminate strip for the slab SL08S

Figure 3(b) shows that up to crack initiation, the CFRP laminates were not yet mobilized. In general, at service load, corresponding to a deflection of 7.2 mm at mid-span, the bond stress did not exceed 1.0 MPa. Despite the fact that in some strengthened slabs it was not possible to evaluate the average bond stress variation up to the ultimate load, it can be noticed, however, that the bond stress did not surpass 5.0 MPa. This maximum value is much lower than the bond stress limit value (12 MPa), registered in pullout-bending tests[11].

3. NUMERICAL ANALYSIS

Previous works[3] have shown that, using a cross-section layered model that takes into account the constitutive laws of the intervening materials and the kinematic and the equilibrium conditions, the deformational behavior of structural elements failing in bending can be predicted from the moment-curvature relation, M - χ, of the representative sections of these elements, using the algorithm described in a former paper[12]. To evaluate the M - χ relationship, the slab cross section was discretized in layers of 0.5 mm thickness. The slab tangential stiffness matrix was determined evaluating the tangential stiffness matrix of the two nodes Euler-Bernoulli beam elements discretizing the slab (a mesh of 60 elements). The values of the parameters for the models defining the behavior of plain concrete, SFRC, steel bars and CFRP laminates are given elsewhere[12]. Figure 4 shows that the developed numerical strategy is able of fitting with enough accuracy the registered experimental load-central deflection curves of the tested slabs.

As a consequence of the increase of the post-cracking stiffness, provided by CFRP laminates, the service load has increased 54% for the slabs strengthened with NSM (see inset of Figure 4). The hybrid strengthening lead to an increase of about 212% in the service load with respect to the reference (see inset of Figure 4).

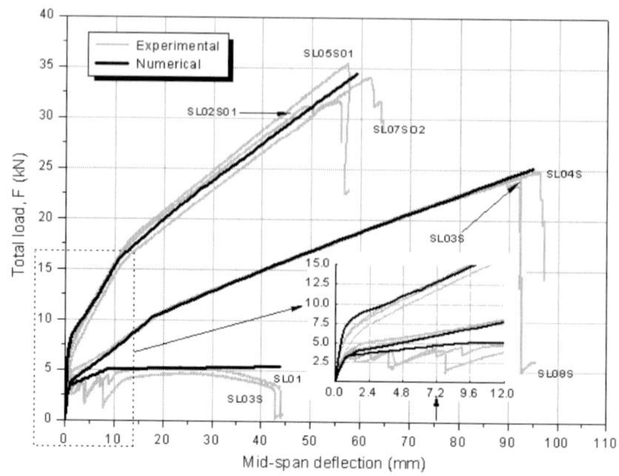

Figure 4. Experimental versus numerical load-central deflection curves

4. CONCLUSIONS

The testing program carried out demonstrated that the hybrid strengthening technique has great potential application towards flexural strengthening of RC slabs. A percentage of 0.12% of CFRP laminates (CFRP reinforcement to the conventional steel reinforcement ratio close to 50%) has increased about 54% the service load of the 1.8 m RC slabs with a steel reinforcement ratio of 0.24%. However, the slabs strengthened with NSM technique and SFRC showed an increase of approximately 212% in the service load, with respect to the reference slabs. Comparatively to the NSM technique, the hybrid strengthening strategy has lead to an increase of about 103% in load at the service load level. The hybrid strengthening system has also lead to an increase of about 570% in the RC slab maximum load carrying capacity with respect to the reference slabs and an increase of about 40% in comparison with the slabs strengthened only with NSM technique. When compared to the reference case, about 390% of increase in the load carrying capacity was attained by the strengthening with NSM technique. The cracking spacing calculations and crack features observations, at the bottom of the slabs, indicate that a significant improvement in the crack behavior of RC slabs can be achieved with the NSM technique. When compared with the bond stress limit recorded in pullout-bending tests, a very low bond stress profile was observed through the interfaces CFRP laminate-epoxy adhesive-concrete, along the laminate strips in slabs where NSM strengthening was applied. The NSM strengthening system has also provided a significant increase in the stiffness and deformation at failure, which are consistent with the high stress redistribution owing to prominent composite action between the CFRP reinforcement and concrete. Since the hybrid strengthening system has lead to substantial increase in flexural load, the shear capacity of the composite slabs has limited their deformability; however, the stiffness of the slabs has strongly increased and high ductility was maintained. The numerical model developed to simulate the load-deflection relationship of RC elements reinforced strengthened with CFRP laminate strips has reproduced with high accuracy the force-mid span deflection of the carried out tests.

Acknowledgments

The authors acknowledge the Portuguese Science and Technology Foundation (FCT) for the PhD grant number SFRH / BD / 11232 / 2002. Thanks also for the companies "Companhia Geral de Cal e Cimento S.A. (SECIL)", Sika S.A., "Central do Pego", "Pedreiras Bezerras", Bekaert NV, "Degussa Construction Chemicals Portugal S.A.", S&P® Reinforcement, which have supplied cement; overlay bond product; fly ash; aggregates; steel fibres; superplasticizer and CFRP adhesive; and CFRP laminate, respectively.

5. REFERENCES

1. L. De Lorenzis, A. Nanni, and A. La Tegola, Flexural and shear strengthening of reinforced concrete structures with near surface mounted FRP rods, In Proceeding of the 3rd Int. Conf. on Advanced Composite Materials in Bridges and Structures.J. Humar and AG Razaqpur, Editors, Ottawa, Canada, pp.521-528, 15-18 Aug. (2000).
2. A. Carolin, Carbon fibre reinforced polymers for strengthening of structural elements, Doctoral Thesis, Division of Structural Engineering, Luleå University of Technology, Luleå, Sweden, p. 190, Jun. (2003).
3. J. A .O. Barros, and A. S. Fortes, Flexural strengthening of concrete beams with CFRP laminates bonded into slits, *Cement. Concr. Compos.* **27**(4), 471-480 (2005).
4. R. Kotynia, Strengthening of reinforced concrete structures with near surface mounted FRP reinforcement, 5th International Conference - Analytical models and new concepts in concrete and masonry structures AMCM 2005, p. 8, Gliwice - Ustron, 12-14 Jun. (2005).
5. J. A. O. Barros, and S. J. E. Dias, Shear strengthening of reinforced concrete beams with laminate strips of CFRP, International Conference Composites in Constructions - CCC2003, Cosenza, Italy, pp. 289-294, 16-19 Sept. (2003).
6. J. A. O. Barros and J. M. Sena-Cruz, Strengthening a prestressed concrete slab by epoxy - bonded FRP composites and SFRC overlayer, 7th Int. Conf. on Inspection Appraisal Repairs & Maintenance of Buildings & Structures, Nottingham Trent University, UK, 11-13 Sept. (2001).
7. E. Bonaldo, J. A. O. Barros, and P. B. Lourenço, Bond characterization between concrete base and repairing SFRC by pull-off tests, Report 04-DEC/E-13, May (2004).
8. E. Bonaldo, J. A. O. Barros, and P. B. Lourenço, Bond characterization between concrete substrate and repairing SFRC using pull-off testing, *Int. J. Adhes. Adhes.* **25**(6), 463-474 (2005).
9. E. Bonaldo, J. A. O. Barros, and P. B. Lourenço, Steel fibre reinforced concrete and CFRP laminate strips for high effective flexural strengthening of RC slabs, Report 05-DEC/E-14, Oct. (2005).
10. J. A. O. Barros, V. M. C. F. Cunha, A. F. Ribeiro, and J. A. B. Antunes, Post-cracking behaviour of steel fibre reinforced concrete, *Mater. Struct.* **38** (275), 47-56 (2005).
11. J. M. Sena-Cruz, and J. A. O. Barros, Bond between near-surface mounted CFRP laminate strips and concrete in structural strengthening, *J. Compos. Construct.* **8**(6), 519-527 (2004).
12. E. Bonaldo, J. A. O. Barros, and P. B. Lourenço, Concrete slabs strips reinforced with epoxy-bonded carbon laminates into slits, 3rd Int. Conf. on Construction Materials: Performance, Innovations and Structural Implications, Vancouver, CA, 22-24 Aug. (2005).

BENDING PERFORMANCE OF HIGH STRENGTH STEEL FIBRE REINFORCED CONCRETE

Static and fatigue loading conditions

E.S. Lappa, C.R. Braam and J.C. Walraven

Delft University of Technology

Abstract: Four point bending tests on 125/125/1000 mm beams at a 750 mm span were performed under both static and fatigue loading conditions. The results of the static tests were used to determine the chosen fatigue sinusoidal loading at two load levels. Three different concrete mixtures were tested: one ultra high strength mixture and two high strength ones that had small differences in their matrix composition and fibre type and content. The static peak load depends on the amount of fibres in the mixture. The scatter in the fatigue result can be reduced by the use of a good workable and flow-able mixture in the fresh state.

Key words: Steel fibres, Fatigue, S-N lines, Fibre distribution, Ultra high strength concrete

1. INTRODUCTION

Concrete as a construction material is often the choice for heavily loaded structures that are expected to resist millions of repeated loading cycles from traffic or other loads during their service lives. Such structures can be road and railroad bridges, airport runways, and offshore structures, only to name some examples. For these applications, not only the static (bending) behaviour is of importance in design, but also appropriate fatigue verifications are necessary in order to prevent fatigue failure.

Recent developments in concrete production led to the introduction of high strength concretes with strengths exceeding 100 MPa in compression, and even ultra high strength concretes (also called reactive powder concretes) with even higher strengths, of 200 MPa and higher. These high-tech materials combine a carefully chosen, fine and dense cementitious matrix and specially selected aggregates and fillers to reach the high compressive strength. Steel or other types of fibres are added to achieve a higher deformation capacity and improve the tensile and flexural tensile strengths. As a consequence, thin, slender structures can be designed with a reduced total amount of reinforcement due to the higher

M.S. Konsta-Gdoutos, (ed.), Measuring, Monitoring and Modeling Concrete Properties, 133–138.

strengths and fibre reinforcement. Since these materials are relatively new, there is still a need for experimental determination and verification of material properties in studies such as the one presented here.

2. MATERIALS AND TEST SET-UP

Three different materials were used in the experimental series: one ultra high strength fibre reinforced concrete and two high strength fibre reinforced concretes. The mixture composition and procedure will be briefly given here, as well as a short description of the testing series. The material description given here is mainly intended to distinguish the mixtures among each other in the results and discussions section; a more detailed description of the mixtures can be found in the listed references.

The Ultra High Strength Fibre Concrete is a commercially available mixture made out of the premix CERACEM and will be referred here as BSI/CERACEM[1,2]. It contains straight steel fibres with a length of 20 mm and a diameter of 0.3 mm at a volume fraction of 2.5%. The mean value of the compressive strength, as measured on 100 mm cubes was 217 MPa. It is a self-compacting mixture, however its fresh state characteristics with regard to the flow ability are less favourable compared to the other two high strength mixtures.

Two high strength fibre reinforced concretes (HSFRCs) were used, both developed and produced in the Delft University laboratory. They are comparable by means of their ultimate compressive, splitting tensile and flexural tensile strengths. Their differences lie mainly in the fibre types and post-peak bearing capacity. The first one, which will be referred to as HSFRC[3,4], contains 13 mm long steel fibres with a diameter of 0.16 mm at a fibre volume fraction of 1.6%. The second one, the hybrid HSFRC[5], is called hybrid since it contains a combination of two different types of steel fibres: 0.5% by volume 13 mm long fibres with a diameter of 0.2 mm and 1% by volume of 60 mm long fibres with a diameter of 0.75 mm. The average compressive strength as tested on 100 mm long were 146 MPa for the HSFRC and 131 MPa for the hybrid HSFRC. Both mixtures are self-compacting.

The experimental program consisted of four-point bending tests on 125/125/1000 mm beams, loaded at their third-points at a 750 mm span. The same set-up was used for static and fatigue tests. The test set-up and specimen preparation is described in more detail elsewhere[2]. The results given in this paper are focusing on the fatigue test series. The fatigue tests were performed load-controlled, with a sinusoidal load between two pre-set load levels at a frequency of 10 Hz. The upper load level was set to a percentage of the previously determined average static peak load for each mixture, and the lower load level was fixed at 20% of the upper load in all tests.

3. RESULTS AND DISCUSSION

3.1 Static tests

The results of the static tests are given in Table 1. The flexural tensile stress given in the table is only an indication, since it is determined according to the elastic beam formulation and is not corresponding to the actual, non linear elastic stress at the bottom fibre of the beam. Figure 1 shows that the peak load can be related to the amount of fibres in one mixture, when these are expressed with a 'fibre factor' in order to be able to compare fibres of different geometry. This factor is the amount of fibres by volume contained in the mixture, multiplied by the fibre aspect ratio, the latter being the fibre length divided by its diameter.

The form of typical static load deflection curves can be taken out of Figure 5, where normalised curves of the HSFRC mixture are shown. All three mixtures show a 'deflection hardening' stage before the peak load is reached. In that stage, one major crack starts to develop which will evolve into the crack leading to failure. For the BSI/CERACEM and HSFRC, this crack is the only visible one, however a number of microcracks are formed. This is indicated by the presence of the hardening stage after the linear elastic part of the curves. The hybrid mixture, which contained the smallest amount of short fibres which contribute to the bridging of microcracks and therefore can level the peak load, showed more than one visible crack before the localization in one major crack took place. This mixture also had the least steep descending part in the load-deflection curve, due to the presence of longer fibres that are able to bridge wider cracks.

Table 1. Static test results of all three mixtures

	BSI/CERACEM	HSFRC	Hybrid HSFRC
Number of tested beams	6	8	8
Peak load [kN]	77.7	62.1	50.4
Flexural tensile strength [MPa]	29.9	23.9	19.3
Std deviation of strength [MPa]	3.2	2.0	1.4
Coefficient of variation	11%	8%	7%

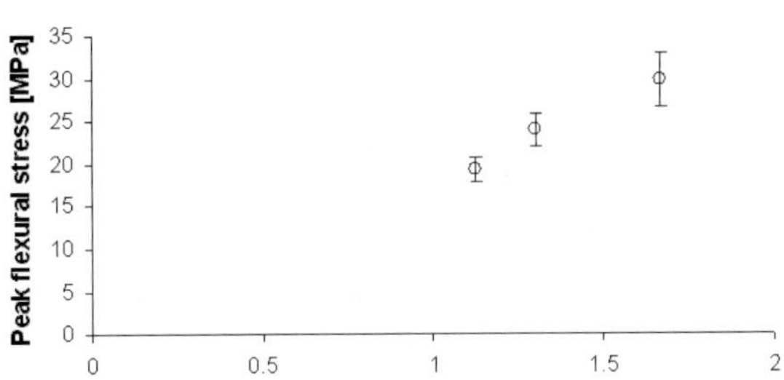

Figure 1. Static tests: Peak load vs. fibre factor

3.2 Fatigue tests

At least four different load levels were applied for each mixture, with 4-6 beams tested per load level. As is often the case in fatigue experiments, the scatter in the results was rather high, and more test specimens would have been needed for a sound statistical interpretation of the results. Still, as will be shown in the following, some conclusions and findings can be drawn from the conducted experiments.

The results of all three mixtures with respect to the upper load level during the fatigue test are given in Figure 2. The average lines shown here are the linear least square regression lines from the average number of load repetitions until failure for each mixture. It has to be noted that the largest scatter was observed for the BSI/CERACEM specimens. The line did not have a good correlation, therefore a linear fit is not appropriate for this mixture and the line is shown only for indicative reasons. For the other two mixtures, the fit is better, with a value for the correlation coefficient R^2=0.8. For all three mixtures, the run-out specimens, that is, specimens that did not fail up to ten million load cycles, were not included in the regression. If these are included, a better fit is obtained. In the graph, the individual results of the BSI/CERACEM mixture are marked with squares, HSFRC with triangles and hybrid HSFRC with circles. It can be seen that in absolute terms, the BSI/CERACEM and the HSFRC have about the same fatigue performance, at least when comparing the linear fit of the results; the individual data points show that the BSI/CERACEM, being the mixture with the highest strength, can survive more load cycles at higher loads, it was the only mixture where upper loads above 55 kN could be subjected.

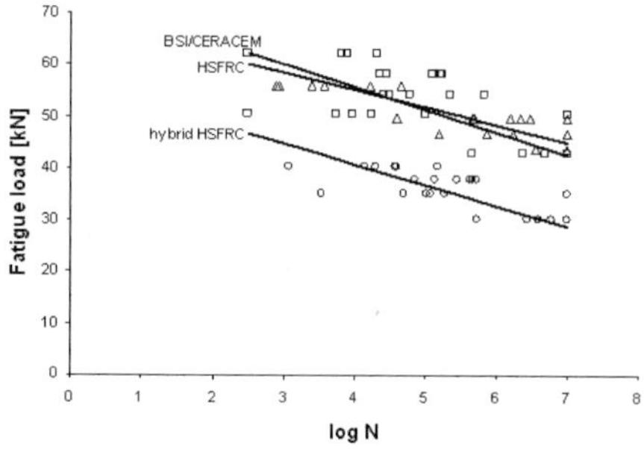

Figure 2. Fatigue load vs. cycles to failure. The BSI/CERACEM results are marked with squares, HSFRC with triangles and hybrid HSFRC with circles

Figure 3 shows the results of Figure 2 with the upper load level normalised with respect to each mixture's static peak load, which is a common way to present fatigue test results. Such curves are also known as S-N curves or Wöhler curves of a material. When presenting the results in this way, the BSI/CERACEM has a worse fatigue performance while the HSFRC is the superior of all three mixtures. In fact, almost all HSFRC beams tested at 70% of the static peak load did not fail up to ten million load cycles, while with BSI/CERACEM the load has to be lowered to 50% in order to obtain a fatigue life of ten million load cycles for most beams. This explains that the average number of cycles to failure of these two mixtures was in the same range at absolute loads of around 55 kN.

More aspects of the fatigue results are given in the Figures 4 and 5. Figure 4 shows the average evolution of deflections at each upper load level for the BSI/CERACEM. In order to compare deflections at midspan of individual beams that had a very different number of cycles to failure, the fatigue lives were normalised with respect to their number of cycles to failure. It can be seen that a higher upper load level leads to higher deflections. The same trend is observed for the two other mixtures.

Figure 5 compares the deflection of the static tests to the deflection at fatigue failure. It can be seen that the static curves in deflection can not serve as an envelope for the fatigue tests, especially not for lower load levels. Subramanian et al.[6] and Suthiwarapirak et al.[7] support this finding. The first authors[6] performed fatigue three point bending tests with plain concrete beams, and concluded that the static deflection cannot serve as an envelope, but the effective crack lengths were better suitable. The second authors[7] observed similar results as in this study, with the deflections at higher stress levels better suitable and significantly lower ones at lower levels. They performed four point bending tests on Engineered Cementitious Composites, ECC, polypropylene fibre reinforced concrete that show multiple cracking in tension and therefore a very ductile behaviour for concrete. At high fatigue load levels, most cracks were formed during the first load repetitions, and fewer new cracks formed during testing up to failure. At these load levels, the fatigue crack growth is the dominant failure mechanism opposed to fatigue crack initiation. At lower load levels, where the applied load does not provoke extensive cracking in the initial load repetitions, multiple cracking was much less pronounced and in fact fatigue crack initiation becomes responsible for failure. They conclude that multiple cracking characteristics are less pronounced at lower load levels, which implies a more brittle type of failure and can explain the smaller deflections compared to the corresponding static ones on the descending branch at the same load, as is also the case for the materials tested in this study.

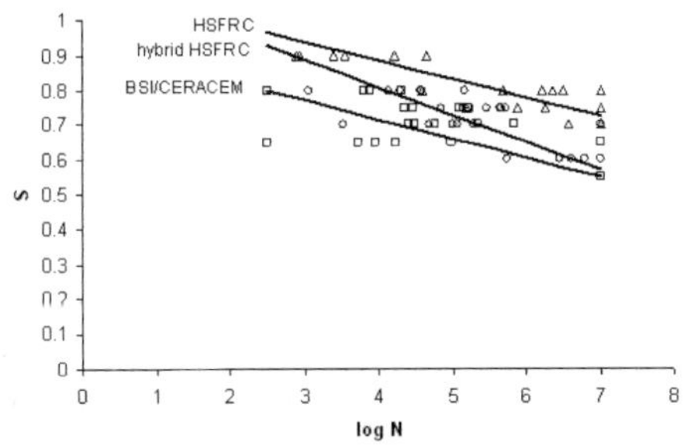

Figure 3. S-N lines. BSI/CERACEM are marked with squares, HSFRC with triangles and hybrid HSFRC with circles

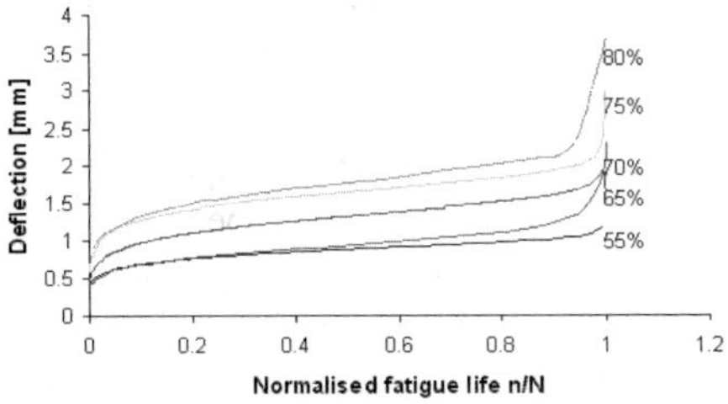

Figure 4. Average deflection during fatigue experiment of BSI/CERACEM

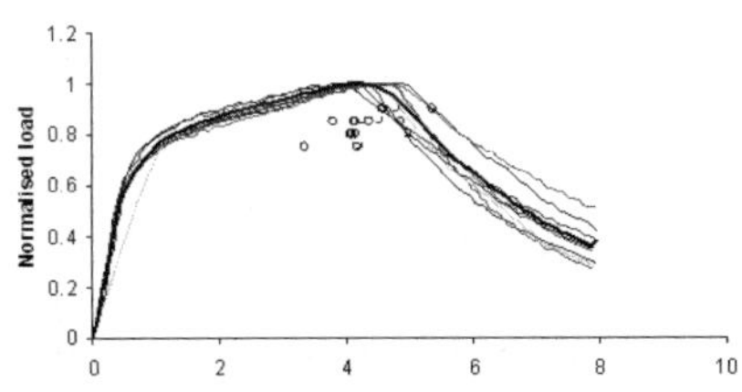

Figure 5. Load-deflection curves for static loading of the HSFRC in comparison with the maximum deflection reached at the fatigue loading (marked with circles)

4. CONCLUSIONS

The main conclusions, that can be drawn from the results presented before, are the following:

1 The fibres present in the mixtures influence the load bearing capacity and ductility in the static experiments.
2 The scatter in the fatigue experiments was the highest for the BSI/CERACEM and lower for the two HSFRCs. The better the workability of a mixture in the fresh state, the lower the scatter, not only with regard to the static but also to the fatigue performance of the mixture.
3 The deflections depend on the upper load level, lower deflections were observed at lower fatigue load levels. The static load-deflection curves cannot serve as envelope curves, especially not for lower load levels.
4 Multiple cracking characteristics, as observed under static load conditions, can only occur at higher fatigue levels, while a single crack failure behaviour is responsible for fatigue failure at lower stress levels.

Acknowledgements

The research at Delft University of Technology is supported by the Technology Foundation STW, applied science division of NWO and the technology programme of the Ministry of Economic Affairs.

5. REFERENCES

1. T. Thibaux, Z. Hajar, A. Simon, and S. Chanut,. Construction of an ultra-high-performance fibre-reinforced concrete thin-shell structure over the Millau viaduct toll gates, in: Proc. of 6th Int. RILEM Symposium on Fibre Reinforced Concrete (FRC), BEFIB (RILEM, 2004), pp. 1183-1192.
2. E.S. Lappa, C.R Braam. and J.C. Walraven, Static and fatigue bending tests of UHPC, in: Proc. of the Int. Symposium on Ultra High Performance Concrete (Kassel University Press GmbH, 2004) pp. 449-459.
3. S. Grünewald, Performance based design of self-compacting, fibre reinforced concrete, PhD Thesis, Delft University Press, 2004.
4. E.S. Lappa, C. van der Veen and J.C. Walraven, Self-compacting, high strength fibre reinforced mortar for pre-cast sheet piles", In proc. of , 3rd Int. Symposium on Self Compacting Concrete, Reykjavik, (RILEM 2003), pp. 732-740.
5. I. Markovic, J.C. Walraven and J.G.M. van Mier, Development and utilization of high performance hybrid-fibre concrete, in: proc. of 5th Int. PhD Symposium in civ. eng., Delft (Balkema publishers, 2004), pp. 1039-1047.
6. K.V. Subramanian, E. O'Neil, J.S. Popovics and S.P. Shah, Crack propagation in flexural fatigue of concrete, J. of eng. Mechanics, Sept. 2000, pp. 891-898.
7. P. Suthiwarapirak, T. Matsumoto and T. Kanda, Multiple cracking and fiber bridging characteristics of engineered cementitious composites under fatigue flexure, J. of materials in civ. Engineering, ASCE, Sept/Oct 2004, pp. 433-443.

AXIAL SYMMETRY ANALYSES OF PUNCHING SHEAR IN REINFORCED FLAT SLABS

L. Trautwein[1], T. Bittencourt[1], R. Faria[2], J.A. Figueiras [2]and R. Gomes [3]
[1]Structural and Foundation Engineering Department, Polytechnic School of São Paulo University, São Paulo, BRAZIL; [2]Faculty of Engineering of the University of Porto, Laboratory for the Concrete Technology and Structural Behaviour, Rua Dr. Roberto Frias, s/n, 4200-465 Porto, PORTUGAL; [3]Civil Engineering Department, Federal University of Goiás, Goiânia, BRAZIL

Abstract: In this paper the results of axial symmetry finite element analyses of punching failure in concrete slabs with or without shear reinforcement are presented. Results from experimentally tested slabs are compared with the numerical simulations. All the slabs failed by punching, either with the failure surface crossing the shear reinforcement (if any), or being formed outside it. The numerical simulations were performed with the software package Diana, adopting the smeared crack approach to reproduce concrete behaviour under tension.

Key words: slabs; punching failure; FE analyses; experimental validation

1. INTRODUCTION

Nowadays flat slabs are a widely adopted solution for buildings, because of their economic and functional advantages. In addition, they simplify and speed up site operations, allow easy and flexible partition of space and reduce the overall height of buildings. However, flat slabs cause structural problems, because of the high shear stress around supporting columns, which can lead to failure due to punching shear at loads well below the flexural capacity. The punching resistance can be increased by a larger column cross section, a higher depth of the slab, more flexural reinforcement, and by increasing the concrete compressive strength, but these solutions in many cases are expensive, less effective or impractical. A solution to increase the punching resistance and ductility of flat slabs that is economic and practical is by providing shear reinforcement.

In previous years several numerical investigations were carried out towards modelling punching failure by the FE method [1], adopting axial symmetry or 3D discretizations. Since experimental punching tests are expensive and the monitored points are restricted,

M.S. Konsta-Gdoutos, (ed.), Measuring, Monitoring and Modeling Concrete Properties, 139–144.

numerical analyses provide an economic and helpful tool to understand the punching phenomenon, predicting the stress and strains of all the model.

The aim of this paper is to reproduce some experimental results reported in [2], validating the numerical modelling strategy. An axial symmetry model will be used to predict the punching shear failure of three flat slabs, with or without shear reinforcement.

2. NON-LINEAR NUMERICAL ANALYSES

The constitutive models for concrete and steel available in the FE package Diana[3] were used for the intended analyses. For the concrete in tension the smeared crack concept and the Multi Fixed Directional Crack approach was adopted, whereas for compression the Mohr-Coulomb criterion with isotropic hardening was selected. Behaviour of concrete in tension prior to cracking was taken as linear elastic, and after cracking a bilinear descending stress-strain diagram was assumed, like the one reproduced in Figure 1 for the tension stiffening phenomenon. The tension stiffening effect is usually referred to in the literature as the ability to gradually redistribute the load in structure from concrete to steel under formation of primary and secondary cracks[4].

The fracture energy according to the CEB/MC90[5] depends on the maximum aggregate size and on tensile strength. The ultimate strain in diagram (Fig.1) ε_u=0,005, corresponding to a fracture energy G_f = 92,5 Nm/m² over unit crack band with h =10mm.

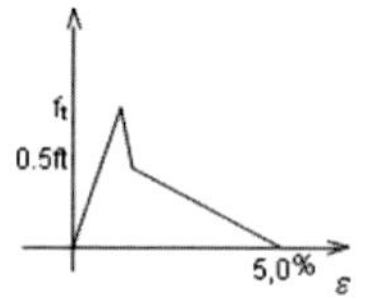

Figure 1. Diagram of tension stiffening

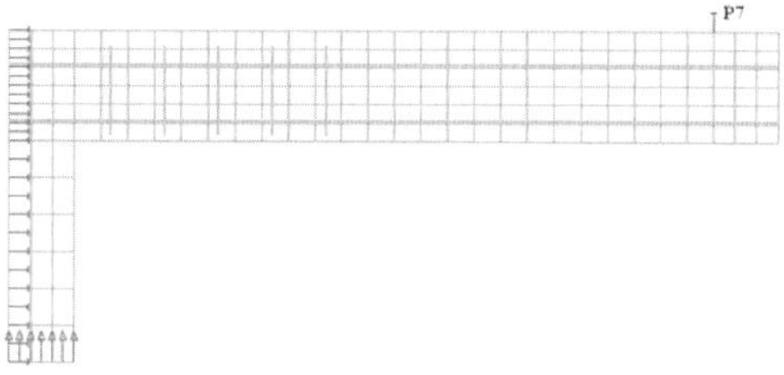

Figure 2. Axial symmetry model and FE mesh

Reinforcement was assumed as embedded, and perfectly adherent to the concrete, following an elastic-perfectly plastic constitutive law. It was taken a thickness of steel equivalent to the total area of rebars in each direction to simulate the flexural reinforcement.

Taking advantage of the geometry particularities, an axial symmetry model was generated with 8-nodded isoparametric FE. Reporting to Figure 2, axis of symmetry along the left hand side of the model has the nodes fixed in the horizontal directions, and point P7, which simulates the ties, is fixed in the vertical direction. In the experimental campaign loads were imposed by a square steel plate centred with the axis of symmetry, which in the FE model was replaced by a fictitious concrete column with a length of two times the height of the slab, with minor interferences on results [6].

3. EXPERIMENTAL TESTS OF MUSSE (2004)

Musse (2004) [2] tested up to failure several square 1.8×1.8m^2 reinforced concrete flat slabs, with a thickness of 0.13m. Each slab was supported by four ties fixed to a strong floor, and loaded upward at the centre by a 0.15×0.15m^2 square steel plate. The first slab (M1) had no shear reinforcement, whereas the other two (M2 and M3) contained headed studs.

The material properties, characteristics of the slabs and the arrangement of the shear reinforcement are summarized in Table 1. The top flexural reinforcement consisted of 19 ϕ12.5mm rebars (f_y = 839MPa and E_s = 200GPa) in each direction, and the bottom one was formed by 11 ϕ6.3mm rebars (f_y = 680MPa and E_s = 210GPa) in each direction.

All models failed by punching (in slabs M2 and M3 the failure surface formed outside the shear reinforcement region): Figure 3 presents a transversal section of the slabs, and it indicates the corresponding failure load and the observed mode of rupture.

Table 1. Characteristics of the slabs tested by Musse (2004)

Slab	f_c (MPa)	f_t (MPa)	E_c (GPa)	V_u (kN)	Shear Reinforcement				
					f (mm)	f_y (MPa)	E_s (GPa)	N. of layers	$S^{(1)}$ (mm)
M1	41.5	3.7	25.3	309	–				
M2	42.0	3.8	25.5	460	10	839	210	3	42
M3	42.5	3.8	25.8	472	10	839	210	5	63

(1) S = radial spacing of studs (8 studs per layer, with depth (d) =90mm).
- Concrete Poisson's ratio υ=0.2; β=0.9.

3.1 Slab M1

The failure load predicted numerically for slab M1, without shear reinforcement, was 304kN, approximately 2% lower than the experimental value (309kN). Figure 4 shows the load-deflection curves derived from the FE analysis and obtained in the experiment in the middle of the slab. The behaviour predicted numerically is quite the same as the one observed in the experimental test; however, after the first cracks the experimental displacements are higher than the ones in the numerical model for the same load.

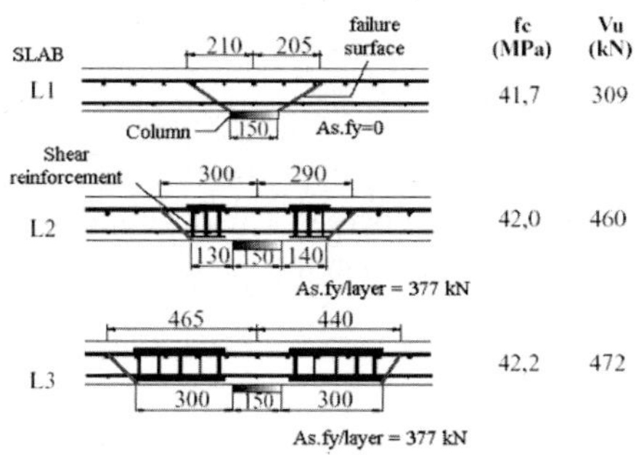

Figure 3. Mode of failure and failure loads of all slabs

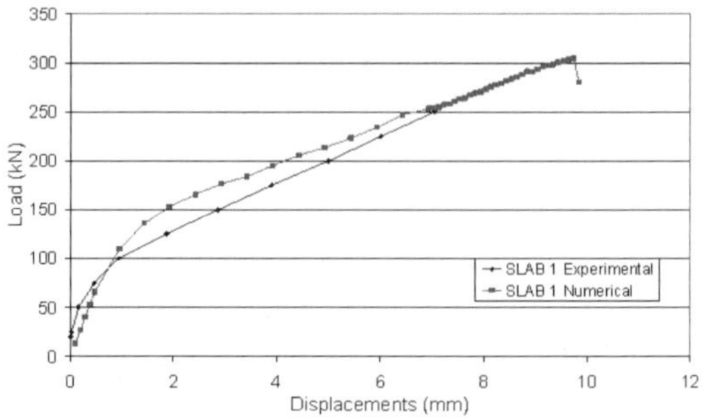

Figure 4. Load-deflection curves for slab M1

A characteristic punching cone was visualized in the numerical analysis, which was very close to the experimental observation: Figure 5 illustrates the punching cone and the crack pattern for the peak load. The angle between the surface of failure and the average plan of the slab obtained numerically was approximately 24.5º, and in the test it was 27º.

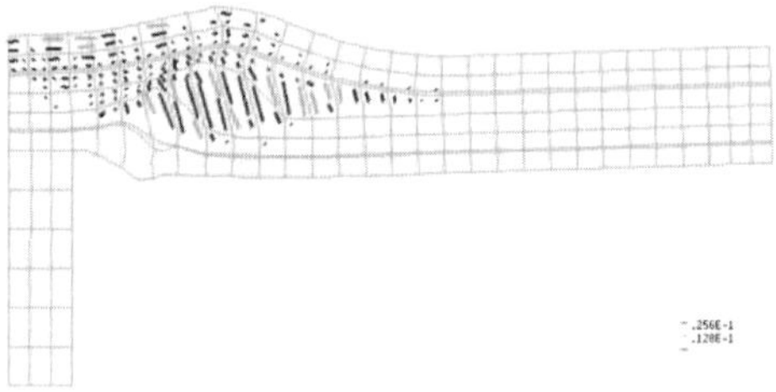

Figure 5. Crack pattern and punching cone for slab M1

3.2 Slab M2

As in the experimental campaign, failure obtained for slab M2 in the numerical analysis was due to punching shear. Comparison with the test results shows that the numerical simulation predicts the same mode of failure as in the experiment, with a shear cone being formed outside the shear reinforcement. Figure 6 shows the crack pattern in the post-peak regime. The calculated and the experimentally measured load-deflection curves at the centre of the slab are plotted in Figure 7. As it can be seen, the agreement between the numerical predictions and the measured data is reasonably good. However, the experimental curve exhibits a more ductile behaviour than the numerical one. This is probably due to the ultimate strain after cracking adopted in the tension stiffening diagram, or to the fracture energy assumed for the concrete. The experimentally measured failure load was 460kN, approximately 6% higher than the value obtained in the numerical analysis (432kN).

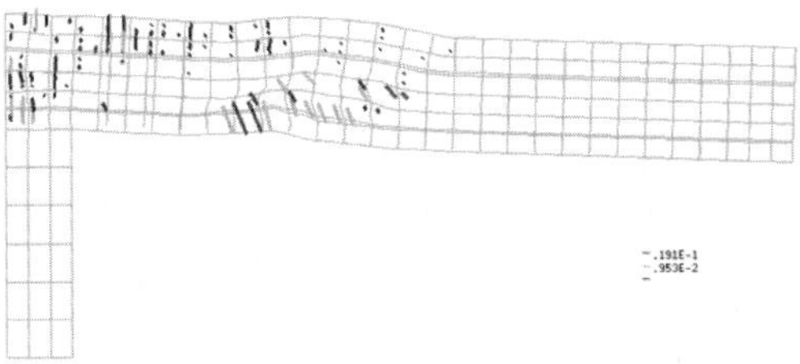

Figure 6. Crack pattern and punching cone for slab M2

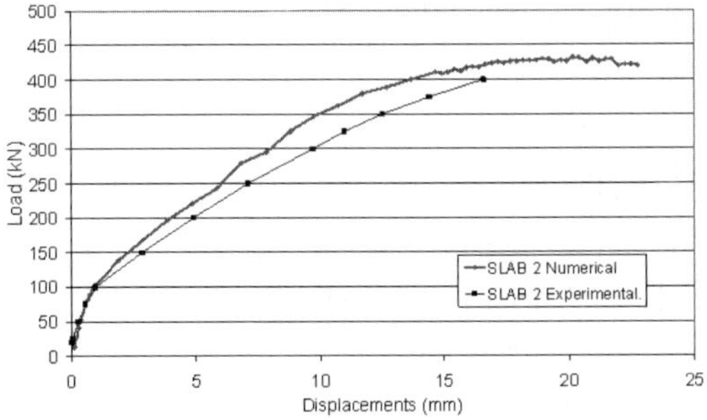

Figure 7. Load-deflection curves for slab M2

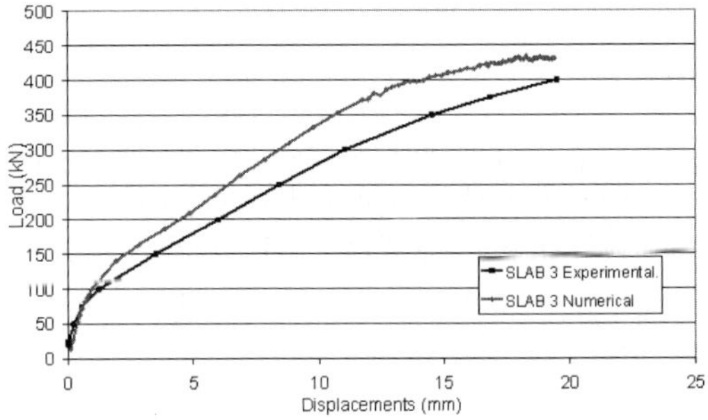

Figure 8. Load-deflection curves for slab M3

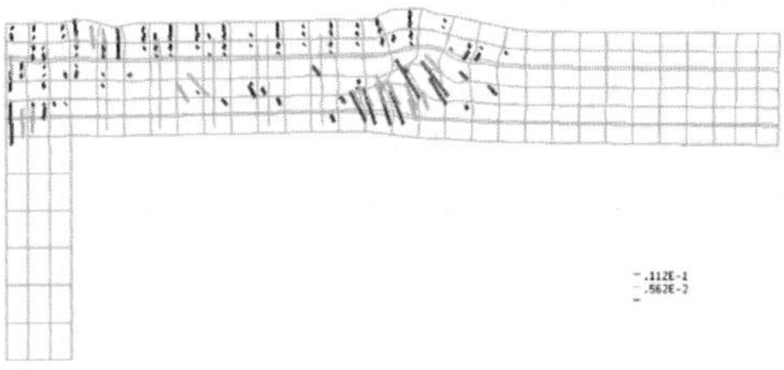

Figure 9. Crack pattern and punching cone for slab M3

3.3 Slab M3

Slab M3 contains five layers of shear reinforcement, constituted by ϕ6.3mm elements with a 63mm radial spacing. Figure 8 shows the load-deflection curves obtained in the test and in the numerical simulation at the centre of the slab. During the initial linear elastic regimen the numerical curve is very similar to the test one, but for serviceability loads higher than the one that induces cracking the numerical response is stiffer than the experimental one. The numerical failure occurs when the applied load reaches 432kN, 91% of the test failure load (472kN).

Illustration of the crack pattern in Figure 9 shows the tangential cracks at the end of the last increment: the punching cone can be easily recognised, which is in rather close agreement with the experiment one, taking place also outside the area with steel shear reinforcement.

4. CONCLUSIONS

In this paper it was put into evidence that the smeared crack approach provided realistic predictions of the behaviour of three RC flat slabs failing by punching shear during experimental tests. The presented comparisons between the numerical predictions and the test results showed good agreement, both in what concerns the load-deflection curves and the failure mechanisms of the slabs.

The failure loads obtained in the FE analyses performed for the slabs with the software Diana were very close to the ones observed in the experimental tests: for slab M3 the difference was 9%, whereas for the others the differences were lower than 4%.

Acknowledgments

The first author wishes to thank the University of Porto and the LABEST research unit for their special contribution and support. Thanks to FAPESP and CAPES for the financial support to the PhD Programme of the first author is also acknowledged.

5. REFERENCES

1. CEB–FIP. Punching of Structural Concrete Slabs. Technical report – Bulletin 12, 2001, 307p
2. Musse, T. H., "Punching in Flat Slabs: Steel fibers and shear reinforcement". Federal University of Goiás, MSc Thesis, 189p (2004).
3. Diana User's Manual Release 8.1. Delft, TNO Building and Construction Research (2002).
4. de Borst, R., Feenstra, P. H. Aspects of robust computational modelling for plain and reinforced concrete. Heron, 38 (4), 1-73 (1993)
5. Comité Euro-International du Béton (CEB). (1990). "CEB – FIP Model Code, Design Code."
6. Trautwein, L. M., Faria, R., Figueiras, J. A., Bittencourt, T., "Numerical Simulation on Punching Shear of Reinforced Concrete Slabs". Report. University of Porto. Porto, Portugal, 76 p (2004).

BOND-SLIP BEHAVIOR OF REINFORCEMENT IN NSC AND HSC WITH AND WITHOUT STEEL FIBERS

A. Dancygier, A. Katz and U. Wexler

The Faculty of Civil and Environmental Engineering, The Department of Structural Engineering and Construction Management, National Building Research Institute, Technion – Israel Institute of Technology, Technion City, Haifa 32000, Israel

Abstract: Application of High Strength Concrete (HSC) in RC structures requires the knowledge of its mechanical properties and of its design implications. In certain cases the brittle nature of HSC requires the inclusion of steel fibers in the mixture, which also influence other material properties. The current study focuses on bond-slip behavior of deformed steel bars in normal and in high strength concrete (NSC and HSC), with and without fibers. This paper describes the first stage of the study, which consists of pullout tests of deformed steel bars of different diameters from large cylindrical specimens made of NSC and HSC. Half of the NSC and HSC specimens included 30-mm hooked-end steel fibers that were given in a constant volume ratio of 0.75%. The paper describes the tests and their results, and points out the conclusions regarding the influence of concrete strength and the inclusion of fibers on the bond-slip behavior of deformed steel.

Key words: Bond, High Strength Concrete, Steel fibers

1. INTRODUCTION

Concrete-reinforcement bond is one of the main properties that affect the material and structural behavior of reinforced concrete. The bond strength as well as the bond-slip behavior are important for various design aspects, such as anchorage, development length or structural ductility. These are affected by several parameters: the type and diameter and the relative location and orientation of the reinforcing bar, the concrete density (related to its compressive strength), the concrete cover, aggregate type, transverse reinforcement, and mix additives such as silica fume or fibers. Pullout tests conducted by Plizzari[1] showed higher displacement ductility of specimens that included 0.38% of steel fibers, which also increased the bond strength of these specimens. It was concluded that

M.S. Konsta-Gdoutos, (ed.), Measuring, Monitoring and Modeling Concrete Properties, 145–150.

this was a result of a confinement action caused by fibers, which also limited radial cracking near the reinforcement bar. It was also found that the action or effect of the fibers is more efficient in HSC. The effect of the fibers on the bond strength was investigated by Harajli et al.[2] and Hamad et al.[3] Azizinamini et al. [4]performed bending experiments from which they concluded that bond stresses in HSC of deformed bars are highly non-uniform along the bar's length and that they are higher near the pulled end of the bar and reduce rapidly away from this point. Based on their results they hypothesized that reinforcement-concrete bond stresses develop in HSC only along few ribs of the deformed bar, located near the applied tension force, and that bond failure may occur prior to the development of uniform stress distribution along the bar's length (as it is commonly assumed in NSC or for smooth bars).

According to the works of Elfgren and Noghabai[5] and Hamad et al.[3] there is no evident conclusion regarding the influence of the bar diameter on its bond strength, although some of the experimental results indicate higher bond strength of smaller bar diameters.

Application of High Strength Concrete (HSC) in RC structures requires the knowledge of its mechanical properties and of its design implications. In certain cases the brittle nature of HSC requires the inclusion of steel fibers in the mixture, which also influence other material properties. The bond-slip behavior of the reinforcing steel bars is one of the properties, which is influenced by concrete strength. This property may also be affected by the inclusion of steel fibers. The bond-slip behavior of the reinforcement is important for understanding the structural RC behavior and for its proper design. The current study focuses on bond-slip behavior of deformed steel bars in normal and in high strength concrete (NSC and HSC), with and without fibers.

2. EXPERIMENTAL PROGRAM

The first stage of the current research includes pullout tests of deformed steel bars. The parameters that were studied here are the influence of the concrete strength, the application of steel fibers and the bar diameter on the bond strength and bond stress-slip relation.

2.1 Test specimens

The tests comprised of 190-mm diameter cylinders made of two types of concrete, NSC and HSC, with characteristic compressive strengths of 35 and 110 MPa (cube strength, respectively). Relatively large diameter concrete cylinder was chosen in order to prevent failure by concrete splitting. Half of the specimens included steel fibers. The volume of fibers in these specimens was kept constant throughout the test at 0.75% (60 kg/m^3). In addition, three 100x100x100 mm cubes were cast for compressive strength tests for each specimen type, and flexural and splitting tensile strength tests were made on beam specimens.

The specimens were denoted N/H-0/1-L_d, where *N* and *H* denote NSC and HSC, *1* and *0* denote specimens that did and did not include fibers and L_d was the adhesion length (mm). Bond breaker was applied at length of 5ϕ (ϕ – bar diameter) at the loaded end of the bar, and ~1ϕ at the unloaded end, following RILEM recommendations for pullout tests[6] with modifications required for high strength concrete as follows. The height of the specimens, thus, was equal to 130+L_d mm (see Fig. 1). Because the HSC specimens were expected to yield high bond strength, the pullout tests were performed in two stages: in the first stage 12-mm deformed steel bars were tested at two adhesion lengths – 60 and 30 mm (5ϕ and 2.5ϕ). The results of these tests showed that using the standard 5ϕ adhesion length causes the steel bar to yield before it is pulled out. Hence, the following stage was performed with 8-mm and 20-mm bars at adhesion lengths that were 25 and 65 mm, respectively (i.e., about 3ϕ), to ensure pullout before yielding of the bar.

2.2 Materials

Hooked-end 35-mm long and 0.55-mm diameter steel fibers were used. These fibers have an aspect ratio of 64 and according to the manufacturer their minimum tensile strength is 1000 MPa. The HSC and NSC mixes included basalt and dolomite coarse aggregates (respectively) of 22 mm maximum size, ordinary Portland cement type CEM I-52.5N and natural quartz sand. Details of mix proportions are listed in Table 1. All the mixes were prepared in a forced pan mixer. The concrete compressive strengths at 28 days are given in Table 2.

2.3 Mechanical properties

Compressive strength was determined by means of 100 mm cubes and 150-mm diameter cylinders. Flexural strength was measured from 280x70x70 beams and splitting tensile strength was evaluated by testing from their prismatic remains. The strength values of the specimens at 28 days are given in Table 2,

Table 1. Mix ingredients

Type	Cement (kg/m^3)	Sand (kg/m^3)	Aggregate (kg/m^3)	Water (kg/m^3)	W/C Ratio	Silica fume (kg/m^3)	WRA[(1)] (Kg/m^3)
NSC	255	721	1162	191	0.75	-	1.27
HSC	494	703	1066	158	0.32	64	5.3

[(1)]Water Reducing Agent and High range water reducing agent

2.4 Pullout tests

The tests were performed with a controlled stroke at the following displacement rates: 0.005 mm/min during the initial 4 mm, 0.01 mm/min during 4 - 8 mm and 0.02 mm/min for stroke displacements larger than 8 mm. The tests were terminated when the post-peak load dropped to ~20% of the peak load.

3. RESULTS

The loads and strains that were measured indicated that the 12-mm bars with 60-mm (5ϕ) anchorage length and the 20-mm bars of the HSC specimens reached their yield strains despite the pre-designed shorter anchorage length of the latter. The bars of the other specimens remained in their elastic range. The bond strengths are given in Table 2 and plotted in Figure 2. These strengths are the average stresses $\tau_{avg,max}$ calculated over the anchorage length (L_d) under the maximum load T_{max} [i.e., $\tau_{avg,max} = T_{max} / (L_d \cdot \pi\phi)$]. Note that the 12-mm results that are plotted in Figure 2 are of the specimens with the 30-mm anchorage length, so that the anchorage length of the specimens plotted in Figure 2 is ~3ϕ (2.5 to 3.25 the bars' diameters).

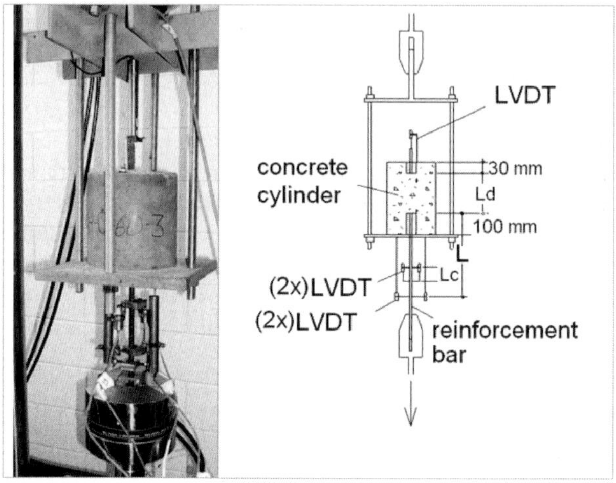

Figure 1. Experimental setup, test specimen and measurements

Table 2. Concrete strengths at 28 days and average bond strength (MPa)

Specimen / bar diameter [mm]	Compressive strength 100/100 cube	Compressive strength 150/300 cylinder	Flexural strength	Splitting tensile strength	Bond strength $\tau_{avg,\ max}$(1)	$\frac{\tau_{HSC}}{\tau_{NSC}}$
N-0-60 / 12	35	29.1	4.3	3.2	6.8 (0.9)	4.33
N-1-60 / 12	34.7	28.6	4.9	3.8	6.6 (0.3)	
H-0-60 / 12	116.1	86.0	11.3	116.1	28.7 (0.8)	
H-1-60 / 12	118.9	92.9	11.2	118.9	29.3 (0.9)	
N-0-30 / 12	35	29.1	4.3	3.2	6.8 (1.1)	4.28
N-1-30 / 12	34.7	28.6	4.9	3.8	5.8 (0.4)	
H-0-30 / 12	116.1	86.0	11.3	116.1	28.6 (1.6)	
H-1-30 / 12	118.9	92.9	11.2	118.9	25.3 (6.8)	
N-0-25 / 8	35.9	27.7	4.8	3.3	18.4 (2.1)	1.58
N-1-25 / 8	37.7	27.8	5.3	3.9	12.6 (1.9)	
H-0-25 / 8	107.4	74.1	10.2	6.8	26.0 (3.1)	
H-1-25 / 8	119.5	82.5	10.7	8.8	23.0 (1.4)	
N-0-65 / 20	35.9	27.7	4.8	3.3	13.2 (1.7)	2.89
N-1-65 / 20	37.7	27.8	5.3	3.9	12.3 (1.4)	
H-0-65 / 20	107.4	74.1	10.2	6.8	35.2 (2.6)	
H-1-65 / 20	119.5	82.5	10.7	8.8	38.4 (0.7)	

(1) values in brackets denote standard deviation

3.2 Fiber effect

Figure 2a, which shows the absolute bond strengths, indicates that except for the HSC specimens with the 20-mm bars, addition of fibers either decreased or did not influence the bond strength. This trend is even more pronounced in all specimens, when the bond strengths are normalized with respect to the concrete (cylinder) strength (Figure 2b). Bond-slip curves of the NSC and HSC specimens, with and without fibers, are shown in Figure 3, where bond values are normalized with respect to the concrete compressive strength. The figure shows that the influence of the fibers in reducing the bond was similar in both pre and post-peak ranges of the bond-slip curves. It is possible that the fibers prevented good compaction of the concrete close to the reinforcing bars, thus affecting the bond strength, mainly in normal strength concrete with the small bar diameter (Table 2).

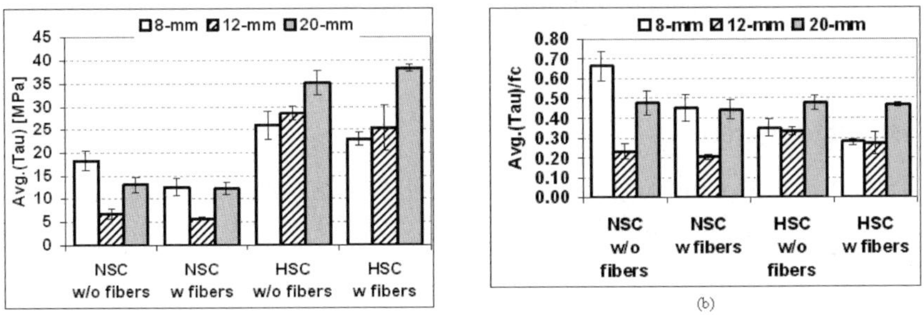

Figure 2. Average bond strengths (a) absolute values and (b) normalized with respect to the concrete cylinder compressive strength f_c (vertical lines indicate standard deviations)

3.2 Concrete strength effect

Clearly (and as expected) the bond that was developed in the HSC specimens was higher than that of the NSC specimens (Table 2 and Figure 2). However, when normalized with respect to the concrete compressive strength the 8-mm and 12-mm bars had in HSC smaller and higher normalized bond strengths (respectively) while the specimens with the 20-mm bars had similar normalized strengths (Figures 2b and 3). These results indicate that the effect of concrete strength on the bond behavior depends also on the bar geometry.

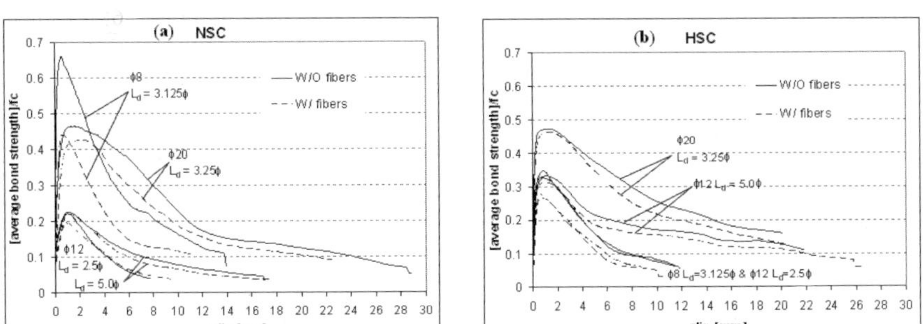

Figure 3. Bond-slip curves for (a) NSC and (b) HSC (Average bond strengths are normalized with respect to the concrete compressive strength)

4. CONCLUDING REMARKS

Pullout tests of deformed reinforcing bars in NSC and HSC specimens, with and without hooked-end steel fibers, were performed. At this stage of the research it was found that addition of fibers either decreased or did not influence the bond strength. The influence of the fibers in reducing the bond was similar in both pre and post-peak ranges of the bond-slip curves. As expected, the bond that was developed in the HSC specimens was higher than that of the NSC specimens. However, as Table 2 shows, this increase in strength is not directly related to concrete strength indicating that it depends not only on the concrete strength but also on the bar geometry (diameter and rib geometry). It is further noted that due to the relatively wide concrete cover in the current tests, pullout was developed without visible external cracking, which caused the negative to no effect of the fibers. Further tests of pullout in bending specimens (with smaller concrete cover) should complete the picture regarding the influence of steel fibers on the bond-slip of deformed bars is NSC and HSC.

Acknowledgement

This study is partially supported by the Israeli Ministry of Construction and housing.

5. REFERENCES

1. G.A. Plizzari, Bond and splitting crack development in normal and high strength fiber reinforced concrete, 13th Engrg. Mechanics Division Conference – EMD99, Baltimore (MD,USA), on CD (1999).
2. M.H. Harajli, A. Ahmad Rteil, Effect of confinement of bond strength between steel bars and concrete, *ACI Str. J.*, **101**(5), 593-603 2004.
3. S. Bilal Hamad, H. Harajli, Ghaida Jumaa, Effect of fiber reinforcement on bond strength of tension lap splices in high-strength concrete, *ACI Str. J.*, **98**(5), 638-647 (2001).
4. A. Azizinamini, M. Stark, J.J. Roller, S.K. Ghosh, Bond Performance of Reinforcing Bars Embedded in High-Strength Concrete, *ACI Str. J.*, **90**(5), 554-561 (1993).
5. L. Elfgren and K. Noghabai, Tension of reinforced concrete prisms. Bond properties of reinforcement bars embedded in concrete tie elements, *Materials and Structures*, **35**, 318-325 (2002).
6. Rilem/CEB/FIP, Bond test for reinforcing steel: 2. Pullout Test, *Materials and Structures*, **3**(15), 175-178 (1970).

APPLICATION OF INVERSE ANALYSIS TO SHRINKAGE AND CREEP MODELS

L.C. de Almeida[1], J.L.A. de Oliveira e Sousa[1] and J. de Azevedo Figueiras[2]
[1] *Faculdade de Engenharia Civil, Arquitetura e Urbanismo da Universidade Estadual de Campinas, Av Albert Einstein, 951, 13083-852 Campinas, SP - Brasil,* [2]*Faculdade de Engenharia da Universidade do Porto R. Dr. Roberto Frias, 4200-465 Porto - Portugal*

Abstract: The identification of adequate parameters for Engineering design, as well as their confirmation from in situ measurements, is a relevant problem in Structural Engineering. One strategy usually applied is the inverse analysis of monitoring data from the actual structure. This paper describes the application of an inverse analysis approach to fit theoretical models, available in the literature, for shrinkage and creep, to data acquired in long term tests on concrete specimens performed at the LABEST – Laboratory of Concrete Technology and Structural Behavior, Faculdade de Engenharia da Universidade do Porto. The following models were considered in this paper: Eurocode 2, ACI 209R-92, Brazilian Code NBR 6118-2003, the model B3 by Bazant and Baweja, and the model GL2000 by Gardner and Lockman. A minimization algorithm developed by Hendriks was implemented in MATLAB for estimating parameters corresponding to each model.

Key words: inverse analysis, shrinkage, creep, concrete, monitoring.

1. INTRODUCTION

The identification of adequate parameters for Engineering design, as well as their confirmation from in situ measurements, is a relevant problem in Structural Engineering. One strategy usually applied is the inverse analysis, performed on data from laboratory tests or monitoring data from actual structures. Example applications of this methodology are the identification of geotechnical parameters for tunnels excavation[1], identification of mechanical behavior of solid materials[2], determination of Young's modulus and active earth pressures coefficient[3,4].

In order to estimate the long-term behavior of reinforced and prestressed concrete structures, equations to predict shrinkage and creep of concrete are required. The North American current practice is based on ACI 209-82[5] recommendations. Since 1970, the

M.S. Konsta-Gdoutos, (ed.), Measuring, Monitoring and Modeling Concrete Properties, 151–160.

Comité European du Béton (CEB) developed successive recommendations of a model code, often modified and adopted in national codes. Eurocode 2/1992 (draft) adopted the proposals of model code CEB 1990 for shrinkage and creep[6]. In 1999, CEB updated the prediction method for concrete shrinkage and creep. Version 2002 of Eurocode 2 recommends the use of shrinkage and creep equations from version 1999 (FIB 1999) of the code model CEB/1990, MC1990-99. Bazant and Baweja[7,8] and Gardner and Lockman[9], respectively, proposed prediction methods in substitution to the ACI 209-82[5].

The objective of this paper is the application of an inverse analysis methodology to adjust theoretical predictions models for concrete shrinkage and creep to strains measured in long term tests. The procedure for estimating parameters is based on an algorithm developed by Hendriks[2], which was implemented in a MatLab[10] routine. This study is limited to serviceability stresses, which are up to 40% of the strength of the concrete, for which creep strains are assumed to have linear dependency on the corresponding stresses. The developed methodology is described, as well as its application to concrete specimens[11], and to a prestressed concrete frame[12]. The results corresponding to the different models are presented and compared with experimental values.

2. IDENTIFICATION METHODS

According to the identification method proposed by Hendriks[2], there are two basic requirements:

- Availability of adequate numerical tools for modeling the mechanical behavior of the target material. The problem is the quantitative determination of material constitutive parameters.
- Availability of an efficient and accurate computational algorithm to solve the error minimization problem.

There is an important difference between the identification methods and the traditional direct material parameters determination methods. For the identification methods it is not necessary that the strain field be homogeneous in some part of the loaded specimen under investigation. It is preferable that the strain field be inhomogeneous. According to Hendriks[2], the inhomogeneous strain fields contain much more information on the material properties that homogeneous strain fields. However, inhomogeneous strain fields come along with three problems:

- Experimental attainment of the inhomogeneous strain field for different applied loadings.
- Need for numerical treatment of experimental because of its complexity.
- Establishment of criteria to compare numerical and experimental criteria results, and then determine material parameters.

The solution of these three problems involves the identification method, as shown in figure 1. The strain distribution, in the specimen, is obtained experimentally. Using a numerical model (e. g., finite elements), the strain field is determined as a function of a given set of parameters. A minimization algorithm is then used to find, the set of parameters that minimize the errors between experimental measurements and numerical model results.

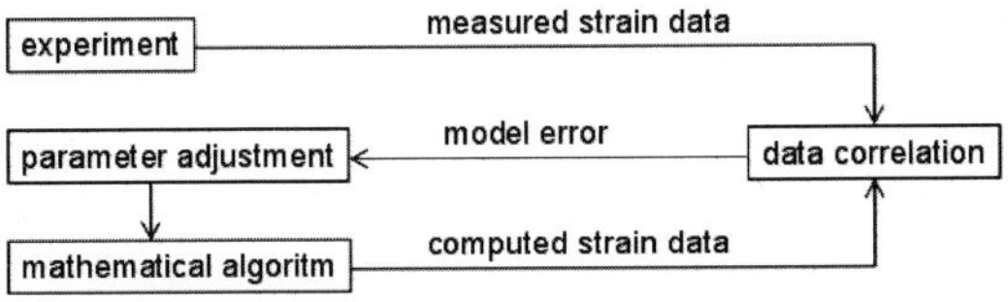

Figure 1. Scheme of identification of parameters

The finite elements method is suitable for this type of problem. A numerical analysis can be performed if a set of parameters is available. This means that initial values of must be estimated for the parameters and, in an iterative procedure, these values are adjusted in order to reproduce, within a predefined accuracy, the observed results.

The comparison between the measured strain field and the computed strain field leads to a quantitative determination of the unknown parameters. Algorithms with fast convergence are required, which should have confidence evaluation. In the systems identification area this is an estimate or reconstruction problem[2]. The estimation problem deals with the determination of physical amounts that cannot be measured directly from those that can.

The identification method is based on the combination of three elements:

- A sufficient amount of measurement data to characterize the strain field on a significant area of the specimen;
- Modeling by the finite element method;
- A technique that adjusts the material parameters supplied to the finite elements model through the comparison between numerical results and corresponding experimental data.

3.1 Parameter estimation

This third element comprises the comparison between experimental values and the values resulting from the finite element model, followed by the update of the material parameters estimate. This estimation problem can be approximated, in a deterministic way, by the least square method[2].

3.2 Problem statement

Assuming that the experimental observations consist of a finite set of vectors y_k, $k = 1,..., N$, a simple way of sorting the complete set is to interpret *k* as a discrete time parameter. However, *k* indicates a loading case for the specimen under investigation. A vector $y_k = (y_1,......y_m)_k^T$ may contain components of material point displacements and other measured quantities, such as force, velocity and pressure. The material behavior is represented quantitatively by a finite set of unknown quantities $x_i, i = 1,..., n$. These parameters define a vector $\boldsymbol{x}$, whose elements are the material parameters to be determined. Assume that an algorithm exists to compute y_k when $\boldsymbol{x}$ has a known initial value. This algorithm, based on the finite elements method, is symbolized by a function $h_k(x)$.

The function $h_k(x)$ describe the dependence of *k*-ths observations on $\boldsymbol{x}$ if there were no observation errors. These errors are represented by a vector v_k.

$$y_k = h_k(x) + v_k \tag{1}$$

where:

$\boldsymbol{x}=(x_1, \ldots, x_n)^T$: is a vector with the material parameters.

v_k : is a vector of observation errors.

Vector $\boldsymbol{x}$ can be called "parameters vector" and may contain, for example, Young´s modulus, Poisson´s ratio, time dependent constants, or a nonlinear function of these material properties.

For the case of linear dependence of x, the equation (1) becomes:

$$y_k = H_k(x) + v_k \tag{2}$$

where: H_k : is a matrix of directions.

There are two relevant aspects in these parameter estimation problems: the first is the use of measured values of displacements y_k to estimate the parameters vector $\boldsymbol{x}$. The estimate can be performed from the mathematical model, equations (1) or (2), an error vector v_k and the previous knowledge of vector $\boldsymbol{x}$. Another problem is to determine how close the estimated vector x_{est} is to the true vector x_{est} . The numerical value of the error $(x_{true}\text{-}x_{est})$ is not known, so the problem is to develop a model of estimate of $(x_{true}\text{-}x_{est})^2$.

2.2.1 Generalized least square method

A generalized procedure of the least squares to estimate the parameters of vector **x** from *k*-ths experimental data is defined through a nonnegative function S_k, defined by:

$$S_k = (y_k - h_k(x))^T W_k (y_k - h_k(x)) \tag{3}$$

where: W_k: is a positive definite symmetric weighting matrix.

Then, by the definition, a procedure to get x_k is to minimize the function S_k with respect to $\boldsymbol{x}$.

The generalized term is used because, in the classic procedure of the least squares, W_k is used as a diagonal matrix, while in this case, although W_k must be symmetrical, it is not necessary to be a diagonal matrix.

For the linear case described by the equation (2):

$$S_k = (y_k - H_k x)^T W_k (y_k - H_k x) \tag{4}$$

The operator of the least squares is also linear, that is, x_k is a linear function of y_k, and can be obtained in closed form by:

$$x_k = P_k H_k^T W_k y_k \qquad where: \; P_k = (H_k^T W_k H_k)^{-1} \tag{5}$$

The dimensions of x_k and y_k are, respectively, n (number of material parameters to be determined) and m (number of observations). If m < n, there are less equations than unknowns and, thus, an indeterminate system of equations that will not supply a unique solution x. If m = n, the numbers of equations and unknowns are the same. Consequently, H_k is a square matrix and, as proven by Hendriks[2], invertible. Therefore, $\boldsymbol{x}$ can be obtained directly to:

$$x_k = H_k^{-1} y_k \tag{6}$$

If m > n, there are more equations than unknowns, and the system is over determined. This is the case of real interest for the estimate by the generalized least square method.

2.2.2 Sequential estimation of parameter

Let it be:

- x_0 , P_0 the initial conditions;
- P_k : covariance matrices of the estimate x_k
- Q_k , R_k : weighting matrices
- *k=1...,N,* the ordering variable for the observations

The following updating equations are used:

$$x_{k+1} = x_k + K_{k+1}(y_{k+1} - h_{k+1}(x_k)) \tag{7}$$

The difference $(y_{k+1} - h_{k+1}(x_k))$ represents new information. This difference is multiplied by the gain matrix K_{k+1}, given by:

$$K_{k+1} = (P_k + Q_k)H_{k+1}^T(R_{k+1} + H_{k+1}(P_k + Q_k)H_{k+1}^T)^{-1} \tag{8}$$

The P_k matrix of the equation (8) is updated by:

$$P_{k+1} = (I - K_{k+1}H_{k+1})P_k \tag{9}$$

The H_{k+1} matrix is given by the expression:

$$H_{k+1} = \left(\frac{\partial h_{k+1}(x)}{\partial x}\right)_{x=x_k} \tag{10}$$

3. SHRINKAGE AND CREEP

3.1 Introduction and definitions

Shrinkage and creep of Portland cement concrete are complex phenomena governed by several physico-chemical processes. The basic process originates within the paste fraction, as a result of the properties of hydration products and the microstructure of the paste, modified by the composed nature of the concrete.[13]

The total observed creep is the sum of basic creep and drying creep. In practice, however, this distinction is not always made. Also, the way to separate creep strain from the "elastic" (i.e. instantaneous strain) is somewhat arbitrary. Creep starts occurring as the load is applied. However, a finite time interval is required for the measurement of the elastic strain.[13]

Basic creep refers to the time-dependent strains that occur when the concrete is loaded in a sealed condition, so that moisture cannot escape. Drying creep (sometimes called the Picket effect) refers to the additional creep. In practice, however, these distinctions are not always taken into account.

The creep coefficient expresses the creep strain as a fraction of the elastic strain, typically, the creep coefficient is in the range 2.0 – 6.0 for ultimate creep (the maximum creep attainable by the system[13]).

Compliance creep (specific creep) is the amount of creep per unit of applied stress, and is used in order to compare the concrete creep potential loaded at different stress levels.

Several shrinkage components, can be defined but, for the models studied here, only two are used: the drying shrinkage and the autogenous shrinkage. Drying shrinkage is the deformation associated with the loss of humidity from the concrete under drying conditions. Autogenous shrinkage (hydration or chemical) occurs when the water is removed internally by chemical combination during the hydration in a moisture-sealed state.

3.2 Prediction models

In the expressions for shrinkage and creep prediction available in Eurocode 2[14], ACI 209[5], NBR 6118[15], the Bazant's B3 model[6,7,16] and the GL2000 of Gardner[8,16], coefficients are introduced to allow fitting to the experimental data. In shrinkage strain prediction models the coefficients C_1 and C_2 are introduced, and for the case of creep, coefficients C_3 and C_4 are introduced. These coefficients are intended to assure a proper adjustment of creep and shrinkage evolution at any time, resulting in the equations to follow.

- Eurocode 2[14]:

$$\varepsilon_{cs}(t) = \beta_{as}(t).\varepsilon_{ca}(\infty) + [\beta_{ds}(t,t_s)]^{C_2} k_h.\varepsilon_{cd,0}.C_1 \tag{11}$$

$$\varphi(t,t_0) = \varphi_0 \beta_c (t,t_0)^{C_4} C_3 \tag{12}$$

- ACI 209[5]

$$\varepsilon(t-t_s) = \left[\frac{(t-t_s)}{35+(t-t_s)}\right]^{C_2} \varepsilon_{su} C_1 \qquad \varphi(t,t_0) = \left[\frac{(t-t_0)^{0,6}}{10+(t-t_0)^{0,6}}\right]^{C_4} \varphi_u C_3 \tag{13, 14}$$

- NBR 6118[15] Model

$$\varepsilon_{cs}(t,t_0) = \varepsilon_{cs\infty}[\beta_s(t) - \beta_s(t_0)]^{C_2} C_1 \tag{15}$$

$$\varphi(t,t_0) = \varphi_a + \varphi_{f\infty}[\beta_f(t) - \beta_f(t_0)]^{C_4} C_3 + \varphi_{d\infty}\beta_d \tag{16}$$

- B3 of Bazant e Baweja[7,8]

$$\varepsilon_{sh}(t,t_0) = -\varepsilon_{sh\infty} k_h \left[S(t)\right]^{C_2} C_1 \tag{17}$$

$$\varphi(t,t') = E_c(t')\left[C_0(t,t')\right]^{C_4} C_3 + E_c(t')C_d(t,t',t_0) - E_c(t')/E_{c28} \tag{18}$$

- Gardner e Lockman's GL2000[8,16]

$$\varepsilon_{sh} = \varepsilon_{shu}\beta(h)\left[\beta(t)\right]^{C_2} C_1 \tag{19}$$

$$\varphi_{28} = \varphi(t_c)\left[\begin{array}{l} 2\left(\dfrac{(t-t_0)^{0.3}}{(t-t_0)^{0.3}+14}\right)+\left(\dfrac{7}{t_0}\right)^{0.5}\left(\dfrac{t-t_0}{t-t_0+7}\right)^{0.5}\ldots\ldots \\ \ldots\ldots\ldots + 2.5(1-1.086h^2)\left(\dfrac{t-t_0}{t-t_0+0.15(v/s)^2}\right)^{0.5} \end{array}\right]^{C_4} C_3 \tag{20}$$

4. DESCRIPTION OF THE LABORATORY TESTS

The experiments were performed at LABEST - Laboratory of Concrete Technology and Structural Behavior, Faculdade de Engenharia Civil da Universidade do Porto, Portugal, and reported in Felix.[17]

The shrinkage and creep tests were performed in concrete specimens, with the characteristics described in table 1, with cross section (15 cm × 15 cm) cm and height 50 cm, manufactured with a regular hardening cement (class N) and regular aggregates.

The specimens were kept in cure, sealed until the eighth day. The specimen intended to the creep study was placed in the creep frame, as shown in the Figure 2, subject to a constant stress 8.80 MPa, and kept in the laboratory environment at an average humidity of 54%, all the time.

Table 1. Characteristic of the concrete

Time (days)	f_{em} (MPa)	f_{ake} (MPa)	E_{ke} (GPa)
7	27.3	22.3	29.7
28	31.3	26.3	29.9

Figure 2. Creep frame

5. RESULTS

The described inverse analysis methodology, implemented in a MatLab program[10], was applied to the experimental data, leading to the adjusted estimates of shrinkage and creep coefficients presented herein.

5.1 Shrinkage

Table 2 presents the coefficients C_1 and C_2 that best adjust the shrinkage models to the experimental data. The curves for drying creep strain corresponding to the chosen models and the creep strain obtained in the laboratorial test are presented in Figure 3. Some models present significant differences, B3[7,8] and ACI 209[5], while others present good match to the experimental results, Eurocode 2[14] and NBR 6118[15]. The curves of the drying shrinkage strain, for the chosen models, adjusted with the coefficients C_1 and C_2, from inverse analysis (Table 2), are presented in Figure 4. A good match between the adjusted models and the experimental results can be observed, even for the models that presented significant variations in the beginning of the process.

Table 2. Coefficients obtained in the shrinkage models fitting

Models	C_1	C_2
Eurocode 2	0.9320	0.5031
GL2000	0.7839	0.5390
ACI 209	3.1587	0.4321
B3	1.2539	0.4615
NBR 6118	0.6119	0.3653

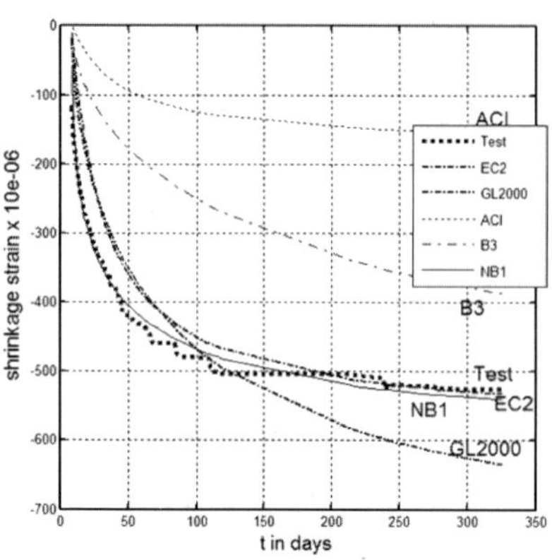

Figure 3. Shrinkage models and experimental results

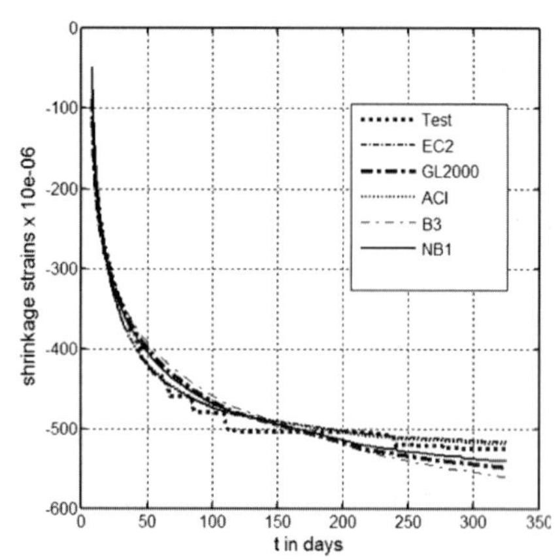

Figure 4. Identification of parameters from 227 experimental data

5.2 Creep

Table 3 presents the coefficients C_3 and C_4 that best fit the creep models to the experimental data.The creep coefficient curves for the models chosen for this study, and the creep coefficient obtained by inverse analysis on the test data are presented in Figure 5. Significant differences with respect to the experimental data can be observed, varying from 25% to 50% for the final creep result, indicating the need for calibration.

Creep coefficient curves for the chosen models are presented in Figure 6, adjusted with the coefficients C_3 and C_4, presented in Table 3, obtained through the inverse analysis.

Table 3. Coefficients obtained in the adjustment of the creep models

Models	C_3	C_4
Eurocode 2	0.7375	1.2152
GL2000	0.5247	1.3731
ACI 209	0.8132	1.0522
B3	0.6080	1.0137
NBR 6118	0.5742	1.1134

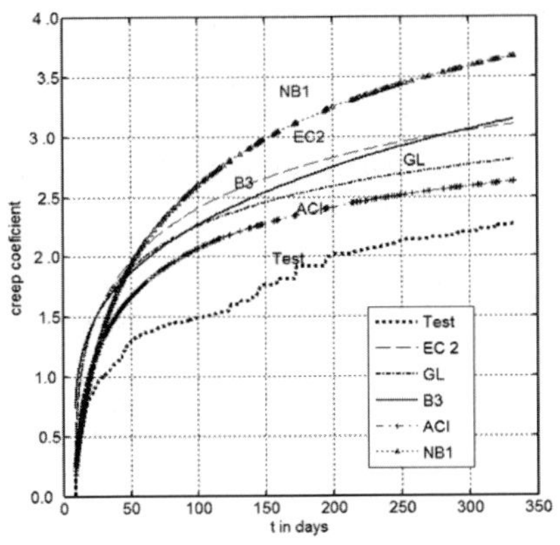

Figure 5. Creep models and results experimental

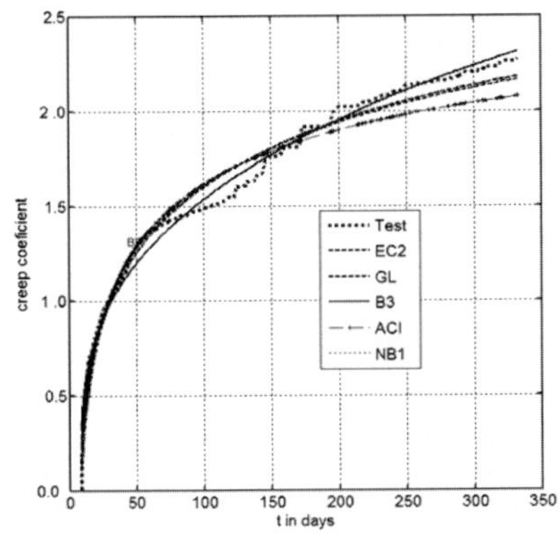

Figure 6. Identification of parameters from 363 experimental data

6. CONCLUSIONS

The methodology has proven efficient for adjusting creep coefficients and shrinkage strain curves, corresponding to theoretical models, through inverse analysis on data obtained from long-term tests performed under laboratory conditions.

This work is part of a project intended to study the time dependant behavior of concrete, using different mixes, comparing results from long-term tests performed in laboratory with results obtained at first from prototypes tested in laboratory and latter from monitoring data acquired from the actual structure.

Acknowledgments

The authors are grateful to CAPES – Coordenação de Aperfeiçoamento de Pessoal de Nível Superior, CNPq – Conselho Nacional de Desenvolvimento Científico e Tecnológico, FEC-UNICAMP – Faculdade de Engenharia Civil, Arquitetura e Urbanismo da Universidade Estadual de Campinas (Brazil), and to LABEST/FEUP – Faculdade de Engenharia da Universidade do Porto (Portugal) for the support to this work. They are also grateful to Dr. Carlos Manuel da Silva Félix, for providing shrinkage and creep test results.

7. REFERENCES

1. Ledesma, A. Identificación de parámetros en geotecnia: aplicación a la excavación de túneles. PhD. Thesis, Escuela Técnica Superior de Ingenieros de Caminos, Canales y Puertos de Barcelona, Universitat Politècnica de Catalunya, Barcelona, Espanha, 1987. (in Spanish)
2. Hendriks, M.A.N. Identification of the mechanical behavior of solid materials. PhD. Thesis, Technische Universiteit Eindhoven, Netherlands, 1991.
3. Ledesma, A., Gens, A., Alonso, E., Parameter and variance estimation in geotechnical back analysis using prior information. International Journal for Numerical and Analytical Methods in Geomechanics, v.20, n.2, p.119-41, 1996.
4. Ledesma, A., Gens, A., Alonso, E.E. Estimation of parameters in geotechnical back analysis - I. Maximum likelihood approach. Computers and Geotechnics. Vol.18, No.1, pp. 1-27, 1996.
5. ACI Committee 209. Prediction of Creep, Shrinkage and Temperature Effects in Concrete Structures. American Concrete Institute, Farmington Hills. Mich, 1982.
6. Bazant, Z.P. Prediction of concrete creep and shrinkage: past, present and future. Nuclear Engineering and Design, 203 p. 27-38, 2001.
7. Bazant, Z.P., Baweja, S. Creep and shrinkage prediction model for analysis and design of concrete structures – model B3. Material Structures, 28, p. 357-365, 1995.
8. Bazant, Z.P., Baweja, S. Justification and refinements of model B3 for concrete creep and shrinkage 2 Updating and theorical basis –. Material Structures, 28, p. 488-495, 1995.
9. Gardner, N.J. Comparison of predictions for drying shrinkage and creep of normal-strength concrete. Can. J. Civ. Eng. 31, p. 767-775, 2004.
10. Mathworks, Inc. MatLab – The language of technical computing. Version 6.5 R13, Mathworks, INC, 2002.
11. Almeida, L.C., Sousa, J.L.A.O., Figueiras, J.A., Determinação de parâmetros elásticos para o concreto por análise inversa de resultados de ensaios de compressão diametral. Anais do 47º Congresso Brasileiro do Concreto, Olinda, PE, Brasil, 2005. (in Portuguese).
12. Almeida, L.C., Sousa, J.L.A.O., Figueiras, J.A., Aplicação de técnicas de análise inversa na determinação de cargas aplicadas em uma estrutura a partir de deslocamentos medidos. Anais do XXVI CILAMCE, Guarapari, ES, Brasil, 2005a. (in Portuguese).
13. Bazant, Z.P. Mathematical modeling of creep and shrinkage concrete. Wiley, Chichester, 1988.
14. EC2 - Eurocódigo 2. Projeto de estruturas de betão - parte 1: Regras gerais e regras para edifícios, 1991.
15. ABNT – Associação Brasileira De Normas Técnicas. Projeto de estruturas de concreto - Procedimento: NBR 6118. Rio de Janeiro, 2003.
16. Gardner, N.J., Lockman, M.J. Design provisions for drying shrinkage and creep of normal-strength concrete. ACI Materials Journal, 98(2), p.159-167, 2001.
17. Felix, C. M. S., Monitorização e análise do comportamento de obras de arte. Tese de doutorado, Faculdade de Engenharia da Universidade do Porto, Porto, Portugal, 2005. (in Portuguese).

Fracture and Deformation of Cement Based Composites

EFFECTS OF LIGHTWEIGHT AGGREGATES ON AUTOGENOUS DEFORMATION IN CONCRETE

B. Akcay and M.A. Tasdemir
Istanbul Technical University, Civil Engineering Faculty 34469, Maslak, Istanbul, Turkey

Abstract: To reduce autogenous deformations in high performance concrete at early ages, dispersed saturated lightweight aggregates (LWAs) are used as water reservoirs. For this purpose, in a concrete with low water cement ratio, normal aggregates have been replaced by natural LWAs at size fractions of 2-4mm or 4-8mm at three different volume fractions such as 10, 20 and 30 percent of total aggregate volume of concrete prepared at a constant low water/cement ratio. Effects of volume fraction and average particle size of LWAs on the load-displacement at mid-span curve are investigated by measuring the fracture energy, the characteristic length and final displacement. The results indicate that the inclusion of fine fraction of LWAs in the concrete reduces the autogenous deformation significantly compared to that of the coarse LWA fraction. It is also shown that the fracture energy, final displacement at the mid-span curve of the beam, splitting tensile strength and compressive strength of concrete with fine fraction are higher than those of the concrete with coarse fraction. Increasing the replacement ratio of LWAs mitigates autogenous deformation, while having an unfavorable effect on fracture and mechanical properties of concrete for both fine and coarse fraction replacements.

Keywords: autogenous deformation; lightweight aggregate; fracture mechanics

1. INTRODUCTION

High range water reducers (HRWRAs) and ultra-fine pozzolanic materials such as silica fume are commonly used to reduce water/cement ratio and to produce high performance/ultra high performance concretes (HP/UHPCs). Since HP/UHPCs are produced using rich mixes with high cement content and low water/cement ratio, the material becomes very dense and more homogeneous than normal strength concrete. Hence, there will be insufficient amount of water in concrete to complete hydration process and also the water income will be prevented because of the impermeable character of concrete (Weber and Reinhardt, 1997). With production of hydrates, after de-moulding (also before) time dependent volume changes of cement paste begins to produce (Jensen and

M.S. Konsta-Gdoutos, (ed.), Measuring, Monitoring and Modeling Concrete Properties, 163–170.

Hansen, 2001). Conventional curing techniques are not effective in mitigating this loss of relative humidity (RH) in the media (Lura et al., 2004, Akcay et al., 2005). Using the pre-soaked fine lightweight aggregate for preventing chemical shrinkage was first introduced by Phileo (1991). The main objective of this study is to determine the effect of the variations in both the volume fractions and the size of saturated LWAs, which alter the fracture and mechanical properties of concrete.

2. EXPERIMENTAL DETAILS

2.1 Materials and mix design

Seven concrete mixtures were prepared using the same cement and silica fume. The mix design, codes and some of fresh properties of the concrete are given in Table 1.

Table 1. Mix proportion and some properties of fresh concretes

Mix code	CREF	CV10L24	CV20L24	CV30L24	CV10L48	CV20L48	CV30L48
Cement, kg/m^3	497	496	497	496	497	496	496
Silica fume, kg/m^3	50	50	50	50	50	50	50
Water, kg/m^3	153	153	153	153	153	153	153
Fine sand (0-0.25 mm), kg/m^3	169	169	169	169	169	169	168
Natural sand (0-2 mm), kg/m^3	514	513	513	513	514	513	512
Crushed limestone sand (2-4 mm), kg/m^3	520	344	173	0	520	519	518
Crushed limestone No:I (4-8 mm), kg/m^3	529	529	529	529	353	177	0
Pumice LWAs (2-4 mm), kg/m^3	0	48	97	145	0	0	0
Pumice LWAs (4-8 mm), kg/m^3	0	0	0	0	47	94	141
HRWRA, kg/m^3	10	10	10	10	10	10	10
Pre-soaked water, kg/m^3	0	17	34	51	16	31	47
Air, %	1.4	1.5	1.3	1.2	1.2	1.3	1.4
Unit weight, kg/m^3	2441	2328	2224	2115	2328	2211	2095

The partial replacements of normal weight aggregate by natural pumice LWAs were chosen as 10, 20 and 30 percent for two different LWA fractions such as 2-4 mm and 4-8 mm. Pre-soaked water content of LWAs was based on the absorption for 30 min. The code of concrete mixtures was designated as a basis of substituted LWAs volume and size fraction. The numbers following the letter V and L in the designation code denote replacement volume and the replaced size fraction of LWAs, respectively. Exception, however, is the reference concrete (CREF) that contains normal weight aggregates only.

2.2. Test procedure

2.2.1 Autogenous deformation

Autogenous deformation can be determined by measuring either the volumetric deformation or the linear deformation. Both methods of measurement have advantages and disadvantages; however the inconsistencies between the two methods are still debatable (Jensen and Hansen, 2001, Bjontegaard et al., 2004). In this study, autogenous deformation of concretes was measured on four sealed prism specimens of 70 x 70 x 280 mm. Stainless steel gage screws were settled on samples after 4 hours and the measurements were taken over a 200 mm gage length.

2.2.2 Mechanical tests

At least five specimens of each concrete mix were tested under each type of loading conditions. The beams prepared for the fracture energy tests were 500 mm in length and 100 mm x 100 mm in cross-section. Modulus of elasticity, compressive strength and splitting tensile tests were also conducted. The fracture properties were based on the fracture energy (G_F) obtained via a three point bending test. For all of the beams, the tests for determining the fracture energy (G_F) were performed according to the recommendation of RILEM 50-FMC Technical Committee (1985). The characteristic length of the material (l_{ch}), which is the measure of brittleness, is given by the expression (Hillerborg, 1985):

$$l_{ch} = \frac{EG_F}{f_t^2} \tag{1}$$

where f_t is the tensile strength obtained from the disc splitting tests.

Some previous studies have shown that the characteristic length from fictitious crack model can be used for determining the brittleness (Tasdemir and Karihaloo, 2001, Lange-Kornbak and Karihaloo, 1998). The characteristic length of a semi-lightweight concrete, however, is higher than that of HSC with normal aggregate (Tasdemir et al., 2002), which means that the material becomes less brittle.

3. RESULTS AND DISCUSSION

3.1 Autogenous deformation

The linear autogenous deformation measurements of concretes for 28 days and 200 days are shown in Figure 1.a and 1.b, respectively. The results obtained indicate that while the linear autogenous deformation of reference concrete (CREF) was -537D ($1D=1\times10^{-6}$) at the age of 28 days, substituting the 2-4 mm size fraction of normal aggregate concentration by LWAs in the rate of 10 percent reduced the autogenous deformation to the value of -410D. In the concretes for which the substitution was done using the same volume but different size such as 4-8 mm, the autogenous deformation was measured as -480 D. When the normal aggregate is replaced by pumice LWAs with 2-4 mm linear autogenous deformation of concrete for 28 days was mitigated by 24, 47 and 92%, for volume fractions of 10, 20 and 30%, respectively. The use of coarse LWAs in 4-8 mm size fractions, however, prevents autogenous deformation of concrete by 11%, 32% and 65% at the same volume concentrations. Thus, the results have shown that the fine LWAs are more effective than the coarser ones to reduce autogenous deformation of concrete. A similar behavior is also seen in 200 days autogenous deformation results.

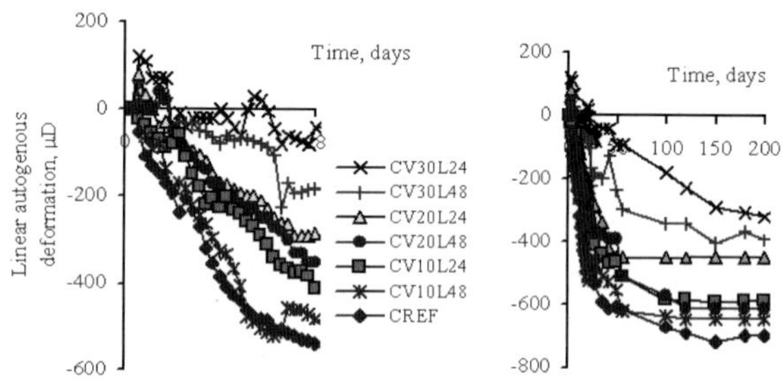

(a) (b)

Figure 1. Linear autogenous deformation vs. time diagram for (a) 28 days and (b) 200 days age of concrete specimens

Because of the fact that the autogenous deformation is an event occurred in cement paste, the coarse mortar (MREF), fine mortar (FMREF) and paste (PREF) phases of the reference concrete (CREF) were casted. In Figure 2, the variations in linear autogenous deformation with time for all these phases are shown. When the autogenous deformation of cement paste at 28 days of age was -760 D, the deformation in fine mortar, coarse mortar and concrete phases were measured as -705, -600 and -537 D, respectively. Both the restraining effect of aggregates and diminishing the cement paste content in a unit volume as a result of addition of aggregate reduce the autogenous deformation.

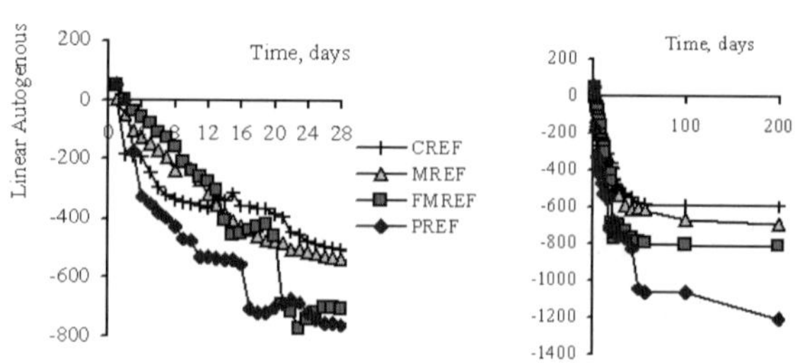

(a) (b)

Figure 2. (a) 28 days old and (b) 200 days old autogenous deformation diagrams for the coarse mortar (MREF), fine mortar (FMREF) and paste (PREF) phases of the reference concrete (CREF)

3.2 Mechanical properties

Table 2 summarizes the mechanical test results of produced concrete along with their fracture parameters.

Table 2. Fracture and strength properties of hardened concretes

Mix code	CREF	CV10L24	CV20L24	CV30L24	CV10L48	CV20L48	CV30L48
Compressive strength, f_c, MPa	93	79	62	50	68	61	48
Modulus of elasticity, E, GPa	46	40	37	32	40	37	27
Area under load-disp. curve, W_0, Nmm	508	563	551	432	451	422	322
Fracture energy, G_F, N/m	85	101	95	77	84	77	59
Final disp.at midspan, δ_0, mm	0.313	0.609	0.687	0.618	0.699	0.671	0.475
Net bending strength, f_{net}, MPa	8.6	7.1	6.0	4.4	6.1	5.0	4.8
Splitting tensile strength, f_{st}, MPa	6.7	6.4	4.8	4.6	5.6	5.0	4.1
Characteristic length, l_{ch}, mm	86	98	155	118	107	115	95

3.2.1 Compressive strength and modulus of elasticity (f_c and E)

A significant decrease in compressive strength for almost all series with LWAs is observed with respect to the reference concrete; because, the modulus of elasticity of LWA is lower than that of the normal aggregate. The compressive strength also decreases with increasing the particle size and volume of the substituted LWA. A slight decrease in the modulus of elasticity was recorded with increasing the replacement volume of the LWAs, although the change in the size of LWA has no significant effect on the modulus of elasticity of concrete

3.2.2 Bending and splitting tensile strengths (f_{net} and f_{st})

As seen in Table 2, the size and volume concentration of LWAs have effects on bending and splitting tensile strengths in a way similar to that observed on the compressive strength results. The use of fine LWA to prevent autogenous deformation of concrete seems more effective in terms of the mechanical properties of concrete. The effect of the size of LWA on bending and also compression can be modeled in that LWA behave in concrete phase as a defect (Tasdemir et al., 2002, Akcay et al., 2005).

3.2.3 Fracture energy (G_F)

The test results have shown that the most significant effect of LWAs is observed on the fracture energy of concrete. The results given in this study are obtained by using the area under the load-displacements curve. As seen in Table 2, in cases of using both fine and coarse fraction, the area under the load displacement curve increases with increasing the replacement volume of LWA, although the peak load decreases with especially the use of fine LWA. Because the LWAs produce more microcracks, the concrete containing these aggregates behaves more ductile (i.e. less brittle) than the reference concrete. Although the slope of the ascending branch of the curves remains almost constant, the measured final displacement at midspan increases significantly with using the LWAs. In particular, replacement of normal aggregate by fine LWAs by volume of 10% causes the fracture energy to have increased by 19% with respect to the reference concrete. However using the coarse fraction of the same volume, although causes a decrease in the peak load, does not change the fracture energy significantly when compared to the reference concrete.

The fracture energy also decreases with increasing the volume of the LWAs. For substitution of normal aggregates with LWAs by 10, 20 and 30 volume percent, concretes with fine fraction of LWAs have greater peak load and also area under the curve increased with respect to that of the coarser fractions. Although the ultimate load does not vary significantly, longer tail is an indication of mean particle size of LWA.

It can be seen that substituting the normal aggregates by LWAs by up to 30 volume percent prevents the autogenous deformation of reference concrete significantly, although such a practice causes also a marked loose in the mechanical properties. However, the negative effect of this application on the mechanical properties of concrete is shown to be smaller in case of using fine LWAs. It should, therefore, be noted that, in order to mitigate the autogenous deformation and cracking in early ages, the volume and size of the LWAs should also be optimized with taking the mechanical properties in to consideration.

3.2.4 Characteristic length (l_{ch})

The most fundamental variations are observed to have occurred on both the fracture energy and the characteristic length of concretes. The characteristic length, which is calculated using the Eq. (1), increases with the use of LWA, indicating that the material becomes more ductile. With decreasing the substituted mean particle size of LWA, the brittleness of concrete diminished. The replacement of the 2-4 mm size of normal aggregate by LWAs by 10 and 20 volume percent increased the fracture energy, while a replacement by 30 percent decreased. On the other hand, no significant effect of replacement on fracture energy is observed when using the size fraction of aggregate of 4-8 mm and the replacement ratio of 10 percent, although a decrease in the fracture energy is observed with increasing the volume of the substitution at ratios greater than 10 percent. The use of LWAs in concrete has an effect of increasing the characteristic length of the material in all size and volume fraction of LWAs. The decrease in the brittleness is more significant in concretes with fine LWAs. It can be shown that replacement of normal aggregate by LWA even by 10 volume percent can play a significant role in decreasing the linear autogenous deformation of concrete at 28 days of age.

4. CONCLUSIONS

Based on experimental studies, the following conclusions can be drawn:

- To mitigate autogenous deformation of concrete, the use of pre soaked pumice LWAs has been found to be effective. Particularly, the replacement of normal aggregate by fine fraction of LWA is more effective in reducing the autogenous deformation compared to the use of the coarse fraction.
- The compressive strength of concrete decreased with increasing the substituted volume of LWA. The compressive strength of cement paste with the fine fraction of LWA is greater than that of the coarser one.
- The use of LWAs in concrete has an effect of increasing the brittleness of the material which can also be clearly observed at the load versus displacement graphs. As the compressive strength decreases, the descending branch at the graph gets with longer tail which means that the material is getting more ductile. As the mean particle size of the LWA decreases, the specific fracture energy, flexural strength and the displacement at mid-span curve increase. It can be concluded that the amount of micro cracks is governed by the low particle size of aggregate substituted and a longer tail is an indication of longer micro cracks.

Acknowledgements

This research was carried out in the Faculty of Civil Engineering at Istanbul Technical University. The authors acknowledge the grant of DPT (State Planning Organization, Project: 2003K120630).

5. REFERENCES

1. Akcay, B., Pekmezci, B.Y., Tasdemir, M.A., 2005, "Utilization of artificial lightweight aggregates in hardened cement paste for internal water curing", Proc. of *fib* Keep Concrete Attractive, Budapest, 23-25 May, Eds., Balazs and Borosnyoi, Vol 1, pp 374-380.
2. Bjøntegaard, Ø., Hammer T.A., Sellevold, E.J., 2004 "On the measurement of free deformation of early age cement paste and concrete", Cem. Conc. Com., Vol.**26**, pp. 427-435,
3. Hillerborg, A., 1985 "Theoretical basis of method to determine fracture energy G_F of concrete", Materials and Structures, Vol. **18**, pp.291-296.
4. Jensen, O.M., Hansen, P.F., 2001 "Autogenous deformation and RH-change in perspective", Cem. Concr. Res. **31**, pp.1859-1865.
5. Lange-Kornbak, D., Karihaloo, B.L., 1998 "Design of fiber reinforced DSP mixes for minimum brittleness", Adv. Cem. Based Mater, **7**, pp. 89-101.
6. Lura, P., Bentz, D.P., Lange, D.A., Kovler, K., Bentur, A., 2004 "Pumice aggregates for internal water curing", Proceedings of the Advances in Concrete Through Science and Engineering Rilem Spring, Northwestern, Illinois, March 23-26.

7. Philleo, R., 1991 "Concrete science and reality", Material Science of Concrete II, American Ceramic Society, Westerville, pp.1-8.
8. RILEM 50-FMC, 1985 "Determination of fracture energy of mortar and concrete by means of three-point bend tests on notched beams", Materials and Structures, Vol. **18**, pp. 285-290.
9. Tasdemir, M.A. and Karihaloo, B.L., 2001 "Effect of aggregate volume fraction on the fracture parameters of concrete: A meso-mechanical approach", Magazine of Concrete Research, Volume **53**, 2001, pp. 405-415.
10. Tademir, M.A., Tademir, C., Grimm, R., König, G., 2002 "Role of aggregate fraction in the fracture of semi-lightweight high strength concrete", Proc. of the 6th Inter. Symp. on Utilization of High Strength/High Performance Concrete, Leipzig, pp.1453-1466.
11. Weber, S., Reinhardt, H.W., 1997 "A new generation of high performance concrete: Concrete with autogenous curing", Adv. Cem-Based Mat. **6**, pp. 59-68.

FRACTURE BEHAVIOR OF HIGH PERFORMANCE FIBER REINFORCED SELF COMPACTING CONCRETE

C. Sengul, Y. Akkaya and M.A. Tasdemir
Istanbul Technical University, Civil Engineering Faculty 34469, Maslak, Istanbul, Turkey

Abstract: The mixture design, workability and mechanical properties of the steel fiber reinforced self compacting concrete mixtures were studied. Nine mixtures were cast without fibers and with fibers, of low and high strength. The compressive and splitting tensile strengths and energy absorption capacities of self compacting concretes with different cementitious material contents, fiber contents and fiber strengths were compared. According to the experimental results, at the same volumetric water/fine material ratio, the compressive strength of specimens without steel fibers did not show significant differences as the cement dosage was decreased from 900 kg/m^3 to 650 kg/m^3 by replacing the cement with finely ground silica powder. With the addition of fibers, the fracture energy and the ductility of the self compacting concrete mixtures were increased compared to those of plain concrete.

Key words: fiber strength, fracture energy, self-compacting concrete, steel fiber, water-powder ratio.

1. INTRODUCTION

The mixture design of self compacting concrete includes fine materials such as cement, fine aggregates and limestone powder, as well as pozzolanic materials such as fly ash and silica fume. Viscosity modifying agents and plasticizers, based on polycarboxylate ether complex, naphthalene sulphonates or melamine sulphonates, are further added to the mixtures, depending on the properties of the targeted workability. The aim of the mixture design is to obtain the desired workability and segregation resistance. This mixture should be able to flow around the steel reinforcement and should not segregate or clump. For this reason, the water/powder ratio and aggregate gradation should be controlled, and effective admixtures should be used during the production of self compacting concrete.

M.S. Konsta-Gdoutos, (ed.), Measuring, Monitoring and Modeling Concrete Properties, 171–177.

Steel fibers and micro-filler materials are widely used in the construction industry. These materials enhance the performance of self compacting concrete, consisting of very fine powder. Studies proved that these materials improve the quality of the concrete both in fresh and hardened states. As the volume of the micro-filler materials increases, the distance between the large size aggregates also increases, reducing the internal friction of the concrete. As the blockage of the large aggregates is prevented, the flow and workability properties of the fresh concrete are improved. The increase in the mechanical properties of the hardened concrete, along with the improved fresh concrete properties, is the subject of this study. The developed volumetric water-to-powder ratio method enables the use of binding materials effectively and provides a tool for optimization, as well as new areas for research on the interaction between the microstructure and mechanical properties of the concrete [1-4].

2. EXPERIMENTAL

Within the scope of this experimental study nine self compacting concrete mixtures were cast of which, three were plain concrete, three were reinforced with low-strength fibers and three were reinforced with high-strength fibers. In these mixtures, the volumetric water-to-cement ratio was kept constant at 0.51. The fine material content of the concrete was composed of cement, silica fume and silica powder. Since the water contents of the mixtures were kept constant, the water-to-cement ratios of the mixtures with the 900 kg/m^3, 650 kg/m^3 and 350 kg/m^3 cement contents were 0.22, 0.33 and 0.61, respectively. The specimens were first cured in 20°C ±2°C water bath for seven days, and then were kept in 85°C ±2°C water bath for three days. At the remaining days until testing at twenty-eighth day, specimens were stored in 20°C ±2°C water bath. Table 1 presents the mixture designs of the concrete.

The steel fibers used were 30 mm long and 0.55 mm in diameter. The strengths of the fibers were 1100 MPa for the low-strength fibers and 2250 MPa for the high-strength fibers.

The unit weight, flow (cm) and t_{50} (s) values of the fresh concrete were measured. In order to determine the properties under restricted conditions, a modified J-ring was also used to measure the flow and t_{50}. The results are presented in Table 2.

Compression test was performed on cylinders with 200 mm height and 100 mm diameter. Modulus of elasticity was determined based on the initial elastic part of the stress – strain curve. Split tension test was performed on 60 mm long cylinders with 150 mm diameter. The results are presented in Table 3.

Deformation controlled bending tests were also performed to obtain the load- deformation curves and to determine the effect of fiber reinforcement on the flexural strength and fracture toughness of the concrete. Three point bending test was employed on prismatic samples with a cross-section of 100x50 mm^2 and 400 mm span. Deformation was measured by an LVDT placed at the mid span. Typical examples of the load - deflection curves are presented in Figures 1-3.

Table 1. Concrete mixture designs, where the mixtures are denoted by specimen codes

(kg/m^3)	V0900	V0650	V0350	V1900	V1650	V1350	V2900	V2650	V2350
Cement	900	650	350	900	650	350	900	650	350
Silica powder	150	350	610	150	350	610	150	350	610
Silica fume	120	120	120	120	120	120	120	120	120
Water	200	200	200	200	200	200	200	200	200
Crushed agg. 1	310	310	310	310	310	310	310	310	310
Crushed agg. powder	352	352	352	352	352	352	352	352	352
Natural sand	334	334	334	334	334	334	334	334	334
Superplasticizer	19	14	8	19	14	8	19	14	8
Steel fiber	0	0	0	40	40	40	40	40	40

Table 2. Fresh concrete properties

	V0900	V0650	V0350	V1900	V1650	V1350	V2900	V2650	V1350
Unit weight (kg/m^3)	2381	2351	2271	2413	2383	2317	2419	2377	2329
Flow (cm)	80	84	77	78	83	68	77	82	69
Restricted flow(cm)	81	84	77	81	82	68	82	82	70
t_{50} (sn)	20	13	13	19	13	26	20	14	23
Restricted t_{50} (sn)	19	14	14	20	14	27	21	15	23

Table 3. Mechanical properties of the hardened concrete

	Compressive strength (MPa)	Modulus of Elasticity (GPa)	Splitting tensile strength (MPa)
V0350	75.5	36.4	7.2
V0650	123.8	44.8	9.6
V0900	126.3	45.8	7.5
V1350	86.0	36.9	8.0
V1650	110.2	43.6	8.6
V1900	124.2	43.9	9.4
V2350	94.9	38.4	8.3
V2650	123.7	42.0	9.5
V2900	138.0	43.8	10.9

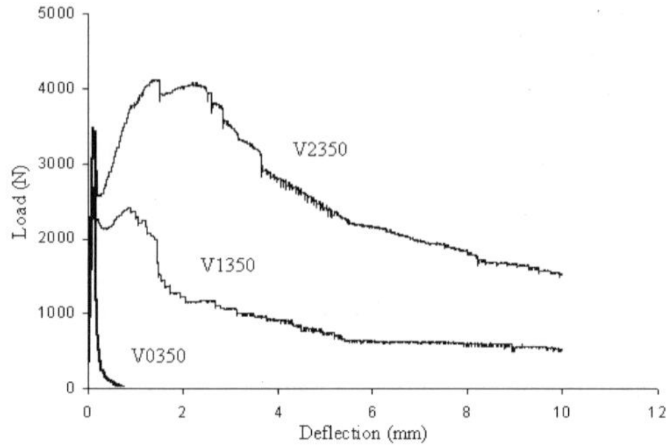

Figure 1. Effect of fiber content and strength for mixtures with 350 kg/m^3 cement content

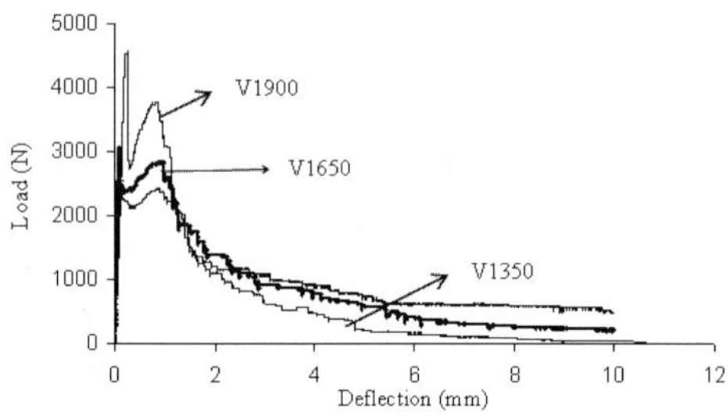

Figure 2. Effect of cement content for mixtures with low-strength fibers

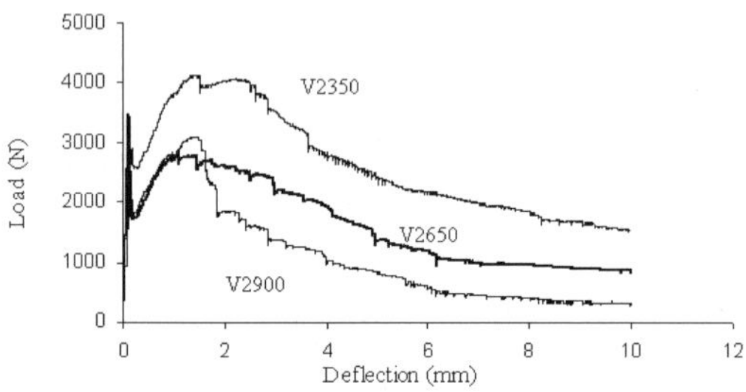

Figure 3. Effect of cement content for mixtures with high-strength fibers

The fracture energy was calculated based on the area under these curves by

$$G_F = (W_o + mg\delta_o) / A_{lig}$$

where W_o: area under the load – deflection curve (N.m), m: weight of sample (kg), g: gravitational acceleration, δ_o: deformation of plain concrete at failure (m) (for the fiber reinforced concrete, δ_o=10 mm), A_{lig}: effective cross-sectional area (m^2). Table 4 presents the experimental results.

Table 4. Fracture energy and the flexural strength of the concrete mixtures

	Final deflection (mm)	Fracture energy (J/m^2)	Flexural strength (MPa)
V0350	0.67	100	7.19
V0650	0.60	77	8.13
V0900	0.60	73	7.41
V1350	10.0	1877	7.29
V1650	10.0	1741	7.66
V1900	9.9	1277	7.34
V2350	10.0	5335	9.14
V2650	10.0	3286	8.27
V2900	9.8	2346	7.58

3. RESULTS AND EVALUATION

The t_{50} values, in Table 2, present that only the concrete with the 900 kg/m^3 cement content exhibits an increase, mainly due to the high cement content and low water-to-cement ratio. There is no significant change in the flow properties and the concrete fulfills the requirements for self compaction.

The t_{50} values of the fiber reinforced concrete mixtures were similar to the plain concrete mixtures, except for the concrete with 350 kg/m^3 cement content. This lower cement content causes a relatively low viscosity compared to the concrete mixtures with higher cement contents. Therefore, fiber addition increased the t_{50} value and decreased the flow diameter of this concrete significantly. The concrete with the 650 kg/m^3 cement content and 0.33 water-to-cement ratio provided the optimum test results in terms of flow and t_{50} values. The results from the restricted flow and t_{50} tests also indicated the same trends. The fiber geometry and content, used in this experimental work, did not affect the flow properties adversely, and the concrete with the 650 kg/m^3 cement content and 0.33 water-to-cement ratio provided the optimum test results.

Figure 1 presents the brittle behavior of a typical plain concrete. It can also be seen that the use of high strength steel fibers improves the mechanical performance and increases the fracture energy of the concrete. Especially after the first crack, the formation of strain hardening is a typical indication of high performance. As it can be seen in Table 4, the concrete with a low cement content, and therefore high water-to-cement ratio, exhibits a more ductile behavior, compared to the high strength concrete with high cement content. With the use of high-strength steel fibers, the crack development can be kept under control and a concrete with strain hardening property could be realized.

Table 3 presents the effect of water-to-cement ratio on the compressive strength of the plain and fiber reinforced concrete mixtures. When the water-to-powder ratio was kept constant, the decrease in cement content to 350 kg/m^3 and the increase in water-to-cement ratio from 0.22 to 0.33, did not affect the compressive strength substantially. However, as the water-to-cement ratio was decreased to 0.61, the compressive strength was reduced 40%.

It can also be seen in Table 3 that the elasticity modulus of the concrete mixtures with 650 kg/m^3 and 900 kg/m^3 cement contents decreased about 5%, with the addition of fibers. The elasticity modulus of the concrete with 350 kg/m^3 cement content did not change with the low-strength fibers, but increased 7% with the high strength fibers.

From Table 3, it can be seen that the splitting tensile strength of the fiber reinforced concrete mixtures decreased with the exchange of cement with silica powder. The decrease was 13% when the cement content was decreased from 900 kg/m^3 to 650 kg/m^3 for high-strength fiber concrete, and 9% for the low-strength fiber concrete. The high-strength fibers provided better performance in split tension test, compared to the low-strength fibers.

The highest fracture energy was obtained for the plain concrete with 350 kg/m^3 cement content and 0.61 water-to-cement ratio (Table 4). This behavior is the result of an increase in toughness due to reduced strength. When the steel fibers were used, the fracture energy of the concrete mixtures increased significantly. Similarly, lower cement content provided higher fracture energy. High-strength fibers exhibited higher performance when compared with low-strength fibers. The trend presented by the flexural strength results was different from the splitting tensile strength results (Table 4). When the low-strength fibers were used, the flexural strengths of the concrete mixtures were similar, irrespective of cement content and water-to-cement ratio.

4. CONCLUSIONS

The effects of steel fiber strength and water-to-powder ratio on the fresh and hardened self compacting concrete properties were investigated. The results of this study can be summarized as follows;

- Increasing the cement content and decreasing the water-to-cement ratio increased the flow time of concrete, due to the increased viscosity. The fiber addition increased flow time of the concrete for only concrete mixtures with low cement content.
- The concrete mixture with 650 kg/m^3 cement content and 0.33 water-to-cement ratio had the optimum fresh concrete properties.
- When the cement-to-powder ratio was kept constant at 0.51, the compressive strength of the plain concrete did not decrease although the water-to-cement ratio was increased from 0.22 to 0.33 and the cement content was reduced from 900 kg/m^3 to 650 kg/m^3. However, the compressive strength of the concrete with 350 kg/m^3 cement content reduced 40 and 30 percent for the plain and fiber reinforced concretes, respectively, compared to the concrete with 900 kg/m^3 cement content.
- The splitting tensile strength of the high-strength fiber reinforced concrete was higher than the concrete reinforced with low-strength fibers.
- Fiber addition increased the ductility significantly. The fracture energy was affected more than the compressive strength. The fracture energy values, up to 10 mm deflection, of the fiber reinforced concrete were 15 to 50 times higher, greater than those of plain concrete.
- Fiber strength also affected the toughness of the concrete. This effect was more pronounced for the mixtures with low cement content, due to the pull-out

resistance of the fibers. As the strength of the concrete increased, the fibers were ruptured, rather than pulling out. Thus, the ductility was less for high strength concrete mixtures.

5. REFERENCES

1. Bornemann, R., Schmidt, M., (2002) "The role of powders in concrete", 6th Internationally symposium on utilization of high strength/high performance concrete, Leipzig, pp 863-872.
2. Grünewald, S., Walraven, J. C., (2001) "Parameter-study on the influence of steel fibers and coarse aggregate content on the fresh properties of self-compacting concrete", Cement and Concrete Research 31 (12), pp 1793-1798.
3. Okamura, H., (1996) "Development of Self-Compacting High-Performance Concrete", Self-Compacting High-Performance Concrete Ferguson Lecture.
4. Sengul, C., (2005) "Effects of water/powder ratio and fiber strength on the mechanicl behavior of steel fiber reinforced self compacting concrete", MSc. Thesis, Istanbul Technical University, Institute of Science and Technology, 64pp.

DETERMINING THE TENSILE SOFTENING DIAGRAM OF CONCRETE-LIKE MATERIALS USING HYBRID OPTIMISATION

J. Hannawald
Institute of Building Materials Research RWTH Aachen University

Abstract: This contribution aims at determining the material parameters tensile strength, fracture energy and the shape of the tensile softening diagram (TSD) for some types of concrete used in Civil Engineering by an inverse finite element (FE) analysis of three point bending (TPB) tests on notched specimens. As an extension to the bilinear approximation of the TSD a multilinear approximation is adopted as the basis of the analysis.

Key words: tensile strength, fracture energy, softening, concrete, finite element analysis, three point bending test, genetic algorithm, optimisation

1. INTRODUCTION

Many non-metallic engineering materials as well as natural materials exhibit under certain load scenarios a strain-softening behaviour. Typical examples are concrete and rock under tension or unconfined compression, or heavily consolidated soils under shear. Modelling such materials as strain-softening continua requires problems of ill-posedness and lack of objectivity resulting in pathological mesh-dependency to be resolved, Jirásek[1]. The simplest remedy, frequently used in engineering applications, is based on an adjustment of the stress-strain diagram depending on the size of the element. Such techniques were proposed for example for the tensile softening of concrete due to smeared cracking by Bazant[2] commonly referred to as crack band model (CBM). The model uses the concept of a cohesive zone, meaning that the formation of a macroscopic stress-free crack is preceded by the development of a fracture process zone, i.e. a region characterised by a highly localised strain and by the development and growth of microcracks, which reduce the cohesion of the material and lead to softening. The idea of the existence of a cohesive zone with regard to concrete has been introduced by Hillerborg[3] and is known as the fictitious crack model (FCM). The FCM lumps the inelastic effects and replaces the process zone by a discontinuity surface across which the displacement

M.S. Konsta-Gdoutos, (ed.), Measuring, Monitoring and Modeling Concrete Properties, 179–187.

field has a jump. The tensile stress transmitted by the discontinuity is considered as a function of the crack opening displacement (COD). The tensile stress-COD law, also called tensile-softening diagram (TSD), provides an objective, simplified description of the decohesion process. The transition to the CBM used for the continuum representation is established by smearing the discontinuity over a certain finite distance, i.e. element size, and transfer it into an inelastic strain. This leads to an inelastic stress-strain law with softening.

The TSD is specified by the tensile strength f_t of the material and a monotonically decreasing function of COD describing the shape, with the area under the curve equal to the Mode-I fracture energy G_F. The shape function has often been approximated by linear, bilinear, exponential or more complex curves such as the model of Reinhardt[4] with the default values specified by Hordijk[5]. It has been pointed out that the shape of the TSD has a significant influence on the numerically determined load-deflection curves in a finite-element (FE) analysis of three point bending (TPB) tests of concrete beams and hence is essential for a realistic material description in numerical analysis, Roelfstra[6]. With the exception of the bilinear approximation the shape of the TSDs mentioned is already completely specified by the two material parameters f_t and G_F. If materials exist, where these shapes are inappropriate the corresponding TSD would lead to an unrealistic material description even if the material parameters f_t and G_F were correct.

This contribution aims at determining the material parameters f_t and G_F and the shape of the TSD for some types of concrete used in Civil Engineering by an inverse FE analysis of TPB tests on notched specimens. As an extension to the bilinear approximation a multilinear approximation of the TSD is adopted as the basis of the analysis.

2. MULTILINEAR TSD

The bilinear TSD is specified by a total of four parameters, i.e. the tensile strength f_t, the ultimate COD w_u and the coordinates (w_1, σ_1) of the point where the slope changes[1]. A straightforward extension of the bilinear approach would consist of adding additional points (w_i, σ_i), where changes of slope occur, in order to arrive at a multilinear curve. But this would require two additional parameters for every intermediate point and would cause great problems in view of the inverse determination of such a large number of parameters. Hence a different approach is adopted.

The basic idea is to generate the multilinear TSD by points (w_i, σ_i) specified by a geometric series, i.e. to let

$$w_i = w_u \cdot p^{i-N}, \quad 1 \le i \le N, \quad w_0 = 0, \quad p > 1 \tag{1}$$

and

$$\sigma_i = f_t \cdot q^{-i}, \quad 0 \le i \le N-1, \quad \sigma_N = 0, \quad q > 1, \tag{2}$$

1. One of the parameters may be replaced by the fracture energy G_F, since the area under the curve must equal G_F.

where N is the number of linear segments of the TSD and p and q may be called the generating parameters of the series. The data points thus defined are listed in Table 1.

Table 1. Displacement/Stress data points of the multilinear TSD

i	w	σ
0	0	f_t
1	$w_u p^{1-N}$	$f_t q^{-1}$
2	$w_u p^{2-N}$	$f_t q^{-2}$
:	:	:
:	:	:
$N-2$	$w_u p^{-2}$	$f_t q^{2-N}$
$N-1$	$w_u p^{-1}$	$f_t q^{1-N}$
N	w_u	0

For arbitrary values of p and $q > 1$ it may happen that the resulting TSD is not a monotonically decreasing function of the COD. The conditions to be obeyed by the parameters p and q are elaborated as follows. Comparing the slopes $S_{1,left}$ and $S_{1,right}$ of the two segments adjacent to point i=1 yields after simple algebra

$$S_{1,right} \geq S_{1,left} \quad \Leftrightarrow \quad p \geq 1 + \frac{1}{q}.$$

Similarly for point i=N-1one obtains

$$S_{N-1,right} \geq S_{N-1,left} \quad \Leftrightarrow \quad p \geq \frac{1}{q-1}$$

whereas for all points $2 \leq i \leq N\text{-}2$ the resulting condition

$$S_{i,right} \geq S_{i,left} \quad \Leftrightarrow \quad p \geq \frac{1}{q}$$

is automatically satisfied since p>1 and q>1. Thus evaluating p to

$$p = \text{Max}\left\{\frac{1}{q-1}, 1 + \frac{1}{q}\right\} \tag{3}$$

ensures that the TSD defined by Eqns. (1) and (2) is a monotonically decreasing function of the COD. The area A under the TSD can be evaluated to

$$A = \frac{w_u \cdot f_t}{2} \cdot B(p,q) \tag{4}$$

where

$$B(p,q)=\frac{1+\frac{1}{q}}{p^{N-1}}+(p-1)(q+1)\sum_{k=2}^{k=N-1}\frac{1}{p^{k}q^{N+1-k}}+\frac{1-\frac{1}{p}}{q^{N-1}} \tag{5}$$

Since A has the physical meaning of the fracture energy G_F, the ultimate COD w_u may be eliminated using Eq. (4), i.e.

$$w_u=\frac{2\cdot G_F}{f_t\cdot B(p,q)}. \tag{6}$$

The TSD defined by Eqns. (1) and (2) with N=10 is compared in Figure 1 with the exponential TSD

$$\sigma(w)=f_t\cdot\exp\left(-\frac{f_t}{G_F}w\right)$$

and the model of Reinhardt[4]

$$\sigma(w)=\begin{cases} f_t\cdot\left[\left(1+a^3\left(\frac{w}{w_u}\right)^3\right)\mathrm{e}^{-b\frac{w}{w_u}}-\frac{w}{w_u}\left(1+a^3\right)\mathrm{e}^{-b}\right] & 0\le w\le w_u, \\ 0 \quad w>w_u \end{cases}$$

where according to Hordijk[5] a=3, b=6.93 and G_F=0.1947 $f_t\, w_u$.
A normalised COD=$w\, f_t/G_F$ and normalised tensile strength =σ/f_t have been introduced so that a choice of particular values for the tensile strength and the fracture energy only results in rescaling of the axis.

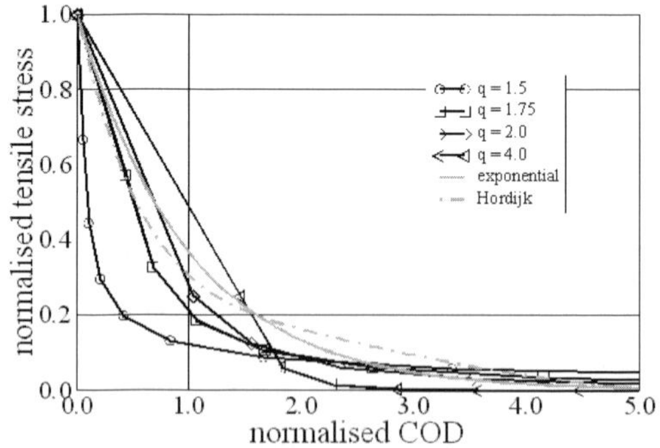

Figure 1. Multilinear TSD defined by Eqns. (1) and (2) in comparison to the exponential model and the model of Hordijk[5]

It may be guessed from Figure 1 that the multilinear TSD approaches the linear approximation for large values of q.

Indeed, in the limit $q \rightarrow \infty$ it may be verified from Eqns. (3) and (5) that $p \rightarrow 1$ and $B(p,q) \rightarrow 1$ and from Eqns. (1) and (2) that $w_i = w_u$ and $\sigma_i = 0$ for i>1. Hence only the first linear segment is reproduced and Eq. (6) yields $G_F = 0.5 f_t\, w_u$ for the triangular area. For $q<2$ the initial slope of the TSD is steeper as compared to the exponential model and the model of Hordijk[5], resulting in a more brittle material behaviour. Hence quite different shapes of the TSD can be produced by varying the generating parameter q resulting in broader spectrum of material behaviour.

The multilinear TSD is completely defined by only three parameters, i.e. the fracture energy G_F, the tensile strength f_t and the generating parameter q, which is even less than the four parameters required for specifying the bilinear TSD.

3. FE-MODEL AND OPTIMISATION SCHEME

TPB tests on notched beams have been performed with several types of concrete. Some details of the tests, i.e. compressive strength (f_c), maximum grain size (d_{max}) of the concrete, beam and notch depth (D and a respectively) and the span of the beam (S) are given in Table 2. For full details refer to references as indicated. The beam thickness was equal to the beam depth in any case.

Table 2. Experimental Details

Concrete type	f_c	d_{max}	D	a	S	Ref.
-	MPa		mm			-
Autoclaved Aerated Concrete	3.7	-	100	50	400	7
C25/30	41.8	8				
C50/60	72.3	32	100	50	500	8
C80/95	116.0	16				
Self-compacting Concrete	77.3	16				
Fine-Grained Concrete	80.0	0.6	40	10	200	9

The two dimensional FE model used for determining the load-displacement curves, Figure 2, is discretised with only 5 quadrilateral elements with quadratic displacement interpolation in the ligament cross-section. Only these elements utilised an inelastic stress strain law in tension specified by the multilinear TSD. The rest of the beam was considered as a linear elastic material. Hence strain localisation was enforced to occur in the ligament cross-section only. The coarse mesh provided sufficient accuracy. This has been checked by increasing the number of elements in ligament cross-section by a factor of six, which resulted in almost the same load-displacement curves after properly adapting the TSD according to the element size. The model has been solved using the FE package DIANA.

A genetic algorithm (GA)[1] was adopted for inversely determining the 3 parameters G_F, f_t and q of the multilinear TSD and Young's modulus E for the elastic material behaviour. Poisson's ratio has been fixed to 0.15 for all types of concrete. In the GA, the population size was 100, an elitist generational replacement was applied, i.e. the most fit string (chromosome) was always copied to the new population.

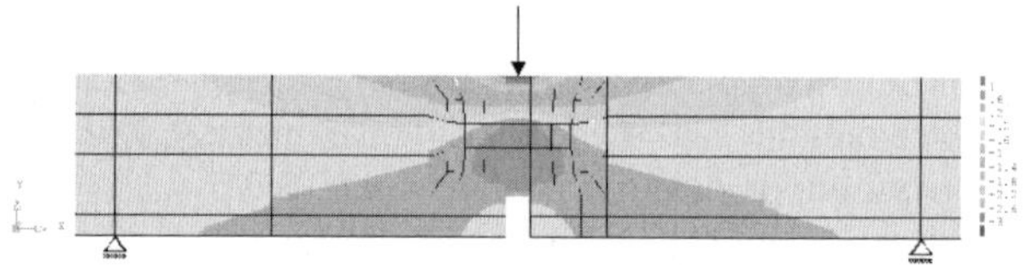

Figure 2. 2D FE model of the TPB test and distribution of bending stress

The remainder of the new population (99 strings) was generated from the two most fit parent strings by crossover and mutation. The number of population generations (iterations) has been limited to 10.

GAs are appropriate for finding promising areas of the search space, but not as suited for fine-tuning within those areas. Methods designed for local optimisation are good at fine-tuning but lack a global perspective. Hence a combination of both methods may outperform either algorithm alone. Therefore a derivative free optimisation method based on a quadratic approximation (QA) of the objective function[11] was incorporated into the GA. Once the fitness of the population has been evaluated, QA is invoked with the most fit string as a starting point. On improvement, the best string is copied to the new population and is being used as one of the parent strings for crossover and mutation.

For both GA and QA open-source code[12, 13] written in C and FORTRAN respectively has been used. The FE package DIANA is started by a C system call from the cost function that evaluates the string. The sum of least squares determined for the numerically calculated and experimental load-deflection curve (LDC) is adopted as string fitness.

4. RESULTS AND DISCUSSION

As an illustration of the operation of the optimisation procedure the total of LDCs and TSDs produced during the search for the best approximation of the LDC for the Autoclaved Aerated Concrete is depicted in Fig. 3. Mean computation time was 30 s for one LDC and approximately 12 h in total on an AMD Opteron machine with 2 GHz.

In Figure 4 the best LDCs are compared with the mean experimental LDC for all types of concrete (left column). The resulting best TSDs are plotted. The best values obtained for the three inelastic parameters are also given. The TSDs are also compared with the exponential model and the model of Hordijk[5] using the best tensile strength and fracture energy obtained in the optimisation procedure (right column). The FE model has been solved again with the inelastic elements being characterised by these TSDs and the resulting LDCs are also shown.

1. See URL[10] for a comprehensive introduction to basic GA terminology and instructive interactive examples of basic GA operation.

It may be seen from Figure 4 that in general the conformance of the numerical LDCs obtained using the multilinear TSD with the experimental curves is very good. From the good conformance it may be concluded that valid material parameters for the tensile strength and fracture energy have been identified by the optimisation procedure. The values obtained span a quite large range from 0.4 MPa (Autoclaved Aerated Concrete) to 4.9 MPa (C80/95) and 9.5 N/m (Autoclaved Aerated Concrete) to 233 N/m (Self-compacting Concrete) for the tensile strength and the fracture energy respectively.

The conformance of the LDCs obtained with the multilinear shape of the TSD is superior to those obtained with the other shapes. In the case of the largest value obtained for the parameter *q*, the initial slope of both exponential TSD and the TSD according to Hordijk[5] is steeper than the initial slope of the multilinear TSD, resulting in an underestimation of the peak load in the LDC, Fig. 4a. For the smallest *q*, the initial slopes of all types of TSD are quite similar. In this case the peak load prediction is the same but the post peak behaviour is different, Figure 4b.

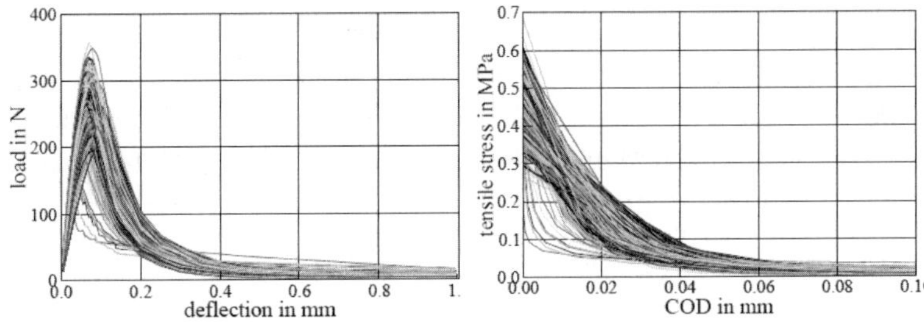

Figure 3. Total of LDCs and TSDs produced by the optimisation procedure for the Autoclaved Aerated Concrete

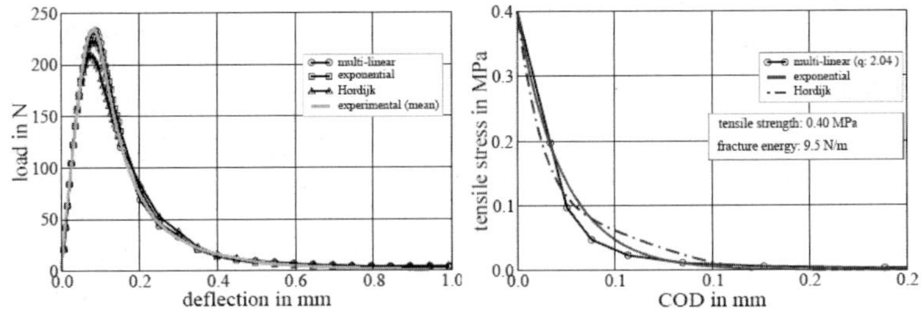

a) Autoclaved Aerated Concrete

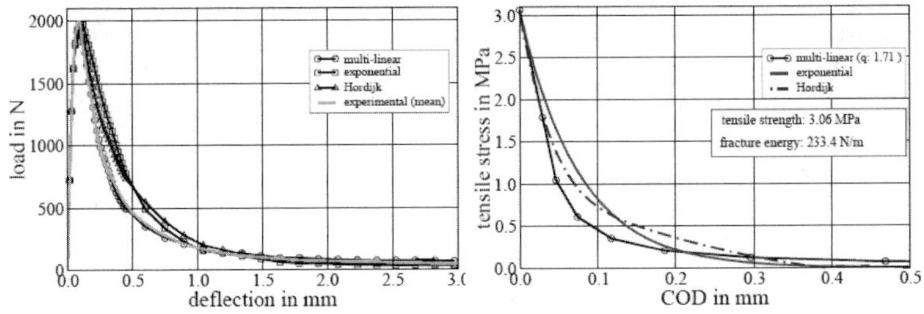

b) Self-compacting Concrete

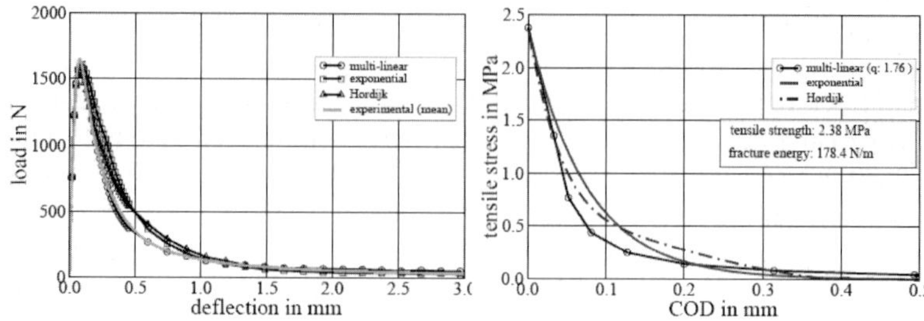

c) Concrete Type C25/30

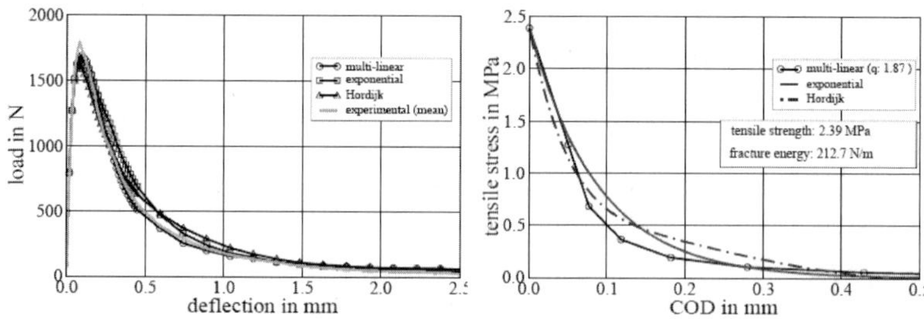

d) Concrete Type C50/60

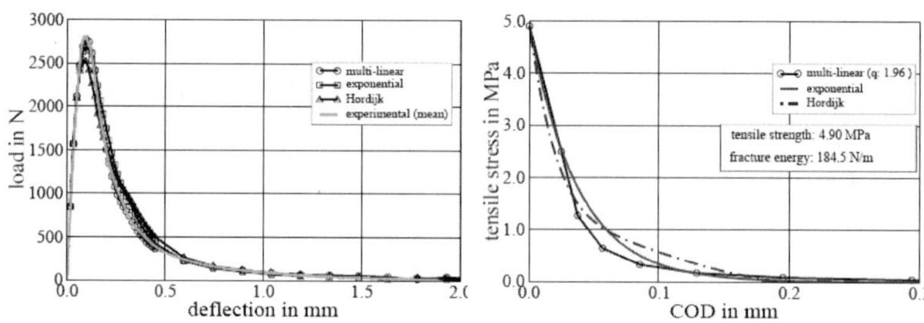

e) Concrete Type C80/95

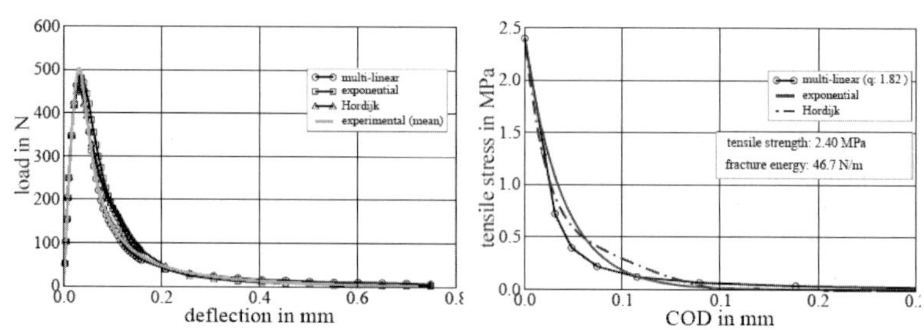

f) Fine-grained Concrete

Figure 4. Comparison of best numerical and experimental LDCs (left) and best TSDs (right)

5. CONCLUSIONS

1. A multilinear tensile softening diagram (TSD) with only three parameters can represent a broader spectrum of material behaviour than existing models.
2. By means of an optimisation scheme based on an genetic algorithm hybridised by a derivative free quadratic approximation of the objective function together with a finite element model the shape of the TSD, the tensile strength and fracture energy can be identified from three point bending tests.

6. REFERENCES

1. Jirásek, M.; Bazant Z. P, 2002, *Inelastic Analysis of Structures*. Chichester : John Wiley & Sons,
2. Bazant, Z.P.; Oh, B.H., 1983, Crack Band Theory for Fracture of Concrete, *Materials and Structures*, vol. **16** no. **93**:155-177
3. Hillerborg, A.; Modeér, M.; Petersson, P.-E , 1976, Analysis of Crack Formation and Crack Growth in Concrete by Means of Fracture Mechanics and Finite Elements, *Cement and Concrete Research*. **6**:773-781
4. Reinhardt, H.W.; Cornelissen, H.A.W.; Hordijk, D.A.,1986, Tensile Tests and Failure Analysis of Concrete, *Journal of Structural Engineering (ASCE)*, **112**, No. **11**:2462-2477
5. Hordijk, D.A., 1991, *Local Approach to Fatigue of Concrete*. Delft, Technical University, PhD
6. Roelfstra, P.E.; Wittmann, F.H., 1986, Numerical Method to link Strain Softening with Failure of Concrete, in: *Fracture Toughness and Fracture Energy of Concrete: Proceedings of the International Conference on Fracture Mechanics of Concrete*, F. H. Wittmann, ed., Elsevier, Amsterdam, pp. 163-175
7. Schmidt, U., Kang, B.-G., Hannawald, J., Brameshuber, W., 2005, Tests on Loadbearing Behaviour of Masonry Shear Walls, in: *Autoclaved Aerated Concrete : Innovation and Development. Proceedings of the 4th International Conference on Autoclaved Aerated Concrete*, Limbachiya, M.C., Roberts, J.J. eds., pp. 375-385
8. Brameshuber, W., Eck. Th., 2006, Application of Fracture Mechanics on Unreinforced Concrete Walls, in: *Proceedings of 16th European Conference of Fracture,* to be published.
9. Brameshuber, W., Brockmann, T., Banholzer, B., 2004, Analytical Evaluation of the Softening Behaviour of Fine Grained Concrete, in: *Proceedings of the fifth International Conference on "Fracture Mechanics of Concrete Structures" (FRAMCOS-5)*, Li, V.C. et al, ed., Vail Colorado, USA Vol. 2, pp. 1145-1153
10. http://cs.felk.cvut.cz/~xobitko/ga/
11. Powell, M.J.D., 2003, On the Use of Quadratic Models in Unconstrained Minimisation, *Report No. 2003/Na03*, www.damtp.cam.ac.uk
12. http://info.mcs.anl.gov/pub/pgapack/
13. http://plato.la.asu.edu/topics/problems/nlounres.html (The package is called NEWUOA)

PERFORMANCE OF PLAIN AND BLENDED CEMENT CONCRETES AGAINST CORROSION CRACKING

E. Güneyisi[1], T. Özturan[2] and M. Gesolu[1]
[1]*Asst. Prof. Dr., Department of Civil Engineering, Gaziantep University, 27310, Gaziantep, Turkey;* [2] *Prof. Dr., Department of Civil Engineering, Boaziçi University, 34342, Bebek, Istanbul, Turkey*

Abstract: An experimental investigation on the resistance of the plain and blended cement concretes against reinforcement corrosion was undertaken. A rapid corrosion testing technique was adopted to compare the corrosion performance of the reinforced concrete specimens. The compressive and splitting tensile strength properties of the concretes have been also evaluated. The blended cements used in this work were in conformity with the relevant European Norm (EN 197-1) and they contained a blend of portland cement clinker, blast furnace slag, natural pozzolans, and limestone powder at different proportions. Ten different concrete mixtures at 0.65 and 0.45 w/c ratios were designed and cast by using a plain and four different blended cements and subjected to three curing procedures namely, uncontrolled, controlled, and wet curing. The specimens from each mixture were tested at the ages of 28 and 180 days. The test results indicated that the specimens with blended cements had superior performance and mostly yielded longer time to corrosion cracking under the condition of well curing practice and latter age. It was also found that data obtained from corrosion test was considerably well correlated with the compressive and especially splitting tensile strength data.

Key words: blended cement; concrete; corrosion; curing; strength.

1. INTRODUCTION

Blended cements have attracted considerable interest, from both scientific and technological points of view, since the beginning of this century[1-4]. In fact, blended cements have now been recognized as a basic material that meets the requirements of the modern society[5]. Not only are they characterized by their performance properties that are absent in ordinary portland cement (OPC), but they also represent gainful utilization of by-prod-

M.S. Konsta-Gdoutos, (ed.), Measuring, Monitoring and Modeling Concrete Properties, 189–198.

ucts from the steel, electric power, and silicon industries. However, the addition of a wide range of blending materials of different chemical composition introduces significant diversity into the cement system. Therefore, it should be recognized that concretes produced with different cements have different properties and performances[6]. Published data on the blended cement concretes suggest that they reduce the chloride-ion penetration into concrete[7-9]. For this reason, a superior corrosion resistance of the steel reinforcement in these concretes is expected[10-12]. However, type of blending material, chemical admixture, water/cementitious material ratio, age, and curing procedure lead to significant variation in performance properties of those concretes.

This paper reports the results of an investigation undertaken to determine the relative corrosion performance of steel reinforcing bars embedded in plain and blended cement concretes. The strength properties of the concretes were also evaluated. In all mixture, a plain portland and four blended cements were used. After casting, the specimens were exposed to three different curing regimes prior to testing. The effect of cement type, curing regime, water-cement ratio, testing age was discussed.

2. EXPERIMENTAL DETAILS

2.1 Materials, mixtures, and curing

A total of five different cements such as portland cement (CEM I), portland composite cements (CEM II/A-M & CEM II/B-M), composite cement (CEM V/A), and blast furnace slag cement (CEM III/A) were used as a cementitious material. The properties of the cements are summarized in Table 1. The blended cements are multicomponent and a blend of portland cement clinker, blast furnace slag, natural pozzolans, and limestone powder at different proportions. Crushed limestone graded to a maximum particle size of 20 mm and having an average fineness modulus of 5.65 was used as coarse aggregate, while a mixture of natural sand and crushed limestone sand having an average fineness modulus of 1.58 was used as fine aggregate. Both aggregates were obtained from local sources. A sulphonated naphthalene formaldehyde-based superplasticizer, having a specific gravity of 1.22, was used to get a workable fresh concrete. Deformed steel bars, 16 mm in diameter, were used as reinforcement.

Two different concrete mixtures, namely plain cement concrete (PCC) and blended cement concrete (BCC) with w/c ratios of 0.65 and 0.45 and cement contents of 300 and 400 kg/m^3, respectively, were designed. The high and low w/c ratio concrete mixtures were denoted by (N) and (H), respectively. Concretes produced with CEM I, CEM II/A-M, CEM II/B-M, CEM V/A, and CEM III/A were designated as (B1), (B2), (B3), (B4), and (B5), respectively. The details of the mixture proportions are given in Table 2. The mixing was performed in a pan mixer in accordance with the ASTM C192 procedures. The fresh mixed concrete was cast into 150x150x150 mm cube and 100x200 mm cylinder moulds for the compressive and splitting tensile strength tests, respectively. The reinforced concrete specimens for the accelerated corrosion test were 100x200 mm cylinders in which a 16 mm diameter steel bar was centrally embedded. From each mixture, nine concrete cubes, nine concrete cylinders, and six reinforced concrete specimens were cast and left in the casting room for 24 hr. They were then demoulded and divided into three equal groups and subjected to the following curing conditions: a) uncontrolled curing (UC): specimens were air cured at uncontrolled temperature and relative humidity until the test age. The variable relative humidity (from 40 to 80%) and temperature (from 8 to 22 °C) of the hall was considered to simulate no curing or uncontrolled curing condition, b) controlled curing (CC): specimens were immersed in 20 ± 2 °C water for 7 days and then air cured in a room at 20 ± 1 °C and 50 ± 5 % relative humidity until the test age, and c) wet curing (WC): specimens were immersed in 20 ± 2 °C water until the test age.

2.2 Test methods

2.2.1 Compressive and splitting tensile strengths

Compressive and splitting tensile strength tests were performed on a 2000 kN capacity testing machine and test procedures followed during the test were in accordance with ASTM C39 and ASTM C496, respectively. Three specimens were tested at the ages of 28 and 180 days for each property.

2.2.2 Accelerated corrosion test

An accelerated corrosion testing technique was used to evaluate the corrosion performance of plain and blended cement concretes. A schematic representation of the experimental setup is shown in Figure 1. As seen from the figure, the reinforced concrete specimens were immersed in a 4% sodium chloride solution and the steel bar (working electrode) was connected to the positive terminal of a DC power source while the negative terminal was connected to stainless steel plates (counter electrode) placed near the specimen. In this circuit, the steel bar is the anode, the plates are the cathode, and the solution is the electrolyte. The corrosion process was initiated by impressing an anodic potential of 30 V. The high impressed voltage was used to accelerate the corrosion process and the shorten test period to practical laboratory conditions. The specimens were monitored periodically to see how long it takes for corrosion cracks to appear on the surface. A data logger was used for recording the current variation with time. The current increased dramatically when the specimen cracked, so it was possible to find out when cracking occurred. The variation of current with time and time to failure of the specimens were determined for all concrete types. The two specimens were tested at the ages of 28 and 180 days for each property.

3. RESULTS AND DISCUSSION

3.1 Compressive and splitting tensile strengths

Compressive strength test results as a function of curing condition and age are given in Figure 2. Strength values varied from 32.5 to 66.7 MPa and from 23.3 to 69.7 MPa for plain and blended cement concretes, respectively, depending on w/c ratio, curing condition, and age at testing. Generally, increasing concrete age from 28 to 180 days, the plain cement concrete exhibited an increase in the strength values from 13 to 23% while that was from 10 to 58% for the blended cement concrete, depending mainly upon the variation in w/c ratio, cement type, and curing condition. The higher strength increment is most probably due to the secondary pozzolanic reaction taking place in the blended cement concretes. Furthermore, it was observed that inadequate curing practice resulted in strength reduction as high as 18% for the plain one but 23% for the blended one. It was also pointed out that concrete with blended cement had considerably greater strength than that with plain cement, especially under the condition of controlled and wet curing and later age.

Table 1. Properties of plain portland and blended cements used

	Cement Type				
Code	B1	B2	B3	B4	B5
Turkish standard	PÇ 42.5	PKÇ	PKÇ	KZÇ	CÇ
European standard	CEM I	CEM	CEM	CEM	CEM
Silicone dioxide SiO_2	20.64	18.38	28.34	25.63	28.81
Aluminum oxide Al_2O_3	5.06	5.05	7.33	5.06	7.2
Ferric oxide Fe_2O_3	3.14	2.89	2.89	3.72	2.31
Calcium oxide CaO	63.98	61.78	52.55	48	49.94
Magnesium oxide MgO	1.2	1.36	2.09	-	4.44
Sulfur trioxide SO_3	2.38	2.34	2.88	2.3	2.41
Sodium oxide Na_2O	0.31	0.28	0.21	-	0.15
Potassium oxide K_2O	0.8	0.73	-	-	0.87
Chloride Cl^-	0.035	0.036	-	0.01	0.027
Insoluble residue	0.46	0.48	7.8	-	0.64
Loss of ignition	1.72	6.44	1.16	-	2.4
Free lime	1.41	1.44	0.35	-	0.83
Specific gravity	3.15	3.12	3.01	3.05	2.94
Specific surface (m^2/kg)	336	334	406	430	464
f_{cc} (2 day) (MPa)	27.5	23.7	23.1	20	13.3
f_{cc} (7 day) (MPa)	41.3	39	35.9	31	24.6
f_{cc} (28 day) (MPa)	51.4	46.2	51.2	45	-
Component fraction in cement (% by weight)					
Clinker, K	95.5	78.7	70.5	57.5	46.7
Blast furnace slag, S	0	2.0	13.0	21.8	48.3
Limestone, L	0	11.9	0	3.0	0
Natural pozzolans, P	0	3.2	13.0	12.6	0
Gypsum	4.5	4.2	3.5	5.1	5.0
Total	100	100	100	100	100

Table 2. Concrete mix proportioning, kg/m^3

Code	W/C	Cement	Water	Coarse Aggregate		Fine Aggregate		SP[a]
				No I	No II	Natural sand	Crushed sand	
N-B1	0.65	308.1	200.3	558.3	616.0	537.6	191.8	0.77
N-B2	0.65	302.8	196.8	547.8	604.4	527.5	188.2	0.76
N-B3	0.65	304.7	198.1	548.6	605.4	528.3	188.5	0.76
N-B4	0.65	306.5	199.2	552.9	610.1	532.4	189.9	0.77
N-B5	0.65	306.4	199.2	549.7	606.5	529.3	188.8	0.77
H-B1	0.45	405.4	182.4	536.2	591.6	516.3	184.2	3.04
H-B2	0.45	399.6	179.8	527.5	582.0	507.9	181.2	3.00
H-B3	0.45	399.5	179.8	523.7	577.9	504.3	179.9	3.00
H-B4	0.45	400.7	180.3	526.0	580.4	506.5	180.7	4.01
H-B5	0.45	399.8	179.9	521.0	574.9	501.7	179.0	4.00

[a] Superplasticizer

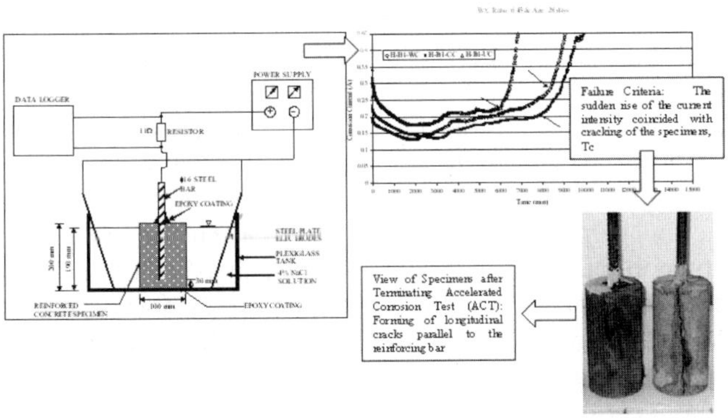

Figure 1. Representation of the experimental setup for the accelerated corrosion test (ACT)

Figure 3 shows the results of the splitting tensile strength tests. The strength increase pattern for the splitting tensile strength is almost similar to that of the compressive strength. However, the rate of the strength increase in the former was lower than that in the latter. Increase in concrete age, the plain cement concrete had an increment from 9 to 24% but from 4 to 43% for the blended one, depending on w/c ratio, cement type, and curing procedure. However, poor curing caused somewhat higher reduction in splitting strength up to 43% and %58 for the plain and blended cement concretes, respectively, as compared to that for compressive strength. It was also noted that in most of the cases the blended cement concretes had relatively higher splitting tensile strength values.

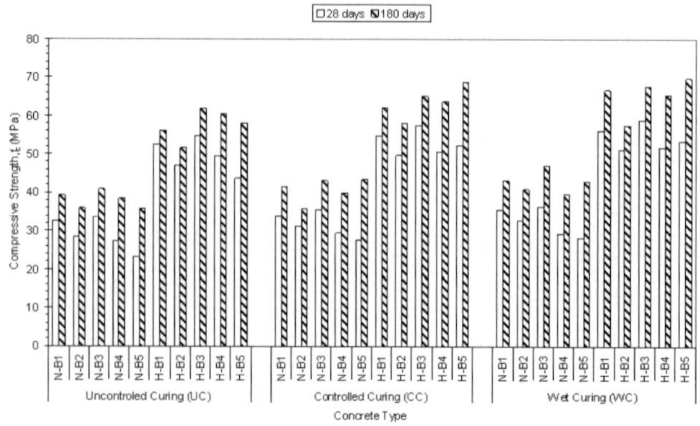

Figure 2. Variation of compressive strengths of plain and blended cement concretes subjected to different curing conditions

3.2 Corrosion resistance

The results of the accelerated corrosion tests are given in Figure 4. Typical curve of corrosion current versus time for the low w/c ratio plain cement concretes cured under different curing conditions is illustrated in Figure 1. The sudden rise of the current intensity coincided with the cracking of the specimen. The formation of the longitudinal crack parallel to the reinforcing bar on the test specimens is also shown in Figure 1. As seen from Figure 4, cement type, w/c ratio, curing condition, and concrete age had significant effects on the time to failure of the specimens. The time to failure for specimens with plain cement was in the range of 67 to 170 hr (3 to 7 days) whereas that with blended cements ranged from 50 to 440 hr (2 to 18 days), depending mainly upon the experimental variables. It was clearly observed that, unlike the strength results, there was higher rate of increment in corrosion time due to increase in age and also greater rate of reduction in corrosion time due the application of inadequate curing to the test specimens. Moreover, blended cement concrete subjected to the proper curing procedure showed noticeably longer time to corrosion than plain one, particularly at later age.

3.3 Relationship between time to failure and compressive strength (or splitting tensile strength)

In order to assess the interdependence between time to failure obtained from corrosion test and compressive and splitting tensile strength of the concretes made with plain and blended cements, the plots given in Figures 5 and 6 are used. Analysis of the results indicated that they had similar tendency and the higher strength resulted in the higher time to failure of the reinforced concrete specimens. It was observed that there was an exponential type of correlation between the measured properties. Moreover, the relevant function parameters (a and b), Pearson correlation coefficient (R^2), and analysis of variance, F statistics, are also summarized in Table 3.

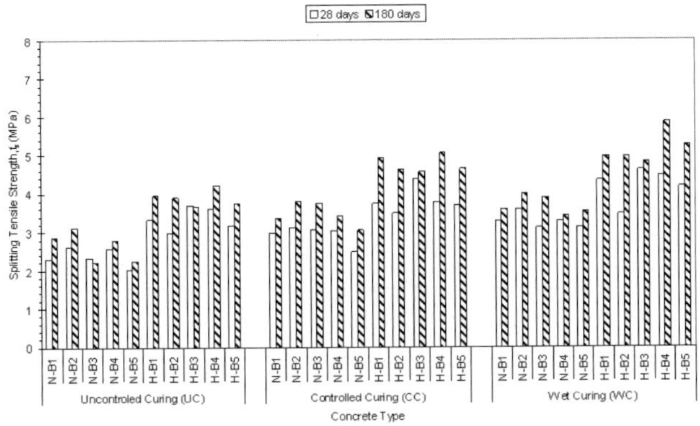

Figure 3. Variation of splitting tensile strengths of plain and blended cement concretes subjected to different curing conditions

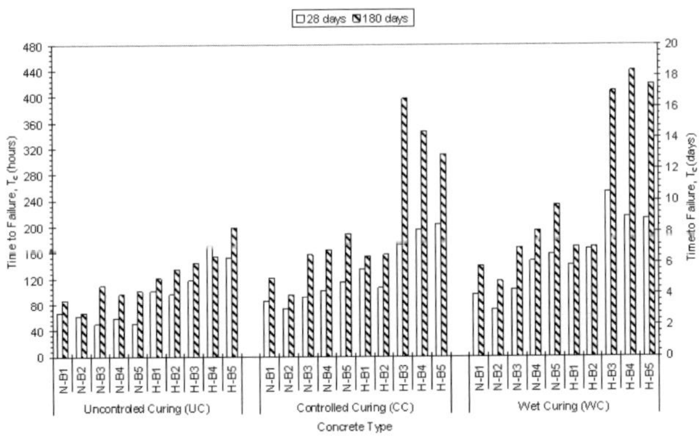

Figure 4. Variation of time to failure of plain and blended cement concrete specimens subjected to different curing conditions

It was noted from the analysis that the function parameters varied for each concrete property. However, test data were fitted with considerably high correlation coefficients. The analysis of variance, *F* statistics, showed that all the relations were significant at a 0.05 level. It was also pointed out that time to failure was well correlated with splitting tensile strength in comparison to compressive strength.

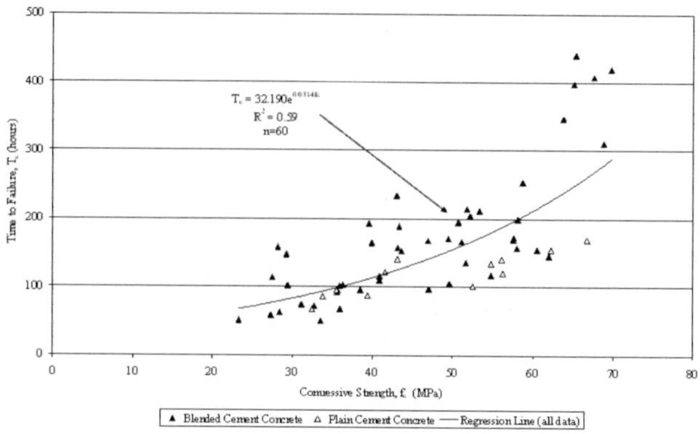

Figure 5. Relationship between the time to failure (Tc) and the compressive strength (fc), representing all data

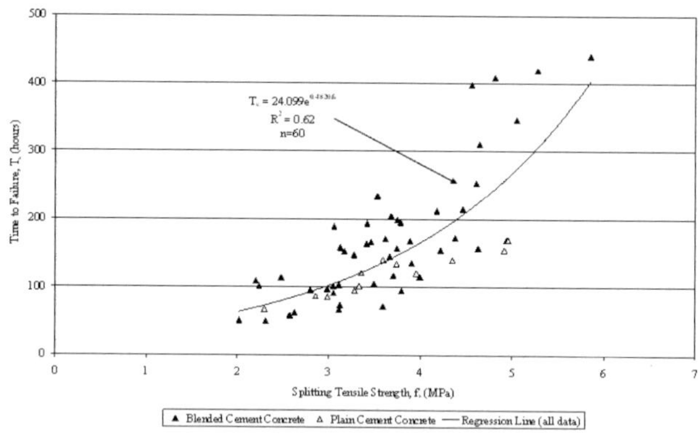

Figure 6. Relationship between the time to failure (Tc) and the splitting tensile strength (fs), representing all data

Table 3. Parameters of the correlations studied for the concrete properties

Model Parameters	Time to Failure vs. Compressive Strength	Time to Failure vs. Splitting Tensile Strength
Trend/Regression type	Exponential, $f(x)=ae^{bx}$	Exponential, $f(x)=ae^{bx}$
a	32.190	24.099
b	0.0341	0.4820
Number of case, n	60	60
Correlation coefficient, R^2	0.59	0.62
$F_{statistics}$	115.3	159.5
$F_{0.05(n-2)}$	4.01	4.01
Significance at 0.05 level	Yes	Yes

4. CONCLUSIONS

Based on the results obtained from the present experimental investigation, the following conclusions can be drawn: Both plain and especially blended cement concrete were more sensitive to curing condition. Poor curing practice resulted in remarkably lower strength and corrosion resistance. Effect of blended cements against reinforcement corrosion was more noticeable for later age, particularly under proper curing procedure. Generally, the resistance of specimens to reinforcement corrosion improved greatly with the use of blended cements, depending mainly on w/c ratio, curing condition, and age at testing. Furthermore, analytical comparison of the measured properties (time to failure, compressive and splitting tensile strengths) revealed that they were well correlated with each other. An exponential type of relationship with fairly good correlation coefficient was observed.

Acknowledgements

The authors are grateful for the financial support provided by Turkish Cement Manufactures' Association (TCMA). Thanks are extended to Mr. lyas Gültekin, Mr. Hasan enel, and Mr. Hamdi Ayar from the Department of Civil Engineering at Boaziçi University for their support during experimental study.

5. REFERENCES

1. P.K. Mehta, Pozzolanic and Cementitious Byproducts as Mineral Admixtures for Concrete–A Critical Review, in: *Fly Ash, Silica Fume, Slag and Other Mineral By-Products in Concrete* (SP-79, American Concrete Institute, Detroit, 1983), pp. 1-46.
2. R. Sersala, Aspects of the Chemistry of Additions, in: *Advances in Cement Technology*, edited by S.N. Ghosh (Pergamon Press, New York, 1983), pp. 537-567.
3. R.F. Feldman, Significance of Porosity Measurements on Blended Cement Performance, in: *Fly Ash, Silica Fume, Slag and Other Mineral By-Products in Concrete* (SP-79, American Concrete Institute, Detroit, 1983), pp. 415-433.
4. P.C. Aitcin, F. Autefage, A. Carles-Gibergues, and A. Vaquier, Comparative Study of the Cementitious Properties of Different Fly Ashes, in: *Fly Ash, Silica Fume, Slag and Other Mineral By-Products in Concrete* (SP-79, American Concrete Institute, Detroit, 1983), pp. 91-113.
5. H. Uchikava and T. Okamura, Binary and Ternary Components Blended Cement, in: *Mineral Admixtures in Cement and Concrete*, edited by S.L. Sarkar (ABI Press, N. Delhi, 1993), pp.1-83.
6. E. Güneyisi, Mechanical and durability performance of plain and blended cement concrete exposed to chlorides and different curing regimes, Boaziçi University, Civil Engineering Department, (March, 2004), PhD.Thesis.
7. C.L. Page, N.R. Short, and W.R. Holden, The Influence of Different Cement on Chloride Induced Corrosion of Reinforcing Steel, *Cement Concrete Research* **16**, 79-86 (1986).
8. K. Byfors, Influence of Silica Fume and Fly Ash on Chloride Diffusion and pH Values in Cement Paste, *Cement Concrete Research* **17**, 115-130 (1987).
9. P.S. Mangat and B.T. Molloy, Influence pfa, Slag and Microsilica on Chloride Induced Corrosion of Reinforcement in Concrete, *Cement Concrete Research* **21**, 819-834 (1991).

10. A. Kumar and D.M. Roy, Pore Structure and Ion Diffusion in Admixture Blended Portland Cement Systems, in: *Proceedings 8th International Conference on the Chemistry of Cement*, (Vol. IV, Rio de Janeiro, 1986), pp. 73-79.
11. C.L. Page and O. Vennesland, Pore Solution Composition and Chloride Binding Capacity of Silica Fume Cement Pastes, *Materials and Structures* **16**(91), 19-25 (1983).
12. N.S. Berke, Resistance of Microsilica Concrete to Steel Corrosion, Erosion, and Chemical Attack, in: *Proceedings of the International Conference on Fly Ash, Silica Fume, Slag and Natural Pozzolans in Concrete*, edited by V.M. Malhotra (SP-114, American Concrete Institute, Farmington Hills, Mich., 1989), pp. 861-886.

MECHANICAL BEHAVIOR AND OPTIMUM DESIGN OF SFRC PLATES

F. Koksal[1], A. Ilki[2], F. Bayramov[2] and M.A. Tasdemir[2]

[1]Erciyes University, Department of Civil Engineering, Yozgat, Turkey; [2]Istanbul Technical University, Faculty of Civil Engineering, Istanbul, Turkey

Abstract: The effects of the aspect ratio and the volume fraction of steel fibers on the mechanical behavior of steel fiber reinforced concrete (SFRC) plates are investigated by using EFNARC testing technique. Optimum solutions for the design parameters (L/d) and (V_f) were found while the brittleness of concrete is minimized. The experimental design is done by using Response Surface Method, which is a promising approach for optimizing SFRCs to meet several performance criteria such as minimum brittleness. According to the experimental results, significant improvement on toughness of concrete is observed by addition of steel fibers into concrete and the longer fibers are more effective in obtaining higher toughness results than the shorter ones.

Keywords: aspect ratio; composite desirability; fiber volume fraction; Response Surface Method; steel fiber; toughness.

1. INTRODUCTION

In recent years, concretes of strength over 100 MPa can be easily produced using ordinary materials and applying conventional mix design methods as a result of rapid developments in concrete technology. The brittleness of concrete, however, increases with an increase in its strength. In other words, the higher the strength of concrete, the lower is its ductility. This inverse relation between strength and ductility is a serious drawback and limits the use of high strength concrete (Balagaru et al.,1992). Therefore, improving the ductility of concrete becomes a major problem for high strength concrete. The use of steel bers in concrete greatly increases its energy absorption and ductility (Wafa and Ashour, 1992). The main contribution of steel bers to concrete can be seen after matrix cracking. If a proper mix is designed, after the matrix cracking, randomly distributed, short bers in the matrix act as crack arresters by bridging mechanism, undergo a pull-out process, delay crack formation and limit crack propagation (Banthia

M.S. Konsta-Gdoutos, (ed.), Measuring, Monitoring and Modeling Concrete Properties, 199–205.

and Trottier, 1995; Kurihara et al., 2000). Debonding and pulling out of bers from matrix require more energy; therefore, a substantial increase in toughness occurs (Barros and Figuerios, 1999). These types of materials are useful if a large amount of energy absorption capacity is required to reduce brittle failure (Tasdemir and Bayramov, 2002). Fiber type, aspect ratio (length/diameter), volume fraction, orientation of fibers in matrix and pull-out resistance of fibers as well as matrix properties inuence the performance of SFRC. SFRC has a wide-range of applications such as; pavements and overlays, industrial floors, precast products, hydraulic and marine structures, repairing and retrofitting of reinforced concrete structures, tunnel linings and slope stabilization works (Balaguru and Shah, 1992).

The main objective in this work is to optimize the toughness of SFRCs to obtain a significantly more ductile behavior than that of plain concrete. The test method used in the experimental part is based on the French Standard known as EFNARC (1996). A multiobjective simultaneous optimization technique and three-level full factorial experimental design by means of Response Surface Method (RSM) are used for optimization and obtaining desirability functions. Optimum solutions are carried out by maximizing toughness and splitting tensile strength while minimizing the cost (i.e. volume fraction of fibers).

2. TOUGHNESS CHARACTERIZATION

For toughness characterization, the test method based on the French standard EFNARC (1996) can be used. In this method, a test plate of 600x600x100 mm is simply supported on its four edges as shown in Figure 1. The test ends, when the midpoint of the plate reaches to the displacement of 25 mm. The area under the load-deflection curve up to 25 mm deflection is taken as toughness or energy absorption capacity of the plate.

A typical load-deflection curve for midpoint of plate specimen made of sprayed SFRC is illustrated in Figure 1. In this figure, OA part of the curve represents the linear elastic region and part AB represents further matrix cracking. Strain hardening behavior occurs between points B and C. Finally, beginning of fiber debonding and fiber pull-out occur in CD region. (Tasdemir et al., 2002). EFNARC recommends the use of the plate test to characterize the energy absorption capacity of fiber reinforced concrete (FRC) for sprayed applications. Toughness classes a, b, or c given in EFNARC require energy absorption capacities of 500, 700, and 1000 Nm, respectively.

3. EXPERIMENTAL WORK

3.1 Specimen characteristics and production

Water/cement ratios of 0.70 and 0.45 were used in the production of concretes. For each water/cement ratio, aspect ratios of fibers were 55, 65, and 80 and volume fractions of steel fibers were 0.26, 0.51 and 0.76 %. Ten concrete batches were made for each water/cement ratio, 0.70 and 0.45. In each series, the volume fractions of cement, siliceous sand (0-0.25 mm), limestone fines (0-4 mm), crushed limestone (4 to 10 mm and 8 to 20 mm), and water were kept constant. Ordinary Portland cement contents in the mixtures with "water/cement" ratios of 0.70 and 0.45 were 300 kg/m^3 and 400 kg/m^3, respectively. For the mixtures with water/cement ratio of 0.45, the amount of high range water reducing admixture varied between 1.3% and 2.0% by weight of cement for different concrete mixtures to maintain approximately the nominal slump between 100 and 130 mm.

The specimens were cast in steel moulds and compacted on a vibration table. All the specimens were demoulded after about 24 hours, stored under wet burlap, at 20^{0}C until 28 days of age and then laboratory air-cured until testing days at 160 days. The dimensions of the plates, prepared for the toughness tests, were 600x600x100 mm. For each mixture, three cylinders, 150 mm in diameter and 300 mm in height, were used for compressive strength and modulus of elasticity tests. Six disc specimens, 150 mm in diameter and 60 mm in height, were prepared for the splitting test. At least three specimens of each concrete mixture were used for each test type.

3.2 Test procedure

Standard strength tests were carried out in accordance with European Standards (EN 206 and EN 12390). Plate test for the determination of the toughness was performed according to EFNARC (1996). During the plate tests, as in Figure 1, a center point load was applied through a contact surface of 100x100 mm. The deflection at the center of plate was recorded simultaneously by using two linear variable displacement transducers (LVDTs). The load was applied by an MTS actuator of 250 kN capacity at a loading rate of 1 mm/min. The load versus center deflection curve for each plate was obtained using the average of two deflection measurements.

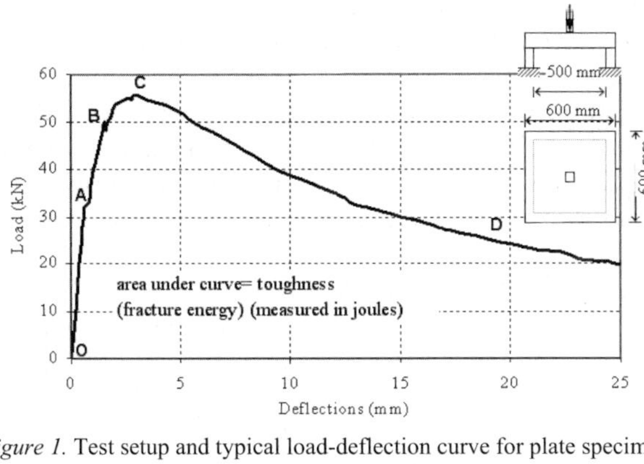

Figure 1. Test setup and typical load-deflection curve for plate specimen

3.3 Mechanical properties

Effects of the fiber volume fraction and the fiber aspect ratio on the compressive strength and modulus of elasticity were found to be inconsistent. The diameter of the steel fibers and possibly their orientation may play a role in compression. On the other hand, the addition of steel fibers into concrete may have an effect of increasing the ductility in the compressive failure rather than the compressive strength itself.

In concretes with water/cement ratios of 0.45, it was shown that the splitting tensile strength increases with increasing steel fiber volume fraction. In concretes with water/cement ratios of 0.45, for the aspect ratio of 65, an increase of the fiber volume fraction from 0 (i.e. normal concrete) to 0.51% has resulted in an increase of 25% in corresponding splitting tensile strength, for the aspect ratios of 55 and 80, this increase was 21% and 15%, respectively. In concretes with water/cement ratio of 0.70, there is no significant effect of the fiber volume fraction on the splitting tensile strength for the aspect ratios of 55 and 80.

It was observed that the opening and propagation of the cracks were controlled by the steel fibers along the fracture plane. After matrix cracking, during the crack propagation, some fibers were broken but some of them were pulled-out of the matrix. After completion of the splitting tests the fracture surfaces were examined, and in most cases, the fibers with the aspect ratio of 65 (L/d= 65) were not broken, but were pulled out of the matrix. However, the fibers with the aspect ratio of 80 (L/d= 80) were broken into two parts. The results obtained for the fibers with L/d=65 might be due to their larger cross sections and smaller bonding length compared to that of fibers with L/d=80. In this study, tensile strength of steel fibers used was 1100 MPa. For high strength concretes, however, the use of high strength steel fibers with a tensile strength of 2000 MPa is suggested (Vandewalle, 1996, Grünewald and Walraven, 2002). The details of experimental results are given by Koksal (2003).

3.4 Toughness

Toughness values of the mixture with the water/cement ratio of 0.45 are given in Figure 2. It can clearly be seen that toughness increases as the fiber volume fraction increases. Toughness test results obtained experimentally can be found in the work of Koksal (2003). The researcher has pointed out that, the SFRC plates made with the water-cement ratio of 0.45 allow obtaining high values of toughness, maximum loads, first crack loads, and as a result a high ductility; depending on volume fractions and aspect ratios of fibers used. The increase in the toughness results in relatively higher energy requirement for fiber pull-out and fiber debonding processes.

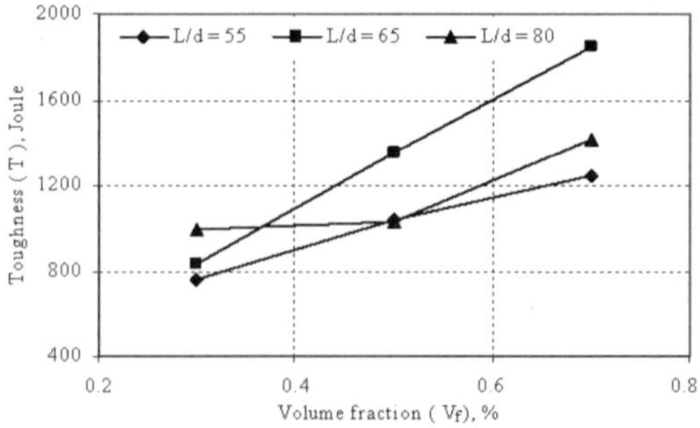

Figure 2. Toughness versus fiber volume fraction at different aspect ratios for w/c=0.45

The reason for the increase in toughness with increasing fiber volume fraction and its aspect ratio stems from a great number of fibers forming a bridge in the tortuous crack propagation. The fracture process of steel fiber reinforced concrete consists of progressive debonding of fiber, during which slow crack propagation occurs. Final failure occurs due to unstable crack propagation when the fibers are pulled out and the interfacial shear stress reaches the ultimate strength. It can be concluded that the results obtained give a clear picture of how a quasi-brittle concrete transforms into a ductile composite with the addition of steel fibers.

4. OPTIMIZATION

In this study, a multi-objective simultaneous optimization technique in which Response Surface Method (RSM), as the basis for finding the best solution, is incorporated. A common response surface experimental plan which can be used to find optimal settings is a two variable (i.e. L/d and V_f), three-level (i.e. L/d=55, 65, and 80; $V_f = 0.26$ %, 0.51 %, and 0.76 %) full factorial experimental design for each water/cement ratio series (i.e., w/c=0.45 and 0.70). In each series, nine experimental data for each response of SFRCs, were fitted to a polynomial type of mathematical model by using analysis of variance (ANOVA). For each mechanical property of SFRCs in each series, the fitted regression models for toughness are given below:

For the mixtures with water/cement ratio of 0.45:

$$T = -7980.5+252.1(L/d)+1245.4(V_f) - 1.81(L/d)^2 +359.4(V_f)^2 -5.18(L/d)(V_f) \quad (1)$$

For the mixtures with water/cement ratio of 0.70:

$$T = -4830.5+151.8(L/d) -1136.4(V_f) -1.08(L/d)^2 +401.6(V_f)^2 +31.01(L/d)(V_f) \quad (2)$$

A numerical optimization technique using desirability functions (d_j), which are defined for each response, can be used to optimize the responses simultaneously (Derringer and Suich, 1980). A desirability function (d_j) varies over the range of $0 \le d_j \le 1$. A multi-objective optimization problem is solved by using the single composite response (D) given in Eq. 3, which is the geometric mean of the individual desirability functions:

$$D = (d_1 \times d_2 \times d_3 \times .....d_4)^{(1/n)} \quad (3)$$

where n is the number of responses included in the optimization. For this experimental work, feasible regions are:

$$55 \le L/d \le 80 \quad (4)$$

$$0.26\% \le V_f \le 0.76\% \quad (5)$$

To attain a more ductile concrete, a concrete with the highest toughness (T) and the highest splitting tensile strength (f_{st}) is to be obtained. The cost of the steel fibers used in the production of composites is also important when the application is of concern. Therefore, it is necessary to maximize T and f_{st}, and minimize V_f, simultaneously. For n=3, Eq. 3 takes the form:

$$D = (d_1 \times d_2 \times d_3)^{(1/3)} \tag{6}$$

where, d_1, d_2 and d_3 are the desirability functions of T, f_{st} and V_f, respectively. For water/cement ratio of 0.45 and 0.70, the solutions of this multi-objective optimization are shown in Table 1. Composite desirability (D) for this multiobjective optimization is shown in Figure 3.

Table 1. Optimum solutions to obtain maximum toughness and maximum splitting strength with minimum fiber volume fraction (water/cement = 0.45)

Water/cement ratio	Aspect ratio (L/d)	Fiber volume fraction, V_f (%)	Splitting tensile strength, f_{st} (MPa)	Toughness (Joule)
0.45	65	0.49	6.1	1324
0.70	80	0.48	4.0	1132

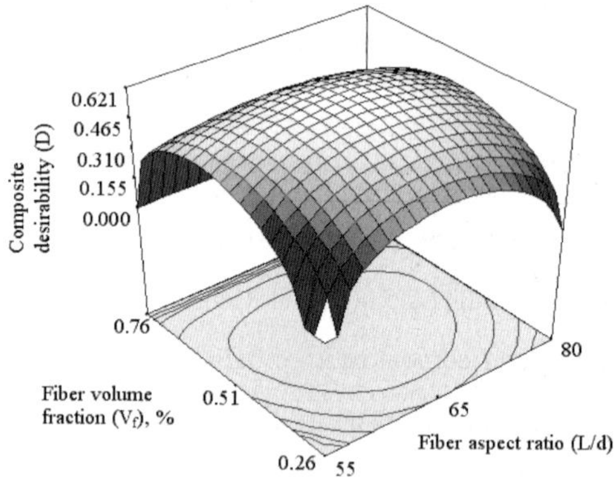

Figure 3. Response surface plot of the composite desirability (D) when T and f_{st} are maximized and steel fiber volume fraction (V_f) is minimized simultaneously (w/c=0.45)

5. CONCLUSIONS

Based on the experimental results obtained on plate test specimens, a multi-objective simultaneous optimization technique is used in this study. It is clearly stated that the toughness of SFRC plates depends on both the fiber volume fraction and the fiber aspect ratio as well as w/c ratio. The longer steel fibers provide better toughness values than the shorter ones. On the other hand, as the steel fiber volume fraction increases, the toughness increases and high values of aspect ratios give higher fracture energies, particularly for lower values of w/c. The experimental design made by using response surface method provides a thorough examination of SFRC properties over the selected ranges of fiber volume fractions and aspect ratios. In order to provide an adequate representation of the responses, fitting quadratic models that are usually assumed to represent each concrete property of interest, can be done in identifying optimal mixes.

6. REFERENCES

1. Balaguru, P.N., Narahari, R., and Patel, M., 1992, Flexural toughness of steel fiber reinforced concrete, *ACI Materials Journal,* Vol. 89, **6** : 541-545.
2. Balaguru, P.N. and Shah, S.P., 1992, Fiber Reinforced Composites, Mc Graw Hill, Inc.
3. Banthia N., Trottier JF.,1995, Concrete reinforced with deformed steel fibres. Part II: Toughness characterization, *ACI Material Journal*, **2**:146–154.
4. Barros, J.A.O. and Figuerios, J.A., 1999, Flexural behaviour of SFRC: testing and modelling. *Journal of Material in Civil Engineering*, Vol. 11, **4**:331-339.
5. Derringer, G. and Suich, R., 1980, Simultaneous optimization of several response variables. *Journal of Quality Technology,* Vol.12, 4:214-219.
6. EFNARC, 1996, *European Specification for Sprayed Concrete*, European Federation of Producers and Applicators of Specialist Products for Structures, EFNARC, 30 pp.
7. Grünewald, S. and Walraven, J.C., 2002, High strength self-compacting fibre reinforced concrete: behaviour in the fresh and hardened state, *In 6th International Symposium on HSC/ HPC*, Leipzig, June, 2: 977-989.
8. Koksal, F., 2003, *Mechanical Behavior and Optimization of Steel Fiber Reinforced Concretes*, Institute of Science and Technology of ITU, Doctoral Thesis, Istanbul, 191 pp. (In Turkish with English abstract)
9. Kurihara N., Kunieda M., Kamada T., Uchida Y., Rokugo K., 2000, Tension softening diagrams and evaluation of properties of steel fibre reinforced concrete, *Journal of Engineering Fracture Mechanics,* 65:235–45.
10. Tasdemir, M.A. and Bayramov, F., 2002, Mechanical behavior of cement based composite materials, *ITU Journal/d Engineering*, October, Vol. 1, **2**:125-144 (In Turkish with English abstract).
11. Tasdemir, M.A., Ilki, A. and Yerlikaya, M., 2002, Mechanical Behaviour of Steel Fibre Reinforced Concrete used in Hydraulic Structures, HYDRO 2002, *International Conference of Hydropower and Dams*, November 4-7, Kiris-Antalya, pp.159-166.
12. TS EN 206, 2002, *Concrete- Part 1: Specification, Performance, Production and Conformity*, Turkish Standard Institution, Ankara, 68 pp.
13. TS EN 12390, 2002, *Testing Hardened Concrete - Part 1: Shape, Dimensions and Other Requirements for Specimens and Moulds*, Turkish Standard Institution, Ankara, 7 pp.
14. Vandewalle, L., 1996, Influence of the yield strength of steel fibres on the toughness of fibre reinforced high strength concrete. *Proc., the CCMS Symposium, Worldwide Advances in Structural Concrete and Masonry*, Chicago, pp. 496-505.
15. Wafa F.F. and Ashour S.A., 1992, Mechanical properties of high-strength fibre reinforced concrete. *ACI Material Journal*, 89, **5**:449–455.

MECHANICAL PROPERTIES OF HYBRID FIBER REINFORCED CONCRETE

A.E. Yurtseven, I.O. Yaman and M. Tokyay
Department of Civil Engineering, Middle East Technical University 06531 Ankara Turkey

Abstract: Fiber reinforcement is commonly used to provide toughness and ductility to brittle cementitious matrices. Reinforcement of concrete with a single type of fiber may improve the desired properties to a limited level. A composite can be termed as *hybrid*, if two or more types of fibers are rationally combined to produce a composite that derives benefits from each of the individual fibers and exhibits a synergetic response. This study aims to characterize and quantify the mechanical properties of hybrid fiber reinforced concrete. For this purpose nine mixes, one plain control mix and eight fiber reinforced mixes were prepared. Six of the mixes were reinforced in a hybrid form. Four different types of fibers were used in combination, two of which were macro steel fibers, and the other two were micro fibers. Volume percentage of fiber inclusion was kept constant at 1.5%. In hybrid reinforced mixes volume percentage of macro fibers was 1.0% whereas the remaining fiber inclusion was composed of micro fibers. 28-day compressive strength, flexural tensile strength, flexural toughness, and impact resistance tests were performed in the hardened state. Various numerical analyses were carried out to quantify the determined mechanical properties and to describe the effects of fiber inclusion on these mechanical properties. It was observed that micro steel fibers contributed to the strength and toughness whereas polypropylene fibers are effective in providing ductility.

Key words: Fiber reinforcement; Hybrid composite; Impact resistance

1. INTRODUCTION

Inherently, concrete is a brittle material and reinforcement in terms of continuous rebars or short discrete fibers are often used to improve this brittle behavior. When short discrete fibers are used, the corresponding material will then be termed as fiber reinforced concrete (FRC), which is defined by ACI Committee 544 as a concrete made of hydraulic cements containing fine or fine and coarse aggregates and discontinuous discrete fibers [1].

M.S. Konsta-Gdoutos, (ed.), Measuring, Monitoring and Modeling Concrete Properties, 207–214.

FRC is often regarded as a composite material with two phases in which concrete represents the matrix phase and the fiber constitutes the inclusion phase. Even though, properties and volume fractions of individual phases affect the composite behavior, those pertaining to fibers are often more important, which include its volume fraction, geometry and mechanical properties. Depending on these factors, reinforcement of concrete with a single type of fiber will improve the properties of the composite to a limited level. However, using the concept of hybridization -a rational combination of two or more types of fibers- more attractive engineering properties of the composite could be achieved from a possible synergetic response of the different fibers.

Various researchers have studied the effects of fiber hybridization on the mechanical properties of concrete and concluded that a composite with superior mechanical characteristics can be obtained through combined use of fibers with different properties[2-5]. According to Benthur and Mindess[6] the advantages of hybrid fiber systems can be listed as follows:

- To provide a system in which one type of fiber is smaller, so that it bridges the micro cracks of which growth can be controlled. This leads to a higher tensile strength of the composite. The second type of fiber is larger, so that it arrest the propagating macro cracks and can substantially improve the toughness of the composite.
- To provide a system in which one type of fiber, which is stronger and stiffer, improves the first crack stress and the ultimate strength, and the second type of fiber, which is more flexible, and ductile leads to improved toughness and strain in the post-cracking zone.
- To provide a system, in which the durability of fiber types is different. The presence of the durable fiber can increase the strength and/or toughness relation after age while the other type is to guarantee the short-term performance during transportation and installation of the composite elements.

The objective of this study is first to develop hybrid fiber reinforced concrete composites and then to characterize and quantify the benefits obtained by the concept of hybridization. Moreover, an attempt will be made to compare the mechanical properties and impact resistance of these composites.

2. EXPERIMENTAL PROGRAM

2.1 Materials and mix proportions

The cement used in all mixes was normal Portland cement, which corresponds to ASTM Type I. As for the aggregates, crushed limestone and crushed sand from the same local source were used with maximum aggregate sizes of 12 and 5 mm, respectively. To obtain sufficient consistency in fiber reinforced mixes a novel polycarboxylic type superplasticizer was also used. Using the abovementioned ingredients, a common concrete matrix was prepared (Table 1) and used for all mixtures. This common matrix was designed to obtain a slump of 15 cm and a 28-day compressive strength of 35 MPa.

Table 1. Mix proportions of the plain concrete mix used as the common matrix

Constituent	Water	Cement	Fine Agg.	Coarse Agg.	Superplasticizer
Amount (kg/m^3)	210	422	784	819	4.22

Four different types of fibers were used in combination, keeping the total volume percentage of fibers at 1.5%. Steel fibers Dramix RC 80/60 and Dramix ZP 305 were used as macro-fibers. Steel fiber OL 6/16 and polypropylene fiber Duomix 20 were used as micro-fibers. Properties of the fibers are presented in Table 2. The volume percentage of fibers was kept constant at 1.5%. Macro steel fibers constituted two thirds of the total fiber content whereas the remaining part was composed of micro fibers in hybrid fiber reinforced mixes. The mixtures were designated such that the label contains the volumetric ratio of the fibers. For example, in Table 3, mixture R1.0-L0.3-D0.2 contains 1.0% R, 0.3% L and 0.2% D type of fibers.

Table 2. Properties of fibers

Fiber Name	Designation	Density (kg/m^3)	Length (mm)	Diameter (mm)	Min f_t * (MPa)	Geometry
Dramix RC 80/60	R	7850	60	0.75	1050	Hooked
Dramix ZP 305	Z	7850	30	0.55	1100	Hooked
OL 6/16	L	7170	6	0.16	2000	Straight
Duomix 20	D	910	20	0.016	400	Fibrillated

* Minimum tensile strength on the wire

2.2 Specimen preparation and test procedures

For each mixture, three 150 mm cubes, two 150x300 mm cylinders, and two 150x150x500 mm beams were cast. Specimens were demolded after one day and then placed in the curing room with 90 ± 5% relative humidity and 20 ± 1 ^{o}C temperature until testing day. Cubes were used for the determination of compressive strength. Cylinder specimens were sawn into 150x60 mm discs which were used in the impact resistance tests. Beam specimens were used for the determination of flexural tensile strength and flexural toughness.

28-day compressive strength of each mixture was determined in accordance with TS 3114 ISO 4012 standard. Average of the test results of three specimens belonging to a mixture was recorded as the 28-day compressive strength. Specimens were tested so that the direction of loading was 90° with the direction of casting.

Tests on beam specimens were carried out in the Turkish Cement Manufacturers Association's laboratory, in a load controlled manner. Thus, post crack portions of the load deflection curves could not be obtained. A single point load was applied at the mid span of the specimens and the deflections were measured from the bottom of the specimens using a mechanical dial gauge with an accuracy of 0.00254 mm. For each load increment corresponding deflection was read and recorded. Area under the load deflection curve was designated as the energy absorbed upto flexural strength, calculated using the trapezoidal rule. Average of the test results of two beam specimens belonging to a mix was accepted as the flexural tensile strength and flexural toughness of that mix. Specimens were tested so that the direction of loading was 90° with the direction of casting.

Impact resistance was determined in accordance with the repeated drop weight method suggested by ACI Committee 544[6]. For this purpose standard Marshall hammer apparatus located in the Transportation Laboratory was modified. Number of blows up to cracking and ultimate failure were determined. In this test eight specimens were used because adopted test method is known to yield highly variable results. Average of the test results of eight specimens belonging to a mix was accepted as first crack strength and ultimate failure strength of that mix. Specimens were tested so that the direction of loading coincided with the direction of casting.

3. TEST RESULTS

3.1 Mechanical properties

Table 3 summarizes the results of all tests that are performed to characterize the mechanical properties. When the compressive and flexural tensile strength examined, it can be seen that for all composites micro steel fiber OL 6/16 increased compressive strength, contrary to micro polypropylene fiber Duomix 20, which caused a decrease in the compressive strength. OL 6/16 fibers are high strength micro steel fibers and they can contribute to the strengthening component of hybrid fiber reinforcement successfully. On the other hand Duomix 20 is weaker than the matrix itself and the decrease in compressive strength of composites with Duomix 20 content could be expected. Apparently fiber inclusion of all types resulted in considerable increases in flexural tensile strength values. In general, macro steel fiber RC 80/60, which has a higher aspect ratio, was more effective in increasing the flexural tensile strength when compared with macro steel fiber ZP 305.

Table 3. Measured Properties

Mix Designation	f_{comp} (MPa)	f_{flex} (MPa)	Energy absorbed upto f_{flex} (J)	Number of blows needed to reach	
				First cracking, FCS	Ultimate strength, UFS
Control	37.21	4.8	3.86	44	48
R1.5	39.91	8.8	51.99	53	512
R1.0-L0.5	42.84	11.0	52.71	51	429
R1.0-L0.3-D0.2	42.35	7.4	41.34	60	522
R1.0-D0.5	37.44	7.8	44.78	61	499
Z1.5	41.22	6.6	29.97	60	238
Z1.0-L0.5	45.37	7.6	45.14	69	182
Z1.0-L0.3-D0.2	41.51	6.4	31.29	61	139
Z1.0-D0.5	37.61	6.8	18.83	55	151

Combination of macro steel fibers with micro steel fiber OL 6/16 gave higher flexural tensile strength than the composites reinforced simply with macro steel fibers. However, similiar synergy was not observed between OL 6/16 and micro polypropylene fiber Duomix 20. Combination of OL 6/16 and Duomix 20 gave the lowest flexural tensile strength among all hybrid fiber reinforced composites.

In addition, the energy absorbed upto flexural strength is also calculated (Table 3) using the load deflection curves shown in Figure 1. Fiber inclusion prevented sudden and brittle failure as seen from the increase in the ultimate displacements. Fiber inclusion of all types greatly enhanced the energy absorption capacity when compared with the plain control mix. Especially, macro steel fiber R, proved to be effective in increasing the absorbed energy. R-type fibers have a high aspect ratio, thus pulling out of these fibers requires more energy. Composite R1.0-L0.5 had the highest absorbed energy. High strength micro steel fiber L, is manufactured from high strength steel wires and has relatively smaller dimensions. Thus, L-type fibers can successfully delay the formation of micro cracks and prevent their propagation up to a certain extent. As a result during pulling out macro steel fibers from a matrix already reinforced with L, more energy is required. Hybridization with R and micro polypropylene fiber D, however, resulted in a decline in the calculated absorbed energy. However, D-type fibers provided ductility due to its high ultimate elongation capacity. For a certain energy level, D-type fiber inclusion shifted the failure deflection upwards for composites with R. Due to its lower aspect ratio, macro steel fiber Z, was not as effective as R in enhancing flexural toughness. Combination of Z with L-type resulted in a considerable increase in the energy absorbed when a comparison is made among composites with Z-type fibers. However hybridization of Z with D adversely affected both energy absorbed and ductility. A synergetic response was not observed between these two fibers.

3.2 Impact Resistance

Obtained impact resistance test results are also presented in Table 3. Main effect of fiber inclusion was on ultimate failure strength (UFS); first crack strength (FCS) was not significantly affected with fiber inclusion. In composites with macro steel fiber, UFS was increased almost 10 times when compared to the plain control mix.

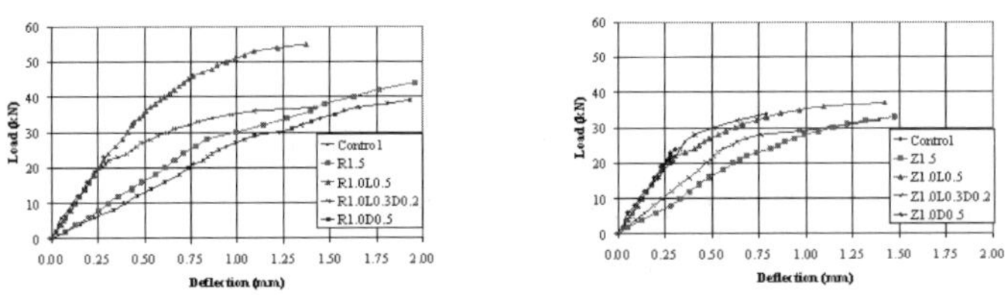

(a) RC 80/60 (b) ZP 305

Figure 1. Load deflection curves for all mixtures

Addition of 0.5% of micro steel fiber L resulted in a 16% decrease in UFS relative to the composite simply reinforced with R. However this decrease was compensated with 0.2% replacement of L with micro polypropylene fiber D; highest UFS was obtained in this composite. D-type fibers proved to be effective under dynamic loading and a synergetic response was observed between R and D-type fibers as well as L and D-type of fibers. Due to its lower aspect ratio macro steel fiber Z, did not enhance UFS as much as R, even though UFS was at least three times increased in composites with macro steel fiber Z.

4. ANALYSIS OF RESULTS

In order to obtain a continuous load-deflection curve the following fitting function is proposed.

$$P = P_u \, \beta (\delta/\delta_u)/(\beta - 1 + (\delta/\delta_u)^{\beta}) \tag{1}$$

where β is the material parameter, P_u is the ultimate flexural load, δ_u is the corresponding ultimate deflection. After performing a non-linear least squares regression analysis, empirical constant, β, was determined. Typical generated load-deflection curves using the above function are presented in Figure 2. Extrapolated portion of the load deflection curves are shown in dashed lines. When extrapolation is carried out, typical load deflection curves common for FRC are obtained, as a consequence this function could be used to estimate the post crack portion of the load-deflection plot.

In order to calculate energy absorptions up to specified multiples of first crack deflection, first crack point should be located. For these calculations, first crack in the fiber reinforced composites was assumed to occur at the ultimate load for plain mix, i.e. effect of fiber reinforcement on the first crack load was neglected. Obtained results are summarized in Table 4.

Flexural toughness tests are relatively hard to carry out and calculation of toughness indexes and other toughness related parameters is a time consuming process. Thus, it would be useful to relate energy absorption under flexural loading to ultimate failure strength suggested in repeated drop weight impact resistance test, as this test is relatively simpler. Figure 3 presents two of such results. As seen from these plots, the macro fiber increased both properties considerably and a distinction for the macro fiber type could be made. However, for each specific macro fiber no followable trend could be observed between the two parameters compared.

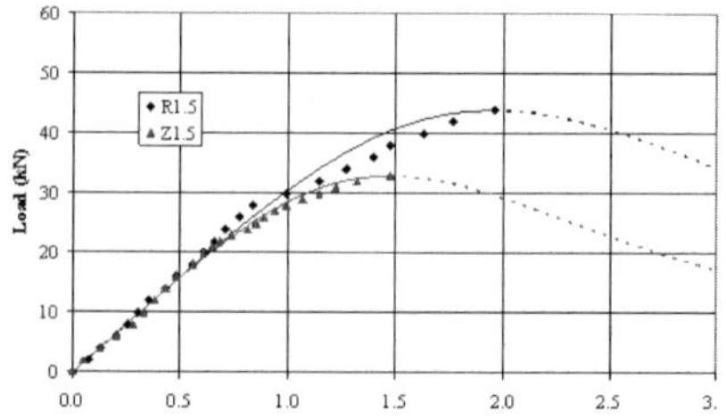

Figure 2. Typical estimated load-deflection curves

Table 4. Calculated energy absorptions up to specified deflections

Mix Designation	Energy upto first crack (J)	Toughness Index	
		I_5	I_{10}
Control	3.65	1.00	1.00
R1.5	8.00	7.63	15.17
R1.0-L0.5	3.48	7.70	19.34
R1.0-L0.3-D0.2	5.84	5.67	12.08
R1.0-D0.5	9.99	6.96	11.19
Z1.5	10.28	5.68	8.48
Z1.0-L0.5	5.84	5.67	12.08
Z1.0-L0.3-D0.2	6.96	5.82	11.53
Z1.0-D0.5	4.14	6.05	11.69

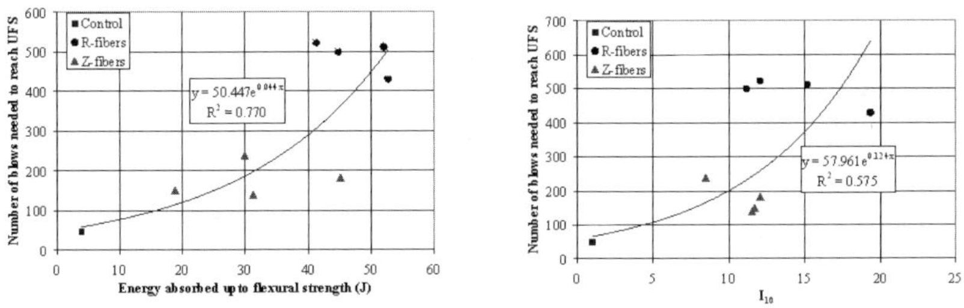

Figure 3. Relation between impact properties and energy absorption

5. CONCLUSIONS

After reviewing the obtained results, following conclusions could be drawn as a result of this experimental study:

- Fiber inclusion of all types increased compressive strength, although this increase was not that significant. High strength micro steel fiber OL 6/16 proved to be efficient in strengthening the matrix. This result can be attributed to its high strength and relatively smaller dimensions.
- Macro steel fiber RC 80/60, which has a higher aspect ratio than the other macro steel fiber ZP 305, was more effective in increasing flexural tensile strength. Combination of both macro steel fibers with micro steel fiber OL 6/16 yielded better tensile strength. However, a synergetic response was not observed between OL 6/16 and Duomix 20 with neither of the macro-fibers.
- Macro steel fiber RC 80/60 was more efficient in enhancing flexural toughness when compared with ZP 305. Combination of macro steel fibers with micro steel fiber OL 6/16 yielded better flexural toughness results. However, combination of ZP 305 with Duomix 20 had an adverse effect on flexural toughness. A synergetic response in neither flexural toughness nor ductility was observed between these fibers.

- First crack strength was not significantly affected with fiber reinforcement, but ultimate failure strength (UFS) was greatly enhanced. Macro steel fiber RC 80/60 was more efficient in increasing UFS when compared with ZP 305.

6. REFERENCES

1. ACI Committee 544, Measurement of Properties of Fiber Reinforced Concrete, ACI 544.2R-89
2. C.X. Quian, P. Stroeven, Development of Hybrid Polypropylene-Steel Fibre Reinforced Concrete, Cement and Concrete Research 30, 63-69 (2000)
3. C.X. Quian, P. Stroeven, Fracture properties of concrete reinforced with steel–polypropylene hybrid fibres, Cement and Concrete Research 22, 343-351 (2000)
4. W. Yao, J. Li, K. Wu, Mechanical Properties of Hybrid Fiber-Reinforced Concrete at Low Fiber Volume Fraction, Cement and Concrete Research 33, 27-30 (2003)
5. N. Banthia, C. Yan, K. Sakai, Impact Resistance of Fiber Reinforced Concrete at Subnormal Temperatures, Cement and Concrete Composites 20, 393-404 (1998)
6. A. Bentur, S. Mindess, Fibre Reinforced Cementitious Composites, Elsevier, London, (1990).

INFLUENCE OF TENSION STIFFENING EFFECT ON DESIGN AND BEHAVIOUR OF REINFORCED CONCRETE STRUCTURES

A. Elenas,[1] L. Vasiliadis,[1] E. Pouliou[2] and N. Emmanouilidou[2]
[1]*Democritus University of Thrace, Department of Civil Engineering, GR-67100 Xanthi, Greece;*
[2] *Institute of Structural Mechanics and Earthquake Engineering,*

Abstract: This paper presents the beneficial act of the tension stiffening effect in the design process according to the rules of the EC2 Eurocode and DIN 1045-1. Different approaches for the modelling of this phenomenon are offered. To describe this effect, two different basic approaches are possible using micro- or macro-elements. The first technique is coupled with a discrete representation of cracking, while the second one is coupled with a smeared representation of cracking. The advantages and the disadvantages of using these models are discussed. A numerical example shows the beneficial act of the tension stiffening effect on a reinforced concrete plate. The economy in reinforcement, when the tension stiffening effect is taken into account during the design procedure, is quantified. Finally, non-linear analyses expose the difference in stiffness and deflection, when tension stiffening is considered.

Key words: tension stiffening; reinforced concrete; numerical methods; design process.

1. INTRODUCTION

It is well known that when the concrete tensile stress in a member reaches the tensile strength, cracking develops. Simultaneously, the load carried by the concrete before cracking is transferred to the reinforcement crossing the crack. Then, the reinforcing bars transfer local tensile stresses to the concrete between the cracks through bond stresses at the steel–concrete interface. Therefore, the intact concrete between cracks continues to carry tensile stresses and offers stiffness. This phenomenon resulting from crack formation and the bond between steel and its surrounding concrete is defined as the tension-stiffening effect[1]. This is a property neither of the reinforcement nor of the concrete. It is a typical property that appears only in the composite material of reinforced concrete, where the two constitutive materials are present. Its ignorance leads to a too soft approx-

M.S. Konsta-Gdoutos, (ed.), Measuring, Monitoring and Modeling Concrete Properties, 215–220.

imation of the structure. Loading progression continues the crack building and with this a drop of concrete stress between the cracks is connected. At last, the cracks are too close for developing tension concrete stress and the concrete has no carrying capacity perpendicular to the crack direction. So in this case the influence of the tension stiffening effect vanishes.

Using the finite element method to describe this behaviour, two different basic approaches are possible utilising micro- or macro-elements[2-5]. The first approach is coupled with a discrete representation of cracking and is used when the size if the finite elements is of the same order if the crack width. The second approach, coupled with the smeared representation of cracking, describes the tension stiffening phenomenon by a modification of the stress-strain relation of the steel or the concrete material.

2. DIRECT APPROACH OF TENSION STIFFENING EFFECT

As mentioned in the introduction section, the direct approach of tension stiffening effect is coupled with the discrete modelling of concrete cracking. This is done by disconnecting the displacement at nodal points for adjoining elements and adding link elements perpendicular and parallel to the reinforcement bar axis[6]. Setting the spring stiffness of the link element both normal and tangential to the crack surface from the initial large values to zero represents the crack initiation. The closing or reopening of a crack is judged on the crack width. When a crack is closed, only the vertical spring stiffness to the crack surface is set equal to the initial large value.

One obvious difficulty in the discrete modelling is that the location and orientation of the cracks are not a priori known. Thus, geometrical restrictions imposed by a preselected finite element mesh cannot be avoided. This can be rectified to some extend by redefinition of element nodes. The use of high-order elements, particularly the isoparametric ones, yields rather poor-quality corner-stress definition which does not blend well with the edge cracking associated with the discrete crack concept. These are two disadvantages of the bond link element.

A second way of modelling bond in a finite element calculation is through the employment of contact elements[7]. These elements connect the nodes of a steel element with the respective of an adjacent concrete element. The contact element has a finite dimension and at least two double nodes and the same length as the connected reinforcement element. Contact elements can also be used in a two dimensional formulation.

Finally, a third group of bond zone elements[8] differs significantly from the two types described previously. The most important difference is their finite dimension. They model the contact surface between steel and concrete as well as the concrete in the immediate vicinity of the reinforcing bar by an adopted material law that considers the special properties of this bond zone.

Such direct techniques of modelling of bond effects are extremely complex. Thus, these approaches can be only used in simulation of laboratory tests and for very simple structures or elements.

3. INDIRECT APPROACH OF TENSION STIFFENING EFFECT

The indirect approach of tension stiffening effect is coupled with the smeared modelling of concrete cracking[2-4]. This assumption is physically reasonable, in view of the material non-homogeneity of concrete. Its representation may be achieved by changing the element stiffness matrix, assuming that the material becomes orthotropic with the modulus for extension normal to crack band reduced to zero. Any direction of crack propagation may be represented equally well, and incremental-iterative solution procedures may account for crack propagation. The crack directions are not affected by the finite element mesh pattern. In connection with the smeared crack model, there are two basically different approaches to represent the tension stiffening effect: it can be attributed either to the concrete itself or to the reinforcement[9]. Figure 1 shows the modified stress-strain diagrams for steel and concrete in tension for numerical modelling of the tension stiffening effect.

The first approach postulates a fictitious unloading branch in the stress-strain curve of concrete instead of the abrupt drop-off of the tensile stress. They have proposed different length and the appearance of the descending branch. Orientation and properties of reinforcement have no influence. If no tension stiffening is taken into account, then the descending branch at tensile concrete stresses vanishes.

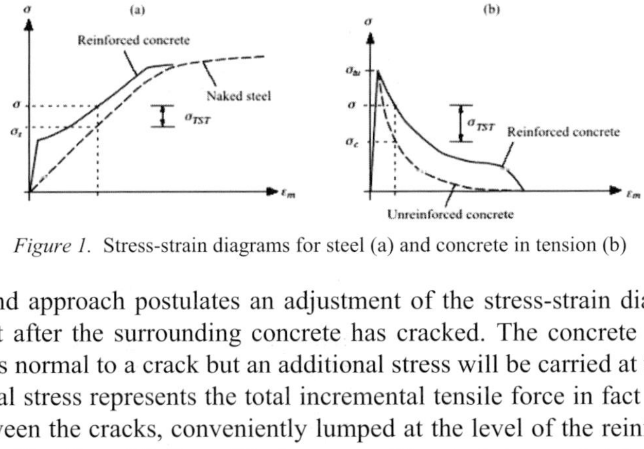

Figure 1. Stress-strain diagrams for steel (a) and concrete in tension (b)

The second approach postulates an adjustment of the stress-strain diagram of steel reinforcement after the surrounding concrete has cracked. The concrete is assumed to carry no stress normal to a crack but an additional stress will be carried at the steel level. This additional stress represents the total incremental tensile force in fact carried by the concrete between the cracks, conveniently lumped at the level of the reinforcement and oriented in the direction of the bars. The increase of the steel stresses is taken into account by an increase of the steel modulus of elasticity. Its magnitude decreases with increasing strains (further development of cracks) until there is no tension stiffening after the formation of the final crack pattern. This concept developed by Gilbert/Warner[10] is popular since tension stiffening as a bond property depends strongly on the orientation of the reinforcing bars. A variation of the second approach is the representation of tension stiffening by a fictive additional reinforcement.

4. TREATMENT OF TENSION STIFFENING IN DESIGN CODES

Modified steel stress-strain laws are frequently used in non-linear analyses and are also implemented in EC2 Eurocode[11] and DIN 1045-1[12] to take into account the tension stiffening effect. In these models the steel strain results numerically from a horizontal move for constant steel stress. The physical meaning is a strain decrease of the steel reinforcement for the same stress. For the horizontal movement there are different approaches in the literature presented. As shown in Figure 2, the EC2 Eurocode specifies a parabolic approach, while the DIN 1045-1 a piece-wise linear one. An important difference between the two codes is that the EC2 model ignores the tension stiffening effect in the yield region of steel.

Moreover, in DIN 1045-1 additional models for the numerical treatment of the tension stiffening effect are given, for different design procedures as it is explained in Figure 4. Thus, the models a, b and d can be used for the evaluation of the internal forces using non-linear numerical procedures. Model b is to be used for the evaluation of the structural displacements. Finally, Model c is to be used for the evaluation of crack widths.

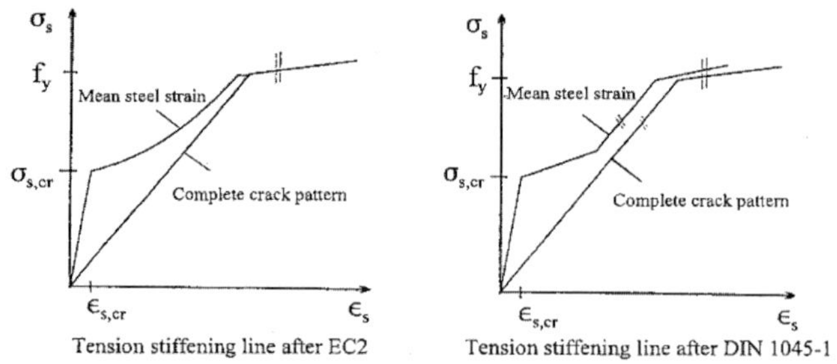

Figure 2. Tension stiffening lines after EC2 and DIN 1045-1

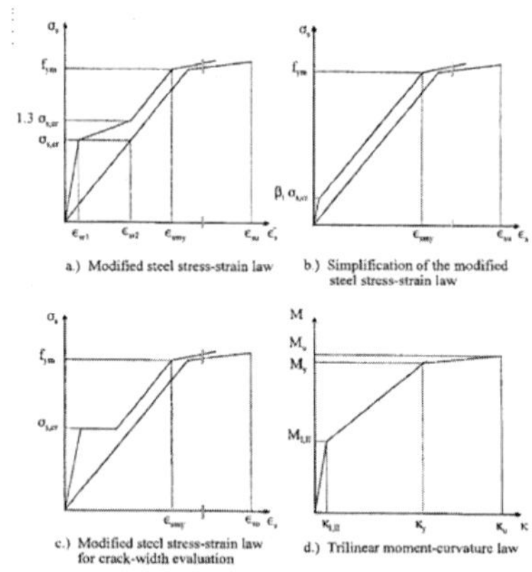

Figure 3. Alternative tension stiffening lines after DIN 1045-1

5. NUMERICAL EXAMPLE AND RESULTS

A simply supported square plate has been designed in accordance to the design rules of the EC2 Eurocode[11] for uniformly applied load of 10 KN/m^2. The dimensions of the plate are 10 m x 10 m and its thickness is 20 cm. The concrete material is C20/25 and the steel material is S500. First the plate is designed for the ultimate limit state which provides 4.404 cm^2/m required longitudinal reinforcement bars in the directions parallel to the plate edges. After that, the plate is designed for the serviceability limit state for different crack widths from 0.05 mm to 0.30 mm. The latter design procedure was carrying once neglecting and once taking into account the tension stiffening effect. This effect was taken into account by considering the average steel strain between the cracks.

The tension stiffening effect decreases the amount of the required longitudinal reinforcement bars, especially for small crack widths. Figure 4 shows the required area of the longitudinal reinforcement bars for different crack widths (0.05 mm up to 0.30 mm). The results are exposed graphically once neglecting the tension stiffening effect (curve 1) and once taking into account the same effect (curve 2). Observing the two curves it can be recognised that curve 2 remains permanently below curve 1. This means that the economic effect is present in all the cases, when the tension stiffening effect is taken into account. The values where the reinforcement is less than statically required (4.404 cm^2/m as mentioned previously), is not of practical interest and is here only presented for completeness reasons.

Next, Figure 5 shows the difference of the required area of longitudinal reinforcement between neglecting and taking into account the tension stiffening in the design procedure. Figure 5 shows the absolute value of the difference, which is about 2.23 cm^2/m for 0.05 mm crack width. This value drops below 0.8 cm^2/m for 0.30 mm crack width.

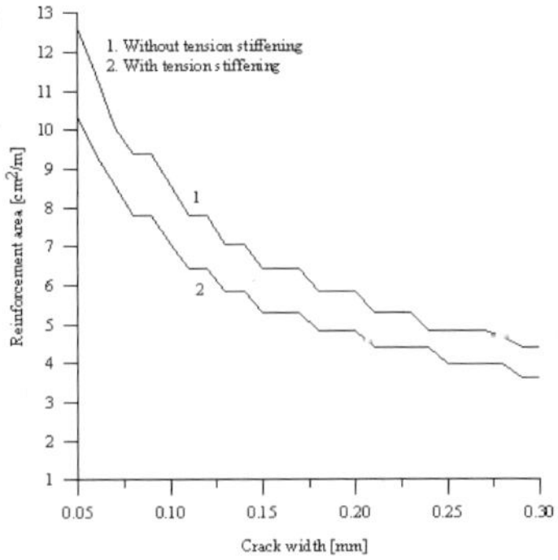

Figure 4. Required longitudinal reinforcement

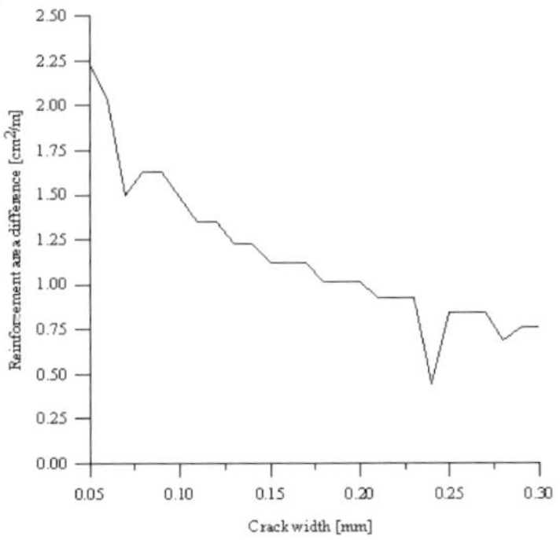

Figure 5. Difference of the required longitudinal reinforcement

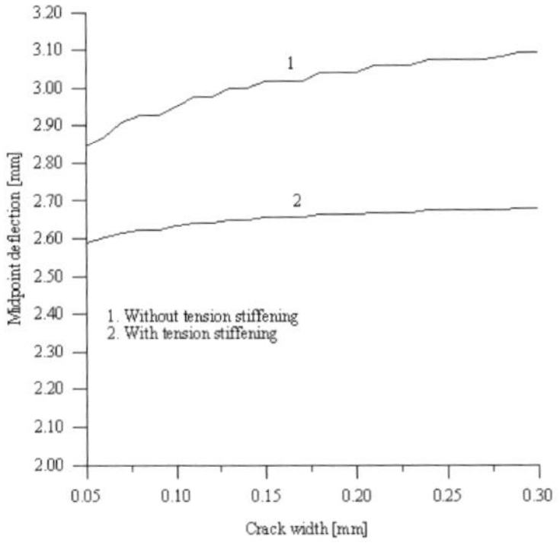

Figure 6. Midpoint deflection of the plate

Non-linear analyses have been carried out for the estimation of the influence of tension stiffening on the maximum plate deflection. Figure 6 presents the midpoint deflection in dependence of the crack width. Two different cases have been considered. The first one is neglecting the tension stiffening effect, while the second one is taking into account the same effect. The midpoint deflection of the plate is about 10% to 15% increased when tension stiffening is neglected.

6. CONCLUSIONS

Several numerical treatments of the tension stiffening effect in reinforced concrete elements have been presented. To describe this behaviour, two different fundamental approaches are possible using micro- or macro-elements. Since non-linear analyses have been introduced in modern design codes for reinforced concrete structures, rules for handling the tension stiffening effect have also been introduced in these codes. Thus, modified steel stress-strain laws are used in EC2 Eurocode and DIN 1045-1 to take into account the tension stiffening effect.

A numerical example demonstrated that neglecting the tension stiffening effect leads to a too soft approximation of the structure. The economy in reinforcement, when the tension stiffening effect is taken into account during the design procedure, has been quantified. Thus, the mean difference percentage in reinforcement for the example used in this investigation was 21.5%. Furthermore, non-linear analyses pointed out the difference in deflection when tension stiffening is considered. Finally, all these comments should be taken into account for the safe and economic design of composite reinforced concrete elements in the civil engineering praxis.

7. REFERENCES

1. Wicke, M., Cracking and deformation in structural concrete, *IABSE reports* **62**, 49-57 (1991).
2. Cauvin A., Influence of tension stiffening on behaviour of structures, *IABSE reports* **62**, 153-158 (1991).
3. Eibl, J., Die FE-Technik im Massivbau zwischen Praxis und Wissenschaft, in: *Finite Elemente in der Baupraxis*, edited by E. Ramm, E. Stein and W. Wunderlich, (Ernst & Sohn, Berlin, 1995), pp. 1-10.
4. Von Grabe, W. and Tworuschka, H., Baupraktische Anwendung nichtlinearer Traglastermittlung - Bemessung im Stahlbetonbau, in : *Finite Elemente in der Baupraxis*, edited by E. Ramm, E. Stein and W. Wunderlich (Ernst & Sohn, Berlin, 1995), pp. 227-286.
5. Parche, S., Querschnittsverhalten und Schädigung stabförmiger Stahlbetonkonstruktionen unter schiefer Biegung und veränderlicher Normalkraft infolge Erdbeben, *Technisch-wissenschaftliche Mitteilung* Nr. 96-10, Institut für konstruktiven Ingenieurbau, (Ruhr-Universität Bochum, Bochum, 1990).
6. Nilson, A.H., Non-linear analysis of reinforced concrete by the finite element method, *ACI Journal* **65**, 757-766 (1968).
7. Schäfer, H., A contribution to the solution of contact problems with the aid of bond elements, *Computer Methods in Applied Mechanics and Engineering* **6**, 335-354 (1975).
8. Dragosavic, M., Modelling of bond, *IABSE reports* **54**, 131-138 (1987).
9. Pravida, J.M., Zur nichtlinearen adaptiven Finite–Element–Analyse von Stahlbetonschei-ben, *Dissertation*, (Technische Universität München, München, 1999).
10. Gilbert, R.I. and Warner, R.F., Tension stiffening in reinforced concrete slabs, *ASCE J. Struct. Div.* **105**, 1885-1990 (1978).
11. EN 1992 Eurocode 2, Design of concrete structures, (1992).
12. DIN 1045-1, Tragwerke aus Beton, Stahlbeton und Spannbeton, (2001).

ASSESSMENT OF MODEL PARAMETERS FOR FRACTURE SIMULATION IN BRITTLE DISORDERED MATERIALS LIKE CONCRETE AND ROCK

J.G.M. van Mier

ETH Zurich, Institute for Building Materials, Materials Research Centre, CH-8093 Zurich, Switzerland

Abstract: Different fracture models for brittle disordered materials require varying empirical content in the form of model parameters that, somehow, must be measured in an experiment. The success of a model to 'predict' situations hitherto unexplored by experiment depends on the correctness of the empirical content of the model, i.e. on the success in capturing material behaviour in the said model parameters. For fracture this seems only possible when the model is capable of simulating fracture *mechanisms* to a high degree of accuracy in the first place. Different models may be applicable at different scale levels, use may be limited by constraints from the roughness of the material structure itself (RVE), but in the end when fracture becomes fatal, specimens/structures are separated into two or more distinct parts separated by localised macrocracks of size similar to the characteristic size of the considered specimen/structure. In particular deriving correct empirical content for the localisation stage is extremely difficult due to intertwining of material and structural effects. In classical fracture mechanics these effects are elegantly taken care of in the structure of the theory, but in models based on continuum principles, the link between material and structure is lost, resulting in tremendous difficulties in determining the empirical content. The solution must be found in physics-based approaches, which are based on correctly simulating fracture *mechanisms* in the first place.

Key words: fracture models; empirical content; material parameters; 4-stage fracture process, fracture mechanism; scaling; fracture strength; pre-critical cracking; representative, volume element (RVE); disordered materials; cement; concrete; rock; ice; geomaterials.

M.S. Konsta-Gdoutos, (ed.), Measuring, Monitoring and Modeling Concrete Properties, 221–232.

1. MODELLING MODE-I FRACTURE

This contribution presents a personal and rather controversial view on the development of fracture models for brittle disordered materials. The single most important question to be asked when developing models simulating fracture in such materials is: 'how can we measure un-biased and size-independent empirical content'? To some, who tend to take a pragmatic attitude towards cohesive crack models this question is considered highly controversial, and may opine that it should not be raised at all. However, the stakes must be set higher, and fracture models should have *predictive qualities* in order to improve current standards of analysis, which often do not get beyond 'post-diction'. We should not be satisfied with less than 'pre-diction', not even for pragmatic reasons.

Since the 1960s much effort has been devoted to finding appropriate fracture models for concrete, rock, ceramics and other brittle disordered materials, also frequently gathered under the name 'geo-materials'. In such undertaking the most common way is to start of with proven models that exist for other materials. Kaplan[1] tried to use principles from Linear Elastic Fracture Mechanics. His work concluded with the finding that LEFM is not applicable to concrete, like many of his contemporaries confirmed. This conclusion is at least correct for small laboratory-scale specimens/structures. LEFM requires knowledge about the size of an initial notch, a crack propagation criterion in terms of a critical energy release rate G_{Ic} or a critical stress intensity factor K_{Ic}, under the assumption that the material can be schematised as an isotropic continuum. Especially the latter assumption is difficult when it comes to extreme disorder as found in the structure of materials like concrete, rock, and ice. In these materials, the size of the largest material structural elements is in many laboratory experiments just 3 to 5 times smaller as the considered specimen/structure itself and one may question whether the sample can be considered as an isotropic continuum. For concrete, I concluded that, when the exact drying conditions are unknown, the size of an RVE should be at least 8 times the maximum aggregate size in order to have a constant scatter band for sample sizes larger than this lower bound[2].

The normal solution, when something does not fit to satisfaction is to add more parameters. The step made for concrete in the 1970s, and some decades later for other materials, was to use cohesive crack models along the line of earlier developments for ductile metals (Dugdale/Barenblatt plastic crack tip models). The change made was in the crack-closing pressure, which was chosen as a softening relation following experimental observations made by Evans & Marathe[3] in 1968. Softening was first demonstrated experimentally for rock and concrete in those years probably due to the extreme disorder in these materials, which makes softening rather pronounced, in contrast to, for example metals and polymers. Hillerborg and co-workers[4] divided the stress-strain diagram, which was the usual representation of mechanical properties, in a pre-peak stress-strain curve and a post-peak stress-displacement curve. The post-peak branch was used as the stress-closing pressure in the cohesive crack model. Continuum mechanics principles remain since for modelling softening stress is used as one of the state variables. One complication that arises is that in plastic metals the frontal tip zone is usually extremely small, whereas using a similar criterion for concrete, the size of the 'plastic' zone, or process zone increases, sometimes beyond the size of the structure or laboratory sample under consideration. The surface under the softening curve is the fracture energy G_f, and can be considered as a crack propagation criterion. The crack initiation criterion is the tensile strength of the material under consideration. For concrete and rock, the tensile strength is in the order of 3-6 MPa, which explains the enormous size of the process zone. The tensile strength is thus the crack (failure) initiation criterion, which is an improvement of LEFM-based analyses that always require an initial notch a_0 as a starter criterion. In LEFM, the determination of the initial notch size is flawed.

Thus, in order to achieve better results, the two-parameter LEFM model (K_{Ic} and a_0) is replaced by a cohesive crack model with many more parameters (E, f_t, $\sigma(w)$, with derived parameters G_f (fracture energy = $\int\sigma(w)dw$) and l_{ch} (characteristic length = EG_f/f_t^2)). The shape of the softening curve must be derived from a displacement-controlled uniaxial tension test between non-rotating loading platens on a representative material sample. The test is hampered by many difficulties, such as alignment problems, glue-problems, and not in the last place the selection of the most appropriate control parameters, see for example in Van Mier & Shi[5]. It appears that the shape of the softening curve is affected by the boundary conditions selected in the experiment, whereas other assumptions seem not very appropriate either when compared to the outcome of experiments[6]. This should be the main worry, not only when it comes to the Fictitious Crack Model (FCM), but to any type of fracture model: can the model parameters be determined in a possibly simple and straightforward manner, and is the fracture *mechanism* captured correctly. Only then may we hope for a wide application of the model.

The application of LEFM requires an experiment for determining the crack propagation parameter K_{Ic} or G_{Ic}, when one limits oneself to mode I crack growth. There seems to be agreement that large ([m]-size) samples should be used. The application of FCM (or other cohesive crack models) requires the determination of the 'yield' strength f_t (tensile strength in the case of geo-materials) and the shape of the crack-closing stress profile. Application of LEFM requires the insert of a sharp initial notch to trigger crack propagation, but the same constraint applies to FCM because of requirements from the displacement-control in the test set-up, which basically asks to trigger crack propagation from a known location unless very advanced solutions for test-control are used[5,7]. Alternatives to the FCM include the crack-band model (CBM)[8], and higher-order continuum models, which – in addition to the above-mentioned parameters – require an internal length scale l_c for modelling the displacement jump in the localization zone. Table 1 gives an overview of required model-parameters for different types of fracture models used in the wide field of geo-materials.

Note that all parameters in FCM are size-dependent, i.e. they depend on the size of the considered specimen/structure[7,9]. In LEFM the size of the crack plays an important role, and results in size-dependent behaviour directly. For FCM, and higher-order continua scaling behaviour is derived as well, but, at least for the latter type of models, depends very much on the characteristics of the strain-softening diagram and the internal length scale l_c[10].

The strong demand for a model *to predict* fracturing of structures made of geo-materials places extreme demands on the reliability of the model parameters. One question to be asked is how accurate the model parameters must be to derive predictions from models.

Table 1. Parameters in selected fracture models applied for geo-materials

Model	Parameters			
LEFM		a_0	$K_{Ic}(a)$ (or $G_{Ic}(a)$)	
FCM	E	f_t	$\sigma(w)$	
	(derived parameters: $G_f = \int \sigma(w)dw$) and $l_{ch} = EG_f/f_t^2$)			
Higher order Continua*	E	f_t	$\sigma(\varepsilon)$	l_c

*) e.g. strain gradient plasticity, non-local damage models, etc

2. PRE-CRITICAL CRACKING

Both in LEFM and FCM it is assumed that the material is homogeneous and can be considered as an isotropic continuum. For materials like concrete and rock this demand usually leads to rather large laboratory specimens in order to comply with these demands. As mentioned, to my opinion the RVE should be at least 8 times the maximum aggregate size for concrete, whereas similar demands can be derived for rock and ice[2]. For an aggregate size of 16-32 mm, which is common for many practical concretes, this would lead to sample sizes in the order of 130-260 mm. The constraint derives simply from strength observations that show that the scatter bar gets the same length after the aforementioned size threshold is exceeded[7]. Thus, below the RVE, one can still consider structures/ specimens to be made up of the continuum material, but in practical laboratory tests the uncertainty about the validity of the model decreases when tests with increasingly larger scatter below the RVE must be used for comparison. Thus, the roughness of the tested material simply decides the limitation of continuum models, where it must be added that this line of reasoning can be followed only up till the strength point. Discussions about the shape of a lower asymptote in size effect laws (see for example the ongoing debate between Bažant[11] and Carpinteri et al.[12]) is in this respect rather debatable, since no experiment can ever discriminate between competing propositions about the lower asymptotic behaviour.

Careful consideration of the generalized fracture process in many engineering materials shows that four distinct stages in the breakdown can be distinguished: (O) linear elastic behaviour, (A) pre-critical non-linear behaviour up till peak stress; (B) post-critical crack growth leading to displacement localisation; and (C) bridging. The process is a convenient way to look to material breakdown under mechanical load, and was recently published[6]. When we limit ourselves to brittle disordered materials subjected to far field uniaxial tension (mode I), during stage (A) small microcracks cause the pre-peak non-linearity as sketched in Figure 1. These microcracks are small cracks in the sense that they can be arrested and delayed by elements in the structure of the disordered material itself. Thus, large aggregates, larger air voids, and fibres are examples of material structural elements that may arrest, deflect or delay the growth of microcracks. For normal concrete it is an accepted fact that the first microcracks appear alongside the interface between aggregates and matrix, and a larger strength threshold (or material toughness) must be overcome in order to extend these microcracks in the surrounding (stronger and tougher matrix) material. This remark must be understood under the premise that concrete can be considered as a 3-phase composite consisting of aggregates embedded in a matrix, with a weak interfacial transition zone in between these two phases. Using simple meso-mechanical models it is relatively straightforward to show pre-critical cracking in stage (A). Important parameters in such models are the amount of aggregate and the thickness of the interfacial transition zone[13,14]. Recent experiments with (hybrid) fibre concrete demonstrate that pre-critical cracking can be delayed further when fibres are added, and good care is taken that the bond between the fibres and the surrounding cement matrix is appropriate, i.e. warrants a control-led and slow pull-out of fibres from the matrix. The pre-peak deformations can thereby increase substantial, as does the maximum tensile strength[15].

The processes happening at the meso-scale, and also at lower size/scale levels (micro-level, i.e. cement in concrete, or other mineral binders like calcium carbonate in sandstone) determine the macroscopic strength of the material. In this view scaling of strength is primarily caused by differences in microcrack processes in specimens/structures of varying size, where not only (1) statistical – Weibull-type – effects are important, but also (2) structural effects like stress-/strain-gradients in the structure under consideration. Interestingly, such structural effects may already be observed in uniaxial tension tests because in the pre-peak regime such tests tend to show non-uniform stress-/strain- distributions, simply caused by the disorder in the material. Van Vliet[16] analysed these effects in detail in the case of uniaxial tension on specimens of varying size in a range of 1:32, whereas for hollow cylinder tests subjected to hydrostatic pressure the same analysis was performed more recently[17,18]. Understanding the structural conditions, plus a proper understanding of the pre-critical crack processes in stage (A) is therefore rather essential for understanding scaling of strength. Strength must be seen as a combined structural and material property, and not just as a property of the material only. This is perhaps not so much visible for uniaxial tension, but when it comes to compressive strength, the enormous influence of boundary restraint, for example, must certainly be considered a struc-tural effect, and as such must be accounted for in any analysis.

The pre-critical cracking stage (A) is the part of the breakdown process in a material specimen before the maximum stress is reached. In the past there has been enormous confusion since softening parameters were used in an attempt to model these micro-crack processes. Softening is, however, a completely differrent part of the breakdown process, as can be seen in Figure 1.

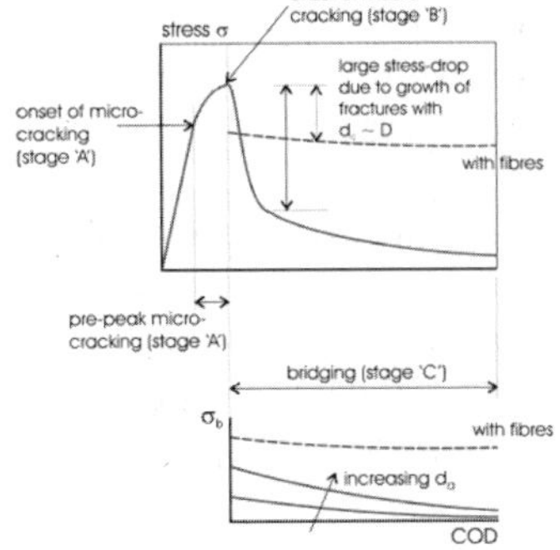

Figure 1. 4-stage fracture model: a convenient concept to assess the breakdown of materials[6]

During softening (*strain*-softening in CBM and higher-order continuum models) large macro-cracks develop in the considered specimen/structure. A characteristic of these large cracks is that they cannot be arrested anymore by elements in the structure of concrete, rock or ice, unless rather extreme measures are taken, for example in the form of fibres that are stiffer and stronger compared to the surrounding matrix (in concrete and ice one could envision this situation). Softening is a consequence of a gradually decreasing effective load-carrying area in the considered specimen/structure, and the only restraint can come from the structure (i.e. stress/strain gradients), or from crack bridging inside the material (stage (C)).

3. SOFTENING

Softening is the final breakdown of a specimen/structure. Since we are dealing with a significant decrease of the effective load-carrying cross-section of a specimen/structure – cracks have sizes equal to the characteristic size of the specimen/structure itself – the boundary conditions and the size of the specimen/structure cannot be ignored anymore. Softening is not a material property, although there is a clear influence of the material: the softening regime may be more or less prominent depending on the actual material structure. This has been sketched in Figure 1. For fine grained materials the role of bridging is very limited; for coarser materials like concrete, granite and ice bridging may be more important[19], whereas in the case of fibre concretes bridging by fibres might even balance the loss of carrying capacity in the matrix[12]. In a tensile experiment these various stages in the breakdown process can be readily identified, especially when one plays around with the allowable rotations and movements of the loading platens during the test. The situation becomes more obscure in bending tests. In recent four-point bending tests of hybrid fibre concrete[20] multiple narrow cracks were observed in the constant moment regime before the maximum load was reached. After passing the peak, localisation in a single large crack occurred, and clearly fibres bridging the crack could be seen. These 'microcracks' (in the pre-peak regime) can be as long as the localized crack in the post-peak regime, but clearly their functioning under mechanical load seems to differ. Most likely, bonding of the fibres to the matrix is still more-or-less intact during the pre-peak regime, whereas this has been totally impaired and limited to frictional restraint along a single critical crack during the post-peak stage. To use a softening diagram to describe the behaviour of the very fine 'microcracks' does not seem very appropriate, and the matter has to be resolved in a different way. In numerical (lattice) analysis of pre-peak fracture process in three-point bending beams, also in the pre-peak regime a wider region with distributed microcracking is observed, and the localised (fatal) macrocrack first appears after the peak has been exceeded[21]. Note that in beams of varying size the effective specimen volume where stresses may exceed the tensile strength will be quite different. Thus, the amount of microcracking in the pre-peak regime may differ substantially for specimens of varying size, and only through a thorough understanding of these microcrack processes it is possible to explain scaling of (flexural) strength. As mentioned in the previous section, scaling of strength can only be understood when the material and structural aspects are separated. The microcrack processes can be understood in detail by applying models at a lower size/scale level. Unavoidably, more parameters enter the model, but this time the situation looks more promising since the new parameters relate all to describing the structure of the material at a pre-selected size/scale level. More about this is to follow in the next section.

There appears to be a fundamental mix-up of different stages in the fracture process, and looking back, it is my impression that the identification of softening with the development of a 'cloud of microcracks in front of a macrocrack-tip' has been the source of quite some confusion. This view has dominated all discussions on fracture in brittle disordered materials in the past three decades, but to my knowledge the existence of such a zone in front of a macrocrack tip has never been shown experimentally. Impregnation experiments clearly show that post-peak behaviour is governed by a single large macrocrack[19]. From a mechanics point of view, a long crack in a specimen of given size will lead to a larger stress-drop compared with a specimen containing multiple short cracks separated by intact ligaments. Bridging, as said, may – to some extent – improve the softening behaviour (i.e. make the material more ductile), and as sketched in Figure 1 must be considered as the real crack closing pressure. The growth of the localised macrocrack can be modelled using LEFM principles. In the view presented in Figure 1, the fracture process is divided into a few distinct stages, that follow the observed fracture process in more detail, rather than using a 'lumped' solution like the cohesive crack model where all fracture behaviour is described by a softening diagram (which is partly a structural property), and where the pre-critical crack regime has been omitted for 'convenience'. With that omission the single most important stage in the fracture process has been removed from cohesive models.

4. MULTI-SCALE MODELLING

Fracture models, like those mentioned in Table 1, can be applied at different size/scale levels. In the foregoing fracture was considered at the macro-level, where concrete, rock and ice are considered as isotropic continua. Descending the size/scale hierarchy more and more detail will have to be taken into account directly in a model representing material behaviour, as already alluded to in the previous section. For cement and concrete the interesting size/scale levels are the meso-, micro- and nano-levels: at all levels important processes take place that contribute to strength and fracture in varying degrees. Numerical concrete[22,23] is an attempt to step-wise understand various processes at the micro-, meso- and macro-level. In an effort to understand strength, fracture and softening my group, first at TU Delft, now at ETH Zurich, has studied various factors contributing to mentioned phenomena using both experiments and numerical (lattice) simulations. This has been limited to meso- and macro-level observation and simulation. The contribution from micro- and nano-level processes cannot be denied, however, and for an in-depth understanding the transition to these scales must be made. One of the major questions is whether one can isolate small samples of material, for example at the micrometer scale, and use the strength and fracture properties determined on those samples as input for models operating at a larger scale. As a matter of fact, this question is completely identical to the one addressed in the previous sections of this paper. As long as the tested or simulated material samples are representative, i.e. have dimensions to warrant that an RVE is considered, and the physical processes in the sample do not lead to localised behaviour, a multi-scale approach must be considered safe. Thus, in the elastic regime (stage (O); the linear portion starting from the origin in Figure 1), it is possible to compute, for example, the elastic properties of a composite by considering the structure and volume fractions of constituting material phases at different size/scale levels. Thus, in cement, one can distinguish un-hydrated cement, different hydrate phases, water, pores, and various interfaces between un-hydrated and hydrated phases to compute effective elastic properties of an RVE of cement. At the meso-scale, one can use the outcome to compute the effective elastic properties of concrete using the data for the cement RVE. Note that at each of these levels the constituting phases are always considered to behave as a continuum. This, in view of the foregoing discussion, directly answers the question if a multi-scale approach may be used for strength and fracture. The answer must be a clear and unequivocal *NO*. Strength is a combined material/structural property, and undeniable this holds for softening too. Both strength and fracture properties cannot be derived from continuum considerations, as this will lead to problems in material parameter identification, at least if one maintains the strong demand that the models must be used for predicting behaviour in situations that have not been explored by experiments before. The converse stands for post-diction, i.e. simulating situations that have been explored by experiment, which is much easier since the outcome is already known. The philosophical problem of multi-scale modelling is intractable and only ab-initio analysis from the (sub-) atomic level to large size/scales seems a feasible approach leading to predictions rather than post-dictions, at least if one can rely on the premise that atomic, or quantum properties are indeed the final truth. There the last word has not been said, and perhaps one has to take a more pragmatic attitude and be satisfied with the possible. Since meso-level analysis of representative material volumes is already rather tedious and time-consuming, ab-initio analysis from the atomic level are impossible considering current computational possibilities.

5. MEASUREMENT ISSUES

In spite of the foregoing discussion interesting experiments and simulations can be conducted, for example the experimental determination of the tensile fracture properties of cement and concrete. For concrete, in the past few decades new and advanced test-methods have been developed, and it was enjoyable to be part of that[5,7]. In a multi-scale approach, even though it does not make sense from a philosophical point of view (in particular when strength and fracture are considered), one also has to determine the same properties, and basically one could, as a first approach attempt to scale down the tensile experiments used at the meso- and macro-scales. In Figure 2 test-rigs used at the macro- and micro-level are depicted. The test set-up in Figure 2a is used for samples at a scale of typically 100 mm. The specimens are glued between steel platens that are pulled in a servo-controlled hydraulic test machine between cables[5]. This test was specifically developed to improve test control systems, in particular in soft-machines (which is the reason to use cables). The loading platens can be considered to rotate freely during the fracture experiment. The test set-up in Figure 2b has been developed for determining the micromechanical properties of cement[24], in particular the micro-strength and to study how cracks propagate through the heterogeneous cement structure at [μm]-scale. Specimens are typically of size 100 μm, i.e. a factor 1000 smaller compared to the experiment depicted in Figure 2a. Since the micro-tensile test was designed as part of an effort to perform 3-dimensional crack detection in the Swiss light source at Villigen (Paul Scherer Institute), i.e. in the tomography test set-up in the synchrotron, the loading frame was specially designed and consists of a glass capillary as can be seen from Figure 2b. The loading is applied by means of a piezoelectric element.

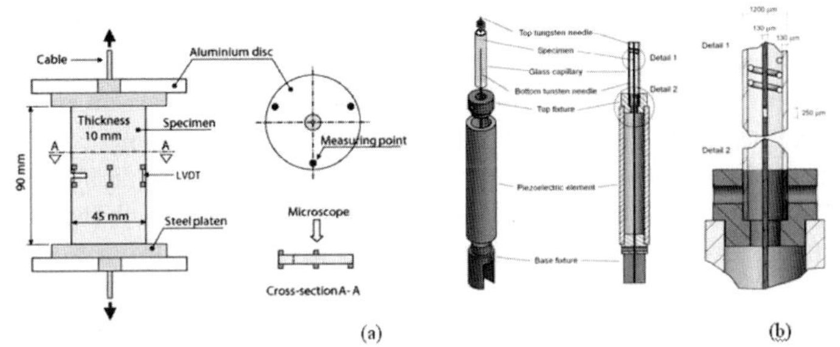

Figure 2. Multi-scale determination of tensile strength and fracture properties: (a) cable tensile test[5] at [mm] scale, and (b) micro-tensile test[24] at [μm]-scale, i.e. a factor 1000 smaller. Typical specimen dimensions are 50 x 100 mm for test set-up (a), and just 100 x 250 μm for the test set-up in Figure (b)

The micro-tensile test must be considered as a tensile test between fixed loading platens: rotations of the specimen ends are almost completely prevented, and, as known from macroscopic experiments, this has a tremendous effect on the test outcome[25]. In principle all phenomena recognised at the macro-level occur here as well, and the ideal boundary conditions and specimen size cannot be specified. The results from either test, in particular when it comes to strength and softening, are valid only for the specimen/machine system used, and cannot be transferred to other situations. This can also be summarised as follows: the technical problem of testing at micro- or macro-scale is different in the sense that microscopic techniques are generally more tedious, yet the philosophical problem remains identical since strength and softening response are still dependent on the experimental context.

This conclusion may appear to be rather negative but the issue must be raised in order to clear the air and search for new approaches that may help to solve the problem of specimen/machine interaction, in particular when it comes to strength and fracture properties.

6. OTHER LOADING CASES

The above can be extended to other loading situations like for example uniaxial compression. In 1984 I found that failure in concrete subjected to uniaxial or multiaxial compression is a localised phenomenon as well[26,27]. This was most clearly shown in a series of uniaxial compression tests on prisms of varying slenderness tested between frictionless loading platens. Later, I found publications in the field of rock mechanics, in particular the experiments by Hudson et al.[28], which were found to show the intertwining of material and structural effects on specimens of varying size too, but because still steel platens were used in their tests, the localisation effect remained hidden. It can be concluded that the same 4-stage fracture model as depicted in Figure 1 is applicable to compressive states of stress too, including multiaxial compression. Under compressive load, microcracks tend to take an orientation parallel to the loading direction. Those cracks are much more stable than microcracks in a tensile stress field, and as a consequence the pre-peak microcrack regime (stage (A)) is much larger in uniaxial compression, and even more extensive when lateral confinement is applied. Restraining microcracks has a positive effect on the pre-peak nonlinearity and leads to larger strains before the maximum stress is reached, and generally the strength is larger too.

When we limit the discussion to uniaxial compression, there is one significant additional effect that has to be considered in testing, namely the frictional restraint between specimen and loading platens. Kotsovos[29] showed that the inclination of the localised failure zone in cylinders loaded in uniaxial compression gradually change from a 20-25 degrees inclination to parallel to the loading axis when the restraint between loading platen decreases. In the context of a round robin test conducted by RILEM technical committee 148SSC, the combined effect of specimen slenderness and boundary restraint was subsequently investigated, and gave the full picture[30]. In compression the complicating factor is friction, not only between loading platens and specimen, but also in the localised crack in the softening regime. The bridging stage in the 4-stage fracture model of Figure 1 must be augmented to include frictional restraint in the localised fracture zone, whereas boundary restraint is an additional specimen/machine interaction that cannot be ignored. For example, applying an LEFM-based model for simulating softening in compression, the boundary restraint effect would imply an initial notch in the specimen with larger or smaller inclination to the loading axis. This is a rather simplified view, and basically a numerical concrete analysis would be needed to understand the full process. At the moment there is still ample experimental information missing for that.

7. CONCLUSION

In this contribution some views on the assessment of model parameters for fracture of brittle disordered materials are discussed. The empirical content of a given model depends very much on the size/scale of operation of the model, as well as on the state variables used. There is a tendency to include more parameters when matters do not fit, as well as a strong movement towards continuum-based state variable like stress and strain. In experiments it has been observed that the fracture process in a specimen/structure subjected to any state of stress below the brittle/ductile transition can be regarded as a four-stage fracture process. The four stages are (O) linear elastic stage, (A) pre-critical (micro-) cracking in the (non-linear) pre-peak regime, (B) critical (macro-) crack growth in the post-peak softening regime, and (C) bridging. The pre-critical (micro-) crack processes define strength, and are important to understand when it comes to scaling of fracture strength. Contrary to the popular view that pre-peak cracking can be modelled with a non-linear fracture mechanics approach including softening as a material parameter, it is argued that softening is a different physical process leading to a mix-up of material and structural effects. The separation of the specimen/machine system is difficult – if not impossible – in the softening regime, and even is already debatable for strength. The difference between crack growth in the pre-peak non-linear stage and the post-peak softening regime (B) is that the former cracks can be stabilized by the material structure, whereas this is impossible for the (macro-) cracks leading to localisation. With this distinction in mind it must even be clear for the largest advocate of non-linear fracture models based on 'softening' that something must be wrong. Non-linear softening models do not follow the physics of fracture, and ultimately this will lead to an in-ability to predict situations hitherto not explored by experiment. Consequently the first test for any fracture model is a check of the model is capable of correctly representing fracture *mechanisms*, the mechanics will follow next.

8. REFERENCES

1. M.F. Kaplan, Crack propagation and the fracture of concrete, *ACI Journal*, **58**(5), 591-610 (1961).
2. J.G.M. van Mier, in: *Scaling Laws in Ice Mechanics and Ice Dynamics*, edited by J. Dempsey and H.H. Shen (Kluwer Academic Publishers, Dordrecht, 2001), pp. 171-182.
3. R.H. Evans and H.S. Marathe, Microcracking and stress-strain curves for concrete in tension, *Mater. Struct. (RILEM)*, **1**, 61-64 (1968)
4. A. Hillerborg, M. Modéer and P.-E.Peterson, Analysis of crack formation and crack growth in concrete by means of fracture mechanics and finite elements, *Cem.Conc.Res.*, **6**(6), 773-782 (1976).
5. J.G.M. van Mier, and Ch. Shi, Stability issues in uniaxial tensile tests on brittle disordered materials, *Int.J.Solids Struct.,* **39**(13/14), 3359-3372 (2002).
6. J.G.M. van Mier, (2004) In: *Proc. 5th Int. Conf. on 'Fracture of Concrete and Concrete Structures' (FraMCoS-V)*, edited by V.C Li, C.K.Y.Leung, K.J. Willam, and S.L. Billington, (IA-FraMCoS, Evanston, IL, 2004), pp. 11-30.
7. M.R.A. van Vliet and J.G.M. van Mier, Experimental investigation of size effect in concrete and sandstone under uniaxial tension, *Engng. Fract. Mech.*, **65**(2/3), 165-188 (2000).
8. Z.P. Bažant and B-H.Oh, Crack band theory for fracture of concrete, *Mater. Struct. (RILEM)*, **16**(93), 155-177 (1983).
9. B. Trunk, Einfluss der Bauteilgrösse auf die Bruchenergie von Beton, *Building Materials Reports*, ETH Zurich, Volume 11, Aedificatio Publishers (2000), p. 155.
10. C. Iacono, L.J. Sluys and J.G.M. van Mier, Estimation of model parameters in non-local damage theories by inverse analysis techniques, *Comput.Methods Appl.Mech.Engrg.*, 2006 (in print).
11. Z.P. Bažant, Size effect in blunt fracture: Concrete, rock, metal, *J.Eng.Mech. (ASCE)*, 518-535 (1984).
12. A. Carpinteri and B. Chiaia, Multifractal nature of concrete fracture surfaces and size-effects on nominal fracture energy, *Mater. Struct.(RILEM)*, **28**, 435-443 (1995).
13. E. Prado and J.G.M. van Mier, Effect of particle structure on mode I fracture process in concrete, *Engng. Fract. Mech.*, 70(14), 1793-1807 (2003).
14. G. Lilliu and J.G.M. van Mier, On the relative use of micro-mechanical lattice analysis of 3-phase particle composites, *Engng. Fract. Mech.*, 2006, accepted.
15. I. Markovic, High Performance Hybrid Fibre Concrete – Development and Utilisation, *PhD-thesis*, Delft University of Technology, January 16, 2006, p.211.
16. M.R.A. van Vliet, Size Effect of Fracture in Concrete and Rock under Uniaxial Tension, *PhD-thesis*, Delft University of Technology, January 31, 2000, p. 192.

17. A.S. Elkadi, Fracture Scaling of Concrete under Multiaxial Compression, *PhD-thesis*, Delft University of Technology, December 20, 2005, p. 179.
18. A.S. Elkadi and J.G.M. van Mier, Experimental investigation of size effect in concrete fracture under multiaxial compression, *Int. J. Fract.*, Special Issue Symposium 34, ICF-11, (2006), accepted.
19. J.G.M. van Mier, Mode I fracture of concrete: Discontinuous crack growth and crack interface grain bridging, *Cem. Conc. Res.*, **21**(1), 1-15 (1991).
20. P. Stähli and J.G.M. van Mier, Manufacturing, fibre anisotropy and fracture of hybrid fibre concrete, *Engng.Fract.Mech.*, 2006 (in print).
21. H.K. Man and J.G.M. van Mier, Analysis of 2D- and 3D- fracture scaling by means of 3D-lattice simulations, In *Proc. ECF-16 'Failure Analysis of Nano and Engineering Materials and Structures'*, Aleaxandroupolis, Greece, July 3-7, 2006 (in press).
22. P.E. Roelfstra, H. Sadouki and F.H. Wittmann, Le Béton Numérique, *Mater Struct. (RILEM)*, **18**(107), 327-335 (1986).
23. E. Schlangen and J.G.M. van Mier, Experimental and numerical analysis of the micro-mechanisms of fracture of cement-based composites, *Cem. Conc. Comp.*, **14**(2), 105-118 (1992).
24. P. Trtik, E.N. Landis, M. Stampanoni, P. Stähli and J.G.M. van Mier, Micro-tensile testing and 3D imaging of hydrated Portland cement, *Mater.Struct (RILEM)* 2006, submitted.
25. J.G.M. van Mier, *Fracture Processes of Concrete (*CRC Press, Boca Raton, FL, USA, 1997).
26. J.G.M. van Mier, Strain-Softening of Concrete under Multiaxial Loading Conditions, *PhD-thesis*, Eindhoven University of Technology, November 1984, p. 349.
27. J.G.M. van Mier, Multiaxial strain-softening of concrete, *Mater. Struct. (RILEM)*, **19**(111), 179-200 (1986).
28. J.A. Hudson, S.L. Crouch and C. Fairhurst, Soft, stiff and servo-controlled testing machines: A review with reference to rock failure, *Eng.Geol.*, **6**, 155-189 (1972).
29. M.D. Kotsovos, Effect of testing techniques on the post-ultimate behaviour of concrete in compression, *Mater Struct. (RILEM)*, **16**(91), 3-12 (1983).
30. J.G.M. van Mier, S.P. Shah, M. Arnaud, J.P. Balayssac, A. Bascoul, S. Choi, D. Dasenbrock, G. Ferrara, C. French, M.E. Gobbi, B.L. Karihaloo, G. König, M.D. Kotsovos, J. Labuz, D. Lange-Kornbak, G. Markeset, M.N. Pavlovic, G. Simsch, K-C. Thienel, A. Turatsinze, M. Ulmer, H.J.G.M. van Geel, M.R.A. van Vliet, D. Zissopoulos, Strain-softening of concrete in uniaxial compression - Report of the Round-Robin test carried out by RILEM TC 148SSC, *Mater. Struct. (RILEM)*, **30**(198), 195-209(1997).

CRACK EXTENSION DUE TO CORROSION BY SIGMA-AE AND BEM

M. Ohtsu and F.A.K.M. Uddin
Graduate School of Science & Technology, Kumamoto University, 2-39-1 Kurokami, Kumamoto 860-8555, Japan

Abstract: Concrete structures suffer from corrosion of reinforcing steel bars due to severe environments. A detailed study on crack extension due to corrosion of reinforcement in concrete is performed. Moment tensor analysis of acoustic emission (AE) can identify cracking kinematics of location, crack-type and crack orientation, by implementing SiGMA code. Visualization system has been developed by using VRML (Virtual Reality Modeling Language). Corrosion cracking was simulated by employing expansive agent in experiments. Nucleation of microcracks was thus identified by SiGMA-AE analysis. By applying the two-domain boundary element method (BEM), crack trajectory in the arbitrary direction was analyzed. Mechanisms of crack extension due to corrosion are investigated, applying the dimensionless stress intensity factors and comparing results by SiGMA-AE and BEM.

Key words : acoustic emission; boundary element method; corrosion; SiGMA-AE; VRML; fracture mechanics; stress intensity factors

1. INTRODUCTION

Concrete structures suffer from corrosion of reinforcing steel bars (rebars) due to severe environments. Consequently, crack extension due to corrosion of reinforcement in concrete has been investigated experimentally and analytically[1]. In respect to experimental techniques, nucleation of a microcrack is readily detected by acoustic emission (AE) technique. Historically, theoretical treatment of AE waves leads to the moment tensor analysis for source kinematics[2] and the deconvolution analysis for kinetics[3]. In order to perform the moment tensor analysis for general AE waves, one powerful technique has been developed as SiGMA (Simplified Green's functions for Moment tensor Analysis)[4], which can determines crack kinematics on locations, types and orientations. Because these kinematical outcomes are obtained as three-dimensional (3-D) locations and vectors, 3-D visualization procedure is developed by using VRML (Virtual Reality Modeling Language)[5].

M.S. Konsta-Gdoutos, (ed.), Measuring, Monitoring and Modeling Concrete Properties, 233–238.

Recently, crack propagation in cementitious materials has been studied on the basis of fracture mechanics[6]. Here, the two-dimensional boundary element method (BEM) is applied to trace crack extensions in arbitrary orientations, based on the linear elastic fracture mechanics (LEFM). Mechanisms of crack extension due to corrosion are investigated by applying the dimensionless stress intensity factors. These microcracking mechanisms are compared with the results of SiGMA-AE analysis.

2. SIGMA ANALYSIS

Defining two vectors **l** of crack motion and **n** of crack normal, moment tensor M_{pq} is introduced,

$$M_{pq} = C_{pqkl} l_k n_l \Delta V \tag{1}$$

M_{pq} is defined by the product of the elastic constants [N/m^2] and the crack volume [m^3], which leads to the moment of physical unit [Nm]. Here, C_{pqkl} are the elastic constants and ΔV is the crack volume. Based on the far-filed term of P wave in the integral representation, one procedure has been developed, which is suitable for a PC-based processor and robust in computation. The procedure is now implemented as a SiGMA (Simplified Green's functions for Moment tensor Analysis) code. By extracting only P wave motion of the far field term of Green's function in an infinite space, taking into account the effect of reflection at the surface and neglecting the source-time function, a mathematical formula of the amplitude of the first motion is derived. Since the moment tensor is symmetric, the number of independent unknowns M_{pq} to be solved is six. Thus, multi-channel observation of the first motions at more than six channels is necessary to determine the moment tensor components.

In the solution procedure, displaying AE waveform on CRT screen, two parameters of the arrival time (P1) and the amplitude of the first motion (P2) are determined. Source location **y** is determined from the arrival time differences. Then, distance R and its direction vector **r** are determined. Substituting the amplitudes of the first motions at more than 6 channels, the components of the moment tensor M_{pq} are determined. Since the SiGMA code requires only relative values of the moment tensor components, the relative calibration of the sensors is sufficient enough.

The classification of a crack is performed by the eigenvalue analysis of the moment tensor. Setting the ratio of the maximum shear contribution as X, three eigenvalues for the shear crack become X, 0, -X. Likewise, the ratio of the maximum deviatoric tensile component is set as Y and the isotropic tensile as Z. It is assumed that the principal axes of the shear crack is identical to those of the tensile crack. Then, the eigenvalues of the moment tensor for a general case are represented by the combination of the shear crack and the tensile crack. Because relative values are determined in the SiGMA, three eigenvalues are normalized and decomposed,

$$1.0 = X + Y + Z$$

$$e_{12} = 0 - \frac{Y}{2} + Z$$
$$e_{13} = -X - \frac{Y}{2} + Z \tag{2}$$

where e_{12} is the ratio of the intermediate eigenvalue to the maximum eigenvalue, and e_{13} is the ratio of the minimum eigenvalue to the maximum. X, Y, and Z denote the shear ratio, the deviatoric tensile ratio, and the isotropic tensile ratio, respectively. In the present SiGMA code[4], AE sources of which the shear ratios are less than 40% are classified into tensile cracks. The sources of X > 60% are classified into shear cracks. In between 40% and 60%, cracks are referred to as mixed mode. In the eigenvalue analysis, three eigenvectors are also determined. Thus, vectors **l** and **n**, which are interchangeable, are readily recovered, and a visualized procedure has been developed by using VRML (Virtual Reality Modeling Language)[5].

3. BEM ANALYSIS

According to LEFM, the angle of crack extension θ is obtained from the maximum circumferential stress[7],

$$K_I \sin\theta + K_{II}(3\cos\theta - 1) = 0\,. \tag{3}$$

Here K_I and K_{II} are the stress intensity factors of mode I and mode II, which can be computed from the displacements on the crack-tip elements in BEM. Introducing the critical stress intensity factor K_{IC}, the initiation of crack extension is governed by,

$$\cos\frac{\theta}{2}\left(K_I \cos^2\frac{\theta}{2} - \frac{3}{2}K_{II}\sin\theta\right) = K_{IC} \tag{4}$$

From Eqs. (3) and (4), contribution K_I and K_{II} to K_{IC} can be obtained as the normalized stress intensity factors K_I/K_{IC} and K_{II}/K_{IC}, K_{IC} is the critical stress intensity factor and is determined in the previously studies[8]. By testing the same concrete mixture, it is found to be equal to 0.723MPa $m^{1/2}$.

4. EXPERIMENT AND ANALYTICAL MODEL

Concrete specimens of dimensions 25 cm x 25 cm x 10 cm with a hole of 3 cm diameter were tested. The hole corresponds to rebar location with 4 cm cover-thickness as shown in Figure 1. For concrete, mixture proportion of water: cement: sand: gravel was 0.5: 1.0:2.41:2.95 by weight. The maximum size of gravel was 20 mm. The compressive strength at 28-day standard curing was 37.9 MPa. The velocity of P wave was 4730 m/s and the modulus of elasticity was 29.7 GPa. P-wave velocity was applied to AE-SiGMA analysis and Young's modulus was employed in BEM analysis.

To simulate corrosion cracking in concrete, hydrostatic pressure was introduced by employing expansive agent. Dolomite paste (water:dolomite=1:2 by weight) was inserted into the hole of the specimen. Stress distribution around rebar, $\sigma_{\theta\theta}$, can be approximated, by assuming an annular elastic-body of the diameter, a, and the outer diameter, b. Setting the outer diameter, b, to the distance from the center of the hole to boundary surfaces of the specimen, stress distributions around the hole were calculated.

Results are shown in Figure 2. Expansive pressure is set to as 10 MPa and the cover-thickness was varied from 2 cm to 4 cm. As it is expected, the maximum stress is always observed at the shortest cover-thickness (at the top in the figure). This suggests that the surface crack is always nucleated first.

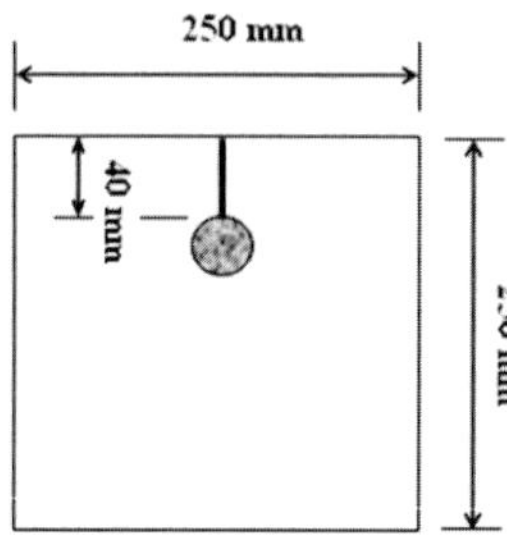

Figure 1. Specimen configuration

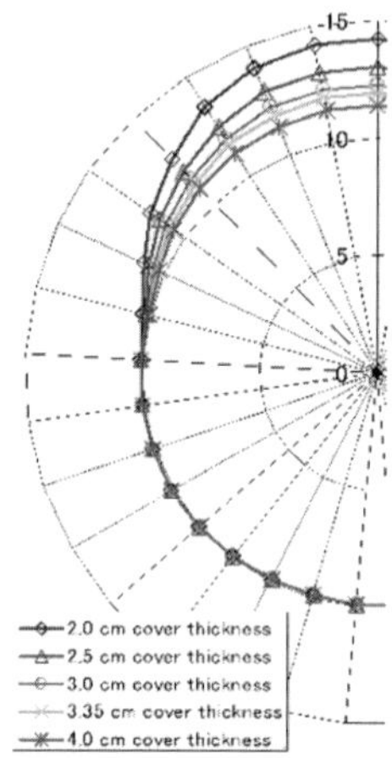

Figure 2. Stesss distribution (MPa)

Thus, one specimen with a hole (specimen A) and the other with a notch connecting the hole with the top surface as shown in Figure 1 (specimen B) were made.

During the test, the expansive pressure was measured by using a pressure gauge embedded inside the hole. AE events were detected by six-channel AE system. Six AE sensors (PAC, UT1000) were attached to the concrete specimen by using silicone grease. AE waves detected bt the sensors were amplified by 50dB with a preamplifier, and the waveforms and parameters of those waves exceeding the threshold level 50dB were recorded and analyzed by TRA212 AE system (PAC). Sampling time for recording AE waves was 1 μsec.

An analytical model to trace an extension of a diagonal crack is illustrated in Figure 3. The model consists of two domains stitched by the interface. At the crack tip elements, crack orientation is calculated by Eq. (3) and pressure which could satisfy with Eq. (4) is calculated. Then, along the crack orientation, two new crack-tip elements are created, separating the interface. The figure shows the initial model. Solving Eq. (3), new boundaries of crack surfaces are created in an arbitrary orientation automatically[9].

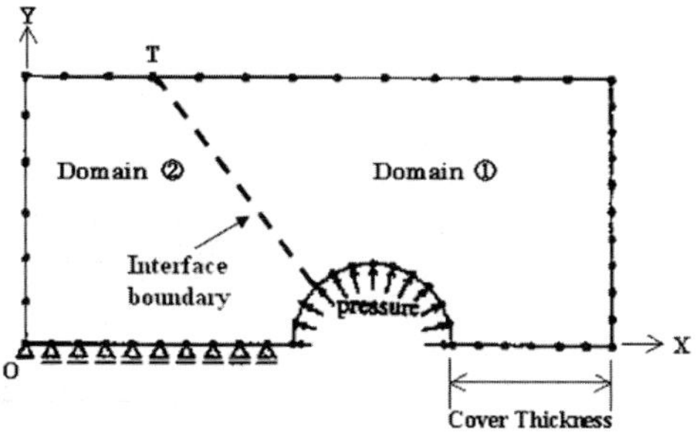

Figure 3. Two-domain BEM model

5. RESULTS AND DISCUSSION

Although a variety of crack patterns were reported in the literature[1], two diagonal (inter) cracks were observed in the both specimens. It is noted that the surface crack in specimen A was first observed and then two diagonal cracks extended almost simultaneously. These results suggest that the diagonal cracks could be generated following the surface crack. When the surface cracks are found and repair work is conducted removing the cover concrete, it should be careful for the presence of diagonal cracks inside rebars.

In order to clarify the mechanisms for nucleating diagonal cracks, BEM analysis was conducted. As illustrated in Figure 3, the model in which the surface crack is already nucleated was analyzed. Assuming three crack orientations of 45°, 90° and 135°, crack trajectories were computed. Three pressure distributions were taken into account, because stress distributions may vary, depending on expansion of corrosive products[1]. Neglecting either horizontal or vertical component of uniform pressure, the effect of pressure distribution was studied. It was observed that crack trajectories were not much different, depending on the pressure distribution. Following nucleation of the surface crack, vertical pressure in Figure 3 could be the most plausible. Consequently, a relation between pressure and crack extension was obtained. Results of specimen B are given in Figure 4. In the figure, case 1 indicates the case crack propagates toward 45°, case 2 toward 90° and case 3 toward 135°.

It is clearly observed that the lowest pressure at the beginning is observed in the 135° orientation, suggesting nucleation of the diagonal crack. This result clarifies a reason why the diagonal cracks were generated, following the surface crack in the tests. It is noted that crack extension in the 90° orientation is often derived in the finite element (FEM) analysis. But, this may result from the effect of mesh division, according to stress concentration in this orientation. Microscopically, all the types of tensile, mixed-mode and shear cracks were observed as AE sources from SiGMA-AE analysis. These imply that microcracks are accumulated and thus macroscopic cracks are nucleated as the surface crack and the diagonal crack. Thus, initiation mechanisms of the diagonal crack was studied, computing the dimensionless stress intensity factors from BEM analysis. Crack orientations observed and results of SiGMA-AE analysis in specimen B were also applied to compute the factors. Results are shown in Figure 5.

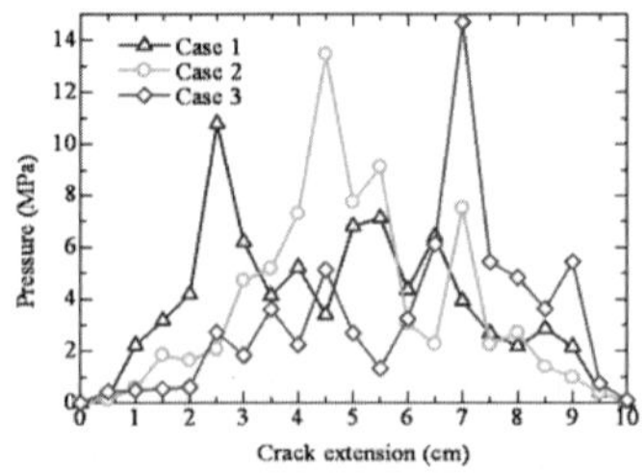

Figure 4. Pressure v.s. crack extension (Case 1:45°, Case 2:90°, and Case3 135°)

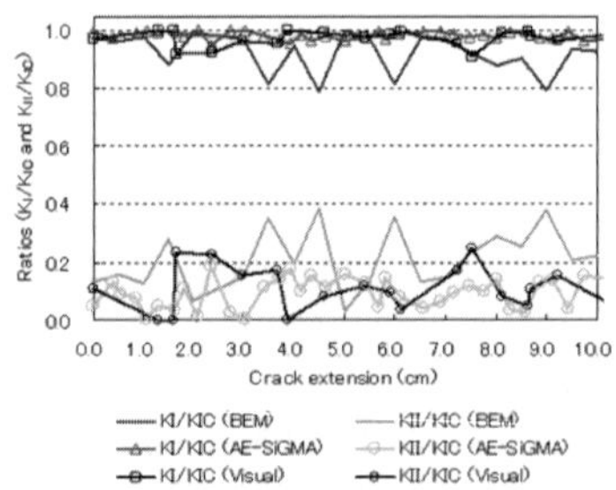

Figure 5. Dimensionless stress intensity factors

Results of SiGMA-AE analysis are in remarkable agreement with those of visual observation. They are in fairly good agreement with results of BEM analysis. In allthe cases, it is found that the dimensionless stress intensity factors K_I/K_{IC} are always dominant and almost equal to 1.0. In contrast, the factors K_{II}/K_{IC} are smaller than 0.4. These results confirm that the mode I cracks are dominantly generated during the extension of the diagonal crack.

6. REFERENCES

1. K. Toongoenthong and K. Maekawa, Simulation of Coupled Corrosive Product Formation, Migration into Crack and Propagation in Reinforced Concrete Sections, *J. Advanced Concrete Technology*, **3**(2), 253-265 (2005).
2. K. Y. Kim and W. Sachse, Characterization of AE Signals from Indentation Cracks in Glass, *Progress in Acoustic Emission II*, JSNDI, 163-172 (1984).
3. N. N. Hsu and S. C. Hardy, Experiments in AE Waveform Analysis for Characterization of AE Sources, Sensors and Structures, *Elastic Waves and Nondestructive Testing of Materials*, AMD-29, 85-106 (1978).
4. M.Ohtsu, Simplified Moment Tensor Analysis and Unified Decomposition of Acoustic Emission Source : Application to In Situ Hydrofracturing Test," *Journal of Geophysical Research*, **96**(B4), 6211-6221 (1991).
5. M. Ohtsu and M. Shigeishi, Three-Dimensional Visualization of Moment Tensor Analysis by SiGMA-AE, *e-Journal of Nondestructive Testing,* **7**(9) (2002).
6. A. Carpinteri,*Mechanical Damage and Crack Growth in Concrete* (Martinus Nijhoff Pub., 1986).
7. F. Erdogan and G. C. Sih, On the Crack Extension in Plates under Plane Loading and Transverse Shear, *J. of Basic Eng.*, 12, 519-527 (1963).
8. F. A. K. M. Uddin and M. Ohtsu, Application of AE to Fracture Toughness and Crack Analysis by BEM in Concrete, *e-Journal of Nondestructive Testing*, **7**(9) (2002).
9. A. H. Chahrour and M. Ohtsu,Crack Growth Prediction in Scaled –Down Model of Concrete Gravity Dam, *Theoretical and Applied Fracture Mechanics,* 21, 29-40 (1994).

SIZE EFFECT ON CONCRETE SPLITTING TENSILE STRENGTH AND MODULUS OF ELASTICITY

A. Kanos, A.E. Giannakopoulos and P.C. Perdikaris

University of Thessaly, Department of Civil Engineering, Pedion Areos, 38334 Volos, GREECE

Abstract: The work presented in this paper is part of an experimental investigation program regarding the tensile behavior of intermediate strength concrete. In total, ninety-one concrete cylinder specimens were casted and eighty-two of them were tested in uniaxial compression and indirect tension. The concrete specimens were systematically varied in size. The splitting tensile strength and modulus of elasticity of concrete exhibited a definite size effect. It was attempted to rationalize the size effect following a size effect law of the specific elastic fracture energy. A tensile concrete strain invariance is supported by the experimental findings.

Key Words: compressive strength; concrete; cylinder specimens; fracture energy; size effect, splitting tensile strength.

1. INTRODUCTION

The splitting tensile test for concrete (Carneiro and Barcellos[1], Akazawa[2]) is also used for the testing of other materials such as ceramics and rocks. The applied compressive force along two anti-diametrical generators results in the development of a largely homogenous and uniform tensile stress field perpendicular to the loading plane. The tensile stress, $_x$, perpendicular to the loading plane (y) resulting in the splitting failure of the cylinder specimen has been determined by Hondros[3] based on elasticity theory as:

$$\sigma_x = \frac{2P}{\pi LD}, \tag{1}$$

where P is the applied concentrated anti-diametrical compressive force, acting on the y-plane, L the length and D the diameter of the cylinder specimens.

According to existing testing standards the compressive force P is distributed along a loading strip, of a width equal to d, on both sides of the cylinder specimen. The testing standards suggest a constant d/D ratio, which means different widths for the loading strip

M.S. Konsta-Gdoutos, (ed.), Measuring, Monitoring and Modeling Concrete Properties, 239–246.

depending on the size and shape of the specimen used. For d/D <0.16, Eq. (1) holds true along almost the entire diameter of the specimen. Therefore, except for the regions close to the loading areas, a nearly homogeneous stress field along the loading plane is established.

2. EXPERIMENTAL INVESTIGATION.

The experimental work is based on the results of the splitting cylinder tests on a series of concrete specimens of three different diameters, D = 76 mm, 101 mm and 152 mm. The cylinder specimens had a diameter to height ratio of ½. Two series of concrete specimens were casted which differed as to the maximum aggregate size, d_{max} = 5 mm and 15 mm. For each concrete mix 4÷5 cylinder specimens were tested. The typical microstructure of specimen series 1 and 2 is shown in Figs. 1 and 2.

Cement type I (42.5 N, Standard ELOT–EN 197 - 1) manufactured by Hercules-Lafarge in Volos, Greece was used. The weight ratios for the two concrete mixes were set at cement (c) / aggregates (a) / sand (s) / water (w) = 1 / 2 / 1 / 0.4. Superplasticizer was used in both mixes to increase their workability. Because of the low w/c ratio of 0.4, higher than normal strength concrete was produced. Plastic molds of different sizes were used for casting the concrete cylinder specimens.

The concrete specimens where cured for three months by submerging them in a water tank at a steady temperature of about 20° C. Upon exiting the tank, the specimens where left to dry before they were weighed and their dimensions recorded. This was done in order to ensure that the density of the material did not differ with specimen size. Large fluctuations of material density with respect to specimen size may severely affect the accuracy of the experimental procedure when examining size effect. The fluctuation of the measured concrete density in this study did not exceed ±1% and can be seen in Table 1.

The meridians of each specimen were marked for proper location of the loading strips. All loading strips had a width of d = 1 cm, which did not differ with specimen size. Although this led to different ratios of d/D for each specimen size, the accuracy of the experimental results was not compromised. The ratios d/D were 0.066, 0.099 and 0.123, respectively for each specimen size, well within the proper range of 0.04 to 0.16 proposed by Rocco et al.[4, 5] According to the elastic theory, the expected deviation of the measured splitting tensile strength for concrete compared to the direct tensile strength, f_t, is less than 2%.

A total of thirty-one (31) specimens where tested using a DMG hydraulic compression testing machine digitally controlled by a computer (stroke control). A steel setup was manufactured for the splitting cylinder tests to uniformly distribute the applied compressive force, anti-diametrically on each cylinder specimen along the specimen's meridional. The recorded response during each test included the applied load and the stroke of the actuator. As expected, the load displacement curve was found to be almost linear up to the maximum applied load. Cracking was sudden and progressed extremely fast.

Uniaxial compression tests were also carried out on cylinder specimens of the same size for both concrete mixes[6]. These specimens were instrumented with three (3) equally spaced strain gages at midheight of the specimen to measure the axial compressive concrete strain. The results of the modulus of elasticity of the concrete specimens tested in uniaxial compression are also discussed in this paper.

Figure 1. Representative microstructure for the cylinder specimens Series 1 with D =152 mm (dmax=15 mm).

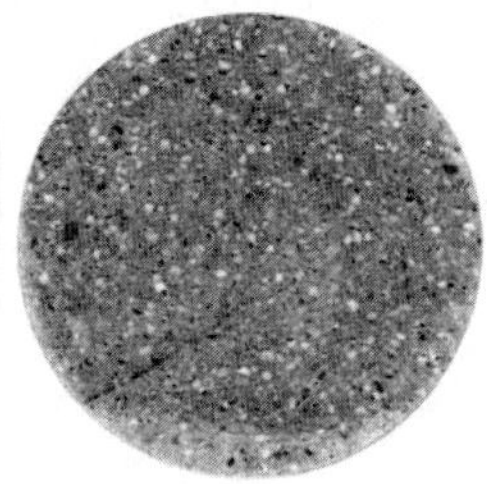

Figure 2. Representative microstructure for the cylinder specimens Series 2 with D =152 mm (dmax=5 mm).

Table 1. Measured density of the cylinder concrete specimens

Series 1 (d_{max} = 15 mm)			**Series 2 (d_{max} = 5 mm)**	
Specimen	Concrete Density (Kg/m^3)		Specimen	Concrete Density (Kg/m^3)
008	2350.35		060	2328,21
009	2327.63		061	2339,67
010	2343.15		062	2335,19
012	2365.05		063	2327,93
013	2357.75		074	2313,65
021	2364.11		075	2325,41
022	2349.46		076	2321,72
023	2340.28		077	2312,75
024	2353.22		086	2299,83
030	2320.79		087	2287,14
031	2309.50		088	2306,18
032	2309.16		089	2281,17
033	2334.35		090	2301,07
034	2338.59			

3. EXPERIMENTAL RESULTS

The experimental results for the concrete splitting tensile strength, f_{st}, based on Eq. (1) yielded clear signs of size effect, following the general trends reported in the literature. Table 2 and Fig. 3 include the results of the splitting tensile tests for the concrete specimens of the 1st series (concrete mix with a maximum aggregate size of 15 mm). The results of the splitting tensile tests for the concrete specimens of the 2nd series (concrete mix with a maximum aggregate size of 5 mm) are presented in Table 3 and Fig. 4.

The modulus of elasticity, E, was measured by strain gages mounted on at least three specimens of each size and concrete batch in the uniaxial compression tests. These measured values for the modulus of elasticity of concrete on selected specimens are presented in Figs. 5 and 6 for the concrete mix with a maximum aggregate size of 5 mm and 15 mm, respectively.

4. MODULUS OF ELASTICITY

The modulus of elasticity is an important material property and it is considered to be independent of specimen size. The experimental results of the present investigation seem to indicate otherwise. The results of the measured modulus of elasticity for the cylinder concrete specimens exhibit a size effect, shadowing the well known size effect of tensile strength. The tensile strength results from the splitting tests and the experimental results for the modulus of elasticity from the uniaxial compression tests showed a similar size effect. The behavior of a quasibrittle material such as concrete under tension is generally linear up to the peak load,whereas in compression linearity does not exist all the way up to the peak load but only initially.

Table 2. Experimental tensile splitting strength results obtained for specimens of Series 1

Series 1 Maximum Aggregate size, $d_{\max} = 15\ mm$		
Specimen No.	Average Specimen Diameter D	Splitting Tensile Strength f_{st} (MPa)
008	76 mm	4.13
009		4.80
010		5.15
012		5.24
013		3.54
021	101 mm	3.62
022		4.67
023		4.70
024		4.56
030	152 mm	3.13
031		2.61
032		4.27
033		3.35
034		3.24

Table 3. Experimental tensile splitting strength results obtained for specimens of Series 2

Series 2 Maximum Aggregate size, $d_{max} = 5\ mm$			
Specimen No.	Average Specimen Diameter D	Splitting Tensile Strength f_{st} (MPa)	
060	76 mm	3.80	
061		3.88	
062		4.24	
063		4.64	
074	101 mm	3.37	
075		4.19	
076		4.26	
077		3.84	
086	152 mm	4.42	
087		3.62	
088		3.80	
089		3.55	
090		3.23	

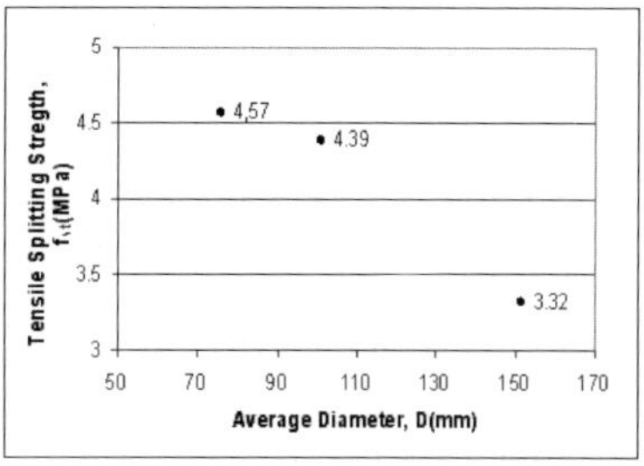

Figure 3. Average experimental tensile splitting strength results for Series 1

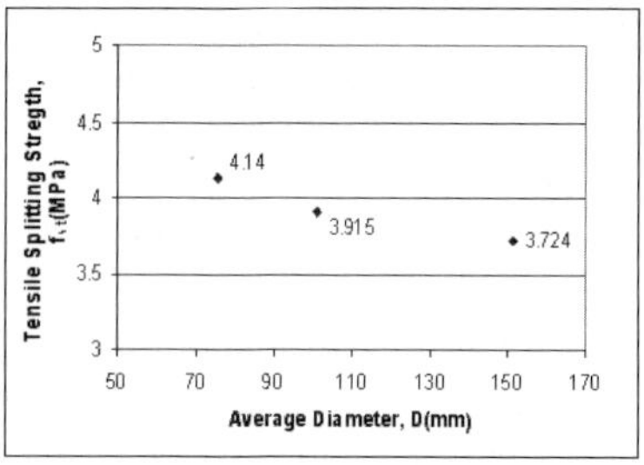

Figure 4. Average experimental tensile splitting strength results for Series 2

This is true because material failure in tension occurs with the propagation of a single crack and the subsequent seperation of the specimen, while in compression the mechanism of failure is far more complex and occurs through the propagation of multiple cracking.

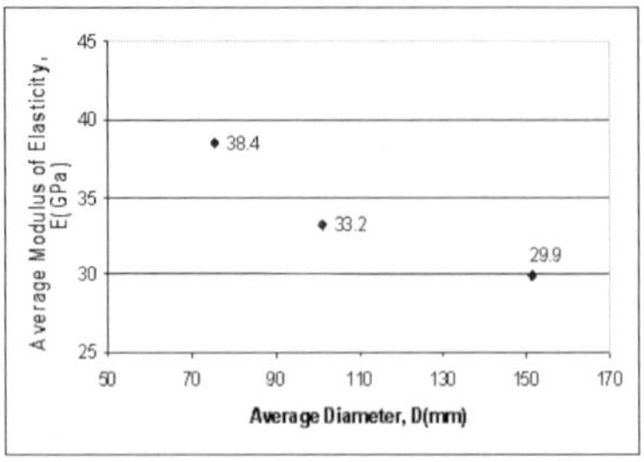

Figure 5. Measured average modulus of elasticity of Series 1

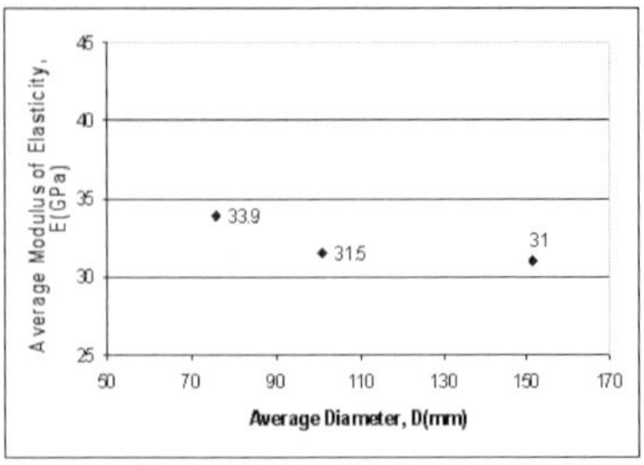

Figure 6. Measured average modulus of elasticity of Series 2

Assuming that the modulus of elasticity in compression, E_c, is practically identical to the modulus of elasticity of tension, E_t ($E_c = E_t$), a characteristic tensile strain at peak load, $_f$, can be defined simply as:

$$\varepsilon_f = f_{st}/E_c \ , \tag{2}$$

where ε_f is the tensile strain at peak load.

It is evident from Figs. 7 and 8 that the average concrete characteristic tensile strain at the peak load, ε_f, is more or less constant and independent of specimen size. The average concrete characteristic tensile strain at peak load, ε_f, appears to be about 0.12 x 10^{-3} for both concrete mixes. This indicates that a major source for the size effect of tensile strength for intermediate strength concrete may actually be the material's modulus of elasticity. Strength being a macroscopic measure for the material is probably influenced by the fluctuation of the modulus of elasticity, which in turn is influenced by the aggregate size to specimen diameter ratio. Microscopically, the concrete material of each concrete mix is identical, independent of specimen size. Thus, the characteristic concrete tensile strain at peak load should also be independent of specimen size, as it appears to be. This strain invariance in combination with the variance of the modulus of elasticity with specimen size appears to be a possible explanation to the so-called size effect of material strength. This characteristic tensile strain depends on the overall stress field, as described in this case by Eq. (1)

Implying that linear elasticity holds up to the peak applied load it is assumed that the specific elastic fracture energy, E_f, is proportional to the splitting tensile strength, f_{st}. Instead of considering a constant modulus of elasticity independent of specimen size it is implied that a characteristic concrete tensile stain at peak load exists retaining an invariance to specimen size.

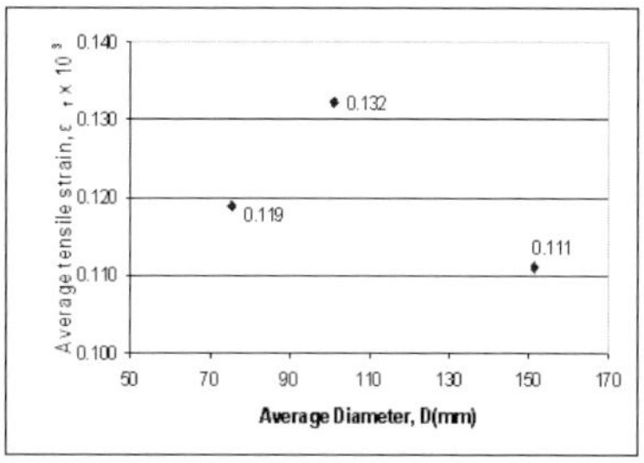

Figure 7. Average calculated characteristic concrete tensile strain at peak load for Series 1 specimens

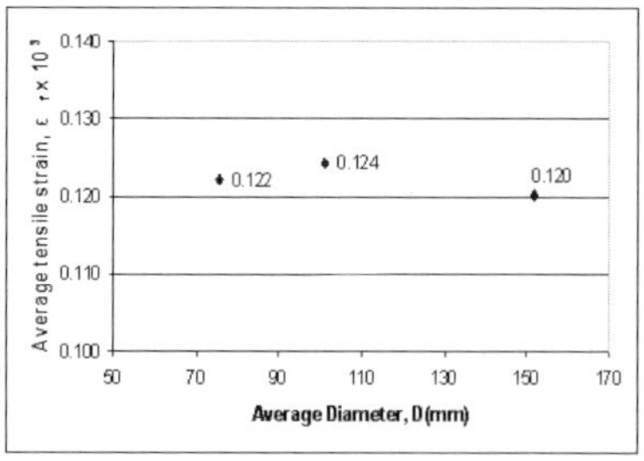

Figure 8. Average calculated characteristic concrete tensile strain at peak load for Series 2 specimens

5. CONCLUSIONS

The present work presents evidence of size effect in the splitting tensile strength of intermediate strength concrete. Two concrete mixes, differing only in the maximum aggregate size are investigated. Measurements on the modulus of elasticity of concrete in compression, which showed a similar size effect with the tensile strength, are included. It is true that the characteristic concrete strain at peak tensile strength is nearly independent of specimen size.It can be suggested that the size effect on the tensile strength of concrete can be explained because of the observed size effect of its modulus of elasticity.

Acknowledgements

The authors wish to thank the Department of Civil Engineering of the University of Thessaly for providing financial support for this investigation. Appreciation is expressed towards Associate Proffesor Dr. Michael Petrou and Dr. George Efremidis for their help and guidance during the setup of the experiments. Also appreciation is expressed towards Mr. Alekos Koutselinis and Mr. Demetris Karamberopoulos for their technical support in the Laboratory of Concrete Technology and Reinforced Concrete Structures.

6. REFERENCES

1. Carneiro, F.L.L., and Barcellos, Aquinaldo, "Tensile Strength of Concrete" RILEM Bulletin No. 13, Union of Testing and Research Laboratories for Materials and Structures, Paris, France, Mar. 1953, pp. 97-123.
2. Akazawa T., "Tensile Test Method for Concretes", International Association of Testing and Research Laboratories for Materials and Structures , Paris, Bulletin No. 16, 1953, pp 11-73.

3. Hondros, G., "The Evaluation of Poisson's Ratio and the Modulus of Materials of a Low Tensile Resistance by the Brazillian (indirect tensile) Test with Particular Reference to Concrete." Aust. J. Appl. Sci., 1959, 10, 243-264.
4. Rocco, C., Guinea, G.V., Planas, J., and Elices, M. "Mechanisms of Rupture in Splitting Tests", ACI Materials J., Vol. 96, No. 1, January-Febuary, 1999, pp. 52 – 60.
5. Rocco, C., Guinea, G.V., Planas, J., and Elices, M., "Review of the Splitting-test Standards from a Fracture Mechanics Point of View", Cement and Concrete Research 31 (Elsevier), 2001, pp. 73 -82.
6. Kanos, A., "Experimental Investigation on the Size Effect of Tensile and Compression Behavior of Concrete and Theoretical Modeling of Experimental Data" Undergraduate Diploma Thesis, Department of Civil Engineering, University of Thessaly, Volos, Greece (in Greek), October 2004.

NOTATIONS

D = cylinder diameter
d = width of loading strip
d_{max} = maximum aggregate size
E_c = modulus of elasticity in compression
E_f = specific elastic fracture energy
E_t = modulus of elasticity in tension
f_{st} = splitting tensile strength
f_t = direct tensile strength
L = specimen length
P = applied force
ε_f = characteristic tensile strain at peak load
σ_x= normal stress in x direction for splitting tests

MIXED-MODE CRACK PROPAGATION THROUGH REINFORCED CONCRETE

An experimental study

J.R. Carmona, G. Ruiz, and J.R. del Viso

ETSI Caminos, C. y P., Universidad de Castilla-La Mancha Avda. Camilo Jose Cela s/n, 13071 Ciudad Real, Spain

Abstract: This paper presents the results of a very recent experimental research program aimed at investigating mixed-mode fracture of reinforced concrete. The tests were designed so that only one single mixed mode crack generates and propagates through the specimen, as opposed to the usual dense crack pattern found in most of the tests in scientific literature. The specimens were three-point-bend beams of three different sizes. They were notched asymmetrically and reinforced with various ratios of longitudinal and of inclined reinforcement. These experiments may help to understand the mechanisms of crack initiation and propagation through reinforcing bars under mixed load conditions.

Keywords: Reinforced concrete, mixed-mode fracture, size effect.

1. INTRODUCTION

This paper presents some very recent results of an experimental program aimed at disclosing some aspects of the propagation of mixed-mode cracks through reinforced concrete. Specifically, the program was designed to investigate the influence of the size of the specimen and of reinforcement detailing on mixed-mode crack propagation. This research is an extension of previous works on the nucleation and propagation of mode I cracks in reinforced concrete (Ruiz et al., 1998, Ruiz and Carmona, 2005). By focusing on mixedmode cracks we aim at completing the study of the generation and development of the different types of cracks that may appear in reinforced concrete.

Mixed-mode crack propagation in plain concrete has been thoroughly studied (Shah et al., 1995). In reinforced concrete the topic is mainly addressed from a technological standpoint. The dense crack pattern that results from theusual reinforcement detailing and element geometry may somehow make it difficult to induce direct relations between

M.S. Konsta-Gdoutos, (ed.), Measuring, Monitoring and Modeling Concrete Properties, 247–255.

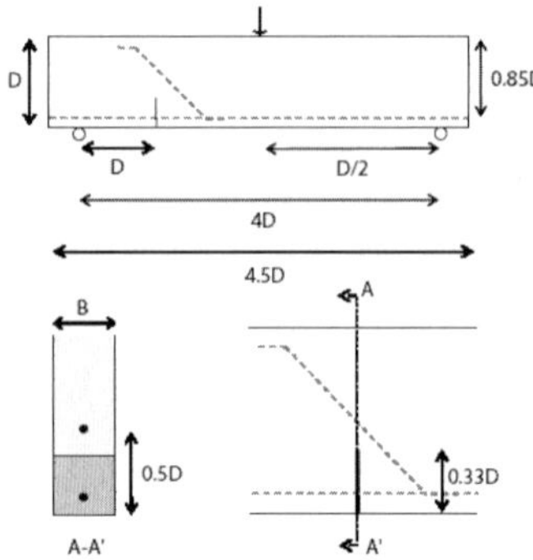

Figure 1. Beam Geometry

causes and effects. That is why we focus on the propagation of one single mixed-mode crack.Of course, there are some other excellent studies sharing our metodology. They addressed problems related to the shear resistance of reinforced elements, like the study on fracture by diagonal tension performed by Bazant and Kazemi (Bazant and Kazemi, 1991), or the work by Kim and White (Kimand White, 1999) on the generation of extense shear-damaged areas in reinforced concrete.

The organization of the article is as follows. A brief outline of the experimental program is given in Section 2. In Section 3 we describe the characterization and control tests performed on the materials used to make the reinforced beams and list the mechanical properties obtained from them. Section 4 deals about the experimental set-up for the mixed-mode tests and lists some selected results. Finally we present a brief discussion on the results and some concluding remarks, Sections 5 and 6 respectively.

2. OVERVIEW OF THE EXPERIMENTAL PROGRAM

The program was designed to study the propagation of mixed-mode cracks through reinforced concrete. Specifically, we wanted to disclose the influence of the location and the amount of reinforcement on the crack propagation. We also intended to analyze the variations in the crack pattern and in the mechan-ical behavior due to the size of the specimens.

With these intentions in mind we chose the beam sketched in Fig. 1 as a convenient specimen for this research. Our choice revisits the geometry tested by Jenq and Shah to study mixed-mode crack propagation in plain concrete (Jenq and Shah, 1988). It is a notched beam that exhibits a single mixed-mode crack when subjected to bending at three points. In their work, Jenq and Shah provide plenty of insights on the generation and propagation of the crack which are of use here. We reinforce the beams with several ratios of longitudinal and of inclined bars (shear reinforcement). The reinforcement provokes changes in the orientation of the main crack and in the global mechanical response, but the presence of a notch avoids a dense crack pattern that would blur our perception of the changes. At most, some reinforcement configurations generate a secondary flexural crack that competes with the one that starts from the notch tip. In such cases it may happen that the mixed-mode crack arrests and the flexural secondary crack develops completely.

Regarding the size of the beams, we wanted even the largest beams to be reasonably easy to handle and test. At the same time, the behavior of the laboratory beams should be representative of the behavior of beams of a normal size made of ordinary concrete. In order to fulfill both requirements, Hillerborg's brittleness number β_H (Bazant and Planas, 1998) was used as the comparison parameter. β_H is the ratio between the size of the beams -represented by their depth D and the characteristic length of the concrete, l_{ch} (Petersson, 1981). As a first approximation, two geometrically similar structures display a similar fracture behavior if their brittleness numbers are equal. According to this, a relatively brittle micro-concrete was selected with a characteristic length of approximately l_{ch}=90 mm (the details of the micro-concrete are given in the next section), while the beams were made to be 75, 150 and 300 mm in depth. Since l_{ch} of ordinary concrete is 300 mm on average, our 150 mm depth laboratory beams are expected to simulate the behavior of ordinary concrete beams 500 mm in depth, which is considered as a reasonable size for the study.

3. CHARACTERIZATION AND CONTROL TESTS

Standard characterization and control tests were performed to determine the compressive strength, tensile strength, elastic modulus and fracture energy of the concrete. The mechanical parameters of the rebars and of the steel to con-crete interface were also measured in our laboratory.

As mentioned above, a single micro-concrete mix was used throughout the experimentation, made with a lime aggregate of 5 mm maximum size and ASTM type II/A cement. The mix proportions by weight were 3.2:0.45:1 (aggregate water cement). We made characterization specimens out of all batches.

Compression tests were carried out according to ASTM C 39 and C 469 on 75 x 150 mm cylinders (diameter x height). Brazilian tests were also carried out on these kind of cylinders following the procedures recommended by ASTM C 496. Stable three-point bend tests on 75 x 50 x 337.5 mm notched beams were carried out to obtain the fracture properties of concrete. We followed the procedures devised by Elices, Guinea and Planas (that are minutely explained in Bazant and Planas, 1998). Particularly, during the tests the beams rested on anti-torsion devices. They consist of two rigid-steel semi-cylinders laid on two supports permitting rotation out of the plane of the beam and rolling along the longitudinal axis of the beam with negligible friction. Table 1 shows the mechanical parameters of the micro-concrete determined in the various characterization and control tests.

For the beam dimensions selected, and the desired steel ratios, the diameter of the steel bars had to be smaller than that of standard reinforcing bars, so commercial ribbed wires with a nominal diameter of 2.5 mm were used to achieve the desired reinforced configuration for different specimens. In order to measure the strength of the interface,

bars in beams without inclined bars. Finally the combined influence of both types of reinforcement is disclosed. The effect of the size on the experimental results permeates the discussion of all the aforementioned categories.

Effect of inclined reinforcement

Figure 2 shows experimental results given by beams that are reinforced only with inclined bars. Plain beams are also included as a limit case of null reinforcement. Figs. 2a and b depict P-δ curves for small and medium beams respectively. They show that the inclined bars provide stability to the crack propagation. Indeed, the plain beams break in an unstable way which is char- acteristic of tests performed under position control. As the amount of inclined bars increases, the behavior after the peak load turns to be more stable and stronger. Nevertheless, it is surprising that the peak load diminishes as the amount of inclined reinforcement increases. In fact, the maximum load withstood by the plain beams is higher than the beams with one bar and, again, the beams with two bars resist even less maximum load. Such trend can be clearly observed in Fig. 2c, which represents the maximum loads of this kind of beams (large beams also included) versus the size of the specimen in a nondimensional way. Such apparently abnormal behavior may be explained by the changes in the crack trajectory induced by the reinforcement, which are shown clearly in Fig. 2d. The presence of inclined bars modifies the almost straight trajectory of the crack in plain concrete beams. The crack is more vertical and even shows a turning point and changes its direction of growth as the reinforcement ratio increases.

Regarding the scale effect, plain beams follow the universal law proposed by Bazant (Bazant, 1984; please, see the solid curve and the formula inserted in Fig. 2c). Reinforced beams follow a similar trend, i.e. larger beams resist less in terms of stress than smaller ones.

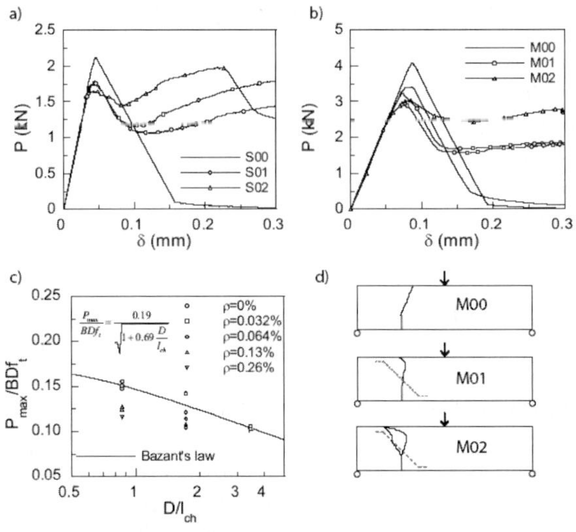

Figure 2. Experimental results given by beams with various ratios of 45'-inclined bars and no longitudinal bars: (a) P-δ curves of the small beams; (b) P-δ curves of the medium beams; (c) size effect on the maximum load; and (d) crack pattern observed in the case of medium beams

Effect of longitudinal reinforcement

Figure 3 shows experimental results given by beams reinforced solely with longitudinal bars. Plain beams are, again, a limit case of this category of beams and, consequently, they are also considered in the figure. In this kind of beams the reinforcing bars are far from the tip of the notch. The crack starting from the notch should behave like a crack that has already crossed the flexural reinforcement layer and goes on progressing under mixed-mode conditions.

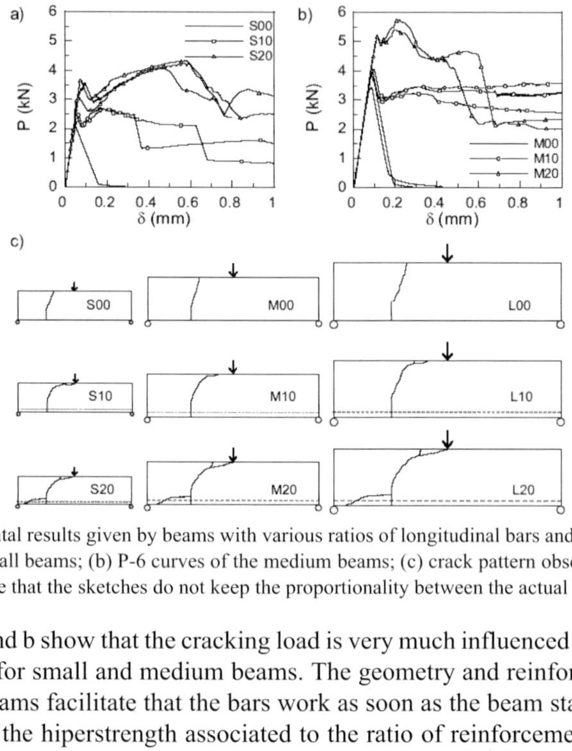

Figure 3. Experimental results given by beams with various ratios of longitudinal bars and no inclined bars: (a) P-6 curves of the small beams; (b) P-6 curves of the medium beams; (c) crack pattern observed for this kind of beams (please, notice that the sketches do not keep the proportionality between the actual beams)

Figures 3a and b show that the cracking load is very much influenced by the reinforcement ratio both for small and medium beams. The geometry and reinforcement arrangement in these beams facilitate that the bars work as soon as the beam starts to be loaded, which provokes the hiperstrength associated to the ratio of reinforcement. Size effect is also evident since M20 beams do not reach twice the cracking load of their small counterparts, S 1 0.

The behavior of the beams with higher ratios of reinforcement as the crack propagates is specially noticeable. S20 and M20 beams start to soften as the crack turns and penetrates the compressed zone of the beam. During this process the reinforcement unloads, since the global displacement of the beam is due mainly to the opening of the crack as it turns towards the load point rather than to the rotation at the notched section. Once the crack tip reaches the loading area, a compressive hinge starts to develop and, from then on, the response of the beam depends only on the steel that sews the notch. The trajectory of the crack is sensitive to the presence and amount of flexural reinforcement, as Fig. 3c clearly shows. The angle at which the crack starts propagating is almost independent of the number of bars, but as the reinforcement ratio increases the crack gets inclined so as to reach the loading point. In this case, crack trajectories for different beam sizes are alike.

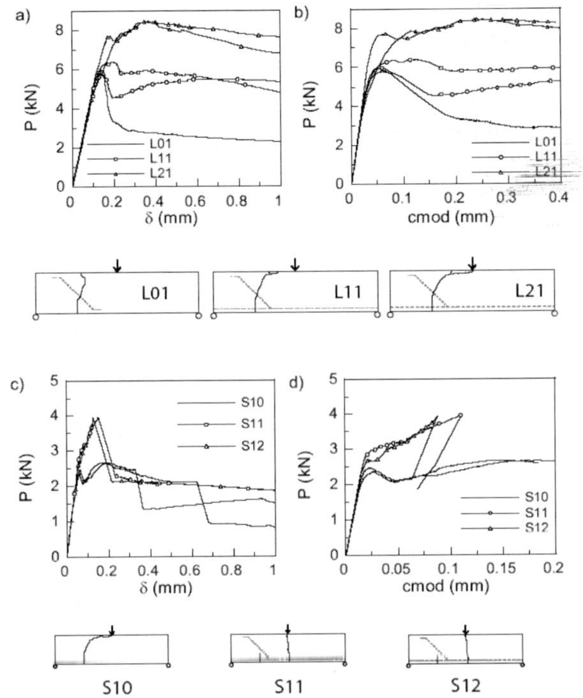

Figure 4. Effect of the amount of reinforcement: (a) P-δ and P-CMOD curves and crack pattern of large beams with a fixed ratio of longitudinal reinforcement but increasing number of inclined bars; (b) P-δ curves of the medium beams; (c) crack pattern observed for this kind of beams (please, notice that the sketches do not keep the proportionality between the actual beams)

Combined effect of longitudinal and inclined reinforcement

Figure 4 shows the mechanical behavior and the crack pattern of beams reiforced with both longitudinal and inclined bars. Specifically, Figs. 4a and b show the results given by large beams reinforced with just one 45'-inclined bar and with zero, one or two longitudinal bars. The comparison between the results given by L01 and L11 beams indicate that small ratios of longitudinal reinforcement modify the crack trajectory ostensibly while the associated hiperstrength is almost negligible. The post-peak behavior is sensitive to the longitudinal reinforcement though. The more longitudinal reinforcement, the more maximum load the beam withstands, as can be induced by comparing the L11 and L21 curves. With only two longitudinal bars, L21 beams even harden after the cracking load and get to another maximum. From then on the beams soften mildly.

Figures 4c and d show results given by small beams in which the longitudinal reinforcement is kept constant while it is the inclined reinforcement that changes. The sketches of the crack trajectories prove that shear reinforcement produces a change in the collapse mechanism of the beam. The presence of inclined bars produces a big increase in the beam carrying capacity because they are located in the area where the crack nucleates.

Indeed, in the two examples shown in the figures, S11 and S12 beams, the inclined reinforcement arrests the propagation of the crack that starts from the notch and provokes the formation of a flexural (mode I) crack in the middle of the beam. The flexural crack grows unstably and subsequently the load capacity of the beam is reduced. Fig. 4d shows that the unloading of the beam causes the closing of the mixed-mode crack. It is interesting to notice that the mechanical behavior of S11 and S12 beams is almost identical in spite that the latter has twice as inclined bars that the former. A slight wider opening of the S11 beam (clearly noticeable in Fig. 4d) shows that the inclined reinforcement is weaker.

6. CONCLUSIONS

This article presents very recent experimental results on the propagation of mixed-mode cracks through reinforced concrete. The tests were designed so that only one single mixed mode crack generates and propagates through the specimen, as opposed to the usual dense crack pattern found in most of the tests in scientific literature. The specimens were three-point-bend beams with an asymmetrical notch of three different sizes reinforced with various ratios of longitudinal (flexural) and of inclined (shear) reinforcement.

Some beams had shear reinforcement only. In these tests the reinforcement provoked changes in the trajectory of the crack so that the reinforced beam ended up withstanding less load than a plain one. Complementarily, we also tested beams reinforced solely with longitudinal bars. The cracking load of such beams was very sensitive to the amount of reinforcement. The crack propagated towards the point where the load was applied. The final stretch of the propagation process induced a drop in the carrying capacity of the beam, specially in case the amount of reinforcing bars was high. Finally, we also tested beams with both shear and flexural reinforcement. Small ratios of inclined reinforcement provoked, in such cases, that the collapse mechanism of the beam changed. The mixed-mode crack developing from the notch got arrested, while a mode I crack in the middle of the span nucleated and eventually progressed, leading to the failure of the beam.

The effect of the size of the beams is noticeable in our tests. On the one hand, large beams resisted less load in terms of stress. On the other hand the larger the beam, the more leaned towards the load point the crack trajectory was.

Acknowledgments

The authors gratefully acknowledge financial support for this research provided by the Ministerio de Educaci?n y Ciencia, Spain, under grant MAT2003-00843, and by the Ministerio de Fomento, Spain, under grant BOE305/2003. Jacinto R. Carmona and Javier R. del Viso also thank the Junta de Comunidades de Castilla-La Mancha, Spain, and the Fondo Social Europeo for the fellowships that support their research activity.

7. REFERENCES

1. Bazant, Z. P. (1984). Size effect in blunt fracture: Concrete, rock, metal. Journal of Engineering Mechanics-ASCE, 110:518-535.

2. Bazant, Z. P. and Kazemi, M. P. (1991). Size effect in diagonal shear failure. ACI Structural Journal, 88(3):268-276.
3. Bazant, Z. P. and Planas, J. (1998). Fracture Size Effect in Concrete and other Quasibrittle Materials. CRC Press, Boca Raton.
4. Jenq, Y. S. and Shah, S. P. (1988). Mixed mode fracture of concrete. International Journal of Fracture, 38:123-142.
5. Kim, W. and White, R. N. (1999). Shear-critical cracking in slender reinforced concrete beams.ACI Structural Journal, 96(5):757-765.
6. Petersson, P. E. (1981). Crack Growth andDevelopment of Fracture Zone in Plain Concrete and Similar Materials. Report No. TVBM-1006, Division of Building Materials, Lund Institute of Technology, Lund, Sweden.
7. Ruiz, G. and Carmona, J. R. (2005). Experimental study on the influence of the shape of the cross-section and of the rebar arrangement on the fracture of lightly reinforced beams. Materials and Structures. In press.
8. Ruiz, G., Elices, M., and Planas, J. (1998). Experimental study of fracture of lightly reinforced concrete beams. Materials and Structures, 31:683-691.
9. Shah, S. P., Swartz, S. E., and Ouyang, C. (1995). Fracture Mechanics of Concrete. Wiley, New York.

Quantifying Damage for Early Age Concrete

ADVANCED ANALYSIS OF STRESSES FOR CONTROL OF TRANSVERSE CRACKING IN EARLY-AGE CONCRETE DECKS OF COMPOSITE BRIDGES

B.H. Oh and S.C. Choi

Professor, Dept. of Civil Engineering, Seoul National University, Shinrim-dong, Gwanak-ku,Seoul, Korea; Research Associate, Dept. of Civil Engineering, Seoul National University, Shinrim-dong, Gwanak-ku,Seoul, Korea

Abstract: The purpose of the present study is to assess accurately the stresses and the risk of transverse cracking in the early- age concrete decks of composite bridges. An analytical method that can reasonably predict the stresses in early-age concrete decks was investigated first. Several series of test members that can exhibit the early-age behavior of composite bridge decks were fabricated to measure the actual total strains, stress-independent strains, and stresses in the early-age concrete decks. The results obtained from the numerical models show good agreement with measured data. The present study indicates that several important parameters such as ambient temperature, solar radiation, and differential shrinkage affect greatly the total stresses arising in the concrete decks at early ages. It was shown that the simplified uniform shrinkage without considering moisture diffusion and drying at exposed surface underestimates greatly the tensile stresses at the surface region of concrete decks, which is unconservative in controlling the transverse cracking. The method of analysis in this study can be efficiently used to perform such functions, which are very important for control of cracking at early ages.

Key words: transverse cracking, early-age concrete deck, temperature, relative humidity, thermal stress, shrinkage stress, ambient temperature, solar radiation, hydration heat

1. INTRODUCTION

The risk of transverse cracking in the concrete decks of composite bridges is affected by many factors related to the bridge design, materials, and construction. Among others, the thermal and shrinkage stresses are the most important factors that affect the transverse cracking in early-age concrete decks.[1-3] Several studies[2-5] have been carried out to pre-

M.S. Konsta-Gdoutos, (ed.), Measuring, Monitoring and Modeling Concrete Properties, 259–272.

dict the thermal and shrinkage stresses in composite bridges. Most studies, however, considered the thermal or shrinkage stresses separately. Furthermore, the rapid change of material properties of early-age concrete at initial stages has not been considered properly. Most studies have inadequately considered the differential drying shrinkage, which is caused by moisture diffusion from drying at the exposed surface, by assuming that the cross-section of bridge deck shrinks uniformly. Therefore, to accurately assess the risk of transverse cracking in the early-age concrete decks, an analysis of temperature and relative humidity that affect directly the strains and stresses of early-age concrete must be correctly executed first.[1,6] The temperature and relative humidity which depend on hydration processes and environmental curing conditions will affect the strains and stresses arising in the concrete decks at early ages.

The purpose of the present study is to assess accurately the stresses and the risk of transverse cracking in the early age concrete decks of composite bridges. To this end, an analytical method that can reasonably predict the stresses in the early-age concrete decks, was investigated. Several series of test members that can describe the early-age behavior of concrete decks of composite bridges, were also made to measure the actual total strains, stress-independent strains, and stresses that develop in the early-age concrete decks with time after the concrete deck is cast in place. The results of strains and stresses obtained from the present analysis were compared with measured data of strains and stresses of concrete decks. The effects of various parameters on the thermal and shrinkage stresses in the early-age concrete decks of composite bridges have been quantitatively evaluated using the analytical model.

2. MODELLING OF MATERIAL PROPERTIES

2.1 Deformation of concrete

Time dependent deformation of concrete consists of the stress-dependent strain and the stress-independent strain.[7] The stress-dependent strain (ε_i^σ), which is produced by stresses in the concrete, is the sum of the elastic strain (ε_i^e) and the creep strain (ε_i^c). On the other hand, the stress-independent strain (ε_i^0), which is not related to stresses, is the sum of the thermal strain (ε_i^e) and the shrinkage strain (ε_i^{sh}) (see Fig. 1). Therefore, the total strain increment during time step $\Delta t_r (= t_r - t_{r-1})$ is as follows.

$$\begin{aligned} \Delta\varepsilon_i &= \Delta\varepsilon_i^e + \Delta\varepsilon_i^c + \Delta\varepsilon_i^T + \Delta\varepsilon_i^{sh} \\ &= \Delta\varepsilon_i^\sigma + \Delta\varepsilon_i^0 \end{aligned} \tag{1}$$

where $\Delta\varepsilon_i^t$ is the total strain increment; $\Delta\varepsilon_i^e$ is the elastic strain increment; $\Delta\varepsilon_i^c$ is the creep strain increment; ε_i^T is the thermal strain increment; $\Delta\varepsilon_i^{sh}$ is the shrinkage strain increment.

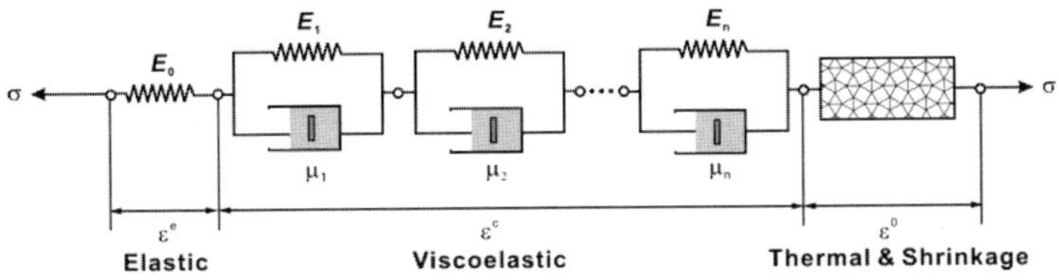

Figure 1. Rheological model describing deformation

2.2 MECHANICAL PROPERTIES OF YOUNG CONCRETE

The material properties of early-age concrete develop with time because of the hydration process and the interaction of material with the environment. To evaluate the risk of early-age cracking in hardening concrete elements, it is necessary to know how the stiffness develops with time. It is not reasonable to describe the material properties in hardening concrete as a function of time only because the development of material properties of each concrete element within a structure is different. The time is not generally an exact parameter to control the chemical and physical processes in concrete, but the degree of hydration is.[1,6,8,9] Therefore, the concept of degree of hydration is used here in the modeling of mechanical properties of concrete for stress analysis.

Replacing the time from adiabatic test with an equivalent time t_{eq}, the degree of hydration α in the adiabatic test may be expressed in terms of maturity-equivalent isothermal condition[8,10]

$$\alpha(t_{eq}) = \frac{Q_{ad}(t_{eq})}{Q^{\max}} \tag{2}$$

in which

$$t_{eq} = \int_0^t \exp\left[\frac{E_A}{R_u}\left(\frac{1}{293} - \frac{1}{T(s)+273}\right)\right] ds \tag{3}$$

where t is the time ; E_A / R_u is the activation temperature (K) ; $T(s)$ is the temperature history, and 20°C is chosen as the reference temperature. The activation temperature has a temperature dependency of the activation energy as follows

$$\frac{E_A}{R_u} = \theta_{ref}\left(\frac{30}{T(s)+10}\right)^{\kappa} \tag{4}$$

where $\theta_{ref} = 4600$ K and k=0.39 for Ordinary Portland Cement.[10]

Using Eq. (3), the degree of hydration of an element in a structure which has different temperature histories, may be evaluated from the corresponding equivalent age.

The mechanical properties of early-age concrete in terms of the degree of hydration may be expressed as follows [8]

$$\frac{f_c'(\alpha)}{f_c'(\alpha=1)} = \left(\frac{\alpha-\alpha_0}{1-\alpha_0}\right)^a \tag{5}$$

$$\frac{E_c(\alpha)}{E_c(\alpha=1)} = \left(\frac{\alpha-\alpha_0}{1-\alpha_0}\right)^b = \left(\frac{\alpha-\alpha_0}{1-\alpha_0}\right)^{a/2} \tag{6}$$

$$\frac{f_t'(\alpha)}{f_t'(\alpha=1)} = \left(\frac{\alpha-\alpha_0}{1-\alpha_0}\right)^c = \left(\frac{\alpha-\alpha_0}{1-\alpha_0}\right)^{\frac{2}{3}a} \tag{7}$$

where f_c' is the compressive strength; E_c is the modulus of elasticity; f_t' is the tensile strength; α_0 is the critical degree of hydration at which the development of stiffness begins[8]; *a, b, c* are material constants. The relations between the compressive strength and the modulus of elasticity and the tensile strength are from ACI 209R-92[11], CEB-FIP MC 90[12], respectively.

2.3 Creep of young concrete

The viscoelastic behavior of concrete can be described by the compliance function. The time-dependent strain $\varepsilon(t)$ arising from arbitrary stress histories $\sigma(t)$ can be obtained by the principle of superposition as follows[9]

$$\varepsilon(t) = \int_0^t J(t,t')d\sigma(t') + \varepsilon^0(t) \tag{8}$$

where $J(t,t')$ is the compliance function which represents the strain at time t caused by the stress imposed at time t'; $d\sigma(t')$ is the stress increment applied at time t!

If integral-type creep law (Eq. 8)) can be converted to rate-type creep law, storing and using the complete history of stresses or strains may not be needed. Such a conversion is possible if the compliance function is expressed as a sum of products of functions of single variables of t and t', i.e., Dirichlet series.[7,13]

$$J(t,t') = \sum_{\mu=0}^{N-1} \frac{1}{C_\mu(t')}\left(1-e^{-\frac{t-t'}{\tau_\mu}}\right) \tag{9}$$

where τ_μ is the retardation time; N is the number of terms in the expansion; $C_\mu(t')$ is determined from the expansion.

$C_\mu(t')$ indicates that the model can be time-dependent, for instance, due to temperature or maturity influence. The expansion can be interpreted by a rheologic model with time-dependent properties, as shown in Fig. 1. Each element in the Kelvin-Chain model represents a term of the expansion. The moduli of elasticity, E_μ, and the viscosities, η_μ, are obtained from the expansion. The relation to the coefficient in the expansion is[9]

$$E_{\mu}(t) = C_{\mu}(t) - \tau_{\mu}\dot{C}_{\mu}(t) \tag{10}$$

and

$$\eta_{\mu}(t) = \tau_{\mu}\dot{C}_{\mu}(t) \tag{11}$$

Several researches have been carried out recently to find the characteristics of the compliance function of early-age concrete.[13]

2.4 Thermal strain

It is necessary to know the coefficient of thermal expansion of concrete to estimate the thermal strains and stresses. Depending on aggregates, constituents and the moisture state of concrete, the coefficient of thermal expansion, α_T, varies within 5-15×10^{-6}/°C.[8] It has been reported that the coefficient of thermal expansion is slightly greater during heating than cooling. Additionally the thermal expansion coefficient of early-age concrete is higher than that of mature concrete.[8] In this study, however, the coefficient is reasonably assumed to be constant, and the thermal strain (ε_i^T) of concrete can be calculated as follows.

$$\Delta\varepsilon_i^T = \alpha_T \Delta T_i \tag{12}$$

where ΔT_i is the temperature increment during time step.

2.5 Shrinkage strain

The shrinkage takes place when the evaporation rate becomes greater than the rate at which bleed water rises to the surface.[15,16] Because concrete has a negligible stiffness before final setting, plastic shrinkage does not induce significant stress and thus has not been considered in this study. The term shrinkage hereafter means the deformation in concrete that has stiffness and thus can carry stress in the matrix.

Autogenous shrinkage is the macroscopic volume reduction of cementitious materials that is not caused by evaporation or temperature change, but by self-desiccation due to chemical reactions.[15] Until recently, many researchers have assumed autogenous volume change of ordinary concrete to be small, which may be ignored in the estimation and control of cracking. Recent researches, however, has demonstrated that the autogenous shrinkage of high-strength concrete with low water/cement ratio can be quite large, and it should be considered in the control of cracking.[6,15]

The relative humidity distribution in a cross-section of early-age concrete is not uniform because of self-desiccation and moisture diffusion due to drying at the exposed surface. Therefore, differential shrinkage is caused by such non-uniform relative humidity distribution.[7,17] A relationship between the volume change of an infinitesimal element and the locally attained relative humidity change is required. In this study, a linear relationship between the change in infinitesimal shrinkage, $\Delta\varepsilon_i^{sh}$, and the relative humidity change, Δh_i , has been assumed as follows[18]

$$\Delta\varepsilon_i^{sh} = \alpha_{sh}\,\Delta h_i \tag{13}$$

where α_{sh} is the coefficient of shrinkage which defines the amount of shrinkage which a material undergoes if it is allowed to dry completely and if it is not constrained in its movement by adjacent elements.

3. TESTS FOR EARLY-AGE BEHAVIOR OF COMPOSITE BRIDGE DECKS

3.1 Test outline

Several series of test members were made to investigate the early-age behavior of composite bridges (See Fig. 2). In order to investigate the actual behavior of composite bridges during construction, a concrete slab was placed on the steel girder and cured in the field. The composite bridges are subject to temperature and relative humidity variations according to the hydration processes and environmental conditions.[2,3,6] The nonlinear distribution of the temperature and relative humidity will generate self-equilibrating stresses in the cross-section.[2,4] To investigate the effect of restraint according to the girder size, different sizes of steel girders were used in test members M1 and M2. Additionally, the steel beam was not used in test member M0. Fig. 2 shows the details of test member M1, and Table 1 summarizes the test variables with test member identification. The average compressive strength of concrete from three cylinders is 26.4 MPa at 28 days.

Table 1. Test variables and test member identification

unit : [mm]

Test member identification	Steel beam				Concrete slab		Span length
	Depth	Web thickness	Flange Width	Flange Thickness	Width	Thickness	
M1	488	12	300	18	1,000	250	8,000
M2	594	14	300	12			
M0*	-	-	-	-			

Note: There is no steel girder in test member M0

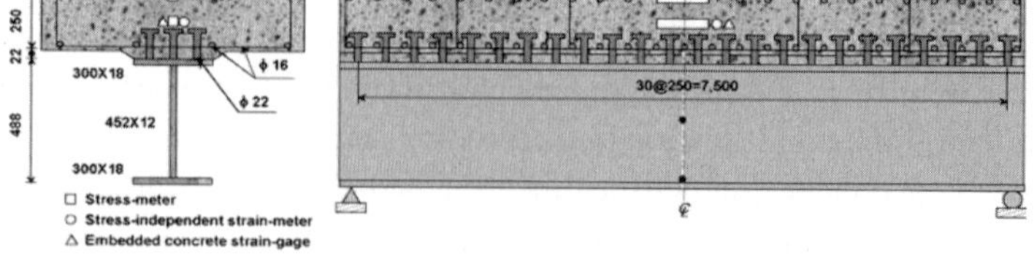

Figure 2. Details of test members (M1) and measurement scheme

3.2 Test method and measurement

Many researchers have reported that strains and stresses in early-age concrete decks are mainly caused by the vertical gradients of temperature and relative humidity along the cross-section of composite bridges. Therefore, embedded-type concrete strain gages, stress-independent strain-meters, and stress-meters were embedded in the central portion of concrete deck in test members to measure the total strains, stress-independent strains, and stresses, respectively. To evaluate the risk of cracking in the surface of deck at early ages, a stress-meter was embedded additionally in the surface portion of the concrete deck in test member M1. The measurement of strains and stresses were started in the test members right after the slab concrete was placed.

The embedded-type concrete strain gage measures the total strains in concrete that consist of stress-dependent strains and stress-independent strains. The stress-independent strain-meters can measure the stress-independent part of the concrete strains by installing the strain gages inside the prefabricated cylinder filled with concrete. The concrete inside the cylinder is separated from surrounding concrete by the cylinder. Using Eq. (1), one can calculate the stress-dependent strain (ε_i^σ) from the measured total strain (ε_i) and stress-independent strain (ε_i^0).

The stress-meters measure directly the stresses arising in concrete by using the load cell in the meters. The stress-meter consists of a load cell in series with the concrete cast inside the stress-meter (hereinafter referred to as a concrete prism), and the surrounding concrete. The concrete prism, load cell, and the surrounding concrete are monolithic through the anchor. The test method using the stress-meter ensures that the properties of concrete prism are the same as the surrounding concrete by allowing free exchange of water. The stress in the concrete prism is obtained by dividing the force value measured with the load cell by the cross sectional area of the prism.[1,18]

4. RESULTS OF TESTS AND ANALYSIS

4.1 Material properties

Table 2 shows the materials properties used in the analysis of thermal and shrinkage stresses of test members. The value of material constant α in Eq. (6), which represents the development of compressive strength, is 1.25, obtained from the compressive strength test conducted with respect to age of concrete. The value of critical degree of hydration α_0 is 0.15, obtained from the measured degree of hydration at which the stress-meter in the test members begins to develop stresses (See Eq. (5), (6), and (7)). To consider the effect of creep on the stress variation in test members, the compliance function in CEB-FIP MC90[12] and five Kelvin-elements are used in a rheological model (See Fig. 1).

Table 2. Material properties used in the analysis of thermal and shrinkage stresses

Material properties	Concrete	Steel
Coefficient of thermal expansion (10^{-6}/°C)	11.0	12.0
Coefficient of shrinkage (10^{-6}/unit)	1,300	-
Modulus of elasticity (MPa)	2.97Þ10^{4} *	2.05Þ10^{5}

Note: Modulus of elasticity at degree of hydration α=1

4.2 Total, stress-independent, and stress-dependent strains

Figures. 3, 4, and 5 show the variation of measured total strains, stress-independent strains, stress-dependent strains using Eq. (1) and calculated stress-dependent strains in the analysis at the central portions of test members M1, M2, and M0, after placement of concrete deck. In these figures, total strains and stress-independent strains were measured from embedded concrete strain gages and stress-independent strain meters, respectively. The stress-dependent strains were calculated using Eq. (1). As shown in Figs. 3, 4, and 5, the variations of stress-independent strains are very similar with that of temperatures. These similarities mean that thermal strain is a major part of stress-independent strains at the central portions of early-age concrete deck. Because the moisture diffusion coefficient is small and the period of drying is short, the relative humidity changes caused by drying at the exposed surface may hardly occur at the central portion of concrete deck. The stress-dependent strain increments are compressive during the period of temperature rise due to the hydration heat, and then changes to tensile or compressive depending on the environmental conditions. Figs. 3, 4, and 5 also indicate that the increments of stress-dependent strain in test member M0 were slightly smaller than those of M1, and M2. This result is due to the difference of the restraining effect by the steel girder. Figs. 4, 5, and 6 also show that the calculated stress-dependent strains agree well with measured strains.

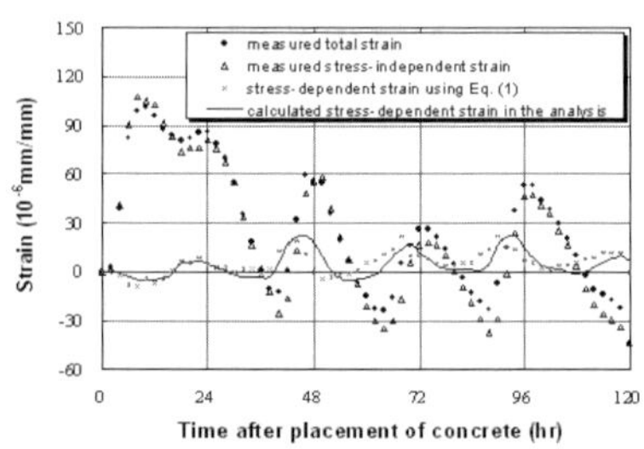

Figure 3. Measured total and stress-independent strains and calculated stress-dependent strain at the central portion of concrete deck (M1)

Figure. 6 depicts the calculated strains at the surface portions of the concrete deck in test member M1. In this figure, the stress-independent strains are affected by the thermal strains as well as drying shrinkage strains at the exposed surface. It is also seen that the stress-dependent strains are tensile at the surface during the initial temperature rise period, which may cause tensile cracking at very early ages.

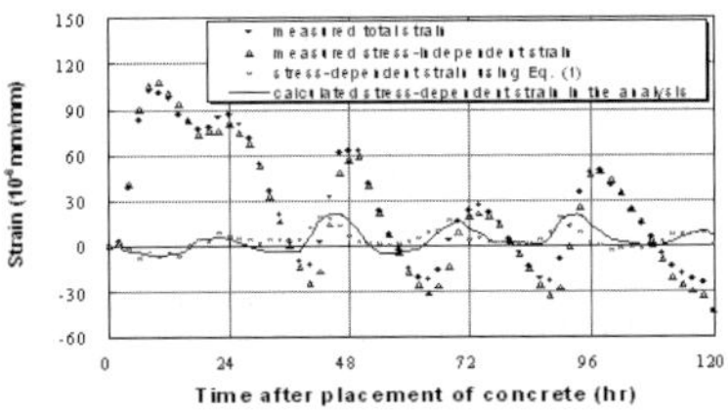

Figure 4. Measured total and stress-independent strains and calculated stress-dependent strain

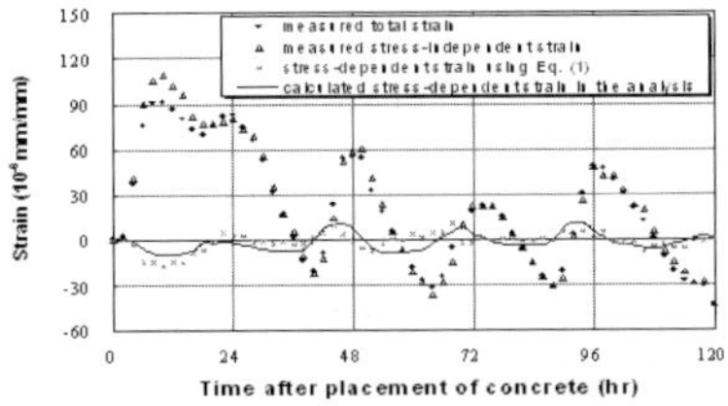

Figure 5. Measured total and stress-independent strains and calculated stress-dependent strain at the central portion of concrete deck (M0)

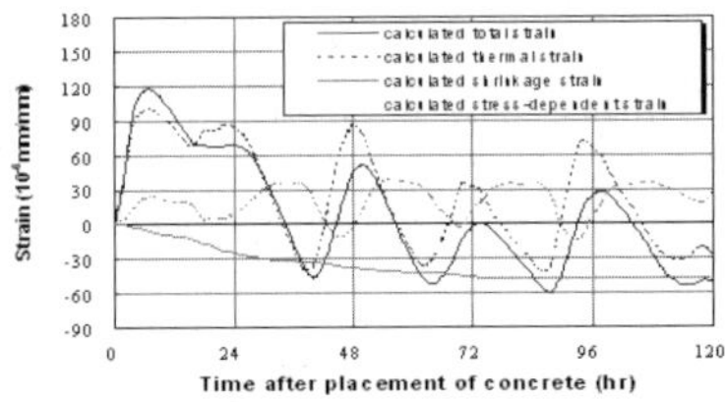

Figure 6. Calculated total and stress-independent and stress-dependent strainsat the surface portion of concrete deck (M1)

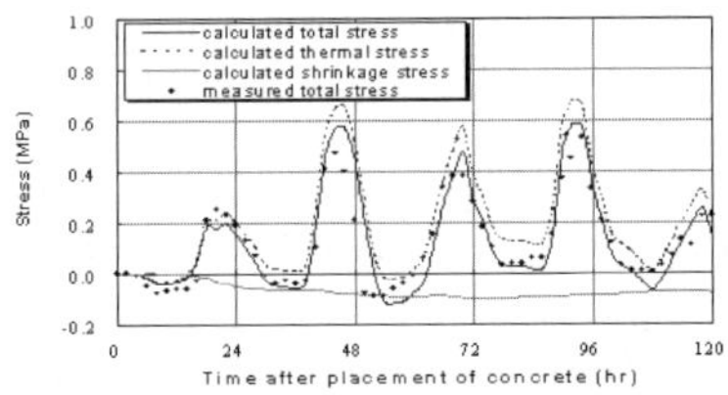

Figure 7. Measured and calculated stresses at the central portion of concrete

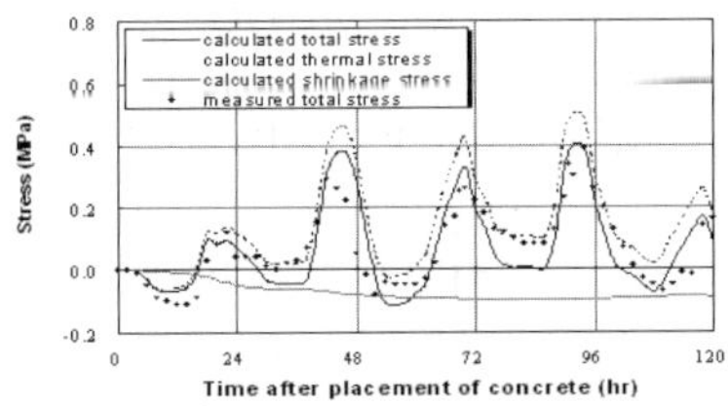

Figure 8. Measured and calculated stresses at the surface portion of concrete deck (M1)

4.3 Stresses

Figure. 7 shows the variations of measured and calculated stresses at the central portion of concrete deck in the test member M1. As shown in Fig. 7, the thermal stress is a major part of the total stress at the central portion of concrete deck, and the shrinkage stress causes small compressive stresses in the central point. The stresses caused by the hydration heat exhibit initially the compressive stresses during early stages. After these stages, the stresses in the concrete decks fluctuate depending on the environmental temperature histories.

Figure. 8 shows the variations of the measured and calculated stresses at the surface portion of concrete deck in the test member M1. Fig. 8 indicates that not only the temperature, but also the shrinkage governs the stress development at the surface of bridge decks.

The shrinkage strains cause larger tensile stresses at the surface as can be seen in Fig. 14, which may increase the probability of cracking at the surface. It is interesting to note here from Fig. 7 and Fig. 8 that the stress patterns of the surface portion (Fig. 7) are almost the opposite to the stress patterns of the central portion of the deck (Fig. 7).

5. ANALYSES OF VARIOUS FACTORS AFFECTING CRACKING

5.1 Effect of hydration heat, ambient temperature, and solar radiation

Figures. 9 and 10 show the effect of hydration heat and environmental conditions on the thermal stresses at the central and surface portions, respectively, of the concrete deck in the test member M1. As shown in Figs. 9 and 10, the thermal stresses at the central portion are affected by both ambient temperature and solar radiation. It can be seen that the solar radiation greatly affects the thermal stresses especially at the surface portion of concrete decks (See Fig. 10).

To investigate the factors affecting the thermal stresses of early-age concrete decks, a parametric study was carried out with the test member M1 (See Fig. 2). To consider the effect of hydration heat, an increase of 5°C in maximum adiabatic temperature rise was compared with that obtained in the present study. The measured ambient temperatures at first and second day, respectively, were repeatedly used to investigate the effects of the ambient temperature. Furthermore, the measured solar radiations at first and second day, respectively, were also used to analyze the effects of the solar radiation. The geographic location, orientation and material properties except the hydration heat remained unchanged. Fig. 11 shows the effects of various parameters in the parametric study on the thermal stresses at central and surface portions, respectively, of the early-age concrete deck in the test member M1. Fig. 11 indicates that the increase of hydration heat causes larger tensile stresses at the central portion after several days. On the other hand, the solar radiation is a major factor that increases the tensile stresses at the surface portion of the deck as shown in Fig. 11.

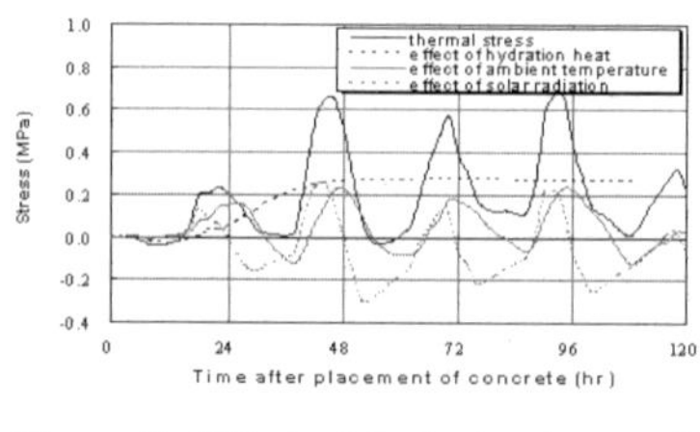

Figure 9. Effect of hydration heat and environmental factors on the thermal stressat the central portion of concrete deck

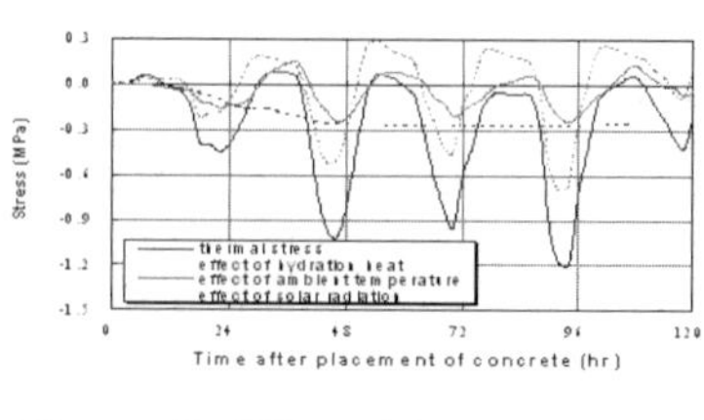

Figure 10. Effect of hydration heat and environmental factors on the thermal stress at the surface portion of concrete deck

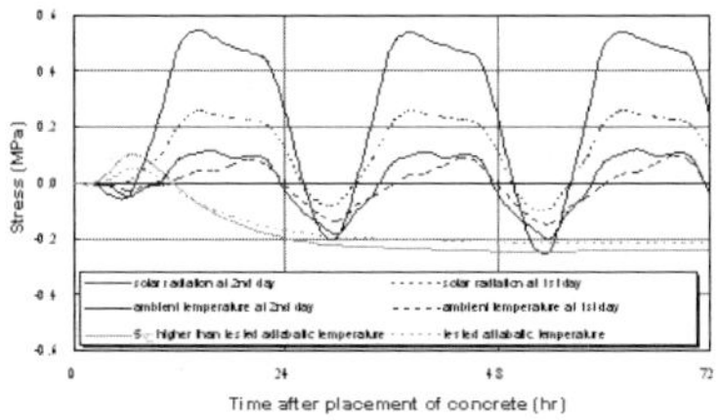

Figure 11. Effects of hydration heat and various parameters on the thermal stresses at the surface portion of concrete deck in the test member M1

5.2 Effects of differential and uniform shrinkage

Figure. 12 shows the comparison of calculated total stresses with measured ones at the surface portion of the concrete deck in the test member M1. Fig. 12 indicates that the stresses assuming the uniform shrinkage strain in the cross-section of the deck underestimate the actual tensile stresses occurring at the surface portion of concrete decks, which is unconservative to control the early-age cracking. Therefore, the differential drying shrinkage strain caused by moisture diffusion and drying at the exposed surface must be considered to obtain realistic results. Fig. 13 shows the effect of autogenous and drying shrinkages on the total shrinkage stresses of concrete deck in the test member M1. As shown in Fig. 13, the autogenous shrinkage does not have significant effect on the total shrinkage stresses at the surface portion of concrete decks. However, the drying differential shrinkage causes large tensile stresses at the surface portion of a concrete deck.

Figure 14 shows the effect of ambient relative humidity on the shrinkage stress at the surface portion of concrete deck in the test member M1. The ambient relative humidity was assumed to be 90 %, 75 %, and 60 %, respectively. Fig. 14 indicates that the shrinkage stress at the surface portion of the early-age concrete decks increases significantly as the relative humidity decreases. This will also increase the possibility of cracking in concrete deck slabs.

5.3 Effect of creep and steel girder size

Figure. 15 represents the effect of creep on the calculated total, thermal and shrinkage stresses at the surface portion of a concrete deck. The comparison of calculated and measured stresses agrees well each other. Fig. 19 indicates that creep reduces slightly the stresses in the concrete decks. It can be seen that the tensile stresses of concrete deck with larger size of girder (M2) are larger than those in smaller girder (M1), which comes from the restraining effect depending upon the girder size.

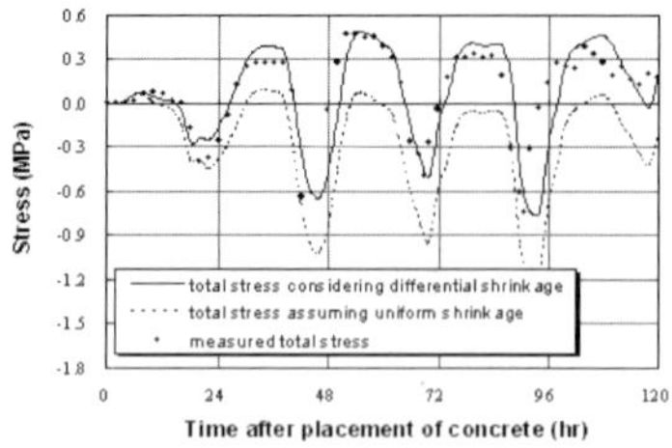

Figure 12. Comparison of calculated total stresses with measured ones under differential and uniform shrinkages

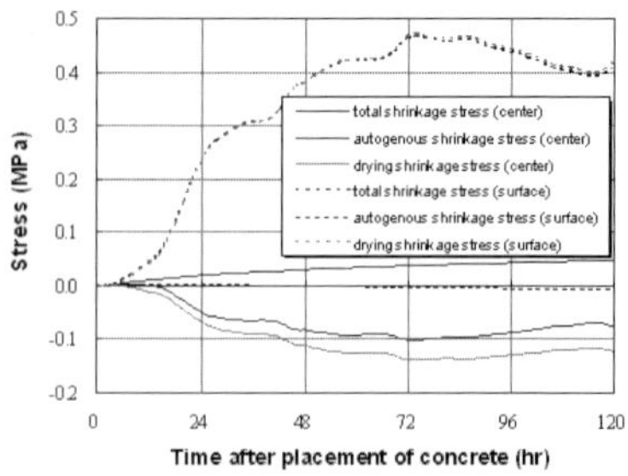

Figure 13. Effects of autogenous and drying shrinkages on the shrinkage stresses of concrete deck in the test member M1

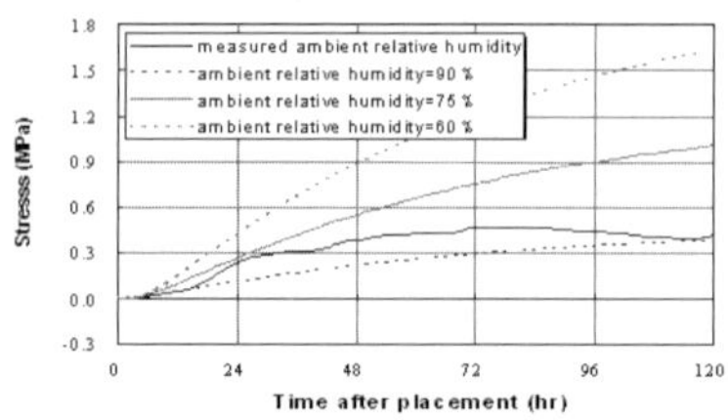

Figure 14. Effect of ambient relative humidity on the shrinkage stress at the surface portion of concrete deck in the test member M1

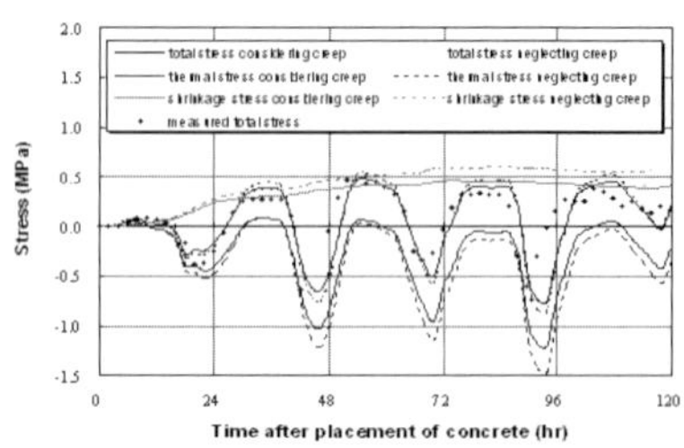

Figure 15. Effect of creep on the stress at the surface portion of concrete deck

6. CONCLUSIONS

The purpose of the present study is to assess accurately the stresses and the risk of transverse cracking in the concrete decks of composite bridges at early ages. The following conclusions were drawn from the present study.

1. The calculated strains and stresses based on the investigated method agree very well with measured strains and stresses of concrete deck in test members. It has been demonstrated that the evolution of material properties must be realistically considered to obtain more reliable and accurate results for strains and stresses arising in the concrete bridge decks at early ages.

2. The present study indicates that the thermal stress is a major part of the total stress at the central portion of a concrete deck and is affected by both ambient temperature and solar radiation. However, the thermal stress at the surface portion is greatly influenced by solar radiation.

3. It was shown that the differential shrinkage causes large tensile stresses at the surface portion of concrete deck. The uniform shrinkage without considering the moisture diffusion and drying at exposed surface, underestimates the tensile stress at the surface portion of a concrete deck. Therefore, the simplification by uniform shrinkage gives

unconservative results in controlling the tensile cracking in the concrete decks at early ages.

4. The present study also indicates that the creep of concrete reduces slightly the stresses arising in the early-age concrete decks.

5. It was shown that larger size of steel girder induces larger tensile stresses at the surface portion of concrete deck, due to the increase of restraining effect of larger steel girder.

6. The analytical method in this study can be efficiently used to predict the strains and stresses more accurately arising in the concrete decks especially at early ages. This will allow efficient control of cracking, otherwise frequently observed in the early-age concrete decks.

Acknowledgment

The financial support from the National Research Laboratory (NRL) Program of Korea is gratefully acknowledged.

7. REFERENCES

1. R. Springenschmid, (ed.), *Prevention of Thermal Cracking in Concrete at Early Ages*, State of the Art Report by RILEM TC 119, E&FN Spon, 348 pp. (1998)
2. H. Dilger, A. Ghali, M. Chan, M. S. Cheung., and M. A. Maes, Temperature Stresses in Composite Box Girder Bridges, *Journal of Structural Engineering*, ASCE, V. 109, No. 6, pp. 1460-1478 (1983)
3. Elbadry, and A. Ghali, Temperature Variations in Concrete Bridges, *Journal of Structural Engineering*, ASCE, V. 109, No. 6, pp. 2355-2374 (1983)
4. Ghali, and R. Favre, *Concrete Structures Stresses and Deformation*, 2nd Edition, E&FN Spon, 444 pp. (1994)
5. D. Cusson, and W. L. Repette, Early-age Cracking in Reconstructed Bridge Barrier Wall, *ACI Materials Journal*, V. 97, No. 4, pp. 438-446 (2000)
6. B. H. Oh, and S. W. Cha, Nonlinear Analysis of Temperature and Moisture Distributions in Early-age Concrete Structures Based on Degree of Hydration," *ACI Materials Journal*, V. 100, No. 5, , pp. 361-370 (2003)
7. Z. P. Bažant, *Mathematical Modeling of Creep and Shrinkage of Concrete*, John Wiley and Sons, 459 pp. (1988)
8. J.Byfors, *Plain Concrete at Early Ages*, CBI Report FO 3:8, Sweden, 345 pp. (1980)
9. G. de Schutter, and L. Taerwe, Degree of Hydration-based Description of Mechanical Properties of Early age Concrete," *Materials and Structures*, V. 29, pp. 335-344 (1996)
10. F.S. Rostasy, and M. Laube, Experimental and Analytical Planning Tools to Minimize Thermal Cracking of Young Concrete," *International RILEM Symposium on Test during Concrete Construction*, Reinhardt, H. W. (eds.), Chapman and Hall, pp. 207-223 (1991)
11. ACI 209R-92, "Prediction of Creep, Shrinkage and Temperature Effect in Concrete Structures", *ACI Manual of Concrete Practice, Part 1*, pp. 47 (1997)
12. CEB-FIP Model Code, *CEB Bulletin* d'information No. 213/214, 437 pp. (1993)
13. Z. P. Bažant, and S. T. Wu, Dirichlet Series Creep Function for Aging Concrete," *Journal of Engineering Mechanics*, ASCE, V. 99, No. 2, pp. 367-387 (1973)
14. G. de Schutter, Degree of Hydration Based Kelvin Model for the Basic Creep of Early age Concrete," *Materials and Structures*, V. 32, pp. 260-265 (1999)

15. JCI, Technical Committee on Autogenous Shrinkage of Concrete, Part 1, Committee Report, *Autogenous Shrinkage of Concrete*, Proceeding of the International Workshop organized by JCI, Tazawa, E., (Ed.), E&FN Spon, pp. 3-67 (1998)
16. O. M. Jensen, and P. F. Hansen, Autogenous Deformation and RH-Change in Perspective," *Cement and Concrete Research*, V. 31, pp. 1859-1865 (2001)
17. A. M. Alvaredo, *Drying Shrinkage and Crack Formation*, Building Materials Reports, No. 5, Lab. For Building Materials, Swiss Federal Institute of Technology, Zurich (Switzerland), 102 pp. (1994)
18. T. Kawaguchi, and S. Nakane, Investigation on Determining Thermal Stress in Massive Concrete Structures," *ACI Materials Journal*, V. 93, N. 1, pp. 96-101 (1996)

CRACK HEALING OF EARLY AGE CRACKS IN CONCRETE

E. Schlangen,[1,3] N. ter Heide[2] and K. van Breugel[3]
[1] *Delft University of Technology, CiTG, Microlab, P.O. Box 5048, 2600 GA Delft, The Netherlands;* [2] *BAM-DMC, P.O. Box 268, 2800 AG Gouda, The Netherlands;* [3] *Intron-Femmasse, P.O. Box 267, 4100 AG Culemborg, The Netherlands*

Abstract: A combined experimental and numerical study is performed on crack healing in hydrating concrete. The aim of the research is to investigate under which conditions cracks that are formed in concrete at very early age can heal again when the cement hydrates further. To study this phenomenon, three-point-bending tests are performed on prismatic concrete specimens at early age to create cracks with a specified crack opening. After the test the specimens are matured further under water with and without compressive loading to close the cracks. Several weeks after the first test the specimens are tested again to investigate the amount of healing and further development of mechanical properties. From the tests it can be concluded that the mechanical properties of the concrete are recovered to a large extent for cracks made after 1 day of hydration, loaded in compression afterwards and healed under water. This is the case for concrete with both BFSC and OPC. If the degree of hydration of the specimens is higher at the moment of cracking, the strength gain is less. A numerical study is performed to investigate the mechanism of self-healing. The results of the simulations indicate that ongoing hydration is responsible for the crack healing observed in the tests.

1. INTRODUCTION

Self-healing is a phenomenon that can "repair" cracked concrete. Although reinforced concrete is designed to crack, in some circumstances cracking should be avoided. Major reasons for requiring crack-free structures are liquid tightness and enhanced durability of concrete structures. Self-healing can help to realize structures without cracks. Within self-healing we actually can distinguish two mechanisms:

- Self tightening: the crack is blocked with small particles from the crack faces or small parts present in fluids that flow through the crack.

M.S. Konsta-Gdoutos, (ed.), Measuring, Monitoring and Modeling Concrete Properties, 273–284.

- Self-healing: a chemical reaction takes place connecting the two crack surfaces. The reaction can be continuing hydration of the cement or a chemical reaction that occurs after hydration (like the formation of calcium carbonate).

An overview of research on self-healing mechanisms is given in [1]. All of the studies described in [1] focus actually on self tightening of old cracks and the formation of calcium carbonate in these cracks. Continuation of the hydration process, because of unhydrated cement particles present, is believed not to happen. Reason for this is that the distance between the two crack faces is generally too large to be bridged by hydration products. Recently a study of temperature influence on crack healing was reported in [2]. From this research it was found that with lower crack widths and at higher temperatures, crack healing is more likely to take place. In self-healing research attention is always focused on blocking of the cracks and increasing the permeability of the material. The recovering of mechanical properties of the material is almost never investigated. An exception on this is given in [3], where an experimental study is described of self-healing of a cementitious composite containing hollow glass fibres carrying air-curing chemicals. The self-healing effectiveness is confirmed by measurement of the elastic modulus of the composite. The elastic modulus is found to regain its original value in a repeat loading subsequent to damage in a first load cycle. An other way to actively create a crack healing environment in a crack is by using micro-organisms [4,5]. These organisms have to be injected into an already formed crack [4] or mixed in the concrete [5] and are able to induce calcite precipitation in the cracks, and in such a way realize blocking of the crack path and increase stiffness and strength.

In massive concrete structures often surface cracks develop already during the first day of hydration [6,7]. With continuation of the hydration process and cooling down of the interior of the structure the surface is loaded in compression. The cracks are closed due to this compressive stress. The question is whether the cracks can heal in this situation. Hydration is still in process and the crack faces are touching each other again. Probably an environment is created in which the cracks can heal. To study this an experimental investigation is started which is described in this paper. The next paragraph describes the tests performed and the variables studied. This will be followed by results of the experiments. A numerical study is performed to get more insight into the mechanism of healing observed in the tests. The last paragraph gives a discussion and concluding remarks of the research.

2. DESCRIPTION OF TEST SET-UP

In order to study the effect of crack healing in early age concrete, first cracks have to be made in the concrete in a controlled way. For this a three-point-bending test is chosen on prismatic concrete specimens. The specimens are cast in steel moulds and have dimensions of 40x40x160mm. The distance between the loading points is 105 mm (see figure 1a). The reaction force is in the centre. The deformation is measured with two LVDT's (at front and back) fixed at the bottom of the specimen (in the centre). The measuring length is 55 mm. The deformation measured with these LVDT's give a bending strain at the bottom of the specimen. If the crack localizes this value is a measure (although not exact) for the crack opening. The specimen is loaded in a special 3-point-bending frame (see figure 1b) in an Instron 8872 servo-hydraulic loading device. In the 3-point-bending frame the load is applied to the concrete via pendulum bars to minimize friction at the loading points.

Figure 1. Specimen dimension (a), 3-point bending loading device (b) in Instron 8872 (c) and application of compression (d)

The test are performed in deformation control using the average signal of the two LVDT's as feed back signal. The speed of loading was set to 0.02 μm/s at the start of the test. In the descending branch of the measured softening curve (at a crack mouth opening of 20 μm) the speed was increased by a factor of two. In the test the loading is stopped at a crack mouth opening of 50 μm. The specimen is unloaded and taken out of the machine. The crack opening decreases after unloading but the crack will not completely close. At this moment the specimen is put in a compression loading device to apply a compressive force to close the crack. This force is measured by means of the deformation of calibrated steel spring. The amount of compressive force is varied in the tests. In figure 1d the compressive loading devices are shown with specimens subjected to 0.5, 1 and 2 N/mm2 compressive stress. The specimens are then stored for a specific period at a certain temperature and relative humidity (or under water) to undergo crack healing. Due to creep of the concrete, the applied load will decrease in time. This is checked during the healing process by measuring the deformation of the spring. It turned out that the decrease in load was smaller than 1 %, so no adjustments were needed. Next to healing of the crack the concrete will also have further hydration. This means that the mechanical properties of the material itself will also improve.

To test the mechanical properties of the healed crack the specimens are, after a certain period of healing, again tested in three-point-bending. To be able to judge the recovering of mechanical properties of the crack, the results have to be compared with cracks that are not healed and with cracks that are made after the healing process. Therefore virgin specimen stored at the same environment and with the same age as the healed specimens were tested. These specimens are first loaded in deformation control up to a displacement of 50 μm. Then the specimen is unloaded and again tested. The maximum load reached in the second stage of the test can be seen as a maximum flexural stress for a specimen with an unhealed crack at an age of 2 weeks. Question is however whether the crack has the same length in a specimen tested at an age of 1 day and at 2 weeks when in both cases the crack opening has reached a value of 50 μm. To be able to answer this, the specimens have been vacuum impregnated with a fluorescence epoxy after the test. The cracks are then visualised under UV-light. The scatter in the crack length that is observed is rather large, but it turned out that in all the situations at a crack width of 50 μm, the crack tip was at about half height of the specimen.

3. DESCRIPTION OF TESTED VARIABLES

In the experiments several parameters are varied. The parameters that are discussed in this paper are:

- the amount of compressive force applied during healing. The variation of this parameter is 0.0, 0.5, 1.0 and 2.0 N/mm2.
- the type of cement on the concrete mix, both a BFSC and an OPC is tested.
- the moment of creating the first crack in the specimen. The cracks are made at an age of 20, 27.5 (further named 1 day), 48 and 72 hours. The age at loading has some variation, since the specimens are cast at the same time, but for testing only one machine is available. Each test, including preparation, takes about 45 minutes.
- the crack (mouth) opening of the crack. Initial crack openings of 20, 50, 100 and 150 μm are discussed in this paper.
- the influence of the Relative Humidity (RH) during healing. Specimens are stored under water and in a climate chamber with 95% and 60 % RH respectively.

Always one parameter is varied. The default parameters in the tests are 1.0 N/mm2 compressive stress, concrete made with BFSC, crack made at age of 1 day, crack opening of 50 μm and healing under water. All the tests are performed at least three times. In table 1 the mix composition of the concrete is given. In the where the concrete mix is changed the CEM III (BFSC) is replaced by the same amount of CEM I 52,5 R (OPC) .

Table 1. Mix composition of the concrete used in the tests

Cement	CEM III/b 42,5 LH HS	375 kg/m^3
Water		187.5 kg/m^3
River gravel	8 - 4 mm	540 kg/m^3
	2 - 4 mm	363 kg/m^3
	1 - 2 mm	272 kg/m^3
	0.5 - 1 mm	272 kg/m^3
	0.25 - 0.5 mm	234 kg/m^3
	0.125 - 0.25 mm	127 kg/m^3

4. EXPERIMENTAL FINDINGS

In this paragraph the results of the experimental findings are described briefly. More detailed results can be found in [8,9].

4.1 Influence of compressifve stress

The first parameter that is investigated is the compressive stress on the specimen during the crack healing. The specimens are tested at an age of 1 day up to a crack opening of 50 μm. Then a compressive force is applied to the specimens and they are stored for 2 weeks under water and tested again. In figure 2 the flexural stress is plotted versus displacement for the reloading test after 2 weeks healing for a test with (1 N/mm2) and without (0 N/mm2) compressive stress. Furthermore the graph is shown of the specimen without healing. The latter is obtained after reloading a specimen tested at an age of 2 weeks. The graph shows that when the crack is not closed (the compressive stress is 0 N/mm2) the recovery of strength is minor. The same observations were done recently for similar tests on HPC [10]. However with a compressive stress of 1 N/mm2 both the stiffness and the strength of the specimen is recovered and shows values close to the reference specimen.

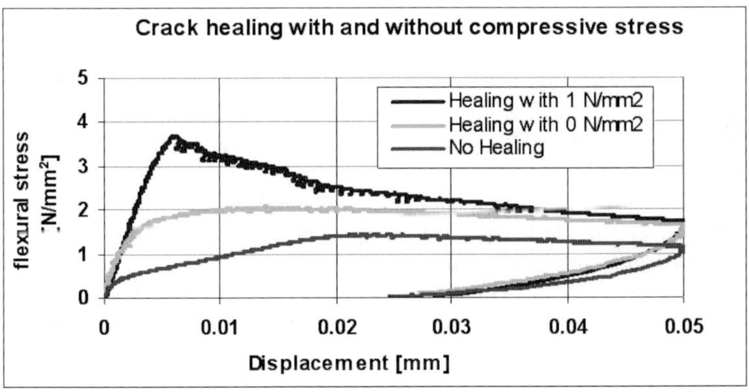

Figure 2. Stress-displacement curve of specimen with and without compressive stress during healing, compared with unhealed cracked specimen

Figure 3 gives the relative strength of the specimen after healing for different amount of compressive stress applied during healing. The relative strength is given in percentage of the strength of the un-cracked virgin specimen tested at an age of 2 weeks (see Figure 4, peak in first loading part). In the figure also a line is shown (with vertical bars showing the scatter) which represents the strength of the material of the unhealed specimen. The figure shows clearly that almost no increase in strength is obtained when the specimen is not loaded in compression (0 N/mm^2). Furthermore it can be seen that, in case a compressive loading is applied to close the crack, the amount of compressive stress is not really influencing the strength gain.

4.2 Influence of cement type

The influence of the cement type in the concrete mix is presented in figures 4 and 5. In Figure 4a the flexural stress is plotted versus the crack opening displacement for the test performed on 1 day old specimen. It can be seen that the specimens with the faster PC have a much higher strength. The concrete with PC is further matured.

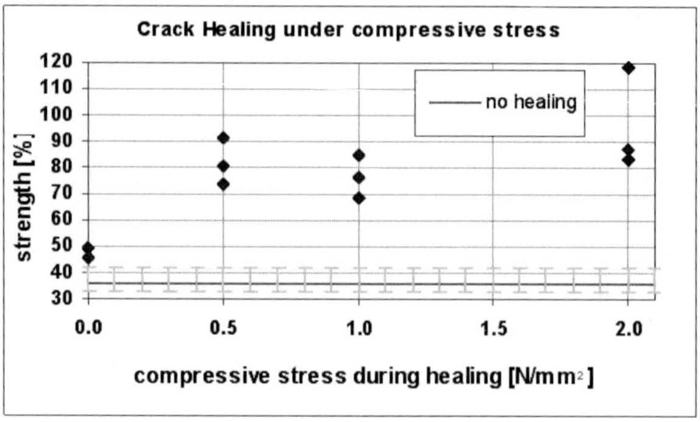

Figure 3. Resulting relative strength after crack healing for various compressive stress values

In Figure 4b the result is plotted for the tests after healing on both materials. Now the strength for both materials is almost equal. Also for the specimen tested for the first time after 15 days (Figure 5a) it can be seen that the strength of both materials is almost equal. Figure 5b gives the plots for the reloading test, which represents the strength of the unhealed specimens. From these tests it is clear that in the BFSC-concrete the strength after 1 day is just minimal. Healing of the crack can then take place easily because still a lot of hydration of the cement has to take place. However in the PC-concrete the result that is obtained is remarkable. Also here the strength is almost fully recovered. Probably this concrete has after hydration still a lot of unhydrated cement left. Thus the potential for crack healing is much higher in PC-concrete.

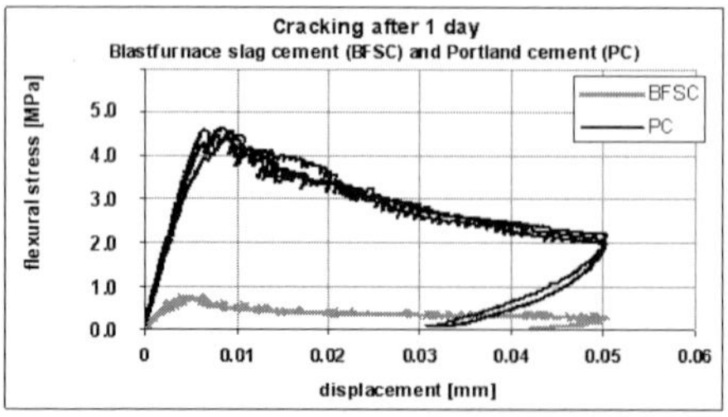

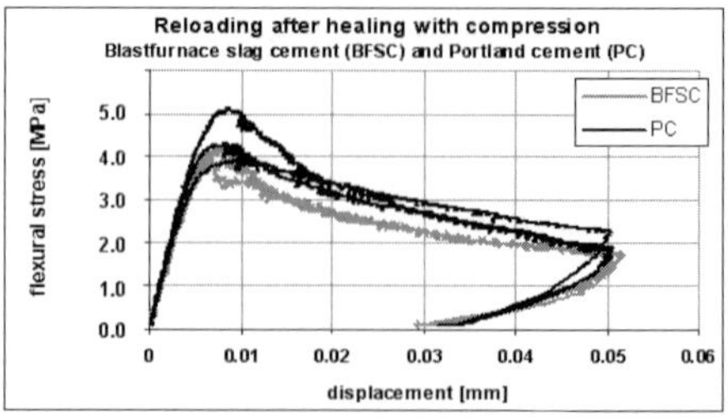

Figure 4. Flexural stress versus crack opening for tests after 1 day and after healing for BFSC and PC concrete

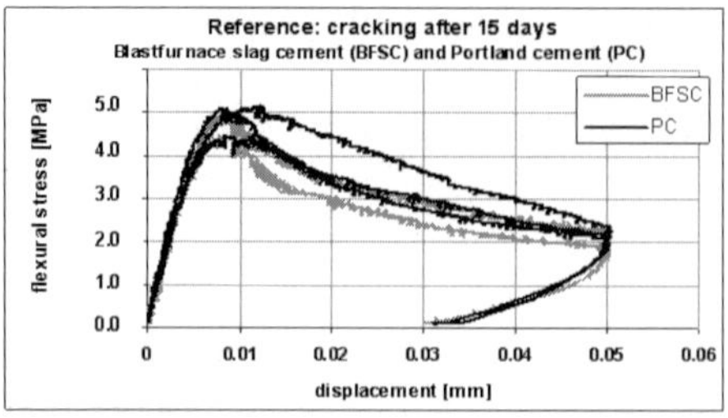

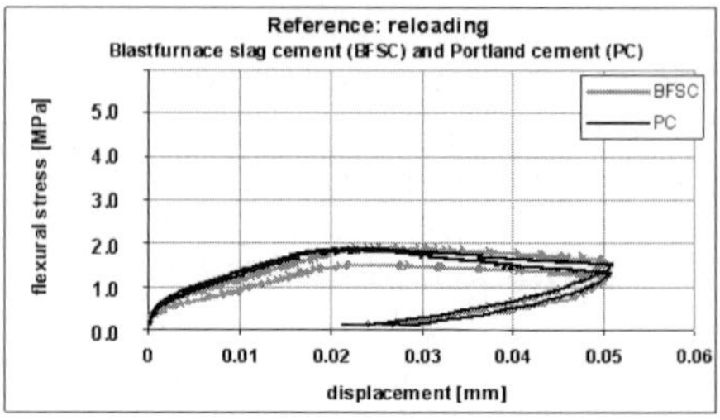

Figure 5. Flexural stress versus crack opening for first test and reloading after 15 days for BFSC and PC concrete

4.3 Influence of age when the first crack is produced

The third parameter that is investigated is the moment of cracking or the age of the specimen when the first crack is produced. In figure 8 the stress-displacement curves are shown for the specimens tested at different age. In these test the crack is opened up to a crack mouth opening of 50 m. Subsequently, the specimens are loaded in compression with a compressive stress of 1 N/mm2 and stored under water for 2 weeks. The strength after healing (relative to the strength of the virgin specimen) is plotted in figure 6 for the various ages of making the first crack. The reference test is for each series always performed at the same age. So this means that for instance the strength of the specimen tested for the first time at 72 hours and subsequently healed for 2 weeks is compared with the strength of a specimen loaded for the first time at an age of 17 days. Although the difference in strength of the virgin material between an age of 14 and 17 days is very small. A clear decrease in strength recovery is observed with increasing age of the specimen when making the first crack.

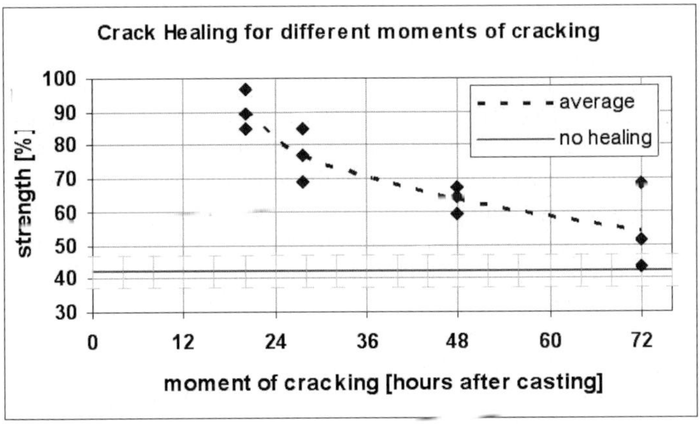

Figure 6. Strength recovery as function of age of specimen when the first crack is made

4.4 Influence of crack width on strength recovery

The fourth parameter discussed in this paper is the influence of the width of the crack that is made in the specimen on the healing mechanism. In these tests the specimens are loaded at an age of 1 day and the compressive stress during healing is equal to 1.0 N/mm2. A larger crack mouth opening will result in a longer crack which has propagated further into the specimen. The load that can be carried at a larger crack opening will be smaller. There is quite some scatter, but there seems to be no influence of crack opening on the strength recovery after healing.

4.5 Influence of relative humidity

The last parameter studied is the influence of the relative humidity on the crack healing. It turned out that only in the case the specimens were stored under water during the healing period recovery of strength is possible. In the case of storing the specimens in an environment of 95% or 60% RH almost no increase in strength was observed. In the case of 95% RH the specimens were even stored for a period of 3 months. Also in that case no crack healing was observed.

5. SIMULATION OF CRACK HEALING

For the simulation of crack healing the module MLS of FEMMASSE [11] is used. MLS is a finite element model based on the state parameter concept. This means that the material properties are a function of the state of the material. The state can be maturity, degree of hydration, temperature or moisture potential. The simulations with the model are performed in order to check the hypothesis that the healing observed in the experiments is due to ongoing hydration of the cement. The ageing of the concrete and the development of the properties in time is taken into account in the analysis. In the analysis a concrete with BFSC is used. The purpose of the analysis was not to exactly match the behaviour as found in the experiment, but more to be able to simulate the mechanism. The properties of the material used in the simulation correspond with an average concrete with BFSC, a w/c of 0.5 and a compressive strength of about 45 MPa. The development of the E-modulus and the tensile strength is shown in Figure 7b. In the analysis the 3-point bending test is simulated in 2D using plane stress conditions. In the middle, under the loading point, a discrete crack is created in the mesh using an interface element with a stress-crack opening relation. The strength of the interface element is taken equal to the strength of the neighbouring concrete at that specific age when the 3-point-bending test is performed. The values used for the different simulations are shown in Table 2. The softening relation is taken bi-linear in all the cases, having the bending point in the descending branch at 20 % of the strength and a crack opening of 0.03 mm and a final crack opening (at zero load) of 0.44 mm. In Figure 7b the bi-linear softening curve for a concrete of 1 and 15 days old is shown. Also a simulation is performed of a specimen which is cracked at age of 1 day and then healed for a period of 14 days. It is assumed that the bottom half of the specimen is cracked and that the strength gain of the material in the crack is equal to the strength gain of the concrete itself in these 14 days. This means that the strength of the interface is equal to strength at 15 days minus the strength at 1 day. The softening relation is then equal to the grey area in the graph in Figure 7b. In the experiments the cracked specimens are stored under water during healing. This means there is sufficient water available for the crack to heal. The unhydrated cement particles present in the crack are then possibly able to fully hydrate and thus are able to generate a higher strength then similar particles in the bulk of the specimen. To check the maximum load that could be reached in such case also a simulation is performed in which the strength of the interface is taken equal to the final strength of the specimen after full hydration minus the strength at one day.

The results of the simulations are shown in Figures 8-10. Figure 8 gives a typical deformed mesh in which the open discrete crack is visible. Figure 9 shows a stress-contour plot when the crack-tip is at half height of the specimen. It also can be seen from this Figure that the crack is not completely stress free, but stress can be transferred by the interface when it is in the descending branch.

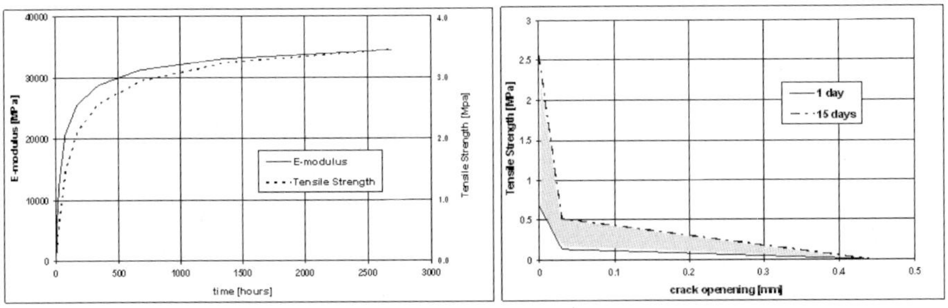

Figure 7. Development of E-modulus and tensile strength in time (a) and softening behaviour in interface element (b)

Table 2. Tensile strength of interface elements used in different simulations

simulation	Tensile strength [MPa]
1 day	0.67
15 days	2.56
1-15 days	1.89
1-final	2.79
(final strength)	(3.46)

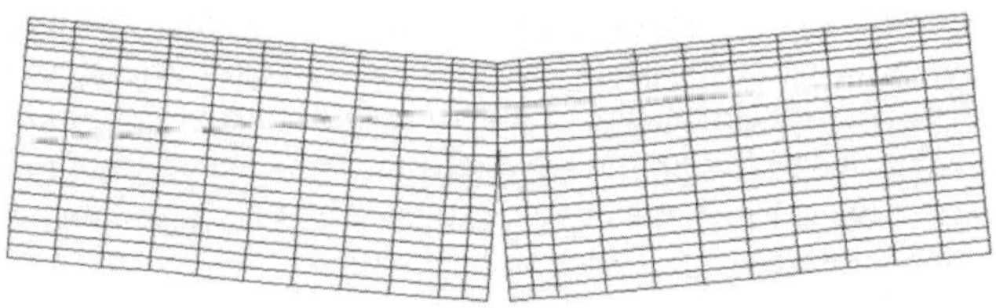

Figure 8. Deformed finite element mesh showing cracked interface element

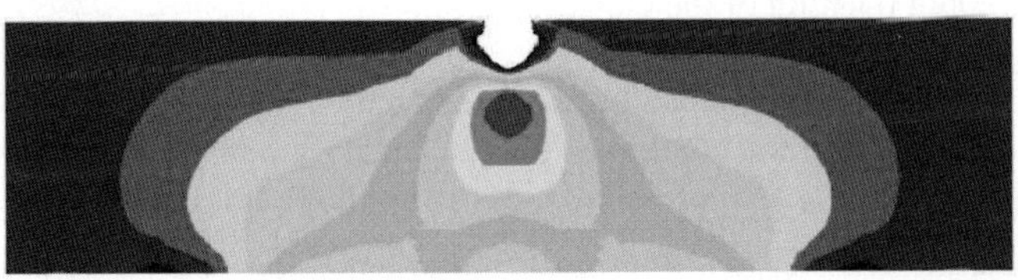

Figure 9. Stress-contour-plot of cracked specimen; crack tip is at half height of specimen

In Figure 10 the flexural stress is plotted versus the displacement (same measuring length as in the experiments (55 mm)) for the different simulations. From this graph the following observations can be made:

- The flexural stress as well as the stiffness found from the specimen tested at 1 day is lower then the specimen tested at 15 days.
- The specimen which is pre-cracked after 1 day and tested after 14 days of healing (1-15 days) has the same stiffness as the un-cracked specimen tested at 15 days. The strength of this specimen is about 77% of the un-cracked specimen.

- If the strength gain in the 14 days of healing is equal to the final strength minus the strength at day 1 then the obtained flexural stress is higher (about 106%) than the strength of the un-cracked specimen tested at 15 days.

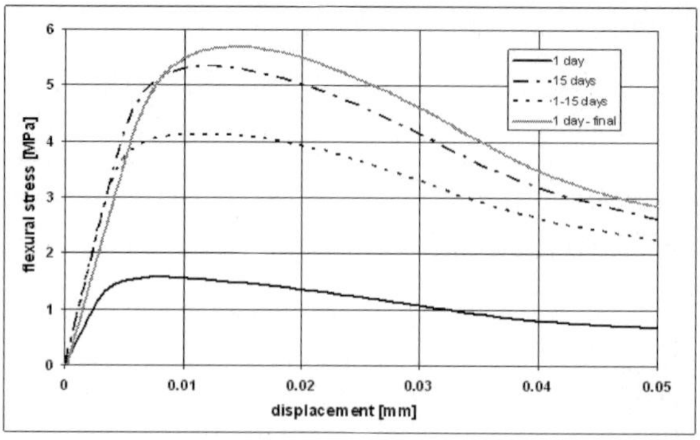

Figure 10. Flexural stress versus crack opening displacement obtained from simulations

6. DISCUSSION AND CONCLUDING REMARKS

This paper discusses the outcome of a combined experimental and numerical investigation performed at the Microlab of Delft University on crack healing of cracks in early age concrete. Prismatic specimens are at a certain age cracked in 3-point-bending after which they are subjected to a compressive load and stored under water to heal.

The variables in the research that are investigated and discussed in this paper are:

- amount of compressive stress during healing.
- type of cement in the concrete
- age of specimen when making the first crack.
- crack mouth opening of the crack.
- RH during the dealing

From the experimental results it can be concluded that:

- Cracks do heal under the conditions that the cracks are made at an early age and the cracks are closed again (a compressive stress is applied) and the specimens are stored under water.

- The amount of compressive stress does not seem to influence the strength recovery. The results indicate that a compressive stress is needed to close the crack, but once the two crack faces tough each other again, or the distance between the crack faces is small enough, crack healing can happen.

- Both for concrete made with BFSC and OPC crack healing takes place. Most probably in the case of OPC still a lot of unhydrated cement is left in the crack. Storing the specimens in water probably opens the way to further hydrate this cement in the crack. In the bulk material water can not reach these unhydrated particles. This means that concrete with OPC probably has some more capacity for crack healing at a later stage compared to BFSC-concrete.

- With increasing age of the specimen at the moment the first crack is made, a decrease in strength recovery is found. The age of the specimen when the first crack is made determines the degree of hydration. With that also the amount of hydration that still can take place is fixed. The amount of strength recovery is therefore also limited when the concrete has already reached a higher degree of hydration when the crack is made.
- The width of the crack does not seem to influence the strength recovery due to healing. The tests with different crack mouth openings show all a similar amount of strength recovery.
- Crack healing is only observed when the cracked specimens are stored under water.

The authors believe that ongoing hydration is the mechanism for crack healing that leads to the strength recovery in this investigation. This mechanism only works when the crack is closed again. It has been shown that the crack healing does take place when enough humidity is present. The simulations that are performed strengthen this hypothesis. It was shown that the increase in strength in the crack due to further hydration could be sufficient to observe the recovery of flexural strength found in the experiments. With the simulations it was also shown that higher strengths can be obtained in the crack compared to the bulk material when it is assumed that due to the water in the crack the final degree of hydration is reached faster in this zone.

For the practical situation of early age surface cracks in (massive) concrete structures, which are a concern from a durability point of view, this investigation shows some promising results. It indicates that these surface cracks can disappear again, at least under the right conditions as discussed in this paper.

Acknowledgments

The authors would like to thank mr. G. Timmers for assisting with the experiment. Furthermore, the authors acknowledge in depth discussions on the subject with Dr. M.R. de Rooij, Dr. E.A.B. Koenders and Ms. J. Bouwmeester.

7. REFERENCES

1. Edvardsen, C.K., Water permeability and self-healing of through cracks in concrete, Deutscher Ausschuss für Stahlbeton, Heft 455, 1996 (in German).
2. Reinhardt, H-W. & Joos, M., Permeability and self-healing of cracked concrete as a function of temperature and crack width, C&CR 33, 2003, 981-985.
3. Li, V.C., Lim, Y.M. and Chan, Y.W., Feasibility study of a passive smart self-healing cementitious composite. Composites Part B 29B (1998) 819-827.
4. Bang, S.S., Galinat, J.K. & Ramakrishnan, V., Calcite precipitation induced by polyurethane-immobilized Bacillus pasteurii, Enzyme and Microbial Technology 28 (2001) 404–409.
5. Ramakrishnan, V., Panchalan, R.K. and Bang, S.S., "Improvement of concrete durability by bacterial mineral precipitation", in Proceedings ICF11, Turin, Italy, March 20-25, 2005.
6. Salet, T. and Schlangen, E., Early-age crack control in tunnels. In Proc. Euromat 97, Sarton, L. and Zeedijk, H. (eds), Maastricht, 1997, Vol 4, 367-377.
7. Thermal Cracking in Concrete at Early-Ages, Proceedings of the International Rilem symposium, no.25, E and FN Spon, London, 1994.

8. Ter Heide, N., Schlangen, E. & van Breugel, K., "Experimental Study of Crack Healing of Early Age Cracks", In Proceedings Knud Højgaard conference on Advanced Cement-Based Materials, Technical University of Denmark, June 2005.
9. Ter Heide, N., Crack healing in hydrating concrete. MSc-thesis, Delft University of Technology, The Netherlands, 2005.
10. Granger, S., Loukili, A., Pijaudier-Cabot, G. and Chanvillard, G. "Mechanical characterization of the self-healing effect of cracks in Ultra High Performance Concrete (UHPC)" in Proceedings Third International Conference on Construction Materials, Performance, Innovations and Structural Implications, ConMat'05, Vancouver, Canada, August 22-24, 2005.
11. Schlangen, E., Lemmens, T. & Beek, T., Simulation of Physical and Mechanical processes in Concrete Floors and Slabs, Proceedings of the international Seminar held at University of Dundee, Scotland, UK on 5-6 September 2002, edited by R. K. Dhir et. Al., pp 45-56.

NON-DESTRUCTIVE MONITORING OF FIBER DISPERSION IN FRCS USING AC-IMPEDANCE SPECTROSCOPY

N. Ozyurt[1], T.O. Mason [2]and S.P. Shah [3]

[1] *Faculty of Civil Engineering, Istanbul Technical Univ., Istanbul, 34469, Turkey;* [2] *Department of Materials Science and Engineering, Northwestern Univ., 2145 Sheridan Road., Evanston, IL., 60208, USA;* [3] *Center for Advanced Cement Based Materials, Northwestern Univ., 2145 Sheridan Road, Evanston, IL., 60208, USA*

Abstract: Alternating Current-Impedance Spectroscopy (AC-IS) was used to study fiber dispersion characteristics in large-scale fiber-reinforced composite (FRC) specimens to understand the feasibility of AC-IS to be used to monitor fiber dispersion in industrial-scale specimens. Results showed that AC-IS can be developed as a non-destructive quality control technique. Fiber orientation and fiber segregation were also studied using alternate methods to verify the results of AC-IS experiments. Strong correlations were found between AC-IS and the alternate methods.

Keywords: Fiber dispersion; AC-Impedance Spectroscopy; non-destructive monitoring

1. INTRODUCTION

AC-IS is an electrical characterization method that can be used to study various features of composite materials such as hydration development, pore structure[1], microcracking (damage evolution)[2] and chloride ion diffusivity[3]. Recently, Torrents and coworkers showed that the use of AC-IS can be extended to obtain the fiber dispersion characteristics of conductive fiber-reinforced composites owing to its unique dual-arc behavior[4]. Previous work on the lab-scale specimens showed that by making use of the dual-arc behavior, fiber dispersion can be characterized in both the fresh and hardened states of cement-based materials[5, 6]. In this study, fiber segregation and fiber orientation are studied using larger-scale specimens to better understand the feasibility of the method to be used on various type and sizes of specimens.

M.S. Konsta-Gdoutos, (ed.), Measuring, Monitoring and Modeling Concrete Properties, 285–290.

2. AC-IS AND EXPERIMENTAL METHODS

The use of AC-IS for fiber dispersion monitoring involves application of an alternating voltage to a specimen and measurement of current response. Then, the experimental data is shown on a complex plane with real part of impedance on the x axis and imaginary part on the y axis. Figure 1 shows typical impedance curves for neat and FRC specimens. As seen in the figure, the dual-arc behavior occurs when conductive fibers are present. The reason for the occurrence of the dual arc behavior is the frequency dependent behavior of steel fibers. Conductive fibers are insulating under the low frequencies (Hz) of AC due to either an oxide film or a polarization layer forming at the fiber-electrolyte interface, while they are conductive under high frequencies (MHz) owing to displacement currents short-circuiting the oxide/polarization layer. Further explanation of the dual-arc behavior is available in the literature [4, 7]. Matrix-resistance (R_m) and composite resistance (R) can be obtained from the experimental data as seen in Fig. 1. The matrix (σ_m) and composite conductivity (σ) are then calculated using the equations below:

$$\sigma_m = \frac{1}{R_{DC}} \qquad \sigma = \frac{1}{R} \tag{1}$$

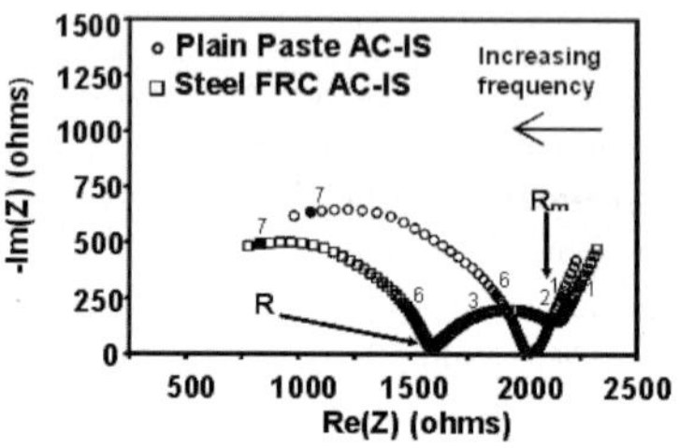

Figure 1. Typical impedance curves for neat and fiber-reinforced cement-based specimens

In this study, matrix-normalized conductivity (σ/σ_m) profiles of the specimens were obtained and used to define fiber orientation or fiber segregation.

Figure 2 (a) and (b) shows the experimental configuration for the beam specimen and the cylinder specimens, respectively. For the beam, circular stainless steel electrodes with radii of 64 mm were positioned side by side in reservoirs that were filled with 1 M NaCl solution. For the cylinders, stainless steel square electrodes (25 mm x 25 mm) were pressed against to specimen. Wet sponges were placed under the electrodes to provide good contact of the electrodes and the specimens. A Solartron 1260 impedance/gain-phase analyzer was used for AC-IS measurements, and the frequency was stepped from 100 mHz to 11 MHz under a voltage excitation of 1V.

3. RESULTS AND DISCUSSION

3.1 Fiber orientation - SCC Beam

A pre-cast fiber-reinforced SCC beam was used to study the orientation of the fibers using AC-IS. CON/SPAN Bridge Systems supplied a FRC beam that had been manufactured at their plant. The beam was cast and tested under bending by the company prior to be sent to the university lab. CON/SPAN Bridge Systems provided the information regarding the materials. The fiber-reinforced beam was cast with self-compacting concrete using 1 % Dramix RC 65/60 (L=60mm, AR=65) steel fibers. The dimensions of the full beam were approximately 15 x 25 x 400 cm. A part of the beam, with the dimensions of 15 x 25 x 70 cm, was used for the tests.

The beam part was divided into 7 parts, as shown in Fig. 3. AC-IS measurements were obtained from all parts of the beam in the two directions to obtain matrix-normalized conductivity profiles. Figure 3 shows the dimensions of the beam, the electrode positions for the X and Z directions and the approximate current paths. AC-IS measurements were obtained along the beam by moving the electrodes to each of the regions. Matrix-normalized conductivities (σ/σ_m) were calculated from the experimental data to obtain the profiles for the X and Z directions and are given in Fig. 4. As can be seen in Fig. 4. σ/σ_m values are higher in the X direction than Z direction, indicating that the fibers are mostly aligned in the plane perpendicular to the casting direction (Z).

Image analysis was done to verify the results obtained using AC-IS and to better understand the extent to which AC-IS predicts fiber orientation. A straightforward method was used to define fiber orientation by means of the number of fibers on the cross-sections cut from the beam. The beam was cut into 7 parts and the number of fibers on the XZ and XY cross-sections was counted. Orientation factors were calculated using the equation below for each part and the orientation factor profiles were plotted (Figure 5).

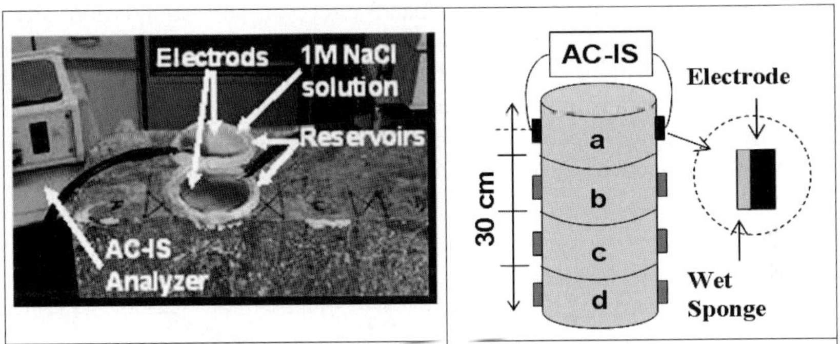

(a) (b)

Figure 2. Experimental set-up for a) SCC beam b) cylinder specimens

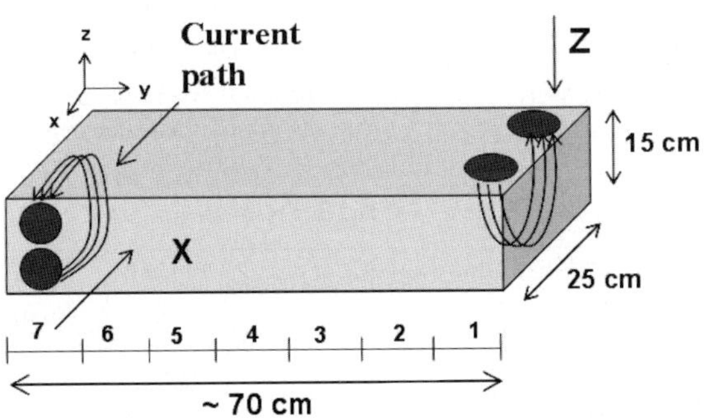

Figure 3. Measurement directions and approximate current paths for X and Z directions

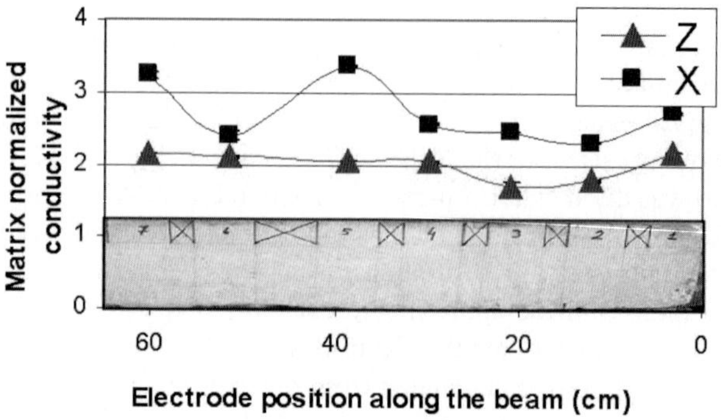

Figure 4. Matrix-normalized conductivity profiles for X and Z directions

$$\eta_\varphi = \frac{N_f * A_f}{V_f} \tag{2}$$

In Eq. (2), η_φ is the orientation number, N_f is the number of fibers per unit area, (1/mm^2), A_f and V_f are the cross-sectional area (mm^2) and fiber volume fraction (vol. %), of a single fiber, respectively. Figure 5 shows the orientation number profiles for the X and Z directions. Fiber orientation was found to be higher on the ZY plane, suggesting that the fibers are more oriented in the X direction than Z direction. This result is consistent with the results of AC-IS measurements, confirming the amenability of AC-IS to be used on larger-scale specimens (Figure 6). As can be seen in Fig. 4., σ/σ_m values are higher in the X direction than Z direction, indicating that the fibers are mostly aligned in the plane perpendicular to the casting direction (Z). Figure 6 shows the orientation number versus matrix-normalized conductivity relation, indicating a linear relationship, with an R^2 value of approximately 0.83.

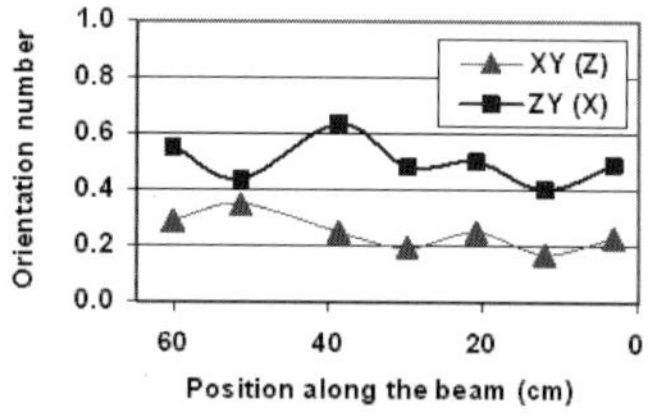

Figure 5. Orientation numbers for X and Z directions

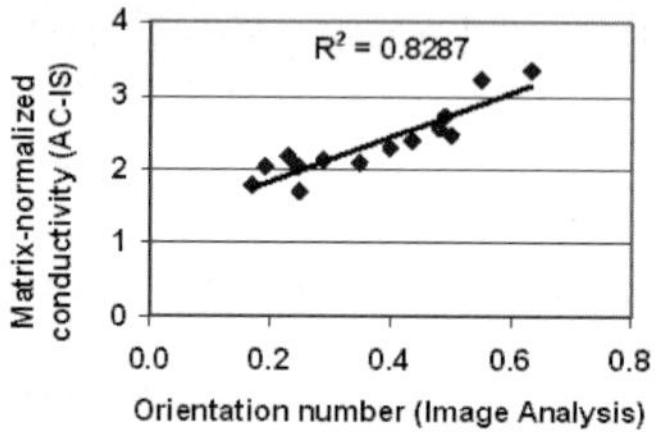

Figure 6. Matrix-normalized conductivity versus orientation number

Table 1. CC and SCC mix designs

	Cement	Water	Fine agg.	Coarse agg.	Fly ash	Sp. (% binder weight)
CC	1	0.4	2	2	-	1
SCC	1	0.4	1.56	1.9	0.25	0.7

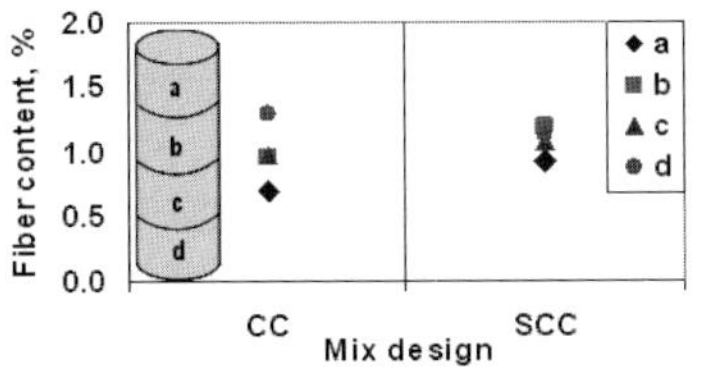

Figure 7. Fiber content distribution for CC and SCC specimens

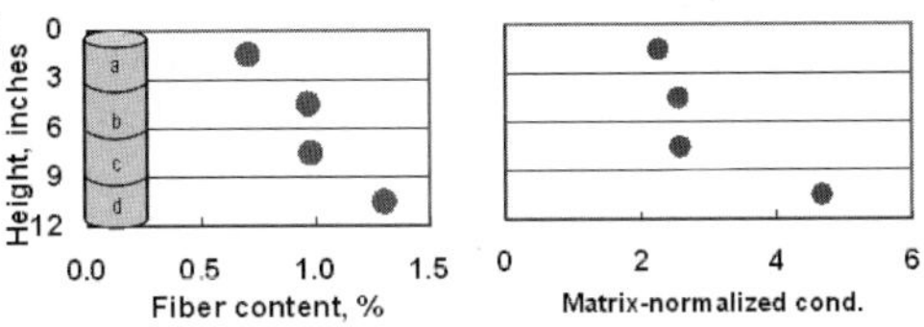

Figure 8. Fiber content and matrix-normalized conductivity profiles of CC mix

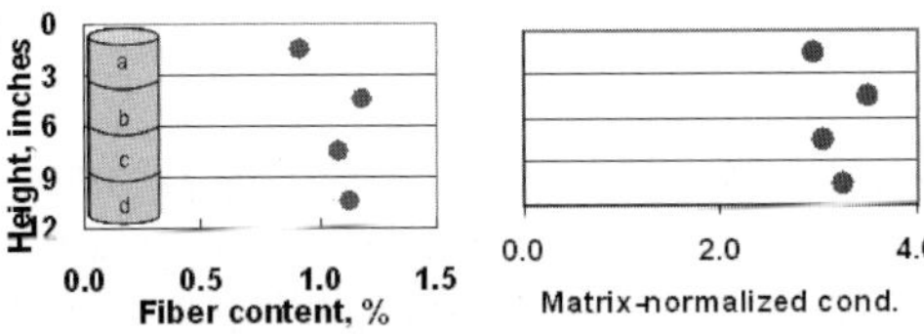

Figure 9. Fiber content and matrix-normalized conductivity profiles of SCC mix

3.2 Fiber segregation - SCCcylinder

In this part of the study, fiber segregation of a self-compacting concrete (SCC) and conventional concrete (CC) are compared. AC-IS was employed to monitor fiber segregation. Fiber segregation was also evaluated by cutting fresh specimens (immediately after initial set) into slices, washing the fibers out and then quantifying the amount of fibers in each section. Table 1 shows mix designs for CC and SCC. Dramix RC-65/40 steel fibers were used (40 mm long and 0.62 mm diameter) for both CC and SCC and the fiber content was kept constant at 1 %. A naphthalene-based superplasticizer was used for CC and a polycarboxylate-based superplasticizer was used for SCC. The water-to-cement ratio was kept constant at 0.40.

Cylinder specimens were cast for each mix design. CC specimens were vibrated for 2 minutes to ensure a good placement of concrete. One of the cylindrical specimens was cut into 4 slices immediately after initial set and fibers were washed out. Next, the amount of fibers in each slice was calculated and fiber segregation was evaluated. Figure 7 shows fiber content distribution for CC and SCC specimens. The remaining specimens were stored under 100 % relative humidity until testing. AC-IS measurements were done 3 days later using the experimental set-up given in Figure 2(b). Electrodes were positioned alongside the cylinder specimens and 4 measurements were obtained. Then, the matrix-normalized conductivity profiles of the cylinders were plotted. Figures 8 and 9 show the fiber content and matrix-normalized conductivity profiles of CC and SCC mixes, respectively. As seen in Figures 8 and 9, fiber content distribution and σ/σ_m profiles of the specimens show very similar tendencies, which confirms that the fiber segregation can be non-destructively monitored using AC-IS. From these figures it can also be concluded that SCC mix in this study is more resistant to the segregation of fibers than CC mix. To mathematically express this result, statistical dispersion (standard deviation) of the fiber contents throughout the specimens was calculated. The standard deviation in fiber content distribution of the CC specimen was found to be two times that of SCC mix, meaning a higher segregation of fibers.

4. CONCLUSIONS

AC-IS was employed together with conventional methods to study fiber orientation and fiber segregation in FRC specimens. The following conclusions can be drawn from the results of this study. 1) Various fiber dispersion issues can be characterized using AC-IS, 2) Experimental set-ups can be modified to extend the use of the method for specimens of various shapes/sizes, 3) The AC-IS results are as reliable as those obtained from conventional methods and can be obtained quickly and non-destructively 4) The potential exists for AC-IS to be utilized as a non-destructive technique for quality control purposes.

Acknowledgements

The authors acknowledge CON/SPAN Bridge Systems and Dave Brodowski for their contribution to this work. The first author also would like to acknowledge the financial support of TUBITAK (The Scientific and Technical Research Council of Turkey), Istanbul Technical University and Northwestern University. The authors are grateful to Dr. Leta Woo for helpful discussions.

5. REFERENCES

1. B. J. Christensen, R. T. Coverdale, R. A. Olson, S. J. Ford, E. J. Garboczi, H. M. Jennings and T. O. Mason, Impedance spectroscopy of hydrating cement-based materials: Measurement, interpretation, and application, *Journal of American Ceramic Society*, **77**(11), 2789-804, (1994).
2. A. Peled, J. M. Torrents, T. O. Mason, S. P. Shah and E. J. Garboczi, Electrical impedance spectra to monitor damage during tensile loading of cement composites, *ACI Materials Journal*, **98**, 313-322, (2001).
3. J. M. Loche, A. Ammar and P. Dumargue, Influence of the migration of chloride ions on the electrochemical impedance spectroscopy of mortar phase, *Cement and Concrete Composites*, **35**(9), 1797-1803, (2005).
4. J. M. Torrents, T. O. Mason, A. Peled, S. P. Shah and E. J. Garboczi, Analysis of the impedance spectra of short conductive fiber-reinforced composites, *Journal of Materials Science,*, **36**, 4003-4012, (2001).
5. N. Ozyurt, L. Y. Woo, B. Mu, S. P. Shah and T. O. Mason, Detection of fiber dispersion in fresh and hardened cement composites, in: *Advances in Concrete Through Science and Engineering, An International Symposium During the RILEM Spring Meeting*, Northwestern University, Evanston, (2004).
6. L. Y. Woo, S. Wansom, N. Ozyurt, B. Mu, S. P. Shah and T. O. Mason, Characterizing fiber dispersion in cement composites using AC-Impedance spectroscopy, *Cement and Concrete Composites*, **27**(6), 627-636, (2005).
7. T. O. Mason, M. A. Campo, A. D. Hixson and L. Y. Woo, Impedance Spectroscopy of Fiber-Reinforced Cement Composites, *Cement and Concrete Composites*, **24**, 457-465, (2002).

EXPERIMENTAL METHODOLOGY TO STUDY PLASTIC SHRINKAGE CRACKS IN HIGH STRENGTH CONCRETE

A. Sivakumar and M. Santhanam
Research Scholar; Assistant Professor, Department of Civil Engineering, Indian Institute of Technology, Madras, Chennai – 600036, India

Abstract: Plastic shrinkage cracks occur primarily in concrete elements having large surface to volume ratios. These cracks develop at early ages of concrete due to the rapid loss of water from the surface of concrete before it is set and has still not attained sufficient tensile strength. High strength concrete incorporating silica fume is more susceptible to plastic shrinkage cracks because of the minuteness of the capillary pores, which causes higher strains on drying. Often, the cracks that develop are so fine that they cannot easily be distinguished from the surface; this is especially true in the case of fibre concretes. This paper describes an experimental methodology to monitor cracking due to plastic shrinkage using a simple and reproducible technique. Rectangular concrete slabs, with a stress riser in the center, and adequate base and end restraints, are subjected to drying in adverse conditions in an environmental chamber. Cracking along the tip of the riser is then studied using a combination of photographs and hand-held microscope. The various crack measurements are then made using image analysis software. A statistically accurate measure of the crack area is then obtained. This parameter is used to compare different types of concrete. This methodology was found to be successful even for the characterization of very fine cracks in hybrid fibre concrete systems.

Keyword: Shrinkage; silica fume; cracking; fibre; stress riser.

1. INTRODUCTION

Concrete undergoes significant volumetric changes during the course of its lifetime, typically from expansion and contraction due to variation in surface temperature and moisture movement from concrete to atmosphere through evaporation. Also, concrete tends to shrink both during and after curing upon drying. Plastic shrinkage occurs at the early ages of concrete as soon as it is placed, and manifests in the form of surface cracks. This is severe in silica fume concrete due to minute pores that result in high capillary

M.S. Konsta-Gdoutos, (ed.), Measuring, Monitoring and Modeling Concrete Properties, 291–296.

pressures. When the bleed water from the surface of concrete is evaporated at a rapid rate (when evaporation rate is more than bleeding rate), a complex meniscus is formed in the pores[1-2]. This results in the formation of negative capillary pressure, and subsequently, cracks, if restrained. In other words, these volumetric changes can cause internal stresses in the concrete matrix, which cracks because it has not attained adequate tensile strength[3]. This type of cracking is more prevalent in the case of highway concrete pavements and industrial concrete floors where inadequate curing and extreme temperature conditions are more likely to cause cracking.

As of now there are no standard methods available to evaluate quantitatively the plastic shrinkage cracks arising in fresh concrete. In this paper an attempt is made to measure cracks arising out of early age shrinkage using a simple reproducible technique. A rectangular slab with a stress riser at the center was chosen for this study, in order to avoid delayed cracking, as well as to have an exact location of the crack.

2. EXPERIMENTAL METHODOLOGY

2.1 Materials used

Ordinary Portland cement conforming to IS 12269[4] was used for the concrete mixtures. Silica fume, obtained from Elkem Materials, India, was also used for the high strength concrete mixtures. River sand with a specific gravity of 2.65 and fineness modulus of 2.64 was used as the fine aggregate, while crushed granite of specific gravity 2.82 was used as coarse aggregate. A naphthalene sulphonate based superplasticizer was used to obtain the desired workability. The fibres used in this study consist of hooked steel and polypropylene, and the properties of these fibres are given in Table 1.

2.2 Mixture proportioning

Trial mixtures were prepared to obtain a target strength of 60 MPa at 28 days, along with a workability of 75 – 120 mm. The detailed mixture proportions for the concrete are presented in Table 2.

Fibres were added at a dosage level of 0.3% by volume of the concrete. In order to obtain the desired workability, only the superplasticizer dosage was varied. Apart from the control mixture (without fibres), one steel fibre concrete mixture and one hybrid fibre concrete (with steel and polypropylene fibres) were prepared.

2.3 Mixing and casting details

The coarse aggregate, fine aggregate, cement, and silica fume, and fibre (where applicable) were first mixed dry in a pan mixer of capacity 100 kg for a period of 2 minutes. The superplasticizer was then mixed thoroughly with the mixing water and added to the mixer and the concrete was mixed for a total of 4 minutes. The fresh concrete was placed in the laminated wooden slab moulds and compacted on a vibrating table. The surface finishing was done very carefully to obtain a uniform perfectly smooth surface.

Table 1. Properties of the different fibres used

Property	Hooked steel	Polypropylene
Length (mm)	30	20
Diameter (mm)	0.5	0.12
Aspect ratio (l/d)	60	166
Density (kg/m^3)	7800	900
Tensile strength (GPa)	1.7	0.45
Elastic modulus (GPa)	200	5
Failure strain (%)	3.5	18

Table 2. Concrete mixture proportions (kg/m^3) used in the study

Cement	Silica fume	Fine Aggregate	Coarse Aggregate		Water	Superplasticizer
			10 mm	20mm		
372	28	750	570	570	160	8

2.4 Specimen geometry and environmental conditions

In this study a simple and modified version of the design by Berke and Dalliare[5], was used to study plastic shrinkage cracking. The slab specimen adopted was 500mm*250mm*110mm and was made of laminated board. A thin polyethylene sheet was covered on the base to eliminate friction between the concrete and laminated board. The slab was provided with a stress riser of 75 mm height at the center and two base restraints of 35 mm height at 35 mm from the ends, both laid along the transverse direction as shown in Figure 1. A bolt and nut arrangement was provided at the ends to minimize the longitudinal movement of slab from edges and thus increase the potential to crack at the notch. The slabs were stored after casting in an environmental chamber with dimensions of 2.75×0.9×0.9 m, as shown in Figure 2. Apart from controlling temperature and humidity, this chamber was equipped with a high-speed fan capable of 1800 rev/sec fixed on the walls of the chamber to accelerate drying of concrete. The slabs were exposed to a constant temperature of 35±1°C, a relative humidity of 40±1%, a wind velocity of 24 km/hr, and the evaporation rate measured was 0.31 kg/m^2/hr.

Figure 1. Moulds fabricated for plastic shrinkage crack study

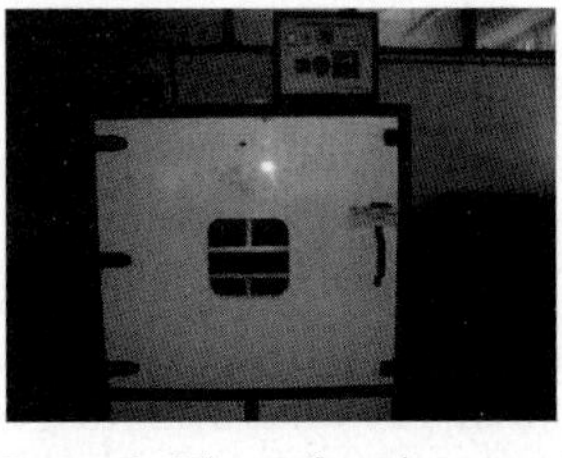

Figure 2. View of environmental chamber used for the study

3. CRACK MEASUREMENT TECHNIQUE

It is a well-known fact that as the cement paste shrinks, it has propensity to develop cracks when restrained by aggregate particles. This causes multiple hairline cracks on the surface of concrete elements having large surface areas and of small thickness. Shrinkage studies on large restrained slabs result in multiple cracking and involve cumbersome calculations to map the entire crack area. Crack measurements become easier as the shrinkage cracks occur in concrete as a result of settlement along transverse direction and longitudinal surface drying[6]. This has necessitated the use of the test set up with a stress riser at the center, which acts as the cracking location. In addition to this, two small risers provided near the ends offer sufficient restraint at the base of the slab and also eliminate biaxial stress. Cracking occurs above the central stress riser across the entire width of the specimen. The cracking at the centre of slab is as a result of the three-dimensional volume change.

In this study, the crack measurements were made after 24 hours in order to ensure that cracks got fully developed and stabilized. The cracked area was then photographed using a 4x optical zoom digital camera, and the image was processed using image analysis software. In order to calibrate the original size of the image that was captured, a measuring scale was placed on the concrete specimen and the entire area was photographed. The distance between two points on the scale was calibrated in terms of pixels and the total pixels were converted to the desired unit. The digitized image was binarized and geometrical operations such as cleaning and filtering were then carried out to get a clean crack profile. The crack parameters such as total length, total crack area, maximum and average crack width were then obtained using the software. In the case of thin hairline cracks with width of less than 1 mm it was difficult to use the crack image (photograph) for further analysis. These thin cracks were studied manually using a crack width microscope, which can measure accurately the crack widths in the range of 0.025 to 2.5 mm. To calculate the mean crack width a total of 30 arbitrary points was chosen along the crack profile. Using this crack width information, and the crack length obtained from image analysis of the binary image, the crack area was obtained.

4. TEST RESULTS AND DISCUSSIONS

Shrinkage cracks were observed in less than about 4.5 hours after casting for the control concrete and 5.5 and 7 hours for the steel fibre and hybrid fibre concretes respectively. Experimental results of the measured/calculated crack parameters are presented in Table 3. For the control concrete, the maximum crack width obtained was 0.5 mm; this was reduced to 0.371 mm (by 25%) in the case of steel fibre concrete (SFRC), and to 0.225 mm (by 55%). This corresponds to a reduced crack area of 57.6 % compared to control concrete and 43% compared to steel fibre concrete. The full width of the cracked slab for control concrete, and smaller areas of cracking from the fibre concretes are shown in Figures 3 to 5. The binarized images for the full width of the slab, for all three concretes, are shown in Figures 6 to 8.

Table 3. Experimental results

Mix ID	Type of fibre	Volume fraction of fibres (%)	Total crack length (mm)	Max crack width (mm)	Mean crack width (mm)	Standard deviation (mm)	Total crack area (mm^2)	% reduction in crack area
Control	-	-	329	0.50	0.456	0.014	150.81	-
SFRC	Steel	0.3	302	0.371	0.314	0.021	112.04	25.7
HFRC	Steel + PP	0.15 : 0.15	284	0.225	0.187	0.028	63.9	57.6

In the case of control concrete, a single crack formed at the center and was continuous across the width of the specimen, and traversed almost in a straight path. This crack was rather discontinuous in the case of fibre concretes, which is essentially because of the presence of fibres along the crack propagation path; this subsequently causes sub-parallel cracks (as shown in Figures 7 and 8). The total crack lengths of both fibre concrete mixtures were not significantly lower than control concrete since sub parallel cracks were formed in addition to the main crack. The formation of sub parallel cracks could be due to the localization and subsequent transfer of shrinkage stresses developed during drying to the neighboring point. These sub parallel cracks were thin and found to be more in the case of hybrid fibre concrete as compared to the steel fibre system. The reason could be increased fibre availability (as a result of substituting steel with polypropylene fibre, which has a much lower density) in the case of hybrid fibre concrete compared to steel fibre concrete. In general the enhancement in the case of hybrid fibre concrete systems could be attributed to the presence of polypropylene fibres.

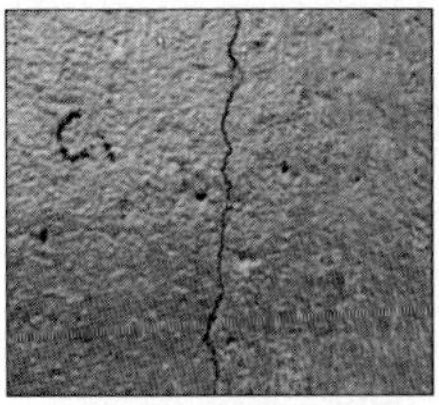

Figure 3. Control concrete (250*250mm)

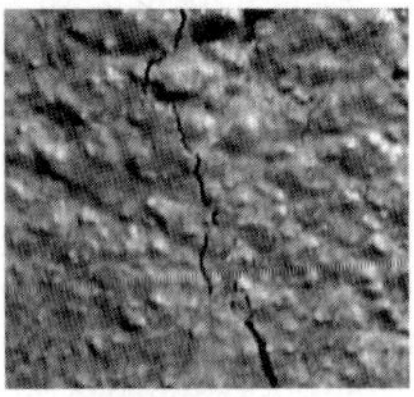

Figure 4. Steel fibre concrete (20*20mm)

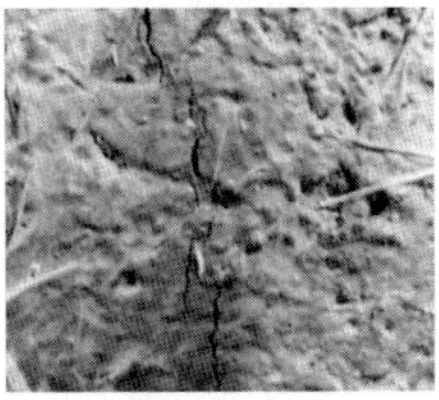

Figure 5. Hybrid fibre concrete (15*15mm)

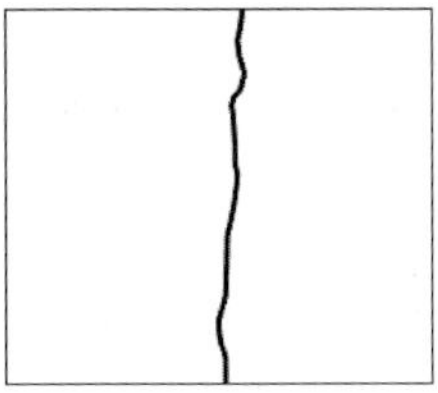

Figure 6. Cleaned binary image showing crack profile of control concrete

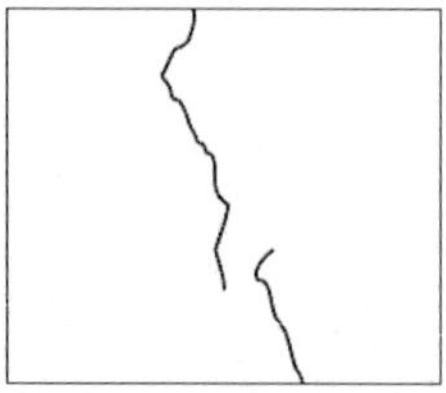

Figure 7. Cleaned binary image showing crack profile of steel fibre concrete

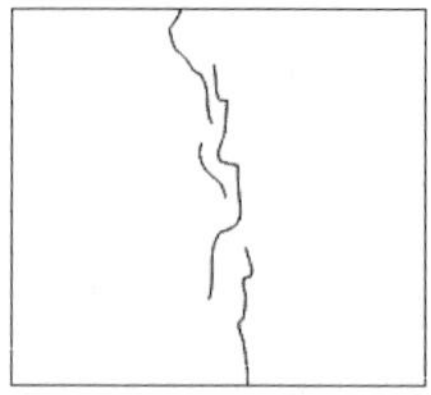

Figure 8. Cleaned binary image showing crack profile of hybrid fibre concrete

5. CONCLUSIONS

This article described the use of a simple technique based on image analysis of photographs and manual crack measurement in order to study cracking due to plastic shrinkage in high strength concrete systems. The technique proved effective in quantifying even the hairline cracks arising in hybrid fibre concrete mixtures.

As for the performance of the mixtures in a hot and dry environment, it was seen that the hybrid fibre combination of steel and polypropylene resulted in the least level of cracking, possibly as a result of the increased availability of fibres (by the use of polypropylene).

6. REFERENCES

1. Cohen M.D., Olek J., and Dolch W.L., 1990, Mechanism of plastic shrinkage cracking in Portland-cement and Portland cement-silica fume paste and mortar, *Cem Concr Res* **20**: 103-119.
2. Cabrera J., Cusens G., and Brookes- Wang Y., 1992, Effect of superplasticizers on the plastic shrinkage of concrete, *Mag Concre Res* **44**:149-155.
3. Soroushian P. and Ravanbaksh S., 1998, Control of plastic shrinkage cracking with specialty cellulose fibers, *ACI Mater J* **89**: 535– 540.
4. Indian Standard Designation, IS 12269:1987, Specification for 53 grade Ordinary Portland Cement, Bureau of Indian Standards, New Delhi, India.
5. Berke N.S. and Dalliare M.P., 1994, The effect of low addition rate of polypropylene fibers on plastic shrinkage cracking and mechanical properties of concrete, *Fiber reinforced Concrete: development and innovations*, ACI SP 142, Detroit, pp.19-41.
6. Weiss W.J., Yang K., and Shah S.P., 2001, Factors influencing durability and early-age cracking in high-strength concrete structures, ACI SP 189, Detroit, pp.387-409.

INVESTIGATION OF THE VISCOELASTIC PROPERTIES OF FRESH PORTLAND CEMENT PASTES WITH AN ULTRASONIC WAVE REFLECTION METHOD

Z. Sun and S.P. Shah

Department of Civil & Environmental Engineering, 112 W. S. Speed Hall, University of Louisville, Louisville, KY 40292, USA; Center for Advanced Cement-Based Materials, Northwestern University, 2145 Sheridan Road, Suite A130, Evanston, IL 60208, USA

Abstract: This study will mainly focus on the viscoelastic properties of Portland cement pastes at very early age, namely before the initial setting time. A method, called ultrasonic wave reflection (UWR-) method, was used to detect the material properties of fresh cast cement pastes. In this technique, a fused quartz plate with a thickness of 10 mm was used as a buffer material that sited next to the cement paste. Two transducers with 2.25 MHz central frequency were coupled to the quartz plate. Ultrasonic shear waves were used in the measurement. Besides measuring the wave reflection coefficient at the quartz-paste interface, the phase shift of the incident and the reflected waves was also measured, which allows the separation of the elastic and viscous properties of a cement paste at very early age by observing the evolution of dynamic storage shear modulus (G') and loss modulus (G") with the hydration of cement. Three cement pastes with water/cement-ratios equal to 0.4, 0.5 and 0.6 cured under water at a constant temperature of 25°C were studied. It was found that the development trend of the storage and loss shear moduli were closely related to the hydration of cement particles.

Key words: viscoelastic properties; ultrasonic wave reflection; storage modulus; loss modulus.

1. INTRODUCTION

This chapter will focus on the viscoelastic properties of Portland cement pastes at very early age, namely before the initial setting time. The methods used in the research were the ultrasonic wave reflection (UWR) method and the oscillatory rheometric (OR) method, both of which are based on the theory of dynamic rheology[1].

M.S. Konsta-Gdoutos, (ed.), Measuring, Monitoring and Modeling Concrete Properties, 297–303.

The ultrasonic wave reflection technique was first applied to the area of cementitious materials by Stepišnik et al. in 1981[2]. This method can be used to monitor the setting and hardening behavior[3] and can also be adopted to mimic the strength development of cementitious materials at early age. In a recent study, more attention was paid to the application of this method to monitor the viscoelastic properties of cement pastes[4]. It was shown that when cement paste is in solid state, the material behavior can be approximated by neglecting the phase shift measured by the wave reflection technique[5]. But few results on fresh cement paste in flowable status were reported.

In this study, the application of the ultrasonic shear wave reflection technique was explored. The technique was used to monitor the viscoelastic properties of cement pastes at very early age by observing the evolution of the storage shear modulus and the viscosity. The calculated storage shear modulus, which represents the elastic properties of the materials were compared to the results measured directly from the oscillatory rheometric method, which is from a microstructural point of view also a non-destructive testing method. In this oscillating rheometric method[6], the storage and loss shear moduli of the cement paste can directly be measured by applying oscillating shear strain according to a sine function and measuring the corresponding shear stress. By controlling the value of oscillatory shear strain and the frequency within the linear viscoelastic region of the material, the microstructure of the cement paste will not be destroyed during the oscillations, and the evolution of the material properties during hydration can be observed.

2. THEORIES AND METHODOLOGIES

2.1 Material behavior

When a linear viscoelastic body is subjected to stress varying sinusoidally with time at a certain frequency, the corresponding strain is not in the same phase as the applied stress, which results in a phase lag between strain and stress[1]. The applied stress can be separated into two independent components: one is exactly in phase with the strain, and the other is π/2 out of phase. According to the decomposed stress components, the relationship between stress and strain of a viscoelastic material can be established by using the modulus of rigidity in a complex format. If a material is subjected to a shear deformation, the shear modulus can be expressed as follows,

$$G^* = \frac{\tau}{\gamma} = G' + iG'' \qquad (1)$$

$$G' = \frac{\tau_0}{\gamma_0} \cos\delta \qquad (2)$$

$$G'' = \frac{\tau_0}{\gamma_0} \sin\delta = \omega\eta \qquad (3)$$

where G^* is the complex shear modulus, G' is the storage shear modulus, which represents the elastic behavior or the energy storage of the material, and G'' is the loss shear modulus, which represents the viscous behavior or energy dissipation of the material. ω is the angular frequency of the applied oscillating strain (stress). The phase lag (δ) between stress and strain, which changes between 0 and $\pi/2$, can also be used to describe the material behavior. When a material is ideally elastic, the phase lag δ equals to zero; when a material is ideally viscous, δ equals to $\pi/2$.

2.2 Wave reflection at an elastic-viscous boundary

In the UWR method, the wave reflection coefficients at the boundary between the tested cement paste and the buffer material were measured. Details about the equipment that used can be found in Ref 3. To enhance the sensitivity of the wave reflection measurement, instead of using a steel plate, a fused quartz plate with 12mm thickness was used as the buffer material in the study presented in this paper.

The real and imaginary part of the shear impedance of the viscoelastic material can be obtained by following equations:

$$R = Z_1 \frac{1 - r_0^2}{1 - 2r_0 \cos\theta + r_0^2} \tag{4}$$

$$X = Z_1 \frac{2r_0 \sin\theta}{1 - 2r_0 \cos\theta + r_0^2} \tag{5}$$

where r_0 is the magnitude of the obtained wave reflection coefficient, and θ is defined as the phase shift between the incident and the reflected waves. According to Whorlow7, the storage shear modulus and the viscosity of the measured cement paste can be calculated by Eq. (4) and (5):

$$G' = Z_1^2 \frac{(1 - r_0^2) - 4r_0^2 \sin^2\theta}{\rho(1 - 2r_0 \cos\theta + r_0^2)^2} \tag{6}$$

$$G'' = Z_1^2 \frac{\omega 4r_0 (1 - r_0^2) \sin\theta}{\omega\rho(1 - 2r_0 \cos\theta + r_0^2)^2} \tag{7}$$

2.3 The oscillatory rheometric method

In this study, a HAAKE RS150 rheometer was used in the OR method. Co-axial cylinders with a gap between the outer and inner cylinder of 0.8mm were used. This allows us to assume that the shear stress is uniformly distributed across the gap[6]. A sample of about 25 g of cement paste was mixed by hand for about 90 seconds. Hand mixing can be helpful to eliminate the effect of pre-shearing on the measurement. After transferring the cement paste into the rheometer, a high shear rate at 600 s^{-1} was used for 30 seconds to let the cement particles go through irreversible structural breakdown[8]. Then the cement paste was allowed to rest for 600 seconds to let each particle achieve its structural equilibrium and by this, to reach the same stress status before the oscillating testing. During the oscillatory testing, the strain-control mode was used. To account for the dependency of the linear viscoelastic region on the w/c ratio of the cement paste, the strain amplitudes used were $1x10^{-4}$ for w/c=0.4 and 0.5, and $7x10^{-5}$ for w/c=0.6. The frequency used was 1 Hz for all the cement pastes. The in-phase and out-of-phase shear stresses can be traced during the measurements. The temperature of the specimens was maintained at 25°C by controlling the circulating water in the water cup in which the outer cylinder is embedded. All the samples were covered by non-seepage oil during the oscillating test to prevent moisture loss.

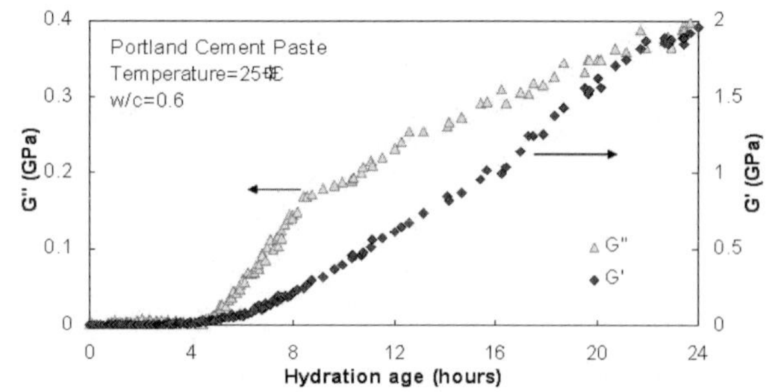

Figure 1. Storage and Loss Shear Moduli Calculated from UWR Measurement

3. RESULTS AND DISCUSSIONS

3.1 Wave reflection measurements

The measured phase change ($\pi - \theta$) increases from a value, that is less than 1° to about 10°. An example of the calculated storage and loss shear moduli from the UWR measurement of a cement paste with w/c=0.6 is given in Fig. 1. It can be noted that both the storage and loss moduli begin to increase with a steeper rate at a very similar hydration age. However, the loss modulus tends to converge to a certain value while the storage modulus keeps increasing. This is a hint that the mechanical behavior of the cement paste gets more and more close to an elastic solid during hydration.

3.2 Oscillatory rheometric measurements

An example of the measured continuous development of storage and loss shear moduli of plain cement paste with w/c=0.4 during the first 3 hours of hydration is given in Fig. 2. The storage modulus increases with time from an initial value of 1 kPa to above 1300 kPa at about 2.5 hours. Different from storage modulus, the loss modulus increases from 10 kPa to around 70 kPa during the first hour and then decreases. Struble et al.[6] reported similar curves for the storage and loss moduli. The reason for the decrease of the loss modulus was not discussed.

Nachbaur et al.[9,10] also reported oscillatory rheometric measurements on cement pastes. Similar results were obtained for the storage modulus, while those for the loss modulus were different. Parallel plates were used during the oscillating test, and both the storage and loss moduli increased with time. This indicates that the development of the storage modulus measured by the oscillatory rheometer is independent of the instrument, which agrees with the results shown by Saak[11], where it is reported that the storage moduli measured by the co-axial cylinder, the parallel plate and the vane rheometer correspond to each other very well. However, the loss modulus development curve was found to be affected by the instrument type. The loss modulus tends to be very sensitive to the stress state and stress distribution applied to the material. The stress distribution in the cement paste measured with a co-axial rheometer and a parallel-plate rheometer are very likely to lead to the different shape of the loss modulus curve due to the geometry of the rheometer.

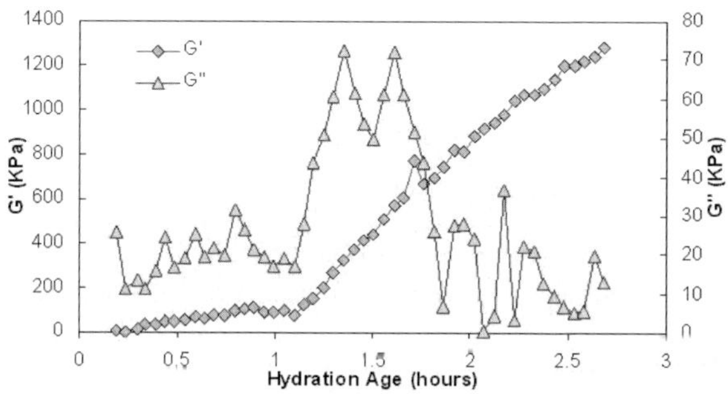

Figure 2. Storage and Loss Modulus as a Function of Hydration Age (w/c=0.4, T=25°C)

3.3 Comparison of the storage modulus

A comparison of the storage shear moduli for cement paste with w/c=0.4 determined with the UWR- and OR- methods is shown in Fig. 3. It can be seen that at a given hydration age, the absolute value of the storage modulus calculated from UWR measurement is somewhat larger than that obtained from the OR-method. This systematic error must due to the different methods used. However, it can also be noticed that both moduli develop after a very similar trend.

In Fig. 4, the evolution of G' for all three cement pastes is plotted. In the figure, the moduli obtained from OR- and UWR methods are normalized with respect to their values at the maximum duration of the OR-measurements. The moduli first exhibit a moderate increase followed by a steeply increasing part. The onset of the sharp increase in the moduli can be considered as the end of the dormant period in the hydration process[12]. It is obvious that the storage shear moduli obtained with the two different methods develop after very similar trends for all tested w/c ratios, which shows that the UWR method can monitor the elastic behavior of cementitious materials in the fresh state.

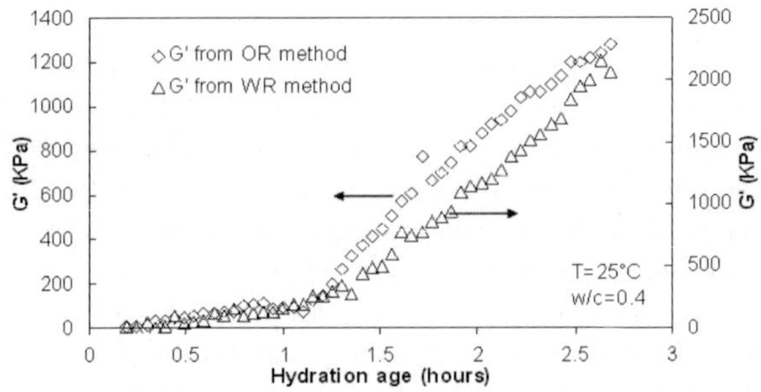

Figure 3. Comparison of Storage Shear Moduli from UWR- and OR- Methods

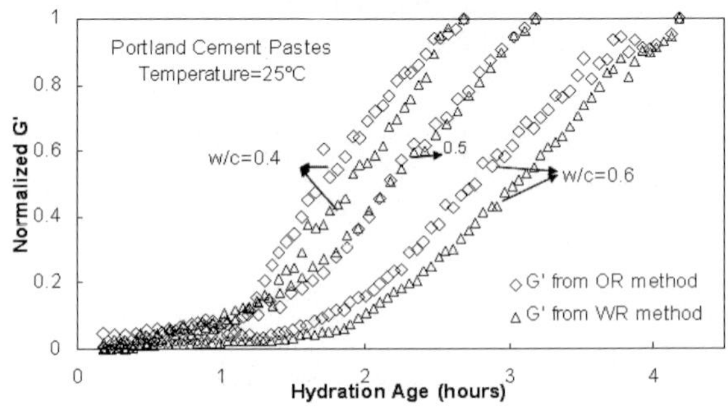

Figure 4. Comparison of Normalized Storage Moduli from UWR- and OR- Methods

3.4 Comparison of the loss modulus

A comparison of the loss shear moduli obtained with two methods is given in Fig. 5. It is obvious that neither the magnitude nor the shapes of the two curves are similar. The reason is still unknown. One explanation could be the different mechanism of the two methods. In OR-method, the frequency of the applied shear strain was 1Hz, while 2.25MHz was used in the UWR measurement. The suspension liquid system can behave very differently when subjected to external forces with various frequencies. G'' may represent different material properties in OR- and UWR-methods.

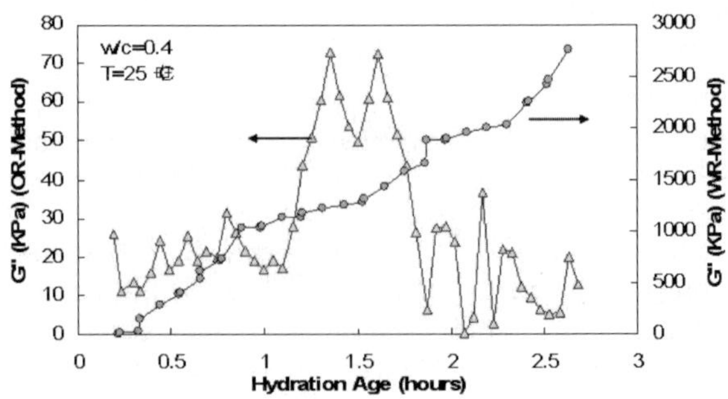

Figure 5. Comparison of Loss Shear Moduli from UWR- and OR-Methods

4. CONCLUSIONS

Based on the research, the following conclusions can be drawn:

- By considering the wave reflection coefficient and the phase shift between the incident and reflected waves, the storage modulus and the viscosity can be calculated with the UWR method.
- The calculated storage moduli from wave reflection measurements have similar trends as those measured with the oscillatory rheology method and can reflect the evolution of solid phases in the microstructure of the cement paste.
- The loss moduli obtained with the two methods do not match to each other. The discrepancy can not be illustrated yet. Further studies on both the physical meaning and the measurement of the loss modulus are necessary.

ACKNOWLEDGEMENTS

The research presented in this paper was funded by the Infrastructure Technology Institute of Northwestern University, the Center for Advanced Cement-Based Materials, and the National Science Foundation (CMS-0408427). Their financial supports are gratefully acknowledged.

5. REFERENCES

1. Eirich, F.R., 1956, *Theology: Theory and Applications*, Vol. I, Academic Press Inc., New York
2. Stepišnik, J., Luka, M. and Kocuvan, I., 1981, Measurement of cement hydration by ultrasonics, *Ceramic Bulletin*, 60(4), pp. 481-483.
3. T Öztürk, J. Rapoport, J.S. Popvics, and S.P. Shah, 1999, Monitoring the setting and hardening of cement-based materials with ultrasound, Concrete Science and Engineering 1(2), pp. 83-91.
4. Labouret, S., Looten-Baquet, I., Bruneel, C. and Frohly, J., 1998, Ultrasound method for monitoring rheology properties evolution of cement, Ultrasonics 36, pp. 205-208.
5. Vali, M.I., 2000, Hydration of cementitious materials by pulse echo USWR method, apparatus and application examples, Cement and Concrete Research 30, pp. 1633-1640.
6. Schultz, M.A. and Struble, L.J., 1993, Use of oscillatory shear to study flow behavior of fresh cement paste, Cement and Concrete Research 23(2), pp. 273-282.
7. Whorlow, R.W., 1992, *Rheological Techniques*, Second Edition, Ellis Horwood Limited, Chichester, England, pp. 335-395.
8. Tattersall, G.H. and Banfill P.F.G., 1983, *The Rheology of Fresh Concrete*, Pitman Publishing Inc. pp. 49-52.
9. Nachbaur, L., Mutin, J.C., Nonat, A., and Choplin, L., 2001, Dynamic mode rheology of cement and tricalcium silicate pastes from mixing to setting, Cement and Concrete Research, 31(2), pp. 183-192.
10. Nachbaur, L., Nonat, A., Mutin, J.C. and Choplin, L., 1997, Dynamic mode rheology of cement pastes, in *Why does Cement Set?*, *Proceedings of the Second RILEM Workshop on Hydration and Setting*, University of Bourgogne, Dijon, France, pp. 281-293.
11. Saak, A.W., 2000, Characterization and Modeling of the Rheology of Cement Paste: with Applications towards Self-Flowing Materials, *PhD Thesis*, Northwestern University.
12. Banfill, P.F.G., Carter, R.E. and Weaver, P.J., 1991, Simultaneous rheological and kinetic measurements on cement pastes, *Cement and Concrete Research*, 21(6), 485-489.

TEMPERATURE AND RELATIVE HUMIDITY ANALYSIS IN EARLY-AGE CONCRETE DECKS OF COMPOSITE BRIDGES

B.H. Oh,[1] S.C. Choi [2]and S.W. Cha [2]

[1]*Professor, Dept. of Civil Engineering, Seoul National University, Shinrim-dong, Gwanak-ku,Seoul, Korea;* [2]*Research Associates, Dept. of Civil Engineering, Seoul National University, Shinrim-dong, Gwanak-ku,Seoul, Korea*

Abstract: The purpose of the present study is to accurately assess the variation of temperature and relative humidity in early-age concrete decks of composite bridges. An analytical method that can reasonably predict the temperature and relative humidity changes was investigated. To examine the early-age behavior of composite bridges, several series of actual bridge deck members were tested, and temperatures were measured directly from the test members. The adiabatic temperature rise, ambient temperature, solar radiation, and ambient relative humidity were also measured and used in the analytical method, The present study indicates that the solar radiation has a pronounced effect on the temperature of the deck surface. The combination of solar radiation and relative humidity difference along the depth of the deck slab may induce larger stresses, which may cause cracking of deck. The present study allows more realistic assessment of temperature and relative humidity distribution of bridge decks which can be effectively used for evaluation of stresses and cracking at early ages.

Key words: early-age concrete deck, temperature, relative humidity, solar radiation, hydration heat, ambient temperature, ambient humidity

1. INTRODUCTION

Even though the causes of transverse cracking in early-age concrete decks have not been fully clarified, the thermal and shrinkage strains have been recognized as the major causes.[1-5]In this regard, several studies[1,5-7] have been carried out to predict the temperature and relative humidity variations that occur in bridge deck structures. In most studies, however, the characteristics of early-age concrete, that is, the rapid variation of temperature and internal relative humidity according to hydration process and curing conditions,

M.S. Konsta-Gdoutos, (ed.), Measuring, Monitoring and Modeling Concrete Properties, 305–316.

have not been properly considered.[8] The methods[6,7] that consider the effect of solar radiation on the temperature variation in bridge structures, are difficult to apply because the turbidity factor, subject to local climatic interaction, is difficult to estimate. Furthermore, when the cross-section of a bridge deck is assumed to shrink uniformly, differential drying shrinkage, which occurs in early-age concrete due to the drying at the surface exposed to outside, may be neglected.[5]

Therefore, the purpose of the present study is to accurately predict the actual variations of temperature and relative humidity, which are used to calculate the stresses and assess the risk of transverse cracking in the early-age concrete decks. To this end, an analytical method that can predict the actual temperature and relative humidity changes in the early-age concrete decks, was investigated. In particular, the method estimating the effect of solar radiation on the temperature of bridges from measured data in the near meteorological station was considered in the analytical method. Several series of test members, which can exhibit early-age behavior of composite bridge, have been made and the actual temperature variations according to elapsed time after concrete placement have been measured. The calculated temperatures based on the analytical method have been compared with the measured temperatures of test members. Furthermore, the variations of relative humidity in early-age concrete were analyzed by using the measured ambient relative humidity.

2. ANALYTICAL MODEL FOR TEMPERATURE AND RELATIVE HUMIDITY DISTRIBUTIONS

2.1 Heat transfer

The temperature within a structure at any point and time is found by applying established principles of heat transfer. As shown in Fig. 1, the temperature is affected by heat transfer within the material by conduction; heat generated within the material, e.g. hydration heat; heat transfer at the surface of the structure by convection. Therefore, heat transfer in a structure can be described by the Fourier equation of flow, relating the temperature T at each point of the cross section to the time t[9]

$$\rho c \frac{\partial T}{\partial t} = k\left(\frac{\partial^2 T}{\partial x^2} + \frac{\partial^2 T}{\partial y^2} + \frac{\partial^2 T}{\partial z^2}\right) + q_v \tag{1}$$

in which k is the thermal conductivity; ρ is the density; c is the specific heat; and q_v is the rate of heat generated per unit volume (exothermic heat of hydration).

The boundary condition associated with Eq. (1) is expressed as follows.

$$k\left(\frac{\partial T}{\partial x} n_x + \frac{\partial T}{\partial y} n_y + \frac{\partial T}{\partial z} n_z\right) + q = 0 \tag{2}$$

where n_x, n_y, n_z are the direction cosines of the unit outward vector normal to the boundary surfaces and q is the rate of energy transferred between the boundary and the environment per unit area. Numerical solution of the Fourier equation may be accomplished by finite element analysis.

2.2 Environmental interaction and solar radiation

The rate of energy transfer, q, between the boundary and the environment is the sum of convection q_c, thermal irradiation q_r, and solar radiation q_s (See Fig. 1).

$$q = q_c + q_r + q_s \tag{3}$$

Temperature difference between the bridge surface and the air causes a bridge deck to lose or gain heat by convection. Convection is expressed by Newton's Law of cooling as

$$q_c = h_c(T_s - T_a) \tag{4}$$

where h_c is the convection heat transfer coefficient; T_s is the temperature of the surface; and T_a is the air temperature [^{0}C].

The convection heat transfer coefficient, h_c, in Eq. (4) is a function of many variables such as wind speed, surface roughness, and geometric configuration of the exposed structure. The following relationship between heat transfer coefficient and wind velocity can be used.[10]

$$h_c = 5.6 + 4.0v \tag{5}$$

where v is the wind speed.

The heat transfer between the bridge surface and the surrounding atmosphere due to long wave radiation, i.e., thermal irradiation, produces a nonlinear boundary condition which can be modeled by the Stefan-Boltzmann radiation law as[6,7]

$$q_r = C_s e[(T_s + T^*) - (T_a + T^*)] \tag{6}$$

where C_s is Stephan-Boltzman constant; e is the emissivity coefficient relating the radiation of the bridge surface to that of an ideal black body ($0 \le e \le 1$); and T^* is the constant used to convert temperature in degree Celsius to degree Kelvin.

Dilger et al.[6], Elbadry and Ghali[7] suggested an equation to calculate the solar radiation on a bridge surface. This equation is a function of the turbidity factor. However, it is difficult to estimate the turbidity factor because it is affected by local climatic condition and air pollution. In the present study, therefore, the total solar radiation on the inclined surface of structures has been estimated from total solar radiation on the horizontal surface, measured hourly in the near meteorological stations.

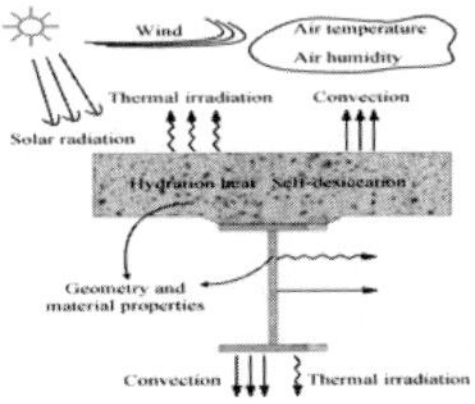

Figure 1. Heat transfer and relative humidity diffusion in a bridge exposed to environment

The rate of heat gain by solar radiation can be expressed as follows.[11]

$$q_s = aI_t \tag{7}$$

where a is the solar radiation absorptivity and I_t is the hourly total solar radiation on inclined surface.

Hourly total solar radiation on the inclined surface, I_t, consists of three components : beam(direct) radiation, diffuse radiation, and solar radiation diffusely reflected from the ground (See Eq. (8)).[11,12]

$$I_t = I_b \frac{\cos\theta}{\cos\theta_z} + I_d\left(\frac{1+\cos\beta}{2}\right) + (I_b + I_d)\rho'\left(\frac{1-\cos\beta}{2}\right) \tag{8}$$

where I_b is the beam(direct) radiation on a horizontal surface; I_d is the diffuse radiation on a horizontal surface; ρ' is the reflectance of the ground; θ is the angle of incidence; θ_z is the zenith angle; β is the slope.

The angle θ can be described in terms of several angles defining the positions of the sun relative to an observer on the earth and the orientation of the surface relative to the surface of the earth as follows (See Fig. 2).[11]

To estimate the hourly total solar radiation on the inclined surface I_t, it is necessary to distinguish between the beam(direct) and diffuse components of solar radiation. For horizontal surface, $\beta = 0°$ and the angle of incidence, θ, is the zenith angle of the sun, θ_z. Eq. (8) becomes

$$I = I_b + I_d \tag{9}$$

where I is the hourly total radiation on a horizontal surface, which is a measured value in the near meteorological stations.

To calculate the total radiation on the surfaces of other orientations from data on a horizontal surface, beam and diffuse radiation must be treated separately. Orgill and Hollands[11] have correlated I_d/I with hourly clearness index, k_T, the ratio of total radiation to extraterrestrial radiation for the hour. The equations for the correlation are

$$\frac{I_d}{I} = \begin{cases} 1.0 - 0.249k_T & for \quad k_T < 0.35 \\ 1.557 - 1.84k_T & for \quad 0.35 < k_T < 0.75 \\ 0.177 & for \quad k_T > 0.75 \end{cases} \tag{10}$$

where an hourly clearness index, k_T, can be defined as Eq. (13)

$$k_T = \frac{I}{I_0} \tag{11}$$

where I_0 is the extraterrestrial radiation on a horizontal surface for an hour (See Eq. (12)).

$$\begin{aligned} I_0 = {} & \frac{12 \times 3600}{\pi} G_{sc}\left[1 + 0.033\cos\frac{360d}{365}\right] \\ & \times \left[\cos\phi\cos\delta(\sin\tau_2 - \sin\tau_1) + \frac{2\pi(\tau_2 - \tau_1)}{360}\sin\phi\sin\delta\right] \end{aligned} \tag{12}$$

where G_{sc} is the solar constant.

Using Eq. (10), (11), and (12), diffuse radiation can be calculated from the measured horizontal surface radiation.

2.3 Relative humidity diffusion

Concrete self-desiccates internally due to cement hydration (See Fig.1). Thus, variations of internal relative humidity in concrete may be expressed as the variation of relative humidity due to moisture diffusion and self-desiccation of concrete, as shown in Eq. (15)[13-15]

$$\frac{\partial h}{\partial t}=\frac{\partial}{\partial x}\left(D_x\frac{\partial h}{\partial x}\right)+\frac{\partial}{\partial y}\left(D_y\frac{\partial h}{\partial y}\right)+\frac{\partial}{\partial z}\left(D_z\frac{\partial h}{\partial z}\right)+\frac{\partial h_{as}}{\partial t} \tag{15}$$

where h is the internal relative humidity; ∂h_{as} is the internal relative humidity change due to self-desiccation; D_x, D_y, D_z are diffusion coefficients. It is reasonably assumed that diffusion coefficients are homogenous, and can be described as a function of $D(h)$.

The humidity drop due to self-desiccation can be evaluated from the autogenous shrinkage strain ε_{cas} and the coefficient of shrinkage α_{sh} (See Eq. (18), (19)).[16]

$$\varepsilon_{cas}(t)=-\alpha_{as}\left(\frac{f_{cm}/f_{cm0}}{6+f_{cm}/f_{cm0}}\right)^{2.5}\left[1-\exp\left\{-0.2(t/t_1)^{0.5}\right\}\right]\cdot 10^{-6} \tag{16}$$

$$\frac{\partial h_{as}}{\partial t}=\frac{1}{\alpha_{sh}}\frac{\partial \varepsilon_{cas}}{\partial t} \tag{17}$$

where α_{as} is the material constant depending on cement type; f_{cm} is the average of the compressive strength of concrete at 28days ; f_{cm0} is 10MPa ; t is age ; t_1 is 1 day.

In the CEB-FIP 90 model code[17], the moisture diffusion coefficient for isothermal condition is expressed as a function of the relative humidity as seen in Eq. (20).

$$D(h)=D_1\left[m+\frac{1-m}{1+[(1-h)/(1-h_c)]^n}\right] \tag{18}$$

where D_1 is the maximum of $D(h)$ and function of compressive strength; $m=0.05$; $h_c=0.80$; $n=15$;

The boundary condition associated with Eq. (15) are expressed as

$$D\left(\frac{\partial h}{\partial x}n_x+\frac{\partial h}{\partial y}n_y+\frac{\partial h}{\partial z}n_z\right)=h_F(h_a-h_s) \tag{19}$$

where h_F is the coefficient of surface humidity transfer, h_a is the environmental relative humidity, and h_s is the relative humidity on the exposed surface.

3. MEMBER TESTS FOR EARLY-AGE BEHAVIOR OF CONCRETE DECK

3.1 TEST OUTLINE

To simulate the actual behavior of composite bridges, a concrete deck was placed on the steel girder and cured in the field. The deck experienced the temperature and relative humidity variation according to the hydration process and environmental conditions, e.g., hydration heat, self-desiccation, ambient temperature and relative humidity, solar radiation.[6-8,18] The temperature of a steel girder would be also changed by the heat conduction of concrete, and environmental conditions.

This nonlinear distribution will also cause self-equilibrating stresses in the cross-section.[6,7] To investigate the effect of restraint according to girder size, differently-sized steel beams were selected for test members M1, M2. In addition to the test members M1, and M2, the deck slab without steel girder M0 was also made. The length of test members is 8,000 mm and the dimension of cross-section of concrete deck is 1,000 mm′250 mm, as shown in Fig. 3. Each test member has a reinforcing bar of D16 (diameter 16 mm), and the stirrups were symmetrically placed with spacing of 500 mm. To develop a full composite behavior between the deck and steel girder, three shear connectors of diameter 22 mm were placed with the longitudinal spacing of 250 mm. A wood formwork was stripped 1 day after placement of concrete. The base and each end plates of the formwork were lined with 3 mm polystyrene sheet, so that the formwork would not restrain the free movement of test members. Table 1 shows the test variables and test member identification. The average compressive strength of concrete from three cylinders is 26.4 MPa at 28 days.

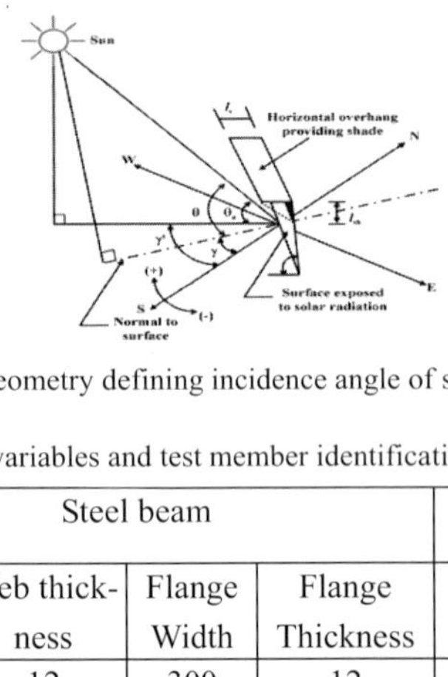

Figure 2. Geometry defining incidence angle of solar radiation

Table 1 Test variables and test member identification, unit : [mm]

Test member identification	Steel beam				Concrete slab		Span length
	Depth	Web thickness	Flange Width	Flange Thickness	Width	Thickness	
M1	488	12	300	12	1,000	250	8,000
M2	594	14	302	18			
M0*	-	-	-	-			

* : There is no steel girder in test member M0

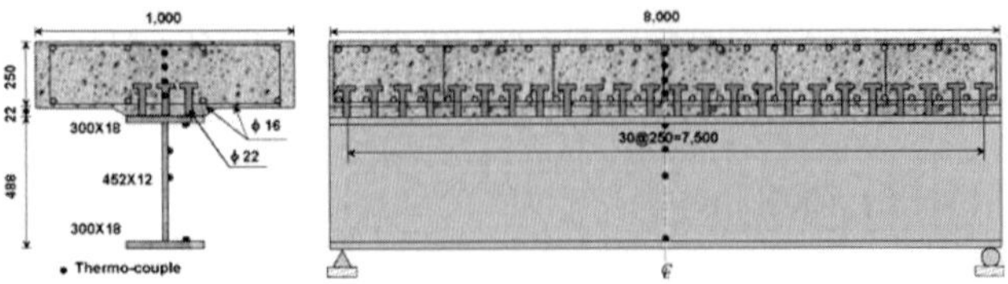

Figure 3. Details of test members (M1) and measurement scheme

3.2 TEST METHOD AND MEASUREMENT

It has been reported that the nonlinearity of temperature distribution in the depth-direction of a composite bridge has higher than those in the length- or width-direction. Therefore, four thermocouples were installed in the concrete deck and steel beam, respectively, at different levels along the depth to measure the temperature distribution of a composite bridge.

4. RESULTS OF TESTS AND ANALYSIS

4.1 Adiabatic temperature rise test

To evaluate quantitatively the effect of hydration heat on the temperature of concrete decks in its early ages, adiabatic test has been conducted. The adiabatic temperature of concrete was measured under conditions where there was neither gain nor loss of heat from or to the environment. Fig. 4 shows the profile of adiabatic temperature of the concrete used in the test member. Concrete hydrates very rapidly during the first few hours and reaches its maximum rate of temperature rise. However, the rate of increase decreases significantly thereafter. From adiabatic temperature $T_{ad}(t_{ad})$, the degree of hydration, α, which indicates the progress of hydration process, may be expressed as follows.[19]

$$\alpha(t_{ad}) - \frac{Q_{ad}(t_{ad})}{Q^{\max}} = \frac{\Delta T_{ad}(t_{ad})}{\Delta T_{ad}^{\max}} \tag{20}$$

where t_{ad} is the time in adiabatic test; Q_{ad} is the liberated heat in adiabatic test; $Q^{\max}$ is the maximum liberated heat; ΔT_{ad} is the temperature rise in adiabatic test ; $\Delta T_{ad}^{\max}$ is the maximum temperature rise.

Replacing the time t_{ad} with the equivalent time t_{eq} [19] in Eq. (20), one can express the degree of hydration in the adiabatic test in maturity-equivalent isothermal condition.[20]

$$\alpha(t_{eq}) = \frac{Q_{ad}(t_{eq})}{Q^{\max}} \tag{21}$$

in which

$$t_{eq} = \int_0^t \exp\left[\frac{E_A}{R_u}\left(\frac{1}{293} - \frac{1}{T(s)+273}\right)\right] ds \tag{22}$$

where E_A / R_u is activation temperature (K) ; $T(s)$ is temperature history, and 20°C is chosen as the reference temperature.

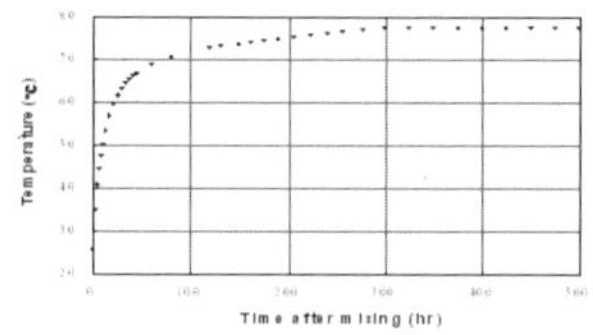

Figure 4. Result of adiabatic temperature rise test

4.2 MATERIAL PROPERTIES AND MEASURED ENVIRONMENTAL CONDITIONS

To analyze realistically the variations of temperature and relative humidity variation in composite bridges after placement of concrete, one needs to know the material properties as well as environmental conditions which are related to the analysis of temperature and relative humidity. Table 2 shows the material properties used in the 3-D finite element analysis of temperature and relative humidity. Fig. 5 indicates the measured ambient temperature and relative humidity histories, respectively. Table 3 shows the material properties, geographic locations, and the orientation of test members, which are related to the analysis of solar radiation. Fig. 6 shows the measured hourly total radiation on the horizontal surface in the near meteorological station. In order to complete the formulation of heat transfer and moisture diffusion, the measured initial temperatures were used and the initial relative humidity of concrete was assumed to be saturation.[16]

Table 2. Material properties used in the analysis of temperature and relative humidity

Material properties	Concrete	Steel
Density (kg/m^3)	2,500	7,800
Specific heat (J/kg °C)	1,047	419
Conductivity of heat (W/m °C)	2.7	51.9
Convection coefficient of heat (W/m^2 °C)	$5.6{+}4.0\nu^{*}$ (8.1^{**})	
Max. diffusion coefficient of humidity (m^2/s)	3.7×10^{-10}	-
Convection coefficient of humidity (m/s)	5.8×10^{-8}	-
Coefficient of shrinkage (10^{-6}/unit)	1,300	-

* : Average velocity ν=1.9 m/s
** : Wood formwork

4.3 ANALYSIS OF TEMPERATURES

Figures. 7 and 8 show measured and calculated temperatures of the slab (at a depth of 50 mm and 150 mm) and steel girder (top flange and web at mid-height) in the test member M1 (See Fig. 3). As shown in Figs. 7 and 8, the temperatures of a concrete deck during the early stages increased in spite of the decrease of ambient temperature and total solar radiation (See Fig. 5 and Fig. 6). This increase is due to the hydration heat of concrete. Furthermore, the increase of temperature in the steel girder was also caused by the conduction of hydration heat during the early stages. After this period, the temperatures of the test member varied according to the environmental conditions. Similar variations of temperature occurred in the test members M2 and M0.

Table 3. Material properties, geometry, and orientation of test membersrelated to solar radiation

Properties and Characteristics	Concrete	Steel
Stefan-Boltzmann constant	5.77×10^{-8} W/m^2 K	
Emissivity	0.88	0.80
Solar radiation absorptivity	0.50	0.685
Diffuse ground reflection	0.2	
Latitude	37.5 N	
Longitude	126.5 E	
Surface azimuth angle	-57.3°	

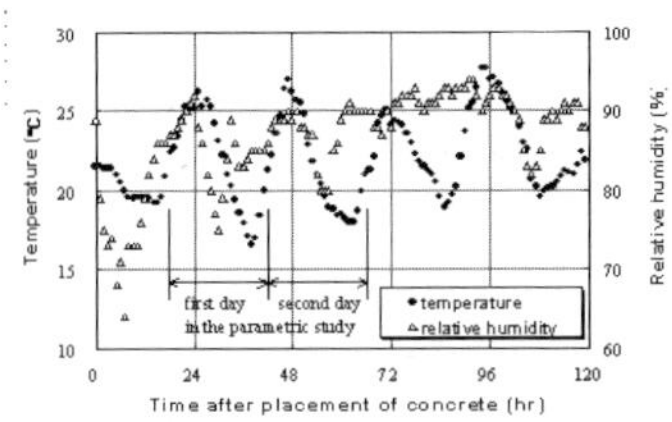

Figure 5. Measured ambient temperature and relative humidity

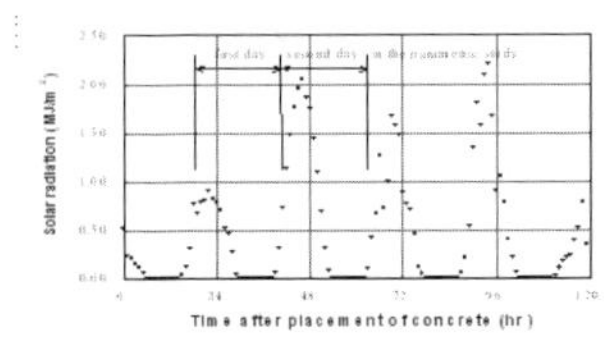

Figure 6. Measured hourly total radiation on horizontal surface in the near meteorological station

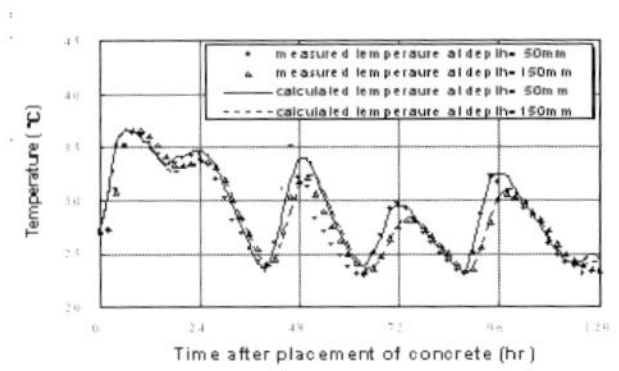

Figure 7. Measured and calculated temperature history of concrete deck in test member M1

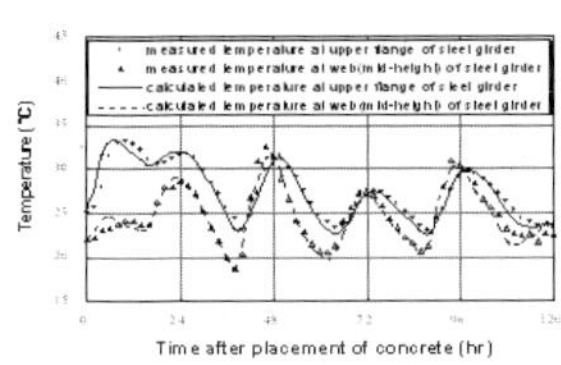

Figure 8. Measured and calculated temperature history of steel girder in test member M1

Figure. 9 shows the effect of thermal factors on the measured temperature changes at the surface (depth= 50 mm) of the concrete deck in test member M1. As mentioned above, the hydration heat increased the initial temperature. On the other hand, the ambient temperature causes the decrease of temperature during the early stages. This decrease is due to the fact that the temperature of concrete at placement was higher than the ambient temperature. The solar radiation has little effect on temperature of concrete deck right after placement because the intensity of radiation was insignificant. As the time after placement elapses, the effect of hydration heat disappeared. Solar radiation mainly affected temperature variations of concrete deck.

Figure. 10 shows the effect of thermal factors on the measured temperature changes at the top-flange in the steel girder in test member M1. The temperatures of the upper portion of the steel girder are directly affected by the heat of hydration at early ages. Therefore, the temperatures of this upper portion become higher at early ages. After this period, the temperature variations of the steel girder were affected by the solar radiation as well as the ambient temperature. However, the magnitude of the temperature changes in the steel girder was smaller than those of the deck because the solar radiation on the steel girder is changed by the shadow of the overhanging cantilever slab.

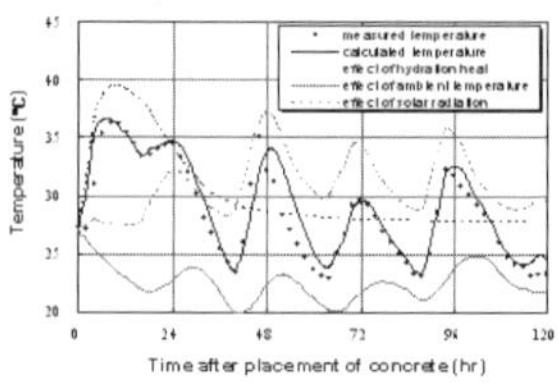

Figure 9. Effect of hydration heat and environment on the temperature of concrete deck at the depth of d=50mm in test member M1

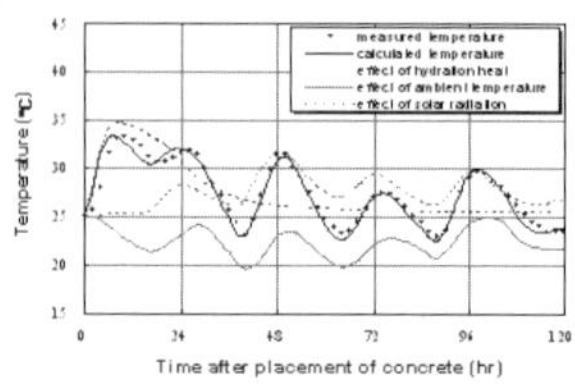

Figure 10. Effect of hydration heat and environment on the temperature of upper flange in test member M1

4.4 ANALYSIS OF RELATIVE HUMIDITY

The internal relative humidity of a concrete deck significantly varies according to the depth from the exposed surface. Furthermore, the changes in the relative humidity were greater at the depth close to exposed surface than at the inner region of the concrete deck, as expected.

5. PARAMETERS AFFECTING TEMPERATURE AND RELATIVE HUMIDITY

5.1 Effect of hydration heat, ambient temperature, and solar radiation

To investigate the effects of various factors affecting the temperature of the early-age concrete decks, a parametric study was carried out with the test member M1. To consider the effect of hydration heat, an increase of 5°C in maximum adiabatic temperature rise than the present test values was assumed and compared (See Fig. 4). The measured ambient temperatures at first and second day, respectively, were repeatedly used to investigate the effects of the ambient temperature (See Fig. 5). Furthermore, the measured solar radiations at first and second day, respectively, were also used to analyze the effects of the solar radiation (See Fig. 6). The time of placement was assumed to be 8:00 a.m., at which the measured ambient temperature on the second day was the same as that on the first day. The geographic location, orientation and material properties except the hydration heat remained unchanged.

The increase of adiabatic temperature causes the rise of temperature during the early stages. However, the variation of ambient temperature does not significantly affect the temperature of a deck. The solar radiation serves as the main parameter affecting the temperature of an early-age concrete deck. Therefore, the solar radiation must be properly considered to control the temperature of early-age concrete decks in composite bridges.

5.2 EFFECT OF AMBIENT RELATIVE HUMIDITY

The ambient relative humidity was assumed to be 90 %, 75 %, and 60 % to investigate the variation of internal relative humidity in the early-age concrete decks. Fig. 11 shows the variations of internal relative humidity at a depth of 25 mm of a concrete deck in test member M1. Fig. 11 indicates that as the ambient air becomes dry, the internal humidity decreases, which may also cause the large tensile stresses and thus increase the probability of cracking.

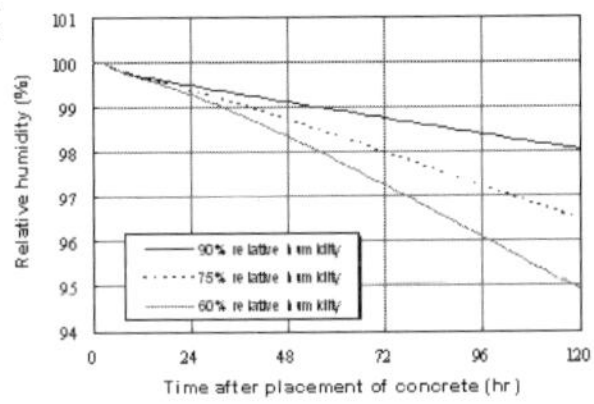

Figure 11. Effect of ambient relative humidity on the internal relative humidity at a depth of 25 mm of concrete deck in test member M1

6. CONCLUSIONS

The purpose of the present study is to assess accurately the variations of temperature and relative humidity in early-age concrete decks of composite bridges. To this end, an analytical method that can reasonably predict the change of temperature and relative humidity was investigated. In particular, the method estimating the effect of solar radiation on the temperature of bridges from measured data in the near meteorological station was considered in the analytical method. The following conclusions were drawn from the present study.

1. The temperatures obtained from the analytical method in this study agree very well with measured temperatures of test members. Therefore, in order to accurately predict the temperature variations of composite bridges at the early ages, various factors, e.g., hydration heat, ambient temperature, and solar radiation, must be realistically considered.

2. The hydration heat of concrete at an early age increases the temperatures of the deck and steel girder during the early stages. However, solar radiation has a greater effect on the temperature of the deck at a depth close to exposed surface.

3. The temperature of steel girder is increased by the conduction of deck heat. However, the changes of temperature in the steel girder are smaller than those of the deck because the solar radiation on the steel is reduced by the shadow of overhanging cantilever slab.

4. The internal relative humidity significantly varies according to the depth from the exposed surface, and the change of relative humidity was greater at the depth close to exposed surface than at an inner region of the concrete. This may cause large tensile stresses near the deck surface which may cause serious cracking.

5. The changes of the relative humidity are mainly affected by self-desiccation in the central portion of the early-age concrete deck.

6. The present study allows more realistic assessment of the temperature and relative humidity distribution in composite bridges at early ages which can be effectively used for predicting the stress distributions and the risk of transverse cracking in early-age bridge decks.

Acknowledgment

The financial support from the National Research Laboratory (NRL) Program of Korea is gratefully acknowledged.

7. REFERENCES

1. P.D. Krauss, and E.A. Rogalla, Transverse Cracking in Newly Constructed Bridge Decks, *NCHRP Report 380*, Transportation Research Board, National Research Council, Washington, D.C., 59 pp.
2. G. E. Ramey, A. R. Wolff, and R. L. Wright, Structural Design Actions to Mitigate Bridge Deck Cracking, *Practical Periodical on Structural Design and Construction*, ASCE, V. 2, No. 3, pp. 118-124 (1997)
3. R. Springenschmid, (ed.), *Prevention of Thermal Cracking in Concrete at Early Ages*, State of the Art Report by RILEM TC 119, E&FN Spon, 348 pp. (1998)
4. A. I. Mohsen, Investigation of Cracking in Concrete Bridge Decks at Early Ages," *Journal of Bridge Engineering*, ASCE, V. 4, No. 2, pp. 116-124 (1999)
5. D. Cusson, W. L. Repette, Early-age Cracking in Reconstructed Bridge Barrier Wall, *ACI Materials Journal*, V. 97, No. 4, pp. 438-446 (2000)
6. W. H. Dilger, A. Ghali, M. Chan, M. S. Cheung, and M. A. Maes, Temperature Stresses in Composite Box Girder Bridges, *Journal of Structural Engineering*, ASCE, V. 109, No. 6, pp. 1460-1478 (1983)
7. M. Elbadry, and A. Ghali, Temperature Variations in Concrete Bridges, *Journal of Structural Engineering*, ASCE, V. 109, No. 6, 1983, pp. 2355-2374 (1983)
8. B. H. Oh, and S. W. Cha, Nonlinear Analysis of Temperature and Moisture Distributions in Early-age Concrete Structures Based on Degree of Hydration, *ACI Materials Journal*, V. 100, No. 5, pp. 361-370 (2003)
9. R. D. Cook, D. S. Malkus, and M. E. Plesha, *Concepts and Applications of Finite Element Analysis*, John Wiley & Sons, 630 pp. (1988)
10. CEB, Thermal Effects in Concrete Structures, Bulletin d'information No. 167, 122 pp. (1985)
11. J. A. Duffie, and W. A. Beckman, *Solar Engineering of Thermal Processes*, John Wiley & Sons, 762 pp. (1980)
12. J. F. Orgill, and K. G. T. Hollands, Correlation Equation for Hourly Diffuse Radiation on a Horizontal Surface," *Solar Energy*, Vol. 19, pp. 357-359 (1977)
13. Z. P. Bazant, and L. J. Najjar, Nonlinear Water Diffusion in Nonsaturated Concrete, *Materials and Structures*, V. 5, No. 25, pp. 3-20 (1972)
14. JCI, Technical Committee on Autogenous Shrinkage of Concrete, Part 1, Committee Report, *Autogenous Shrinkage of Concrete*, Proceeding of the International Workshop organized by JCI, Tazawa, E., (Ed.), E&FN Spon, pp. 3-67 (1998)
15. O. M. Jensen, and P. F. Hansen, Autogenous Deformation and RH-Change in Perspective, *Cement and Concrete Research*, V. 31, pp. 1859-1865 (2001)
16. A. M. Alvaredo, *Drying Shrinkage and Crack Formation*, Building Materials Report, No. 5, Laboratory For Building Materials, Swiss Federal Institute of Technology, Zurich (Switzerland), 102 pp. (1994)
17. CEB-FIP Model Code 1990, *CEB Bulletin* d'information No. 213/214, 437 pp. (1993)
18. F. A. Branco, P. A. Mendes, and E. Mirabell, Heat of Hydration Effects in Concrete Structures, *ACI Materials Journal*, V. 89, No. 2, pp. 139-145 (1992)
19. J. Byfors, *Plain Concrete at Early Ages*, CBI Report FO 3:8, Sweden, 345 pp. (1980)
20. F. S. Rostasy, and M. Laube, Experimental and Analytical Planning Tools to Minimize Thermal Cracking of Young Concrete, *International RILEM Symposium on Test during Concrete Construction*, H. W. Reinhardt, (eds.), Chapman and Hall, pp. 207-223 (1991)
21. J. E. Jonasson, *Modelling of Temperature, Moisture and Stresses in Young Concrete*, Division of Structural Engineering, Luleå University of Technology, Doctoral Thesis, 153D, 225 pp. (1994)

PRELIMINARY NUMERICAL ASSESSMENT OF MICROCRACKING CAUSED BY AUTOGENOUS SHRINKAGE IN A HETEROGENEOUS SYSTEM

J.H. Moon, J. Couch and J. Weiss

[1]*Ph.D. Student, Purdue University;* [2]*Undergraduate Research Assistant, Purdue University;* [3*]*Corresponding Author, Associate Professor, Assistant Head of Research, Purdue University*

Abstract: The majority of research that has been performed on the restrained shrinkage cracking behavior of concrete has assumed that concrete acts as a homogeneous material. While this assumption enables the potential for through cracking to be assessed, it does not provide realistic information on the damage that may develop as a result of the heterogenous nature of the composite (i.e., microcracking). This research uses an object oriented finite element analysis (OOF) to simulate the behavior of the concrete on a meso-scale by considering it as a two-phase system of 'inert aggregates' and 'shrinking paste'. The meso-scale composite images used in the simulations were directly obtained from mortar samples. The procedure of acquiring the images, analyzing the images to enable phase separation, meshing the finite element model, performing the simulation, and analyzing the results is discussed. This paper illustrates how concrete composites could be analyzed on the meso-scale as well as how this information can be used to improve the design of concrete.

Keywords: Shrinkage; Cracking; Finite element analysis; Image analysis; Microcracking

1. INTRODUCTION

Concrete changes volume in response to chemical reactions, moisture changes, and temperature variations. Many documents highlight the importance of considering the influence of volume change on the potential for cracking (ACI 209R-92; Kosmatka et al., 2003); however, the majority of these documents and simulation models (Bernard et al., 2002; Foster, 2000; Weiss et al. 1998) treat the concrete as a homogenous material and focus only on the development of large, visible cracks. Currently, few engineers explicitly consider the heterogeneous nature of concrete in designing for shrinkage cracking. Researchers (Grzybowski and Shah, 1990; Bisschop and van Mier, 2001; Kim and Weiss, 2003; Moon et al. 2005) have shown that the combination of the external restraint from

M.S. Konsta-Gdoutos, (ed.), Measuring, Monitoring and Modeling Concrete Properties, 317–323.

the structure surrounding the concrete and the internal restraint provided by the aggregate can lead to the development of microcracks. These microcracks may play a key role in relaxing stresses (Pease et al., 2005), increasing transport (Yang et al., 2005) and serving as an initiation point for visible crack formation (Chariton et al., 2002). This research describes preliminary results from an effort to utilize a simulation technique to assess the potential for microcrack development in concrete. The behavior of realistic two dimensional concrete composites system were simulated using an Object Oriented Finite Element (OOF) code which has been developed by NIST (National Institute of Standards and Technology, Langer et al., 2001).

2. SIMULATION PROCEDURES

The simulation procedures can be divided into three sections (pre-processing, processing and post-processing) each of which is described in greater detail below.

2.1 Pre-Processing

This research used two-dimensional images directly obtained from mortar and concrete samples. To obtain the images in this paper, a mortar specimen was saw-cut and polished (Moon, 2006). The polished surface was stained using phenolphthalein to change the color of the paste phase to pink thereby enabling each phase to be more clearly identified. The stained surface was then scanned using a scanner with a resolution of 600 dpi.

The scanned image was saved as a TIFF format file that recorded the initial color of each pixel. The saved image file is re-colored through the use of a color histogram to identify areas of similar phase (i.e., areas of similar color) that can be converted to a gray scale image (i.e., black, gray, and white) to represent either phase (aggregate, interfacial zone, or paste) with the assistance of Paint Shop Pro®. In this research, each phase was assumed to have the same material properties throughout the phase. The re-colored image was then saved as a portable pixmap image file and transferred to PPM2OOF code (complementary program of OOF code for pre-processing) for meshing. The mesh was successively refined until the boundaries between phases have a shape that was reasonably similar to the original image. The refinement of the mesh at the boundaries was performed using the PPM2OOF code (Carter et al. 2000). During the meshing procedures, the total number of meshed elements was maintained to be less than 200,000 which was found to be consistent with the computer (RAM-1.5 GB) used for this research. For this research, 3-node triangular elements were used and perfectly bonded condition was applied to the interface between paste and aggregates. Figure 1 shows the pre-processing procedure for the OOF simulation (Moon 2006).

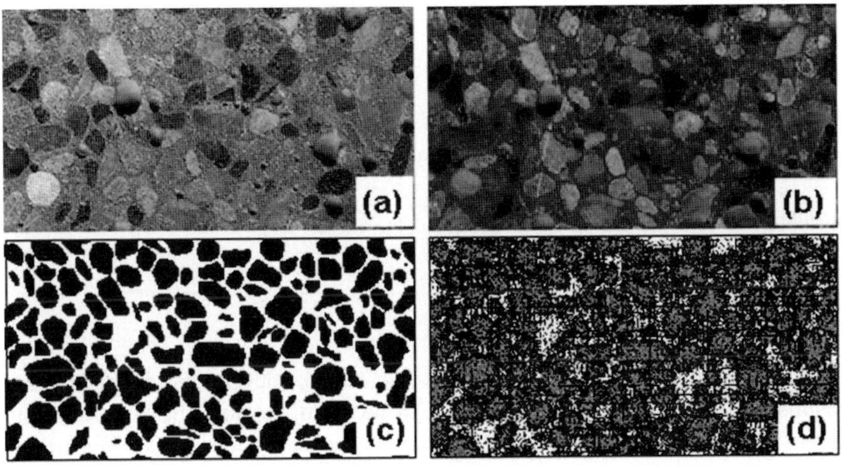

Figure 1. Pre-processing procedures: a) saw-cut surface, b) polished surface treated with phenolphthalein, c) the re-colored surface after using a histogram to convert it to a two phase image, and d) the final meshed image (Size of original image was 2.54 cm × 1.27 cm)

2.2 FINITE ELEMENT SIMULATION

Two different boundary conditions were considered in this study. The unrestrained specimen corresponds to a free shrinkage specimen while the restrained specimen considers restraint in one direction (i.e., the x-direction) similar to what may be expected in a bridge deck or pavement. The loading consisted of the application of a uniform volume change in the paste and no volume change in the aggregate using a temperature substitution analogy (Moon et al., 2005, Moon 2006).

Concrete behaves as a quasi-brittle material, while the behavior of each phase of the concrete (paste and aggregates) more closely resembles a brittle material. In this simulation, each phase was assumed to be perfectly brittle using the 'damisotropic' function in OOF code (Carter et al., 2000) that decreases the elastic modulus of an element to zero when those elements reach to the maximum stress (tensile strength). Simulations were performed by successively increasing the shrinkage strain of the paste and iterating until the composite system reached equilibrium.

To observe the influence of autogenous shrinkage on microcracking and cracking, three different types of paste are chosen and used for the simulations (Table 1). Table 1 shows the material properties of aggregates (sand) assuming aggregates as silicious quartz (West, 1995). For the studies shown in this paper the material properties of cement paste were not considered to vary with time due to hydration and viscoelasticity was not considered.

A mortar specimen was used in this investigation consisting of sieved fine aggregates that were retained between the #16 and #8 sieves (size distribution of fine aggregates: 1.19 ~ 2.36 mm). The mixture had an aggregate volume of 55 % (the image was observed to contain 55.7 % of fine aggregate). After casting, the specimen was cured for two days before pre-processing (saw-cutting, polishing, chemical treatment, scanning, re-coloring, and meshing) as shown in Figure 1.

Table 1. Material properties of paste and aggregates

	Paste-1	Paste-2	Paste-3	Aggregate
Elastic modulus (GPa)	20	10	5	50
Poisson's ratio	0.18	0.18	0.18	0.20
Tensile strength (MPa)	5.00	2.50	1.25	20.00
Compressive strength (MPa)	20	10	5	200

2.3 POST-PROCESSING

The post-processing portion of this work consisted of developing methods to assess the results of the finite element analyses. In addition to using the stresses obtained directly at the nodes from the finite element analysis, image analysis was performed using the color contour images obtained from OOF simulations where each color represents a specific level of stress or strain energy density. The area fraction of each color level was obtained using Image-Pro® Plus™ program.

In this analysis, consistent maximum and the minimum limits for the color scale needed to be determined. The majority of the analysis is only concerned with tensile stresses, and as such the minimum limit of the scale for the color contour of stress distribution is fixed to zero. Therefore, compressive stress is represented along with the zero stress elements. The maximum stress was determined to be 6 MPa.

After the determination of the maximum and minimum limits of scale, type of color scale was chosen. In this research, the 'Tequila_sunrise' option in OOF was chosen (Figure 2). The 'Tequila_sunrise' color scale varies color from red (minimum magnitude of scale) to yellow (maximum magnitude of scale) changing RGB (red, green, blue) values from 255,0,0 to 255,221,0. The value of red is always 255 and blue is always zero in the scale. Therefore, each color level can be defined by observing the change of green color value.

In this research, 80 scale intervals were used. The stress distribution and the energy density distribution were saved as ppm format file for each magnitude of the shrinkage strain in the cement paste. The cracked images were obtained from the OOF code that showed changes in the color of elements to black when the elements reach to the maximum stress. All saved image files were then transferred to Image-Pro® Plus™ program to obtain the area fraction of each color level.

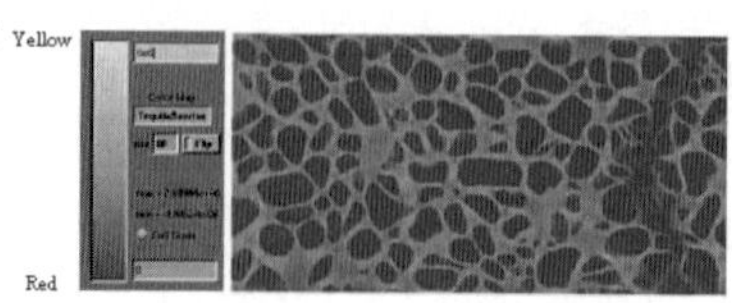

Figure 2. Color scaled image of stress distribution in a composite (Paste-1, red is zero stress, yellow is a stress of 6 MPa one directionally restrained, ε_{paste}= -140 με)

3. RESULTS AND DISCUSSIONS

Two different specimen boundary conditions were considered in the simulations (i.e., restrained and free). Figures 3 (a) and (b) show images of the cement composites (paste-1) where the white areas represent the paste, the gray areas represent the aggregates, and the black areas in the specimen represent cracking. The specimens restrained in the x-direction (Figure 3a), showed localized cracking (Figure 3 (a)) when the shrinkage of the paste reached -140 με for paste-1, -125 με for paste-2, and -120 με for paste-3. The unrestrained specimens (i.e., free) were observed to begin cracking at a slightly higher paste shrinkage strain at first cracking as compared to the restrained specimen (-150 με for paste-1, -140 με for paste-2 and paste-3). In addition, the unrestrained specimen (i.e., free) demonstrated distributed cracking that increased until the simulations were stopped at -300 με (Figure 3 (b)).

The average section stress at failure was determined to be 1.7 MPa, 0.9 MPa, and 0.5 MPa for paste-1, paste-2, and paste-3, respectively for the restrained specimens. These results are consistent with the idea that the composite could have a lower average failure stress (equivalent strength) than the strength of either phase due to stress concentrations. Figures 3 (c) and (d) show the cumulative area fraction of the specimen with principal stresses higher than a specified stress level in a composite (paste-1). The residual stress level in Figure 3 (c) shows that only a small volume fraction (0.5 %) of the specimen exceeds the tensile strength of the paste (if cracking is not permitted) at the time of failure. Figure 3 (c) shows that, after failure, (i.e., through cracking) the restrained specimen had a similar stress distribution as the free specimen as one may expect due to internal restraint from the aggregate.

Figure 3 (e) and (f) show the relationships between the shrinkage strain of paste and the cumulative normalized stored energy (the product of the strain energy density and its volume in the sample). In the case of the unrestrained (i.e., free) specimen, cracks began to initiate when the shrinkage strain of the paste reached -150 με for paste-1, -140 με for paste-2, and paste-3. The normalized stored energy in the unrestrained specimen increased as microcracking continued at the external boundaries of the aggregate (until the shrinkage strain reaches -220 με for paste-1, -210 με for paste-2, and -200 με for paste-3). As the shrinkage of the paste continues to increase, the microcracks begin to coalesce resulting in a decrease in the stored strain energy.

The simulations described in this paper did not consider an interfacial transition zone. Current simulation results indicate that the addition of a transition zone can permit more microcracking before the restrained specimen fails. Further work is needed to correlate the results of the simulations with experimental observations. Once this is done simulations can be performed to assess the influence of the aggregate volume fraction, aggregate shape, aggregate size distribution, and bond conditions.

4. SUMMARY

This paper presented the preliminary application of a finite element modeling approach for assessing the potential for microcracks to develop in a concrete composite undergoing autogenous (i.e., uniform) shrinkage. The use of the NIST object oriented finite element code enables scanned images to be used to develop a finite element mesh which can be used to analyze materials on the meso-scale. The results were analyzed using distributions of stress and stored energy. Different behavior was observed in specimens that were unrestrained (i.e., free) and specimens that were restrained. The unrestrained specimens permitted substantial microcracking while the restrained specimen permitted less microcracking before through cracking occurred. Further research being performed to quantify the influence of the aggregate volume fraction, size distribution, shapes, and bond conditions.

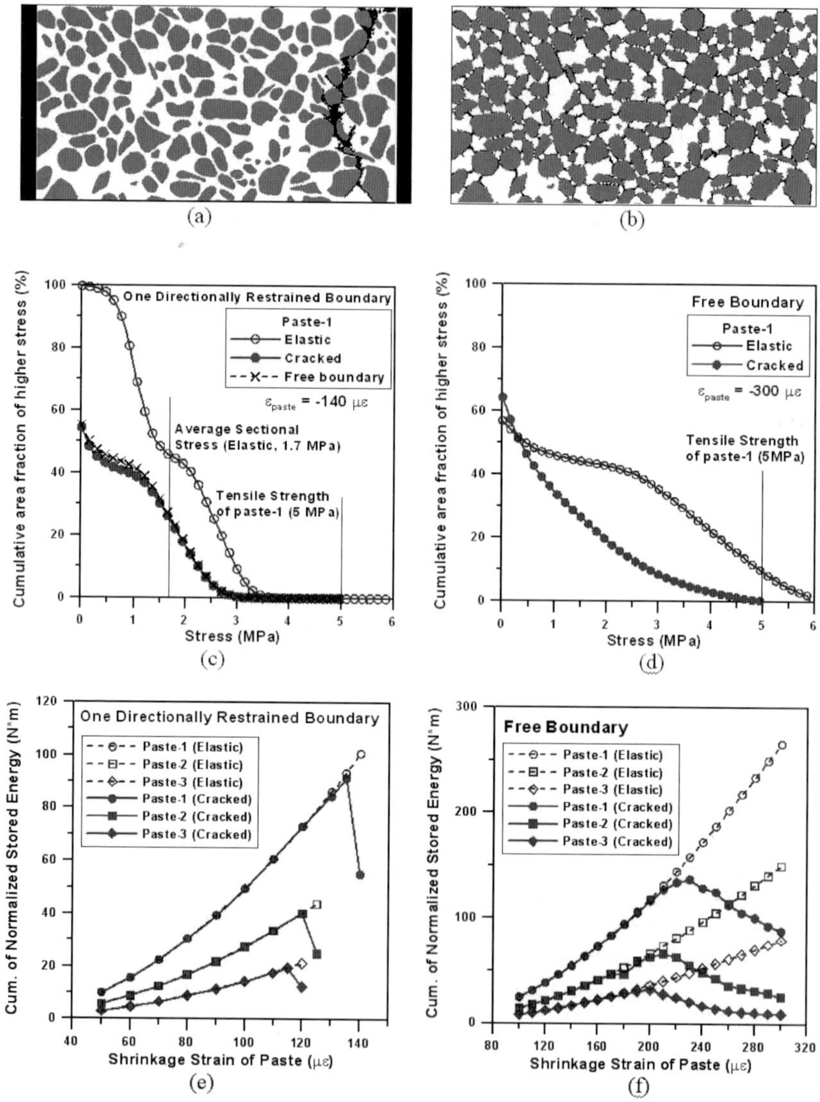

Figure 3. Simulation results: cracked image for a (a) restrained and (b) free specimen, cumulative stress distribution for a (c) restrained and (d) free specimen, and cumulative normalized stored energy for a (e) restrained and (f) free specimen

Acknowledgements

The authors gratefully acknowledge support received from the National Science Foundation (NSF) through Grant No. 0134272. Any opinions, findings and conclusions or recommendations expressed in this material are those of the authors and do not necessarily reflect the views of the National Science Foundation (NSF).

6. REFERENCES

1. ACI 209R-92, 1997, Prediction of Creep, Shrinkage, and Temperature Effects in Concrete Structures, Manual of Concrete Practice, *American Concrete Institute*, Farmington Hills, MI
2. Bernard, O. and Bruhwiler, E., 2002, Influence of autogenous shrinkage on early age behavior of structural elements consisting of concretes of different ages, *Materials and Structures*, Vol. 35, No. 235, pp. 550-556
3. Bisschop, J., and van Mier, J.G.M., 2001, Shrinkage microcracking in cement-based materials with low water-cement ratio, *RILEM International Conference on Early Age Cracking in Cementitious Systems*, EAC'01, ed. K., Kovler, and A. Bentur
4. Carter, W.C., Langer, S.A., and Fuller, E.R., 2000, The OOF Manual: Version 1.0.8.6, *National Institute of Standards and Technology*, (http://www.ctcms.nist.gov/oof/download/index.html#Manual)
5. Chariton, T., Kim, B., and Weiss, W.J., 2002, Using passive acoustic energy to quantify cracking in volumetrically restrained cementitious systems, 15th ASCE EMD Conference, New York, (http://www.civil.columbia.edu/em2002/)
6. Foster, S.W., 2000, HIPERPAV-Guidance to avoid early-age cracking in concrete pavements, ACI Special Publication, Vol. 192, pp. 1109-1122
7. Grzybowski, M., and Shah, S.P., 1990, Shrinkage cracking of fiber reinforced concrete, *ACI Materials Journal*, Vol. 87, no.2, pp. 138-148
8. Kim, B., and Weiss, W.J., 2003, Using acoustic emission to quantify damage in restrained fiber reinforced cement mortars, *Cement and Concrete Research*, Vol. 33, no.2, pp. 207-214
9. Kosmatka, S.H., Kerkhoff, B., and Panarese, W.C., 2002, *Design and Control of Concrete Mixtures*, 14th ed., Portland Cement Association, Skokie, Illinois
10. Langer, S. A., Fuller, E. R., and Carter, W. C., 2001, OOF: An Inage-Based Finite-Element Analysis of Material Microstructures, *Computing in Science and Engineering*, pp. 15~23
11. Moon, J.H., Rajabipour, F., Pease, B., and Weiss, J., 2005, Autogenous shrinkage, residual stress, and cracking in cementitious composites: The influence of internal and external restraint, *Fourth international seminar on self-desiccation and its importance in concrete technology*, Maryland, USA
12. Moon, J.H., 2006, Shrinkage, restraint, residual stress, and cracking in heterogeneous materials, Ph.D. Thesis, *Purdue University*
13. Pease, B.J., Shah, H.R., and Weiss, W.J., (2005), Shrinkage behavior and residual stress development in mortar containing shrinkage reducing admixture (SRA's), *ACI-Special Publication on Concrete Admixtures*, Vol. 227, pp. 285-302
14. Weiss, W.J., Yang, W., and Shah, S.P., Shrinkage Cracking of Restrained Concrete Slabs, *Journal of Engineering, Mechanics Div., ASCE*, 124(7), pp. 765-774
15. West, Terry R., 1995, *Geology Applied to Engineering*, Prentice Hall, Inc, Simon/Schuster Company, Englewood Cliffs, New Jersey
16. Yang, Z., Weiss, W.J., and Olek, J., 2005, Using acoustic emission for the detection of damage caused by tensile loading and its impact on the freeze-thaw resistance of concrete, *International Conference on Construction Materials: ConMat'05*, Vancouver, Canada

FINITE ELEMENT MODELING OF EARLY-AGE CRACKING IN RESTRAINED CONCRETE RING SPECIMENS

O.G. Stavropoulou[1], M.S. Konsta-Gdoutos[2] and G.E. Papakaliatakis[2]
[1]Graduate Student, National Technical University of Athens; [2]Associate Professors, Department of Civil Engineering, School of Engineering, Democritus University of Thrace

Abstract: High strength–high performance concrete (HPC) is an engineered product that provides specific performance advantages in terms of strength and conductivity in comparison with the conventional strength concrete[1,2]. As the use of high performance concrete has increased, problems with early-age cracking have become prominent. The objective of this paper is to present a numerical analytical study using a finite element analysis, to predict cracking of high performance concrete in restrained shrinkage. The numerical method estimates the residual stress development and, accordingly, quantifies the material's potential for cracking.

Key words: early-age cracking; high performance concrete; ring test; restrained shrinkage; finite element analysis.

1. INTRODUCTION

Early age cracking sensitivity is a well known phenomenon, which is associated with mechanisms that are related to microstructural changes and chemical reactions, taking place during the first few days after mixing the binder with water[3]. High strength concretes appear to be liable to cracking in early ages, due to the extremely low water to cement ratio used. Several studies attribute early age cracking of high performance cementitious systems to autogenous shrinkage and the resulting self-desiccation of concrete.

Historically, many test procedures have been used to determine the early age behaviour of concrete under restrained shrinkage. Among them, the ring test is perhaps the most common method applied to assess early age cracking in HPC's. In this test a concrete ring is cast around a restrained core (usually steel) and is allowed to shrink against it. As the concrete dries, shrinkage is prevented by the steel ring resulting in the development of tensile stresses in the concrete. Early age cracking is observed as the residual tensile

M.S. Konsta-Gdoutos, (ed.), Measuring, Monitoring and Modeling Concrete Properties, 325–330.

stresses developed surpass the tensile strength of the concrete. In other words, cracking is the result of the combination of the high tensile stresses developed and low fracture resistance of the young concrete.

Although this experimental method is simple and versatile in use, the fact that is limited with respect to providing quantitative information on early age stress development and the exact conditions under which cracking is taking place is referred as a disadvantage of the method. The work presented here describes the finite element analysis that was used in order to acquire quantitive information on early age stress development and early age cracking, using experimental data acquired from the ring test.

2. NUMERICAL MODELING

The FE modeling of the ring test was performed using the pre-processor of Abaqus. The dimensions of the ring setup, as well as the boundary conditions are shown in Figure 1[4,5]. According to the test, a 75 mm thick concrete annulus was cast around a steel ring. The steel wall thickness was 19.05 mm. The steel ring had 4 strain gages attached at the mid-height of the inner surface of the steel ring. As the concrete shrinks, the steel ring was pressurized at the outer surface. Steel strain was monitored over time. The average strain information monitored by the stain gages was used as an input for the FE analysis.

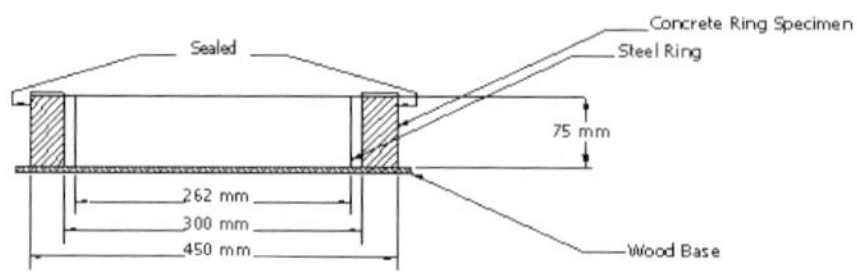

Figure 1. Ring specimen geometry

2.1 Material Properties

A number of difficulties arose when trying to simulate the restrained concrete specimen using FE analysis, mainly because of the dynamic nature of the young cement paste. For the purpose of this study it was decided that the numerical modeling of the ring test should be based on the material properties and the corresponding experimental values of the fifth day (after casting). The mechanical properties of the two materials used, concrete and steel respectively, are presented in Table 1. The concrete material properties were experimentally determined and also compared with published data[4,5]. In particular, the experimental values presented in Table 1 are referring to a high performance concrete mixture with a 28-day compressive strength of 85 MPa, corresponding to a 5-day compressive strength of 57 MPa. The modulus of elasticity of concrete in compression was taken equal to 28.77 GPa. Abaqus pro-processor used the same value for the modulus of elasticity (28.77 GPa) to plot the stress-strain curve in tension. The tensile strength of concrete was taken equal to 4.6 MPa.

With respect to the ring steel, material properties of ordinary steel were chosen from reference, after variations in the steel mechanical properties is shown to have only minor effect on the overall behaviour of the test procedure. On the contrary, the steel ring wall thickness appears to have a major effect on the strain monitored by the strain gages. It is important to note that thinner steel rings demonstrate a much higher strain level than the thicker rings[4].

Table 1. Material properties of concrete and steel

Material	Concrete	Steel
Modulus of elasticity, GPa	28,77	200
Compressive strength, MPa	57	624
Tensile strength, MPa	4,6	-
Poisson's ratio	0,2	0,33

The stress-strain relation of the two materials under compression was also required in the FE modeling. As no experimental data were available with respect to the stress-strain relation of early age concrete (namely the stress-strain plot referring to the age of five days), it was decided that this should plotted using Thorenfeldt's analytical equation. Abaqus allows the non-linearity of materials to be defined by the user in the form of solution-dependent variables, for example strain. Namely, the CONCRETE command within Abaqus allowed the compressive stress to be paired with the plastic strain. The same solution-depended methodology used with the concrete properties was undertaken for steel, using the command PLASTIC (although the behaviour of the steel ring was proved to be linear).

2.2 Modeling of the Ring Test

Abaqus pre-processor was used to generate the solid geometry of the intended model and to mesh it with the desired elements. In order to simulate the exact conditions of the experiment two separate FE models were developed, the first simulating the steel ring and the second the concrete ring respectively. The two FE models are shown in Figure 2. According to the test set up, the two rings are constantly in contact and as the concrete ring dries, strain gages monitor the steel ring strain. In the FE analysis the strain information was used as input using the assumption of a constrained radial displacement of the nodes of the outer circumference of the steel ring. The average strain of the steel ring was taken equal to -63 $\mu\varepsilon$ (experimental value corresponding to the average steel strain of the fifth day). The results of the numerical modeling of the steel ring were used as the input data for the modelling of the concrete ring, based on the assumption that the outer circumference of the steel ring was in fact the inner circumference of the concrete ring, and conditions of equal radial stress component development existed in the interface.

3. RESULTS

The results of the numerical modeling of the concrete ring were automatically plotted, using Abaqus post-processing. They are presented in Figures 3-4. As it was generally expected stress and strain presented a uniform distribution along the ring's circumference. In discussing stresses in circular rings it is advantageous to use polar coordinates.

The concrete ring subjected to a uniform radial pressure presented a uniform stress distribution along the ring's circumference, as stated in the preceding paragraph. In Figures 5-6 the stresses developed along the concrete wall thickness were expressed in polar coordinates.

The radial, σ_r, and circumferential, σ_θ, stress produced a uniform extension in the direction of the axis of the ring. It is interesting to note that the sum $\sigma_r + \sigma_\theta$ is constant through the thickness of the wall of the ring, as theory of elasticity proposes[6]. As expected, the maximum circumferential stress was developed at the inner circumference of the concrete ring (at the interface between concrete and ring).

As soon as the stress field of the concrete ring was estimated, appropriate failure criterions were applied, in order to predict the concrete's potential for cracking. The circumferencial stress distribution along the concrete wall was compared to the concrete's tensile strength, according to the maximum circumferential strength criterion. The result is shown in Figure 6. As expected, Figure 6 indicates that according to the criterion, the ring specimen will be cracked, after the tensile strength of concrete has been exceeded.

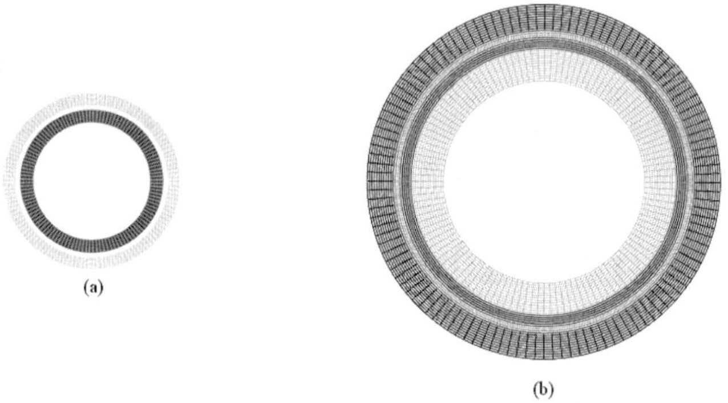

Figure 2. FE models used for the analysis of (a) the steel ring before (grey) and after (black) the applied node displacement, and (b) the concrete ring specimen before (grey) and after (black) the applied load

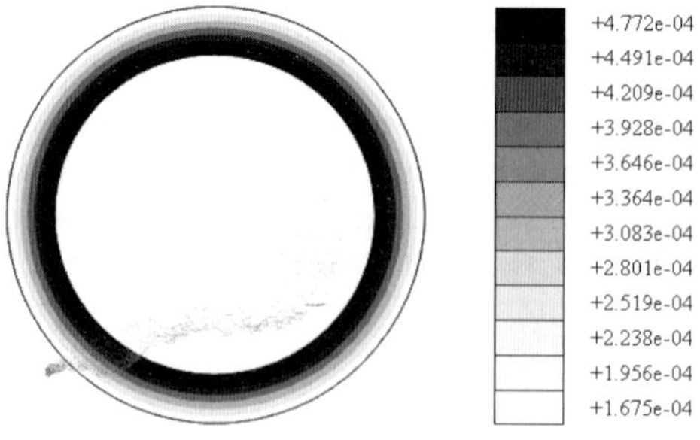

Figure 3. Strain energy density distribution, $\frac{dw}{dV}$, (given in Nmm/mm^2) in the concrete ring

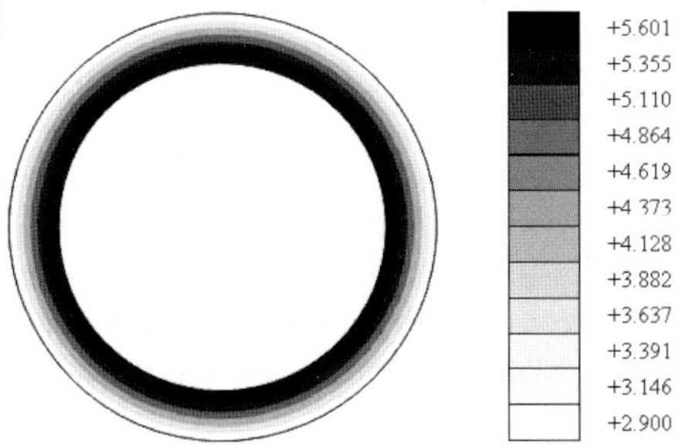

Figure 4. Von-Misses stress distribution (given in MPa) in the concrete ring

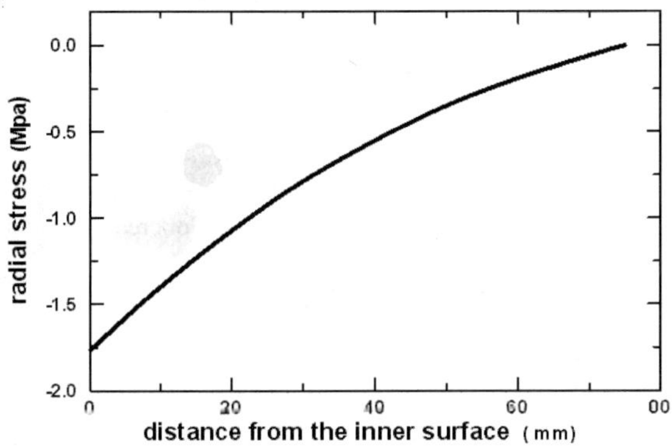

Figure 5. Variation of the stress component in the radial direction

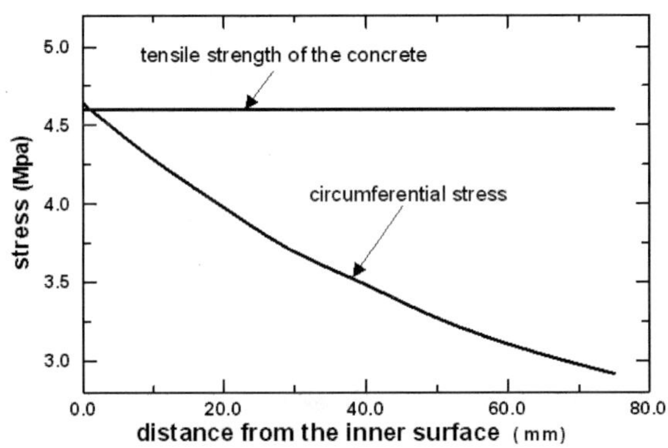

Figure 6. Variation of the stress component in the circumferential direction

Taking the stress distribution along the ring's wall into consideration, it is interesting to note that the first macro crack is expected to grow at any point along the inner circumference of the concrete ring.

4. CONCLUSIONS

This paper concluded that the ring test setup could be effectively modelled using a FE analysis. It has been demonstrated that the restrained ring test can be used to provide quantitive information on early-age stress development and early age cracking. On the other hand, the FE analysis procedure outlined in the preceding paragraphs could be generally used for the modeling of the behaviour of concrete under restrained shrinkage. It is important to note that the results of the numerical modeling were in agreement with experimental results of ring tests carried out by the same authors.

5. REFERENCES

1. Konsta-Gdoutos, M.S., Dattatraya, J.K., and Shah, S.P., "The Influence of Mineral Admixtures on the Autogenous Shrinkage and Porosity of High Performance Concrete," Sixth CANMET/ACI International Conference on Durability of Concrete, Special Publication (SP 212-15) June 1-7, 2003, Thessaloniki, Greece, pp. 227-238, 2003.
2. Konsta-Gdoutos, M.S., Shah, S.P., and. Dattatraya, J.K., "Relationships Between Engineering Characteristics and Material Properties of High Strength-High Performance Concrete", Celebrating Concrete: People and Practice, in Role of Concrete in sustainable Development, edited by R.V. Dhir, M.D. Newlands, pp. 37-46, 2003.
3. Bentur, A., "Early-Age Shrinkage and Cracking in Cementitious Systems", Concrete Science and Engineering, Vol. 3, March 2001, pp.3-12.
4. Hossain, A., and Weiss, J., "Assessing Residual Stress Development and Stress Relaxation in Restrained Concrete Ring Specimens", Vol. 26, Issue 5, July 2004, pp. 531-540.
5. See, H. T., Attiogbe, E.K., and Miltenberger, M.A., "Shrinkage Cracking Characteristics of Concrete Using Ring Specimens", ACI Materials Journal, Vol. 100, No.3, May-June 2003, pp.239-245.
6. Timoshenko S.P., Theory of Elasticity, McGraw-Hill Inc; 1987.

CHEMICAL SHRINKAGE AND CALCIUM HYDROXIDE CONTENT OF EARLY AGE PORTLAND CEMENT MONITORED WITH ULTRASONIC SHEAR WAVE REFLECTIONS

T. Voigt and S.P. Shah
USG Corporation, 700 North Highway 45, Libertyville, IL 6004, Northwestern University, 2145 Sheridan Road, Suite A130, Evanston, IL 60208

Abstract: The ultrasonic wave reflection method, which measures the reflection coefficient (or reflection loss) of high frequency shear waves at an interface between a steel plate and a cementitious material, was used to monitor the early hydration process of Portland cement mortars with water-cement ratios of 0.35, 0.5 and 0.6. To obtain independent measures of cement hydration, chemical shrinkage and calcium hydroxide content of cement pastes with the same water-cement ratios were determined. The obtained results show that, at ages up to three days, the reflection loss measured on the mortars has a strong linear relationship to the chemical shrinkage and the calcium hydroxide content of the appropriate cement paste.

Keywords: nondestructive testing; ultrasonic testing; cement hydration; chemical shrinkage; calcium hydroxide; hydration products

1. INTRODUCTION

Traditional test methods to characterize the setting and hardening of cementitious materials are either arbitrary (e.g. Vicat needle test), discontinuous (e.g. pull-out test) or not based on truly mechanical parameters of the test material (e.g. maturity method). New techniques are needed that have the ability to continuously monitor the evolution of fundamental material properties. These methods should be nondestructive in nature, applicable in-situ on the structure and should not require multiple access points to perform the measurements. An ultrasonic wave reflection method based on shear waves has been introduced previously and shown to be a sensitive indicator of the setting and hardening process of cement-based materials[1-3]. In this paper it is investigated how measurements conducted with the wave reflection (WR)-method are related to the hydration process of cement-based materials as indicated by different direct measures of cement hydration.

M.S. Konsta-Gdoutos, (ed.), Measuring, Monitoring and Modeling Concrete Properties, 331–337.

This analysis is essential for developing an understanding of the physical relationships between the WR-measurements and basic material parameters of cement paste, mortar and concrete. Several methods have been established and proven to be a direct measure of cement hydration. In this paper the amount of calcium hydroxide and the chemical shrinkage of Portland cement paste will be investigated and related to the corresponding WR-measurements.

2. EXPERIMENTAL PROGRAM

2.1 Reflection Loss

The wave reflection method as used for the experiments described in this paper was introduced by Öztürk et al.[1] and Rapoport et al.[2] The technique monitors the reflection loss of transverse or shear waves (S-waves) at an interface between a steel plate and a cementitious material over time. The amount of the lost wave amplitude depends on the reflection coefficient, which in turn is a function of the acoustical properties of the materials that form the interface. The reflection loss measured on the three mortars is given in Fig. 1.

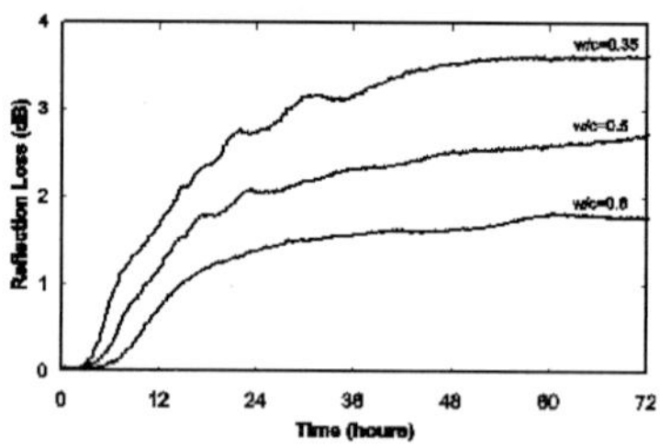

Figure 1. Development of reflection loss measured on mortars with w/c = 0.35, 05, and 0.6

2.2 Amount of calcium hydroxide

Besides the calcium silicate hydrates (C-S-H), calcium hydroxide ($Ca(OH)_2$ or CH), which is also referred to as Portlandite, is a major component of the hydration products and occupies about 20% to 25% of the solid volume of cement paste (Mindess et al.[4]). The amount of CH held in cement paste at a given time is widely regarded as a measure of the degree of hydration of that material. El-Jazairi and Illston[5] have shown that the content of calcium hydroxide and non-evaporable water follow very similar trends during the course of hydration. In a study conducted by Mounanga et al.[6] on cement pastes with various w/c-ratios at a range of different temperatures it was found that the CH content is uniquely related to the degree of hydration determined from the non-evaporable water content. Further investigations that use the content of free CH to characterize the hydration of Portland cement were conducted by Bhatty and Reid[7], Bhatty et al.[8] and Midgley[9]. Given this background, it will be of interest how the development of the amount of calcium hydroxide over time is related to corresponding WR-measurements.

The amount of the free CH in cement pastes with w/c-ratios of 0.35, 0.5 and 0.6 was determined with thermogravimetric (TG) measurements. A representative example of the hydrothermal phases that occur during the TG-measurements is given Fig. 2. The amount of the free CH is derived from the weight loss associated with the dehydroxylation as shown in the inset of Fig. 2. More details about the used procedure are given in Voigt and Shah[10] and Voigt[3].

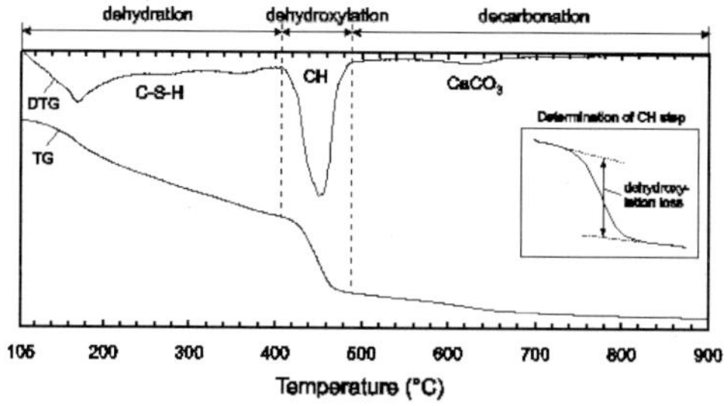

Figure 2. Major decomposition reactions during heating of Portland cement paste measured by TG (w/c = 0.5, age: 120 hours)

Free CH has a crystalline morphology and describes that part of the CH present in the paste that has not reacted with carbon dioxide (CO_2) to calcium carbonate ($CaCO_3$). Evidence of free CH can be found from TG-measurements through a distinct drop of the weight loss curve in the temperature range of about 380°C to 500°C. Derivative thermogravimetric (DTG) curves for cement paste with a w/c-ratio of 0.5 for different ages are given in Fig. 3. These curves clearly show the increasing amount of free CH with time that is decomposing during the specified temperature range (dehydroxylation).

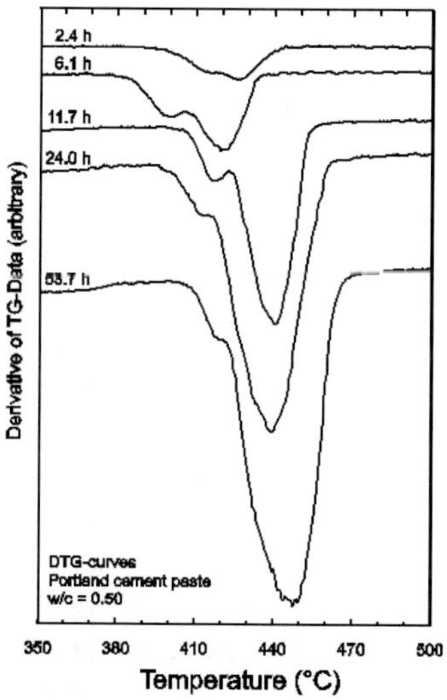

Figure 3. Development of the decomposition of calcium hydroxide characterized by the derivative of TG-curves

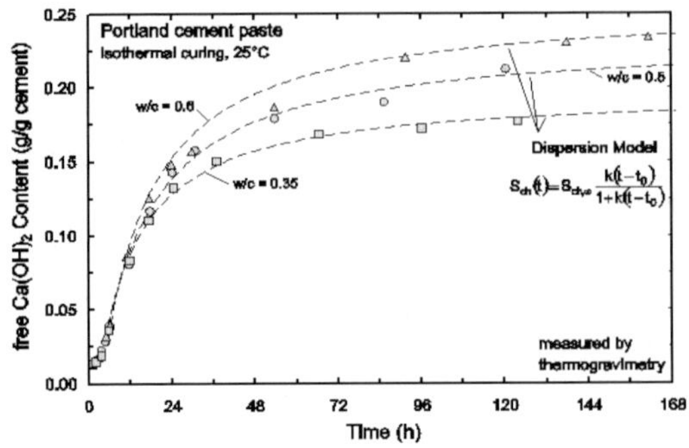

Figure 4. Evolution of $Ca(OH)_2$-content of cement paste with different w/c-ratios determined with thermogravimetry

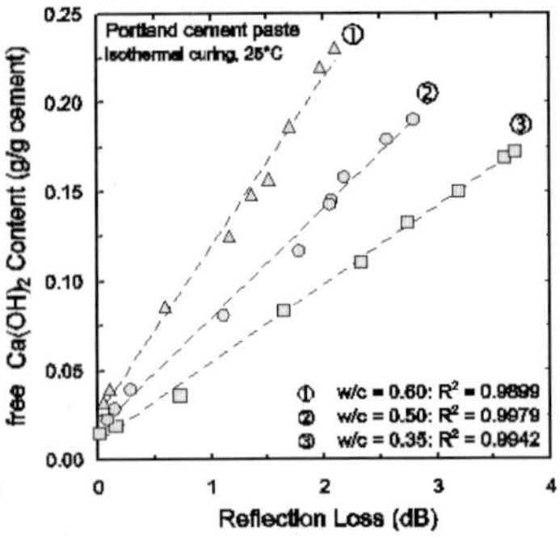

Figure 5. Relationship between $Ca(OH)_2$ content (paste) and reflection loss (mortar)

The quantitative development of free CH in gram per gram of original cement in time for the three tested pastes is given in Fig. 4. The plotted values were calculated from the results of the TG-measurements using the weight loss of the paste during the dehydroxylation period only. The data show that due to the higher availability of water the CH content for higher w/c-ratios ranges on higher values, indicating a faster progress of cement hydration in these mixtures.

The relationship between the free CH content of the cement pastes and the reflection loss measured on mortars with the corresponding w/c-ratios is given in Fig. 5. It can be seen that the free CH content has a linear relationship to the reflection loss for early ages (3-4 days). This indicates a high sensitivity of the WR-measurements to microstructural changes of cementitious materials during hydration.

2.3 CHEMICAL SHRINKAGE

The chemical shrinkage of Portland cement has been shown to be a sensitive indicator of the progress of the cement hydration. In various studies it was found that the chemical shrinkage of Portland cement develops in close correlation with the compressive strength, the heat of hydration, and the non-evaporable water content of cement paste and mortar (Jung[11]; Knudsen and Geiker[12]; Geiker[13]). In studies comparing the sensitivity of different methods to the cement hydration it was also found that the chemical shrinkage is in linear relationship to the degree of hydration determined by quantitative X-ray diffraction analysis (Parrott et al.[14]).

As part of this study, the chemical shrinkage was measured with the dilatometric method. Details about the experimental procedure are given in Voigt et al.[15]. The development of the chemical shrinkage of the cement pastes with w/c-ratios of 0.35, 0.5 and 0.6 is given in Fig. 6. The figure shows that the general development of the shrinkage data obtained from the conducted experiments can be explained by the dispersion model, which was already shown by Geiker[13]. By further analyzing the curves for the different w/c-ratios at ages greater than 24 hours, it can be observed that cement paste with a lower w/c-ratio exhibits also a lower chemical shrinkage. This agrees with results published for example by Powers[16], Czernin[17], and Geiker and Knudsen[18].

Several references can be found that have investigated the influence of the sample size (height) of the tested cement paste on the final value of the chemical shrinkage (Geiker[13]; Tazawa et al.[19]; Boivin et al.[20]. It was found that for low w/c-ratios (w/c < 0.3) the chemical shrinkage decreases with increasing sample height. This effect is due to a lower permeability of the paste preventing the water to permeate through the sample and fill the pores created by the self-desiccation. This effect was not observed for w/c-ratios equal or greater than 0.4. Based on these investigations, the paste with w/c = 0.35 tested here has the potential to be influenced by this size effect. However, the data given in Geiker[13] indicate that the decrease of the chemical shrinkage is most significant at ages greater than three days and for sample heights greater than 3 cm. In the study presented here, the sample height was always 2.5 cm and chemical shrinkage data of the first three days will be used for comparison with the reflection loss data. Under these conditions, the influence of the sample height on the development of the chemical shrinkage can assumed to be minor.

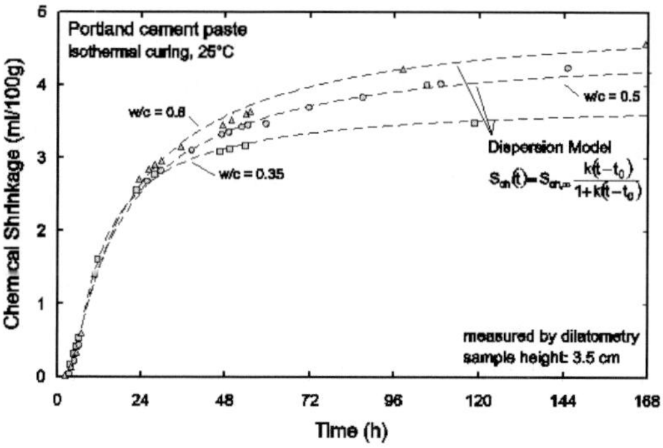

Figure 6. Development of chemical shrinkage of Portland cement pastes with different w/c-ratios

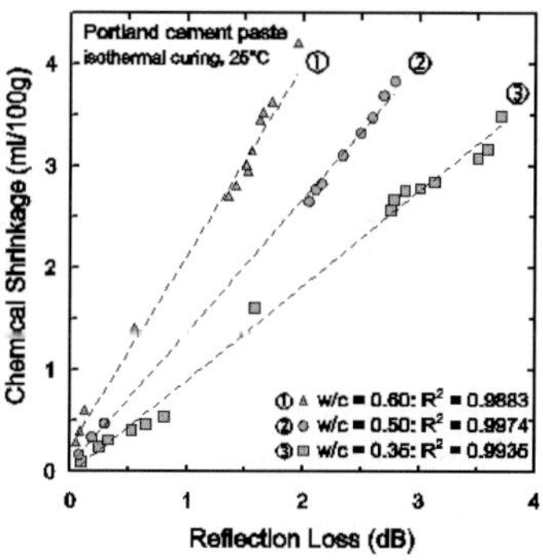

Figure 7. Relationship between chemical shrinkage (paste) and reflection loss (mortar)

The relationships between the chemical shrinkage of the cement pastes and the reflection loss measured on mortars containing the corresponding paste is given in Fig. 7. The figure shows that both parameters are related by a strong linear trend that applies to the entire period that was investigated. The R^2-values for the plotted trend lines are given in the figure to illustrate the statistical significance of the trends. The close correlation between the chemical shrinkage and the reflection loss, expressed by the linear trends, shows that the WR-measurements are governed by basic parameters that are intimately related to the hydration process.

3. CONCLUSION

The presented investigations have found that the shear wave reflection loss measured on the three mortars is intimately related to the chemical shrinkage and calcium hydroxide content of the cement paste phase of these materials. The linear relationship that was found between reflection loss and the two parameters indicates that the WR-measurements are sensitive to the same physical phenomenon, which also manifests itself in the development of chemical shrinkage and calcium hydroxide. In other words, the process of cement hydration is affecting the reflection loss in the same way as it does affect the two measured parameters.

The results also show that the w/c-ratio influences the slope of the linear trends, where a low w/c-ratio corresponds to a small slope. This observation indicates that the reflection loss is influenced by characteristics of the cementitious microstructure that cannot be described by chemical shrinkage or calcium hydroxide only. Additional results relating to the materials described in this paper have been reported elsewhere and show that capillary porosity, describing the cementitious microstructure from a different perspective, is similarly important for explaining the reflection loss development (Voigt[3]). Ultimately, it was found that the gel-space ratio, which describes the properties of the microstructure as a result of degree of hydration and porosity, is uniquely related to the reflection loss development (Voigt[3], Voigt and Shah[10]). A change in w/c-ratio did not affect the singularity of this relationship for the group of the investigated mortars.

4. REFERENCES

1. T. Öztürk, J. Rapoport, J. S. Popovics, S. P. Shah, Monitoring the setting and hardening of cement-based materials with ultrasound, *Concr. Sci. Eng.* **1**(2), 83–91 (1999).
2. J. Rapoport, J. S. Popovics, K. V. Subramaniam, S. P. Shah, The use of ultrasound to monitor the stiffening process of Portland cement concrete with admixtures, *ACI Mat. J.* **97**(6), 675–683 (2000).
3. T. Voigt, *The Application of an Ultrasonic Shear Wave Reflection Method for Nondestructive Testing of Cement-Based Materials at Early Ages: An Experimental and Numerical Analysis* (PhD Thesis, University of Leipzig, Germany, Books on Demand, Norderstedt, 244 p., 2005)
4. S. Mindess, J. F. Young, D. Darwin, *Concrete* (Prentice Hall, Upper Saddle River, NJ, 2nd ed., 644 pp., 2003)
5. B. El-Jazairi, J. M. Illston, Hydration of cement paste using the semi-isothermal method of derivative Thermogravimetry, *Cem. Con. Res.* **10**(3), 361–366 (1980).
6. P. Mounanga, A. Khelidj, A. Loukili, V. Baroghel-Bouny, Predicting Ca(OH)2 content and chemical shrinkage of hydrating cement pastes using analytical approach, Cem. Con. Res. **34**(2), 255–265 (2004).
7. J. I., Bhatty, K. J. Reid, Use of thermal analysis in the hydration studies of a type I Portland cement produced from mineral tailings, *Therm. Acta* **91**(1), 95–105 (1985).
8. J. I. Bhatty, D. Dollimore, G. A. Gamlen, R. J. Mangabhai, H. Olmez, Estimation of calcium hydroxide in OPC, OPC/PFA and OPC/PFA/polymer modified systems, *Therm. Acta* **106**, 115–123 (1986).
9. H. G. Midgley, Determination of calcium hydroxide in set Portland cements, *Cem. Con. Res.* **9**(1), 77–82 (1979).
10. T. Voigt, S.P. Shah, Properties of early age Portland cement mortar monitored with a shear wave reflection method, *ACI Mat. J.* **101**(6), 473–482 (2004).
11. F. Jung, Ein einfaches Verfahren zur zerstörungsfreien Prüfung der Festigkeit von Beton (A simple method for nondestructive testing of the strength of concrete), *Beton- und Stahlbeton* **69**(1), 16–19 (1974) (in German).

12. T. Knudsen, M. Geiker, Chemical shrinkage as an indicator of the stage of hardening. In RILEM (Ed.), International Conference on Concrete of Early Ages. Editions Anciens ENPC, Paris, vol. 1, pp. 163–165 (1982).
13. M. Geiker, *Studies of Portland cement hydration by measurements of chemical shrinkage and a systematic evaluation of hydration curves by means of the dispersion model*, (PhD thesis, Institute of Mineral Industry, Technical University of Denmark, Denmark, 1983)
14. L. J. Parrott, M. Geiker, W. A. Gutteridge, D. Killoh, Monitoring Portland cement hydration: Comparison of methods, *Cem. Conc. Res.* **20**(6), 919–926 (1990).
15. T. Voigt, C.U. Grosse, Z. Sun, S. P. Shah, H.-W. Reinhardt, Comparison of ultrasonic wave transmission and reflection measurements with P- and S-waves on early age mortar and concrete, *Mat. Struct.* **38**(282), 729–738 (2004).
16. T. C. Powers, Absorption of water by Portland cement paste during the hardening process. *Ind. Eng. Chem.* **27**(7), 790–794 (1935).
17. W. Czernin, *Cement chemistry and physics for civil engineers* (Chemical Publishing Co., Inc, New York, 139 pp., 1962).
18. M. Geiker, T. Knudsen, Chemical shrinkage of Portland cement pastes, *Cem. Con. Res.* **12**(5), 603–610 (1982).
19. E. Tazawa, S. Miyazawa, T. Kasai, Chemical shrinkage and autogenous shrinkage of hydrating cement paste, *Cem. Con. Res.* **25**(2), 288–292 (1995).
20. S. Boivin, P. Acker, S. Rigaud, B. Clavaud, Experimental assessment of chemical shrinkage of hydrating cement pastes. In Tazawa, E. (Ed.), Autogenous shrinkage of concrete. Spon Press, Proceedings of the International Workshop organized by the JCI (Japan Concrete Institute), pp. 81–92 (1999).

Development of Innovative Cementitious Materials

A COMPARISON OF HBC & MHC MASSIVE CONCRETES FOR THREE GORGES PROJECT IN CHINA

T. Sui,* J. Li,** X. Peng,** W. Li,*** Z. Wen,* J. Wang* and L. Fan*
**China Building Materials Academy, Beijing 100024, PRC, **China Institute of Water Resource & Hydropower Research, Beijing 100038, PRC, ***China Yangtze Three Gorges Project Development Co., Yichang 443133, PRC*

Abstract: Performance of two massive concretes for the Three Gorges Dam Project (TGP) made by High Belite Portland Cement (HBC) and Moderate Heat Portland Cement (MHC) are reviewed. Comparison tests reveal that both HBC and MHC massive concretes used in the TGP project possess excellent performance in terms of workability, mechanical performance, and durability, such as the resistance to freezing-thawing, permeability and carbonation, and volume stability etc. In particularly, HBC dam concrete exhibits better performance in the aspects of mechanical behavior, thermal properties as well as cracking resistance when compared with MHC dam concrete.

Key words: High Belite Portland Cement, Moderate Heat Portland Cement, massive concrete performance, high cracking resistance

1. INTRODUCTION

Recently there has been a revival of intensive interests especially in Japan, India, and China in exploring high belite Portland cement (HBC) due to the requirement of energy conservation and environment protection for cement manufacturing and the rapid development of high performance concrete[1-3]. C_3S, as the most important constituent of traditional Portland cement (PC), accounts for 50-65%, while C_2S constitutes only 15-30%. High belite cement presented in this paper, however, contains reversal quantity in the composition of C_3S and C_2S, i.e. more than 50% of C_2S and about 25% of C_3S. It has been proved [4-6] that HBC not only possesses the lower heat evolution and higher late age strength, but also exhibits excellent workability, physical mechanical properties and durability.

M.S. Konsta-Gdoutos, (ed.), Measuring, Monitoring and Modeling Concrete Properties, 341–346.

It is well known that the technical feature for both low heat and high cracking resistance is always the most important and difficult in dam concrete design and application. Many approaches such as blending with secondary cementitious materials in dam concrete mix design, strictly controlling the temperature of fresh concrete prior to concreting by cooling the raw materials, and placing cooling pipe inside the dam concrete etc. have been applied to control the temperature rise of the massive concrete in order to avoid or reduce the possibility of concrete cracking especially during the cooling stage of the dam. Yet the results are far from satisfactory based on the practical dam construction, cracking of dam concrete has been frequently seen in many hydropower project. It is therefore crucial to seek for a fundamental solution for this problem. HBC, as previous study indicated[5], might be a better option in meeting the technical requirement of dam concrete in particularly for cracking resistance based on the latest progress of HBC and it sapplication.

2. CEMENT PERFORMANCE

The chemical analysis and mineral composition of HBC clinker are listed in Table 1 in comparison with MHC and PC. The comparison data of the mortar (cement : sand=1:2.5, W/C=0.44) compressive strength development of these three types of cement under standard curing temperature (20°C) and elevated curing temperature (38°C) is shown in Figures 1 and 2. The hydration heat measurement data is also given in Table 2, in which MHC-1, MHC-2 and MHC-3 are moderate heat Portland cement from three main cement suppliers for TGP project.

It can be seen that:

1 Under the standard curing conditions (temperature 20°C and relative humidity exceeding 90%), though the early strength prior to 28-day age is lower for HBC, the strength at the 28-day is equivalent to that of PC and MHC, while the strength at the 90-day age is much higher than that of PC and MHC, which demonstrates the excellent strength development of HBC especially at later hydration age when compared with C_3S-based MHC and PC.

2 Better strength development for HBC can also be proved in Figure 2 when curing the three cements at elevated temperature which is the normal condition especially for massive concrete with hydration heat accumulation inside the concrete.

3 The hydration heat for HBC is obviously lower than that of PC and MHC, for example, the hydration heat of HBC at the 3-day age is approximately 20% lower than that of MHC, while at the 7-day it is lower by 15-20% than that of MHC.

It can therefore be seen that the HBC is a kind for low heat and high strength cement which has great potential for the dam concrete making.

3. CONCRETE PERFORMANCE

Comparison evaluation of the performance of HBC dam concrete with the existing MHC dam concrete presently used for TGP were made on the basis of present Chinese national standards. All the parameters except the cement used for comparison tests were maintained the same. The mix proportion of HBC and MHC dam concretes are listed in

Table 3. It can be seen, through the testing results, that the air content of HBC dam concrete is basically similar to that of MHC dam concrete. However, the slump for HBC dam concrete is slightly larger than that of MHC dam concrete, which might be due to the lower water demand and better compatibility with concrete admixture of the HBC.

3.1 Mechanical performance

The testing results in mechanical behavior upon HBC and MHC dam concretes are listed as Table 4.

1 Again the strength development of HBC dam concrete compared with MHC dam concrete is in good agreement with the mortar strength comparison results.
2 The axial tensile strength and ultimate tensile strain of HBC dam concrete are higher than that of MHC concrete.

3.2 Deformation performance

The deformation performance mainly including the drying shrinkage, creepage, autogenous volumetric deformation, and testing results is shown in figure 3-6.

1 The creep and drying shrinkage deformation for HBC dam concrete is similar to that of MHC dam concrete accordingly for 28 days or 90 days loading period.
2 The autogeneous volume deformation for HBC dam concrete behaves as micro-expansion type, MHC dam concrete, however, shows a micro-shrinkage type.

3.3 Freezing and permeability resistance

The anti-freezing and thawing and permeability resistance of both concretes were evaluated. The testing results, as shown in Table 5, indicate that both HBC and MHC dam concretes can satisfy the technical requirement of TGP concrete with the capability of resisting more than 300 times of quick freezing and thawing cycles and 1 Mpa of water pressure.

3.4 Adiabatic temperature rising

The adiabatic temperature rising of HBC and MHC dam concrete whose mixes are the same with TGP interior concrete is shown in Figure 7. It can be seen that the adiabatic temperature rising of HBC dam concrete is average 3~5°C lower than that of MHC dam concrete, which indicates that HBC is an ideal cementitious material for massive concrete. This result has been well demonstrated by the application of the HBC dam concrete in TGP construction later on.

3.5 Cracking resistance analysis

In the temperature controlling design for dam, the most visual and simple evaluation upon the cracking resistance capability is to apply the ultimate tensile strength of the concrete, σ, as a criteria. Higher σ value represents better capability of the concrete to resist cracking. This can be simply derived from the following equation:

$$\sigma=\varepsilon p*E \quad (1)$$

where: σ-ultimate tensile strength in concrete (MPa)

ε_p-ultimate tensile deformation in concrete

E-tensile elastic modulus of concrete (MPa)

Table 6 gives the cracking resistance capability evaluation results for HBC and MHC dam concretes.Through the test calculating results in the table above, whether the fiducial concrete without any fly ash, or the dam concrete with 20-40% fly ash, the crack resistance abilities of HBC concretes are larger than that of MHC concrete.

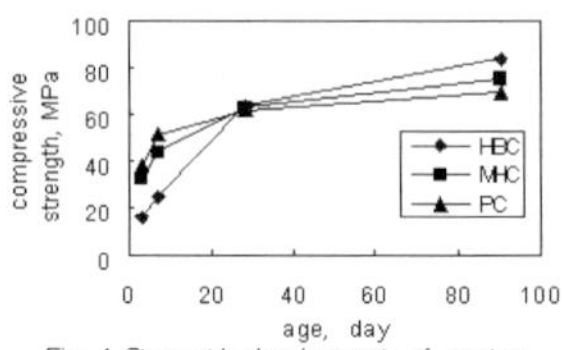

Fig. 1 Strength development of mortar under standard curing

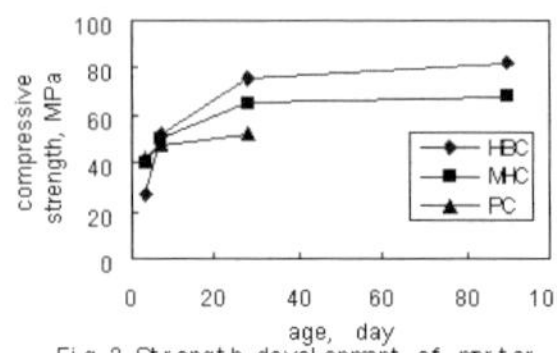

Fig. 2 Strength development of mortar under 38°C curing

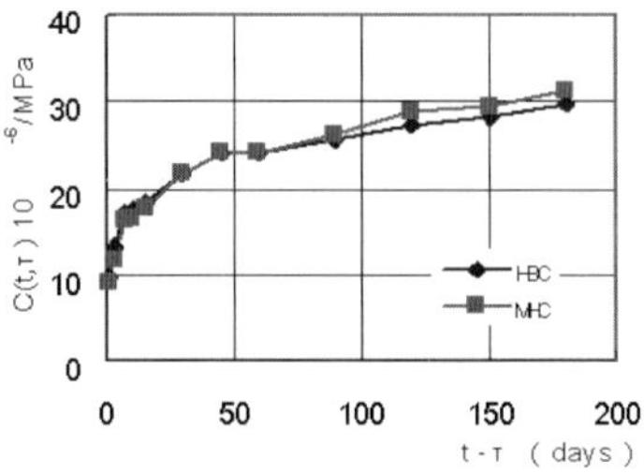

Fig. 3 Creep deformation of HBC & MHC concrete, 28-day

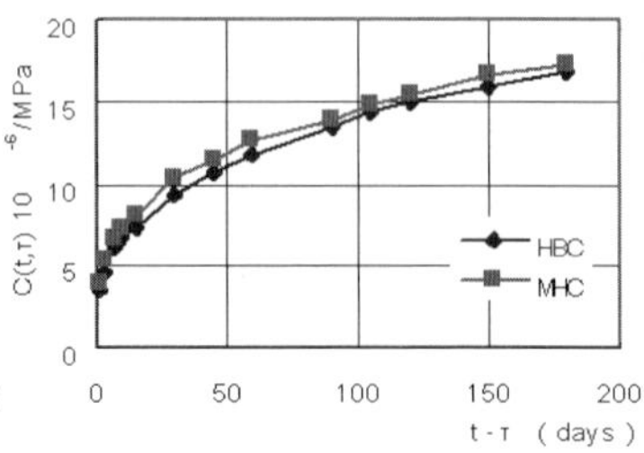

Fig. 4 Creep deformation of HBC & MHC concrete, 90-day

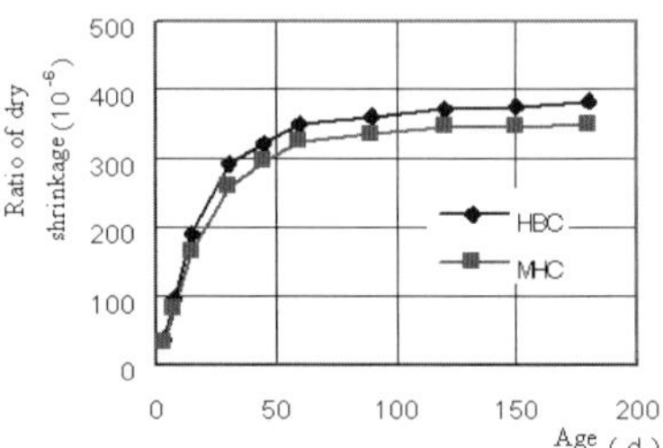

Fig.5 Dry shrinkage of dam concrete

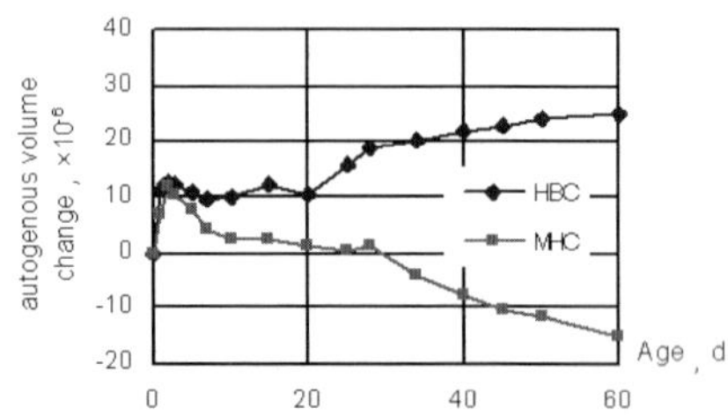

Fig.6 Autogenous volume change of MHC and HBC dam concrete

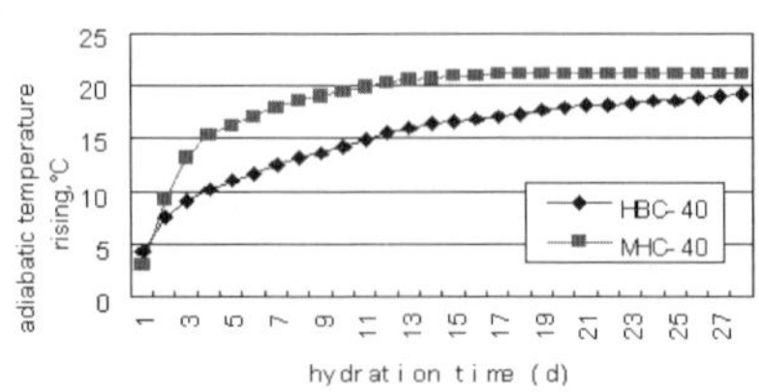

Figure 7. Adiabatic Temperature Rising for HBC and MHC Concretes

Table 1. Chemical analysis and mineral composition of HBC, PC and MHC

Cement	Chemical composition, %						Silicate mineral, %	
	SiO_2	Al_2O_3	Fe_2O_3	CaO	MgO	R_2O	C_3S	C_2S
HBC	23.06	4.59	4.57	59.88	1.39	0.54	20.65	50.53
PC	22.66	5.31	3.29	65.75	1.21	-	55.10	23.40
MHC	21.22	4.68	4.13	62.95	3.95	0.62	53.50	21.64

Table 2. Hydration Heat for HBC, PC and MHC Unit: KJ/kg

Sorts of cement	1d	2d	3d	4d	5d	6d	7d
HBC	159	181	196	208	219	227	234
PC	-	-	247	-	-	-	289
MHC-1	201	233	248	257	265	272	278
MHC-2	180	220	238	248	256	262	267
MHC-3	193	228	245	256	264	271	277

Table 3. Mix design of HBC and MHC Concretes

Concrete No.	Concrete mix design (Kg/ m^3)							Slump (cm)	Air content (%)
	Cement+Pfa	Water	Sand	Aggregate					
				80~150mm	40~80mm	20~40mm	5~20mm		
MHC-00	231.0+0	127.0	621.5	455.0	455.0	300.0	300.0	4.7	1.1
HBC-00	231.0+0	127.0	621.5	455.0	455.0	300.0	300.0	7.5	1.1
MHC-20	290.5+72.5	127.0	653.0	-	-	718.5	588.5	5.5	4.7
HBC-20	290.5+72.5	127.0	653.0	-	-	718.5	588.5	7.5	4.6
MHC-30	133.8+57.2	86.0	555.0	497.5	497.5	330.0	330.5	4.7	4.5
HBC-30	133.8+57.2	86.0	555.0	497.5	497.5	330.0	330.5	6.4	5.4
MHC-40	98.2+61.7	85.0	607.5	490.0	490.0	325.0	325.0	4.5	5.1
HBC-40	98.2+61.7	85.0	607.5	490.0	490.0	325.0	325.0	6.7	5.6

Notes: The serial numbers, such as –00, -20, -30 and –40 are represented for the fly ash replacement percentage.

Table 4. Mechanical physical properties of HBC and MHC dam concrete

Cement	Compressive strength (MPa)			Axial tensile strength (MPa)		Elastic modulus ($\times 10^3$MPa)		Ultimate tensile strain ($\times 10^{-4}$)	
	7d	28d	90d	28d	90d	28d	90d	28d	90d
MHC	26.0/100	35.4/100	46.2/100	2.30/100	3.02/100	29.1/100	38.8/100	0.99/100	1.03/100
HBC	9.2/35	36.1/102	53.1/115	2.58/112	3.18/105	32.3/111	37.5/97	1.01/102	1.08/105
HBC-30*	18.7/72	38.7/109	58.1/126						

*:The curing temperature is 38C, and the concrete mixing proportion is shown in the table 5.

Table 5. Freezing and permeability resistance of HBC and MHC Dam Concretes

	Grade of freezing resistance	Anti-permeation gradekg/cm^2
HBC-30	>D300	>W10
MBC-30	>D300	>W10

Table 6. Cracking Resistance capability for HBC and MHC concretes

Concrete number	Crack resistance ability for concrete $\sigma=\varepsilon_p$ x E (MPa)	
	28d	90d
MHC-00	3.4	4.3
HBC-00	4.5	4.8
MHC-20	3.4	4.5
HBC-20	3.7	5.0
MHC-30	2.9	3.9
HBC-30	3.3	4.2
MHC-40	2.2	3.2
HBC-40	2.3	3.7

4. CONCLUSIONS

1 Both HBC and MHC massive concretes used in the TGP project possess excellent performance in terms of workability, mechanical performance, and durability such as the resistance to freezing-thawing, permeability and carbonation, and volume stability etc.
2 HBC dam concrete exhibits better performance in the aspects of mechanical behavior beyond the 28-day hydration age, such as higher flexural and tensile strength, better strength gain especially at elevated curing temperature.
3 Better thermal properties were also proved for HBC dam concrete by its lower hydration heat evolution and adiabatic temperature rise while higher strength gain at later age.
4 The main advantage of HBC for massive concrete making was also identified as its excellent cracking resistance, which is higher than the mostly used MHC dam concrete for TGP.

5. REFERENCES

1. A. K. Chatterjee, Cement and Concrete Research, Vol.26, No.8, 1996, pp1213-1237.
2. Sui Tongbo, Guo Suihua, Liu Kezhong, et al., *Research on High Belite Cement, Part I;* 4th, Beijing International Symposium on Cement and Concrete.
3. TONGBO SUI, LEI FAN, ZHAIJUN WEN, et al. Journal of Advanced Concrete Technology. Vol.2, No.2, June 2004. pp 201-206.
4. Sui Tongbo, Liu Kezhong.: *Study on the Properties of High Belite Cement*; Journal of Chinese Ceramic Society, No.4; pp.488-492; 1999.
5. Sui Tongbo, Wang Jing, Wen Zhaijun, et al.: *Strength and Pore Structure of High Belite Cement;* Proceeding of the 5th International Symposium on Cement & Concrete; pp. 261-265; 2002.

INFLUENCE OF FLY ASH ON THE PROPERTIES OF MAGNESIUM OXYCHLORIDE CEMENT

J. Chan and Z. Li

Department of Civil Engineering, Hong Kong University of Science and Technology, Clear Water Bay, Kowloon, Hong Kong, P R China

Abstract: The effects of fly ash on the properties of Magnesium Oxychloride Cement (MOC) have been studied. The experimental results indicated that the addition of fly ash changes flow property, setting times, compressive strength development, water resistance, volume stability and microstructures of MOC. It might be due to the change of dissolution rate of magnesium oxide and chemical reaction between MOC and fly ash. MOC containing a certain amount of fly ash has a potential to be utilized in the construction industry.

Key words: Magnesium Oxychloride cement (MOC), fly ash and water resistance

1. INTRODUCTION

Magnesium Oxychloride Cement (MOC) is an air-hardened and expansive material with chemical reaction between light-burned magnesium oxide and magnesium chloride solution. Its main reaction products at ambient temperature are $5Mg(OH)_2MgCl_28H_2O$ and $3Mg(OH)_2MgCl_28H_2O$ (5-1-8 and 3-1-8 phases). Its mechanical properties are mainly based on these phases. Although it has high compressive strength and good bonding with others substances, it deteriorates significantly under moist climate due to the leaching of magnesium chloride. Its water resistance can be altered with the addition of small amount of additive, such as water-soluble phosphate[1, 2] and copper powder[3].

However, the utilization of these additives increases the cost of MOC based products. Therefore, using waste materials to improve the water resistance and reduce the cost of MOC should be further investigated. Fly ash is a promising waste material because it has a strong absorption of chloride ion[4] and prohibits halogenation frosting[5].

Although the feasibility of using fly ash to impart water resistance of MOC had been reported[1], its mechanical properties were still not well understood. In this paper, the effect of fly ash on properties of MOC was studied. Its flow property, setting times, compressive strength development, water resistance and volume stability were well discussed.

M.S. Konsta-Gdoutos, (ed.), Measuring, Monitoring and Modeling Concrete Properties, 347–352.

2. EXPERIMENTAL PROGRAM

2.1 Materials

Light-burned magnesium oxide (MgO), magnesium chloride crystal ($MgCl_26H_2O$) and fly ash were used for this study. Their chemical compositions were listed in Table 1. Magnesium oxide was manufactured by calcination in the temperature of 800°C. Natural river sand was used by putting into oven for 24 hours and passing through 1.18mm sieve.

Table 1. Chemical composition of raw materials

Chemical composition wt. %	Light-burned MgO	Fly Ash	$MgCl_26H_2O$
MgO	96.2	2.9	0
Al_2O_3	0	20.1	0
SiO_2	0.8	49.9	0
SO_3	0.1	1.1	0
K_2O	0	2.0	0
CaO	1.6	8.5	0
Fe_2O_3	0.7	11.8	0
$MgCl_2$	0	0	46.5
H_2O	0	0	50.5
Total	99.4	96.3	97.0

2.2 Specimen preparation

The effect of MOC with fly ash on fluidity, setting times, compressive strength development, water resistance and drying shrinkage were studied by incorporating fly ash ranging from 0% to 30% by weight of magnesium oxide. Four mixes used in this study were given in Table 2. Morphology of MOC was examined using MOC paste at the fixed water-to-powder ratio (W/P) of 0.465. Magnesium chloride crystals were first dissolved into water before blending with magnesium oxide, fly ash and natural river sand (if applicable). The molar ratio of magnesium oxide to magnesium chloride in the mixes was 9.3 in order to have a complete reaction of magnesium chloride.

The fluidity of mortar was tested by using the flow table test (GB2419) that measured the diameter of a conical frustum of mortar in 100mm diameter at its base, after the flow table had been moved up and down through a height of 25mm and 30 times in 30seconds. The setting times of mortar were determined using Vicat apparatus by recording both initial and final setting times.

The compressive strength development of the mortars was investigated by crashing mortar cubes (40x40x40mm) at different ages. All cubes were cured in open air in the laboratory (23 ± 2°C, 60%R.H.). The strength retention coefficients of various compositions of fly ash were used to indicate the water-resistance of MOC.

Table 2. Design mix proportions

Reference No.	Molar ratio		W/P	Fly ash (wt. % of MgO)	Sand/Powder
	MgO	$MgCl_2$			
Control	9.3	1	0.465	0	1.5
FA10	9.3	1	0.465	10	1.5
FA20	9.3	1	0.465	20	1.5
FA30	9.3	1	0.465	30	1.5

Mortar cubes were first cured in air for 14 days and then soaked into distilled water for 28days. Their corresponding strength retention coefficient (W_{28}) was determined by the following equation:

$$W_{28} = R_{c28}/R_c \quad (1)$$

where R_{c28} is the compressive strength of specimen immersing in distilled water for 28 days and R_c is the compressive strength of specimen curing in air for 42 days.

The volume stability of MOC mortar was evaluated by drying shrinkage test. The specimen of size of 25x25x285mm was placed in curing room under condition of 28°C and 50%R.H. The length changes of the specimen were measured by a micrometer.

The morphology of the MOC paste was examined by Scanning Electron Microscope (SEM, Model JEOL-6300F). Specimens were dried in oven for 48 hours at a low temperature of 40°C since oven-drying at a high temperature that might cause a severe error due to change of the microstructure of cement-based materials[6].

3. RESULTS AND DISCUSSION

3.1 Fluidity and setting times of the MOC mortars

The results of the flow table tests are shown in Figure 1 indicated that the fluidity of MOC mortar increased as the fly ash content increased. This might be attributed to the decrease of water demand because of the spherical shape of fly ash. Figure 2 showed both the initial and final setting times of MOC mortars. A larger amount of fly ash retarded the setting process. Since the initial set of MOC was characterized as formation of amorphous gel phases between magnesium oxide and magnesium chloride solutions[7], the utilization of fly ash may yield the retardation of the formation of amorphous gel of MOC through dilution effect.

3.2 Compressive strength development

The compressive strengths of MOC mortars curing in air to different ages were given in Figure 3. From this figure, it was clear that the compressive strength of the MOC mortar decreased with the amount of fly ash. Since the strength of MOC was mainly contributed by the hydrated phases (5-1-8 or 3-1-8 phase) on magnesium oxide crystals[3], higher consumption of fly ash might yield the low formation of hydrated phases. Also, at fixed water to powder ratio (W/P), the addition of fly ash may lead to an increase of free water.

As a result, the amount of un-reacted water in the mortar increased and lowered its strength. On the other hand, the compressive strength development of the MOC mortar with fly ash requires a longer period of time than that of mortar without fly ash. It might be caused by the low dissolution rate of magnesium oxide due to the existence of fly ash.

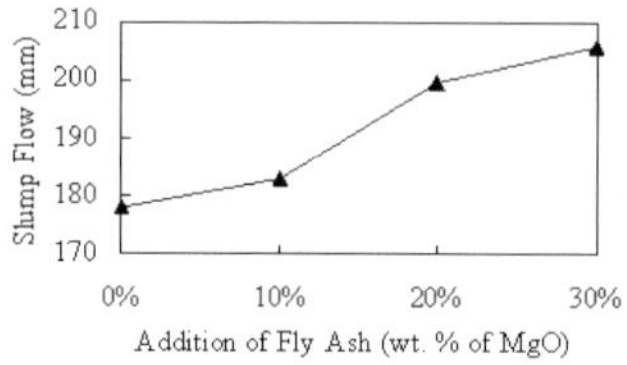

Figure 1. Effect of dosages of fly ash on slump flow of MOC mortar

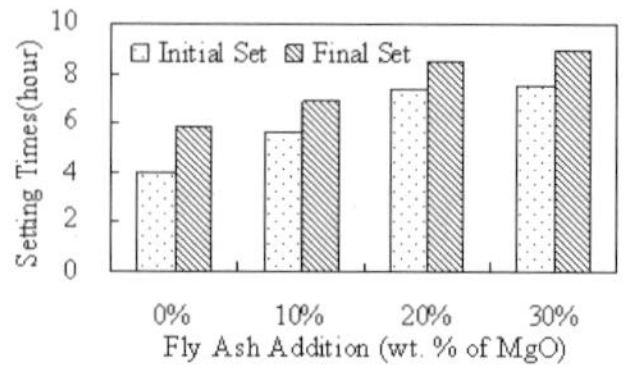

Figure 2. Effect of dosages of fly ash on setting times of MOC mortar

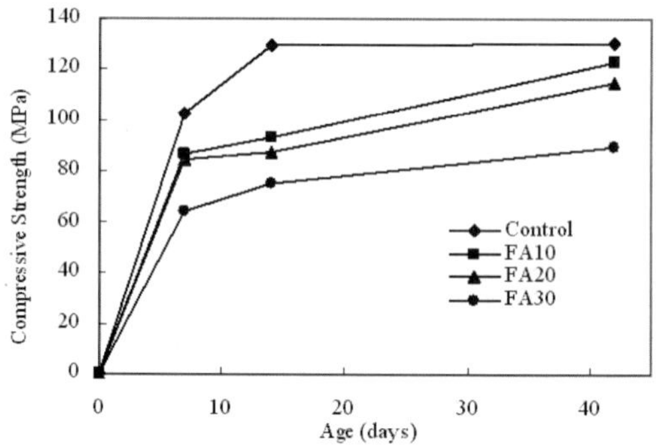

Figure 3. Compressive strength developments of MOC mortars with fly ash addition

3.3 Water resistance

The results of water-resistance tests in Figure 4 revealed that the strength retention coefficients of the MOC mortar were greatly increased with the increase of the fly ash dosage. The SEM micrographs in Figure 5 showed that the well crystallized needles were closely packed with the addition of fly ash. The existence of fly ash in MOC might promote the interpenetration of hydrated phases, and therefore increase the resistance of MOC in water.

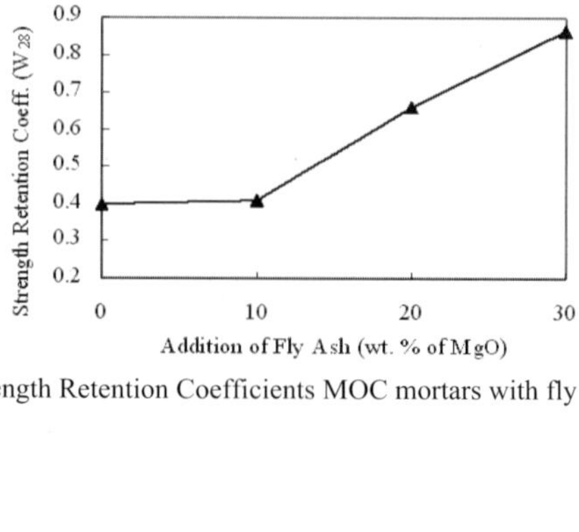

Figure 4. Strength Retention Coefficients MOC mortars with fly ash addition

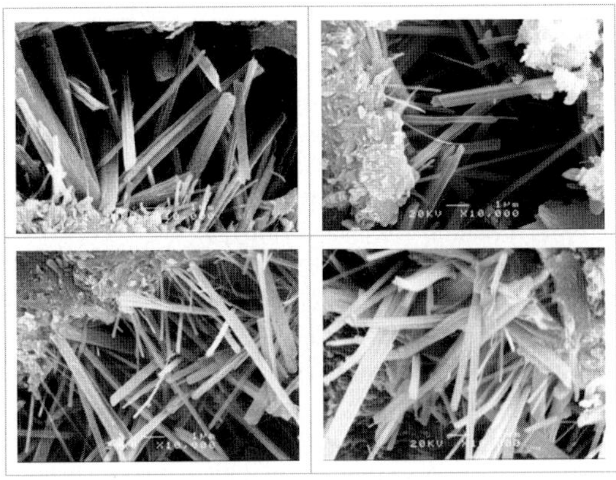

Figure 5. SEM micrographs of MOC pastes (Top left: Control; Top right: FA10; Bottom left: FA20; Bottom right: FA30)

3.4 Volume stability

The drying shrinkage test results illustrated in Figure 6 showed that MOC mortar expanded in nature. Since the amorphous gel phase initially occurred right after mixing and crystalline hydrate phases of the mortar were converted from amorphous gel phase over several days or weeks[8], the expansion of MOC might be caused by the formation of the well crystallized hydrate phases after the mortar has already attained certain rigidity. For mixes containing fly ash, a reduction of expansion occurred and increased as the dosage increased. It might be attributed to slow reaction of fly ash.

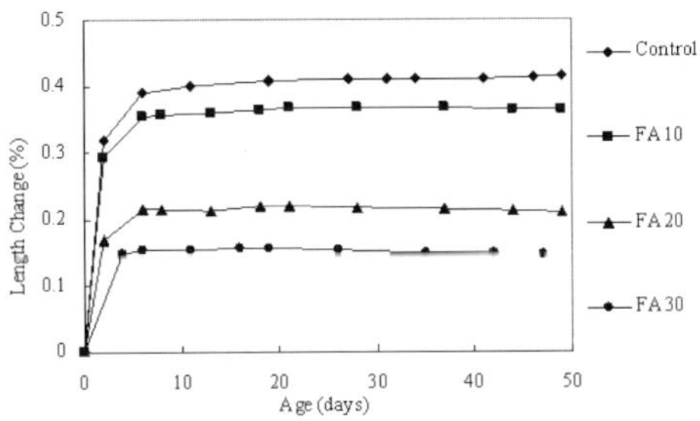

Figure 6. Length changes of MOC mortars with fly ash under 28°C and 50%R.H.

4. CONCLUSIONS

The addition of fly ash in MOC can delay its setting and compressive strength development. It can enhance its fluidity and improve its resistance to moist environment. It also reduces the ultimate compressive strength and degree of expansion. The closely packed hydrated phases of MOC with fly ash may be the mechanism to increase its water resistance.

5. REFERENCES

1. H. P. Lu, P. L. Wang, N. X. Jiang, Design of Additives for Water-resistant Magnesium Oxychloride Cement Using Pattern Recognition, Mater. Let. **20**, 217-223 (1994).
2. X.Y. He and J. B. Wang, Surface Property of Modified Magnesium Concrete, J. of Inner Mongolia Polytechnic University **22**(4), 308-310 (2003).
3. V. S. Ramachandran, R. F. Feldman, J. J. Beaudoin, Concrete Science: Treatise on Current Research (Heyden Press, London, 1981), pp. 317-319.
4. B. Ma, M. Qi, J. Peng, and Z. Li, The Compositions, Surface Texture, Absorption and Binding Properties of Fly Ash in China, Environ. Inter. **25**(4), 423-432 (1999).
5. L. Xu, N. Yang, H. Tao, Y. Yin, Effect of Chemical Activity of MgO on Cracking and Water-resistance of Magnesium Oxychloride Materials, J. of The Chinese Ceramic Society **31**(8), 759-762 (2003).
6. C. Galle, Effect of Drying on Cement-based Materials Pore Structure as Identified by Mercury Intrusion Porosimetry: A Comparative Study between Oven-, Vacuum-, and Freeze-drying, Cem. Concr. Res. **31**(10), 1467-1477 (2001).
7. D. Deng, The Mechanism for Soluble Phosphates to Improve the Water Resistance of Magnesium Oxychloride Cement, Cem. Concr. Res. **33**(9), 1311-1317 (2003).
8. D. Deng and C. Zhang, The Formation Mechanism of the Hydrate Phases in Magnesium Oxychloride Cement, Cem. Concr. Res. **29**(9), 1365-1371 (1999).

CARBONATED CEMENTITIOUS MATERIALS AND THEIR ROLE IN CO_2 SEQUESTRATION

Y. Shao and S. Monkman
Department of Civil Engineering, McGill University, 817 Sherbrooke Street West, Montreal, Quebec H3A 2K6, Canada

Abstract: Carbonation behaviour of four cementitious materials, CSA Type 10 cement, CSA Type 30 cement, fly ash and lime, were studied for their applicability to CO_2 sequestration. It was found that the cements and fly ash could each show CO_2 uptake on the order of 12% while the lime achieved nearly 24%. The two-hour carbonation produced high early age strength in cements and lime, while the strength gain in fly ash was nevertheless not sufficient. XRD analysis determined that the primary carbonation product formed was calcite while C_3S, C_2S, CSH and CH were the phases consumed.

Keywords: carbonation; cement; lime; fly ash; calcium carbonates; CO_2 sequestration.

1. INTRODUCTION

Carbon dioxide is the dominant greenhouse gas resulting from human activities. A significant fraction of the CO_2 discharged to the atmosphere comes from industry point sources. Cement production alone contributes approximately 5% of global CO_2 emissions. In 1995, global cement production was estimated to be 1453 million tons. The demand for cement is growing at an average annual rate of 1.5% (Worrel et al., 2001). Cement and concrete industry is facing challenge to reduce the carbon dioxide emissions. One of the effective ways that is currently practiced is the use of supplementary cementitious materials, such as fly ash and ground granulated blast furnace slag, to reduce the consumption of cement and thus the emission from the production (Malhotra, 1999).

Carbonation of concrete through its early age curing provides an alternative means to directly sequester carbon by converting the carbon to calcium carbonates. For applications without reinforcing steel, the carbonated concrete products can perform better in achieving strength, durability and dimensional stability due to the near-complete depletion of calcium hydroxide. The rapid carbonation reaction with cementitious binders

M.S. Konsta-Gdoutos, (ed.), Measuring, Monitoring and Modeling Concrete Properties, 353–359.

accelerates the strength development and shortens the time required for the production. It is best suited for manufactured concrete products, such as bricks and blocks, cement boards and precast concrete with non-metallic reinforcement.

The mechanisms of early strength development in cement and lime exposed to carbon dioxide were investigated by Berger et al. (1972), Young et al. (1974), Goodbrake et al. (1979) and Moorehead (1986). It was found that the carbonation products were primarily calcium carbonates and silica gel. This paper is to examine the carbonation behavior of the commonly used cementitious materials and the possibility of using the carbonation technology for CO_2 sequestration. The pure CO_2 of 99.5% purity was employed to simulate the recovered CO_2 from exhaust gas at point sources. The carbonation time was fixed at two hours and the properties of carbonated products were evaluated by their carbon uptake and strength development.

2. EXPERIMENTAL PROGRAM

The four materials used in the testing were CSA Type 10 cement, CSA Type 30 cement, fly ash and lime ($Ca(OH)_2$). The chemical compositions of these materials are given in Table 1. Water to cementitious ratio was kept constant at 0.15. Rectangular plate specimens of 14 mm thick, 76 mm wide and 127 mm long were prepared by a press forming under a pressure of 8 MPa. No aggregates were used.

Table 1. Typical chemical compositions of cementitious materials

	CaO	SiO_2	Al_2O_3	Fe_2O_3	MgO	LOI
Type 10 cement	63.1	19.8	4.9	2.0	2.0	2.8
Type 30 cement	62.9	19.6	4.9	2.0	2.0	3.0
Fly ash	29.6	27.6	14.6	2.0	0.55	23.4
Lime	61.4	1.1	0.5	0.2	0.7	34.9

Figure. 1 shows the carbonation curing set-up. A carbon dioxide gas of 99.5% purity was considered as a gas of 100% concentration and employed to simulate a recovered CO_2 from flue gas at point sources. The pressure inside the chamber and the temperature inside one sample were recorded using Measurement Group data acquisition system (System 5000). The CO_2 gas cylinder was regulated to a constant pressure of 0.5 MPa to ensure a continuous supply of carbon dioxide for two hours and prevent the CO_2 starvation due to its absorption by concrete. Vacuum was applied to the chamber before the CO_2 injection. Three plate specimens were placed inside the chamber immediately after press-formed.

CO_2 uptake is characterized as mass gain which was obtained based on the difference between the mass measured before carbonation and the mass after carbonation, in duration of two hours, as expressed by Eq. (1):

$$\%Mass\ gain = \frac{(Mass)_{aft,CO2} - (Mass)_{bef,CO2} + (Mass)_{lost\ water}}{(Mass)_{dry\ binder}} \tag{1}$$

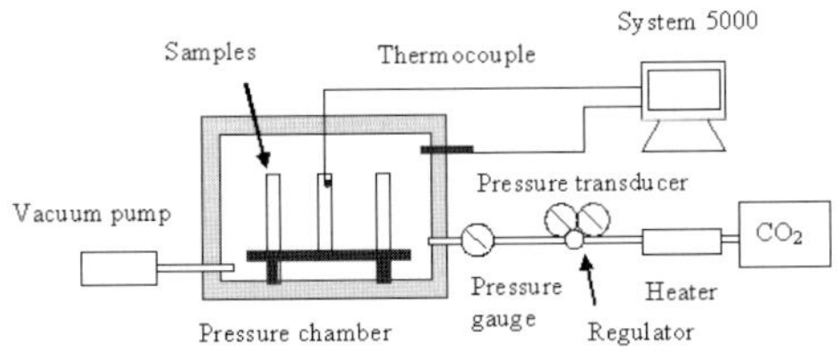

Figure 1. Experimental setup

Water loss was observed during the carbonation curing and the lost water was collected and added to the total mass as the water loss correction. The mass gain defined by Eq. (1) is based on a dry cement binder. The mass gain was also quantified by the powder analysis using infrared based CO_2 analyzer. The CO_2 content determined by the carbon analyzer was based on total mass after carbonation, including dry binder, water, and calcium carbonates. Eq. (1) represents an average measurement while the infrared technique is a point analysis. Since infrared technology requires only 0.5 g powder for analysis, the results can still serve as a comparison. Powders were collected both on surface and at the core to examine the uniformity of carbonation. Three-point bending tests at a span of 101 mm and compressive strength tests were performed right after two-hour carbonation to determine the modulus of rupture (MOR) and compressive strength of carbonated concrete specimens.

Carbonation was repeated using the material in bulk powder form using the equivalent to one compact sample. It was assumed that the porosity of the samples in a loose powder form would be much greater than that of the compact samples and thus display a higher CO_2 uptake if porosity was a controlling factor.

The modulus of rupture (MOR) and compressive strength of the carbonated specimens were tested immediately upon completion of the two-hour carbonation. The two cements were also used to make hydrated compact samples for reference. The hydrated samples were tested at an age of 7 days. For each batch, at least three specimens were tested for average.

3. RESULTS

The specimen temperature increased as soon as the gas was introduced into the chamber. The temperature was observed to have risen rapidly by the time the chamber had reached the desired pressure of 500 kPa which was reached within first few minutes of the test and slowly decreased thereafter. Table 2 summarizes the maximum temperatures reached during the carbonation of both the powder and compact specimens and the water loss during the carbonation of the compact specimens. The carbonation heats of reaction for the main calcium silicate phases are 347 kJ/mol for C_3S and 184 kJ/mol for β-C_2S (Goodbrake et al. 1979) while 74 kJ/mol is released during carbonation of calcium hydroxide (Moorehead 1986). The highly exothermic nature of the carbonation reaction leads to the formation of heat which contributes to the fast solidification. In each case, the maximum temperature measured in the carbonation of the powdered samples was significantly higher than the temperature rise detected in the compact sample. The greater temperature rise could be a reflection of the higher porosity of the powder providing more sites for reaction which might result in a more vigorous exothermic release.

Table 2. Carbonation temperature and percent water loss

Materials	Peak Temperature (°C)		% water lost from compacts[a]
	Powder	Compact	
Type 10 cement	116	70	26.3
Type 30 cement	131	68	26.7
Fly ash	131.7	70.1	10.8
Lime	117	60	92.4

a – expressed as a percentage of original mix water

Table 3 shows the mass gain and the percent carbonation degree of carbonated powder and carbonated compacts. The two cements had a final CO_2 content, as measured by infrared, of about 12.5% for the Type 10 and over 12.7% for the Type 30. Both fly ash and lime showed a significant carbon content in their as-received forms. It was verified by their high percentage of loss on ignition (LOI) in XRF tests (Table 1). The high carbon content in as-received fly ash and lime was likely attributed to the free carbon in fly ash and the uncalcined calcium carbonates in lime. To determine the net CO_2 uptake by the two materials through carbonation curing, the initial carbon was subtracted from the carbonated fly ash and lime samples. Table 3 summarizes the net CO_2 content and the carbonation degree. The calculations suggested that over 25% of the CO_2 reactive materials reacted in the two cements. Significantly higher reaction was seen in the fly ash and lime tests with over 45% and 72%, respectively, of the reactive materials reacted.

Table 3. CO_2 content determined by infrared analyzer

Materials	CO_2 content (%)		Carbonation degree (%)	
	carbonated powder	carbonated compact	carbonated powder	carbonated compact
Type 10 cement	10.9	12.5	22.1	25.4
Type 30 cement	12.3	12.7	25.9	26.8
Fly ash	12.9	11.3	51.5	45.1
Lime	18.5	23.8	56.2	72.3

When comparing the carbonation results reported in Table 3 to the temperature rise in terms of the sample form shown in Table 2, it is seen that, although the carbonation consistently produced higher maximum temperatures for materials in the powder form, the final CO_2 contents as measured by infrared do not show any consistent difference in the level of carbonation. Some samples showed greater carbonation in the powder form while others showed greater carbonation in the compacted from. Thus the reduced porosity of the compacted sample did not have an adverse affect on the overall carbon uptake. The higher carbonation degree of the fly ash and lime reflects the fact that the carbonation reaction of the $Ca(OH)_2$ in these specimens produced water and thus reduced or eliminated the reaction limiting effects associated with water starvation.

The results of the strength testing are given in Table 4. It can be seen that the strength achieved by the cement samples that were carbonated two hours exceeded the strength of comparable specimens allowed to hydrate for 7 days. The strength developed by the fly ash was not high enough to be considered for use as structural materials. However, fly ash can be used as supplementary cementitious materials in concrete products to assist CO_2 uptake during carbonation curing and promote pozzolanic reactions in the subsequent hydration. Lime had also developed sufficient strength with two hours of carbonation and could be combined with other materials to make special building products.

Table 4. Strength of carbonated cementitious materials

Material	Curing	Thickness	Age at test	MOR (MPa)	f_c' (MPa)
Type 10 cement	hydrated	14.3 mm	7-day	4.0	39.7
Type 10 cement	carbonated	14.3 mm	2-hour	8.8	56.8
Type 30 cement	hydrated	14.5 mm	7-day	4.3	41.3
Type 30 cement	carbonated	14.3 mm	2-hour	9.6	56.0
Fly ash	carbonated	20.5 mm	2-hour	1.3	3.5
Lime	carbonated	21.4 mm	2-hour	3.3	24.0

The fracture surface of carbonated cementitious materials was examined using a scanning electron microscope. Figure 2 shows a typical micrograph, obtained for carbonated Type 10 cement. Coatings of granular grains can be seen on the surface of cement particles and inside the pores with grain sizes ranging from1 to 5 microns. These are possibly the calcium carbonate crystals. The carbonated cement demonstrated a dense structure with reduced porosity. An EDX analysis was performed to verify the existence of carbon in carbonated samples. A typical result is also presented in Figure 2. The spectrum demonstrates a strong peak for carbon together with calcium, silicon, magnesium and oxygen, suggesting the presence of carbonates, although the quantification of carbon by spot analysis was difficult due to the use of fracture surface.

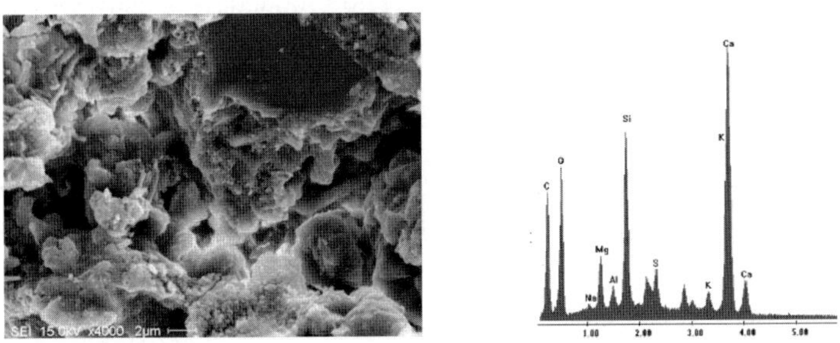

Figure 2. Microstructure and X-ray spectrum of carbonated Type 10 cement

The XRD patterns of carbonated Type 10 and Type 30 cement are displayed in Figure 3 together with natural calcite spectrum as reference. For both carbonated cements, the presence of calcium carbonate in the form of calcite is clear. The diffraction peaks coincide perfectly with that of the calcite reference at 2θ (CuKα) = 29.4° and 39.4°. Since the samples were preserved in 100% alcohol immediately after the 2-hour carbonation and mechanical testing, the hydration of the cement had probably ceased. It is evident from the absence of the calcium hydroxide peak at 2θ = 18° in the XRD patterns. The XRD analysis of carbonated cement confirms that the carbonation process converts the gaseous carbon dioxide to solid calcium carbonates in the preferred form of geologically stable calcite. Chemically unstable calcium hydroxide, a typical hydration product, disappears totally after 2-hour carbonation. It is, therefore, conclusive that the carbonation curing of concrete can produce a chemically and geologically stable product.

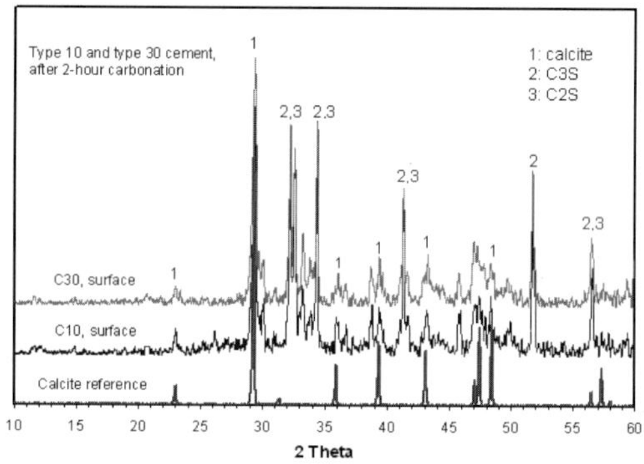

Figure 3. XRD Spectra of carbonated Type 10 and Type 30 cements

4. CONCLUSION

The possibility of using cementitious materials to sequester carbon dioxide is studied. The calcium-carrying materials are found to be able to convert gaseous CO_2 to solid calcium carbonates in a preferred form of calcite. The sequestration is permanent in a geologically stable form. The uptake is readily achievable at a CO_2 to binder ratio of 12.6% in cement, 11.3% in fly ash and 23.8% in lime in a press-formed compact product. The scale of CO_2 sequestration through concrete curing is comparable to that of small geologic storage projects. Nevertheless, the concrete production is daily-based, profit-oriented and can be carried out anywhere in the world. The production line is best suited to set near the cement plant, using the cement onsite and treating the concrete products by CO_2 collected from the stack gas. The accelerated strength development provides an incentive for industry to adopt the technology. A 12.6% CO_2 uptake in cement represents only 25% carbonation efficiency. There exists a large window of opportunity for improvement of the technology. The use of industrial wastes, such as fly ashes, promotes more net gain in global CO_2 sequestration activities.

Acknowledgment

The continued support by Natural Science and Engineering Research Council of Canada (NSERC), St. Lawrence Cement and CJS Technology is gratefully acknowledged.

5. REFERENCES

1. Berger, R.L., Young, J.F., and Leung, K., 1972, Accleration of hydration of calcium silicates by carbon-dioxide treatment. Nature Physical Science, 240: 16-18.
2. Goodbrake, C.J., Young, J.F., and Berger, R.L. 1979, Reaction of beta-dicalcium silicate and tricalcium silicate with carbon dioxide and water vapour, J. American Ceramic Society, March: 168-171.
3. Malhotra, V.M., 1999, Making concrete greener with fly ash. Concrete Int., 21: 61-66.
4. Moorehead, D. R., 1986, Cementation by the carbonation of hydrated lime, *Cem. Conc. Res.*, 16: 700-708.
5. Worrel, E., Price, L., and Martin, N., 2001, Carbon dioxide emissions from the global cement industry, Annual Review of Energy and Environment, 26: 303-329.
6. Young, J.F., Berger, R.L. and Breese, J., 1974, Accelerated curing of compacted calcium silicate mortars on exposure to CO_2, J. American Ceramic Society, 57: 394-397.

MORTAR BASED ON ALKALI-ACTIVATED BLAST FURNACE SLAG

Microcracks formation as a result of the drying shrinkage process

D. Krizan and M. Komljenovic

Dr. Chem. Eng., General Manager, Holcim, Serbia; Dr. Chem. Eng., Higher Scientific Associate, Center for Multidisciplinary Studies, University of Belgrade, Kneza Viseslava 1a, Belgrade, Serbia

Abstract: Blast furnace slag was activated with water glass of different modulus **n** (mass ratio between SiO_2 and Na_2O) and concentration. Linear deformations of alkali-activated slag mortar, i.e. mortar bars drying shrinkage, were examined as a function of the water glass modulus (**n**=0,6; 0,9; 1,2 and 1,5), water glass concentration (3 and 4% m/m calculated as Na_2O) and slag specific surface area (430, 500 and 600m^2/kg). Linear deformations of alkali-activated slag mortar were also compared with linear deformations of ordinary Portland cements mortar (CEM I 42.5).

Key words: Blast furnace slag, alkali activation, water glass, linear deformations, microcracks

1. INTRODUCTION

It is well known that blast furnace slag has latent hydraulic, i.e. binding properties so activator has to be present to enable good binding properties of slag. Alkali activation of slag has lately been the research subject of many scientists. However, due to numerous influencing factors, such as chemical and mineral composition of the slag investigated, the amount of glassy phase, slag specific surface area, then the type and concentration of activator, curing conditions etc., alkali activation of slag is still an insufficiently defined process requiring additional research. The influence of water glass ($Na_2O.nSiO_2$) modulus and concentration, as well as slag specific surface area, on the drying shrinkage process of alkali activated slag was investigated in this paper.

2. MATERIALS

Chemical and physical characteristcs of slag and ordinary Portland cement are given in Table 1.

M.S. Konsta-Gdoutos, (ed.), Measuring, Monitoring and Modeling Concrete Properties, 361–366.

Table 1. Chemical composition (%, m/m) and physical characteristics of slag (GBFS) and ordinary Portland cement (CEM I 42.5)

Composition / characteristic	GBFS	CEM I 42.5
Loss on ignition at 1000°C	0,42	3,12
SiO_2	38,26	21,67
Al_2O_3	9,32	5,42
Fe_2O_3	0,60	2,44
CaO	40,49	60,72
MgO	8,70	2,22
SO_3	0,47	2,86
S	0,42	-
MnO	0,81	0,05
FeO	-	0,24
P_2O_5	-	0,05
Na_2O	0,40	0,52
K_2O	0,39	0,31
Total	99,86	99,62
Density (kg/m^3)	2920	3150
Specific surface area - Blaine (m^2/kg)	430	350

3. SAMPLE PREPARATION AND TESTING

Samples of alkali-activated slag mortar were prepared in accordance with the standard JUS EN 196-1, with water/slag ratio 0,43, while Portland cement mortars were prepared with water/cement ratio 0,5. Water glass was previously mixed with water and then with slag. The mass ratio between binder and sand was 1:3. Dimensions of mortar bars were 40x40x160mm.

Mortar bars were kept in the mold during first 24 hours, in the air-conditioned room (temperature 20±2°C and relative humidity 50±5%), and after demolding under same conditions untill the moment of testing. Linear deformations – shrinkage testing was performed in accordance with the standard JUS BC.8.029/79. Linear deformations before demolding were not investigated.

4. RESULTS AND DISCUSSION

Linear deformations results of the alkali-activated slag mortar (for different water glass modulus **n** and concentrations, and slag specific surface area), as well as portland cement mortar are given in Figures1-3.

These figures show that linear deformations of alkali-activated slag mortar, regardless of the slag specific surface area, grow with the increase in waterglass concentration and modulus. The most significant dimension change occurs during first 21 days, and after that period the rate of linear deformations shrinkage is much slower. In all cases linear deformations of alkali-activated slag mortar are higher than linear deformations of reference Portland cement (CEM I 42.5). A certain reduction of linear deformations of alkali-activated slag mortar occured with the increase of the specific surface. This is more noticeable when the waterglass modulus **n** and concentration are higher (Fig. 4).

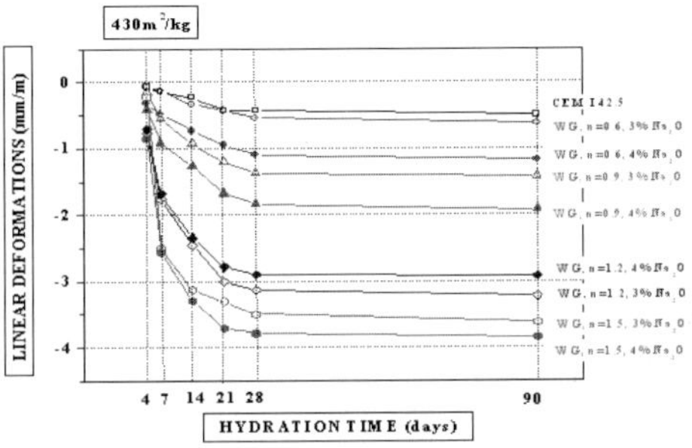

Figure 1. Mortar linear deformations of slag alkali-activated with waterglass (WG) and of ordinary Portland cement (CEM I 42.5) as a function of hydration time (for the slag specific surface area of 430m^2/kg and different values of waterglass modulus **n** and activator concentration – Na_2O content)

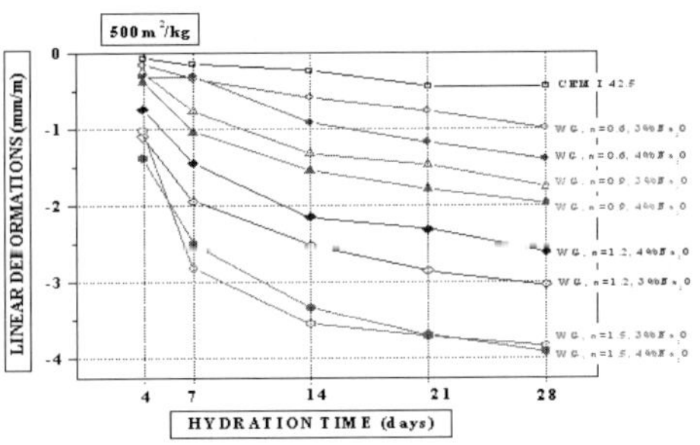

Figure 2. Mortar linear deformations of slag alkali-activated with waterglass (WG) and of ordinary Portland cement (CEM I 42.5) as a function of hydration time (for the slag specific surface area of 500m^2/kg and different values of waterglass modulus **n** and activator concentration – Na_2O content)

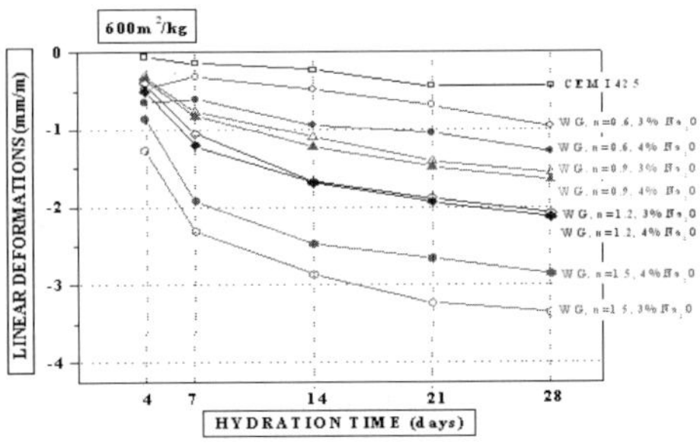

Figure 3. Mortar linear deformations of slag alkali-activated with waterglass (WG) and of ordinary Portland cement (CEM I 42.5) as a function of hydration time (for the slag specific surface area of 600m^2/kg and different values of waterglass modulus **n** and activator concentration – Na_2O content)

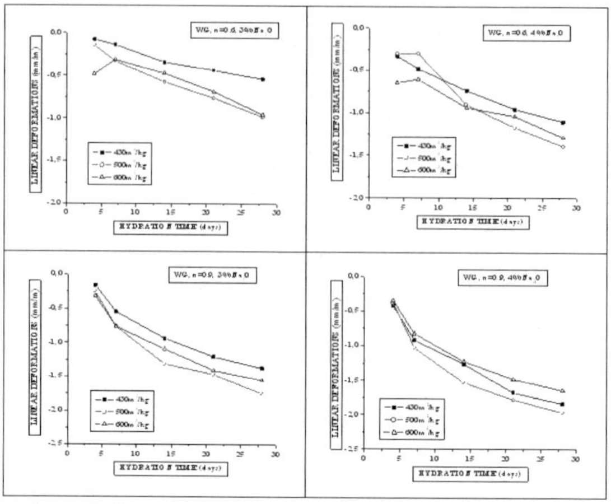

Figure 4. Linear deformations of alkali-activated slag mortar as a function of hydration time (for different values of the specific surface area of slag and waterglass modulus **n** and concentration – Na_2O content)

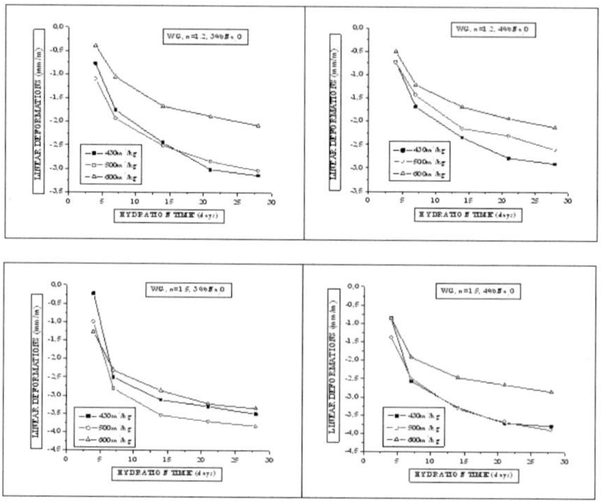

Figure 4. (continued) Linear deformations of alkali-activated slag mortar as a function of hydration time (for different values of the specific surface area of slag and waterglass modulus **n** and concentration – Na_2O content)

Krizan et al.[1] have found that poorly crystallized low base calcium silicate hydrate C-S-H (I) is the main hydration product of slag activated with water glass, which can be noted already after the first day of hydration. This is the reason for rapid strength growth of alkali-activated slag and also very high strength values. During drying process the structure of C-S-H gel is altering, which causes linear deformations – shrinkage. As a result of drying shrinkage process microcracks in the structure are forming, in this case the length is about 10-20μm and width less than 1μm (Fig.5.). Thus, the same factor that has a positive influence on compressive strength of alkali-activated slag mortar has a negative influence on linear deformations.

Figure 5. Microcracks in the structure of alkali-activated slag after 28 days of hydration

Other authors[2-4] also have reported increased shrinkage of mortar or concrete based on slag activated with water glass. They considered it as the consequence of the formation of silica or silica-rich gel during hydration. Glukhovsky[5] considered that the initial phase forming by alkali-activation of slag is rich in hydrosilicates and silica acid. Silica acid can polymerize into a silica gel. This tendency grows with the increase of the pH value. In an alkali medium silica acid is either in a molecular form or weakly dissociated and polymerization occurs within only a few minutes. Silica gel contains a large amount of water. The shrinkage begins with the syneresis process i.e. when free water enmeshed in the gel is spontaneously expelled. Syneresis comes to an end when the water content of the silica gel is still high (about 90%). During further drying the silica gel continues to shrink and number of particle-to-particle bonds in the structure gradually break.

5. CONCLUSION

Linear deformations of alkali-activated slag mortar, i.e. mortar bars drying shrinkage, were examined as a function of the water glass modulus (**n**=0,6; 0,9; 1,2 and 1,5), water glass concentration (3 and 4% m/m calculated as Na_2O) and slag specific surface area (430, 500 and 600m^2/kg). Linear deformations of alkali-activated slag mortar were also compared with linear deformations of ordinary Portland cements mortar (CEM I 42.5).

It was established that linear deformations of alkali-activated slag mortar, regardless of the slag specific surface area, grow with the increase in waterglass concentration and modulus. In all cases linear deformations of alkali-activated slag mortar are higher than linear deformations of reference Portland cement.

A certain reduction of linear deformations of alkali-activated slag mortar occured with the increase of the specific surface. This is more noticeable when the waterglass modulus **n** and concentration are higher.

During drying process the structure of C-S-H gel, the main hydration product of slag activated with water glass, is altering, which causes linear deformations – shrinkage. As a result of drying shrinkage process microcracks in the structure are forming.

6. REFERENCES

1. D. Krizan, M. Komljenovic and B. Zivanovic, The influence of different parameters on the hydration process of binders based on alkali activated slag, J. of the Serbian Chem. Soc., **70** (1), 97-105, 2005.
2. C. Shi, R.L. Day, X. Wu and M. Tang, Comparison of the microstructure and performances of alkali-slag and Portland cement pastes, Proc. of the 9th Int. Cong. on the Chemistry of Cements, New Delhi, India, vol. **3**, 298–304, (1992).
3. S.-D. Wang, X.-C. Pu, K. Scrivener and P.L. Pratt, Alkali-activated slag cement and concrete: A review of properties and problems, Advances in Cement Research, **27** (7), 93–107, (1995).
4. T. Kutti, Hydration products of alkali activated slag, Proc. of the 9th Int. Cong. on the Chemistry of Cements, New Delhi, India, vol. **3,** 468– 474, (1992).
5. V. Glukhovsky, Yy. Zaitsev and V. Pakhomov, Slag–alkaline cements and concretes - structure, properties, technological and economical aspects of the use, Silicate Industrials, **48** (10), 197–200, (1983).

INFLUENCE OF TEMPERATURE AND CHEMISTRY ACTIVED ON THE CEMENTING PROPERTIES OF COAL GANGUE

W. Zhang, S. Zhou, J. Ye, D. Li,* and Y. Chen

*China Building Materials Academy, Beijing 100024, P R China; * Nanjing University of Technoly, Nanjing 210009, P R China*

Abstract: Coal gangues, which are gotten from different six coal mines, were used as raw materials in the paper. The chemical compositions and the minerals of the original coal gangue have been tested. The results state that there is the different optimum temperature for the different coal gangue. For most coal gangues, the optimum activated temperature is 800. The 28days compressive strength of mortar with activated coal gangue can be increased when calcining coal gangue with a certain limestone. The structure of calcining coal gangue and calcining with limestone is tested by the method of NMR. The related mechanism of the calcining coal gangue is discussed in the paper.

Key words: Coal gangue; Calcining temperature; Cementitious properties; NMR method

1. INTRODUCTION

Coal gangue is waste residue during the coal mine production process of coal mining and coal washing, is the one of present biggest industrial solid junk of our country of discharge capacity. Generally coal gangue comprehensive discharge capacity takes the 15%-20% of the output of raw coal. Statistics according to national state economic & trade commission shows that the annual emission of coal gangue in our country is near to one-hundred millions tons[1-2], which is now amounted to accumulation of more than 3000 millions tons, covering an area of 1.20000 ha.

Because of mineral contents of the coal in every place differs, its chemical composition is relatively complex, which is about 10 kinds of element probably[3]. Its major composition normally is silicon and aluminum. Most of the coal gangue is a clay rock; its main mineral, for example, kaolin and quartz. Numerous researches show[4] the gel role of fresh coal gangue is very weak, because this kind of coal gangue have steady crystal structure, its atom, ion and member etc. are put according to certain law by order. After

M.S. Konsta-Gdoutos, (ed.), Measuring, Monitoring and Modeling Concrete Properties, 367–372.

calcinations under high temperature of 800-1000, the clay mineral is dewatered and disintegrated, the carbon component is removed with the deteriorative impurity burned out. The crystal is disintegrated and transformed into amorphous non-crystal; this makes the coal gangue active. The active SiO_2 and Al_2O_3 in it can react with $Ca(OH)_2$ and produce CSH gel, calcium aluminate and calcium sulfoaluminate hydrates[5-6].

2. MATERIALS AND TEST

Coal gangue was obtained from six coal mine, a,b,c,d,e,f. The 52.5R Portland cement produced by Huaxin Cement plant in Hubei Province was used in all the mortar mixtures. The chemical composition characteristics of them are described in Table. 1.

Table 1. Chemical composition of six regions coal gangue

Composition	SiO_2 %	Al_2O_3 %	CaO %	MgO %	Fe_2O_3 %	SO_3 %
a coal gangue	54.34	13.75	9.53	1.11	3.95	2.57
b coal gangue	43.97	27.72	1.92	0.6	5.02	0.45
c coal gangue	49.09	27.37	1.79	0.13	1.95	0.13
d coal gangue	54.52	17.64	2.49	1.62	6.5	1.67
e coal gangue	57.54	19.21	3.08	1.21	4.92	1.67
f coal gangue	57.95	19.02	3.16	0.82	5.32	0.64
Portland cement	21.06	6.04	63.98	2.67	3.63	2.25

a,b,c,c,e,f, stands for different regions, a-Xinzhi, b-Yangquan, c-Huozhou, d-Yanzhou, e-zhenzhou, f-Xuzhou

The mortar specimens were prepared according to "The Standard for Cement Testing in China", which specifies a water to cementitious material ratio of 0.5 and a sand to cementitious material ratio of 2.5 by weight.

The specimens were stored in water with temperature of 20±2 for curing one day after being cast. Compressive strengths of the mortar specimens were determined at 3,28days. The compressive strength of cement pastes was tested after the thermal activated coal gangue to Portland cement with a mass fraction of 30 percent.

2.1 X-ray diffraction

Coal gangue was obtained from six coal mine, they are Yangquan coal mine, Huozhou coal mine, Xinzhi coal mine, Yanzhou coal mine, Zhenzhou coal mine, Xuzhou coal mine. Its main mineral composition consists of kaolin, illite, quartz, mica, anorthite, siderite and calcite, etc.

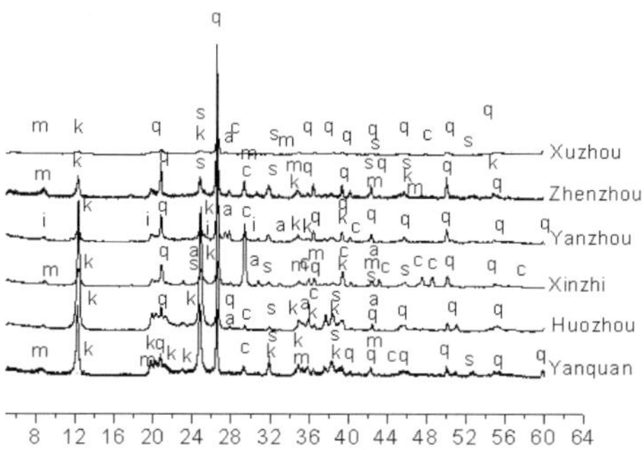

Figure 1. XRD patterns for six regions coal gangue

2.2 Effect of calcined temperature of coal gangue

The calcining temperature of clay affects the pozzolanic reactivity of the resulting product. The clay is in its most reactive state when the calcining temperature leads to loss of hydroxyls and results in a collapsed and disarranged clay structure. The calcining temperature producing the active state is usually in the range of 600-1000. In this paper, the calcining temperature of coal gangue in turn is 800900and 1000. 3d and 28d compressive strengths of different coal gangue calcining at different temperature are shown in Figures 2 and 3.

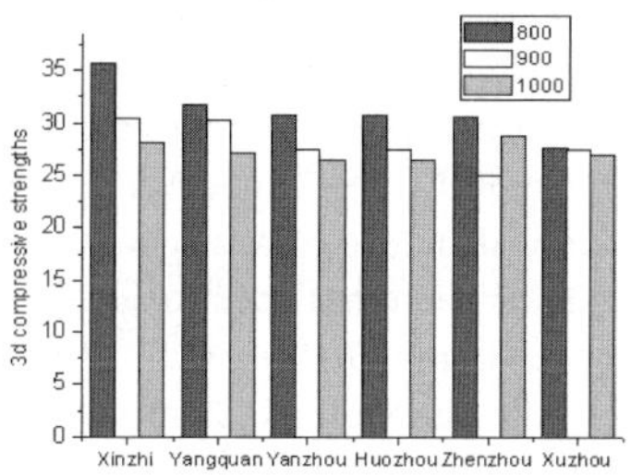

Figure 2. 3d compressive strength of different coal gangue calcining at different temperature

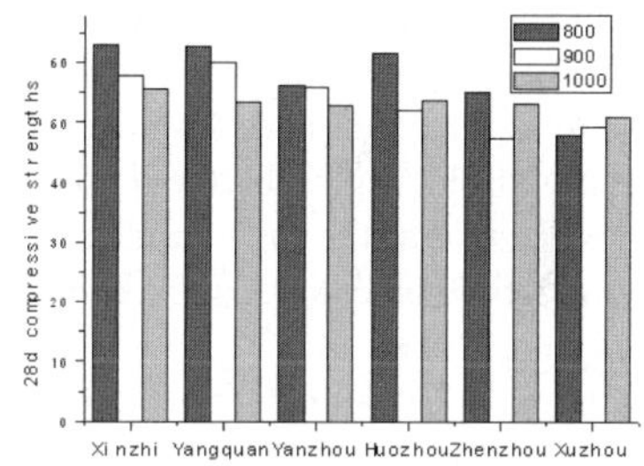

Figure 3. 28 d compressive strength of different coal gangue calcining at different temperature

2.3 Effect of calcining coal gangue with limestone

Because of CaO content of most coal gangue is very low, ordinarily below 3 percent, the way of chemistry increase calcium was adopted. One of CaO content is from 8 percent to 10 percent. One is around 15 percent, another is around 25 percent. When the CaO content is exceeding 25 percent, the cementitious properties will be declined according to the reference[4]. So the way of calcining coal gangue with limestone was chosen. The calcining temperature is 820. The compressive strength of mortar was tested with a mass fraction of 30 percent activated coal gangue.

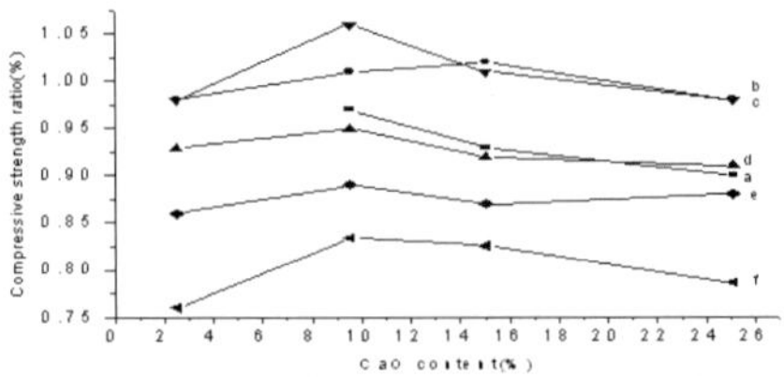

Figure 4. Compressive strengths of different CaO contents coal gangue blended cement

Figure 4 shows that changes of 28 days compressive strengths of thermal activated coal gangue with different CaO content. From the Fig.4, it can be seen that the compressive strengths of different region coal gangue blended cement are increased when calcining with a certain limestone at 820.

2.4 NMR analysis of temperature activated and Chemistry increased calcium

Figure 5 present the Si^{29} spectra of the samples of Xuzhou original coal gangue(a)calcining at 800°C,(b)900°C,(c)1000°C,(d) and calcining at 820°Cwith limestone(e). The results show that the silica and oxide tetrahedron structure of of coal gangue have much changed when calcining at different temperture and calcining with limestone at 820°C. Fig. 5 (a) show that silica and oxide tetrahedron of Xuzhou original coal gangue are unsymmetrical polymeric state. It behave like spectra apex of unsymmetrical structure as well as chemistry displacement situated at $-85*10^{-6}$. After calcining at the temperature of 800-1000°C, the chemistry displacement is not lie at former place, but move at the direction of high field. Fig. 5 (b) show that when calcining at 800°C, Si^{29} spectra apex was split up. It explains that the unsymmetrical silica and oxide structure of coal gangue was discompounded when calcining. At the same time, it can be seen that spectra apex moved at low polymeric state. But Fig. 5 (d) shows that when calcining at 1000°C Si^{29} spectra was aculeate and spectra apex moved at high polymeric state as well as chemistry displacement situated at $-110*10^{-6}$. Fig. 5 (e) shows that calcining at 820 °C with limestone the Si^{29} spectra apex was split up and spectra apex become wide. Silica and oxide turned into three kinds of separated silica and oxide tetrahedron. There were Q^0Q^1 Q^2 forms. This phenomenon reflects the mechanics properties, namely the compressive strength is increased.

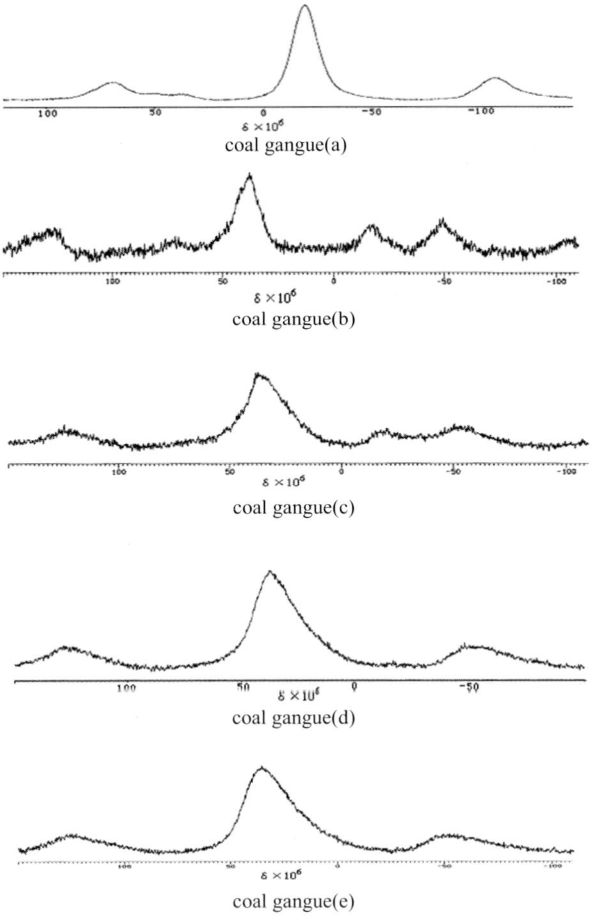

Figure 5. Si^{29} NMR spectra of different coal gangue

3. CONCLUSIONS

The following conclusions can be drawn from the present study:

1 Most of the coal gangue is a clay rock; its main mineral composition consists of kaolin, illite, quartz, mica, anorthite, siderite and calcite.

2 There is the different optimum temperature for the different coal gangue. For most coal gangue, the optimum activated temperature is 800.

3 28days compressive strength of mortar with activated coal gangue can be increased when calcining coal gangue with a certain limestone.

4 The silica and oxide tetrahedron structure of of coal gangue have much changed when calcining at different temperture and calcining with a certain limestone at 820.

Acknowledgments

The authors wish to thank the National Key Fundamental Research and Development Program of China for their financial support (2001CB610703)

4. REFERENCES

1. "Policy point of coal gangue utilization put out by national state economic & trade commission". Coal Clean Technology, 2000(1): 56
2. Wu Zhongwei and Lian Huizhen. High Performance Concrete. China Railway Press, 1999
3. Chen Jianxiong et al. "Investigation on Durability of Buildings under Severe Acid Rain." Concrete, 2001(11),44-47
4. Dongxu Li, Xuyan Song, Chenchen Gong, et al. "Research on cementitious behavior and mechanism of composite cement with coal gangue. Journal of the Chinese Ceramic Society, 2004,32(3): 358-363
5. Li Min, and Xu Yuyan. "Utilization of Aggregate from Self-combusted Gangue." New Building Materials, 2001(3):51-56
6. B.B. Sabir, S. Wild, J. Bai. "Metakaolin and calcined clays as pozzolans for concrete: a review". Cement &Concrete Composites 23 (2001):441-454

PERFORMANCE CRITERIA FOR THE USE OF FGD GYPSUM IN CEMENT AND CONCRETE PRODUCTION

G. Tzouvalas, G. Rantis and S. Tsimas
National Technical University of Athens, School of Chemical Engineering, 9 Heroon Polytechniou Str., Zografu Campus, 15773, Athens

Abstract: FGD gypsum is evaluated as a possible partial or total substitute of natural gypsum for the control of cement setting. Extended laboratory and industrial scale trials in cement and concrete mixes were carried out. FGD gypsum after been dried presents an excellent performance compared with natural gypsum.

Keywords: FGD gypsum; cement setting time; solubility; dihydrate; compressive strength; moisture; durability.

1. INTRODUCTION

In the cement industry, natural gypsum ($CaSO_4{\cdot}2H_2O$) is added during grinding of Portland cement in order to delay the rapid reaction between C_3A ($3CaO{\cdot}Al_2O_3$) and water and to regulate cement setting properties. However its continuous use has led to the reduction of its high purity stock in some countries. In certain cases anhydrite ($CaSO_4$) has proved particularly reliable for the partial replacement of natural gypsum[1, 2]. FGD gypsum, a waste material of the desulphurisation process in coal burning power plants, where flue gases with high SO_2 emmissions, are converted into calcium sulfate dihydrate ($CaSO_4{\cdot}2H_2O$), is an important alternative source of natural gypsum. Strong efforts have been made to utilize FGD gypsum from the technical and economic point of view[3-8]. At the same time FGD gypsum has gained recognition in the European Union and in the OECD as being a product. There are two major methods of producing FGD gypsum: the wet-type scrubbing process, which is the most common (90%) and thus applied in Greek power plants (Megalopolis and Florina plants), where mixture of $CaO/CaCO_3$ with water (slurry) is used, and the dry or semidry scrubbing process. The chemical reactions that can take place are the following: $SO_2 + H_2O \rightarrow H^+ + HSO_3^-$, $H^+ + HSO_3 + 1/2O_2 \rightarrow 2H^+$

M.S. Konsta-Gdoutos, (ed.), Measuring, Monitoring and Modeling Concrete Properties, 373–378.

+ SO_4^{-2} and $2H^+ + SO_4^{-2} + Ca(OH)_2 \rightarrow CaSO_4 \cdot 2H_2O$ (>95%w/w content in $CaSO_4 \cdot 2H_2O$). FGD gypsum annual production in Greece approaches 1.000.000tn which is totally rejected to the environment after been mixed with fly ash[6, 8, 9].

2. EXPERIMENTAL

2.1 Laboratory and industrial cement production

FGD gypsum was selected from Megalopolis PPC thermal plant and had a moisture about 40%, so it was first dried in an oven at the temperature of 45^0C in order not to been partially or totally dehydrated forming hemihyrate or anhydrite. The dry process is necessary, because according to industrial scale estimations, the high moisture content of FGD gypsum (~10%) could cause handling and feeding problems when it is used in large percentages. Natural gypsum was selected from Titan mine quarries. The chemical analysis of these calcium sulfate bearing materials (CSBMs), according to ASTM C 471M – 95, the hydrated forms of calcium sulfate, as well as their solubility at 25^0C, are shown in Table 1.

Table 1. Chemical analysis, hydrated forms determination and solubility of CSBMs

	Gypsum	FGD Gypsum
Combined water, %	19,30	18,05
SO_3, %	43,41	42,19
SiO_2, %	0,65	0,30
CO_2, %	2,51	1,58
CaO, %	32,40	33,40
$Fe_2O_3 + Al_2O_3$, %	0,03	0,10
MgO, %	0,92	0,10
$CaSO_4 \cdot 2H_2O$, %	90,97	87,96
$CaSO_4$, %	1,87	2,17
Solubility (g/100g H_2O)	0,260	0,273

These CSBMs were interground for an hour in a lab ball mill in different proportions (4–7,5%) with clinker produced in Titan Cement Plant (Kamari). SO_3 content in the cement produced was ranging from 2,3 to 4,4% and the specific surface obtained was in the range of 3700 to 3900 cm^2/g. The optimum percentage of SO_3 relating to the compressive strength, is 3,5% for cements with both CSBMs. It was also shown that the results of compressive strength of cements with FGD and natural gypsum are in the same order, especially around the optimum SO_3 addition of 3,5%[8, 9].

In order to achieve a SO_3 value about to 3.5%, which is the optimum addition of the CSBMs, cement mixtures using blends of gypsum and FGD were also prepared. The mixtures of CSBMs expressed as natural gypsum/FGD gypsum ratio were: 0/100 (CGF1), 20/80 (CGF2), 40/60 (CGF3), 50/50 (CGF4), 60/40 (CGF5), 70/30(CGF6), 100/0 (CGF7)[8, 9].

During industrial production of cement, in the cement mill, gypsum undergoes partial dehydration at 110-130^{0}C to form hemihydrate ($CaSO_4 \cdot 0.5H_2O$) and in some cases total dehydration, at 170-190^{0}C, to form soluble anhydrite (-$CaSO_4$)[2, 7]. In order to apply the laboratory findings in industrial trial, clinker of various temperature (130-250^{0}C) was fed into a ball mill with mixtures of CSBMs [100% gypsum (CG), 75% gypsum/25% FGD (CGF)] and filler (limestone) to produce cement (CEM I, specific surface:3500 to 3650 cm^2/g). The temperature at the exit of the mill was maintained below 118^{0}C by spraying the necessary quantity of water at the inlet and/or the middle of the mill. The quantity of sprayed water and the mill dew point for each trial were calculated through the construction of mill heat balances. All these in addition to clinker temperature, SO_3 values obtained for cements and the admixing ratios of dihydrate and anhydrite form to each CSBM are shown in Table 2. In the same time physico-mechanical properties, such as setting time and compressive strength of the produced cement were determined based on EN-norms[10].

Table 2. Clinker temperature, admixing ratios of dihydrate and anhydrite to CSBM, water sprayed and mill dew point

	CG1	CG2	CGF1	CGF2	CGF3
Clinker temperature (^{0}C)	250	192	250	189	133
% dihydrate	3,70	3,70	4,73	4,54	4,71
% anhydrite	0,64	0,64	0,36	0,43	0,35
H_2O (kg/h)	5894	4146	6252	3966	2142
Mill production (t/h)	95.5	97	99	99	94
Dew point (^{0}C)	73	66	76	65	53

2.2 Concrete production

Cements (CEM I) with natural gypsum and FGD as setting regulator in additions aiming to a SO_3 addition in cement around 3.5%, with two types of aggregates crushed fine and coarse limestone and sand at total cement to aggregate ratios of 1:0.38, 1:2.62, and 1:3.26 respectively were used to prepare concrete mixes of category C20/25 and of slump class S3 (10-11cm) according to EN 206-1 at a w/c ratio of 0.68. Compressive strength measurements were carried out according to Greek standards. The resistence of concrete mortars to sulfate attack was estimated through the determination of strength and mass loss after an exposing period of 50 days in a solution of H_2SO_4 3%w/w. The capillary absorption was testified according the method: Determination of the capillary absorption of water of hardened concrete (Materials and Structures, Vol. 32, April 1999). The uptake of water by capillary absorption was measured through the weight of the specimens at time intervals of 10, 30 minutes, 1, 4 and 24 hours[11].

3. RESULTS AND DISCUSSION

The 28-day compressive strength results for cements with mixtures of CSBMs are shown in Figure 1. No significant differences, except the 20/80 mixture, are observed. The compressive strengths of CGF cements showed the same values for all examined ratios[8-10].

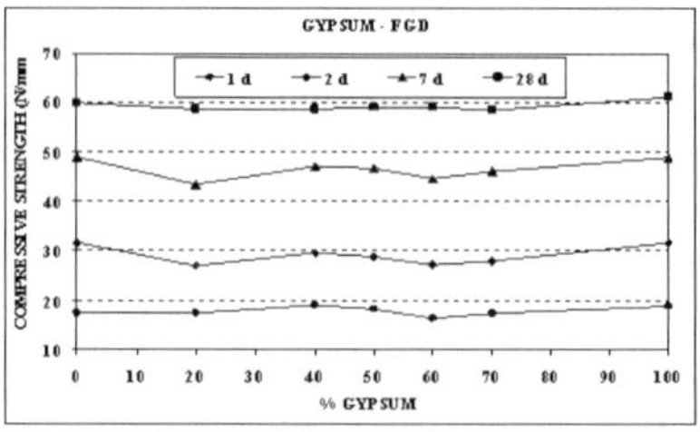

Figure 1. Compressive strengths of CGF cements vs. the ratio of gypsum in the mixture of CSBM

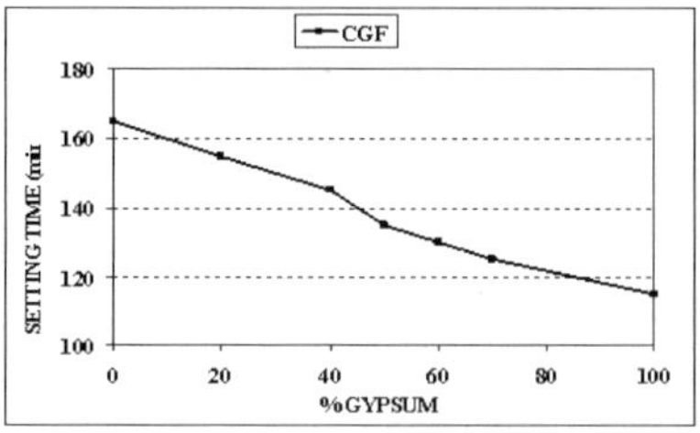

Figure 2. Initial setting times of CGF cements vs. the ratio of gypsum in the mixture of CSBM

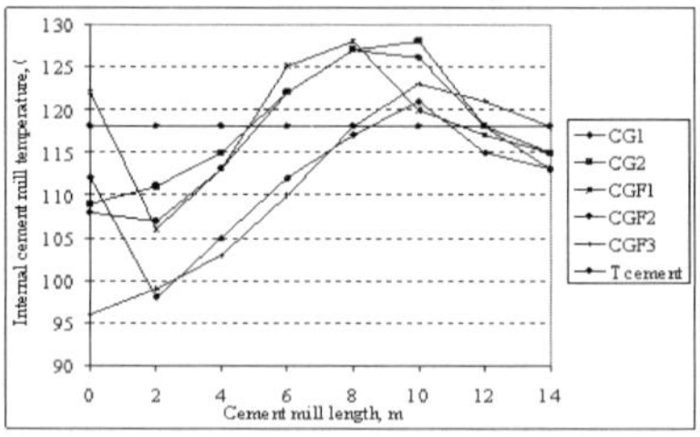

Figure 3. Cement with various admixtures of gypsum and FGD gypsum produced in industrial scale: Internal cement mill temperature vs. cement mill length

In Figure 2 the initial setting times of all cements versus the percent content of gypsum in the CSBM mixture are plotted. The addition of FGD leads to slightly higher setting times. It is also extracted that FGD is a very good controller of setting, as it can give, in mixtures with gypsum, a large range of setting time (115–170 min). Although they both contain the same chemical compound, i.e. $CaSO_4 \cdot 2H_2O$, the setting time increases as more dihydrate of FGD gypsum replaces dihydrate of gypsum in the admixture[8-10].

In Figure 3 the profile of the internal mill temperature for each trial is plotted. Temperatures did not exceed 130^0C and gypsum dehydration was limited to the formation of hemihydrate and not γ-$CaSO_4$. The behavior of each CSBM in the cement setting and the compressive strength performance was confirmed from the industrial samples as well. It must be pointed out that the mill production was increased when FGD was used as setting regulator. Therefore, it was obvious that FGD gypsum increased the setting time without deteriorarting the process of the industrial production[10].

The compressive strengths of 2, 7, 28 and 90 days for concrete mixes with CSBM addition 6% are presented in Figure 4. When the addition of CSBM was 4 and 5% the results were also in a similar trend. In general the results are in great accordance with those in cement pastes, indicating the suitability of FGD as alternative cement setting retarders[11]. Capillary sorptivity is a serious indication for the durability of concrete with different CSBM as setting retarders, since water presence is an obligatory requirement for any chemical external attack to mass concrete. Concrete with FGD as setting retarder shows less capillary sorptivity for all additions in cement (Figure 5).

The results of the resistance of concrete with different CSBMs in sulfate attack are presented in Figure 6. Concrete with FGD gypsum show a more vigorous resistance compared with natural gypsum. Concrete with NG are appeared with a strength loss ranging from 12 to 19% and mass loss about to 6%. Deterioration of compressive strength in FGD is eliminated below 8.5% for all additions. Mass loss in concrete with FGD gypsum does not exceed the 3% of the total concrete mass[11].

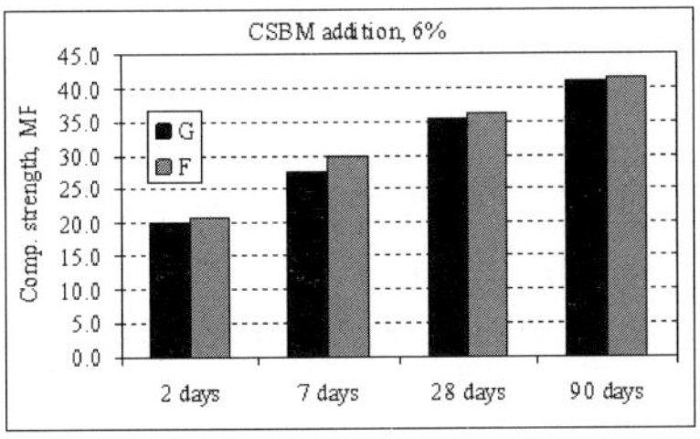

Figure 4. Compressive strength of 2, 7, 28 and 90 days of concrete containing 6% CSBM addition

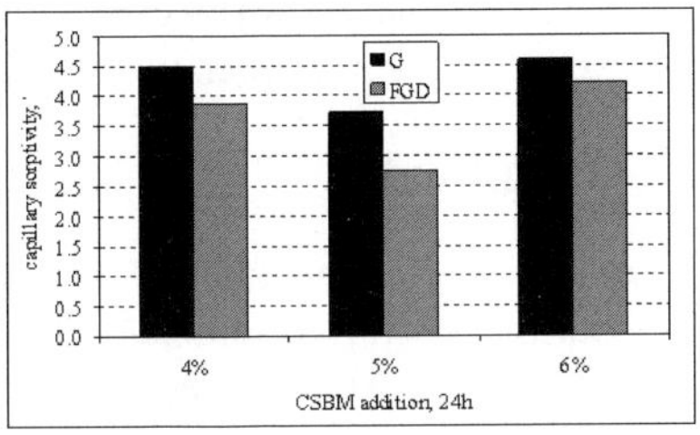

Figure 5. Capillary sorptivity of concrete at 24 hours for different aditions of CSBM. addition

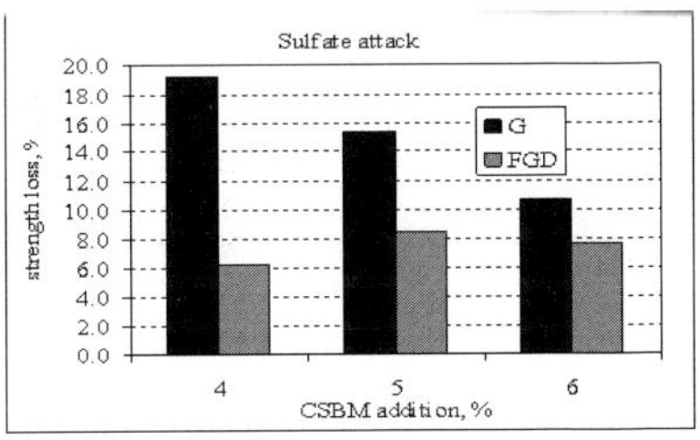

Figure 6. Strength loss of concrete containing different CSBMs due to sulfate attack

4. CONCLUDING REMARKS

Laboratory and industrial scale trials where natural gypsum and FGD gypsum have been used in various proportions in the cement production have shown that the addition of FGD gypsum increases setting time. This must be attributed to the higher solubility of FGD gypsum compared with natural gypsum (Table 1) and consequently to the different quantities of soluble SO_3 which exist in cement after its mixture with water and react with C_3A retarding its hydration. The more FGD gypsum is added in the mixture of FGD / NG, the more soluble SO_3 in cement is available and consequently the higher setting times are obtained[8, 9]. FGD gypsum showed similar to gypsum behaviour in compressive strength development, especially in the range of the optimum SO_3 addition, which was found to be about 3,5%. Additionally the durability of concrete seems to be improved with the use of chemical by-product (FGD gypsum) as setting regulator instead of NG.

However, the high moisture content of FGD gypsum (~10%) could cause handling and feeding problems when it is used in large percentages. FGD is produced from PPC in Megalopolis (700000t/y) and in Florina (300000t/y). The Florina Plant has eliminated moisture content under 10%. In the same time PPC is trying to develop a dry process in Megalopolis Plant in order to reduce moisture content below 5%. If this happens the total production of FGD can be directly used for the cement production with very satisfactory environmental and economical benefits.

In concluding FGD gypsum is a suitable alternative to gypsum for the cement setting regulation when it satisfies the following performance criteria: i) its moisture is less than 5% ii) it has more than 95%w/w content in $CaSO_4 \cdot 2H_2O$ and iii) its addition leads to the optimum SO_3 content in cement which is about 3,5%.

Acknowledgments

The authors wish to thank the "Titan Cement Company" and especially the director of R & D Department Mr. E. Chaniotakis and Dr. A. Papageorgiou for the technical and scientific support at the experimental scale trials.

5. REFERENCES

1. Locher F. G., Richartz W., Sprung S., Setting of cement – Effect of adding calcium sulfate, ZKG, vol. 6, 271-277, 1980.
2. Theisen K., Relationship between gypsum dehydration and strength development in Portland Cement, ZKG, vol. 10, 571-577, 1983.
3. Ozkul Hulusi M., Utilization of citro- and desulphogypsum as set retarders in Portland cement, Cement and Concrete Research, vol. 30, 1755-1758, 2000.
4. Manjit Singh, Influence of blended gypsum on the properties of Portland cement and Portland slag cement, Cement and Concrete Research, vol. 30, 1185-1188, 2000.
5. Hamm H., Coping with FGD gypsum, ZKG, vol. 8, 443-449, 1994.
6. Roulia M., Flue gas desulphurization technologies with calcium compounds, Chimika Chronika, vol. 3, 73-77, 1999.
7. Roy S., Ghosh S. N., Case study of rising cement mill temperatures – incidence of false set, ZKG, vol.54 (No 4), 206 – 212, 2001.
8. Tsimas S., Delagrammatikas G., Papageorgiou A., Tzouvalas G., Evaluation of industrial byproducts for the control of setting time of cements, Cement and Concrete World, vol.6, 54-62, 2001.
9. Tzouvalas G., Rantis G., Tsimas S., Alternative calcium sulfate bearing materials as cement retarders. Part II: FGD gypsum, Cement and Concrete Research, vol 34 (11), 2119-2125, 2004.
10. Tzouvalas G., Tsimas S., Papageorgiou A., Mill heat balances. A key for the industrial production of cements with different CSBMs, The 6th Intern. Conf. on concrete technology for developing countries, Amman-Jordan, vol.1, 81–90, 2002.
11. Tzouvalas G., Efstratiou K., Chaniotakis E., Tsimas S., Effect of calcium sulphate bearing materials on concrete properties and durability, The 7th Intern. Conf. on concrete technology for developing countries / Innovations and emerging technology in concrete industry, Kuala Lumpur - Malaysia, 285 – 294, 2004.

CEMENT-BASED NANOPIEZO 0–3 COMPOSITES

Z. Li and H. Gong
Department of Civil Engineering, Hong Kong University of Science and Technology, Clear Water Bay, Kowloon, Hong Kong, People's Republic of China; [2]School of Materials Science and Technology, Shandong University, Jinan 250061, People's Republic of China

Abstract: In this paper, study on a new 0-3 type cement-based NanoPiezo composites is presented. The nanopowder was prepared by a sol-gel process. The combined results of the DTA/TGA, XRD, laser particle size analysis and SEM studies indicated that the PZT powder calcined at 700°C obtained perovskite structure with an average crystallite diameter of 26.4 nm. Using a pressing process, up to 80vol% NanoPiezo powder could be incorporated into cement-based composites. The behaviors of the composites under different polarizing conditions are investigated. The composites were compacted and had good response behaviors in impedance spectra. The variations of piezoelectric properties of the nanocomposites to PZT contents were studied. It shows that cement-based NanoPiezo composites have some advantage to the cement-based PZT composites. There is a good potential for application of 0-3 type cement-based NanoPiezo composites in civil engineering

Key words: cement; PZT; nanopowder; piezoelectric properties; composites.

1. Introduction

Lead zirconate titanate ($Pb(Zr_xTi_{1-x})O_3$ or PZT) piezoelectric ceramics have been used successfully in industry applications, such as transducers, micro-positioners, rotary actuators and sensors.[1-2] In recent years, the piezoelectric composites containing PZT ceramics were attracted more interests for a wider variety of applications. The general goal of the research was to tailor the properties of composites by combining particular properties of the individual components. For example, the combination of a piezoelectric ceramic with a polymer provides a flexible or shapeable material with good piezoelectric properties.[3-4] In order to suit the requirements of civil engineering applications, cement-based 0-3 piezoelectric composites were designed by mixing PZT powders and cements. The piezoelectric properties of the materials were slightly higher than that of 0-3 polymer

M.S. Konsta-Gdoutos, (ed.), Measuring, Monitoring and Modeling Concrete Properties, 379–384.

matrix piezoelectric composites.[5] However, the basic properties of the cement based piezoelectric materials, such as responsibility, sensitivity and stability, were expected to be further improved for the application as sensors or actuators in civil engineering.

The dispersed PZT phases play a critically important role in the 0-3 type cement based piezoelectric compositesWhile, the quality of PZT powder strongly depends on synthesis process. Compared to other methods, the sol-gel process offers many advantages, including high purity, chemical homogeneity, and controlled particle size.

In this work, $Pb(Zr_{0.52}Ti_{0.48})O_3$ nanopowders were prepared by an alkoxide-based sol-gel process. Cement based NanoPiezo 0-3 composites were fabricated by pressing the mixture of cement and PZT powders. The dielectric properties of the composites were determined by using HP4294A. It was found that sol-gel generated powder had superior piezoelectric properties to that made by conventional solid reaction method.

2. EXPERIMENTAL PROCEDURES

$Pb(Zr_{0.52}Ti_{0.48})O_3$ nanopowders were prepared by sol-gel process. Firstly, the lead acetate (AR, Nacalai tesque inc, Japan) was initially dissolved in acetic acid. The associated water was removed during a period of distillation at 110°C for 1h. After cooling down to 80°C, the stoichiometric amount of zirconium n-propoxide (70 wt% in 1-propanol, Aldrich, USA) was added to the solution and the mixture was refluxed at 110°C for 2h. Then the stoichiometric amount of titanium iso-propoxide (98 %, Acros, USA) was added and refluxed at 110°C for 2h. Deionized water was added to the solution and continued to heat at 110°C. After 1h, a clear slight-yellow gel was obtained. Then the gel was dried until solidified completely.

The crystal phase and composites of the PZT powders were analyzed by XRD (PW1830, Philips) and XRF (JSX-3201z, JEOL), respectively. The particle size distribution was measured by an laser particle size analyzer (Model LS230, Coulter). The morphology of powders and PZT/cement composites was observed by SEM (JSM 6300, JEOL) connected with a Energy Dispersive X-ray Analysis (EDAX).

To prepare the cement/PZT composites with 0-3 connectivity, calcined PZT powder and white cement were mixed thoroughly, then about 10 wt. % water was added into the mixed powders. The mixture was pressed under a pressure of 100 MPa to form discs with diameter of 14mm and thickness of about 1.5 mm. The samples were then cured at 60°C in a relative humidity of 100% for 24 h. Silver paste was coated on sides of the discs as electrodes. The poling was carried out for 1h at a rate of 5kV/mm at 140°C in silicon oil.

The piezoelectric strain factor, d_{33}, was measured by using a d_{33} meter (Model ZJ-3B, Institute of Acoustics Academia Sinica, China). The impedance spectra of the samples were measured by an impedance/gain-phase analyzer (Model 4294A, Hewlett Packard, Tokyo, Japan). The dielectric constant was measured using the same apparatus at 1 kHz.

3. RESULTS AND DISCUSSIONS

3.1 Characterization of the nanoPZT powder

The XRD patterns of the precursor gel calcined for 1 hour at various temperatures are shown in Figure 1. The sample shows an amorphous structure at the calcined temperature of 80°C. When the calcined temperature increases to 300°C, small amounts of pyrochlore phase is observed with a characteristic broad bump near 30° 2 in the XRD pattern. With the increase of the calcined temperature, the pyrochlore phase is gradually reduced and the perovskite phase increases. At the temperature of 700°C, XRD patterns shows the crystallization of perovskite phase only and no pyrochlore phase is observed.

The average crystallite size in the PZT powder prepared from the annealed gel can be calculated using Scherrer's equation.[6] The crystallite diameters of the PZT powders calcined at 700°C, 800°C and 900°C are calculated to be 23.6nm, 26.4nm and 30.7nm, respectively. The particle size distributions of PZT gel calcined at 700°C and 800°C are shown in Figure 2. It can be seen that the as-calcined powders have a bimodal distribution. It is well known that aggregation of nano powder is very easy to occur due to its high activity. As a result, the measured average particle size of the nano PZT powder at 700°C is about 200 nm, while the calculated average crystalline size from XRD measurement is only around 26 nm. The particle size distribution shows that the size of some aggregated particles can even up to 2 m, which is due to the further aggregation of the PZT particles. Compared to the powders calcined at 700°C, the powders calcined at 800°C have more volume percentage of the large particles, which indicate that the aggregation easily occur at a higher calcined temperature.

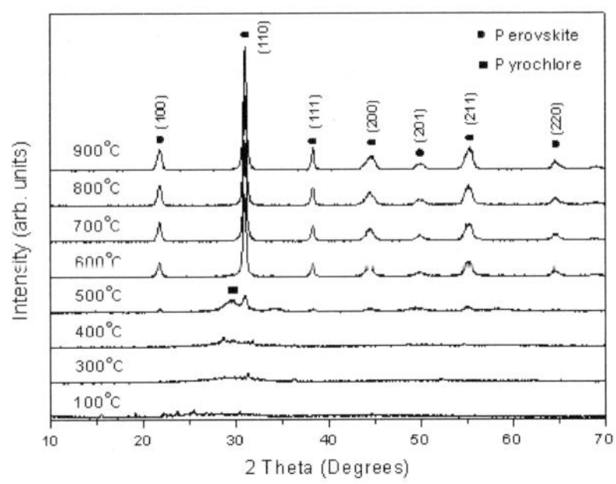

Figure 1. XRD patterns of the PZT powders annealed at different temperatures for 1h

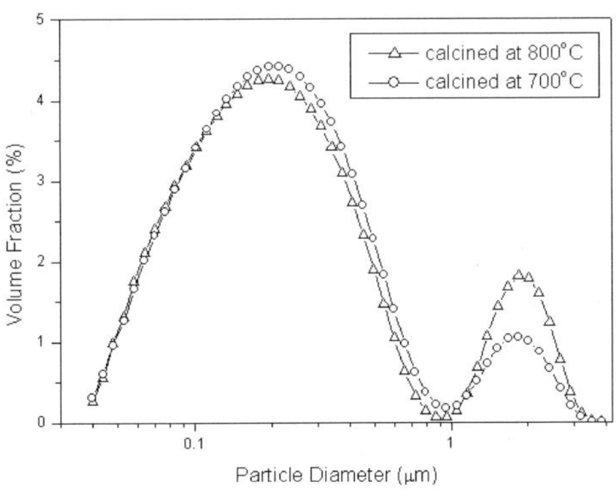

Figure 2. The particle size distributions of PZT gels calcined at 700°C and 800°C

The SEM photographs of PZT powder calcined at 700°C are shown in Figure 3. It can be seen that the particles are almost spherical and have an average size of about 200 nm, which is consistent with the results measured by laser particle size analyzer. When observed in higher resolution, each particle consisting of a number of smaller crystallites can be observed (Figure 3b). The morphology of the calcined powders observed by SEM are consistent with the results measured by laser particle size analyzer.

According to the XRF results of the PZT powder sintered at 700°C for 1h, the composites have a Zr:Ti ratio of 51.84:48.16, which is in good agreement with the stoichiometric ratio of Zr:Ti is 52:48. The measured ratio of Pb to total of Zr and Ti is 0.99, indicating no significant loss of Pb during the process.

(a) ×12,000 (a) ×12,000

Figure 3. SEM photographs of the PZT powders calcined at 700°C at different magnification levels: (a) ×12,000; (b) ×30,000

3.2 The properties of the NanoPiezo composites

The cement-based NanoPiezo 0-3 composites were fabricated by pressing the mixture of white cement and nano PZT powders. The pressing process could incorporate a high volumetric percentage of the ceramic phase and lead to a denser microstructure. There are total four different mixture proportions. The adding amounts of PZT powders were 35, 50, 65 and 80 percent by volume, named NP 35, NP 50, NP 65 and NP 80, respectively.

The impedance spectra of the composites were measured by HP4294A, and the electromechanical coupling factors (K_t) can be calculated using the resonant and antiresonant frequencies of the composites in the impedance spectra.[5, 7]

The piezoelectric and dielectric properties of the cement based NanoPiezo 0-3 composites are shown in Table 1. It can be observed that the piezoelectric properties of the composites increase with increasing of the volume percentage of nano PZT particles. The highest values of $k_{t,\ r}$ and d_{33} are determined to be 18.1%, 130.5, and 53.7 pC/N, respectively, which obtained from the composite containing 80 vol. % nano PZT powder.

In the previous studies,[5] the average particle size of the PZT powder of the 0-3 cement based piezoelectric composites are about 4 m and 83 m, respectively. For the composites with 50 % PZT by volume, corresponding d_{33} values are 9.5 and 12.5 pC/N, which are lower than that of the cement/ nano PZT composite (18.5 pC/N). In addition, the relative dielectric constants ($_r$) of the corresponding cement/ PZT composites are 63.9 and 94.2, which are quite close to that of cement/ nano PZT composite (83.4). However, the PZT powders used in the previous studies were ground from a commercial product. The d_{33}

and $_r$ values of the PZT ceramic were 513 pC/N and 3643, respectively. While that of $Pb(Zr_{0.52}Ti_{0.48})O_3$ ceramic are normally around 300 pC/N and 1000 only.[8, 9] As the piezoelectricity of the cement can be ignored, it is believed that the piezoelectric properties of the cement/PZT composites come from the embedded PZT particles. The piezoelectricity of the nano PZT used in the current studies is lower than that of the commercial PZT used in the previous studies. The overall piezoelectricity of the cement/PZT composites containing the two kinds of PZT powders are, however, about the same. It is reasonably to believe that the nano PZT powder is more effectively utilized in the cement based composites.

There are two possible reasons for such phenomenon. First one is the denser structure of the cement-based NanoPiezo 0-3 composites. In the current studies, the composites were fabricated by pressing, which can reasonably form a more compacted microstructure than the casting method used in the previous studies. The second is due to the better connectivity among the PZT particles. The microstructure of the cement based composites with 50 vol. % nano PZT powders is shown in Figure 4. Combining the results of EDAX, the large black grains are determined to bealcium hydroxide, which is the main product of cement hydration; the black matrix is cement; and the white phases are PZT particles. It can be seen that the composite is well compacted and the PZT powders are dispersed evenly, though some large PZT aggregated particles are observed. The good connectivity among the nano PZT particles and denser microstructure of the composites are probably the main reasons for the good piezoelectric properties of the cement-based NanoPiezo 0-3 composites.

Table 1. The piezoelectric and dielectric properties of the composites

Samples	PZT particle content (vol%)	Piezoelectric strain factor d_{33} (pC/N)	Electromechanical coupling coefficient k_t (%)	Relative dielectric constant $_r$
NP 35	35	7.3	10.1	21.2
NP 50	50	18.5	10.5	83.4
NP 65	65	29.7	12.6	112.8
NP 80	80	53.7	18.1	130.5

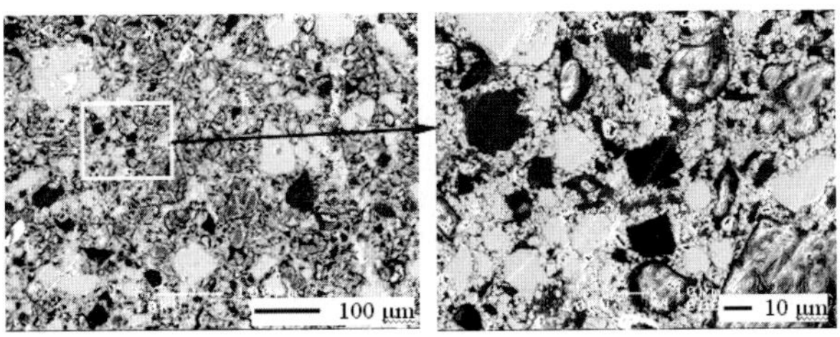

Figure 4. SEM photos of the NanoPiezo composites with 50 vol.% PZT nanopowders

4. CONCLUSIONS

Nano PZT powder was prepared by a sol-gel process using metal alkoxides. The X-ray diffraction studies confirmed that only perovskite phase left for the PZT powders sintered at above 700°C. The average particle size of the PZT powder sintered at 700°C was about 200nm while the calculated crystallite size is 26.4 nm. The cement based 0-3 piezoelectric nanocomposites were fabricated by pressing the mixture of white cement and PZT powders into discs under 100MPa and curing at 60°C in relative humidity of 100% for 24 h. Up to 80 vol. % PZT powder could be incorporated. The piezoelectric properties, such as ε_r, k_t and d_{33}, increased with the increase of volumetric percentage of nano PZT powders. The better properties of the composites with nano PZT powders than that of the composites with coarse PZT powders were probably due to the good connectivity among the nano PZT particles and the compacted microstructure of the nanocomposites.

5. REFERENCES

1. A. J. Moulson and J. M. Herbert, *Electroceramics* (2nd Edition, John Wiley & Sons press, New York, 2003), pp. 339-411.
2. C. Z. Rosen, B. V. Hiremath, and R. Newnham, *Piezoelectricity* (American Institute of Physics, New York, 1992) pp. 284-469.
3. C. R. Bowen and V. Yu Topolov, Piezoelectric sensitivity of $PbTiO_3$-based ceramic/polymer composites with 0-3 and 3-3 connectivity, *Acta Materialia* **51**, 4965-4976 (2003).
4. W. Nhuapeng and T. Tunkasiri, Properties of 0-3 lead zirconate titanate-polymer composites prepared in a centrifuge, *J Am Ceram Soc* **85**(3), 700-702 (2002).
5. Z. J. Li, D. Zhang and K. R. Wu, Cement-based 0-3 piezoelectric composites, *J Am Ceram Soc* **85**(2), 305-313 (2002).
6. C. Suryanarayana and M. G. Norton, *X-ray Diffraction: a Practical Approach* (Plenum Press, New York and London, 1998), pp. 207-223.
7. IEEE standard on piezoelectricity, *ANSI/IEEE Std.* (IEEE, New York, 1987), pp. 176-188.
8. K. Rama Mohana Rao, A. U. Prasada Rao and S. Komarneni, Reactive PZT precursor powder by coprecipitation, *Materials Letters* **28**, 463-467 (1996).
9. B. Guiffard and M. Troccaz, Low temperature synthesis of stoichiometric and homogeneous lead zirconate titanate powder by oxalate and hydroxide coprecipitation. *Materials Research Bulletin* **33** (12), 1759-1768 (1998).

CONCRETE STRENGTH PREDICTION IN STRUCTURAL ELEMENTS MADE WITH PULVERISED FUEL ASH

A. Hatzitheodorou and M.N. Soutsos,
University of Liverpool Department of Engineering, Civil Engineering Tower Buuilding, Brownlow Street, Liverpool, L69 3GQ , United Kingdom

Abstract: Two maturity functions, the Nurse-Saul and one based on the Arrhenius equation have been examined to determine their applicability for in-situ strength prediction of concretes made with pulverised fuel ash (PFA). Results from a PFA concrete mix with 30% of its binder being PFA were compared to a Portland cement concrete mix with similar 28-day target mean strength. The strength development of these two concretes in a block (1.5 x 1.5 x 1.5m) and a wall (3 x 2 x 0.3m) was determined.

Key words: concrete strength, fast-track construction, maturity functions, pulverised fuel ash, strength prediction

1. INTRODUCTION

Striking times of formwork, removal of falsework and safe application of post-tensioning on site depend largely on early-age in-situ concrete strength. These times will more or less govern the overall construction schedule and cost and hence influence considerably the feasibility of the project.

The majority of the contractors worldwide are still assessing the early-age concrete strength development by testing cubes or cylinders that are cast from concrete supplied to the site and cured alongside the structure[1] or cured under water at 20 °C (standard cube curing regime BS EN 12390-2:2000[2]). Consequently in most cases the curing regime for the specimens being tested is not representative of the actual on-site conditions. As a result the testing-specimens' temperature profile at early ages is generally significantly lower than the actual in-situ temperature profile.

Concrete strength gain at early ages related to its temperature history. Therefore by testing specimens, which are cured at a notably lower temperature than the concrete on site, contractors are underestimating the in-situ concrete strength and are subsequently

M.S. Konsta-Gdoutos, (ed.), Measuring, Monitoring and Modeling Concrete Properties, 385–394.

overestimating striking times. The strength development of pozzolanic mixes is more sensitive to temperature than it is for Portland cement mixes. Hence the use of such methods penalises the use of mixes containing Pulverised Fuel Ash (PFA) and Ground Granulated Blastfurnace Slag (GGBS) in fast-track construction.

In recent years, contractors and engineers have started realising the importance of using non-destructive methods for monitoring in-situ concrete strength that consider the actual in-situ temperature history. Probably the most widely used technique is the "maturity method". The actual or predicted in-situ temperature history is applied on maturity functions to calculate the concrete maturity development. Maturity is a temperature-time factor that describes/quantifies the combined effect of age and temperature on the concrete strength[3]. The concrete maturity development is converted to strength development, using maturity-strength correlations that have been established under laboratory conditions.

Maturity functions, such as the Nurse-Saul and the Equivalent Age method, which is based on the Arrhenius equation, have mainly been developed for Portland cement and their applicability to pozzolanic mixes must therefore be investigated.

2. BACKGROUND

In 1997, the Concrete Society (UK) on behalf of the Department of the Environment and Transport (DETR) carried out a large-scale research project that aimed to investigate the relationship between strengths obtained from cores and standard cured cubes.

Sixteen different concrete mixes were used to cast blocks, walls and slabs, during winter and summer weather conditions. Concrete strength was monitored by testing cores that were extracted from the structural elements. Standard cured cubes were also tested. Thermocouples were inserted in various locations inside the structural elements and their temperature history was recorded. Temperature matched curing (TMC) tests were carried out on a selection of walls and blocks containing GGBS and PFA[4].

This paper will present the results from TMC testing carried out on two of the DETR mixes for two different types of structural element. The two mixes presented are a Portland cement mix with a strength grade of 50 MPa labelled as PC 50 and a pozzolanic mix with 30% of its total binder being PFA, which also has a strength grade of 50 MPa and is labelled PFA 50. The mix proportions for the two mixes are shown in Table 1.

The two structural elements studied are a 1.5 x 1.5 x 1.5m block and 3 x 2 x 0.3m wall. The dimensions of the block and wall and the location of the thermocouples used to record the in-situ temperature history are shown in Figures 1 and 2 respectively; all dimensions are in mm.

3. RESEARCH SIGNIFICANCE

This research was undertaken to investigate the applicability of the maturity method in predicting the early age in-situ strength of PFA mixes.

Table 1. Concrete Mixture Proportions and Compressive Strengths

	PC 50				PFA 50 (30% PFA)			
Cement	357				278			
PFA	-				118			
GGBS	-				-			
Gravel saturated	1172				1343			
Sand saturated	749				522			
Free Water	170				147			
Free w/b	0.47				0.37			
Total Aggregate	1921				1865			
% of Fines	39				28			
Gravel oven dry	1144				1310			
Sand oven dry	743				516			
Total Water	205				185			
AGE	**Compressive Strengths (MPa)**							
	Block		**Wall**		**Block**		**Wall**	
	20 ^{0}C	TMC	20 ^{0}C	TMC	20 ^{0}C	TMC	20 ^{0}C	TMC
1-day	14.09	25.85	16.49	21.47	12.28	20.76	13.63	18.69
2-day	23.48	34.77	-	-	20.53	32.97	22.91	26.72
3-day	28.41	38.90	28.02	30.90	25.24	41.40	28.14	29.06
7-day	36.70	42.63	36.70	35.53	32.83	49.40	36.17	34.70
14-day	44.53	42.00	43.30	42.60	40.53	49.47	43.67	42.33
28-day	50.40	45.37	48.63	46.87	49.70	49.20	53.90	50.33
42-day	-	-	-	-	52.43	50.27	55.70	53.83

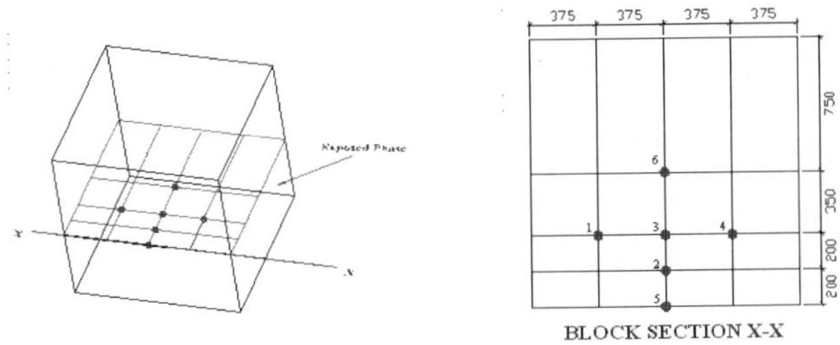

Figure 1. DETR Concrete Block Dimensions and Thermocouple Positions

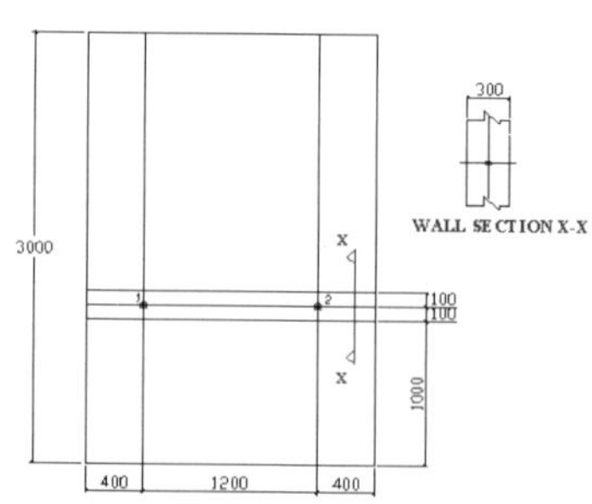

Figure 2. DETR Concrete Wall Dimensions and Thermocouple Positions

4. MATERIALS & EXPERIMENTAL METHODS

4.1 Materials

Materials similar to those used for the DETR blocks and walls were obtained for this project. These included: Portland cement, provided by British Lime Industries (BLI), pulverised fuel ash (PFA) obtained from Scottish Southern and 20-5mm well graded uncrushed round gravel and fine aggregate of medium grading provided by Tarmac

4.2 Mixing, Casting and Testing Procedures

For each simulation a concrete mix was cast in 1 batch using a 0.1 m^3 capacity horizontal pan mixer. The materials were mixed dry for 1 minute, then water was added and mixing continued for a further 5 minutes. In total 48 concrete specimens were cast, using two types of steel moulds; 24 in 100mm cast steel single-cube moulds and 24 of the same dimensions in 8 3-gang stainless steel moulds. All specimens were compacted using a vibrating table. To ensure a satisfactory level of compaction the moulds were initially filled up half way, vibrated and then filled up to the top and vibrated once more.

The concrete specimens inside the single-cube moulds were used as control specimens. They were originally covered with wet Hessian and a polythene sheet and cured under room temperature conditions, approximately 20 ^{0}C. Within 24 hours after casting they were demoulded and placed inside a water tank set at 20 ^{0}C to be cured under standard curing conditions.

The concrete specimens cast in the 3-gang moulds were sealed using cling film and tape and placed inside a computer controlled Temperature Matched Curing (TMC) tank. The next day once the TMC specimens had hardened the cling film was removed to ensure that the specimens would remain moist. The TMC specimens remained inside their moulds, until they were removed from the tank, either to be tested or to be cured under ambient conditions.

The TMC specimens remained in the TMC tank until the concrete temperature dropped close to the average ambient temperature. The specimens cured under the summer block temperature histories were removed from the TMC tank at 14 days, while those cured under the summer wall temperature histories were removed after just 7 days. All were wrapped with wet Hessian and polythene sheets and placed in a curing cabinet set at 17 ^{0}C, which was the average ambient temperature during the summer phase of the DETR project.

The DETR ambient temperature history was compared to averaged summer temperature data obtained from the Met Office web site[5], which proved that it offered a good approximation of the prevailing summer ambient conditions in Northwest England, where the DETR project was carried out. In other words the summer ambient conditions identified in the DETR project were representative of the ambient conditions that are likely to occur on site in Northwest England during the summer period.

The testing ages for both the control and TMC cured specimens were 1, 2, 3, 5, 7, 14, 28, 42. The values in Table 1 are the mean of three specimens.

4.3 Curing Regimes and Apparatus

For each mix two DETR in-situ temperature histories were investigated. The first was based on the recordings taken from thermocouple 4 from an insulated concrete block (Figure 1) that was cured under summer ambient conditions; the formwork was removed after just 18 hours. This thermocouple, was not in the centre, however it was the furthest away from the exposed phase and therefore exhibited the highest in-situ temperature history. It is important to note at this stage that the blocks in the DETR project, with the exception of one exposed phase, unlike the rest of the structural elements, remained insulated with polystyrene after the formwork was removed; furthermore this insulation was only removed for a short period of time when coring took place.

The second in-situ temperature history was based on the recordings taken from thermocouple 1 from a concrete wall that was also cured under summer ambient conditions. The thermocouple in the wall was located 1100mm from its foundation, 400mm from the edge and approximately 122.5mm from the exposed surface as shown in Figure 2. The formwork in the wall was removed permanently after two days.

A computer controlled Temperature Matched Curing (TMC) Tank, Figure 3, was programmed to follow the in-situ temperature history shown in Figure 4.

5. RESULTS AND DISCUSSION

5.1 TMC Temperature Histories

The temperature inside the concrete walls (Figure 4), which are much thinner elements than the concrete blocks and consequently have a much higher surface to volume ratio, is significantly lower than that observed in the blocks. Furthermore it rapidly, after less than 3 days, starts to follow the ambient temperature. The significant difference in temperature is also due to the fact that the blocks remained insulated, while the walls were left exposed to the weather after their formwork was removed.

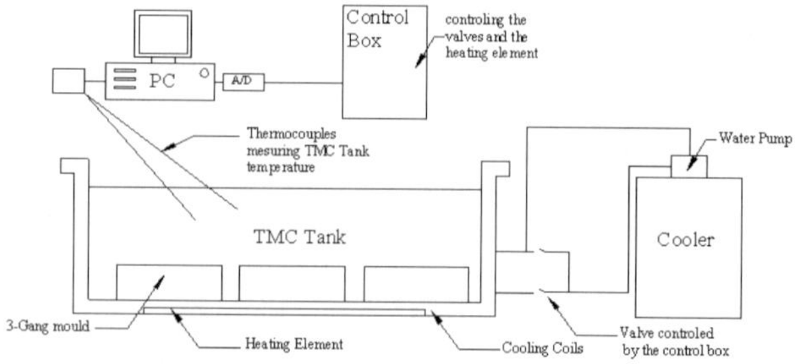

Figure 3. Computer Controlled Temperature Matched Curing Tank Setup

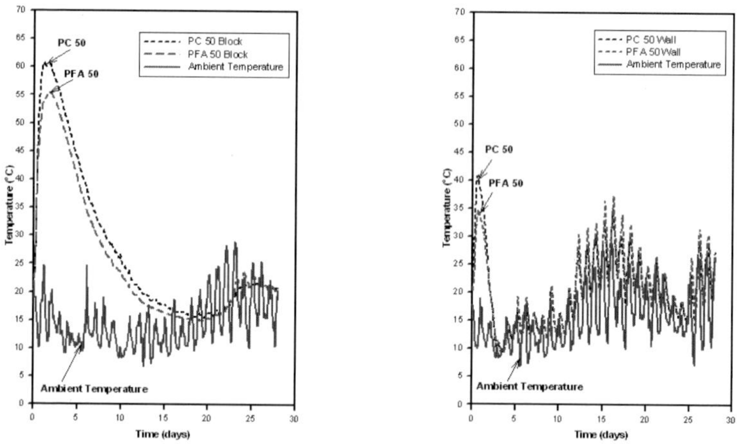

(a) (b)

Figure 4. DETR In-situ Temperature Histories used for TMC Tests, (a) Grade 50 Summer Blocks and (b) Grade 50 Summer Walls

5.2 Strength-Maturity Correlations

The equation used to predict the strength development of the TMC specimens according to the Nurse-Saul method[6], as described by Carino[4], is given below:

$$S = \frac{S_u \times k \times (M - M_0)}{1 + k \times (M - M_0)} \tag{1}$$

Where S is the compressive strength, S_u the limiting compressive strength, M the concrete maturity, M_0 the concrete maturity at which the strength development is initiated and k a rate constant.

The concrete maturity is given by the following equation:

$$M = \sum_{0}^{t} (T - T_0)\Delta t \tag{2}$$

where M is the concrete maturity, t the time interval, T the average temperature during the time interval t and T_0 the datum temperature, assumed to be -11 ^{0}C.

The Arrhenius equation used to predict the strength development of the TMC specimens according to the Equivalent Age method is given below:

$$S = \frac{S_u \times k \times (t_e - t_0)}{1 + k \times (t_e - t_0)} \tag{3}$$

where S is the compressive strength, S_u the limiting compressive strength t_e the concrete equivalent age at the reference curing temperature, t_0 the time at which the strength development is initiated and k a rate constant. The equivalent age t_e was calculated according to Freiesleben, Hansen and Pederson[7], using equation 4.

$$t_e = \sum_{0}^{t} e^{-\frac{E}{R}\left(\frac{1}{T} - \frac{1}{T_r}\right)} \tag{4}$$

where E is the Activation Energy, R the Universal gas constant, T the concrete temperature,T_0 the reference curing temperature and k the rate constant.

The Nurse-Saul and Arrhenius equation coefficients were evaluated by carrying out a regression analysis on the standard cured specimens' test results, using a spreadsheet software called SigmaPlot. Typical values of activation energies, similar to those found in literature were used for the Equivalent Age method. The activation energies were 30000 and 35000 for the PC 50 and PFA 50 mix respectively. The datum temperature was taken to be equal to -11 ^{0}C.

The maturity functions were used to estimate the strength development of the TMC concrete specimens according to their recorded temperature history; the results were compared to their actual concrete strength development as shown in Figure 6. The experimental test results are shown in Table 1. The strength development of the standard and TMC cured specimens is shown in Figures 5 and 6.

5.3 Discussion of Results

In all cases the early-age strength development of the TMC specimens is higher than that of the standard cured specimens. This is more evident for the TMC block specimens, which were cured under a significantly higher than the ambient temperature, and especially for the PFA specimens, which were more sensitive to temperature. The high early-age curing temperatures have a significant negative effect on the 28-day strength of the PC 50 TMC block specimens. On the other hand, for the PFA 50 block specimens that were cured under similar conditions, the effect of the high early-age curing temperature on the 28-day strength was negligible. The strength improvement in the TMC wall specimens for both mixes ends after approximately just three days; after that the TMC strength development becomes nearly the same with that of the standard cured specimens.

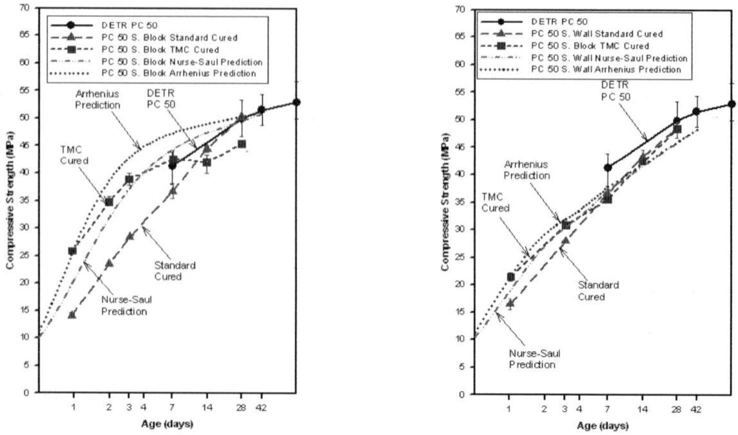

(a) (b)

Figure 5. PC 50 - Comparing Actual to Predicted Strength Development, (a) PC 50 Summer Block and (b) PC 50 Summer Wall

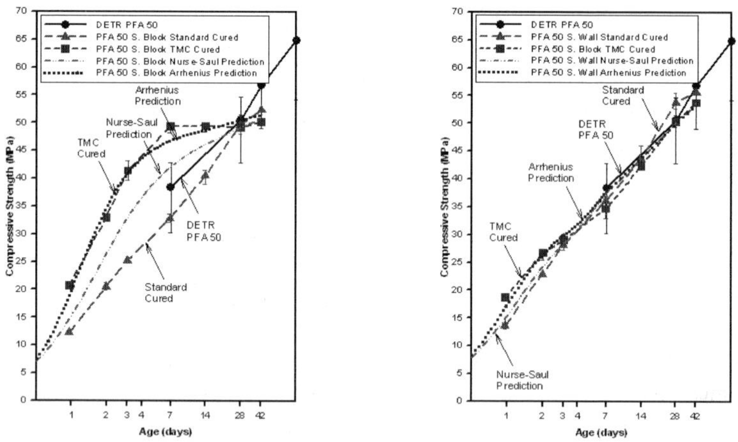

(a) (b)

Figure 6. PFA 50 - Comparing Actual to Predicted Strength Development, (a) PFA 50 Summer Block and (b) PFA 50 Summer Wall

The Nurse-Saul method appears to underestimate the early-age strength development; this is more evident in the TMC block specimens, especially for the PFA mix. On the other hand, at least for the blocks, it still seems to give a much higher and more accurate estimate than that obtained from the standard-cured specimens. For the PC 50 mix the

Nurse-Saul method significantly overestimates the block's later age strength; this is due to the fact that it does not account for the effect that high early-age curing temperatures have on the later age strength development of Portland cement mixes.

Similarly the Arrhenius predictions significantly overestimate the later age strength of the PC 50 TMC block specimens, even after just 3 days. On the other hand for the PFA 50 TMC block specimens this effect takes place at a much later age, after approximately 28 days. This can be explained by the fact that for PFA mixes high early-age curing temperatures have a negative effect on the strength development at a much later age than they have for Portland cement mixes. Overall the Arrhenius predictions are much higher than the Nurse-Saul predictions and in most cases, especially for the PFA 50 specimens are much closer to the actual strength development. The overall degree of accuracy of the Equivalent Age method, as for the Nurse-Saul method, although probably adequate for most applications in the construction industry may mean that they are still not reliable and consistent enough for fast-track construction.

6. CONCLUSIONS

Temperature-matched curing can be an important tool in attempting to simulate in-situ curing conditions and study how different curing regimes affect concrete strength development. However the temperature histories used are not indicative of how temperature changes with location inside the concrete element. In other words they do not offer the temperature profile of the whole concrete over time; they just show the temperature development at certain points inside the concrete where the TMC probes were inserted. The temperature and maturity history in structural elements is not uniform; hence the actual in-situ concrete strength development will vary according to location. The TMC estimated in-situ strength development and the associated striking time are therefore influenced by the location of the TMC probes.

The Nurse-Saul method significantly underestimates the early-age concrete strength for specimens cured under high early-age temperature conditions and the later-age strength is adversely affected. Strength predictions made with the Arrhenius equation, most importantly at early ages, are generally more accurate.

The use of maturity methods to predict/monitor concrete strength on site can benefit the use of PFA mixes, which are definitely more temperature sensitive. However the maturity methods' degree of accuracy suggests that there is scope for improvement for both Portland cement and PFA mixes, especially for concrete elements with a low surface to volume ratio and/or a high degree of insulation. In such structural elements, large quantities of heat are retained for a long period of time leading to in-situ temperatures that are much higher than the ambient and the standard curing temperatures. There is therefore scope for using the maturity methods to account for these high early-age in-situ temperatures in predicting/modelling the in-situ concrete strength development.

Acknowledgements

The authors would like to express their appreciation for Dr. S. Barnett's overall contribution to the work. They also wish to thank Mr. T. Cowan and Mr. M. Jones for their assistance in their laboratory work.

7. REFERENCES

1. S. J. Barnett, M. N. Soutsos, J. H. Bungey and S. G. Millard, *Fast Track Concrete Construction using Cement Replacement Materials*, SP-221-7: Eighth CANMET/ACI International Conference on Fly Ash, Silica Fume, Slag, and Natural Pozzolans in Concrete, 2004
2. British Standards Institution, BS EN 12390-2:2000, *Testing hardened concrete – Part 2: Making and curing specimens for strength tests* , 2000
3. V. M. Malhotra and N. J. Carino *NONDESTRUCTIVE TESTING OF CONCRET" Chapter 5 The Maturity Method*, pp 5-1 – 5- 47
4. DTI, The Concrete Society Project Report 13, *In situ Concrete Strength, Investigation into the relationship between core strength and standard cube strength*, reported by working party of the Concrete Society.
5. The Met Office web site, Climate Averages – long-term averages, http://*www.met-office.gov.uk/climate/uk/averages.*
6. A.G.A Saul, *Principles underlying the steam curing of concrete at atmospheric pressure*, Magazine of Concrete Research, 2(6), 1951, pp. 127
7. Hansen P. F. and Pedersen, J., *Maturity computer for controlled curing and hardening of concrete*, Nordisk Betong, V. 1, 1977, pp. 19-34

STUDY ON PROPERTIES OF RUBBER INCLUDED CONCRETE UNDER WET-DRY CYCLING

Y. Zhang, Sun Wei and Chen Shengxia

Dept. of Materials Science & Engineering, Southeast University, Nanjing, 210096, China, Phone: 86-25-83795618, Fax: 86-25-83795982,

Abstract: In the current paper, the relative dynamic modulus of elasticity, compressive strength, flexural strength and the concentration of chloride in rubber included concrete treated with water or composite salt solution were investigated. The mechanism of the degradation of rubber included concrete was briefly analyzed. The experimental results showed that rubber included concrete performed comparably with reference concrete when concretes were immersed in water or composite salt solution, however, under the action of wet-dry cycling, the properties of rubber included concrete was much inferior to that of reference concrete. Furthermore, the action of composite salt solution accelerated the degradation of rubber included concrete. It is therefore suggested that rubber included concrete should not be used where long-term wet-dry cycling, hot and dry weather occurs or where sulfate attack may happen.

Key words: rubber; concrete; wet-dry cycling; composite salt solution.

1. INTRODUCTION

Concrete in practical engineering usually undergoes complex action from both loading and environment. For concrete at tidal zone of sea structure and piers of bridges over river, the frequent washing of seawater or river water is one of the key factors affecting the durability, besides, wet-dry cycling has important influence as well. Even for buildings and bridges over land, structural concretes always serve under frequently changing weather condition, in particular, the change of temperature and humidity is frequent. Therefore, investigation into the performance of concrete under frequent changing of both temperature and humidity is necessary and of great importance for evaluating the durability of concrete.

In recent years, research interest in rubber included concrete has been increasing[1~6]. Siddique et al.[7] overviewed the properties of concrete containing scrap-tire rubber. The reduction in mechanical strength of concrete manufactured with rubber aggregates may

M.S. Konsta-Gdoutos, (ed.), Measuring, Monitoring and Modeling Concrete Properties, 395–401.

limit its use in some structural applications, but rubberized concrete also has some desirable characteristics such as lower density, higher impact, frost and toughness resistance, enhanced ductility and better sound insulation. Study by Hernardez et al.[8] showed the inclusion of low volumetric fractions of rubber reduced the risk of explosive spalling of high strength concrete at high temperatures because water vapor could exit through the channels left as the polymeric particles get burnt. Although concrete with rubber included has advantages over conventional ordinary concrete, its long-term performance under weathering condition or severe environment is not yet clear.

This paper focuses on the properties of rubber included concrete under the action of wet-dry cycling treated with water and Na_2SO_4/NaCl composite salt solution.

2. RAW MATERIALS AND EXPERIMENTS

Portland cement with specific surface area of 3460 cm^2/griver sand with fineness modulus of 2.3 and crushed basalt with 5~15 mm continuous grading were used to produce concretes. Rubber powder with an average size of 140 μm and coarse rubber particles sized between 3 to 4 mm were applied as partial substitutes for sand. A naphthalene-type plasticizer was used to adjust the workability of concretes. Mix proportion of concretes was given in table 1. C50 was prepared as reference concrete with 28 d compressive strength aiming at 50 MPa. In addition, 1.5% (mass fraction) of NaOH was added to each mixture to improve the bonding between rubber particles and cement paste.

Table 1. Mix proportion of C50 concrete

Code	Mix Proportion / kg/m^3					
	Cement	Sand	Stone	Water	Rubber	
					140 μm	3~4 mm
C50	457	586	1279	155	0	0
C50p7.5s	457	552	1279	155	34	0
C50G7.5s	457	552	1279	155	0	34
C50G10s	457	540	1279	155	0	46
C50G15s	457	519	1279	155	0	67

In this table, p represents 140 μm rubber powder s means sand was replaced by rubber G represents 3~4 mm rubber particles.

The size of specimens was 40 mm×40 mm×160 mm. After molds were removed, specimens were cured in a fog room at 20±1 °C for 28 days. Then, some specimens were immersed in composite solution of 7% NaCl and 10% Na_2SO_4 (NaCl/Na_2SO_4 composite solution) and in tap water, separately. NaCl/Na_2SO_4 composite solution was refreshed every 30 d. Other specimens were used to take wet-dry cycling test. The regime of one wet-dry cycle was as following: first immerse specimens in NaCl/Na_2SO_4 composite solution or water for 56 h, then dry at 55 °C for 16 h. The dynamic modulus of elasticity of specimens was examined periodically. After 180 d, the compressive strength and the flexural strength of specimens were tested, and the cores of specimens were drilled for determining both free and total chloride content.

3. EXPERIMENTAL RESULTS

3.1 Change of relative dynamic modulus of elasticity

Figure 1 shows the variation of the relative dynamic modulus of elasticity (the ratio of dynamic modulus of elasticity after treatment to that without treatment at 28 d) of specimens immersed in water for 180 d. It can be seen the dynamic modulus of all the specimens kept increasing with time and the relative dynamic modulus of elasticity at 180 d was as high as 1.1.

Figure 2 gives the variation of the relative dynamic modulus of elasticity of specimens treated with wet-dry cycling for 180 d. The relative dynamic modulus of elasticity of C50,C50G7.5s and C50p7.5s increased till 90 d, but began to decline after that. The dynamic modulus of elasticity of C50G15s and C50G10s kept decreasing after wet-dry cycling began. C50G15s had the lowest relative dynamic modulus of elasticity close to 0.8 at 180 d.

Figure 3 shows the variation of the relative dynamic modulus of elasticity of specimens immersed in NaCl/Na_2SO_4 composite solution for 180 d. The relative dynamic modulus of elasticity increased with time initially, about 90 d later, the curves leveled off till 150 d, then, they began to decline slightly.

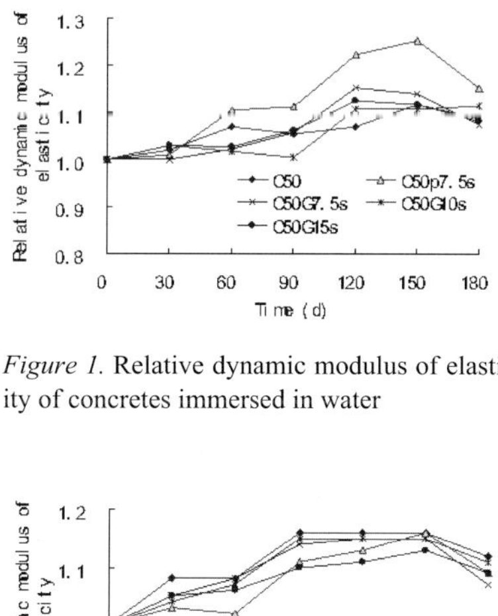

Figure 1. Relative dynamic modulus of elasticity of concretes immersed in water

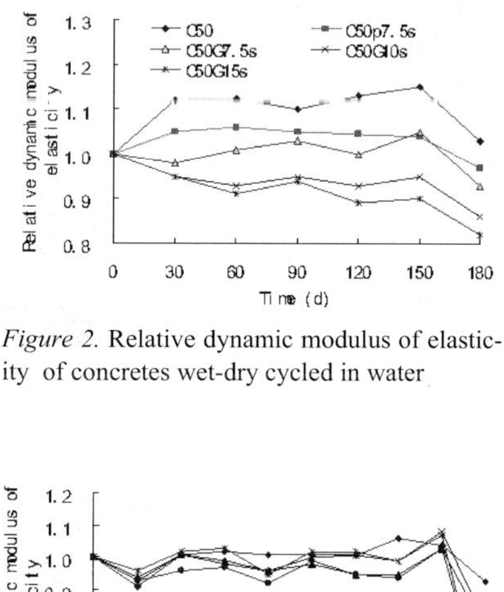

Figure 2. Relative dynamic modulus of elasticity of concretes wet-dry cycled in water

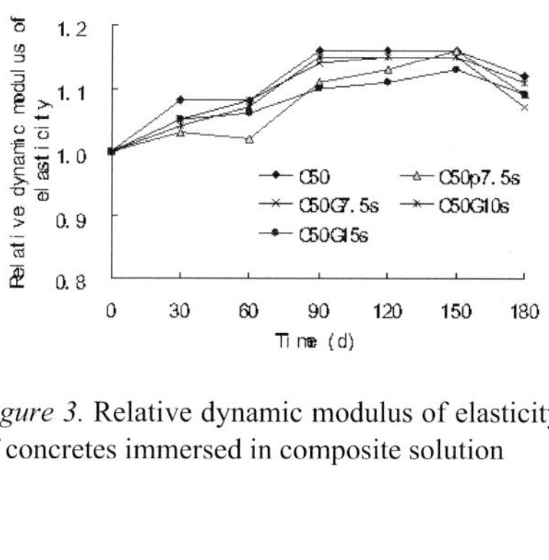

Figure 3. Relative dynamic modulus of elasticity of concretes immersed in composite solution

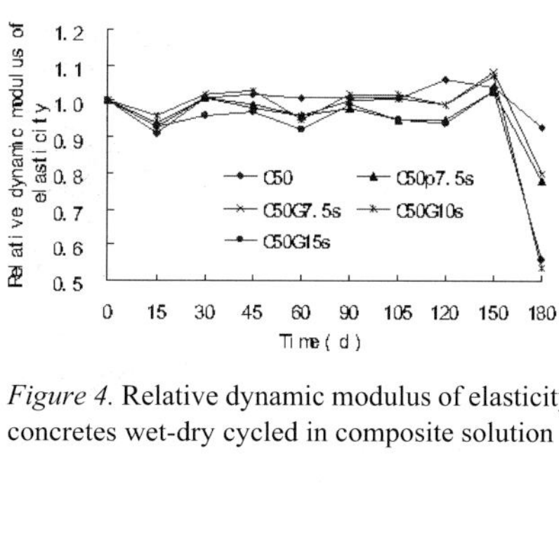

Figure 4. Relative dynamic modulus of elasticity of concretes wet-dry cycled in composite solution

Figure 4 shows the variation of the relative dynamic modulus of elasticity of specimens wet-dry cycled in NaCl/Na_2SO_4 composite solution for 180 d. The relative dynamic modulus of specimens kept at a relatively stable level till 150 d for all the specimens, but decreased markedly at 180 d for rubber included concrete. The relative dynamic modulus of elasticity of C50G10s and C50G15s dropped to below 0.5 at 180 d.

3.2 Change of mechanical strength of specimens treated by wet-dry cycling

To compare the effect of immersion and wet-dry cycling on the mechanical performance of specimens, relative compressive/flexural strength is defined by dividing the compressive/flexural strength after treatment by the compressive/flexural strength without treatment at 28 d of corresponding specimen. The relative strength of specimens after immersed or wet-dry cycled in water for 180 d is shown in figure 5. For reference concrete C50, both the relative compressive strength and the flexural strength were higher than 1.0 no matter what kind of treatment were applied. For rubber included concrete, however, treatment method greatly affected the final mechanical strength, especially the flexural strength. The relative flexural strength of rubber included concrete specimens immersed in water was higher than 1.0. For specimens wet-dry cycled in water, the relative flexural strength of C50p7.5s and C50G7.5 was higher than 1.0, but that of C50G10s and C50G15s dropped to 0.95 and 0.84, respectively.

The relative strength of specimens after immersed or wet-dry cycled in NaCl/Na_2SO_4 composite solution for 180 d is shown in figure 6. For reference concrete, both the compressive strength and the flexural strength increased after treated with composite solution for 180 d. The compressive strength of rubber included concrete immersed in composite solution did not change a lot, but the flexural strength increased. For wet-dry cycled rubber included concrete, however, both the relative compressive strength and flexural strength decreased to below 1.0. The decrease of the relative flexural strength was remarkable. The relative flexural strength of C50p7.5s, C50G7.5s, C50G10s and C50G15s dropped to 0.78, 0.67, 0.45 and 0.29, respectively.

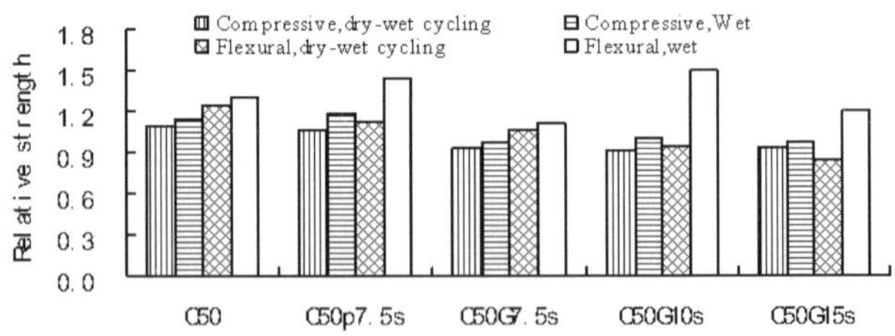

Figure 5. Relative strength of concretes after treated with water for 180 d

3.3 Concentration of chloride in specimens treated with composite solution

Table 2 lists the concentration (mass fraction based on dry samples) of chloride in concrete after treated by immersion or wet-dry cycling in $NaCl/Na_2SO_4$ composite solution for 180 d. Table 2 demonstrates that, for both reference concrete and rubber included concrete, the concentration of both free chloride and total chloride in wet-dry cycled specimens was higher than that immersed in composite solution at the same depth. Furthermore, under wet-dry cycling, the concentration of chloride in rubber included concrete was higher than that of reference concrete. With the increase of rubber in concrete, the concentration of chloride increased. For specimens immersed in composite solution for 180 d, the concentration of chloride in rubber included concrete was however close to that of reference concrete at the same depth.

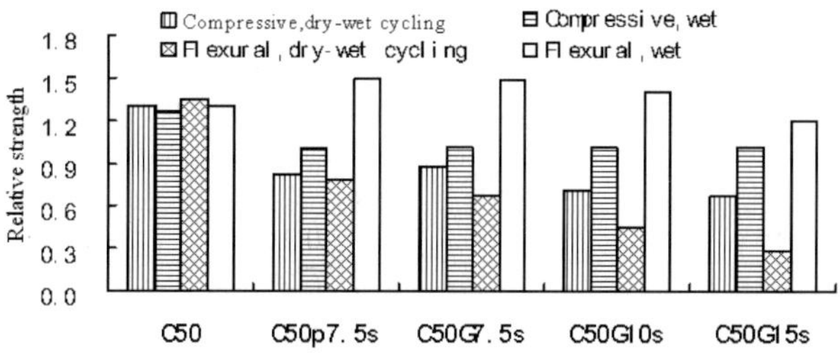

Figure 6. Relative strength of concretes after treated with composite solution for 180 d

4. DISCUSSION

Experimental results demonstrate that rubber included concrete immersed in water or $NaCl/Na_2SO_4$ composite solution exhibited comparable performance to reference concrete. The difference occurred when concrete was attacked by wet-dry cycling. Rubber included concrete suffered decrease of dynamic modulus of elasticity, compressive strength and flexural strength to different extent under wet-dry cycling. The decrease in composite solution was higher than in water, and the decline of flexural strength was greater than that of compressive strength.

The current investigation reveals that wet-dry cycling is one of the key factors causing the degradation of rubber included concretes. Under the action of heat, oxygen and water, irreversible changes in appearance and molecular structure may take place and cause the ageing of rubber. Although the dosage of rubber was low and rubber particles were uniformly distributed in concrete, ageing of rubber happened when concrete was subject to long-term wet-dry cycling. With of the ageing of rubber, the bonding between rubber and cement paste was decreased and defects increased.

Investigation also reveals that the action of $NaCl/Na_2SO_4$ composite solution accelerated the degradation of rubber included concrete under wet-dry cycling. Observations by SEM and EDS showed remarkable quantities of AFt, Freidels salts and minor crystalline products having the morphology similar to that of AFm and Freidels salts. The formation of lots of AFt around rubber particles might be the main reason that caused the worsening of rubber included concrete in composite solution. Furthermore, since flexural strength was more sensitive to cracks and defects than compressive strength, the decline of flexural strength under wet-dry cycling was therefore more obvious than the latter.

Table 2. The concentration of chloride in concrete after treated by immersion or wet-dry cycling in NaCl/ Na_2SO_4 composite solution for 180 d

Code	Depth /mm	Free Cl^- /%		Total Cl^- /%	
		Dry-wet cycling	Wet	Dry-wet cycling	Wet
C50	5	0.67	0.53	1.43	1.14
	10	0.46	0.21	0.97	0.46
	15	0.48	0.17	0.97	0.38
	20	0.48	0.36	1.03	0.80
C50P7.5s	5	0.82	0.49	1.64	0.98
	10	0.62	0.18	1.28	0.44
	15	0.59	0.15	1.21	0.43
	20	0.58	0.08	1.17	0.20
C50G7.5s	5	0.92	0.61	1.88	1.28
	10	0.63	0.32	1.40	0.66
	15	0.61	0.23	1.29	0.51
	20	0.61	0.17	1.29	0.41
C50G10s	5	0.95	0.53	1.96	1.07
	10	0.70	0.32	1.45	0.64
	15	0.67	0.24	1.37	0.49
	20	0.67	0.17	1.35	0.37
C50G15s	5	1.03	0.59	2.07	1.19
	10	0.77	0.34	1.73	0.70
	15	0.68	0.29	1.38	0.60
	20	0.61	0.23	1.25	0.48

5. CONCLUSIONS

Rubber included concrete immersed in water or NaCl/Na_2SO_4 composite solution exhibited comparable performance to reference concrete. Rubber included concrete suffered the decrease of dynamic modulus of elasticity, compressive strength and flexural strength to different extent under wet-dry cycling. The decrease in composite solution was higher than in water, and the decline of flexural strength was greater than that of compressive strength.

Under wet-dry cycling, the concentration of chloride in rubber included concrete was higher than that of reference concrete. With the increase of rubber in concrete, the concentration of chloride increased. For specimens immersed in composite solution, the concentration of chloride in rubber included concrete was however close to that of reference concrete at the same depth. It is therefore suggested that rubber included concrete should not be used where long-term wet-dry cycling, hot and dry weather occurs or where sulfate attack may happen.

6. REFERENCES

1. I.B. Topcu, Collision behaviors of rubberized concrete, *Cem. and Concr. Res.* **27**(12), 1893-1898 (1997).
2. D. Raghavan, H. Huynh and C. F. Ferraris, Workability, mechanical properties and chemical stability of a recycled tire rubber-filled cementitious composite, *J. of Mater. Sci.* **33**, 1475-1752 (1998).
3. N. Segre and I. Joekes, Use of tire rubber particles as addition to concrete paste, *Cem. and Concr. Res.* **30**, 1421-1425 (2000).
4. O. F. Hernardez, G. Barluenga, M. Bollati and B. Witoszek, Static and dynamic behaviour of recycled tyre rubber-filled concrete, *Cem. and Concr. Res.* **32**, 1587-1596 (2002).
5. A. Benazzouk, O. Douzane and M. Queneudec, Transport of fluids in cement-rubber composites, *Cem. & Concr. Comp.* **26**, 21-29 (2004).
6. Y. M. Zhang, S. X. Chen, B. Chen et al. Frost resistance and permeability of rubber included concrete, *Key Eng. Mater.* **302-303**, 120-124 (2006).
7. R. Siddique, T. R. Naik, Properties of concrete containing scrap-tire rubber – an overview, *Waste Management.* **24**, 563–569 (2004).
8. O. F. Hernardez, G. Barluenga, Fire performance of recycled rubber-filled high-strength concrete, *Cem. and Concr. Res.* **34**, 109-117 (2004).

REACTIVE SILICA OF FLY ASH AS AN INDICATOR FOR THE MECHANICAL PERFORMANCE OF BLENDED CEMENTS

S.K. Antiohos and S. Tsimas
National Technical University of Athens, School of Chemical Engineering, 9 Heroon Polytechniou Str., Zografou Campus, 157 73, Athens, Greece

Abstract: Research on fly ash has currently shifted towards exploring the characteristics that determine its activity in the cement paste environment. Reactive silica is the principal parameter that determines the pozzolanic potential of a fly ash, its tendency, that is, to react with available calcium hydroxide to form hydration products with binding properties. Once reactive silica is accurately measured, there exists a good opportunity to relate it with the future strength development of a blended cementitious system. In the work presented herein such a correlation was developed. Its validity was tested for the case of several systems containing various kinds of fly ashes (of different lime content, multicomponent containing both types of fly ash and processed one). Results testify that the developed relationship can be applied to obtain a first approximation of the k-value of the respective blended cements, enabling an almost rapid prediction of the quantity and, most importantly, the quality of the ash used in the mix design.

Key words: amorphous material; blended cements; compressive strength; fly ash; predictive model; reactive silica.

1. INTRODUCTION

Despite having being explored for more than fifty years, the exact behaviour of fly ash in blended cement systems is not clear yet. This is not only because of its inhomogeneity and vast physicochemical variations from country to country (or even from producing site to site), but additionally to its inherent characteristics that can either act beneficially or pose significant threat to the service life-time of a structure[1]. Factors like these create scepticism concerning the wider use of this material and are considered responsible for its low utilization rate (approximately 10-12% in Greece according to a recent report[2]).

M.S. Konsta-Gdoutos, (ed.), Measuring, Monitoring and Modeling Concrete Properties, 403–409.

Research has now moved on towards fully exploring the inherent characteristics of fly ash, especially those that determine its activity in an alkaline environment such as the one created when blended cement is mixed with water. In the literature[3, 4] there is consensus that active silica, which is part of the total silica of the fly ash, is the fraction which is involved in the hydration reactions producing calcium silicate hydrates (CSH) to which the strengthening of cement is attributed. Active silica is non-crystalline silica glass, present in the amorphous and mostly vitreous part of the fly ash[5], which can be combined with the lime giving increased contents of C-S-H gel[6], unlike crystalline silica that exhibits very low reactivity[7]. Efforts to relate this critical parameter with cement and concrete properties are limited in the literature. Papadakis[2, 8] in previous publications had observed that the final strength gain in SCM concrete is proportional to the glass phase content, while Sharma et al.[9] proposed an empirical correlation between the compressive strength of mortars and the fineness and the soluble silica of fly ash.

It total, despite the fact that authors have attempted to connect the pozzolanic effect with a number of parameters such as, fineness[10], curing temperature[11] and alkalinity of the pore solution[12], there seems to be a lack in the literature regarding the effect of the reactive silica fly ash behavior as additive in cement and concrete. The present work aims at filling this gap by developing a relationship between the reactive silica content of various fly ashes utilized in blended cements and their respective k-values. Using such a relationship and once reactive silica is accurately measured, there exists a good opportunity to relate it with the future strength development of a binary (or even ternary) system that incorporates its carrier. By means of a brief review, the work presents the beneficial and delicate aspects of the methods most commonly used for estimating reactive silica in fly ash, while at the same time it rationalizes the use of the European standard EN 450-1 and introduces the need for a faster and less tedious technique.

2. ACTIVE SILICA DETERMINATION PROCEDURES

In the past, soluble silica was expressed as the difference between total and free silica[9], which were determined after fusion by the gravimetric method before and after treating the fly ash with hydrochloric acid. Sivapullaiah et al.[13] also used the gravimetric method to determine the reactive silica indirectly as acid soluble silica, giving surprisingly low values for different fly ashes. Mehta[14] established the concept of the 'silica activity index' meaning the percentage of available silica that is dissolved in an excess of boiling 0.5 M NaOH solution during a 3-min extraction period. Simpler methods have been proposed[15] based on the titration of sample suspensions with methylene blue, providing an index of the amorphousness of the silica contained in the ash. Paya et al.[16] recently proposed a rapid method for the determination of amorphous silica in rice husk ashes, based on bringing the siliceous non-crystalline fraction of the pozzolan into solution as glycerosilicate by treating the test material with glycerol.

In the present research, the European standard EN 450-1[17] was used to determine the active silica contents of all ashes used. This standard specifies that the reactive silica content in fly ashes should be estimated by subtracting from the total silica content, the silica fraction that is contained in the insoluble residue (and thus considered non-reactive). The attack by hydrochloric acid and boiling 4 M potassium hydroxide in a 4-h extraction that is described in the standard has been criticized[16] as too drastic that it dissolves not only the amorphous silica, but also crystalline silica. Nevertheless, and despite that fact that the particular method is considered time consuming and requires increased caution by the analyst, it provides more reliable results when compared with resembling methods.

3. EXPERIMENTAL SECTION

Two different fly ashes from Greek power plants (produced by Puplic Power Corporation) were used, i.e., a high-calcium fly ash (from Ptolemais plant, designated as TF) and a fly ash of relatively lower calcium content (from Megalopolis plant, defined as TM). Both were ground prior to use up to similar fineness in order to deactivate the effect of specific surface on their pozzolanic activity. A normal setting Portland cement was used (CEM I 42.5 according to European Standard EN 197). Based on the initial materials, new fly ash specimens were prepared (i) by mixing TF and TM in several proportions until homogeneity was reached. (i.e. 50% TF and 50% TM for the new ash coded as T1 and 75% TM and 25% TF for the new fly ash coded as T2), and (ii) by separating and futher grinding the coarse fraction (>45 μm) of TF ash (thus obtaining a new ash coded as TFP). In total five ashes (TF, TM, T1, T2, TFP) of similar fineness were investigated.

Oxide analyses for all materials are presented in Table 1. The fraction of SiO_2 which is active for pozzolanic reactions is also given (active silica ratio). The mortar specimens for strength measurements were cast in prisms of 40x40x160 mm^3, vibrated for 20 sec on a vibration table and then covered to minimize water evaporation. A 20% fly ash addition instead of cement was adopted in all cases. A mortar without any fly ash was also prepared (control) for comparison purposes. The molds were stripped after 24 hr, and the specimens were immersed in deionized water at 20°C, until testing. The testing age was after 2, 7, 28, and 90 days. For each age, two specimens of each mixture were tested for compressive strength and the mean value of these measurements is reported.

4. RESULTS AND DISCUSSION

4.1 Compressive strength development

Compressive strength development of the control and fly ash mortars are presented in Table 2. In general it is observed that during the first week of hydration, fly ash specimens develop strength slower than the control since at this stage ash acts mostly as an inert material. However, after the first week pozzolanic systems are starting to develop strength at a faster rate than the control. In fact after 28 days of curing, all fly ash samples are either approaching or outperform the no-fly ash mortar. At this age, the intermixture prepared with equal contributions from TF and TM ashes (T1) is the mixed sample exhibiting maximum strength value, a fact that indicates that synergy between the two types of fly ashes has taken place. The strength performance of cement with TFP samples is even more impressive after only two days of curing. Considering that the particle size distribution of all samples tested is practically the same, the strength superiority of the TFP blended systems could be ascribed to the different chemical composition, and in particular its higher content of active silica.

As hydration evolves, TFP mortars continue to show better performance than the other ash-containing samples, reaching emphatic values of 62.7 MPa after three months of curing. The difference with the mechanical strength of the control sample grows larger with curing time, and becomes noticeable after the first month, since at this curing stage active silica is critically involved into pozzolanic reactions[18].

Table 1. Chemical oxide analyses (%) for cement and fly ashes used

	C	TF	TM	T1	T2	TFP
SiO_2	20.73	36.92	51.36	44.08	48.00	42.79
active SiO_2[a]	-	29.13	31.16	30.36	32.02	32.46
γs[b]	-	78.90	61.05	68.87	66.70	75.86
Al_2O_3	4.78	13.50	16.73	15.70	15.92	14.92
Fe_2O_3	3.87	7.06	8.75	8.75	8.92	5.82
CaO	64.73	29.79	13.80	21.50	17.96	21.02
MgO	2.05	2.69	2.26	2.45	2.36	3.32
K_2O	0.50	0.50	1.52	0.93	1.26	0.56
Na_2O	0.10	0.92	0.77	0.81	0.78	0.74
SO_3	2.47	5.10	1.49	3.18	2.39	3.09

[a] Estimated according to EN 450-1.

[b] Reactive silica ratio (ratio of active: total silica in fly ash)

Table 2. Compressive strength values for control and fly ashes mortars

	Compressive strength (MPa)					
Age (days)	Control	TF	TM	T1	T2	TFP
2	24.7	19.4	21.1	21.3	21.0	25.8
7	39.2	38.0	31.4	34.7	33.2	38.3
28	50.6	49.4	45.9	49.1	45.4	51.7
90	59.7	59.7	57.8	60.5	59.2	62.7

4.2 Strength, reactive silica and the k-value concept

It has been well established[2,8, 18] that in the case of mortars and concrete that incorporate supplementary cementing materials, the *k*-value derives from the following expression for the measured compressive strength (f_c):

$$f_c = K\left(\frac{1}{W/(C + kP)} - a\right) \tag{1}$$

where K is a parameter depending on the cement type, C and P are the cement and fly ash contents respectively in the mortar (kg/m^3), W is the water content (Kg/m^3) kept constant in all the mixes and a, a parameter depending mainly on time and curing. Using the above equation and compression test results, the k-values of the examined systems were calculated and presented in Table 3.

In the frame of previous efforts, Papadakis[8, 19] has demonstrated that the strength of a cementitious system (mortar or concrete) is proportional to the CSH-content, and for compressive strength (f_c) the following equation was developed (for ages >28 days):

$$f_c = 2.85\ m\ (f_{S,C}\ C + \gamma_S f_{S,P}\ P) \tag{2}$$

where m is a parameter depending mainly on water content, aggregate content and type and other compositional parameters of the mortar or concrete, C is the cement content, P is the pozzolan content, $f_{S,C}$ and $f_{S,P}$ are the weight fraction of silica in cement and pozzolan respectively, and finally γ_S is the weight fraction of the oxide SiO_2 in the pozzolan that contributes to the pozzolanic reactions (or else the ratio of active silica to the total silica in the pozzolan).

Table 3. Comparison of calculated by Eq. (3) and measured k-values

	Experimental k-values				
Age (days)	TF	TM	T1	T2	TFP
2	0.67	0.81	0.82	0.80	1.07
7	0.92	0.72	0.94	0.84	0.94
28	0.92	0.88	1.08	0.85	1.07
90	1.00	0.97	1.15	1.06	1.19
	Theoretical k-values (applying Eq. (3))				
2	0.44	0.48	0.46	0.49	0.50
28	0.91	0.97	0.90	0.95	1.05
90	1.08	1.15	1.10	1.16	1.23

For a pozzolanic mortar or concrete, by combining Eqs. (1) - (2) and using the estimated value for m (here equal to 0.32) the following relationship for k-value is obtained:

$$k = (\gamma_S f_{S,P} / f_{S,C})\ (1 - a\ W/C) \tag{3}$$

By applying the above equation for the fly ashes of the present work, k-values for the respective cement systems were calculated for 2, 28 and 90 days and provided in Table 3. The estimated values for the advanced hydration ages (>28 days) are in a good agreement with the experimental ones, given also in Table 3, showing the validity and the predictive power of Eq. (3). It becomes clear that Eq. (3) can be applied for a first approximation of the longer-term k-value not only in the case of typical fly ashes (TF and TM), but additionally in the case of ashes derived from their simultaneous use (T1 and T2) and the processed TFP. On the contrary, the theoretical values calculated for the early hydration stage (2 days) are, in all cases, notably different than the measured ones indicating that Eq. (3) should not be used for predicting early strength values. The poor early-age predictive potential of the developed expression may be attributed to the fact that fly ash acts as an inert material during this stage; subsequently its active silica content is not engaged into pozzolanic reactions therefore using an expression mainly controlled by soluble silica is not suitable.

5. CONCLUSIONS

Nowadays, that the use of a wide range of solid by-products in the construction sector is considered a common practise, a large part of the research is focused on the complete exploration of their properties, mainly those that are governed by inherent characteristics. The clearest challenge in this field is to relate critical factors of such waste materials with the future performance of the systems that incorporate them. Unambiguously, in the case of fly ash, such a factor is the reactive silica content, since this is the fraction involved in the pozzolanic reactions. Having elucidated that the literature lacks a published work directly correlating reactive silica of fly ash with the mechanical performance of blended cements, the authors attempted to fill this gap by introducing a theoretical expression between amorphous silica and k-value of cementitious systems.

In the frame of this work, this expression was tested experimentally and found to be valid for various kinds of fly ashes at elevated (>28 days) hydration stages. This was attributed to the fact that the contribution of active silica in the strength development of FA-cements is significant after the first month of hydration where its liberation from fly ash spheres gradually progresses. On the contrary, during the first stages of curing, amorphous silica is usually confined in the interior part of the fly ash particles, thus it cannot participate critically in the reactions taking place. However, the fact that the credibility of the developed relationship was confirmed, not only in the case of typical ashes (of high and lower calcium content), but also in the case of multicomponent and potentially reject samples reinforces the belief that it can be applied for a rapid prediction of the quantity, but most of all the quality of the fly ash used in the mix design so that the final product will meet certain specified requirements.

6. REFERENCES

1. S. N. Ghosh, L. S. Sarkar, *Mineral Admixtures in Cement and Concrete* (ABI Books Pvt. Ltd., New Delhi, 1993).
2. V. G. Papadakis, S. Tsimas, Supplementary Cementing Materials for Sustainable Building Sector Growth, European Commission DGXII, Project No HPMF-CT-1999-00370, Marie Curie Fellowship, National Technical University of Athens, Technical Report, Greece, 2001.
3. F. Massazza, U. Costa, Factors determining the development of mechanical strength in lime-pozzolana pastes, in: Proceedings of the XII Conference on Silicate Industry and Silicate Science, vol. 1, 537, Budapest, 1997.
4. J. F. Young, S. Mindess, R. J. Gray, A. Bentur, *The science and technology of civil engineering materials*. (Prentice Hall, New Jersey, 1999).
5. J. S. Damtoft, D. Herfort, E. Yde, Concrete binders, mineral additions and chemical admixtures: state of the art and challenges for the 21st century. Proceedings of Creating with Concrete International Congress, R.K. Dhir, T.D. Dyer (eds.), 1, Dundee: Thomas Telford, 1999.

6. A. Weinberg, R. Hemmings, Hydration and weathering reactions in by-products from clean coal technologies: effect on material properties. *Fuel*, **76**(8), 705-709 (1997).
7. M. Enders, Microanalytical characterization (AEM) of glassy spheres and anhydrite from a high-calcium lignite fly ash from Germany, *Cem. Concr. Res.* **25**(6), 1369-1377 (1995).
8. V. G. Papadakis, Experimental investigation and theoretical modeling of silica fume activity in concrete, *Cem Concr Res*. **29**(1), 79-86 (1999).
9. R. C. Sharma, N. K. Jain, S. N. Ghosh, Semi-theoretical method for the assessment of reactivity of fly ashes, *Cem. Concr. Res.* **23**(1), 41-45 (1993).
10. D. Ravina, Optimized determination of PFA (fly ash) fineness with reference to pozzolanic activity. *Cem. Concr. Res.* **10**(4) 573-580 (1980).
11. Y. Maltais, J. Marchand, Influence of curing temperature on cement hydration and mechanical strength development of fly ash mortars. *Cem. Concr. Res.* **27**(7) 1009-1020 (1997).
12. D. Li, Y. Chen, J. Shen, J. Su, X. Wu, The influence of alkalinity on activation and microstructure of fly ash. *Cem. Concr. Res.* **30**(6) 881-886 (2000).
13. P.V. Sivapullaiah, J.P. Prashanth, A. Sridharan and B.V. Narayana, Reactive silica and strength of fly ashes, *Geotech. Geol. Eng.* **16** 239-250 (1998).
14. P.L. Mehta, Belgian patent No 802909, 1973.
15. D.J. Cook, Rice husk ash, Concrete technology and design, Cement replacement materials, Edited by R. N. Swamy, Surrey University Press, London 1986, 3, 171.
16. J. Paya, J. Monzo, M.V. Borrachero and L.M. Ordonez, Determination of amorphous silica in rice husk ash by a rapid analytical method, *Cem. Concr. Res.* **31**(2) 227-231 (2003).
17. European Standard EN 450, Fly ash for concrete - definitions, specifications and conformity criteria. Brussels: CEN, 2000.
18. S. Antiohos and S. Tsimas, Investigating the role of active silica on the hydration mechanisms of high-calcium fly ash/cement systems, *Cem. Concr. Comp* **27**(2) 171-181 (2005).
19. V. G. Papadakis, Effect of fly ash on Portland cement systems. Part I: Low-calcium fly ash. *Cem. Concr. Res.* **29**(11) 1727-1736 (1999).

OPTIMIZATION OF LADLE FURNACE SLAG FOR USE AS A SUPPLEMENTARY CEMENTING MATERIAL

I. Papayianni and E. Anastasiou
Laboratory of Building Materials, Aristotle University of Thessaloniki, Aristotle University Campus, 54124, Thessaloniki, Greece

Abstract: Ladle furnace slag is a by-product of the steel industry, deriving from the second stage of the production process, which presents some cementitious properties, mainly due to its high CaO content. A series of test mortars with 20% by weight of the cement substituted with ladle furnace slag was produced in the laboratory, in order to investigate its potential as a supplementary cementing material, the varying parameter being the fineness of the slag. Thus, following some screening of the material, four mortars were produced with slag of maximum grain size 105 μm, 75 μm, 60 μm and 45 μm and also two more test mortars were produced with groundslag retaining 31% and 21% at the 45 μm sieve, all of which were compared to a test mortar with 100% cement as a binder. The test mortars developed compressive strength as high as 94% of the 100%-cement mortar, which shows that ladle furnace slag can be used as a supplementary cementing material.

Key words: ladle furnace slag; concrete; supplementary cementing materials.

1. INTRODUCTION

According to the global need for the minimization of the consumption of raw materials as well as the reduction of the energy consumption in any production process[1] it is necessary for every country to exploit to the maximum every available industrial by-product. In the case of the concrete industry, a by-product of fine gradation can be used as filler or as fine aggregate[2]. Of course, there is a higher benefit when this can be used as binding agent[3], substituting a percentage of cement. In these terms, according to the European standard of ConcreteEN -206[4] the additions are separated in Type I (practical inert material) and type (with pozzolanic or hydraulic properties).

The slags of the steel industry are well-known by-products in the construction industry, most of which have been successfully inserted in the concrete production, as aggre-

M.S. Konsta-Gdoutos, (ed.), Measuring, Monitoring and Modeling Concrete Properties, 411–417.

gates or even as binding agents[5]. In this frame there has been developed and adopted by other countries the use of ground granulated blast-furnace slag – ggbs, like an alternative binder for concrete[6]. In Greece there is production of electric-arc furnace slag, which is used as aggregate[7]. In the second stage of the steel production process the ladle furnace slag is additionally produced. In this research of the possibility of using the ladle furnace slag of SIDENOR industry as an alternative binder for the concrete production is investigated.

In general, a basic characteristic of the industrial by-products and mainly of the steel industry's slag is that their physico-chemical properties differ very much analogous to the raw materials and the manufacturing process. In this way by-products can be either of low cost for applications of low demands[8], or products of a high added value that can improve the properties of the final concrete product [9]. Based on the chemical composition (Table 1) and the granulometry of the specific furnace slag it was decided to find its capability to substitute cement in a percentage of 20%. The rate of the strength development and the achievement of satisfactory mechanical strength is estimated. The basic parameter of the study is to test the influence of the fineness of the material to the final strength development, as well as the perspective of its grinding in order to increase its effectiveness. Apart form the possibility of furnace slag's use for the partial replacement of cement, it is also tested its capability to the development of strength with lime for the production of mortars. This research work is therefore a basic research, on which the furthermore exploitation of the material will be based on. An estimation of the necessity of treatment of the material in order to improve its properties will also be made.

2. EXPERIMENTAL PART

The furnace slag used for this experimental work comes from the steel industry SIDENOR and is handled by AEIFOROS A.E. It results from the second process stage of the production of steel, when after its removal from the melting area it is firstly frozen by spraying of water. Consecutively, the material is separated from the remnants of metal, which are being returned to the steel production process and what remains is a slag of various gradation. The fraction that has been used for testing was selected from this material by sieving. Its chemical composition, as well as some physico-chemical properties are presented in Table 1.

2.1 Use of furnace slag with cement

The most important role for the reactivity of a specific binder is the fineness and specific surface of its grains. Therefore, in order to make all the tests for the reactivity with cement, it was decided for the slag to be tested in four fragments: 0-105μm, 0-75μm, 0-60μm, 0-45μm. In addition one raw slag sample was grinded and two more fractions were achieved, one with 31% retained in the sieve 45m and another with 21% retained in the sieve 45μm. With these six fractions there were produced six laboratory compositions of mortars, as it is presented in Table 2. For the comparison of the results another synthesis with 100% of cement was manufactured.

All compositions were produced with a stable expansion of 16±0,5, while the demand of water was measured in order to achieve the specific expansion measured by flow table (Table 2). The compositions are presented in Table 3. From these trial mixes resulted cubic specimens 7x7x7 cm, which were preserved in saturated with $Ca(OH)_2$ water, until the time of the tests. The specimens were tested in compressive strength in the age of 7 and 28 days, in order to measure the rate of strength development.

2.2 Use of furnace slag with lime

An alternative application of furnace slag is the manufacture of mortars with lime. The capacity of developing adequate mechanical strength is an obvious indication of the material's pozzolanicity. For this aim there have been manufactured two trial mixes with fractions of 0-75μm (L--75) and 21% (L--21%) retained in sieve 45μm, of the available slag, as it is presented in Table 4. From the mixtures resulted prismatic specimens 4x4x16cm and cylindrical of diameter 5cm and of 10cm height, in which there have been tested the compressive and flexural strength as well as the strength development rate.

Table 1. Chemical analysis of the furnace slag

	Soluble in acids	Soluble in HCl 0.1 N	
Na_2O	0,34	0,03	
K_2O	0,04	0,01	
CaO	54,1	52,7	
MgO	5,55	5,12	
FeO	1,72	1,25	
Al_2O_3	2,50	2,47	
SiO_2	32,5	26,4	
Soluble salts (% p.w.	Cl^-	NO_3^-	SO_4^-
	0,03	0	0,21
Loss of Ignition (950°C) (%)	3,19		
Loss of Ignition (550°C) (%)	2,62		
Specific Gravity (g/cm^3)	2,590		

Table 2. Use of slag of different fineness in the trial mixes

Composition		-105	-75	-60	-45	-31%	-21%
Slag fraction	-	0-105μm	0-75μm	0-60μm	0-45μm	31% retained in sieve 45μm	21% retained in sieve 45μm
Cement replace-ment p.w.	0%	20%	20%	20%	20%	20%	20%
Expansion (cm)	16,5	16,5	16,0	16,5	16,5	16,0	16,5
Water demand (% of)	-	105%	100%	96%	96%	96%	98%

Table 3. Description of the compositions with cement

	Model composition	Compositions of replacement
Cement CEM I42,5 (g)	1000	800
Furnace slag (g)	0	200
Standard sand (g)	2700	2700
Water (g)	484	484*
Composition's specific gravity (g/cm^3)	2,312	2,143
Water binder ratio (W/B)	0,484	0,484*

* there has been an adjustment of water in order to have expansion 16±0,5 cm

3. RESULTS

3.1 Trial mixes with furnace slag and cement

From the results of the compressive strength (Figure 1) it is obvious that the furnace slag‘s fineness influences directly the mechanical strength development and specifically the compressive strength of the 28 day age. The values achieved were 37,1 MPa for the fraction of 0-105μm and 42,6MPa for the fraction of 0-45μm. The more fineness is the slag the highest is the mechanical strength. Grinding also provides better hydraulic properties to the material, which is possibly due to the greater reactive surface of the grains after milling. Furthermore, it is important to point out that the two grinded slag samples (31% and 21%) do not differ very much regarding the final strength.

In Table 2 appears that the addition of furnace slag in the mixture does not increase the demand of water. In the finer samples (<60μm) there is a slight decrease of water demand and therefore the ratio of water to binder does not change. This fact influences positively the final strength of the mixture, as well as its durability. In Table 5 it is presented the percentage of the strength development in relation to the model composition with 100% cement. With the fraction of 0-45μm it is achieved the 87% of the primary strength, while with the grinded mixtures the percentage is increased to 94%. In addition, from the comparison of the 7 and 28 day strength, it seems that the rate of strength development is not influenced by the addition of slag, which is very important for the contemporary constructions. This is due to the fact –according to Table 1- that the furnace slag has a high percentage of calcium and silica oxide and therefore presents cementitious as well as pozzolanic properties.

Table 4. Description of trial mixes with lime

	L-
1 proportion p.w. lime (g)	150
2,4 proportion p.w. furnace slag (g)	360
9 proportion p.w. standard sand (g)	1350
Water demand for expansion 11±0,5cm (g)	240
Composition's specific gravity (g/cm^3)	2,172
Ratio water per binder (W/B)	0,47

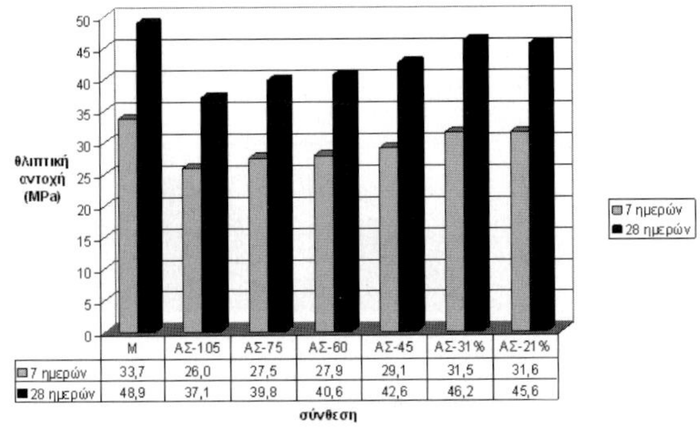

Figure 1. Strength development of mortars with furnace slag and cement

3.2 Trial mixes with furnace slag and lime

In accordance to the previous compositions, trial mixtures of slag with lime have been tested, which mainly shows the pozzolanicity of the material. The compressive and flexural strength development of the trial mortars are presented in Figures 2 and 3. For the better control of the pozzolanic reactivity, which takes place slowly, the strength was measured 7, 28 and 200 days from the manufacture. From these results it is obvious that the furnace slag has remarkable pozzolanic properties, since the rate of strength development is not stopped at 28 days but is continued, resulting in the highest compressive strength at 200 day age. The compressive strength achieved is 5,11 MPa, while the flexural strength is 2,04 MPa. In the composition with the fraction of 0-75m as well as in this with the grinded fraction 21% the mechanical strength is being doubled from 28 to 200-day, which shows that the binder's reactivity is still going on.

Table 5. Comparison of the mechanical strength of the trial mixes with the one with 100% of cement

Composition	7-day strength (Percentage to the model)	28-day strength (Percentage to the model)
	(%)	(%)
	100%	100%
105	77%	76%
75	82%	81%
60	83%	83%
45	86%	87%
31%	94%	93%
21%	93%	94%

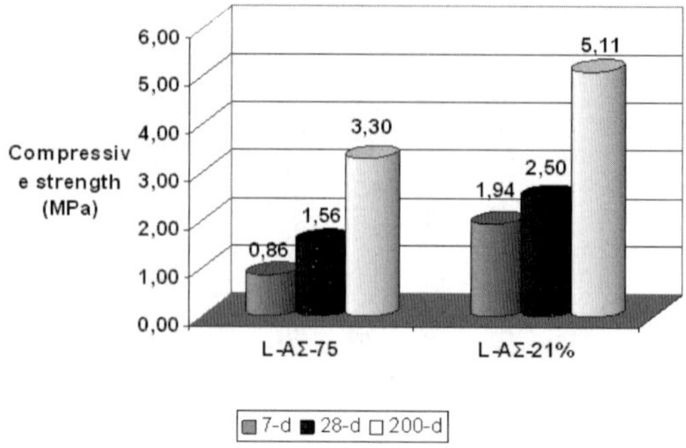

Figure 2. Compressive strength development of mortars with furnace slag and lime

In the case of the trial mixes with lime, as well as in those with cement, the increase of the slag's fineness with grinding gives higher final strength. Comparing the results of the compositions of slag with lime, with those of slag with cement, it is obvious that grinding influences much the pozzolanicity of the material. Moreover, bibliography [10] shows that the furnace slag could be more reactive with the use of an alkali catalyst, which could also give to the material the properties of an extremely powerful binder.

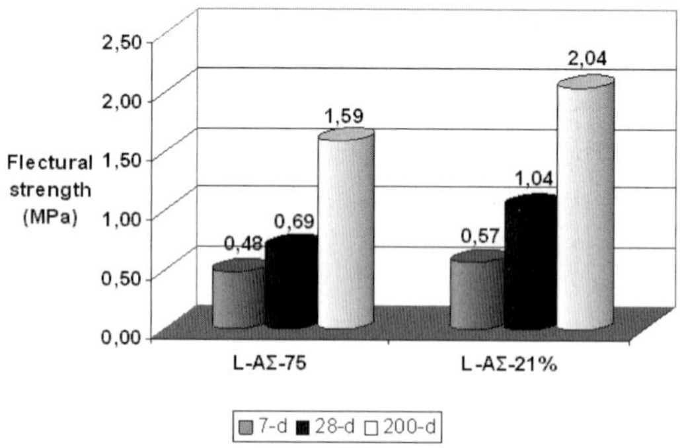

Figure 3. Flexural strength development of mortars with furnace slag and lime

4. CONCLUSIONS

In this research work it has been studied the possibility of using the furnace slag for the production of concrete, by replacement of a 20% of cement. The pozzolanicity of it was investigated with trial mixes with lime. The basic conclusions that derived are:

1. The reactivity of the furnace slag, as an additional binder, increases significantly with its fineness. This increase can be achieved by selecting the finer fragment of the material or by grinding it, although with the latter one a higher final strength is developed.
2. When 20% of the cement is replaced in mortar mixtures by the fine fragment of furnace slag, a 28-d strength level of 42MPa is achieved, equivalent of 87% of the

control cement mortar. In the case of ground furnace slag, the achieved 28-d strength amounts 45MPa and corresponds to 94% of the control cement mixture.

3. The furnace slag does not delay the cement's hydration and does not demand more quantity of water in order to achieve mortars of the same workability.
4. As a binder it has pozzolanic properties. This explains the fact that the strength development continues even until 200 days from the manufacture. However, the rate of strength development is relatively low. The use of activators could accelerate it.

5. REFERENCES

1. Mehta P.K. 'Concrete Technology for Sustainable Development – An Overview of Essential Principles', in Mehta P.K. (ed.) *Proc. of the Int. Symposium on concrete for sustainable development in the twenty-first century.* Hyderabad, India (1999) 1-22.
2. Yamada K., Ishiyama S. 'Maximum dosage of glass cullet as fine aggregate in mortar', in R.K. Dhir et al (eds.) *Achieving Sustainability in Construction, Proc. of the Int. Congress Global construction: ultimate concrete opportunities.* Dundee, Scotland (2005) 185-192.
3. Malhotra V.M., Mehta P.K. *High-performance, high volume fly ash concrete: Materials, Mixture Proportioning, Properties, Construction Practice, and Case Histories.* Ottawa, Canada: Marquardt Printing (2002).
4. EN 206-1 'Concrete - Part 1: Specification, performance, production and conformity' (2000).
5. Motz H., Geiseler J. 'Products of steel slags an opportunity to save natural resources', *Waste Management* **21**: (2001) 285-293.
6. Babu K.G., Kumar V.S.R. 'Efficiency of GGBS in concrete', *Cement and Concrete Research* **27**: (2000) 1031-1036.
7. Papayianni I., Anastasiou E. 'Heavyweight Concrete with Steel Slag Aggregates', in R.K. Dhir et al. (eds.) *Role of Concrete in Nuclear Facilities, Proc. of the Int. Congress Global construction: ultimate concrete opportunities.* Dundee, Scotland (2005) 25-32.
8. Claisse P.A. et al. 'Recycled Materials in Concrete Barriers', in V.M. Malhotra (ed.) *Proc. of the Sixth CANMET/ACI Int. Conference on Durability of Concrete.* Thessaloniki, Greece (2003) 951-972.
9. Shi C., Qian J. 'High performance cementing materials from industrial slags – a review', *Resources, Conservation and Recycling* **29** (2000) 195-207.
10. Xu H., Lukey G.C., van Deventer J.S.J. 'The Activation of Class C-, Class F-Fly Ash and Blast Furnace Slag Using Geopolymerisation', in V.M. Malhotra (ed.) *Proc. of the Eighth CANMET/ACI Int. Conference on Fly Ash, Silica Fume, Slag, and Natural Pozzolans in Concrete.* Las Vegas, USA (2004) 797-820.

CRITERIA FOR THE USE OF STEEL SLAG AGGREGATES IN CONCRETE

E. Anastasiou and I. Papayianni
Laboratory of Building Materials, Aristotle University of Thessaloniki, Aristotle University Campus, 54124, Thessaloniki, Greece

Abstract: The successful incorporation of steel slag as aggregates in construction products requires the consideration of certain issues. Firstly, as steel slag is an industrial by-product until recently disposed in landfills, the question is whether it is suitable for use in construction. Then the technical characteristics of the material are examined because due to its physicochemical properties steel slag requires special care, but can also provide maximum value if used for specific applications. The utilisation of a by-product in suitable applications –mainly where it is advantageous compared to traditional materials, but also where it is most economical– can give a higher added value to the product. Finally, there are a number of economy-related parameters that allow for a new product to enter the construction market like the situation of the local aggregate market or the need to communicate the efficiency of a new product through demonstration projects. Through all the above considerations and knowledge through practice we look at the way steel slag aggregates enter the construction market in Greece.

Key words: industrial by-products; steel slag; standards.

1. INTRODUCTION

The aim of the regulative frame for the use of building materials is basically the safety of construction and generally the avoidance of failures that are due to the material and its cooperation with other construction materials. The prescriptions that govern the use of a material have resulted from multi-schientific cooperation and are mainly based on the accumulated experience. Usually they follow the applications for which the materials have been already used. The coverage that the standards offer for the trading of a building material is necessary for its promotion. However, there are many innovative products, that without having the coverage of a regulative frame are very successfully used in construction, because the properties of the final product are improved.. The criteria of regulations can refer to general or more specific needs. In the case of the use of steel slag as

M.S. Konsta-Gdoutos, (ed.), Measuring, Monitoring and Modeling Concrete Properties, 419–426.

aggregates for the production of concrete, there are the general demands that refer to the aggregates for concrete and the specific ones that refer to concretes of specific applications. Due to the origin of aggregates there can be more controls.

The use of slags, by-products of the iron industry (blast furnace slag), as aggregates for concrete is well-known in Europe for more than 30 years, since relevant regulations have been already published in 1974 (BS 1047: Air cooled blast furnace slag used as coarse aggregate for concrete), where there are presented specific demands concerning the concentration of soluble sulfates and the volume stability. The slags that are produced during the steel production (steel slag) differ to their composition, specifically regarding the concentration of Fe_2O_3, the relation SiO_2 / CaO and are comparatively of a higher specific gravity. They can be characterized -in contradiction to the blast furnace slag aggregates– as aggregates of heavy concretes. Specific regulation of the use of steel slag as aggregates in concrete does not exist. However, the possibility of their use in concrete with technological and ecological benefits, has been globally studied from several researchers [1,2,3].

2. REGULATIONS FOR CONCRETE AGGREGATES

The demands for the use of a grained material in concretes are prescribed in the European standard 12620: Aggregates for Concrete and in the EN 13242: Aggregates for unbound and hydraulically bound materials for use in civil engineering work and road construction. In the standard 206: Manual for concrete Practice there are also included articles that concern aggregates. Relevant to the aggregates' demands for concrete are the regulations ASTM C33: Concrete Aggregates, ASTM C637: Aggregates for Radiation-Shielding Concrete, with the test methods that are prescribed in ASTM Annual Book of ASTM Standards Volume 04.02.

Concerning the regulations for the use of grained materials as aggregates in asphaltic mixtures, such as the standard 13043: Aggregates for bituminous mixtures and surface treatments for roads airfields and other trafficked areas, the difference is that in the standards 12620 prEN 13242 steel slags are embodied to the cement paste, which is of a maximum binding capacity, with adequate durability and stability during time.

In the paragraphs concerning the aim and the determination of the above mentioned standards (EN 12620, EN 13242) are described as aggregates the grained materials that result from the process of natural or industrially produced or even recycled materials , that have dry density of grains bigger than 2,0 kg/m^3 and can be used in concrete of usual applications that are prescribed 206-1 standard, or in roads, floors or other products of prefabricated concrete.

According to the determination, the steel slag aggregates are included in the standards when provide that the tests all the demands that are prescribed in the standards are fulfilled. Regarding the tests by which the demands are fulfilled, there is a catalogue of relevant standards in the annex.

2.1 Granulometry

The grain size of the steel slag aggregates is determined by crushing and sieving of the raw slag. According to the standards, the segregation to coarse and fine aggregates or to a mixture is made with the sieve size of the smallest (d) and the biggest (D) (Tables 1, 2).

Table 1. Granulometry demands of aggregates according to 12620

Aggregate	Size*	Total passing % ..					Category G
		2D	1,4D	D	d	d/2	
Coarse	*D/d ≤ 2 or*	100	98-100	85-99	0-20	0-5	$G_c85/20$
	D ≤ 11,2mm	100	98-100	80-99	0-20	0-5	$G_c80/20$
	D/d > 2 and D > 11,2mm	100	98-100	90-99	0-15	0-5	$G_c90/15$
Fine	*D ≤ 4mm d=0*	100	95-100	85-99	-	-	G_F85
Natural gradation 0-8	*D = 8mm and d = 0*	100	98-100	90-99	-	-	$G_{NG}90$
One gradation (all-in)	*D ≤45mm and d=0*	100	95-100	85-99	-	-	G_F85

*where D and d the biggest and smallest nominal size of aggregate fragment

Table 2. Granulometry demands of aggregates according to 13242

Aggregate	Size* (mm)	Total passing % ..					Category G
		2D	1,4D	D	d	d/2	
Coarse	*d ≥ 1 D > 2*	100	98-100	85-99	0-15	0-5	G_c85-15
		100	98-100	80-99	0-20	0-5	G_c80-20
Fine	*d = 0 D ≤ 6,3*	100	98-100	85-99	-	-	G_F85
		100	98-100	80-99	-	-	G_F80
One gradation (all-in)	*D ≤ 45mm d = 0*	100	98-100	90-99	-	-	G90
		100	98-100	85-99	-	-	G85

*the same as previously

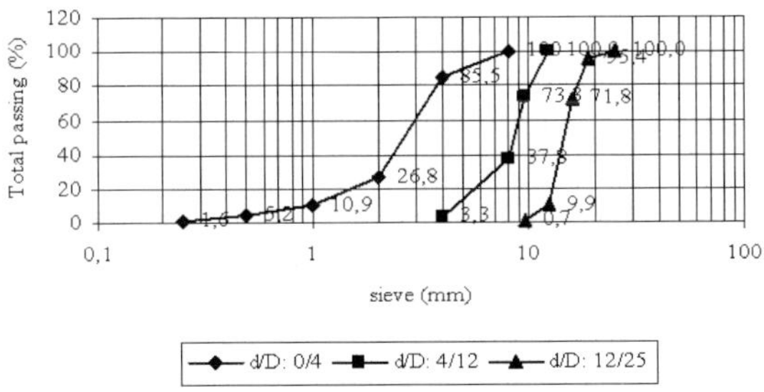

Figure 1. Indicative gradation curves of three fragments of steel slags nominal size d/D 0/4, 4/12 and 12/25

2.2 Morphological characteristics

The standard 12620 prescribes the demands for the flakiness index and for the percentage of fines that passes from the sieve 63m. After laboratory tests, the slag aggregates gave the results of Table 3. It is worth mentioning that in both cases the slag meets the minimum of the relevant categories.

Table 3. Morphological characteristics of slug aggregates

Test	Fragment	Result	Category/Requirement	
			12620	13242
Flakiness index	Coarse	8%	FI_{15}	FI_{20}
Total passing the 63m sieve	Coarse	0,2%	$f_{1,5}$	f_2
	Fine	1,0%	f_3	f_3

2.3 Physical properties of aggregates

These properties are related to the origin, structure, and the surface characteristics of the grains and are:

- Crashing test (Los Angeles)
- Resistance to surface abrasion (AAV)
- Impact strength
- Resistance to abrasion (Micro – Deval)
- Resistance to tire weathering

Table 4. Physical properties of aggregates

		Category/Requirement	
Test	Result	12620	13242
Apparent Specific Gravity of fine	3,330	> 2,0	> 2,0
Apparent Specific Gravity of coarse	3,333		
Bulk density	1,562	-	-
Water permeability	1,3%	WA_{24} 2	WA_{24} 2
Abrasion and Impact (Los Angeles)	13,32%	LA_{15}	LA_{20}
(PSV)	64	PSV_{62}	is not required
Abrasion (AAV)	3	AAV_{10}	is not required

The test specifications are presented in the standards of the annex. For the slag aggregates that have been tested, the results are presented in Table 4. In all these tests, the results are in between the limits of the standards and in the best categories. In the standard 12620 there is also the demand of the test of coarse aggregates in freezing-thawing cycles, alkali-aggregate reaction and drying shrinkage. For the highest resistance to freezing-thawing cycles there are presented in the same standard (tables 18, 19) prices not only for freezing-thawing but even for resistance to sulfate magnesium. According to 13242 paragraph 7.3 that concerns the resistance of aggregates for the category of water absorption $WA_{24}\ 2$ there is not a demand for control and the aggregate is considered to be resistant to freezing-thawing cycles. The rest control tests have been only made to concrete that was produced with these aggregates and it is proposed to be made in the future in the frame of certification of the steel slag aggregates.

2.4 Chemical demands - Performance in concrete mixtures

According to the standards it is necessary to measure the concentration of chlorides and sulfates of the aggregates. The results of the tests are presented in table 5. In both standards it is included the test of the presence of harmful for the hydration and hardening of concrete (paragraph 6.4.1). The demand for the strength development of a test mixture is not to delay the hardening of the testing mixture more than 120minutes and not to reduce the 28-d compressive strength more than 20%. In Figure 2, it is obvious that the 28-d strength is being increased up to 21% with the replacement of typical aggregates with steel slag, while there is no increase in the setting time of concrete mixtures.

Table 5. Chemical properties of aggregates

		Category/Demand	
Control test	Concentration	According to 12620	According to 13242
Concentration in Cl^-	0,039%	-	-
Soluble sulfates	<0,8%	$AS_{0,8}$	$AS_{0,8}$
Total sulfates	<1,0%	<1%	S_1
MgO	4,7%	-	V_5

For the categories of steel slag even though it was not followed the methodology of 1744-1 for the control tests for harmful components, there is a de facto documentation from the rate of the strength development for concrete mixtures with steel slug aggregates (Figure 2). The Standard 13242 regarding volume stability of iron and steel slag in non binding mixtures, paragraph 6.4.2.1, refers to the control tests of the volume stability of steel slag, mainly due to its concentration of MgO, which can be measured according to 196-2. In the Table 14 of the same standard, there are given according to the test value the anticipated categories for steel slag aggregates in non hydraulic binding mixtures. Although, in this article the steel slags are confronted as aggregates of hydraulically bound mixtures, the measurement of the concentration of MgO showed that it does not exceeds the 5%. In addition, all the concrete compositions with these aggregates did not presented any time of problem of volume stability even 5 years form their manufacture. Additionally, as it is shown in Figure 3, the cement-aggregate interface seems to be very dense without cracks or other discontinuities.

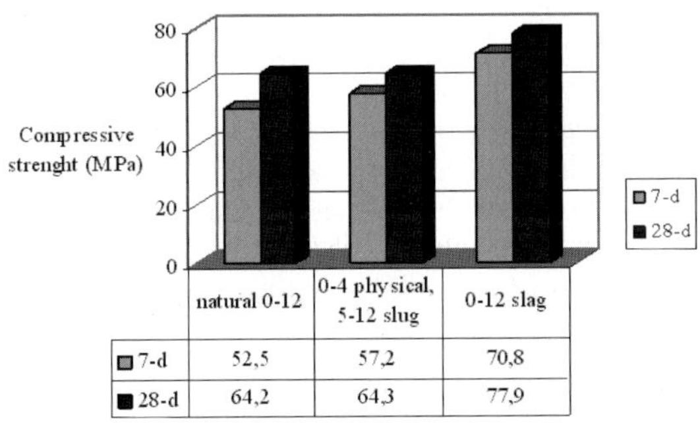

Figure 2. Strength development of concretes with slag aggregates

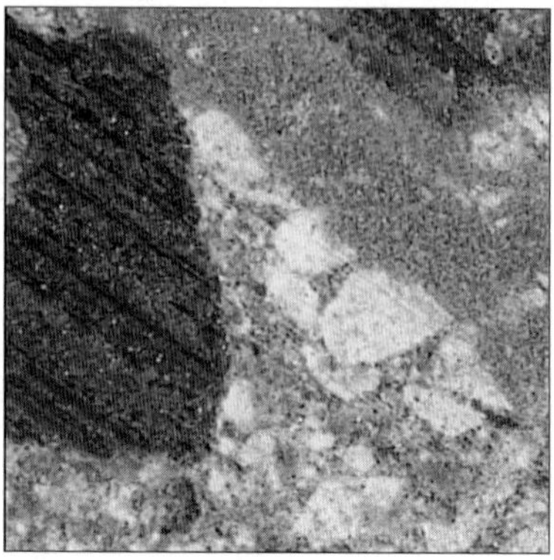

Figure 3. Matrix-aggregate transition zone after wet-dry cycling in seawater – inside section (microscope x12.5)

In paragraphs 6.4.2.2, 6.4.2.3 there are specific control tests for the iron slag aggregates. Regarding the slag aggregates it was tested in concretes the possibility of the iron oxidization and the appearance of stains. After many freezing-thawing cycles there was not any kind discoloration or rust stains in the concrete. In the above mentioned standards there are incorporated instructions for the evaluation of the compliance with the regulations, in order for the results to be reliable.

3. CONCLUSIONS

Although the steel slag aggregates are not certified according to the process which is anticipated from the regulative frame, it can be said that they meet the criteria of regulations as aggregates for hydraulically bound mixtures for road bases. There is more than 5 years of experience in their use in concretes for road bases (pavement, paving blocks, concrete of high density) in laboratory and worksite scale.

It is certified that:

- There is good cooperation of aggregates with cement or with binding systems such as cement+pozzolanic materials and there is an increase of strength of the concrete.
- There can be used in several granulometries or in cooperation with river sand or calcitic aggregates without any problem.
- There no need for specific regulations for the production of concrete. There are followed typical ways of design and production of concrete.
- Concrete presents during its fracture an enhancement of the interfacial transition zone.
- It does not present volume changes through time apart form the anticipated ones for concrete.
- It does not present stain in the surface due to iron oxidization.
- The leaching tests showed a high retention capacity of the metals in the concrete with slag aggregates, a fact that is desirable and it is strengthening the ecological profile of slag concretes

- The concrete that is produced by steel slag aggregates is of a high specific gravity 2,8 tn/m^3 compared to the typical concretes. The specific gravity can be increased or can be reduced proportionally to the granulometry or the combination of aggregate deposits from different orientation.

4. REFERENCES

1. Beshr H., Almusallam A.A., Maslehuddin M. 'Effect of coarse aggregate quality on the mechanical properties of high strength concrete', *Construction and Building Materials* **17** (2003) 97-103.
2. Shi C., Qian J. 'High performance cementing materials from industrial slags – a review', *Resources, Conservation and Recycling*, **29** 3 (2000), 195-207.
3. Matsunaga H. et al. 'Environment-Friendly Block Made From Steel Slag', in V.M. Malhotra (ed.) *Proc. of the Eighth CANMET/ACI Int. Conference on Fly Ash, Silica Fume, Slag, and Natural Pozzolans in Concrete.* Las Vegas, USA (2004) 457-470.
4. Annual Book of ASTM Standards Volume 04.02: Concrete and Aggregates (1996)
5. 12620: Aggregates for Concrete (2002)
6. EN 13242: Aggregates for unbound and hydraulically bound materials for use in civil engineering work and road construction (2002)
7. 13043: Aggregates for bituminous mixtures and surface treatments for roads airfields and other trafficked areas (2002)

ANNEX

Standard methodologies of controlling the suitability of aggregates (EN12620 and EN 13242)

1. EN 196-21:1989, Methods of testing cement — Part 21: Determination of the chloride, carbon dioxide and alkali content of cement.
2. EN 932-3, Tests for general properties of aggregates — Part 3: Procedure and terminology for simplified petrographic description.
3. EN 932-5, Tests for general properties of aggregates — Part 5: Common equipment and calibration.
4. EN 933-1, Tests for geometrical properties of aggregates — Part 1: Determination of particle size distribution – Sieving method.
5. EN 933-3, Tests for geometrical properties of aggregates — Part 3: Determination of particle shape — Flakiness index.
6. EN 933-4, Tests for geometrical properties of aggregates — Part 4: Determination of particle shape — Shape index.
7. EN 933-5, Tests for geometrical properties of aggregates — Part 5: Determination of percentage of crushed and broken surfaces in coarse aggregates particles
8. EN 933-7, Tests for geometrical properties of aggregates — Part 7: Determination of shell content — Percentage of shells in coarse aggregates.
9. EN 933-8, Tests for geometrical properties of aggregates — Part 8: Assessment of fines — Sand equivalent test.
10. EN 933-9, Tests for geometrical properties of aggregates — Part 9: Assessment of fines — Methylene blue test.

11. EN 933-10, Tests for geometrical properties of aggregates — Part 10: Assessment of fines — Grading of fillers (air jet sieving).
12. EN 1097-1, Tests for mechanical and physical properties of aggregates — Part 1: Determination of the resistance to wear (micro-Deval).
13. EN 1097-2:1998, Tests for mechanical and physical properties of aggregates — Part 2: Methods for the determination of resistance to fragmentation.
14. EN 1097-3, Tests for mechanical and physical properties of aggregates — Part 3: Determination of loose bulk density and voids.
15. EN 1097-6, Tests for mechanical and physical properties of aggregates — Part 6: Determination of particle density and water absorption.
16. EN 1097-8:1999, Tests for mechanical and physical properties of aggregates — Part 8: Determination of the polished stone value.
17. EN 1097-9, Tests for mechanical and physical properties of aggregates — Part 9: Determination of the resistance to wear by abrasion from studded tyres — Nordic test.
18. EN 1367-1:1999, Tests for thermal and weathering properties of aggregates — Part 1: Determination of resistance to freezing and thawing.
19. EN 1367-2, Tests for thermal and weathering properties of aggregates — Part 2: Magnesium sulfate test.
20. EN 1367-3, Tests for thermal and weathering properties of aggregates — Part 3: Boiling test for "Sonnenbrand" basalt.
21. EN 1367-4, Tests for thermal and weathering properties of aggregates — Part 4: Determination of drying shrinkage.
22. EN 1744-1:1998, Tests for chemical properties of aggregates — Part 1: Chemical analysis.
23. EN 1744-3, Tests for chemical properties of aggregates — Part 3: Preparation of eluates by leaching of aggregates.
24. EN 196-2, Methods of testing cement — Part 2: Chemical analysis of cement.

Designing Concrete for Unconventional Properties

CONCRETE FOR THE CONSTRUCTION INDUSTRY OF TOMORROW

M. Corradi
Senior Vice President Technology & Development, BU Admixture Systems Europe, Degussa Construction Chemicals

Abstract: The construction industry has made impressive progress since reinforced concrete was introduced as structural material. At the beginning of the 20th century the first technical bulletin on "Tests of Reinforced Concrete Beams" published by Robert Talbot (1), and based on extended experimental work, opened the way to the modern use of reinforced concrete. Reinforced concrete was first used for factories, warehouses and low-rise residential houses. In 1903, the first high-rise in concrete (15 floors), the Ingalls Building in Cincinnati, Ohio, showed the suitability of concrete for tall structures, and the benefit of concrete over steel. Many sceptics believed that the 15-floor building would have collapsed on its own weight, when completed; in fact, it stood firmly on its foundations and opened new horizons to the construction industry. Even the comparison with steel constructions showed the superior safety of concrete in terms of fire resistance and it was a determining success factor for concrete after the "Great Fire" in Chicago in 1873. The technological changes and progress in the construction industry can be recognized in the development of high rises in the past century and the beginning of the currentone. Progress in the construction industry were essentially based on the improvement of materials, structural models and construction methods.

1. MATERIALS

The development of lightweight structural concrete with compressive strengths ranging from 20 to 41 MPa and specific weight between 1440 and 1920 kg/m^3, allowed high multi-storey buildings to be built. The higher cost of this kind of concrete was greatly balanced by the lower concrete weight, the consequent reduction of column sizes and more commercial space available. A 52-storey building, the Shell Plaza, Houston, Texas, fully built in lightweight concrete in 1971, still remains the highest lightweight building in the world.

M.S. Konsta-Gdoutos, (ed.), Measuring, Monitoring and Modeling Concrete Properties, 429–440.

The Ingalls Building, Cincinnati, Ohio

The Shell Plaza, Houston, Texas

High Strength Concrete (HSC) developed in the late sixties and its first structural use was implemented in Norway in 1971. HSC is a new concrete composition where silica fume and superplasticizing admixtures are essential components, together with cement, fine sand and crushed aggregate. Due to the binding property of silica fume and the high packing capability of the mix, this concrete can reach strengths higher than 130 MPa. HSC is more and more frequently used for high rise constructions, with the advantage of reducing the size of columns thanks to the high strengths developed. In addition, it is pumped easily even to very high or long distances.

Completed in 1998, Petronas Towers are a bright example of use of HSC for high-rise buildings. They were the tallest buildings in the world for many years, superseded, only recently, by 101 Building in Taiwan, and by others which are under completion.

Petronas Towers, Kuala Lumpur, Malaysia

Taipei 101, Taiwan

Modern high strength concrete is characterized by long term high strengths and, even more, by very high early strengths, impermeability, chemical resistance and durability. Many of these properties are imparted by continuously improved admixtures. As a matter of fact, the development of new chemistry in the field of admixtures has allowed the extraordinary improvement of admixture performances to be obtained.

Thirty years ago the admixtures state-of-the-art was based on ligninsulphonates, which allowed a 6 to 8% reduction in concrete mixing water, or a 2 to 4-cm slump increase. Superplasticizers based on Naphtalene Sulphonate (NS) or Melamine Sulphonate (MS), developed in the seventies, marked a quantum leap progress in the admixture science by reducing concrete mixing water 15 to 20%, improving workability to very fluid concrete and allowing 25 to 30% higher early and long term strengths to be obtained.

Durability and reliability of placed concrete improved rapidly and superplasticizers became an essential part of modern concrete. More recently, a new chemistry, based on polycarboxylate ethers (PCE), set a new standard in the admixture science, not only allowing the further reduction of concrete mixing water from 30 to 40%, workability improvement up to self compacting concrete and strengths improvement up to 40- 60%, but also new performances, such as workability retention, very early high strength and a higher degree of durability to be achieved. The great innovation of PCE is the chemical structure and its flexibility to obtain desired performances (2).

2. STRUCTURAL MODELS

A structure is designed to sustain loads, the building load first of all, wind or lateral loads, snow and equipment. In the beginning, high-rise buildings used to have a mixed functionality: retail space at street level, offices at the floors immediately above and apartments at the upper levels.

A landmark in this kind of structure is Marina City, by Bertrand Goldberg Associates, in Chicago, built in 1964 – 1967 where, in two sixty-storey towers, a complete self-sufficient structure was created. It includes 18 parking levels, offices, a bank, restaurants, a gym, a theatre, a television studio and a marina. These multiple functionalities are very complex to accommodate and require complex structures according to the different uses. The structural model in this concrete tower structure is based on cylindrical cores carrying the loads of the cantilevered floors. Marina City is an example of how structural models evolved from gravity and lateral load-resisting structures to the most recent Modular Tube structure that allows to build up even higher than 80 storeys.

The structural model evolution goes through more and more complex and greater load bearing models. When reinforced concrete started to be used in building constructions there were limitations on the heights a building could reach. Innovative structural models were developed to allow higher and higher buildings. The Frame Model based on traditional columns and girders allowed buildings up to 20–25 floors; the Shear Wall system, developed in the forties, is constituted of vertical cantilevered beams resisting lateral winds and seismic forces transmitting loads to floors diaphragms. Shear walls can be shaped in circular, oval, box-type or triangular structures and they often contain building services like lifts, heating and conditioning systems.

Marina City Towers, Chicago, Illinois

Further progress was made with the development of Framed Tube Structures and their evolutions, in the sixties, defined by Fazlur Khan, of Skidmore, Owings & Merril (SOM) as a "three dimensional space structure composed of many frames joined at their edges to form a vertical tube like structural system, capable of resisting lateral forces in any direction cantilevering from the foundation." (3). Many high rises built in Chicago used that structural system, as The Onterie Center, Brunswick Building and One Magnificent Mile.

The Onterie Center, Chicago, Illinois

One Magnificent Mile, Chicago, Illinois

The construction models development allowed the height of high rises to be increased from the 15 floors of Ingalls Building (65 mt) to the 452 mt of Petronas Towers. Higher buildings in concrete or composite materials are currently under construction and Taipei 101, 508 mt, under completion, will be the highest in the world. The most advanced model currently in use for structural and fire resistant applications is the core concrete structure bearing cantilevered slabs.

3. CONSTRUCTION METHODS

The development of construction methods has brought many economical, technical and time efficiency improvements to the building industry. Labour productivity is one of the critical - difficult to be measured - factors, but data show that the construction industry has failed to keep the pace with the performance improvements of other sectors. An OECD source indicates that that the labour productivity growth rate was only 1.25 per cent on average in the OECD countries in the period 1985 – 1993 (4). This has contributed to a relative construction cost increase and declining efficiency of the construction process.

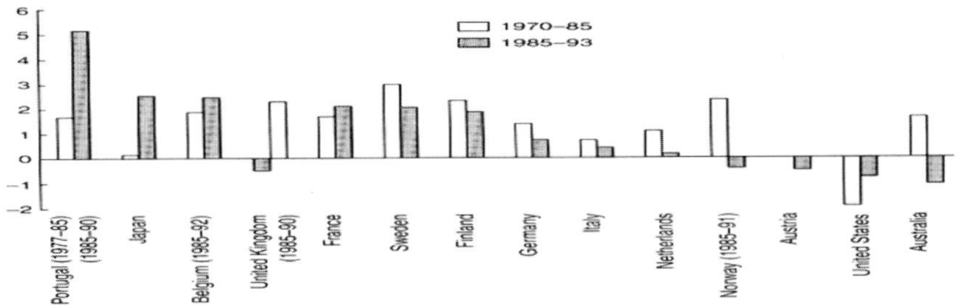

Fig. 1.5. Labour productivity growth in construction, 1970–1985 and 1985–1993. Note: labour productivity is measured by constant price value-added divided by the number employed; the figure shows compound annual growth rates as percentages (source: OECD 1998, p. 153)

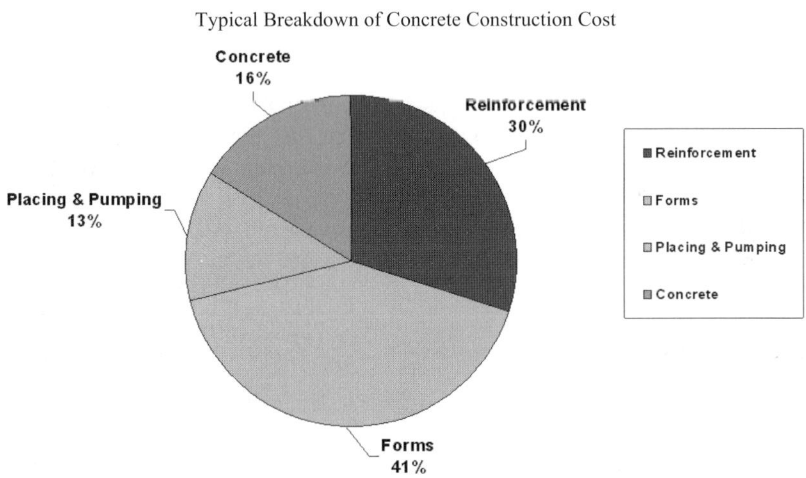

Considering the average breakdown of construction costs, we can see how labour productivity affects the key drivers of the total construction costs (5). The construction process is based on a sequence of activities, such as: installing forms, mixing, transporting, placing and demoulding concrete. Each of these steps has progressed, supported by design and concrete technology improvements. The possibility of larger and larger single concrete pouring, the development of new types of formwork for faster and more intensive usage allowing early striking, the transport of concrete to long distance and height by innovative pumping methods are among the most important improvements achieved.

Admixtures and concrete technology progress, together with mechanical innovations in equipments, have allowed to make concrete competitive with steel in terms of construction and erection speed.

4. CONCRETE FOR THE CONSTRUCTION INDUSTRY OF TOMORROW

The Construction industry has recently made some substantial changes and new drivers have come into the picture of its development, progress and innovation. There are ongoing substantial changes in the design of new buildings showing more focus on environment, safety, aesthetics and economy. The European Community in its "Strategic Research Agenda" (6) has introduced, as to the research programs of the next decade, topics like "Sustainability", which implies: to reduce resources consumption (energy, water, materials) and reduce environmental and man-made impact.

In the next decades, the construction industry is expected to move from technology-driven to demand-driven industry, and from resource-intensive to knowledge-intensive activities. New sciences, such as nanotechnology, are expected to play an important role in the development of new materials and new knowledge-based materials with tailored properties and enhanced process abilities. Continuous innovation is the basis of the new approach to the construction industry progress (ECTP, Plenary Assembly European Construction Technology Platform, Paris, October 5, 2005). Innovation will also be triggered by legislative measures, such as reducing energy consumption, lowering CO_2 emissions, limiting aggregate excavations and disposing of waste from demolition.

New shape of buildings, unthinkable a decade ago, are today in the agenda of the major Architects. New structural models, as the Headquarter of the European Central Bank in Frankfurt, by United Architects Associates, the new office area in Milan, Italy, by Liebeskind and Associates, or the Guggenheim museum in Bilbao by Frank O. Gehry, are examples of unconventional structures no longer based on squared sharp edged boxes.

Plan of the Headquarter of the European Central Bank, Frankfurt, Germany

Guggenheim Museum, Bilbao, Spain

The construction industry is therefore changing to fit the requirements of new designs, new social pressure, environmental demand and more stringent economical parameters. Three drivers will lead the change of the construction industry of tomorrow: 1) New materials and improved performances of concrete, 2) Construction processes and economy improvement, 3) Energy saving and environment protection.

5. NEW MATERIALS AND IMPROVED CONCRETE PERFORMANCES

The concrete technology development will probably show the most spectacular progress of the next decades, pushed by new needs of architects, engineers and contractors. Requirements, such as total durability of structures, and not only of concrete itself, is an area where a lot of work has to be done.

The new European norms EN 206 (7) assure durability, in relation to concrete exposure, but do not assure total durability of structures: water/cement ratio, strengths and minimum cement content do not assure that, once concrete is placed in the structure, it will resist aggressive environment attacks if concrete cracking occurs, or if impermeability is not adequate.

A new holistic approach to durability is necessary, which requires design, construction practice and materials properties to be considered more deeply. As to material properties, the following aspects must be optimized:

- Low water content
- Curing (external / internal)
- Shrinkage elimination
- Low cement content
- Cracks elimination
- Easy placeability
- Concrete-to-reinforcement bond
- Freeze-thaw cycles resistance
- Impermeability
- Alkali aggregate reaction inhibition
- Inhibition of reinforcement corrosion
- Chemical resistance

Admixtures can greatly contribute to solve or minimize the problems related to the above mentioned topics, which the admixture science must approach in a radical way, eliminating the negative effects. Admixture science is at the eve of a radical change.

Nanotechnology has introduced a new approach to the solution of concrete technology problems: the understanding of the relationship between structure of chemicals and its effect on cement behaviour and performances has created the basis for the solution of the above mentioned problems. Much work has still to be done, but the progress made with the development of Polycarboxylic Ether Polymers (PCE) has already shown this potential (8).

We already know a good deal on how:

- length of backbone chain
- length of side chains
- density of side chains
- electrical charges
- functional groups

influence properties, such as adsorption on cement particles, repulsion among particles, hydration kinetic of cement, paste density and paste rheology. We therefore have control on water/cement ratio, workability and workability retention, setting time and strengths development.

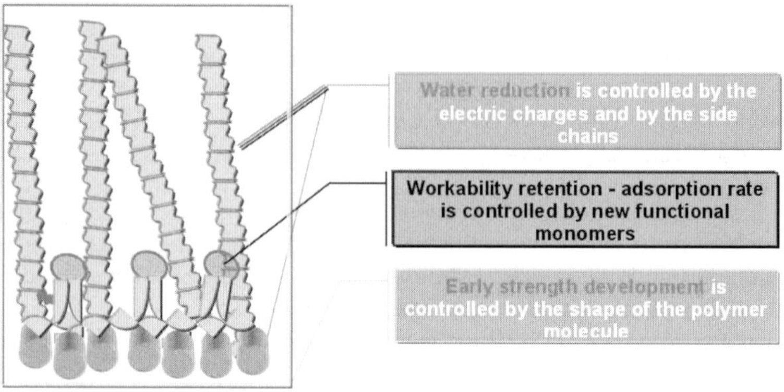

Investigation on interaction between polymers and cement hydration, especially at early age, allowed to understand the role of key components of cement, in particular: C_3A, in orthorombic and cubic phase, $CaSO_4$ in its various hydrated forms and alkalis content (9). This investigation allows cement composition and admixture polymer structure to be matched.

Cement composition becomes a critical factor for the best performances of admixtures and for the achievement of concrete advanced properties. Unfortunately, in recent years, cement has been subject to important changes due to environmental reasons which we will see later on; this creates a serious conflict between desired and achievable performances.

Concrete technology is anyway progressing, in spite of these problems, improving economy and friendliness of use of new, advanced types of materials. According to ERMCO statistics (10), Self Compacting Concrete represents only 0.5% of the total concrete produced by the ready-mixed concrete industry in Europe, in spite of self evident economical and productivity advantages. A US statistics among the construction industry members (11) shows that 18% of the interviewed do not use SCC because too expensive and 13% because too difficult and complex to control. An additional roadblock is the logistic complexity: additional silos, additional materials as fillers, transport organization. A great improvement in the diffusion of SCC would be the possibility of using ordinary sand and aggregate available at the job site, in addition to lower cost due to a reduced amount of cement and fine particles to meet the strengths requested by customers and engineers. Assuring the needed packing factor of the mix through appropriate concrete design, the use of engineered Viscosity Modifying Agent may allow to achieve the simplification and the economy required.

High Strength Concrete (HSC) has already delivered performances that allow to build structures over 400 mt. in height. But HSC still lacks for some key structural properties, like adequate flexural strengths and ductility. Recently developed Ultra High Performance Concrete greatly improved these properties (200 MPa compressive and 50 MPa flexural strengths, improved ductility at fracture), yet, its use still shows some serious drawbacks, due to

- material cost: approx 30 times more expensive than ordinary concrete
- logistic: to assure performances, pre-mixing of selected material at a central plant is needed
- very complex rheology makes concrete mixing extremely difficult by ordinary mixers.

The development of new, specific superplasticizers (the best ones are PCE-based superplasticizers), new packing models assuring full space filling, and engineered VMA will allow to achieve the targeted friendliness of this material. The availability of friendly and economical UHPC will allow to compete with steel in construction.

Material Properties of Steel and Ultra High Performance Concrete (UHPC)

Properties	*Steel (FeB 44K)*	*UHPC*
Behaviour	isotropic	anisotropic
Density	7.85	about 2.40
Coefficient of expansion	12 μm/m/°C	11.8 μm/m
Elastic modulus	210 GMpa	60 GMPa
Yield stress	275 MPa	150-220 MPa
Rupture stress	430 MPa	150-220 MPa
Elongation at failure	>24%	0.5 - 1.5%
Poisson's ratio	1	0.2
Flexural Strength	275 MPa	30-50 MPa
Fire resistance	Poor. It Needs protection	Good. Can be improved with poly propylene fibres

Even if steel shows some better properties with respect to UHPC, it is important to take into consideration the much lower specific weight of UHPC, which allows much lower weight at comparable strength. Considering the market price fluctuation of steel, UHPC has serious possibilities of becoming very competitive with steel in many major structures of today and tomorrow.

6. CONSTRUCTION PROCESS AND ECONOMIC IMPROVEMENtT

The traditional construction practice is heavily dependent upon plentiful building materials supply and labour and does not guarantee sustained business growth in the future. If the financial cost of a project keeps being directly related to the build starting time until the rental and / or sale income time begins, then it is clear that construction speed will continue to be the decisive factor in the choice of building options.

Pre-cast elements are the most effective way to improve building ability and the speed of construction. Admixture and plant producers have greatly improved the productivity of the pre-cast process. However, constructions based on pre-cast elements remain a minority compared to in-situ constructions.

In the large, in-situ constructions, all the process steps, like:

- concrete mixing

- transporting
- delivering
- placing
- form striking

have to be reviewed to reduce the time required by each step. Many of the above mentioned steps are speeded up by enhancement in mechanical equipments and by planning larger and continuous concrete pouring.

New concrete technology development will help to speed up all the steps of the process; in particular, great potential will come from rheology control that allows concrete to be distributed from a central mixing plant to different job sites by pumping and making truck mixers no longer the key transport system. Admixtures allowing extended workability retention to be obtained and setting time and strength development to start soon after placing, are now in the phase of experimental, preliminary testing and, if successful, will make form striking much faster. Admixture science has the potential and the basic knowledge to achieve all those targets.

7. ENERGY SAVING AND ENVIRONMENT PROTECTION

Unquestionably, green house gases production is growing much faster than the bearing capability of the planet eco-system. Comparing the CO_2 concentration worldwide in 2000 with respect to 1986, we clearly observe the dramatic worsening of the atmospheric composition, with a sharp CO_2 concentration growth all over the world.

The construction industry and, in particular, the cement industry signed the " Kyoto Protocol to the United Nations Framework Convention on Climate Change", and committed themselves to comply with the requirements of Article 3 which prescribes: "to reduce the overall emissions of the greenhouse gases by at least 5% below 1990 levels in the commitment period 2008 to 2012."

Considering the forecast that in the next 30 years cement consumption will double, the commitment to Kyoto Protocol means that much lower quantities of clinker should be used for each cubic meter of concrete. And, the same result can be obtained replacing cement with secondary cementing material (SCM) directly in concrete. Cement amounts to replace by SCM in concrete, in order to achieve zero increase in CO2 emission over the next 15 years, has been figured out to range from minimum 2% in North America - already blending fly ash with cement - to 55% in the Russian Federation,.

CO_2 discharge from fossil fuel consumption and cement production

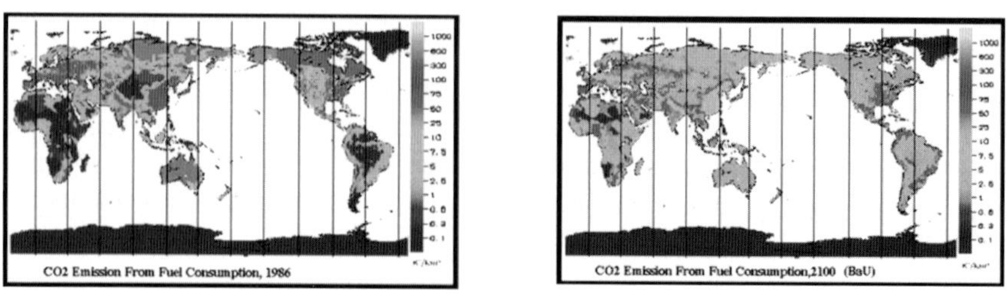

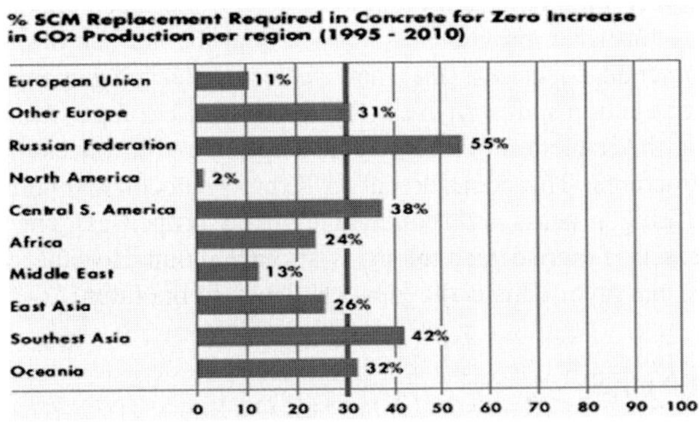

A credit in quantities of CO_2 emissions will be given to cement plants if and when they use a mix of secondary fuel materials. This benefit, being, in the end, a benefit for the whole community, includes the use of waste urban materials and industry by-products. A huge quantity of rubber tyres, domestic garbage, residual meat and bones from the food industry and chemical residues are already fed to the clinker kiln daily.

This great variety of secondary fuel generates temperature fluctuations in the kiln, thus originating a number of problems in the clinker composition consistency. The combined effect of secondary fuel materials, secondary cementing materials, progressive reduction of clinker quantity in concrete makes the quality consistency of concrete difficult to maintain. On the other hand, the very high performances requested by the new construction standards are in conflict with the lower cement quality. Cement performances always comply with National and European standards, but the same standards do not guarantee that many important cement requirements, in addition to setting time, strengths and mixing water quantities, are maintained at adequate levels. For example, the interaction cement / admixtures is very often variable and unpredictable.

Critical inefficiency is found in water reduction and workability retention, particularly when superplasticizers are used. Today, admixtures are recognized as an essential part of any concrete mix like water, cement, sand and aggregate, and therefore the above drawback is particularly serious. A new trend in the cement industry aimed at overcoming the inconvenience is to produce different types /qualities of clinker in order to obtain tailor-made cement according to the specific use it is addressed to. This would be a very effective way to overcome the conflicting trend between cement production with low environmental impact and the high and consistent quality of concrete delivered.

8. CONCLUSION

Unquestionably, the construction process and the materials have made impressive progress in the last 100 years. What has been built in the last decades was unimaginable before, and it was possible thanks to the use of advanced structural models, innovative construction techniques, new concrete properties and, last but not least, the excellent job of engineers, architects and chemists.

The race to higher and higher building, to more and more complex structures is not finished yet; many further improvements are clearly at the horizon of our future. However, the environmental concern is the new factor influencing many aspects of the future of the construction industry. New legislations will compel the search for new solutions allowing the reduction of energy consumption and lower environment impact of man-made structures. The commitment to Kyoto Protocol will heavily affect the cement manufacturing process and concrete ultimate properties. The obtainment of envisaged progress in concrete technology is strongly bound to the contribution the admixture science can give. This is the great challenge of the future.

9. REFERENCES

1. "Tests of Reinforced Concrete Beams" Technical Bulletin, Robert Talbot, University of Illinois, Urbana-Champaign, 1904
2. Corradi M., Khurana R. Magarotto R. "Total Performance Control: An Innovative Technology for Improving the Performances of Fresh and Hardened Ready Mixed Concrete" Proceedings of ERMCO, Helsinki, Finland, June 2004.
3. Khan, F.R. and Sbaronius, J.A. "Interaction of Shear Walls and Frames in Concrete Structures under Lateral Loads", Journal of the American Society of Civil Engineers, June 1964
4. Gann D.M., "Building Innovation" Thomas Telford Publishing, 2000
5. Bennet, D. "Innovations in Concrete", Thomas Telford Publishing, 2002
6. ECTP, Plenary Assembly European Construction Technology Platform, Paris, October 5, 2005.
7. EN 206-1 Standard, "Concrete - Part 1: Specification, performance, production and conformity", CEN European Committee for Standardization.
8. Corradi, M., Magarotto, R. "Chemical Nano Design to Engineer Intelligent Concrete Admixtures", RILEM Symposium on Nano Technology, Bilbao, Spain 2005
9. Flatt, R.J., Houst, Y.F., "A simplified view of effects perturbing the action of superplasticisers", Cement and Concrete Research, 2001, vol. 31, pages 1169-1176.
10. ERMCO Statistics, 2003
11. 4th RILEM Conference on SCC, Chicago, October 2005

MODELLING THE INFLUENCE OF SRA ON PROPERTIES OF HPC

V. López and A. Pacios
Fundación PRODINTEC, Parque Científico y Tecnológico, Gijón, 33203 Asturias, Spain; Departamento de Mecánica Estructural y Construcciones Industriales, ETSI Industriales, Univesidad Politécnica de Madrid, José Gutierrez Abascal 2, 28006 Madrid, Spain

Abstract: The present paper summarizes the study carried out in order to understand the influence of SRA (shrinkage reducing admixtures) on shrinkage and other mechanical properties of HPC (high performance concrete). A revision of different theoretical models for predicting shrinkage of HPC is made and an alternative prediction method based on ANN (artificial neural networks) is proposed. A comparison between results of these two kinds of methods is shown.

Key words: high performance concrete (HPC); shrinkage reducing admixtures (SRA); explicit models; artificial neural networks (ANN).

1. INTRODUCTION

HPC can be defined as concrete specifically designed for specific properties: low density, self compacting properties, special surface finish, high durability, high strength. The higher initial cost of these concretes, due to the use of special component materials and the need of previous testing, can be compensated with an improved durability and a longer service life. To obtain high durability, the matrix of the concrete should be very dense, with low permeability, and it will be also necessary to avoid cracking, especially at early ages, when the evolution of mechanical properties is faster.

It is known the influence of shrinkage of concrete on the formation of cracks. Shrinkage can be partially compensated by using expansive cements, and opening of cracks can also be limited by the addition of fibers. Another possibility is the use of SRA. These admixtures reduce the surface tension of the water in the mix, which causes a decrease of the capillary stresses formed in the porous microstructure, and a global less important reduction of volume.

Existing models for shrinkage and other mechanical properties stimation of HPC, do not normally perform well (ACI 209, 2000; ACI 318, 2000; ACI 362, 2000; Bazant et al.,

M.S. Konsta-Gdoutos, (ed.), Measuring, Monitoring and Modeling Concrete Properties, 441–447.

1991; Bazant et al., 1993; CEB, 1993), since, they do not consider the effect of specific component materials, as for example SRA for example, and they do not take into account the faster evolution of properties at early ages. These models (explicit models), use mathematical formulae to relate known and unknown variables by means of coefficients calculated from statistical studies over a certain population. The characteristics of the population (component materials, dosages, ages, environmental conditions…) will limit the valid range for application of the model.

As an alternative to explicit methods, ANN can be used as implicit models to predict mechanical properties (Konderla et al., 2000). With ANN there is no need to assume any previous mathematical relationship between variables. These networks are formed by a number of simple elements (neurons) that respond to a certain input according to a mathematical function, giving an output that can be used as an input to a new neuron.

To obtain a good model with ANN, a training process must be carried out. In this process, a set of examples (input data) with know outputs (target outputs) is presented to the ANN. The network then uses these relationships between known inputs and outputs, to adjust the weights of the internal connections between neurons. The objective of this training process is to minimize errors between the network outputs and the target outputs. After training, the ANN is able to generalize rules and will be able to respond to previously unseen input data to predict a more or less accurate output (depending on the quality of the training: ranges for input variables, number of iterations,…), within the domain covered by the training examples.

2. EXPERIMENTAL PROGRAM

To understand the influence of SRA on shrinkage and mechanical properties of cement pastes, mortars and concrete, different mixes were tested. Studied properties shows the studied properties along with the test methods used. Component materials are shown in Component material used.

The mortar mixes had a water/cement ratio of 0.35. Cement dosage was always 450 kg/m^3. SRA was tested at dosages of 1, 1.5 and 2%, always with respect to cement weight, and SP was optimised to not modify workability. Sand/cement ratio was always 3:1. Concrete proportions include coarse aggregate amount of 951,8 kg/m^3, and coarse aggregate/sand ratio is 1:1. Dosage of SRA for concretes presented is 1.5%, as this proportion was considered representative of the normal use for industrial production. The slump for all concrete batches was greater than 15 cm and compression strength was 52.5 MPa for concrete without SRA and 49.5 MPa for concrete incorporating SRA (measured at 28 days).

2.1 Shrinkage of mortars and concrete

The specimen's geometry for mortars shrinkage testing was in accordance with ASTM C490 1" x 1" x 11¼"). Steel gage studs were placed in the specimens at the time of casting to perform the measurement with a manual comparator of 0.01 mm of accuracy. After mixing, the mixture was placed in the molds, vibrated, and then sealed with a plastic film.

Table 1. Studied Properties

Study	Properties	Experimental procedure
Shrinkage of mortars	Autogenous and total shrinkage	ASTM C490, (2001)
Shrinkage of concretes	Autogenous and total shrinkage	TI-B 102 (95), (1996)
Microstructure	Total porosity and pore size distribution	Mercury Intrusion Porosimetry
Mechanical performance	Compressive strength Modulus of elasticity	UNE 83304:1984, (2000) UNE 83316:1996, (2000)

Table 2. Component material used

Cement		CEM I/42,5R
SRA		Glycol ether blend (S.R. = 0% = 0.93 g/ml)
Superplasticizer (SP)		Synthetic modified melamine: SR = 40%; = 1.23 g/ml
Sand	Mortar	Standard sand (CEN UNE-EN 196-1)
	Concrete	Siliceous sand (ϕ_{max} = 4.5 mm ; Fms = 3.4)
Gravel		Siliceous gravel (ϕ_{max} = 20 mm; Fmg = 5.7)

A total of eight specimens were cast for each mix (four for measurement of autogenous shrinkage and four for total shrinkage). At approximately 18 hours, they were demoulded and moved to a chamber with constant ambient conditions (20°C y 50% RH) where the first measurements were taken. Prior to the first measurement, the specimens for autogenous shrinkage were immediately wrapped with several layers of plastic film to avoid humidity exchange with environment.

For testing concrete shrinkage, two cylindrical specimens (700 mm length and 140 mm diameter) per batch were cast and moved to a chamber with constant environmental conditions (20 ± 2 ºC and 50% RH). Specimens used for autogenous shrinkage were kept in the mold, starting the measurement from 4 to 6 hours after casting. Specimens used for total shrinkage were demoulded from 20 to 24 hours after casting, and the measurement started at that time. In all cases, the internal temperature in one of the two specimens was registered, from casting time. For each specimen two LVDT's, of 5 mm range and 3 m of accuracy with a gage length of 400 mm, were placed at approximately half of the length of the specimen. Temperatures were measured with embedded T type thermocouples.

2.2 Microstructure of mortars

The microstructure properties were studied on mortar specimens. A small sample of one of the specimens used for the measurement of both, autogenous and total shrinkage, was tested by Mercury Intrusion Porosimetry. Among the results that can be obtained with this technique are total porosity and pore size distribution (PSD).

2.3 Mechanical properties of concrete

The concrete mixes used for measurement of mechanical properties, were the same that the ones used for measurement of shrinkage. Properties were obtained at 1, 3, 7, 28 and 56 days. Testing methods, including curing conditions for specimens, were in accordance to Spanish standards referred in Studied properties.

3. RESULTS

3.1 Shrinkage of mortars and concrete

SRA effectively reduce shrinkage of both mortars and concrete, in different amounts, depending on the environmental conditions to witch specimens have been exposed (isolated or allowing exchange of humidity with ambient). This can be seen from Figures 1 and 2.

3.2 Microstructure of mortars

Figures 3 and 4 represent total volumetric porosity for mortars. Figure 3 includes specimens used for the measurement of autogenous shrinkage, while Figure 4 shows specimens used for measurement of total shrinkage.

Decreases of total volumetric porosity and changes in PSD's are observed when adding SRA. A refinement of the porous microstructure is obtained because of the SRA, as can be observed from the figures. From the total volumetric porosity, the proportion corresponding to smaller diameters increases when SRA is used.

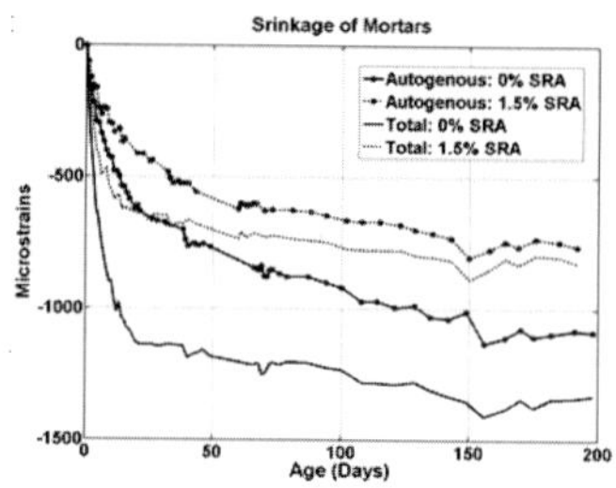

Figure 1. Shrinkage of mortars: effect of SRA on isolated and exposed specimens

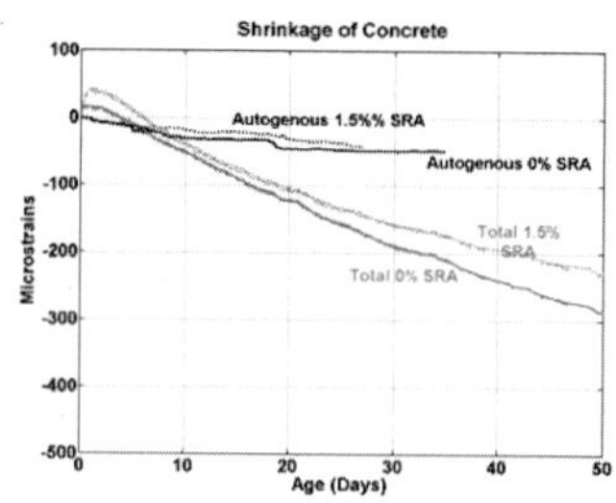

Figure 2. Shrinkage of concrete: effect of SRA on isolated and exposed specimens

3.3 Mechanical properties of concrete

A decrease of compressive strength is observed when adding SRA, from about 2% at 1 day to 10% at 56 days. With regard to modulus of elasticity, the effect of SRA is to increase it, resulting in stiffer concretes (from 6% increase at 1 day to 4.5% increase at 56 days).

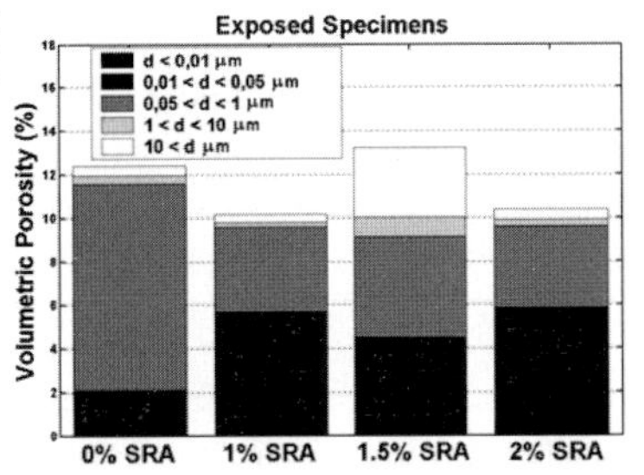

Figure 3. Total volumetric porosity and PSD for mortars sealed from environmen

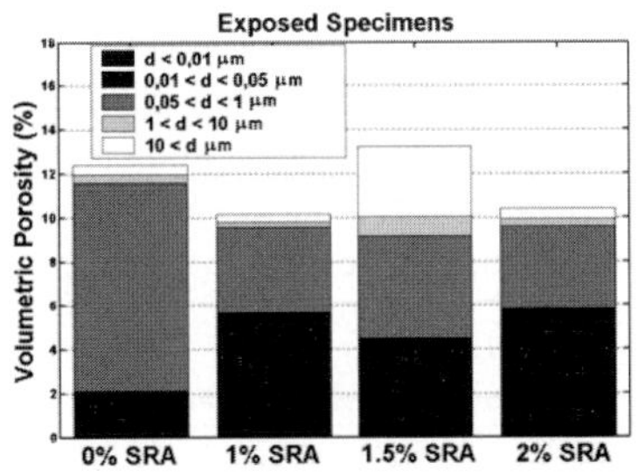

Figure 4. Total volumetric porosity and PSD for mortars exposed to environment

4. COMPARISON OF SHRINKAGE RESULTS WITH SHRINKAGE MODELS AND ANN

To check the ability of shrinkage models to predict shrinkage of HPC, obtained results where compared with ACI model (ACI, 2000), CEB (CEB, 1990), and BP-KX model (Bazant et al, 1991; Bazant et al, 1993). Experimental results and predictions from these models are also compared with predictions from ANN.

For the definition and training of the different ANN tested, the following set of input variables was chosen: 1) Age of specimen (days); 2) Size of the specimen (mm), represented as the mean thickness: 2Ac/u (Ac = cross sectional area; u perimeter); 3) SRA dosage (l/m3); 4) Aggregates (%) (total volume of sand over total volume of the specimen); 5) Type and binder amount (kg/m3); 6) Characteristics of the paste, represented as the amount of water (l/m3) and volume of paste (% of volume of the specimen); 7) Environmental conditions, represented as the RH (%) of drying, since the time and ambient conditions of the curing were the same for all tests.

From the total results of experimental program, 37.5% were used for training, 3.3% for the validation of generalization capability, and 59.2% were finally used for testing the accuracy of the ANN. For each tested training algorithm, the initial ANN was built with only one hidden layer. When, after several tests with different conditions (number of epochs, number of neurons in hidden layer...), a good estimation was not obtained, a second hidden layer was added. Several combinations of number of neurons in the first and second hidden layer were tried, and each of these combinations was tested with the algorithms and training conditions used with the ANN of one hidden layer. Transfer function for all hidden neurons was tansig, and for the output layer the identity function (y=x) was used.

In Figures 5 through 8 comparisons between experimental results, models and predictions from ANN are shown. After training and testing different networks, best results were obtained with a ANN of two hidden layers, with 4 neurons in the first layer, and 2 neurons in the second.

With regard to autogenous shrinkage, closest predictions to experimental results were obtained with BP-KX model. ACI model performed also quite well.

For total shrinkage, the best model was the ACI. The predictions from the ANN were worse than those from the other models for autogenous shrinkage, while for total shrinkage it gave good results, mainly for concrete without SRA. It must be noted that the evolution of shrinkage is best predicted with the ANN (rate and slope). Difference on shrinkage values are due to bad estimation of shrinkage at early ages. To compensate this, more shrinkage results at these ages should be used in the training of ANN.

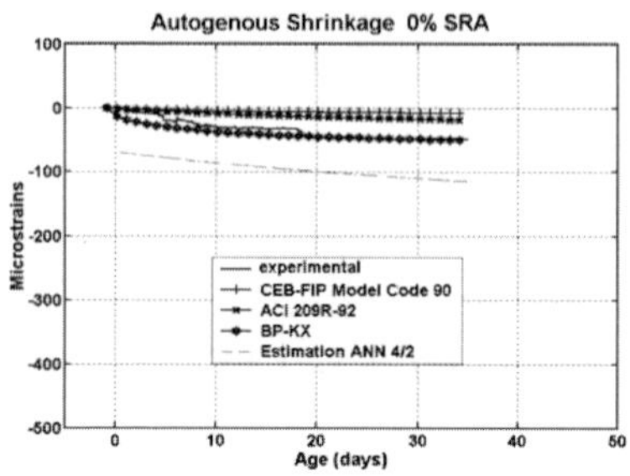

Figure 5. Experimental results, models and ANN (0% SRA, autogenous shrinkage)

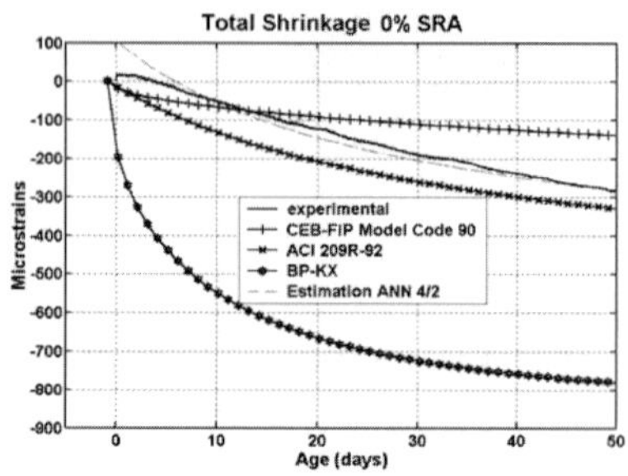

Figure 6. Experimental results, models and ANN (0% SRA, total shrinkage)

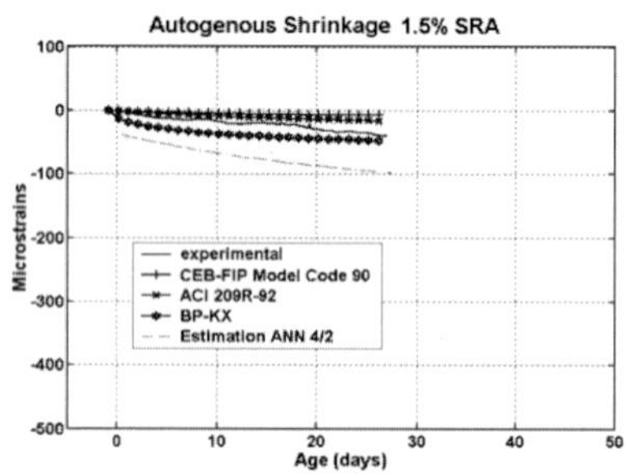

Figure 7. Experimental results, models and ANN (1.5%SRA, autogenous shrinkage)

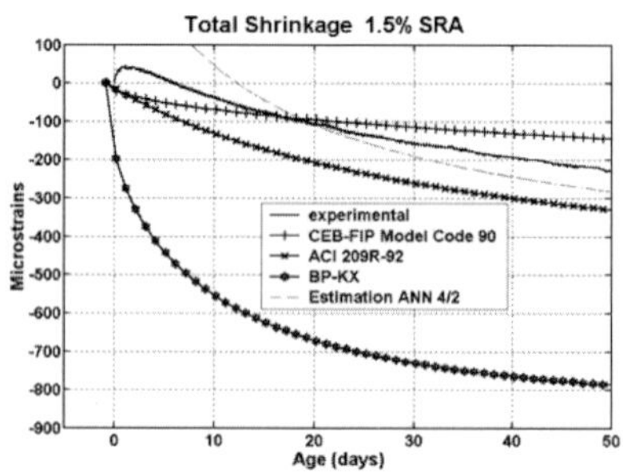

Figure 8. Experimental results, models and ANN (1.5% SRA, total shrinkage)

5. CONCLUSIONS

- SRA reduces total volumetric porosity and refines the microstructure (proportion of smaller diameters is more important).
- SRA reduces compressive strength, and makes concretes stiffer.
- SRA reduce shrinkage of cement based materials from early ages. The total reduction depends, apart from the dosage of SRA, from the relationship between total volume of paste and the ambient conditions.
- ANN can serve as an alternative to predict properties like shrinkage, influenced by several variables, as long as data used for training are representative enough of the mixes and ambient conditions.

6. REFERENCES

1. ACI Committee 209. 2000, Prediction of creep, shrinkage and temperature effects in concrete structures, American Concrete Institute, Detroit.
2. ACI Committee 318. Building code requirements for structural concrete, 2000, American Concrete Institute, Detroit.
3. ACI Committee 362. State-of-the-art report on high strength concrete, 2000, American Concrete Institute, Detroit.
4. Bazant Z.P, Kim J.K, Panula L., 1991, Improved prediction model for time-dependent deformations of concrete: part 1 – shrinkage, *Materials and Structures,* **24**(144): 409-421
5. Bazant Z.P, Xi Y., Carol I., 1993, Preliminary guidelines and recommendations for characterizing concrete creep and shrinkage in structural design codes, in: *5th International RILEM Symposium of Creep and Shrinkage of Concrete*, E & FN Spon, London.
6. CEB-FIP Model code 90, 1993, Comité Euro-International du Béton, Thomas Telford, London.
7. Konderla, P., Mokanek, T., 2000, Comparison of two methods for the analysis of composite material, *Journal of Materials Processing Technology*, **106**.

A STUDY OF THE INTERACTION BETWEEN VISCOSITY MODIFYING AGENT AND HIGH RANGE WATER REDUCER IN SELF COMPACTING CONCRETE

N. Prakash and M. Santhanam

Research Scholar; Assistant Professor, Department of Civil Engineering, Indian Institute of Technology, Madras, Chennai – 600036, India

Abstract: This study pertains to the interactions between a water soluble polysaccharide VMA and two types of HRWRs (one based on sulfonated naphthalene formaldehyde – SNF - and the other on polycarboxylic ether – PCE - technology) in an SCC system. The experimental investigation includes the evaluation of cement paste rheological parameters using a rheometer as well as empirical laboratory tests on flowability of cement paste produced using these admixtures. Results indicate that while both types of HRWRs are compatible with the VMA used (Welan gum), the pseudoplastic nature of the VMA and thixotropy of paste depends on the type of HRWR used.

Keyword: HRWR; VMA; SCC; viscosity; rheology.

1. INTRODUCTION

Self-compacting concrete has to fulfill contradictory requirements of high flowability when it is being cast, and high viscosity when it is at rest, in order to prevent segregation. These requirements make the use of mineral and chemical admixtures essential for SCC. High flowability is achieved using High Range Water Reducer (HRWR), while stability against segregation is achieved either by using a large quantity of fine materials, or by using an appropriate viscosity modifying agent (VMA).

Common VMAs in concrete include microbial polysaccharides (such as Welan gum), cellulose derivatives (methyl cellulose), and acrylic polymers. The mechanism of action in each case is different. In the case of Welan gum and cellulose, the long chain polymer molecules adhere to the periphery of water molecules, thus imbibing and fixing a part of the mixing water. Some other VMAs adsorb on cement particles and increase viscosity by promoting inter-particle attraction. The advantage of using VMA is that the fines content

M.S. Konsta-Gdoutos, (ed.), Measuring, Monitoring and Modeling Concrete Properties, 449–454.

(and paste content) in SCC is at the level of normal concrete (as a result, the potential for creep and shrinkage of SCC is reduced). Current research shows that SCC produced with low powder content and VMA had similar fresh concrete properties as SCC with high powder contents produced without VMA (Lachemi et al., 2003).

The presence of two dissimilar chemicals having different mechanisms of action could lead to a number of issues of incompatibility that affect the properties of the resultant SCC. Because of the complex nature of this system, understanding the exact mechanism of any incompatibility issue is difficult. There have been some attempts by researchers to unravel this problem.

For underwater concrete applications, lignosulfonates and melamine formaldehyde based HRWRs have been known to perform well with VMAs, while naphthalene formaldehyde admixtures show erratic results (Saucier and Neely, 1987). Sometimes superplasticizers are blended with retarders or lignosulfonates (which may have sugar in them), for workability retention in hot weather conditions. When a VMA is used along with such blended superplasticizers, concrete may experience setting problems. Most VMAs tend to entrain air at high dosages because of their inherent surfactant properties (Rixom and Mailvaganam, 1999). In combination with lignosulfonates or SNF, the air entrainment could be substantial, leading to lowered strengths.

Lachemi et al. (2004) reported that viscosity of cement mortar increased while yield stress decreased with the increase in the content of polysaccharide based VMAs (at a fixed dosage of sulfonated naphthalene formaldehyde – SNF - based HRWR). All VMAs were found to have a shear thinning behaviour. The study on self-compacting concrete showed that bleeding and segregation were dependent on combinations of HRWR and VMA, and not just on the VMA alone.

Khayat and Yahia (1997) found that the apparent viscosity (at both low and high shear rates) increased with the dosage of VMA, irrespective of the content of (SNF based) HRWR. Shear thinning behaviour was more pronounced at high quantities of VMA and at low concentration of HRWR. Aligning of VMA chains in the direction of flow at high shear rates was explained as the reason for the shear thinning. Setting time was more affected by VMA dosage. In a rheometric study of SCC equivalent mortars, Ghezal and Khayat (2003) were able to show that the efficiency of VMA depends on the class of superplasticizer used (that is, whether it was a copolymer, polysulphonate, or polycarboxylate). Yammamuro et al. (1997) reported that non-adsorptive viscosity agents provided better compatibility with HRWR, while adsorptive VMAs competed with the HRWR for adsorption sites on the cement, resulting in reduced efficacy of the HRWR.

The current study is based on the evaluation of the combined effects of two types of HRWRs – PCE (polycarboxylic ether) based and SNF based – and Welan gum, a microbial polysaccharide VMA.

2. EXPERIMENTAL METHODOLOGY

2.1 Materials used and mixture proportions

Ordinary Portland cement (specific gravity 3.14 and Blaine 325 m^2/kg) and Type F fly ash (specific gravity 2.13 and Blaine 500 m^2/kg) from Ennore thermal power station (north of Chennai, India) were used for the study. Two HRWRs were used – one based on polycarboxylic ether (PCE – 33 % solids) and the other based on sulfonated naphthalene formaldehyde (SNF – 44% solids). A microbial polysaccharide VMA (Welan gum) was used to study the interaction effects.

The proportions of ingredients for the SCC paste were chosen based on previous investigations in the laboratory. Cement and fly ash were used in the ratio 60:40, while the water to powder ratio by volume was kept at 1.0. The type and dosage of HRWR, and the dosage of Welan gum, are listed in Table 1.

Table 1. Mixture proportions for pastes

Mix ID	HRWR SNF (%)	PCE (%)	Welan Gum (%)	Remarks
SNF0	0.44	-	0	No bleeding
SNF1	0.44	-	0.02	Air bubbles; No bleeding
SNF2	0.44	-	0.03	Air bubbles; No bleeding
SNF3	0.44	-	0.04	Air bubbles; No bleeding
PCE0	-	0.33	0	Bleeding
PCE1	-	0.33	0.02	No bleeding
PCE2	-	0.33	0.03	No bleeding
PCE3	-	0.33	0.04	No bleeding

Note: All admixtures are dosed on % solids by weight of cement basis
SNF and PCE dosage based on previous experience with SCC in the laboratory

2.2 Mixing and Testing

The ingredients for the paste were mixed in a Hobart mixer as per two different sequences, primarily to evaluate the effects of batching the HRWR and VMA in a given order. These are designated as SEQ 1 and SEQ2, respectively. The general procedure for preparing the pastes is as follows.

1. All the mix water was first added to the bowl
2. In SEQ1, the HRWR was added to the mix water and the mixer was run at low speed for 30 sec. The VMA was then added and mixing continued for 30 more seconds. In SEQ2, VMA was added first, followed by HRWR (30 sec mixing each time)
3. Cement and fly ash, mixed together, were then added to the bowl, and the mixing continued for 2 min at low speed

The Marsh cone and mini-slump (Kantro, 1980) tests were immediately performed on the fresh paste (within 5 min of mixing). 500 ml of fresh paste (mixed as per SEQ 1) in a beaker was used for the measurement of rheology using a Brookefield Viscometer. In this test, the shear rate was first ramped up from 10 RPM to 100 RPM, and viscosity was measured every 10 RPM for a period of 15 sec. Following this, the viscosity was again measured as the shear rate was ramped down from 100 to 10 RPM in steps of 10 RPM.

3. RESULTS

The subjective remarks provided in Table 1 indicate that the SNF – Welan gum combination could have generated extra air (although this was not quantitatively evaluated). In the case of PCE – Welan gum combination, no such effect was seen.

Figures 1 and 2 depict the results of the Marsh cone and mini slump tests conducted on different mixes (mixed using two differing batching sequences). A study of Figures 1 and 2 shows that the batching sequence in this study was not found to have any influence on the fresh paste behaviour. As expected, the flow properties were better for PCE pastes, as compared to the SNF pastes. However, the use of VMA produced interesting modifications in the flow behaviour of these two pastes. In the case of pastes with PCE, there was a uniform increase in flow time and decrease in spread, with an increase in VMA dosage. However, in the case of SNF, although the flow time increased in the Marsh cone test, a significant drop in the mini slump spread was not seen with an increase in VMA dosage. This indicates that flow in SNF-Welan gum mixtures would not be greatly affected by the dosage of Welan gum (the effect of SNF dosage would need to be evaluated). In terms of rheology, this would mean that the shear yield stress for mixes with SNF would be less affected by the dosage of VMA.

Even in the case of PCE pastes, there was no significant change in the mini slump spread when the VMA dosage was increased from 0.02 to 0.04%. In both series of pastes, however, there was a uniform increase in the Marsh cone flow time (which corresponds to viscosity).

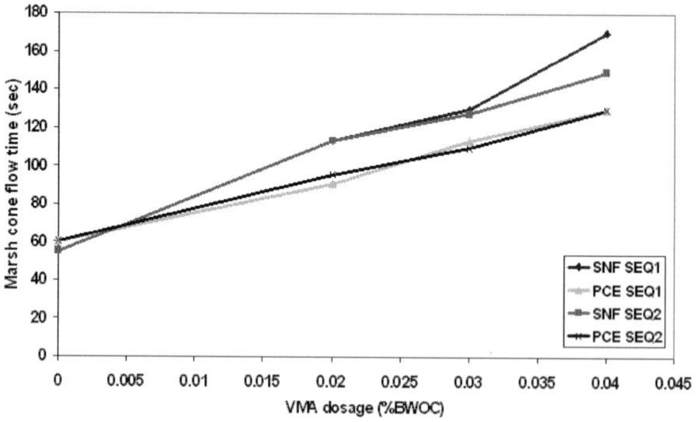

Figure 1. Marsh cone flow time for different paste mixes

Results from rheological measurements are presented in Figures 3 to 6. In the case of pastes with SNF (Figures 3 and 4), it is evident that the ramping up and ramping down curves are similar, indicating the absence of any thixotropic behaviour. The viscosity of the pastes is nearly the same on both the ascending and descending branches. There is clear evidence of a shear thinning (pseudoplastic) behaviour for pastes with VMA (SNF1, SNF2, and SNF3). In the case of pastes with PCE (Figures 5 and 6), the ascending and descending curves are well separated, indicating thixotropy. While a shear thinning response is evident on the ascending branch, there is a shift to a linear (Bingham/Newtonian) response on the descending curve. Except at low shear rates, the viscosity is nearly constant when the rate is ramped down. This is generally expected from most SCC mixes.

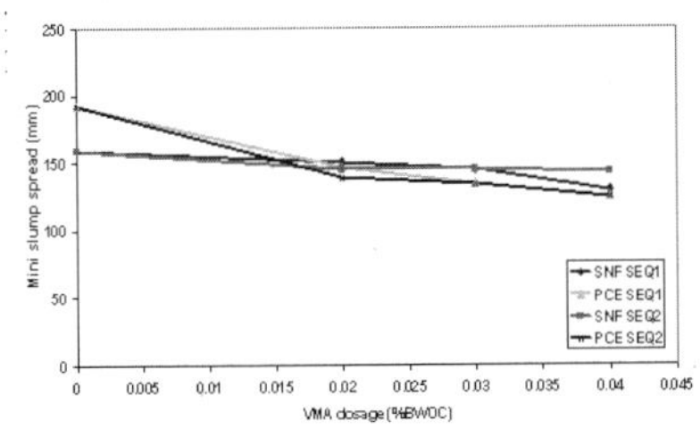

Figure 2. Mini slump spread for different paste mixes

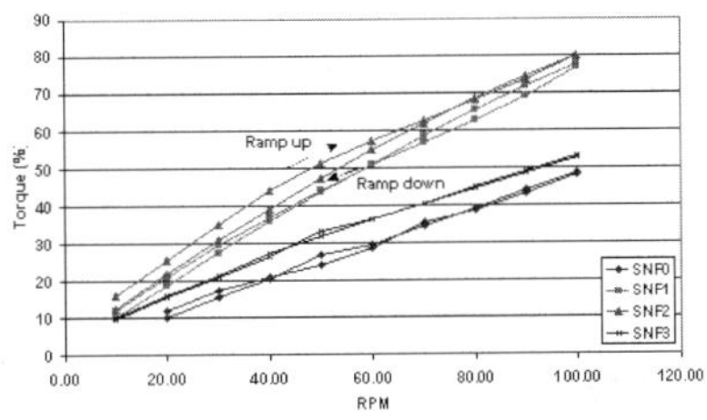

Figure 3. Rheological response of pastes with SNF HRWR

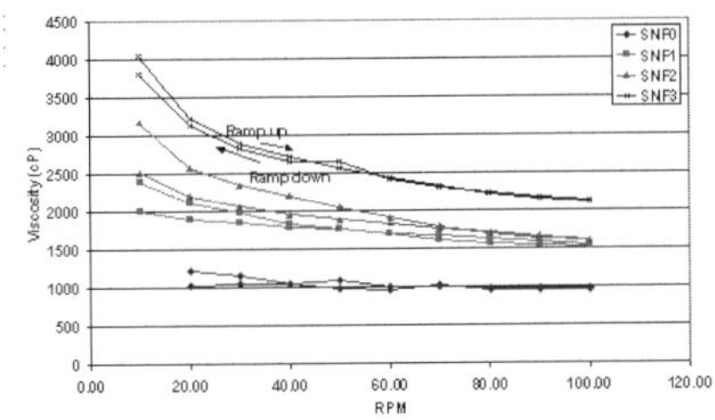

Figure 4. Viscosity measurement of pastes with SNF HRWR

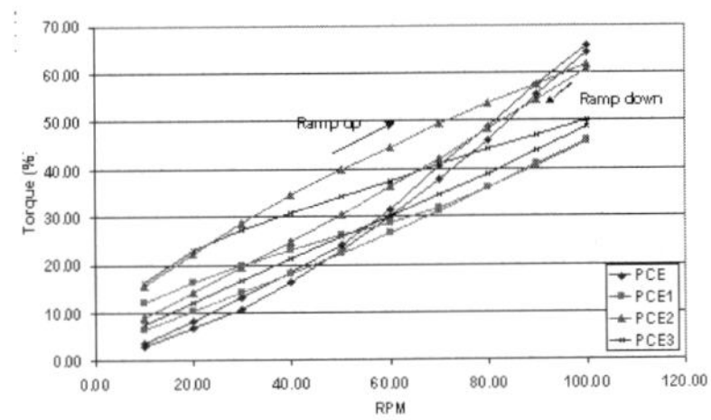

Figure 5. Rheological response of pastes with PCE HRWR

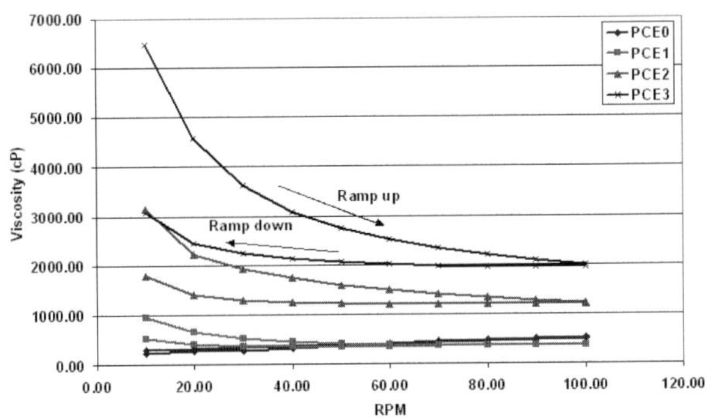

Figure 6. Viscosity measurement of pastes with PCE HRWR

4. CONCLUSIONS

Although some qualitative evidence of incompatibility (air entrainment) was seen for the combination of SNF and Welan gum, it is safe to say that the flow properties of SCC pastes produced with combinations of SNF/PCE and Welan gum are satisfactory. Increasing dosage of Welan gum only results in the increase of viscosity, without compromising on the flowability (within the dosage range tested in this study).

Rheological studies showed that while the PCE-Welan gum combination showed evidence of thixotropy, the same was not observed for SNF-Welan gum combinations.

Acknowledgments

The authors would like to thank the Department of Science and Technology, Government of India, for their financial support under the SERC: Fast Track Proposals program. The authors also express their sincere appreciation for the help of Dr. Murali Krishnan, IIT Madras, for conducting the rheology studies.

5. REFERENCES

Ghezal A.F. and Khayat K.H., 2003, Pseudoplastic and thixotropic properties of SCC equivalent mortars made with various admixtures, *Proceedings of the 3rd International RILEM Symposium on Self-Compacting Concrete,* O. Wallevik and I. Nielsson, Ed., RILEM Publications, Paris, pp. 69 – 83.

Kantro D.L., 1980, Influence of water reducing admixtures on properties of cement pastes - a miniature slump test, *Cement Concrete Aggregates* **2**:95-102.

Khayat K.H. and Yahia A., 1997, Effect of Welan gum – high range water reducer combinations on rheology of cement grout, *ACI Mat J* **94**(5):365 – 372.

Lachemi M., Hossain K.M.A., Lambros V., and Bouzoubaa N., 2003, Development of cost-effective self-consolidating concrete incorporating fly-ash, slag cement, or viscosity-modifying admixtures, *ACI Mat J* **100**(5).

Lachemi M., Hossain K.M.A., Lambros V., Nkinamubanzi P.-C., and Bouzoubaa N., 2004, Self-consolidating concrete incorporating new viscosity modifying admixtures, *Cem Concr Res* **34**:917 – 926.

Rixom R. and Mailvaganam N., 1999, *Chemical Admixtures for Concrete*, E&FN Spon, UK, 437 pp.

Saucier K.L. and Neely B.D., 1987, Antiwashout admixtures in underwater concrete, *Concrete International* **9**(5).

Yammamuro H., Izumi T., and Mizunuma T., 1997, Study of non-adsorptive viscosity agents applied to self-compacting concrete, *5th CANMET/ACI International Conference on Superplasticizers and Other Chemical Admixtures in Concrete*, ACI SP-173, Detroit, pp. 427 – 444.

EARLY HYDRATION OF CLINKER PHASES ANALYZED BY SOFT X-RAY TRANSMISSION MICROSCOPY: EFFECTS OF VISCOSITY MODIFYING AGENTS

D.A. Silva and P.J.M. Monteiro

Grace Brasil Ltda., R&D Cement Additives, Sorocaba, SP, Brazil; Department of Civil and Environmental Engineering, University of California at Berkeley, U.S.A.

Abstract: Along with superplasticizers, viscosity modifying agents (VMA) such as cellulose ethers and natural gums are being used for the production of self-leveling concretes in order to provide stability to the concrete during pumping and casting. Besides affecting the properties of fresh concrete, VMAs affect the early hydration reactions of portland cement. Soft X-ray transmission microscopy is a new technique used for the analysis of wet cement samples. It allows the *in situ* observation of anhydrous phases dissolution and precipitation and growth of hydrates since few minutes after mixing with water, under atmospheric pressure. The effects of the cellulose ether hydroxypropyl methylcellulose (HPMC) and welan gum on the hydration of C_3S and C_3A were analyzed by such technique since 8 minutes until more than 2 hours after mixing with water. The images have shown that both agents delay the dissolution and precipitation steps of cement hydration. Moreover, they change the overall aspect of hydration of both phases, promoting rather the formation of inner than outer products. However, the effect of the VMAs on the hydration of C_3A has shown to be more important, because the formation of ettringite needles was hindered.

Key words: viscosity modifying agents; C_3S; C_3A; hydration; soft X-ray transmission microscopy; HPMC; welan gum.

1. INTRODUCTION

Hydroxypropyl methylcellulose (HPMC) and welan gum are water soluble polymers used for the production of self leveling concretes in order to promote stability to the mixing, to avoid sagging and bleeding, and to improve cohesion. HPMC is a nonionic, semi-synthetic polymer obtained by the partial substitution of the hydroxyls in the cellulose

M.S. Konsta-Gdoutos, (ed.), Measuring, Monitoring and Modeling Concrete Properties, 455–461.

skeleton with hydrophobic groups. Welan gum is an anionic, biologic polymer produced by a controlled aerobic fermentation process. Welan gum has the sugar backbones replaced with sugar side chains (Lachemi et al., 2004). Both are long chained, high molecular weight polymers, which bind to water molecules by hydrogen bonding, reducing the availability of water and increasing the viscosity of the solution. The resulting solutions manifest pseudoplastic behavior (Khayat, 1998).

Because the polymers dramatically change the concrete rheology, techniques that permit the assessment of their effects when the concrete is still plastic are of utmost importance for the understanding of the microstructure and behavior of the hardened concrete.

A new, powerful microscopic technique is being used to analyze the dissolution of cement grains and the formation of hydrates as early as few minutes after mixing with water. The effects of several chemical admixtures on the hydration mechanisms of clinker phases are being assessed (Silva and Monteiro, 2005a, 2005b; Juenger et al., 2003; Juenger et al., 2005). The soft x-ray transmission microscope operates at the E.O. Lawrence Berkeley National Laboratory (LBL, Berkeley, CA, USA) and uses radiation in the wavelength range of soft x-rays generated by a synchrotron source (Advanced Light Source, ALS). The great advantages of the microscope are the low energy of the radiation (less than 543 eV), which does not damage the cement hydrates, and the high resolution and magnification (40nm and 2400x, respectively). Moreover and especially, the samples can be analyzed under atmospheric pressure and room temperature, which means there is no need for drying the sample. Cement particle are thus analyzed when surrounded by water. The images obtained clearly show nucleation and growth of crystals, which are embedded in the aqueous phase.

This research investigated the effects of HPMC and welan gum on the hydration of C_3S and C_3A by soft x-ray transmission microscopy during the first 3 hours of hydration.

2. EXPERIMENTAL

Commercial hydroxypropyl methylcellulose (Methocel 228, Dow Chemical) and welan gum (Kelco-crete®) were used for this investigation. Characteristics of the cellulose ether are presented in Table 1. Both polymers are provided as water-soluble powders.

The hydration of C_3S and C_3A particles embedded in solutions containing HPMC or welan gum was analyzed using the soft x-ray transmission microscope. Instead pure water, we used a solution saturated with respect to calcium hydroxide and gypsum in order to simulate the mixing water of concrete during the first few hours of hydration. The solution was previously prepared with fresh boiled, de-ionized water inside a glove bag filled with N_2 gas to avoid carbonation. Polyethylene and Teflon flasks and test tubes were used instead of glass to avoid alkali-silica reaction.

Table 1. Chemical composition and characteristics of HPMC[a]

Hydroxypropyl methylcellulose	85-99%
Water	1-10%
NaCl	0.5-5%
Proprietary polyglycol, carboxylic acid, and aldehyde	< 5%
Surface tension at 25C, 0.05% solution	43-55 dynes/cm
Specific gravity at 4C, 1% solution	1.0012
Typical viscosity (Brookfield RVT, 20 rpm, 20C, 2% solution)	3,800 mPa.s
pH, 2.5% solution[b]	4.26

[a]characteristics informed by the manufacturer
[b]characteristic determined by pHmeter

The concentrations of cement particles (C_3S or C_3A) and polymers in the solution were 0.2 and 0.002 g/cm^3, respectively, corresponding to a water/cement ratio of 5 and a polymer/cement ratio of 0.01, respectively (weight ratio). HPMC and welan gum were mixed to the solution and allowed to rest for 10 minutes prior the addition of the cementitious particles to assure complete dissolution by the time of mixing.

Because of the sample thickness restriction (10 μm) and the need of enough space to permit the transmission of the radiation, samples should be highly diluted, and solid particles should be not larger than a few micrometers. Centrifugation of the samples for 30 seconds is needed in order to attend such requirements. After centrifugation, a 2μl droplet is taken from the supernatant and squeezed between two silicon nitride windows for the analysis. The hydration was observed from less than 10 minutes to more than 2.5 hours after mixing.

The soft X-ray transmission microscope (XM-1) is operated by the Center of X-rays Optics (CXRO) at the beamline 6.1.2 of the Advanced Light Source (ALS) facility, in the LBL. The energy of the radiation was set in 517 eV (2.4 nm wavelength), which assures the transparency of the water to the x-rays and the contrast of the cementitious phases to the aqueous phase. Kurtis (1998) and Atwood (1999) provide further details about the sample preparation, the optics features of the synchrotron radiation and the x-ray microscope.

3. RESULTS

Figure 1 shows the soft x-ray images of C_3S hydrating in the calcium hydroxide-gypsum saturated solution containing HPMC. During the first minutes of hydration, thin and short hydrated fibers grow outwards from the surface of the particles. The reactions then slow down and burst again at approximately 1 to 1.5 hours after mixing, leading to a fast dissolution of the anhydrous particles and to the development of more hydrated fibers between them. The dissolution is observed to develop inwards the surface of the C_3S particles, followed by the formation of inner products (bright regions developing inside the original boundaries over time).

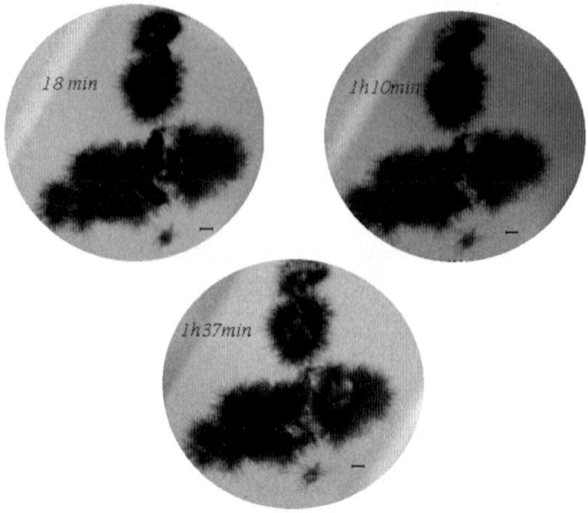

Figure 1. Soft x-ray micrographs of C_3S particles in the solution containing 0.2% HPMC. Hydration time is indicated. Scale bar is 700 nm

The aspect of the C_3A particles hydrating in the saturated solution with HPMC is shown in Figure 2. Fiber-shaped hydrates formed on the surface of C_3A at early contact with the saturated solution. Some growth of hydration products was observed to occur up to around 20 minutes after mixing. From this point, the hydration strongly slowed down and did not develop further by the end of the analysis (3 hours after mixing), differently from the hydrating C_3S particles. For both C_3S and C_3A systems, the hydrates only precipitated near the surface of the dissolving particles, contrarily to pure C_3S and C_3A systems, where many crystals precipitated also far away from the original anhydrous particles.

Figure 2. Hydration of C_3A particles in the saturated solution with 0.2% HPMC. Hydration time is indicated. Scale bar is 1.15μm

Figure 3 shows images of C_3S particles hydrating in the solution with welan gum. The particles are partially or completely covered by a discontinuous layer of a gel-like product. Also interesting is the hydration that occurs mainly inside the boundaries of the original particle. The inner product is brighter than the anhydrous particle, evidencing a much lower density. Very few, thin and short fibers of C_3S hydrates are observed to grow near the surface during the analysis.

The aspect of C_3A hydration in the solution with welan gum is similar to C_3S, i.e., the hydration develops mainly inside the boundaries of the original particle. Hydrates are mostly thin and short fibers with amorphous aspect. Figure 4 shows two images obtained at 31 minutes and 2 hours and 10 minutes after mixing welan gum with the saturated solution and C_3A. Even though the aspect of C_3S and C_3A particles is similar, the kinetics seems to be faster for the latter. Note that the bigger particles in the images seem to be completely hydrated, because the radiation is able to pass through them. However, no signs of well-formed ettringite crystals are evident from the images.

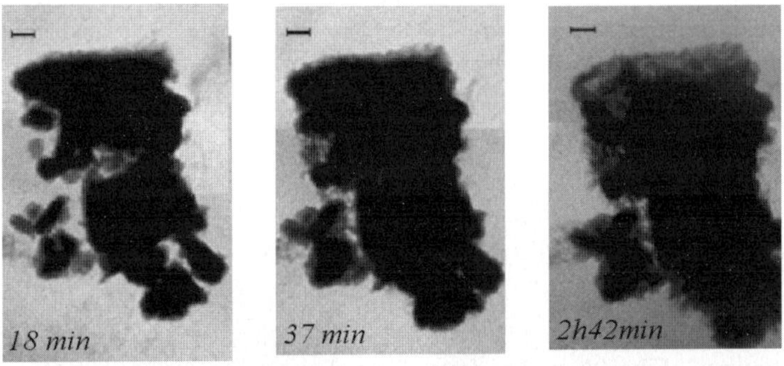

Figure 3. C_3S particles in the saturated solution with 0.2% welan gum. Hydration time is indicated. Scale bar is 700 nm

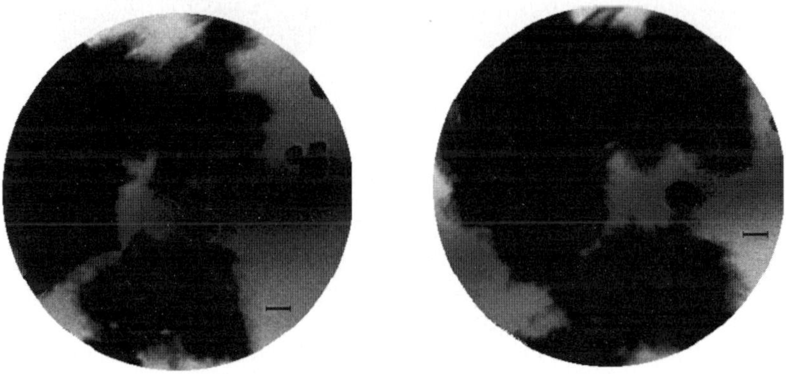

(a) 31 min (b) 2h10min

Figure 4. Images of C_3A particles in the saturated solution with 0.2% welan gum. Hydration time is indicated. Scale bar is 700 nm

4. DISCUSSION

Images of C_3S and C_3A particles hydrating in pure solutions, i.e., without any polysaccharide, are shown in Figure 5. It is evident that welan gum and HPMC strongly influence the hydration of the particles during the first few hours of hydration. The major effects of the polymers are: (i) on the kinetics of hydration, and (ii) on the morphology of the hydrated products.

The effects of the polymers are probably mostly due to changes in the aqueous solution: a gel structure is formed through hydrogen bonds of the hydroxyl groups with water molecules, hindering the motion of the dissolved ions and consequently the precipitation of hydrates. Alesiani et al. (2004) demonstrated the phenomenon using proton relaxation NMR studies on HMEC/water and HMEC/C_3S/water systems.

In addition, polysaccharides are able to adsorb on the surface of the hydrating particles through different mechanisms involving hydrogen bonding, hydrophobic interactions, and chemical complexation (Liu et al., 2000). The adsorption of the polysaccharides on the surface of the cement grains creates a water barrier, resulting in a delayed dissolution.

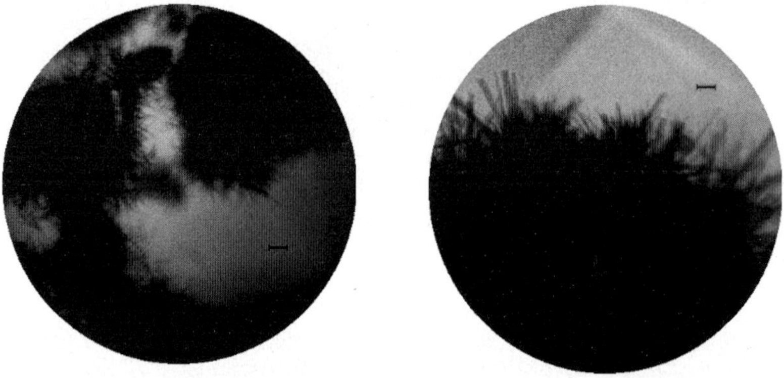

(a) 3h46min (b) 2h59min

Figure 5. Images of (a) C_3S and (b) C_3A particles hydrating in the saturated solution without any chemical admixture. Hydration time is indicated. Scale bar is 700 nm

Furthermore, the different aspect of the hydrates, especially the inner products of C_3S and C_3A particles in solutions with welan gum might be an evidence of structural changes of the hydrates (mainly C-S-H and ettringite). Clearly, more research in this area is needed.

Finally, the authors warn that the images showed herein are valid only for the samples investigated. Any projection to concrete's behavior should be made carefully due to the differences of the systems, especially the high dilution. However, the technique has proven to be powerful to compare the effects of different chemicals on the hydration of cement.

5. CONCLUSION

Soft x-ray images of C_3S and C_3A particles hydrating in solutions containing HPMC or welan gum have evidenced that both polymers retard the kinetics of hydration of the cementitious particles. Welan gum favored the hydration inwards the surface of the particles, leading to the formation of amorphous-like inner products. The polymer mostly hindered the formation of fiber or needle-like hydrates, which are the common morphologies of C-S-H and ettringite crystals. HPMC, on the other hand, allowed the development of very thin and short fiber-like hydrates outwards the surface of hydrating C_3S particles.

Because of the very different aspect of hydration when the polysaccharides are present in the solutions, we conclude that more research on the effects of VMAs on the micro and nanostructure of C-S-H and ettringite is clearly needed in order to establish structure-properties relationships for self-leveling concretes.

Acknowledgements

The authors acknowledge the financial support of the Brazilian government through CAPES (grant BEX 2368/02-1) and CNPq (grant PQ 307930/2003-3). The authors are grateful to Dr. Angelic Pearson, Bob Gunion and Weilun Chao of the Center for X-ray Optics and the Lawrence Berkeley National Laboratory for their assistance in acquiring the images. Research at XM-1 is supported by the United States Department of Energy, Office of Basic Energy Sciences under contract DE-AC 03-76SF00098.

6. REFERENCES

Alesiani, M., Capuani, S., Giorgi, R., Maraviglia, B., Pirazzoli, I., Ridi, F. and Baglioni, P., Influence of cellulosic additives on tricalcium silicate hydration: nuclear magnetic resonance relaxation time analysis, 2004, *J. Phys. Chem. B* **108**: 4869-4874.

Atwood, D., 1999, *Soft X-Rays and Extreme Ultraviolet Radiation: Principles and Applications*, Cambridge Univ Press, UC Berkeley/LBNL, 1999.

Juenger, M.C.G., Monteiro, P.J.M., Gartner, E.M. and Denbeaux, G.P., Using soft x-ray transmission microscopy to examine cement hydration in the presence of retarders, 2003, *Proceedings of the 11th International Congress on the Chemistry of Cement (ICCC)*, Durban (South Africa), 249-258.

Juenger, M.C.G., Monteiro, P.J.M., Gartner, E.M. and Denbeaux, G.P., 2005, A soft x-ray microscope investigation into the effects of calcium chloride on tricalcium silicate hydration, *Cem Conc Res* **35**: 19-25.

Khayat, K.H., 1998, Viscosity-enhancing admixtures for cement-based materials – an overview, *Cem Conc Comp* **20**: 171-188.

Kurtis, K.E., 1998, *Transmission Soft X-Ray Microscopy of the Alkali-Silica Reaction*, UC Berkeley Dissertation.

Lachemi, M., Hossain, K.M.A., Lambros, V., Nkinamubanzi, P.C. and Bouzoubaâ, N., Self-consolidating concrete incorporating new viscosity modifying admixtures, *Cem Conc Res* **34**: 917-926.

Liu, Q., Zhang, Y. and Laskowski, J.S., 2000, The adsorption of polysaccharides onto mineral surfaces: an acid/base interaction, *Int J Min Proc* **60**: 229-245.

Silva, D.A. and Monteiro, P.J.M., 2005a, Hydration evolution of C_3S-EVA composite analyzed by soft x-rays microscopy, *Cem Conc Res* **35**: 351-357.

Silva, D.A. and Monteiro, P.J.M., 2005, Analysis of C3A hydration using soft X-rays transmission microscopy: effect of EVA copolymer, *Cem Conc Res* **35**: 2026-2032.

RHEOLOGICAL PROPERTIES AND SEGREGATION RESISTANCE OF SCC PREPARED BY PORTLAND CEMENT AND FLY ASH

M.H. Ozkul and U.A. Dogan
Istanbul Technical University, Civil Engineering Faculty, Maslak, Istanbul

Abstract: Self-compacting concrete (SCC) provides high flowability, high filling capacity, high passing ability through reinforcing bars, which are owned by using a powerful superplasticizer as well as reducing both the coarse aggregate content and water/powder ratio. It should also expose high segregation resistance, which can be obtained by using high amount of fine material or by adding a viscosity modifying admixture, or both. In this study, the rheological properties and segregation resistance of SCC are examined. The effect of coarse aggregate concentration between 225-375 dm^3/m^3 on both flow behavior and segregation resistance of SCCs are investigated on concretes prepared with cement and fly ash as binder. The total binder content (including both cement and fly ash) varies between 450 and 650 kg/m^3. The maximum aggregate sizes were chosen as 12, 16 and 20 mm. In the experiments a specially designed apparatus, which was expired from the slump-flow and L-shape box tests, has been used. This apparatus also allows the measurement of segregation resistance. The other tests applied were slump-flow and penetration resistance.

Key words: self-compacting concrete; segregation; rheology.

1. INTRODUCTION

Self-compactability is defined as a capability of concrete to be uniformly filled in every corner of a formwork by the gravity force without any vibration during casting. SCC is originally developed in Japan for underwater applications but it can also be utilized in many cases such as repair of voids, production of tall walls and concrete members with complex shape or heavily reinforced.

Self-compactability keeps on increasing with the increasing amount of superplasticizer until it reaches a maximum value and then declines due to the segregation of the concrete[1]. Slump-flow, V-funnel, U-box and L-box tests can be used to measure the flow-

M.S. Konsta-Gdoutos, (ed.), Measuring, Monitoring and Modeling Concrete Properties, 463–468.

ing, filling and passing ability of SCC[2]. Not only high flowability, but also high resistance to segregation are important properties for SCC. However, the most practical test, slump flow test, to detect the deformability of SCC, is not sufficient for evaluating resistance to segregation. Penetration test[3], sieving fresh concrete through 5 mm mesh test[4], determining the coarse aggregate concentration in a L-box[3] and J-ring with the Orimet[5] are some of the proposed methods to measure the segregation resistance of SCC.

In this study, a previously developed apparatus[6,7] by the authors is used to measure rheological properties and segregation resistance of SCC prepared by Portland cement and fly ash. Effects of total binder content, maximum aggregate size and coarse aggregate concentration were chosen as testing parameters. The results obtained by the developed apparatus were compared with those obtained by slump flow and penetration tests.

2. EXPERIMENTAL

2.1 Materials

2.1.1 Aggregates

Natural and crushed stone sand were used as fine aggregates and their specific gravities were 2.62 kg/m^3 and 2.71 kg/m^3, respectively. In order to investigate the effect of maximum size of aggregate on SCC, crushed stone, with a specific gravity of 2.71 kg/m^3 was separated into three different parts having maximum sizes of 12 mm, 16 mm and 20 mm. The grading of fine and coarse aggregate was calculated separately. The reference grading was calculated by using Fuller parabola (1) where d_i is the mesh size of the sieve and P_i is the percentage of the aggregate passed form the ith sieve. In all the batches, fine aggregate grading was same and coarse aggregate grading was changed according to maximum size of aggregate.

$$P_i = \sqrt{\frac{d_i}{D_{max}}} \tag{1}$$

2.1.2 Cement, additive and admixture

An ordinary Portland cement, PC 42.5 (CEM I, in accordance to TS EN 197-1 standard) and a fly ash, maintained from Cayirhan are used. The fly ash/cement ratio is kept as 2.5/3 in all mixes. A polycarboxylate based HRWR admixture was employed in all the mixtures to obtain a sufficient workability.

2.2 Mixing Proportions

The important criteria, fine particle content, coarse aggregate concentration and maximum size of aggregate to obtain a SCC are investigated in this study. Each of these three variables had three levels to be examined. Total binder content was chosen as 450 kg/m^3, 550 kg/m^3 and 650 kg/m^3 with a fly ash/cement ratio of 2.5/3. The effect of coarse aggregate concentration on both flow behavior and segregation resistance of SCC were studied at 225, 300 and 375 dm^3 of coarse aggregate contents. In all mixes, maximum size of aggregate, affecting segregation and passing through the highly congested reinforcement, was changed between 12 mm 16 mm and 20 mm. Finally, 27 different batches composed of 3 variable with 3 levels of each were produced. Water/binder ratio was adjusted to have the best spread of itself at flow test.

2.3 Test Methods

In the experiments a specially designed apparatus (confined-slump flow), which was expired from the slump-flow and L-shape box tests, has been used. It looks like the J-ring apparatus, but was developed unaware of it, and consists of a cylinder with a 15 cm diameter and 30 cm height, surrounded by ϕ12 bars having 35 mm gap between adjacent bars. This apparatus (Fig. 1) also allows the measurement of segregation resistance while the concrete is flowing through the bars. For the latter purpose, the concrete left between the bars is taken out after the flow is completed, wet-sieved through the 4 mm mesh, and the change in the coarse aggregate concentration is determined with respect to the initial state. Final diameter of concrete can be also recorded with this apparatus similar to slump-flow test. The other tests applied were slump-flow and penetration[3].

*Figure 1. Confined-s*lump flow apparatus

3. TEST RESULTS AND DISCUSSION

A summary of the test results is presented in *Table 1,* together with the mix proportions of the concretes. Slump flows of concretes were measured simultaneously by using Abrams cone in inversed position and by the specially designed apparatus, and are shown in *Figure 2. Table 1* exhibits that, it is necessary to increase the superplasticizer content over 2% to obtain slump-flow over 60 cm for the concretes with 450 kg/m^3 binder content. However, in the confined-flow test, the spreads were remained under those obtained from the free slump-flow, as shown in *Figure 2*, although the concrete in the developed apparatus is 35 % larger than that of Abrams cone. This indicates that, flow is prevented by the bars due to the insufficient passing ability of the concrete.

On the other hand, for the concretes with 550 kg/m^3 binder content, a superplasticizer dosage of 1.2% was found sufficient to obtain slump-flows over 60 cm (*Table 1*). Furthermore, the free and confined-slump values obtained on the latter concretes are close to each other and there is no definite difference between the tendencies of results for the two slump-flow methods (*Figure 2*). *Table 1* and *Figure 2* also show that, the level of flows for both methods at 650 kg/m^3 binder content, are higher than those of the former two binder contents.

Table 1. Mix proportions and test results

Binder (kg/m^3)	Coarse Agg. (dm^3)	D_{max} (mm)	W/B	Admix (%)	Slump Flow Confined f_d	Slump Flow Free f_d	Penet. (mm)	Seg. Ratio
450	375	20	0.31	2.2	46	58		
450	300	20	0.31	2.5	49	61		
450	225	20	0.32	2.5	58	63		
450	375	16	0.31	2.5	41	55		
450	300	16	0.32	2.5	58	62		
450	225	16	0.33	2.0	54	62		
450	375	12	0.30	2.5	47	55		
450	300	12	0.31	2.0	44	58		
450	225	12	0.33	2.5	54	61		
550	375	20	0.26	1,2	60	65	8	1.07
550	300	20	0.26	1,2	64	62	8	1,15
550	225	20	0.27	1,2	62	61	8	1.23
550	375	16	0.26	1,2	63	68	6.5	1.02
550	300	16	0.27	1,2	64	63	5	1.10
550	225	16	0.27	1,2	66	63	3	1.33
550	375	12	0.26	1,2	62	60	5	1.08
550	300	12	0.27	1,2	65	62	8	1.00
550	225	12	0.27	1,2	66	66	2	1.15
650	375	20	0.23	1,2	72	69	4	1.11
650	300	20	0.23	1,2	70	65	9	1.12
650	225	20	0.23	1,2	68	65	15	1.34
650	375	16	0.22	1,2	68	63	10	1.14
650	300	16	0.23	1,2	74	68	17	1.23
650	225	16	0.23	1,2	74	68	15	1.04
650	375	12	0.22	1,2	66	67	6	1.02
650	300	12	0.23	1,2	74	68	20	1.21
650	225	12	0.24	1,2	78	73	17	1.05

D_{max}: Max. aggregate size; W/B: Water/binder ratio; f_d: final spread (in cm)

Moreover, for this higher binder content, the spread values obtained in the confined-slump flow test are larger than those of the free one, indicating a high passing ability through the obstacles in general, which can be due to the increased amount of powder material (binder) in the mixtures.

Figure 3 illustrates the relation between segregation ratio and penetration. The segregation resistance was calculated as the ratio of the coarse aggregate content over 4 mm size measured on the sample taken from the concrete left between the bars, to the coarse aggregate content of the original concrete. The results of 450 kg/m^3 binder content were not shown, because they exhibited high segregation. For the binder content of 550 kg/m^3, all the penetrations are equal or smaller than 8 mm, which was given as the penetration limit for an unsegregated SCC[3]. *Figure 3* also shows that the segregation ratio values measured in this study remain under 1.20 (which corresponds a 20% segregation) except the mixtures with maximum aggregate sizes of 16 and 20 mm at a coarse aggregate content of 225 dm^3/m^3. Since a powder content of 550 kg/m^3 seems reasonable for SCC,[6,7] a segregation resistance of 20% can be taken as an upper limit.

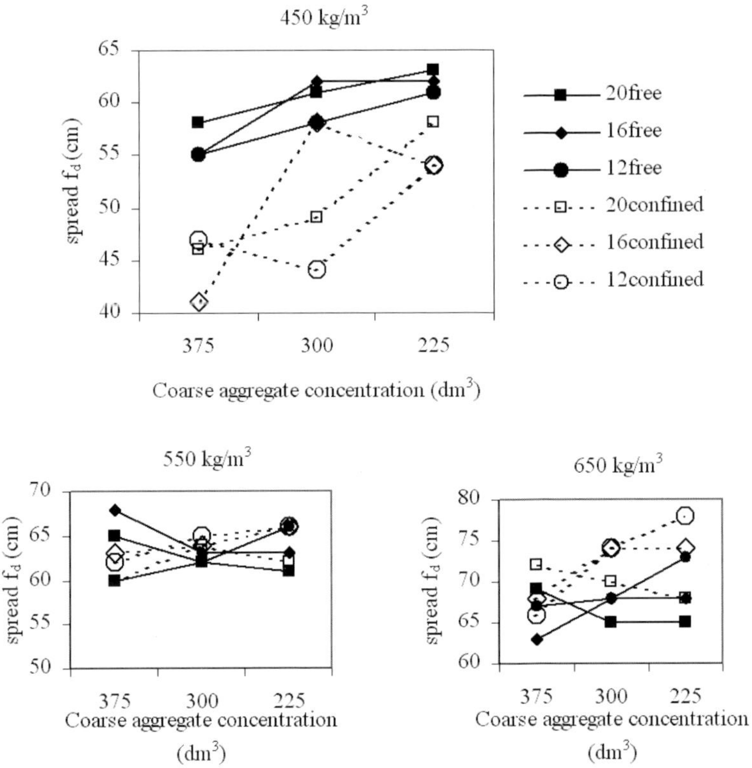

Figure 2. Comparison of confined and free spread methods

Figure 3 exhibits the segregation ratio and penetration relations for the binder content of 650 kg/m^3. All the penetration results except two (belonging to the coarse aggregate content of 375 dm^3/m^3) remain over the limit of 8. However, the segregation resistance of 1.20 is exceeded by only three mixtures, and four mixtures exhibited penetrations over 8. It seems that for a high binder content, when the coarse aggregate content is lower, than the penetration becomes higher. However, for these mixtures, higher the penetration does not mean that lower the segregation tendency.

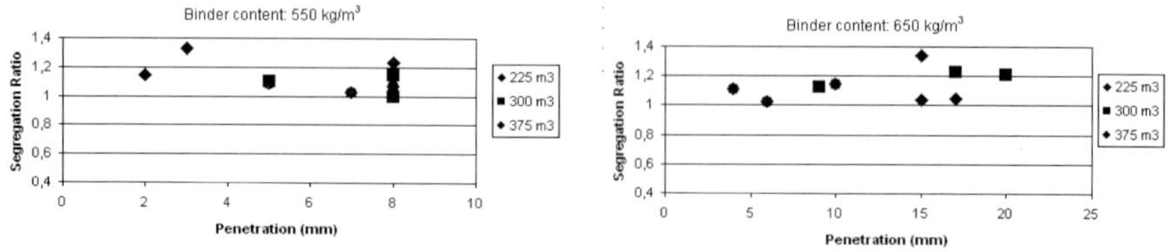

Figure 3. Relation penetration and segregation ratio with respect to binder content and coarse aggregate ratio

4. CONCLUSION

The following conclusions can be drawn based on the results of this study:

1 For the low binder contents (450 kg/m^3), the confined-slump flow values obtained by the developed apparatus remained under those of the free slump-flows, indicating the low passing ability and segregation resistance of these mixtures.

2 For the moderate binder contents (550 kg/m^3), the confined and free slump values were found close to each other and they did not exhibit a definite tendency over each other. The penetrations remained under the limit of 8 for these mixtures, however, two of them showed segregation over 20%.

3 For the high binder contents (650 kg/m^3), some of the mixtures exhibited penetrations over 8, although they showed segregations under 20%, which may be due to the high binder content.

5. REFERENCES

1. T. Shindoh and Y. Matsuoka, Development of combination-type self-compacting concrete and evaluation test methods, *Journal of Advanced Concrete Technology,* **1**(1), 26-36 (2003).
2. Self Compacting Concrete, Ed. By A. Skarendahl and O. Petersson, RILEM Pub., France (2000).
3. V. K. Bui, D. Montgomery, I. Hinczak, and K. Turner, Rapid testing method for segregation resistance of self-compacting concrete, *Cem. Concr. Res.,* 32, 1489-1496 (2002).
4. H. Fujiwara, Fundemental study on self-compacting property of high-fluidity concrete, *Proc. Jpn. Concr. Inst.*, 14 (1), 27-32 (1992).
5. M. Sonebi, Application of statistical models in proportioning medium-strength self-consolidating concrete, *ACI Materials J.,* 101 (5), 339-346 (2004).
6. M. H. Ozkul, U. A. Dogan, Z. Cavdar, A. R. Saglam and N. Parlak, Properties of fresh and hardened concretes prepared by new generation superplasticizers, *Int. Conf. on Modern Concrete Materials: Binders, Additives and Admixtures,* Ed. by R. K. Dhir, Dundee, Scotland, 467-474 (1999).
7. M. H. Ozkul, U. A. Dogan, Z. Cavdar, A. R. Saglam and N. Parlak, Effects of self compacting concrete admixtures on fresh and hardened concrete properties, *2nd Int. Symp. on Cement and Concrete Technology in the 2000s,* Ed. by A. Yeginobali, Istanbul, Turkey, 493-502 (2000).

OPTIMIZATION OF SUPERPLASTICIZER CONTENT IN SELF-COMPACTING CONCRETE

K.A. Melo and W.L. Repette

Universidade Federal de Santa Catarina – UFSC, UFSC - Departamento de Engenharia Civil, Bloco B, Sala 114 CEP 88040-900, Florianópolis-SC, Brazil

Abstract: According to Okamura´s method for self-compacting concrete mix design, the dosage of superplasticizer is first determined in mortar mixtures and it is after adjusted in concrete mix trials. Other SCC mix proportioning methods rely on the definition of the superplasticizer saturation dosage in pastes. These approaches to mix design have advantages over the ones based exclusively on concrete batching because it is less expensive and material demanding to perform tests in mortar and paste than in concrete. Nevertheless, difficulties are being reported which concern to the lack of correlation between the amounts of superplasticizer determined in paste and mortar mixtures and the dosages necessary to produce SCC. This article presents the results of an experimental research carried out to investigate the use of tests performed in paste and mortar to define the dosage of superplasticizer for self-compacting concrete. The materials employed were cement of high initial resistance, limestone filler, fine and coarser sand as fine aggregate, a maximum 10mm diameter coarse aggregate and a policarboxylate based superplasticizer. The saturation point of the superplasticizer was determined in pastes by the use of Marsh cone and a coaxial cylinder viscometer. The amounts of superplasticizer for the mortar mixtures were determined with the use of Slump-flow and V-Funnel tests. The final superplasticizer dosage was determined in concrete, so it the requirements for Slump-flow, V-Funnel and the L-Box tests were satisfied. The results show a comparison of the admixture contents determined for the different phases. Better relationships were found between the amounts of superplasticizer determined for mortar and concrete. Poor relationship was observed between the dosages on paste and concrete. This allowed for the conclusion that in SCC mixture proportioning methods, the determination of the superplasticizer content in the paste phase is dispensable, and the amount of admixture is more reliably determined by testing mortar mixtures. Nevertheless, final adjustment of the superplasticizer dosage was always needed in order to produce SCC.

Key words: superplasticizer, mix design, self-compacting concrete

M.S. Konsta-Gdoutos, (ed.), Measuring, Monitoring and Modeling Concrete Properties, 469–477.

1. INTRODUCTION

Self-compacting concrete (SCC) is characterized by its filling ability and deformability, which is governed by fluidity and cohesion of the mixture. To these properties should also be added requirements for passing ability and resistance to segregation, which are mostly related to the volume, dimension and distribution of the aggregate particles. These properties can be evaluated in SCC with specific tests, such as Slump flow, V-Funnel and L-Box [1, 2, 3].

SCC can also be characterized by its rheological properties, which are a low yield stress (τ_0), to guarantee the increase of the fluidity, and a moderate viscosity (η), to promote the necessary mixture stability. The yield stress relates with the distance between the particles in the paste matrix and governs the mix deformability, while viscosity is governed by the contact between the particles and it can be used to indicate the resistance to the segregation. SCC has been satisfactory described by the Bingham model (Eq. 1).

$$\tau = \tau_0 + \eta \cdot \dot{\gamma} \tag{1}$$

where

τ = shear stress
τ_0 = yield stress
η = plastic viscosity
$\dot{\gamma}$ = shear rate

One of the critical parameters to be determined during SCC mixture proportioning is the superplasticizer content. The amount of superplasticizer affects the cost and the fresh and early-age properties of the SCC. Excess of superplasticizer can cause mixture instability and setting delay; in small amounts, the superplasticizer does not promote the necessary fluidity and problems with fluidity maintenance are verified. Due to its economical and technological importance for the SCC, the definition of an optimum superplasticizer dosage is of extreme importance. Many technologists propose that superplasticizer dosage in concrete should relate well with the saturation dosage determined in studies carried out in cement paste, through the use of viscometers, Marsh-cone or mini-slump testing. Others indicate that a satisfactory definition of the superplasticizer content to be used in SCC can derive from studies performed in mortar. This article investigates the existence of interrelation between the Marsh-cone and the rheological parameters for the definition of the saturation point in pastes, and the admixture contents determined in the paste and mortar with the final dosage necessary to produce SCC.

2. EXPERIMENTAL

The materials used in this work come from the South Region of Brazil. The cement was a high early strength cement, CP V ARI RS (similar to ASTM Type III cement) with Blaine fineness of 511.8 m^2/kg. The limestone filler had specific gravity of 2.87 kg/dm^3 and 60% of its particles smaller than 75 μm. Figure 1 shows the grain-size distribution for the cement and the limestone filler, obtained by laser-diffraction-analysis, including only the passing 75 μm material.

A well-graded fine aggregate was composed by mixing 20% of fine and 80% of coarse sand (by mass) in order to increase the fluidity of the concrete. A maximum 10 mm diameter crushed granite rock was used as coarse aggregate. The main characteristics of the aggregates are summarized in Table 1. The aggregates were used in the saturated condition to avoid the loss of fluidity caused by the penetration of water in the grain voids.

A polycarboxilic acid based superplasticizer (Glenium 51, by Degussa) was used. The water content of the mixtures were adjusted considering the admixture solid content of 30% (mass), and the dosages were defined as the superplasticizer solid content in relation to cement content, by mass.

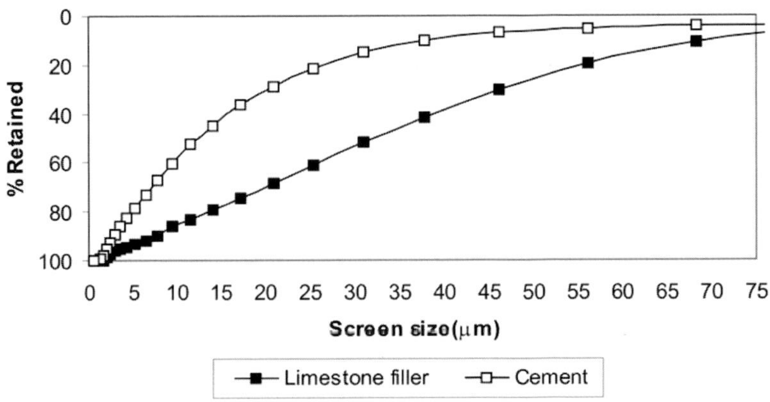

Figure 1. Grain-size distribution of cement and limestone filler (smaller than 75 μm)

The adjustment of admixture dosages was made in paste, mortar and concrete, and was carried out to validate a methodology for SCC mix design. The Repette-Melo Method [4] is based on the optimization of each component of the paste and mortar of the SCC. The water/cement ratio (w/c) is the starting point for the mix and it is determined from Abrams curves (compressive strength versus w/c) of normal concrete made with the same materials (not proportions) being used in the SCC. In the present work, the mixtures were prepared with w/c of 0.82, 0.67 and 0.55, respectively to normal concrete compressive strengths of 20, 30 and 40 MPa. The superplasticizer content was adjusted to maintain the fluidity of the mixtures for at least 30 minutes after the contact between cement and water.

2.1 Pastes

Limestone filler content was determined in the paste phase to guarantee the appropriate cohesion and to avoid bleeding of the mixture. In this phase, only the portion of filler with particle diameter smaller than 75 m was used. Limestone filler was added to the paste in increments of 5%, related to the cement volume. For each w/c, the ideal content was defined as minimum amount necessary to minimize segregation and bleeding of the mix. The volumes of filler, as the percentages of the total volume of solids, are presented in Table 2.

The saturation point of the superplasticizer was determined on pastes with the use of the Marsh-cone test and by the determination of plastic viscosity and yield stress with a coaxial cylinder viscometer (Haake Viscotester VT550), with a cylinder gap of 1 mm. The plastic viscosity and the yield stress were determined considering the Bingham Model over the results obtained for shear rates between 1 and 100 s^{-1}. The temperature of the pastes was maintained at 23 °C. For the determination with the Marsh-cone the admixture was incremented at dosages of 0.2% (solids/cement, by mass), while for the tests performed on the viscosimeter this value was 0.1%.

Table 1. Physical characteristics of the aggregates

Screen size (mm)	Fine sand	Coarse sand	Coarse aggregate
38	0	0	0
25	0	0	0
19.5	0	0	0
12.5	0	0	0
9.5	0	0	1.10
6.3	0	0.08	47.26
4.8	0	0.32	82.06
2.4	0	4.33	99.18
1.2	0	21.65	99.72
0.6	0.02	52.66	99.72
0.3	7.77	77.39	99.72
0.15	96.04	89.70	99.72
0.075	99.93	96.00	99.72
Specific gravity (kg/dm^3)	2.65	2.62	2.68
Absorption (%)	0.15	2.10	1.03

Table 2. Content of limestone filler and nomenclature of the pastes

Mixture	w/c	Limestone filler (%)*
A82-F50	0.82	50
A82-F55		55
A82-F60		60
A67-F35	0.67	35
A67-F40		40
A67-F45		45
A55-F25	0.55	25
A55-F30		30
A55-F35		35

* Portion of filler with particles finer than 75 m

2.2 Mortars

Mortars were obtained by adding sand to the previously determined paste mixtures (cement, limestone filler and water). The fine aggregate volume, expressed as the percentage of the total volume of the mortar, was determined experimentally and corresponded to the maximum amount of material to for which the stability of the mixture was maintained as the volume of superplasticizer was increased to result the necessary fluidity, without compromising the stability of the mixture. Slump flow and V-funnel mortar tests were performed in the mixes A82-F55, A67-F40, A55-F30. The ideal volume of fine aggregate in the mortar was found to be 45% of the mortar volume, which was adopted the other mixtures studied. The coarser portion of limestone filler (particles bigger than 75 m) was incorporated at the fine aggregate content.

The superplasticizer content was adjusted in increments of 0.1% (solids/cement, by mass) for the 9 mixtures listed in Table 2. The saturation dosage defined for the paste was the starting dosage for the adjustment of the superplasticizer content of the mortar. The final value was determined by evaluation of the deformability and fluidity according with flow and funnel tests described by Okamura and Ouchi [2]. The required values used to consider the mortar mix satisfactory to be used in SCC were of 225 to 280 mm for the opening flow diameter and of 3 to 10 s for the flowing time in the funnel, based on experimental investigations and on values proposed by Edamatsu et al apud Gomes [5].

2.3 Concretes

The optimum coarse aggregate volume was determined experimentally as the maximum amount that could be incorporated to the mortar with w/c=0.67 and volume of limestone filler of 40% without compromising the mixture stability and the self-compactability characteristics measured by the Slump flow, V-Funnel and L-Box tests, with the requirements presented in Table 3. Concretes were produced with coarse aggregate volumes of 27, 28.5, 30, 31.5 and 33% to the total volume of concrete and superplasticizer increments of 0.1% (solids/cement, by mass). The results showed that the ideal volume of coarse aggregate was 28.5%, value adopted for the production of the other mixtures.

Table 3. Acceptance limits of SCC according to the test method

Test method	Limits for SCC
Slump flow (d)	600 to 700 mm
V-Funnel (t)	10 s
L-Box (H_2/H_1)	0.8

With the coarse aggregate volume of 28.5%, concretes were prepared for the mortars A82-F55, A67-F35, A67-F40, A67-F45 and A55-F30. The dosage of superplasticizer started with the value determined for the mortar, and was increased in increments of 0.025% to the point that SCC was obtained and complied with the requirements presented in Table 3. It was observed that the superplasticizer content defined in mortars was not enough to guarantee the self-compactability of the concretes.

3. RESULTS AND ANALYSIS

Figure 2 shows, for the paste A82-F60, the results of the tests performed with the Marsh-cone test and the viscosimeter (yield stress (a) and plastic viscosity (b)). Similar curves were obtained for the different pastes tested, and showed to exist better agreement between the results of time to flow (Marsh-cone) and plastic viscosity than between time to flow and yield stress.

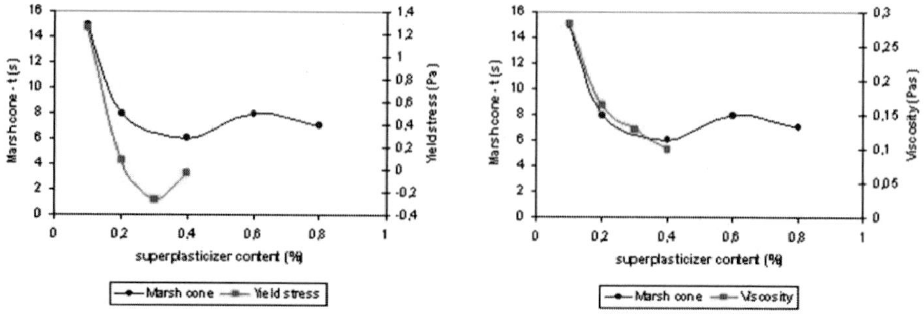

(a) (b)

Figure 2. Results of the tests performed to determine the superplasticizer saturation dosage in the A82-F60 paste

The first two parameters were used for the determination of the admixture saturation dosage, presented for each paste in Figure 3. Figure 3 shows that there was a good agreement between the admixture saturation dosage determined with the Marsh-cone test and plastic viscosity when the w/c was 0.67. On the other hand, the test results were considerable different for pastes of w/c 0.82 and 0.55. For the paste A82-F60, the results of plastic viscosity did not allow for the definition of the saturation dosage despite the fact that the test was repeated three times. The instability of the mix was likely responsible for the described result.

For pastes prepared with w/c of 0.82 and 0.55, the results of the Marsh-cone tests indicate an increase of the admixture saturation dosage that is proportional to the mixture fines content. For w/c of 0.67, the results showed an opposite trend. The analysis of the results obtained with plastic viscosity do not allow for the definition of a coherent relationship between the admixture dosage and the fines content, particularly for the pastes prepared with w/c of 0.55. The definition of superplasticizer saturation dosage was more clearly determined with Marsh-cone than with the rheological parameters as were determined in this study.

Table 4 presents the superplasticizer dosages defined for the production of mortar mixtures suitable for the production of SCC. The results of the Slump flow and V-Funnel tests are also presented. The superplasticizer dosages obtained in mortars were directly proportional to the amount of fines in the mixtures of same w/c. Only the mixture A82-F60 exceeded the established limit for self-compactability, particularly for the V-Funnel flow time, despite the high volume of admixture employed in the mix. It should be considered that this particular mixture had a high volume of fines, and as a consequence, was highly viscous. This demonstrates that if the volume of fines is high in the mixture, bigger dosages of the superplasticizer do not cause the reduction of plastic viscosity, associated with a smaller V-Funnel flow time.

The results obtained for concrete are presented in Table 5. Similarly to what was verified for mortars, the amount of required superplasticizer admixture for the production of SCC increased as the volume of fine particles in the mixture was bigger.

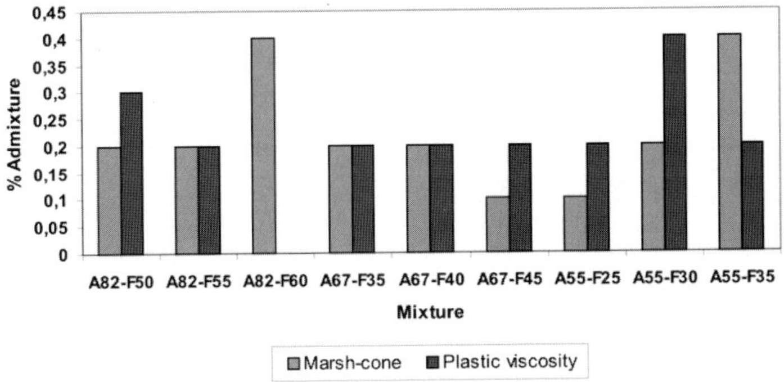

Figure 3. Superplasticizer saturation dosages in the paste

Table 4. Mortar composition and slump flow and V- funnel test results

Mixture	Cement (kg/m^3)	Limestone filler (kg/m^3)	Fine aggregate (kg/m^3)	Admixture (% mass cement)	Slump Flow d (mm)	V-Funnel t (s)
A82-F50	367.72	594.49	706.10	0.20	265	4.005
A82-F55	350.52	689.35	922.74	0.25	262	9.409
A82-F60	330.54	798.80	882.45	0.40	325	13.447
A67-F35	462.71	401.02	1027.82	0.15	257	3.737
A67-F40	446.99	476.79	1000.28	0.15	225	6.073
A67-F45	428.61	564.33	968.00	0.25	287	7.641
A55-F25	550.57	293.64	1066.73	0.15	280	4.371
A55-F30	533.69	364.69	1010.63	0.20	287	5.873
A55-F35	514.43	445.84	1010.87	0.25	280	8.141

Figures 4 and 5 show the differences of admixture dosage obtained for paste, mortar and concrete, respectively to the mass of cement and to the mass of fine particles (not larger than 75m). The admixture saturation dosage determined in pastes is often considered valuable information for the definition of the admixture dosage in SCC and a good relationship should exist between the amounts of admixture determined for paste and concrete. The results do not support this assumption, as the saturation dosage defined in paste can not be used to predict the amount of superplasticizer obtained for the correspondent self-compacting concrete, regardless of whether the amount of admixture is expressed in relation to the cement or the fines content.

The admixture saturation dosage determined for pastes showed a reasonable correlation with the dosages obtained for mortars when the water to cement and, by consequence, the water to fines ratios were the highest, 0.82. Interaction of cement and admixture becomes more critical for mixtures with lower water contents, and this could be the cause for a poor relationship between the dosages defined in paste and mortar for w/c of 0.67 and 0.55.

The amounts of superplasticizer obtained for mortars correlated satisfactory with the values adjusted for the production of SCC. Even though the amounts of superplasticizer determined in mortar mixtures were not equal to the dosages obtained for concrete, the amount of superplasticizer needed to be incorporated to the concrete was always very similar and close to 0.1% in relation to the mass of cement. This implies that the test methods and the evaluation criteria used for the determination of self-compacting mortars are adequate to screen mortar compositions suitable for the production of SCC.

Table 5. SCC composition and tests results

Mixture	Cement (kg/m³)	Limestone filler (kg/m³)	Fine aggregate (kg/m³)	Coarse aggregate (kg/m³)	Admixture (% m.c.)	Slump Flow d (mm)	V-Funnel t (s)	L-Box H_2/H_1
A82-F55	250.82	493.28	660.29	762.38	0.35	663.3	9.97	0.89
A67-F35	331.05	286.91	735.36	762.38	0.25	665.0	4.21	0.84
A67-F40	319.79	341.11	715.64	762.38	0.275	658.3	5.77	0.98
A67-F45	306.76	403.90	692.80	762.38	0.30	610.0	6.95	0.81
A55-F30	381.96	261.01	744.78	762.38	0.30	650.0	9.16	0.84

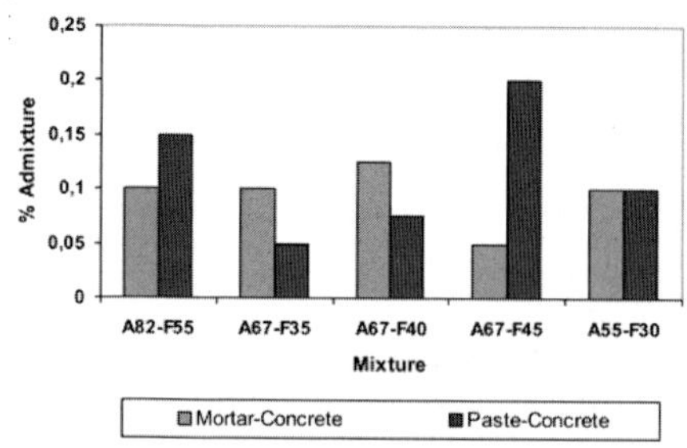

Figure 4. Difference in admixture content between the mixture phases related to the mass of cement

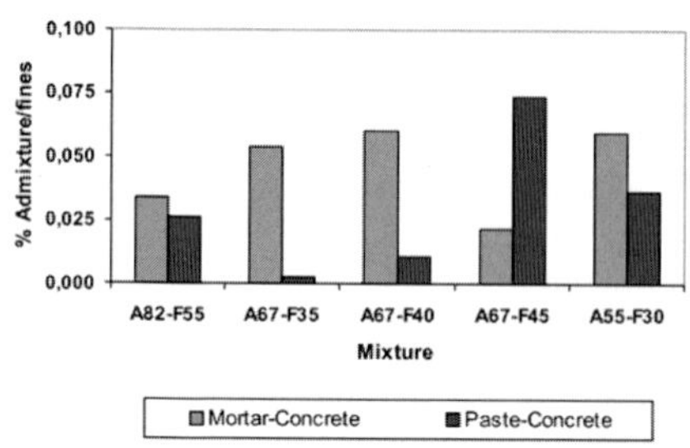

Figure 5. Difference in admixture content between the mixture phases related to the mass of fines

4. CONCLUSIONS

The results of the tests performed on pastes with a coaxial cylinder viscosimeter indicated that the definition of the superplasticizer saturation dosage was more adequate when plastic viscosity, and not yield stress, was use as the parameter affected by the admixture dosage. Notwithstanding, more coherent results were obtained when time to flow measured with the Marsh-cone was used for the determination of the admixture saturation dosage. The admixture saturation dosage determined for paste mixtures showed a week correlation with the dosage necessary for the production of the correspondent self-compacting mortar and concrete. The saturation dosage determined in pastes has limited use for the mixture proportioning of SCC.

The amounts of superplasticizer determined in mortar mixtures related satisfactorily with the admixture dosage necessary for production of SCC. Considering the dosage determined for the mortar mixtures, the amount of admixture needed for production of concrete was almost constant and equal to 0.1% when the superplasticizer content was related to the mass of cement rather than mass of fines.

In general, mixtures with higher w/c demanded higher volumes of filler, necessary for mixture stabilization, and higher amounts of superplasticizer to promote the necessary self-compatibility properties.

Acknowledgments

Authors are thankful to CNPq – Brazilian Research Council for its financial support to the research reported in this paper. Thanks to Votoratin Cement for the cement supply, to Degussa for providing the superplasticizer and to Calfipar for the limestone filler supply.

5. REFERENCES

1. EFNARC, "Specifications and Guidelines for Self-Compacting Concrete", 32p, 2005.
2. Okamura, H. and Ouchi, M., "Self-compacting concrete". Journal of Advanced Concrete Technology, **1**(1), 5-15 (2003).
3. Nunes, S. C. B., Betão Auto-Compactável – Tecnologia e Propriedades, University of Porto, 2001.
4. Melo, K. A., Proposição de método de dosagem de concreto auto-adensável com adição de fíler calcário, Federal University of Santa Catarina, 2005.
5. Gomes, P. C. C., Optimization and characterization of high-strength self-compacting concrete, Polytechnic University of Catalunya, 2002.

CAPILLARY RHEOLOGY OF EXTRUDED CEMENT-BASED MATERIALS

K.G. Kuder and S.P. Shah
Seattle University, 901 12th Ave, Seattle, WA 98122-1090; Northwestern University, 2145 Sheridan Rd., Suite A130, Evanston, IL 60208

Abstract: Extrusion processing is a technique used to produce high-performance fiber-reinforced cement-based composites (HPFRCC), which has shown great promise for manufacturing materials that are strong, ductile, durable, design versatile and environmentally friendly. Despite these advantages, extrusion is still primarily limited to laboratory-scale work. One reason this technology has not been adopted by industry is the high cost of the cellulose ether processing aids that are required for extrusion. In this research, the possibility of partially replacing cellulose ethers with less expensive clay binders is investigated. Extrudable and not extrudable mixes are identified and capillary rheology is used to describe the rheological parameters of the various mixes. The results indicate that clay binders can be used as a partial replacement for cellulose ethers and that capillary rheology can be used to describe extrudability.

Key words: Extrusion, capillary rheology, processing, clay, cellulose ether

1. INTRODUCTION

Extrusion is a special processing technique that is used to produce HPFRCC. HPFRCC exhibit a strain-hardening response, with a significant increase in both strength and toughness when compared to plain and conventional fiber-reinforced composites. Research shows that this high performance can be achieved in a variety of ways, including using micromechanical modeling [1], tailored fiber geometries [2] and advanced processing techniques (such as extrusion) [3-7]. In addition to enhanced mechanical performance, composites demonstrate a significant improvement in durability due to the high density that results from the extrusion technique [5, 8].

Laboratory-scale research has demonstrated great potential for extrusion technology. However, the technology has not been widely adopted by industry. One reason for this limited use is that expensive cellulose ether processing aids are needed to control the

M.S. Konsta-Gdoutos, (ed.), Measuring, Monitoring and Modeling Concrete Properties, 479–484.

fresh state properties of extruded materials. In this research, the use of less-expensive processing aids is examined. The extrudability of mixes containing various amounts of processing aids is evaluated and then capillary rheology is used to describe the rheological parameters of both the extrudable and not extrudable mixes.

2. EXPERIMENTAL PROGRAM

For a material to be extrudable, 1) it must be soft enough to flow through the die, yet rigid enough to maintain it shape upon exit of the die, 2) the pressure required for extrusion must be reasonably low (to control manufacturing costs), 3) phase migration must be avoided and 4) the material should be shape stable. Initial work involved determining mixes that were extrudable. First, the minimum amount of the cellulose ethers needed for extrusion was evaluated. Next, the possibility of replacing the cellulose ethers with clay was systematically examined. Finally, capillary rheology was used to characterize extrudability.

2.1 Sample Preparation

The matrix composition used, by volume, consisted of 33% Class F fly ash (produced by Dynegy Midwest Generation, Inc., mean particle size = 10 μm), 12% silica fume (W.R. Grace Force 10,000), 14% cement (LaFarge Type I), 39% water and 1% high-range-water-reducing admixture (Daracem 19). Two different cellulose ethers, Methocel (D) and Walocel (W), and two different clay types, Concresol (C) and Metamax (M), were studied. Mixes were prepared using a planetary (Hobart) mixer, with the dry and wet ingredients first mixed separately, then combined and mixed by hand, followed by mixing on the slow mixer speed for approximately 5 minutes, and then mixing on the medium speed for 10-15 minutes until a cohesive dough was formed.

The properties of the processing aids are given in Tables 1 and 2. The clay binders are approximately one hundredth the cost of the cellulose ethers and previous research has shown that clay has the potential to enhance the fresh state properties of stiff cementitious materials [9,10]. When clay was added, water weighing 60% of the clay weight was added to account for the water absorptive properties of the clay. The amount of cellulose ethers and clays incorporated was based on the extrudability testing (presented in Section 2.2) and is given as a percentage of the weight of the total binder (cement + fly ash + silica fume). For convenience, a shorthand was used to express the mix designs, with the binders of the mix described by weight percentages. For example, W0.25C0.25, contains 0.25% Walocel and 0.25% Concresol.

2.2 Extrudability

No standardized test method exists to determine whether or not a composition is extrudable. In this work, extrudability was evaluated by extruding open cross-sections, using a cellular die, with two cells, that had a total length of 25.4 mm, a width of 15 mm and a wall thickness of 3.25 mm. Specimens were extruded at a rate of 1 mm/s. If poor shape stability, phase migration, an excessively high extrusion pressure, or surface defects (usually edge tearing) was observed, the material was considered not extrudable.

Table 1. Cellulose ether properties [11, 12]

Trade name	Producer	Chemical designation	Specific gravity g/cm^3	Viscosity Pa*s
Methocel 4fm*	Dow Chemical Company	Hydroxypropyl methylcellulose	1.3	2.3 - 3.8
Walocel M-20678**	Wolff Celulosics	Methylhydroxyethyl cellulose	3.2 - 4.0	10.4 - 12.4

*Viscosity determined for 1% solution, 20° C, Rotovisko rheometer

** Viscosity determined for 1% solution, 20° C, "by rotation"

Table 2. Clay properties [13, 14]

Trade name	Producer	Mineral Composition	Mean particle size (μm)
Concresol	Stephan Schmidt Group	kaolinite (45%), mica (35%)and free silica (20%)	~ 0.5
Metamax HRM	Engelhard	calcined kaolinite	1.2

It is important to note that extrudability is related to extrusion velocity as well as to the shape being extruded. Therefore, the minimum amounts of binders defined here are dependent on the velocity and die used.

Table 3 summarizes the extrudable mixes for D and W. By incrementally adding cellulose ethers to the base mix, the minimum amount needed for extrusion was determined. As Table 3 indicates, half that amount of W was needed compared to D. Once these minimal amounts were determined, the amount of cellulose ether was reduced by half and the clays were added. As Figure 1 demonstrates, once 0.3% of the clay was added, an extrudable mix was achieved. Similar results were found when either C or M was added. However, if all the cellulose ether was removed, the material was no longer extrudable. These results, which are explained in more detail in [15], indicate that it is important to find the most effective type of cellulose ether (here twice as much D is needed, compared to W, while the costs are comparable) and that cellulose ethers can be partially replaced with clay binders.

2.3 Capillary Rheology Theory

The rheological properties of the mixes presented in Table 3 were characterized using the Benbow-Bridgewater model and capillary rheology. The results obtained using the Benbow-Bridgewater model can be found elsewhere [15]. Results from the capillary rheology analysis are presented here.

Capillary rheology can be used to determine fundamental flow properties. Capillary analysis assumes that flow is laminar (Reynolds number < 2000), is fully developed and that there is no slip at the wall. The apparent shear stress (τ_{app}) and shear rate ($\dot{\gamma}_{app}$) are given in Equation (1)) and (2)), respectively.

Table 3. Extrudable mixes

(a) D

Material	Extrudable
D0.5	
D1	x
D2	x
D0.5/C0.15	
D0.5/C0.3	x
D0.5/C3	
D0.5/M0.15	
D0.5/M0.3	x
D0.5/M3	

(b) W

Material	Extrudable
W0.25	
W0.5	x
W1	
W0.25/C0.15	
W0.25/C0.3	x
W0.25/C3	
W0.25/M0.15	
W0.25/M0.3	x
W0.25/M3	

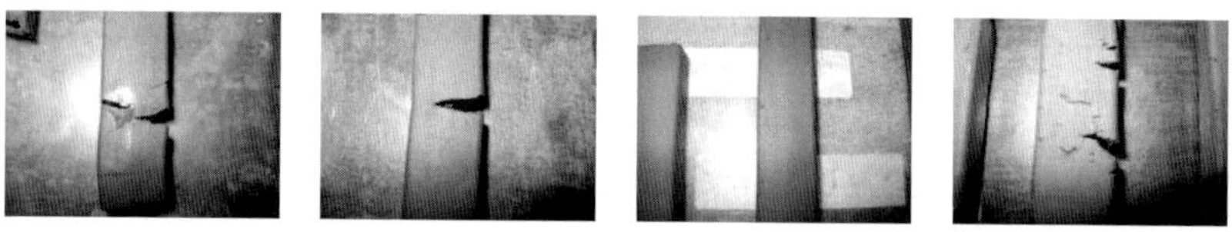

0.25W 0.25W0.15C 0.25W0.3C 0.25W3C

Figure 1. Effect of clay addition on extrudability

$$\tau_{rx} = \tau_{app} = \frac{PD}{4L} \tag{1}$$

$$\dot{\gamma}_{app} = \frac{8V}{D} \tag{2}$$

Where P is the extrusion pressure (kPa), V is the mean extrudate velocity in the capillary (mm/s), L is the capillary length (mm) and D is the capillary diameter (mm). In addition, the end effects that occur when the flow regime is complicated, as is the case for paste systems, can be taken into consideration using Bagley's end correction [16], which determines the true wall shear stress in the capillary, τ_w, by:

$$\tau_w = \frac{PD}{4(L+ND)} \tag{3}$$

Where N is the end correction factor for the imaginary extension of the capillary length.

By using Equations (3) and (2), shear stress versus apparent shear rate curves can be obtained. However, researchers have also found that corrections may be needed to account for the wall slip that occurs in the capillary with highly stiff, concentrated pastes. [17-20] Recent work by Zhou and Li indicates that the Jastrzebski correction should be used when studying the rheology of fiber-reinforced cement pastes to account for wall slip [20]. Since only one capillary diameter was used in this investigation, wall slip corrections cannot be made and flow curves are given as wall shear stress versus apparent shear rate.

2.4 Capillary Analysis

Capillary analysis was conducted by extruding mixes through three different die lengths and at six different velocities. The extruder barrel had a diameter of 38.1 mm and a length of 125 mm, allowing approximately 120 ml of material to be extruded at a time. For each mix that was tested, eighteen experimental runs were made. Three die lengths (giving L/D = 1, 2 and 4) and six piston velocities, 0.2, 0.5, 1, 2, 3 and 5 mm/s, which correspond to extrudate velocities of 1.8, 4.5, 9, 18, 27 and 45 mm/s, respectively, were used. The rheometer was mounted in a closed-loop, MTS testing machine with a 24 kN load cell. Preliminary work showed that repeatable results were obtained from extruding the same mix in two different runs. Therefore, subsequent testing only involved one extrusion run per variable tested. The stiff cementitious dough was forced through the die at a constant displacement rate and the load and piston displacement were recorded.

Figure 2 presents an example of a rheometric curve obtained using capillary analysis. Yield stress (τ_0) was approximated using the lowest two data points and extrapolating to the y-axis. Using the differential viscosity versus apparent shear rate curve, an equilibrium viscosity ($\eta_{equilibrium}$) was defined as the differential viscosity at which the system equilibrated.

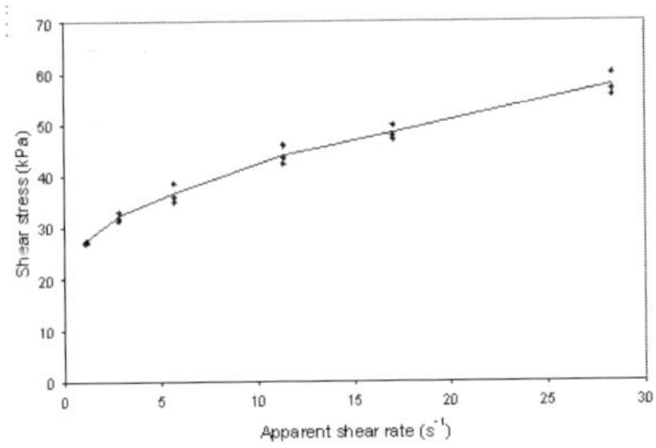

Figure 2. Example of rheometric curve obtained using capillary analysis (shown for W0.5)

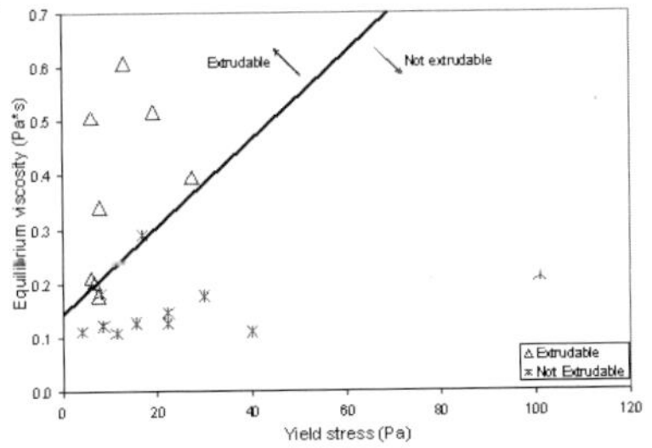

Figure 3. Equilibrium viscosity versus yield stress for extrudable and not extrudable mixes

The two rheological parameters obtained, τ_0 and $\eta_{equilibrium}$, were examined independently to see if either gave an indication of extrudability. However, no trends were observed. Figure 3 presents the two parameters plotted together, for both the extrudable and not extrudable mixes, and demonstrates that, when considered together, τ_0 and $\eta_{equilibrium}$, can be used to evaluate extrudability. Figure 3 suggests that an extrudable mix is one in which the yield stress is reasonably low (facilitating extrusion) and the equilibrium viscosity (probably related to thixotropy) is high.

3. CONCLUSION

This research examines the possibility of replacing cellulose ether with clay binders. Once extrudable and not extruable mixes are identified, capillary rheology is used to describe extrudability. The results indicate that cellulose ethers can be partially replaced by clay binders and, for the two clays examined, the effects are similar for either clay. Furthermore, the research shows that capillary rheology can be used to determine extrudability, with extrudable mixes having a low yield stress and a high equilibrium viscosity.

4. REFERENCES

1. V. C. Li and S. Wang, Tensile Strain-Hardening Behavior of Polyvinyl Alcohol Engineered Cementitious Composites (PVA-ECC), ACI Materials Journal. 98 (6), 483-492 (2001).
2. A. E. Naaman, Engineered Steel Fibers with Optimal Properties for Reinforcement of Cement Composites, Journal of Advanced Concrete Technology. 1 (3), 241-252 (2003).
3. S. H. Li, S. P. Shah, Z. Li, and T. Mura, Micromechanical Analysis of Multiple Fracture and Evaluation of Debonding Behavior for Fiber-Reinforced Composites, International Journal of Solids and Structures. 30 (11), 1429-1459 (1993).
4. Y. Shao, Z. Li, and S. P. Shah, Matrix Cracking and Interface Debonding in Fiber-Reinforced Cement-Matrix Composites, Advanced Cement-Based Materials. 1 (2), 55-66 (1993).
5. A. Peled, M. Cyr, and S. P. Shah, High Content of Fly Ash (Class F) in Extruded Cementitious Composites, ACI Materials Journal. 97 (5), 509-517 (2000).
6. A. Peled and S. P. Shah, Processing Effects in Cementitious Composites: Extrusion and Casting, Journal of Materials in Civil Engineering. 15 (2), 192-199 (2003).
7. B. Mobasher and A. Pivacek, A Filament Winding Technique for Manufacturing Cement Based Cross-Ply Laminates, Cement and Concrete Composites. 20 (5), 405-415 (1998).
8. P. L. Burke and S. P. Shah. "Durability of Extruded Thin Sheet PVA Fiber-Reinforced Cement Composites." ACI SP-190 High Performance Fiber-Reinforced Concrete Thin Sheet Products1999.
9. T. Malonn, K. Hariri, and H. Budelmann, Optimizing the Properties of No-Slump Concrete Products, Betonwerk + Fertigteil-Technik. 71 (4), 20-26 (2005).
10. T. Voigt, T. Malonn, and S. P. Shah, Green and Early Age Compressive Strength of Extruded Cement Mortar Monitored with Compression Tests and Ultrasonic Techniques, Cement and Concrete Research. (accepted).
11. Dow Chemical Company, Methocel Cellulose Ethers Technical Handbook, (2002).
12. Wolff Cellulosics Company, Preliminary Specification Walocel M-20678, (2004).
13. Stephan Schmidte Gruppe, Technisches Datenblatt Concresol 105 (in german), (2004).
14. Engelhard Company, Basic Concrete Materials and Methods Section 03050 - MetaMax, (2002).
15. K. G. Kuder, Extruded Fiber-Reinforced Cementitious Composites for Use in Residential Construction, Thesis in Civil and Environmental Engineering. 200 (2005).
16. E. B. Bagley, End Correction in the Capillary Flow of Polyethylene, Journal of Applied Physics. 28 624-627 (1957).
17. Z. D. Jastrzebski, Enterance Effects and Wall Effects in an Extrusion Rheometer During the Flow of Concentrated Suspensions, Industrial and Engineering Chemistry - Fundamentals. 6 (4), 445-453 (1967).
18. A. U. Khan, B. J. Briscoe, and P. F. Luckham, Evaluation of Slip on Capillary Extrusion of Ceramic Pastes, Journal of European Ceramic Society. 21 (4), 483-491 (2001).
19. P. J. Halliday and A. C. Smith, Estimation of the Wall Slip Velocity in the Capillary Flow of Potato Granule Pastes, Journal of Rheology. 39 (1), 139-149 (1995).
20. X. Zhou and Z. Li, Characterizing Rheology of Fresh Short Fiber Reinforced Cementitious Composites Through Capillary Extrusion, Journal of Materials in Civil Engineering. 17 (1), 28-35 (2005).
21. K. G. Kuder and S. P. Shah, Effects of Pressure on Resistance to Freezing and Thawing of Fiber-Reinforced Cement Board, ACI Materials Journal. 100 (6), 463-468 (2003).
22. K. G. Kuder and S. P. Shah, Freeze-Thaw Durability of Commerical Fiber-Reinforced Cement Board, in ACI SP-224: Thin Reinforced Cement-Based Products and Construction Systems, A. Dubey, Editor. 2004. p. 210.
23. K.G. Kuder, B. Mu, M.F. Cyr, and S.P. Shah. "Extruded Fiber-Reinforced Composites for Building Enclosures." NSF Housing Research Agenda Development and Workshop. Orlando, FL, USA 2004.
24. K. G. Kuder, E. B. Mu, and S. P. Shah, A New Method to Evaluate the Nailing Performance of HPFRCC for Residential Applications, Journal of Materials in Civil Engineering. (accepted 2005).

DESIGN OF HIGH STRENGTH SELF-COMPACTING CONCRETE FOR TUNNEL LININGS

B. Barragán,[1] R. Gettu,[2] X. Pintado[1] and M. Bravo[1]
[1]*Department of Construction Engineering, Universitat Politècnica de Catalunya, Barcelona, Spain;* [2]*Department of Civil Engineering, Indian Institute of Technology Madras, Chennai, India*

Abstract: A case study is presented where high strength self compacting concrete was developed for application in tunnel linings. The high compressive loads on the tunnel, limited lining thickness and heavy reinforcement necessitated the use of such a concrete. Limitations such as the unavailability of fillers led to specific mix optimization for three tunnels in Spain. A limited study on the use of a shrinkage reducing admixture to reduce long-term deformations in the concrete is also reported.

Keywords: Self-compacting concrete, mix design, tunnel lining, shrinkage.

1. INTRODUCTION

The use of chemical admixtures, such as superplasticizers and viscosity modifying agents (VMAs), has led to the development and use of self-compacting concrete (SCC) in various applications. Two such fields of application are the construction of slender elements, and the strengthening of structures, where the dimensions are often limited and the reinforcement density is high. SCC has been successfully used in the construction of thin-walled densely-reinforced structures, such as liquefied gas tanks[1-5] and tunnel linings[6], and in repair[7,8].

The present work reports on the development of high strength SCCs for a heavily reinforced tunnel lining in Spain The material had to fulfil strict requirements in terms of flowability, passing ability, resistance to segregation, shrinkage, early-age strength and long-term deformations. Since SCC is yet not a common construction material in Spain, specifications had to be defined for all relevant aspects of the concrete, and trial mixes had to developed in the laboratory. The details of the application and the mixes are detailed here. Several SCC mixes were designed by means of a four-step mix design method[9,10], which consists of optimizing the superplasticizer and filler dosages, the aggregate proportions and the paste volume. In addition to the mix development, data

M.S. Konsta-Gdoutos, (ed.), Measuring, Monitoring and Modeling Concrete Properties, 485–491.

from tests on SCCs without and with a glycol-based shrinkage reducing admixture are given, demonstrating the benefit of using this type of admixture to improve the long-term structural behaviour.

2. CASE STUDY

2.1 Background

In 2002, cracking and large deformations were observed in three tunnels (denoted as Camp Magre, Puig Cabrer and Lilla, respectively) of a high-speed train line being constructed in Spain. The damage to the tunnel lining was attributed to unforeseen expansion of the soil. In order to repair the tunnels and make them serviceable, a new lining was designed to resist the forces generated by the soil expansion, as well as the other actions to which the lining was subjected. The floor of the tunnel was to be constructed first and subsequently the semi-circular arch lining (see Fig. 1) with a thickness of 76 cm in all three tunnels, except for a section of the Lilla tunnel that had a lining thickness of 46 cm. The design compressive strength of the concrete was 80 MPa.

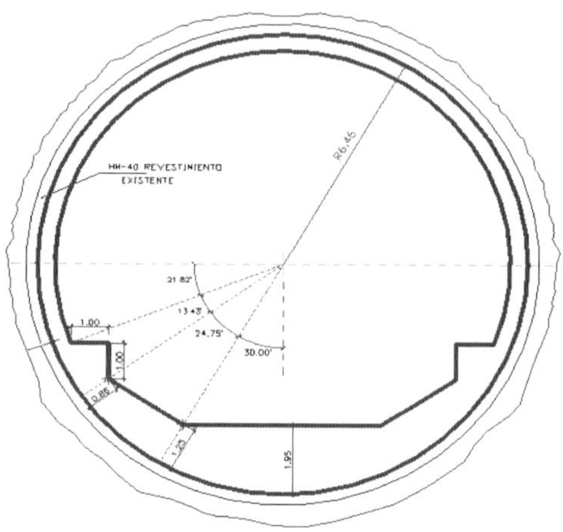

Figure 1. Cross-section of tunnel lining

The structural design specified circumferential reinforcement consisting of 8 bars of 32 mm diameter per meter, which could result in a minimum bar spacing of 30 mm in the splice zone. Also, in some zones, a second layer of 32 mm diameter bars was placed with a separation of 30 mm from the primary layer. Therefore, the concrete had to be able to flow through a 30 mm gap without blocking. The bar spacing in the longitudinal and shear reinforcement were always larger than 70 mm. The minimum clear cover was specified to be 35 mm. In all, the reinforcement density was about 10000 kg of steel per longitudinal meter of tunnel.

The project originally contemplated the use of a high slump concrete and intensive compaction through the use of shutter vibrators. However, considering the small bar spacing and the density of the reinforcement, the use of SCC in the construction of the tunnel linings was emphasized.

Other factors that had to be taken into account in the concrete design included the ambient temperature range of 5 to 35°C during the construction period, difficulty in curing of the lining with water (due to lack of adequate drainage within the tunnel), and the need for a transportation time of at least 30 minutes, high pumping speeds and early strength to facilitate the reuse of the formwork.

2.2 Specifications related to the concrete

Considering the characteristics of the project, structural design and the type of concrete that would be required, specifications such as the following were defined:

- Cement type CEM I 42.5 or 52.5 (according to the European standard EN 197-1:2000) and high sulphate resistance had to be used.
- The maximum aggregate size was limited to 12 mm.
- A silica fume dosage of up to 10% by weight of cement was recommended.
- A fly ash dosage of up to 25% by weight of cement was permitted.
- A limestone filler (with maximum grain size of 63 microns) dosage of 25% by weight of cement was permitted.
- The slump flow test had to be performed on site for each truckload of concrete. The slump flow spread (D_f) had to be in the range of 650±50 mm, and the time for a spread of 500 mm (T_{50}) had to be in the range of 6±3 seconds.
- The concrete had to satisfy the J-ring test in conjunction with the slump flow test, with the difference (D) between the slump flow spread without and with the J-ring not exceeding 50 mm.
- In the V-funnel test of the fresh concrete, the flow time (T_V) had to be in the range of 10±5 seconds.
- Compressive strength tests had to yield characteristic cylinder strengths of at least 12.5 MPa at 24 hours and 80 MPa at 28 days.
- In addition, the mix had to be evaluated previously for ensuring a splitting-tensile strength of at least 8 MPa at 28 days, and an autogenous shrinkage strain (measured in sealed specimens maintained at 20°C) of not more than 300 microstrains at the age of 3 months.
- Water curing for at least the first 7 days was prescribed in order to limit cracking due to early-age shrinkage and thermal stresses.

3. MIX DESIGN

The first phase of the experimental program consisted of the design of SCC mixes following the mix design methodology proposed by Gomes et al.[9,10] Within this procedure, the cement paste is optimized for high fluidity and moderate cohesion using the Marsh cone and mini-slump tests, and the best particle packing of the aggregate skeleton is determined by checking the void content in the dry uncompacted state. Subsequently, the minimum paste volume was decided based on the tests for self-compactability specified above. The mixes were first evaluated in the laboratory and then checked with trials on site.

Several mixes had to be designed considering practical limitations in terms of availability of the components and subsequent changes required by the contractors. In terms of strength and durability, it was decided that the cement content should be in the range of 350-500 kg/m^3, and the maximum water/cement ratio would be 0.35.

The mix proportions used for the laboratory and on-site trials are given in Table 1, along with the fresh concrete properties. All concretes had mean compressive strengths of 95-105 MPa. Mix L1 was the first that was proposed with silica fume and limestone filler; note that the properties were not all within the prescribed range but an improvement was expected in the more efficient plant mixing. However, the readymix plant could not obtain the needed quantities of filler and, therefore, further mixes had to be proposed with less or no filler. Consequently, Mix L2 was proposed with a filler dosage of only 10%, by weight of cement. In Mix L3, the limestone filler of Mix L1 was substituted with fly ash, to explore the possibility of convincing the contractor to use it but again sufficient quantities could not be obtained. Mixes L1 and L3 had a cement content of 498 kg/m^3 and a powder content of 672 kg/m^3, and exhibited similar properties. Mix L2 had about the same cement content but much lower powder content, i.e., 600 kg/m^3. The lower paste content led to a lower V-funnel flow time.

Mixes L4, L5 and L6 had a powder content of 600 kg/m^3 (consisting of cement and silica fume) and no fillers. However, to compensate for the reduced powder content a VMA was incorporated in order to provide the needed cohesiveness and increase the robustness of the concrete. Different combinations of sands were tried in these three mixes to give more options to the construction contractors. The properties are similar in these three mixes. Mix L7 is identical to Mix L6 with the exception that had glycol-based shrinkage reducing admixture (SRA) at a dosage of 5 kg/m^3. This mix was used to evaluate the improvement in the shrinkage due to the incorporation of the SRA.

On site trials were performed with two mixes that were obtained from the compositions of the most feasible laboratory mixes. The Mix S1 is between to Mixes L1 and L2, in terms of the limestone filler dosage. It exhibited good self-compactability and higher slump flow than the laboratory mixes, as expected. The only drawback of this mix was the need to use limestone filler. Mix S2 is a modification of Mix L6, with lower water content, that satisfied the specifications with satisfactory self-compactability. Eventually, a mix composition similar to Mix S2 was used for almost the entire tunnel lining.

4. SHRINKAGE STUDIES

In order to study the shrinkage of some of the SCC mixes, and to evaluate the effect of an SRA on such response, cylindrical specimens of 150×300 mm were cast. The mixes that were tested are L4, L6 and L7. The autogenous shrinkage strains obtained in sealed specimens are shown in Fig. 2. It can be observed that the concrete with the SRA (L6) exhibits much lower shrinkage than the other two. Fig. 3 presents the corresponding curves for specimens subjected to drying in an environment of 20°C temperature and 50% r.h., after a curing period of 105 days. Again, the reduction of total shrinkage (mainly drying shrinkage) due to the SRA is evident. Such trends have been previously reported for normal strength concrete incorporating the same SRA[11].

Table 1. Mix Proportions and self-compactability parameters

	Mix L1	Mix L2	Mix L3	Mix L4	Mix L5	Mix L6	Mix L7	Mix S1	Mix S2
Components (kg/m^3)									
Cement I 52,5 SR	498	500	498	500	500	500	500	492	500
Limestone filler	124	50	-	-	-	-	-	93	-
Silica fume	50	50	50	50	50	50	50	49	50
Fly ash	-	-	124	-	-	-	-	-	-
Water	182	186	182	202	197	193	193	188	165
Crushed limestone sand (0-2 mm)	-	-	-	426	-	-	-	-	-
Natural siliceous sand (0-2 mm)	337	352	337	-	-	-	-	373	-
Crushed limestone sand (0-5 mm)	505	528	505	638	1067	576	576	508	576
Natural siliceous sand (0-5 mm)	-	-	-	-	-	411	411	-	411
Crushed granite gravel (6-12 mm)	688	720	688	573	575	658	658	684	658
Superplasticizer (polycarboxylate)	15.7	13.2	14.5	11.8	10.5	11.8	11.8	12.7	11.0
VMA	-	-	-	1.8	2.5	2.0	2.0	-	2.0
SRA (glycol-based)	-	-	-	-		-	5.0	-	-
Test results									
Slump flow spread, D_f (mm)	630	630	610	640	610	590	590	710	610
Slump flow time, T_{50} (s)	4	3	4	3	3	3	3	2	4
V-funnel time, T_V (s)	23	9	16	11	12	9	9	11	11
J-ring difference, D (mm)	50	45	50	30	40	30	30	0	40

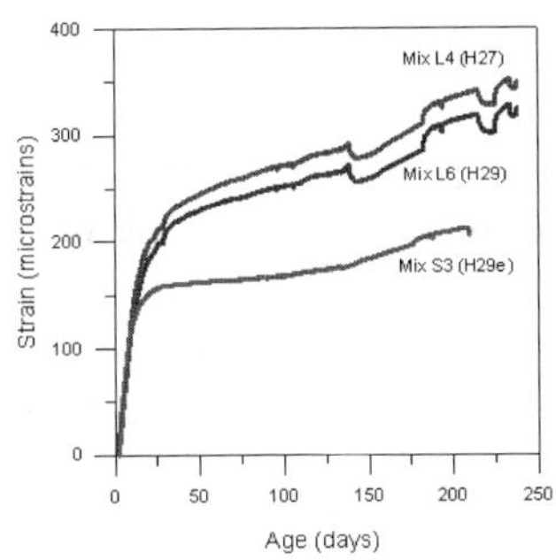

Figure 2. Autogenous shrinkage

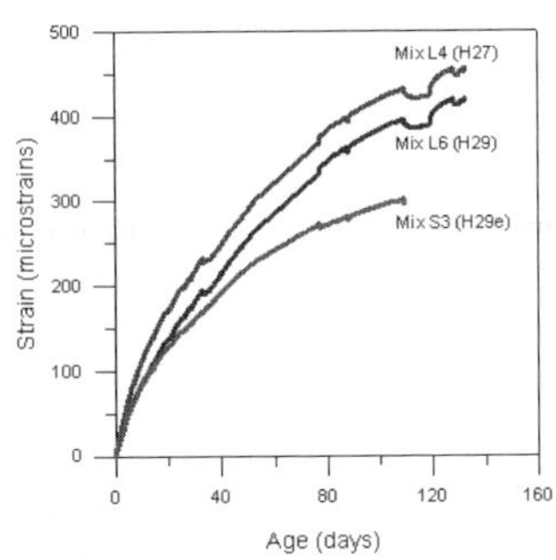

Figure 3. Total shrinkage

5. CONCLUSIONS

In situ placing of concrete in tunnel linings can benefit significantly from the use of high strength self-compacting concrete, especially when the lining thickness is relatively small and the reinforcement density is high.

One of the major restrictions encountered when SCC is used at the job site is the unavailability of fillers and the impossibility of storing and batching fillers in an existing readymix concrete plant. Under such conditions, an adequate combination of a superplasticizer and a viscosity modifying agent helped overcome such limitations in the case study presented here.

The beneficial role of the chemical admixtures can be extended also to the hardened state when a shrinkage reducing admixture is incorporated to reduce the autogenous and total shrinkage values significantly.

Acknowledgments

Partial funding for this work was provided by the Spanish Ministry of Science and Technology grant MAT 2003-5530, and the Spanish Ministry of Education and Science grant PSE 11-2005 to the UPC. The authors thank Corsan-Corviam and Ferrovial Agroman, the principal construction contractors, for the financial and technical support for the work at UPC and on site. The initiative of GIF, the promoter of the project, which made possible the collaboration between the UPC and the constructors, is gratefully appreciated. The help of Getinsa, the technical assistance unit on site, is also acknowledged. The second author is grateful to UPC and IITM for facilitating a stay in Barcelona as visiting professor in 2005, which made the collaboration possible.

6. REFERENCES

1. Okamura, H., "Self-Compacting High-Performance Concrete", *Concr. Intnl.*, V. 19, No. 7, pp. 50-54, 1997.
2. Nishizaki, T., Kamada, F., Chikamatsu, R. and Kawashima, H., "Application of High-Strength Self-Compacting Concrete to Prestressed Concrete Outer Tank for LNG Storage", *Proc. 1st Intnl. RILEM Symp. on Self-Compacting Concrete*, Eds. A. Skarendahl and Ö. Petersson, RILEM Publications S.A.R.L., Cachan, France, pp. 629-638, 1999.
3. Ouchi, M., "Self-Compacting Concrete: Development, Applications and Investigations", *Nordic Concrete Research*, No. 23, http://www.itn.is/ncr/ publications/doc-23-3.pdf, 5 p., 1999.
4. Seto, K., Okada, K., Yanai, S. and Nobuta, Y., "Development and Applications of Self-Compacting Concrete", *Proc. Intnl. Conf. on Engineering Materials* (Ottawa, Canada), Eds. A.Al-Manaseer, S.Nagataki and R.C.Joshi, CSCE/JSCE, Ottawa/Tokyo, Vol. I, pp. 413-429, 1997.
5. Ouchi, M., "Current Conditions of Self-Compacting Concrete in Japan", *Proc. Second Intnl. Symp. on Self Compacting Concrete* (Tokyo), Eds. K.Ozawa and M.Ouchi, COMS Engineering Corp., Kochi, Japan, pp. 63-68, 2001.
6. Takeuchi, H., Higuchi, M. and Nanni, A., "Application of "Flowable" Concrete in a Tunnel Lining", *Concr. Intnl.*, V. 16, No. 4, pp. 26-29, 1994.

7. McLeish, A., "Evaluating the Flow Properties of Flowable Concrete", *Special Concretes: Workability and Mixing* (Proc. Intnl. RILEM Workshop, Paisley, Scotland), Ed. P.J.M.Bartos, E&FN Spon, London, pp. 249-257, 1993.
8. Lacombe, P., Beaupré, D. and Pouliot, N., "Rheology and Bonding Characteristics of Self-Levelling Concrete as a Repair Material", *Proc. Fifth CANMET/ACI Intnl. Conf. on Superplasticizers and Other Chemical Admixtures in Concrete* (Rome), Supplementary Papers, pp. 163-184, 1997.
9. Gomes, P.C.C., Gettu, R., Agulló, L. and Bernad, C., "Experimental Optimization of High-Strength Self-Compacting Concrete", *Proc. Second Intnl. Symp. on Self-Compacting Concrete* (Tokyo), Eds. K.Ozawa and M.Ouchi, COMS Engng. Corp., Kochi, Japan, pp. 377-386, 2001.
10. Gettu, R., Gomes, P.C.C., Agulló, L. and Josa, A., "High-Strength Self-Compacting Concrete with Fly Ash: Development and Utilization", *Proc. Eighth CANMET/ACI Intnl. Conf. on Fly Ash, Silica Fume, Slag, and Natural Pozzolans in Concrete* (Las Vegas, USA), ACI SP-221, Ed. V.M.Malhotra, American Concrete Institute, Farmington Hills, USA, pp. 507-522, 2004.
11. Roncero, J., Gettu, R. and Martín, M.A., "Evaluation of the Influence of a Shrinkage Reducing Admixture on the Microstructure and Long-Term Behavior of Concrete", *Proc. Seventh CANMET/ACI Intnl. Conf. on Superplasticizers and Other Chemical Admixtures in Concrete* (Berlin), Supplementary papers, pp. 207-226, 2003.

Quantitative Image Analysis for Microstructural Characterization of Concrete

CHARACTERISING THE PORE STRUCTURE OF CEMENT-BASED MATERIALS USING BACKSCATTERED ELECTRON AND CONFOCAL MICROSCOPY

H.S. Wong, M.K. Head and N.R. Buenfeld
Concrete Durability Group, Imperial College London, SW7 2AZ London

Abstract: The pore structure of cement-based materials affects their mechanical properties, shrinkage behaviour, molecular/ionic transport properties and durability. This paper presents an overview of our work on imaging of capillary pores in cement-based materials using 2D backscattered electron microscopy and 3D laser scanning confocal microscopy. Our aim is to develop an integrated imaging approach that is able to provide relevant parameters of the pore structure that can be used as input values to transport prediction models. Topics covered in this paper include sample preparation, the importance of epoxy penetration and its relevance to patch microstructure, application of Monte-Carlo methods to simulate electron-solid interactions in cement-based materials, development of image analysis tools for accurate pore segmentation and for obtaining microstructural gradients at interfaces, and preliminary results on transport prediction using data obtained from 2D image analysis. The applicability of laser scanning confocal microscopy for 3D imaging of pores at sub-micron resolution and evidence for interconnectivity of Hadley grains with capillary pores are presented.

Key words: backscattered electron microscopy; Euclidean distance mapping; Hadley grains; image analysis; interfacial transition zone; laser scanning confocal microscopy; Monte-Carlo simulation; patch microstructure; pore structure; transport properties.

1. BACKSCATTERED ELECTRON IMAGING

In the last two decades, backscattered electron imaging (BSEI) has proven to be an important technique for qualitative and quantitative study of the microstructure of cement-based materials[1]. The brightness of various phases in BSEI is a function of the mean atomic number and so the resin-filled pores appear the darkest. The pores are

M.S. Konsta-Gdoutos, (ed.), Measuring, Monitoring and Modeling Concrete Properties, 495–502.

clearly contrasted and can be segmented for subsequent measurements. Considering its applicability across a range of magnifications and its high resolution, BSEI is an ideal method for 2D characterisation of the pore structure.

1.1 Epoxy Penetration And Patch Microstructure

In BSEI, it is essential for the pores to be saturated with epoxy, which upon hardening, supports the delicate microstructure and provides atomic contrast to the pores. Non-resin filled pores are not visible in BSEI. However, grinding could easily exceed the epoxy penetration depth since conventional vacuum impregnation gives only ~100-200 μm epoxy penetration[2]. We have developed a slight modification to the conventional impregnation technique by using toluene to reduce the viscosity of the epoxy and by applying a small over-pressure to force the epoxy into the pores[3]. Penetration of several millimetres has been achieved. This gives more tolerance during grinding and ensures that the imaged surface remains epoxy-saturated. Using this new technique, we showed that the recently proposed 'patch microstructure'[4] is actually an artefact caused by grinding beyond the epoxy depth[3]. Area matching BSE images found that samples that were re-impregnated using the new method no longer displayed the broad dense patches that were originally present.

1.2 Monte-Carlo Simulation

Knowledge of the size of the interaction volume and the sampling volume of various signals within it is important for interpretation of images and analytical results obtained from electron microscopy. To this end, we have been using a Monte Carlo technique[5] to simulate the electron trajectories in order to determine the shape and size of the interaction volume (Figure 1), the spatial and energy distribution of backscattered electrons and characteristic x-rays in cement-based materials. Monte Carlo simulation has also been used to determine the optimal imaging strategy for cement-based materials, to study the signal variation across phase boundaries and to determine the theoretical resolution limit for quantitative imaging of pores.

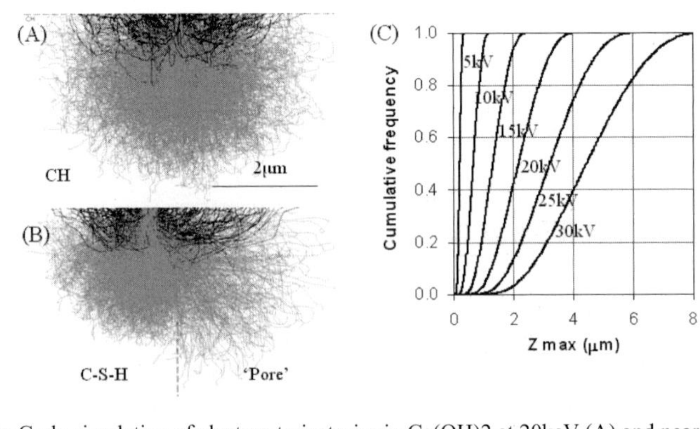

Figure 1. Monte Carlo simulation of electron trajectories in Ca(OH)2 at 20keV (A) and near a hypothetical C-S-H/epoxy-filled pore boundary (B). The electron trajectory is followed until it loses all of its energy (grey lines) or is backscattered (black lines). Figure (C) shows the effect of accelerating voltage on the maximum penetration depth of electrons in Ca(OH)2

1.3 Image Analysis And Transport Prediction

Accurate quantitative microscopy requires analysis of many images for statistical confidence and this presents an obstacle for practical applications. The choice for the pore threshold is crucial. We found that using an objective greyscale thresholding rule could eliminate operator judgement and substantially reduce the number of images necessary. We developed a segmentation technique for pores and cracks[6] and proposed that the threshold be determined from the inflection point of the cumulative brightness histogram (Figure 2). This represents a critical point where a small incremental grey value causes a large increase in segmented area, a condition termed as *overflow*.

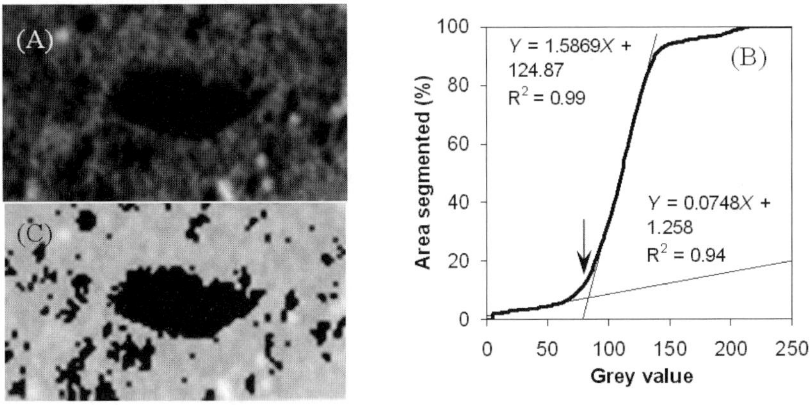

Figure 2. The 'overflow' pore segmentation technique

There have been disputes regarding the conventional ITZ/bulk paste model[7]. We developed a new image analysis routine for computing microstructural gradients that is much faster, achieves greater resolution and is unrestrained by boundary conditions, using Euclidean Distance Mapping[8]. Initial findings confirmed that there is a strong gradient in 'average' porosity at the ITZ, but substantial variation from one image to another according to the presence of $Ca(OH)_2$ deposits on aggregate surfaces and the adopted sampling procedure (Figure 3). The higher sensitivity of the new method enabled it to detect previously unreported effects of $Ca(OH)_2$ on the 'average' porosity gradient. The arrow in Figure 3A shows a sudden drop in average detectable porosity at less than 5μm away from the aggregate-cement paste interface.

Figure 4 shows preliminary results on transport prediction of mortars with a range of pore structure characteristics using porosity and specific surface values obtained from 2D image analysis[9]. Oxygen diffusivity and permeability was predicted from the Van Brakel and Heertjes model, and a modified Kozeny-Carman equation[9] respectively. We assumed that the aggregate particles are impermeable and used the paste tortuosity as a lower bound estimate for the pore tortuosity. The preliminary results (Figure 4) are encouraging and show that despite the limitations of 2D imaging, it is still a viable tool for extracting quantitative information for transport prediction.

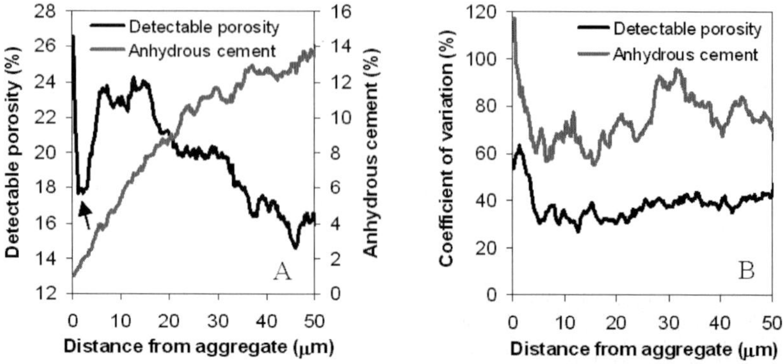

Figure 3. A) Detectable porosity and anhydrous cement gradients from the aggregate surface, measured at single pixel strip width using Euclidean Distance Mapping, and B) Coefficients of variation. Values are average of 30 frames. Sample is an OPC concrete of w/c ratio 0.4

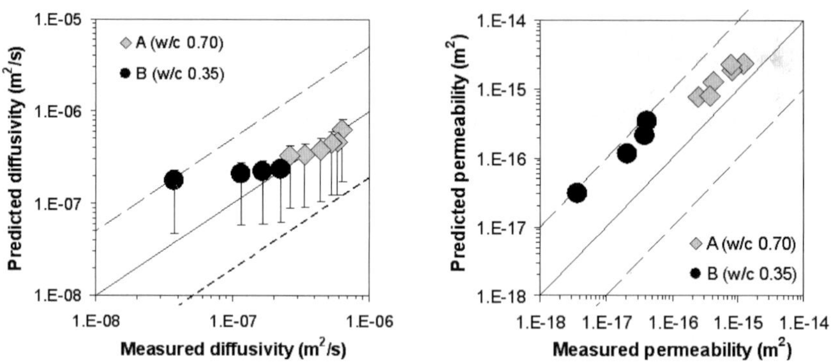

Figure 4. Predicted and measured values for oxygen diffusivity and oxygen permeability on mortars of different w/c ratios, curing age and conditioning regime

2. LASER SCANNING CONFOCAL MICROSCOPY

To investigate the 3D nature of pores and voids in hardened concrete and mortar, various 3D imaging techniques were evaluated to identify the one that most closely meets the high resolution imaging requirements of the work. Laser scanning confocal microscopy (LSCM) was found to provide the best image resolution, when used in epi-fluorescent imaging mode. An argon gas laser emits light at 488 nm and induces excitation of fluorophores in the impregnated resin causing light to be emitted at 505 nm, which is passed to a photo detector via a very small pinhole, blocking out-of-focus light. Confocal microscopy has been previously used to investigate cementitious materials, but work has been based on the use of visible light. This has the disadvantage that the pore structure cannot be contrasted against solid phases, and so images are only of solid surfaces. When fluorescent dye is added to the impregnated epoxy resin, it is possible to reveal structures from below a specimen surface, as the laser probe is scanned across the specimen.

2.1 Image capture

We have examined several mortars using LSCM[10], and BSEI was used to confirm features that were imaged. Figure 5A is a BSEI of a mortar (0.7 w/c) showing two partially reacted cement grains, subsequently area-matched by LSCM (Figure 5B).

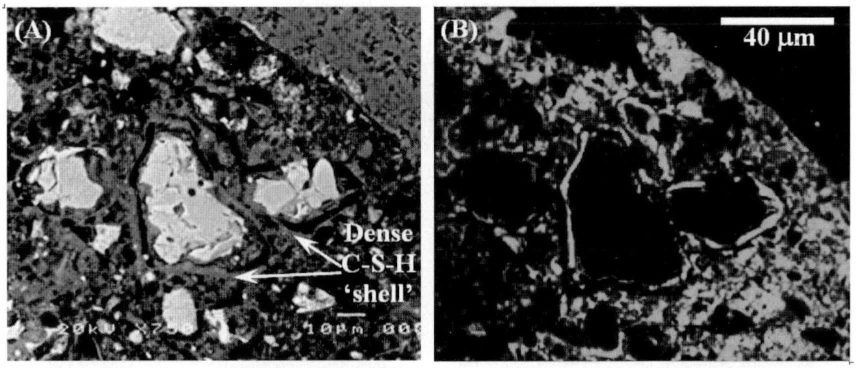

Figure 5. A) BSEI of mortar area-matched by LSCM (B). From Ref [10]

The grains display a porous inner 'zone' that surround an anhydrous core, and is regularly imaged during BSEI investigations. Surrounding this porous zone is a more dense 'shell' of inner product C-S-H (marked in Figure 5A). In the backscattered image (Figure 5A), pixels corresponding to internal pore space appear black due to differences in the atomic numbers of hydration and anhydrous products (brighter), and the epoxy resin (darker). In the confocal image however (Figure 5B), fluorescent resin filled features appear bright green, due to direct detection and imaging of fluorescent light. Microcracks and air voids are not evident in this image, but capillary pores are visible as small fluorescent features in the background.

2.2 Resolution

The resolution of the confocal technique provides the highest currently available by an optical technique and actually exceeds BSE imaging when performed with 'standard' SEMs, which are limited by effects of interaction volumes (see section 1.2). So far features as small as 0.17 μm have been resolved by fluorescent LSCM, but imaging is limited to the upper 10 μm or so of the surface due to signal loss with increased depth. However, this is sufficient to image sub-μm size features.

2.3 Early application of technique to hadley grains

This technique is well suited to the imaging of hollow shell cement or 'Hadley' grains[11], and evidence of possible impregnation routes into Hadley grains was observed.

Figure 6 is a high resolution image of a capillary pore 'p' and hollow shell cement grains 'h' observed in a 0.6 w/c mortar. Two 'links' can be observed to connect the pore to two of the grains. These were interpreted as sub-micrometer connecting channels 'cc', as opposed to micro-cracks, mainly due to their morphology and that they do not extend into any other plane. If this hypothesis is true, hollow cement grains linked by capillary pores may facilitate the flow of fluids and ions.

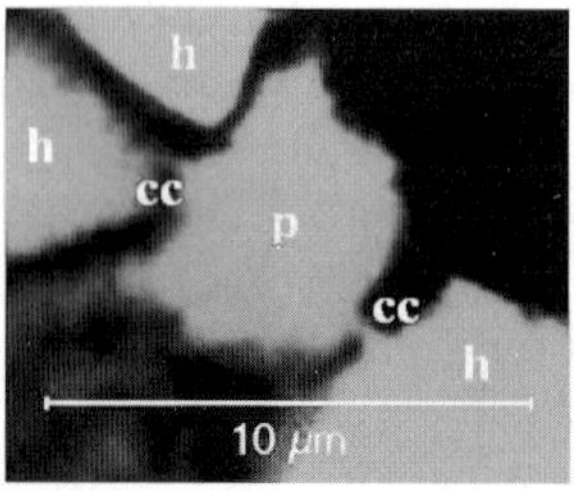

Figure 6. Capillary pore 'p' connected by two small porous channels 'cc' to hollow shell cement grains 'h'. From Ref [13]

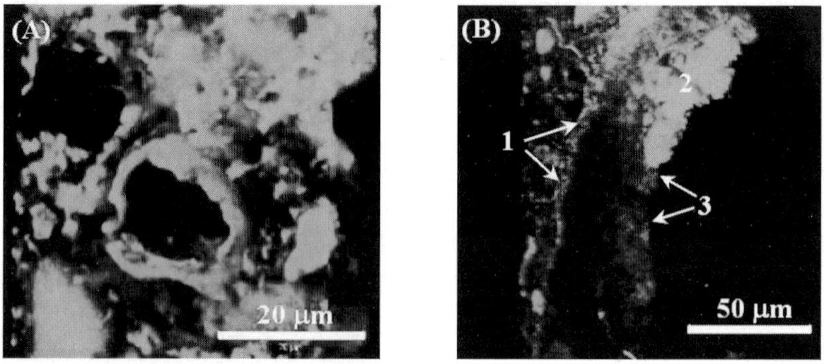

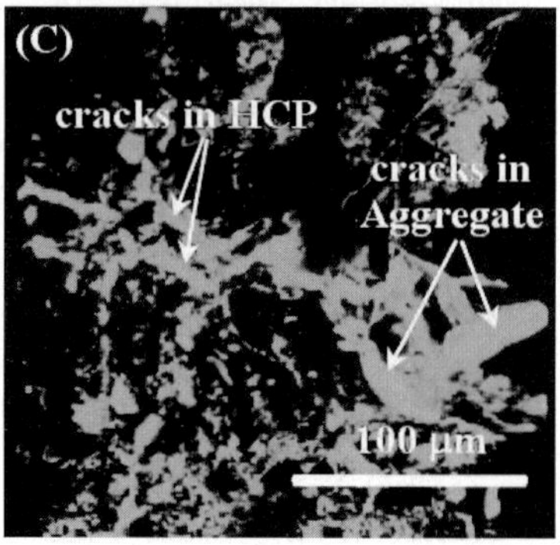

Figure 7. 3D projections of A) a partly reacted cement grain, B) a natural aggregate interface and C) a micro-crack crossing the interface between HCP and natural aggregate

2.4 3D projections of impregnated porous structures

Figure 7A is a 3D projection showing a partly reacted cement grain (image centre), outside of which, other fluorescent features represent capillary pores. A small fully reacted cement grain can be seen in the lower left of the image. Figure 7B is an image of a natural aggregate interface. A bright strip (1) can be seen running from the top centre edge to the lower left edge of the image, which marks the specimen upper surface as it intersects with the aggregate interface. An area of high interfacial porosity can be observed as a bright fluorescent patch in the upper right of the image (2), intersecting the lower bonded edge (3) of the specimen.

Figure 7C is a 3D projection of a micro-crack that cuts across the interface between a natural aggregate particle and an area of hardened cement paste (HCP). The interface between aggregate and HCP is less clear in this image, but is located just to the right of the two arrows that mark a division in the crack as it appears on the surface. Manipulation of the virtual model in 3D space however, reveals that the crack is still linked below the surface, and that different crack geometries are present in both HCP and aggregate.

3. CONCLUSIONS

This paper presents a summary of our on-going work in characterising pores in cement-based materials, using an integrated approach of backscattered electron imaging and laser scanning confocal microscopy. We have made improvements in sample preparation techniques and shown that recently reported patch microstructure is an artefact caused by grinding beyond the epoxy depth. A Monte Carlo technique was used to assess the interaction volume, and sampling volume of backscattered electrons and characteristic x-rays. The Monte Carlo simulations also allowed us to determine the optimal imaging strategy for cement-based materials, to study signal variation across phase boundaries and the theoretical resolution limit for quantitative pore imaging. We have also developed image analysis methods for accurate pore segmentation and computing microstructural gradients at interfaces. Preliminary results on transport prediction using the quantified pore structure were presented.

It is possible to image the surface area of specimens impregnated with fluorescent epoxy resin using LSCM, and an electron microscope can be used in BSE mode to confirm the nature of porous features observed. By scanning the specimen in the z-axis direction, 3D images can be constructed at sub-micrometer resolutions. So far, very small connections between capillary pores and partly hydrated cement grains have been imaged, together with natural aggregate interfaces, and micro-cracks.

Acknowledgments

HSW acknowledges the financial assistance given by Universities UK, via the ORS Awards Scheme. We would also like to acknowledge the EPSRC for support under grant numbers M97206, T25439, and S18175. We are grateful to Mr. R.A. Baxter for his help with the laboratory work.

4. REFERENCES

1. K.L. Scrivener, Backscattered electron imaging of cementitious microstructures: understanding and quantification, *Cem. Concr. Compos.*, 26 (8) 935-945 (2004).
2. K.O. Kjellsen, A. Monsøy, K. Isachsen, R.J. Detwiler, Preparation of flat-polished specimens for SEM-backscattered electron imaging and X-ray microanalysis – importance of epoxy impregnation, *Cem. Concr. Res.*, 33, 611-616 (2003).
3. H.S. Wong, N.R. Buenfeld, Patch microstructure in cement-based materials: Fact or artefact? *Cem. Concr. Res.*, 2005 (In press).
4. S. Diamond, Percolation due to overlapping ITZs in laboratory mortars? A microstructural evaluation, *Cem. Concr. Res.*, 33 (7) 949-955 (2003).
5. H.S. Wong, N.R. Buenfeld, Monte Carlo simulation of electron-solid interactions in cement-based materials, *Cem Concr. Res.*, 2005 (Submitted).
6. H.S. Wong, M.K. Head, N.R. Buenfeld, Pore segmentation of cement-based materials from backscattered electron images, *Cem. Concr. Res.*, 2005 (In press).
7. S. Diamond, J. Huang, The ITZ of concrete – a different view based on image analysis and SEM observations, *Cem. Concr. Compos.*, 23, 179-188 (2001).
8. H.S. Wong, N.R. Buenfeld, Euclidean Distance Mapping for computing microstructural gradients at interfaces in composite materials, *Cem. Concr. Res.*, 2005 (In press).
9. H.S. Wong, N.R. Buenfeld, M.K. Head, Estimating transport properties of mortars using image analysis on backscattered electron images, 10th Euroseminar on Microscopy Applied to Building Materials, June 22-25, 2005, University of Paisley.
10. M.K. Head, N.R. Buenfeld, Confocal imaging of porosity in hardened concrete, *Cem. Conc. Res.*, 2005 (In press).
11. M.K. Head, H.S. Wong, N.R. Buenfeld, Characterisation of 'Hadley' grains by confocal microscopy, *Cem Concr. Res.*, 2005 (Submitted).

FRACTOGRAPHY OF FIBER-CEMENT COMPOSITES VIA LASER SCANNING CONFOCAL MICROSCOPY

B.J. Mohr and K.E. Kurtis
Tennessee Technological University, Department of Civil and Environmental Engineering, Box 5015, 1020 Stadium Drive, Cookeville, TN 38505; Georgia Institute of Technology, School of Civil and Environmental Engineering, 790 Atlantic Drive, Atlanta, GA 30332

Abstract: Fracture surface characteristics of pulp fiber-cement composites have been quantitatively evaluated by laser scanning confocal microscopy (LSCM) to examine the influence of fiber addition rate and matrix composition on mechanical behavior and fracture processes. A strong correlation was found between the fracture surface roughness and the post-cracking toughness in these composites. In addition, while not contributing to toughness, an inherent surface roughness, likely due to the inhomogeneous and porous microstructure of the hydrated cement paste, was apparent in both the roughness number and fractal dimension measurements. The fractal dimension of the fracture surfaces showed that matrix cracking was a contributing factor to increased toughness. Increased toughness of the composites was attributed to increased fiber pull-out, as compared to samples with minimal toughness which primarily failed by fiber fracture. The partial replacement of portland cement with supplementary cementitious materials did not have an observable effect of the fracture surface roughness.

Key words: cement; composite; fractography; fractal dimension; fibers; microscopy; roughness

1. INTRODUCTION

Fractography is a method by which physical features of a material's fracture surface can be related to mechanical properties, primarily toughness or fracture energy. The fracture surface characteristics of fiber-reinforced cement-based materials are of particular importance when evaluating the toughening mechanisms of the fibers. Two fracture parameters that have been subject to previous research include the surface roughness number and fractal dimension[1-4].

M.S. Konsta-Gdoutos, (ed.), Measuring, Monitoring and Modeling Concrete Properties, 503–508.

Several methods are available for the acquisition of a fracture surface image or profile at the microscale including, stereoscopic SEM, profilometry, and confocal microscopy. Confocal microscopy has significant advantages over the prior techniques, though its use has generally been limited to applications in the biological sciences and in materials science, where it has been widely applied to characterize the surface of semiconductors.

However, laser scanning confocal microscopy (LSCM) has emerged as a tool for cement-based materials microcharacterization[5]. LSCM has the capability of generating three-dimensional volumetric fracture surface representations through the acquisition of a series of images obtained at focal planes separated by a user-specified distance. In this research, LSCM has been used to quantitatively evaluate fracture surface characteristics of pulp fiber-cement composites. The aim was to investigate the use of fractography of composite fracture surface as a means to correlate fracture surface characteristics to composite post-cracking toughness, and to use the measures to assess the effects of environmentally-induced composite degradation and matrix composition.

2. METHODOLOGY

LSCM observations were conducted using an argon laser (= 488-514 nm) Leica Confocal TCS NT. Fracture surfaces were observed in reflected light mode using an H PLAN 20x/0.40∞/1.8 Q/B objective lens (magnification = 400X).

Examined pulp-fiber cement composite samples were prepared at a water-to-cement ratio of 0.60 using commercially available Type I cement and softwood kraft pulp fibers, and were tested in flexure by ASTM C 348-97 and C 293-94. Further details regarding sample composition, preparation, and testing are provided in Mohr et al.[6,7]

Each fracture surface had a total projected area of 25.4 x 25.4 mm. For this research, eight 500 x 500 μm regions of interest (ROIs) were imaged for each sample; this number was determined to well-represent the features of the surface. Each area was imaged in the reflective mode and at a voltage of 435 V. For this particular research, a step size of 5 μm between focal planes was chosen to optimize resolution and imaging time.

Upon completion of the series scanning, fractography measurements were made. Quantitative measurements included fracture surface roughness number (RN) and fractal dimension. These factors were determined to evaluate the influence of the fiber degradation due to environmental exposure and matrix composition on the composite fracture behavior. Fracture surface roughness was determined by the actual surface area to the projected surface area ratio[5].

In addition, by evaluating the fractal dimension, the effect of fiber addition on matrix roughness may be isolated. The fractal dimension was determined by acquiring a z-axis depth profile of a user-defined line on the LSCM fracture surface image. The fractal dimension was the determined using the ruler method. The ruler method was most applicable to this research as the profiles were obtained in actual dimensions, as opposed to gray scales values (i.e., 0-255) that are on a different scale from the x- and y-axes.

3. RESULTS AND DISCUSSION

3.1 Fracture Surface Roughness Number

Based on preliminary results, as seen in Figure 1, a strong correlation exists between the fracture surface roughness and the post-cracking toughness in pulp fiber-cement composites. This fractography measurement is an important parameter, as toughness is also related to fracture energy (e.g., K_{Ic}). This correlation appears to be linear in nature. One interesting aspect of this linear behavior is the roughness number y-axis intercept (i.e., no theoretical toughness). In theory, a fracture surface with zero toughness should exhibit a flat fracture surface as would be indicated by a roughness number of 1.0. However, in the observed measurements, the roughness number intercept is significantly greater than 1.0. For these composites, the increased roughness may be related to the imaging technique (i.e., maximum observable resolution), which may be evaluated by performing LSCM on the same fracture surface but at different magnification. However, it is believed that the inherent porosity and heterogeneity of the hydrated cement paste contributes to the roughness of the fracture surface, while not contributing to fracture energy dissipation.

Pulp-fiber cement composites containing varying amounts of supplementary cementitious materials (SCMs), by weight percent of cement, were also examined. Previous research has produced conflicting results regarding the influence silica fume on fracture surface roughness. Wang and Diamond[8] have shown that 15% silica fume had a negligible effect of the roughness of cement paste fracture surfaces. However, Abell and Lange[1] observed a notable increase in the cement paste fracture surface roughness number with silica fume replacement values of 5 and 10%.

This research investigated the effect of silica fume (10, 30, and 50%), slag (10, 30, 50, 70, 90%), and Class C fly ash (10, 30, 50, 70%) on the fracture surface roughness number of fiber-cement composites. As shown in Figure 2, no significant differences can be seen between the three types of SCMs and at varying SCM replacement amounts in agreement with Wang and Diamond[8].

3.2 Fracture Surface Fractal Dimension

Another fractography parameter examined was the surface fractal dimension, which allowed for the isolation of matrix roughness from the net fracture surface roughness. Thus, the effect of fiber addition rate on the composite toughening can be evaluated. In theory, a matrix roughness line with zero slope would indicate that the fibers do not contribute to energy dissipation during fracture due to pure fiber pull-out. An increase in the slope of the net surface line indicates that the fibers have an increasing effect on dissipating energy through matrix crack deflection. In addition, the relative slopes of these lines may be used to quantify the extent of fiber-cement bonding and the degree of fiber pull-out versus fiber fracture during composite failure.

As seen in Figure 3, differences in the roughness of the net fracture surface (including fiber effects) are apparent as compared to the matrix roughness. In these materials, it can be seen the fibers do contribute to the energy dissipation, indicating that composite fracture was a combination of fiber pull-out and fiber fracture.

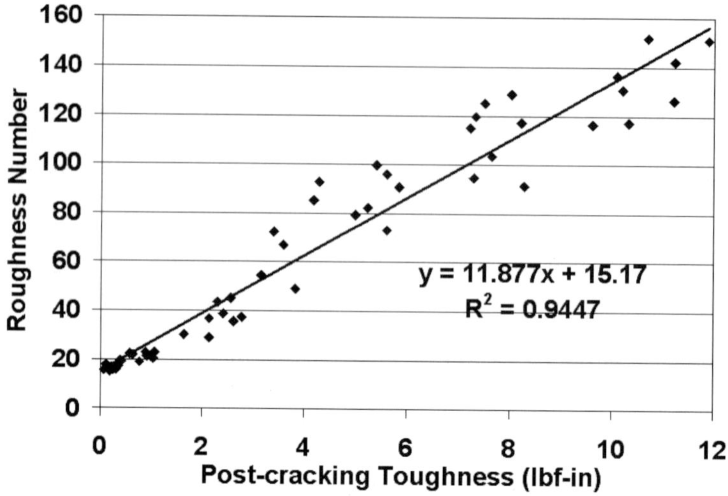

Figure 1. Correlation between fracture surface roughness number and composite post-cracking roughness

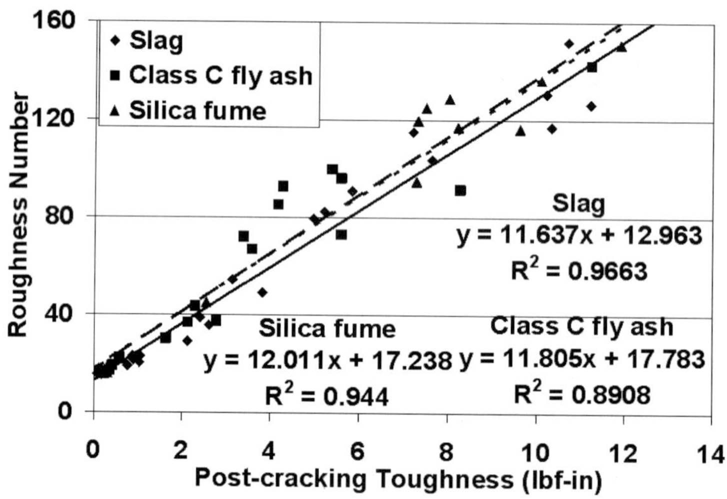

Figure 2. Effect of SCM replacement of portland cement on composite fracture surface roughness number

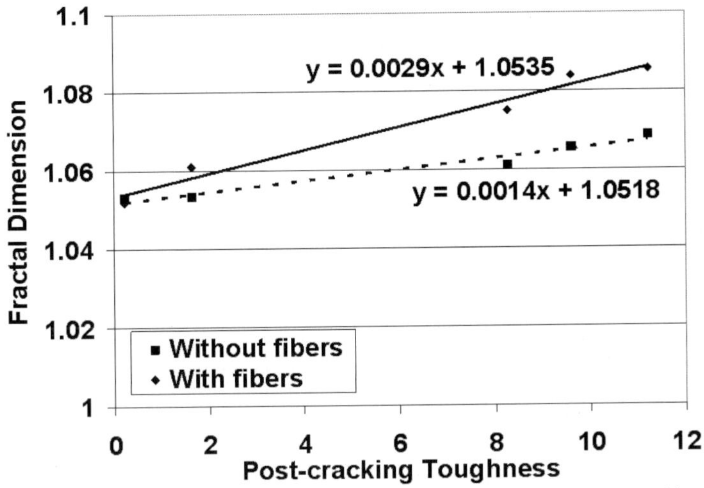

Figure 3. Correlation between fracture surface fractal dimension and composite post-cracking toughness

As with the fracture surface roughness number, the y-intercept of the fractal dimension lines was greater than 1.0, indicating that an inherent degree of roughness exists in these composite even with zero theoretical flexural toughness. In addition, as the post-cracking toughness of the composite decreases, it appears that the tendency for fiber fracture increases as shown by minimal differences between the matrix and net surface fractal dimension curves.

More research is necessary to elucidate the mechanisms of fracture behavior as measured by the fractal dimension. In addition, by measuring the differences in fractal dimension as a function of the number of wet/dry cycles (as in Mohr et al.[6,7]), the influence of environmentally-induced degradation on fracture behavior can be assessed.

4. CONCLUSIONS

In this research, laser scanning confocal microscopy was shown to be a powerful technique for the examination of fracture surface fractography parameters of fiber-cement composites. The fracture surface roughness number and fractal dimension was determined for these composites. From the results obtained, the following conclusions may be drawn:

- Fiber-cement composite fracture surfaces revealed a strong correlation between composite post-cracking toughness and surface roughness.
- A degree of inherent surface roughness, likely due to the inhomogeneous and porous microstructure of the hydrated cement, which does not contribute to toughness, was measured by both the roughness number and fractal dimension.
- Partial replacement of portland cement with silica fume, slag, or Class C fly ash did not have a noticeable influence on the pulp fiber-cement composite fracture behavior as measured by the fracture surface roughness number.

- The fractal dimension of the fracture surfaces showed that matrix cracking was a contributing factor for increased toughness. Composite failure was shown to be a combination of fiber pull-out and fiber fracture.
- Increased toughness of the composites was attributed to increased fiber pull-out, as compared to samples with minimal toughness which primarily failed by fiber fracture

5. REFERENCES

1. A.B. Abell and D.A. Lange, Fracture mechanics modeling using images of fracture sur-faces, *International Journal of Solids and Structures*, **35**, 4025-4033 (1998).
2. D.A. Lange, H.M. Jennings, and S.P. Shah, Analysis of Surface Roughness using Confocal Microscopy, *Journal of Materials Science*, **28**, 3879-84 (1993).
3. D.A. Lange, H.M. Jennings, and S.P. Shah, Relationship between Fracture Surface Roughness and Fracture Behavior of Cement Paste and Mortar, *Journal of the American Ceramic Society*, **76**(3), 589-97 (1993).
4. D. Zampini, H.M. Jennings, and S.P. Shah, Characterization of the Paste-Aggregate Interfacial Zone Surface Roughness and its Relationship to the Fracture Toughness of Concrete, *Journal of Materials Science*, **30**, 3139-54 (1995).
5. K.E. Kurtis, N.H. El-Ashkar, C.L. Collins, and N.N. Naik, Examining Cement-Based Ma-`terials by Laser Scanning Confocal Microscopy, *Cement and Concrete Composites*, **25**(7), 695-701 (2003).
6. B.J. Mohr, H. Nanko, and K.E. Kurtis, Durability of Thermomechanical Pulp Fiber-Cement Composites to Wet/Dry Cycling, *Cement and Concrete Research,* **35**(8), 1646-9 (2005).
7. B.J. Mohr, H. Nanko, and K.E. Kurtis, Durability of Kraft Pulp Fiber-Cement Composites to Wet/Dry Cycling, *Cement and Concrete Composites*, **27**(4), 435-448 (2005).
8. Y.T. Wang and S. Diamond, A fractal study of the fracture surfaces of cement pastes and mortars using a stereoscopic SEM method, *Cement and Concrete Research,* 31(10), 1385-1392 (2001).

QUANTIFICATION OF CAPILLARY PORES AND HADLEY GRAINS IN CEMENT PASTE USING FIB-NANOTOMOGRAPHY

L. Holzer, P. Gasser and B. Muench
Empa Materials Science and Technology, 3D-Mat group, Ueberlandstrasse 129, CH-8600 Dübendorf, Switzerland

Abstract: Based on high resolution 3D-microsturctural data from FIB-nanotomography the pore structure in the sub-m range can now be described quantitatively. This is demonstrated for a 28 days old cement paste. In contrast to the discontinuous pore size distribution (PSD) which results from mercury intrusion porosimetry, the data from FIB-3D-analysis reveals an exponential PSD at radii larger than 50 nm. Using extended image analysis techniques, porosity in hadley grains and the capillary pore network in the interstitial groundmass can be distinguished and the connectivity between them can be quantified. In the 28 days old cement paste, 34 % of the total pore volume are located within the hadley grains. The connection with the capillary pores in the groundmass is dominated by numerous but small pathways (intersections). The average radius of these intersections is below 100 nm. The average intersection density is 0.75 intersections per m^2 of phaenograin surface. The intersections occupy only 2% of the interface between the phaenograins and the groundmass. In summary, the new 3D-microscopy technique not only provides information about pore size distribution of the bulk microstructure, but it also enables to distinguish different types of pores and to characterize the connectivity between them. This information is considered to be a prerequisite for establishing microstructural models that can predict permeability properties in cement pastes.

1. INTRODUCTION

Porosity represents a fundamental microstructural feature in cementitious materials which has a major impact on permeability, durability and also on mechanical properties. Therefore quantitative analysis of the pore structure is of major importance in order to establish relationships between mix parameters, microstructure and final concrete properties. Unfortunately, even after decades of microstructural investigations with cementitious materials, there is still no method available for reliable analysis of the pore

M.S. Konsta-Gdoutos, (ed.), Measuring, Monitoring and Modeling Concrete Properties, 509–516.

structure. Mercury intrusion porosimetry (MIP), which is the most widely used method for this purpose, is now considered to be "unsuitable for pore analysis in cement materials" (Diamond, 2000). More sophisticated methods such as proton relaxation NMR are emerging (Barberon et al., 2003; Plasssais et al., 2005) but at the moment the results are difficult to interpret and no spatially resolved information is provided. Image based pore structure analysis with SEM or TEM is restricted to 2D information, which is not suitable for the investigation of permeability and connectivity, where higher order topological information related to the 3D-structure of the pore network is required. Since a large part of the pore structure has dimensions below the μm-scale, also conventional tomography techniques for 3D-analysis fail because of insufficient resolution.

Recently, a new high resolution 3D-microscopy technique was developed at Empa using a dual beam FIB (focused ion beam) (Holzer et al., 2004), whereby voxel resolutions below 20 nm can be achieved. The so-called FIB-nanotomography (FIB-nt) method thus opens new possibilities for a quantitative characterization of the pore structure in cementitious materials. The aim of this paper is to illustrate the unique potential of FIB-nt for parametrisation of the capillary pores and the so-called hadley grains.

2. THE PORE STRUCTURE OF CEMENT PASTE

Four different types of pores can be distinguished in cement paste: gel-pores (at nm-scale), capillary pores and hadley grain porosity (10 nm up to μm-scale) and air voids (>> 1 μm). Since the pore structure is a result of the hydration process, the different types of pores are associated with distinct types of hydration products and with the corresponding microstructural domains. The hydration products are differerentiated as belonging either to the inner (IP) or to the outer products (OP) (Taplin, 1959). The IP is formed within the initial boundaries of the cement grains, whereas the OP is formed in the interstitial space between the cement grains. The capillary pores represent remnants of interstitial space that was initially filled with water and that was not overgrown by OP. Capillary pores and OP are thus both located in the former interstitial space.

From a phaenomenological point of view the microstructure of cement paste is better described as consisting of phaenograins and undifferentiated groundmass instead of IP and OP (Diamond & Bonen, 1993). The groundmass mainly contains the OP and capillary pores whereas the phaeno-grains mainly include the IP and the unhydrated clinker. However, a certain number of grains, especially small ones that are fully hydrated, can not be identified as phaenograins and are thus included in the undifferentiated groundmass, which therefore also contains an undefined amount of IP. A considerable number of the phaenograins exhibit pores in the same size-range as the capillary pores. These porous particles are called hollow shell hydration grains or hadley grains (Hadley et al., 2000; Kjellsen & Atlassi, 1999; Kjellsen et al., 1997). Hollow shell hydration is strongly influenced by water content, clinker mineralogy, size of the particles and curing regime (Kjellsen & Atlassi, 1999). With respect to permeability, transport and durability properties, the microstructural characterization of the hollow shell porosity and its connectivity with the capillary pores is of major importance. However, so far quantitative investigations of hadley grains are rare because of a lack of suitable methods. Also here, FIB-nt opens new possibilities.

Figure 1. 3D-structure of sample A: 28 days old cement paste (w/c 0.35). Dimensions of cube:46*32*25μm. Voxel resolution: 74nm

Figure 2. 3D-visualization of pore structure in sample A after binarization of data volumes hown in figure 1

3. MATERIALS AND METHODS

3D-analysis and porosimetry was performed from sample A, a 28 days old cement paste (CEM I 42.5, w/c 0.35). Hydration was stopped in a vacuum-oven at 70°C. Pore size distribution was measured with conventional mercury intrusion porosimetry (MIP). For microscopic analysis with FIB, the sample was prepared according to conventional procedures for impregnating and polishing as described elsewhere (Crumbie, 2001; Kjellsen et al., 2003).

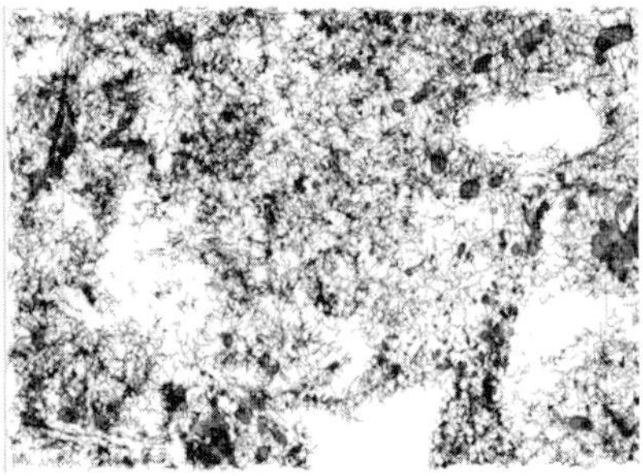

Figure 3. Skeletonization of pore structure from sample A (28 days old cement pasteCube size: 46*32*25μm. Voxel size:74 nm

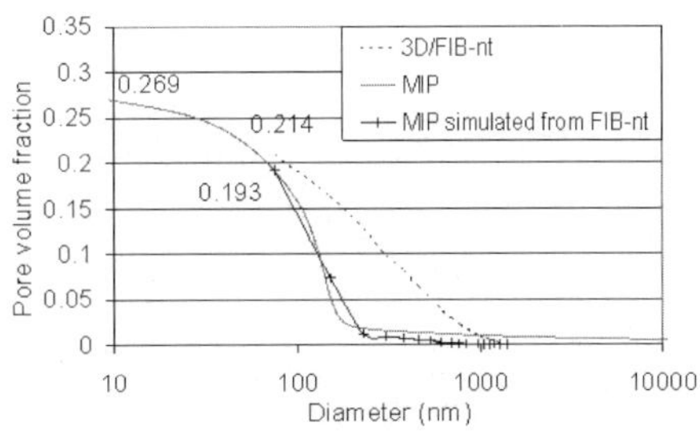

Figure 4. Pore size distribution from sample A measured with mercury intrusion (MIP) and with FIB-nt. In addition the mercury intrusionis simulated based on the 3D data from FIB-nt

FIB-nanotomography (FIB-nt) is based on an automated serial sectioning procedure that was recently implemented on a dual beam FIB (FEI Strata DB235) at Empa (Holzer et al., 2004). Sample A was analysed at different magnifications with a maximum voxel resolution of 20 nm. In this paper, for simplicity we are only discussing one set of data with a moderate voxel resolution of 74 nm. Figure 1 shows a 3D-reconstruction of the raw data, whereby a similar contrast as in backscattered imaging is obtained (dark grey=pore, medium grey=hydration product, bright grey=unhydrated clinker). After segmentation by thresholding, a binary data volume exhibiting the 3D pore structure is obtained (figure 2). For quantitative analysis and topological characterization of the pore network various algorithms for skeletonization and for 3D-distance mapping have been developed in Matlab and in Java. Fig. 3 shows the skeletonized pore structure. Each voxel on the skeleton contains size information, i.e. distance to the nearest pore wall. This information is used for statistical characterization of the pore structure. In addition an algorithm for the simulation of the mercury intrusion process was also implemented.

4. RESULTS

4.1. Pore size distribution: Comparison of FIB-nt vs. MIP

Figure 4 shows pore size distributions (PSD) obtained for sample A from mercury intrusion porosimetry (MIP) and from FIB-nt. The curve from FIB-nt stops at 74 nm, which is caused by the limited voxel resolution. The volume fraction of pores with diameters larger than 74nm is 21.4%.

The corresponding curve follows an exponential PSD. FIB-investigations at 20 nm resolution show that the slope of the curve decreases drastically at diameters below 40-60 nm (unpubl. data). In contrast to FIB-nt, MIP exhibits a discontinuous size distribution with a sharp increase at approx. 200 nm. This apparently discontinuous distribution is attributed to the ink-bottle effect (Diamond, 2000). Thereby the size of the bottle neck is represented by the observed discontinuity at 200 nm, which is also called break-through diameter. Since MIP measurements include pores at the lower nm-scale, a higher total porosity of 26.9 % is measured.

Figure 5. Identification of phaenograins in sample A, based on raw data in Fig. 1. Pores within the phaenograins are shown in dark. The same porosity is isolated in Fig. 7

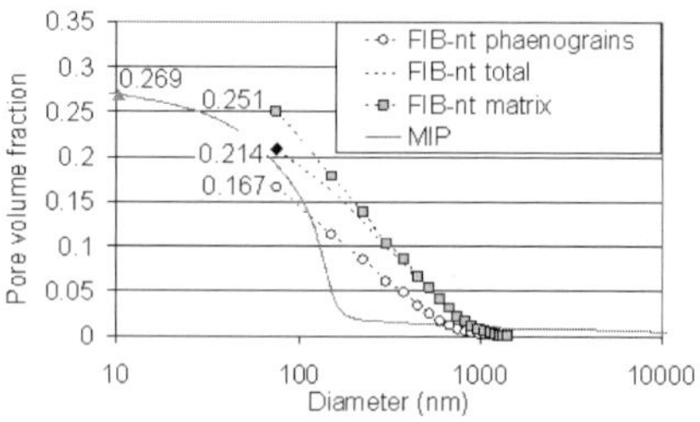

Figure 6. Pore size distribution (PSD) from sample A. Using the phaenograin mask (Fig. 5), porosity from "hollow shell" phaenograins and from groundmass can be analysed seperately

Figure 7. 3D-pore structure from within the hollow phaenograins in sample A

Figure 8. 3D-structure from the capillary pores in the interstitial groundmass of sample A

The difference of 5.5 % (compared to FIB-nt) may represent the porosity with diameters below 74 nm. The third curve in figure 4 is obtained by simulating the mercury intrusion process on the basis of 3D-data from FIB-nt. With the simulation the discontinuous MIP curve and the break-through diameter at 200 nm can be reproduced. This indicates, that the datavolume from FIB-nt is representative for the pore structure in sample A. The difference of 2.1% between total porosity in MIP simulation and in FIB-nt represents the isolated pores that can not be intruded with MIP simulation at the given resolution, i.e. the disconnected pores.

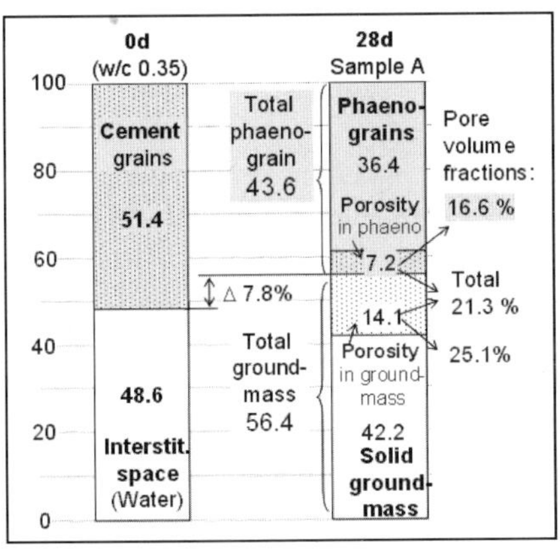

Figure 9. Volume fractions of solid and pore calculated for the initial mixture (0 days) andmeasured from 3D FIB-analysis (28 days)

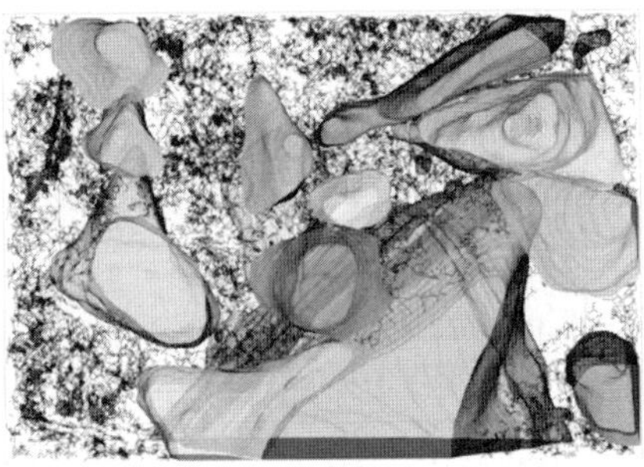

Figure 10. The intersections of the poreskeleton (Figure 3) with the phaenograins (Figure 5) are used to describe the "connectivity" between pores in groundmass and pores in phaenograins

4.2. Porosity in hollow phaenograins and in groundmass

As shown in Figure 5 individual phaenograins can be identified in the raw data volume from sample A (Fig. 1). Thereby, the boundary of the phaenograins is localized in the dense hydration layer which covers the grains. The segmentation is done with a semi-automated interpolation process, whereby the starting points are defined manually. The data for the phaenograins is then used as a mask that can be applied to the total porosity (Fig. 2) in order to distinguish porosity within the hollow phaenograins (Fig. 7) and the capillary pores in the groundmass (Fig. 8). For both domains the pore size distribution follows an exponential curve (fig. 6), whereby the average pore volume fraction in the groundmass (25.1 %) is higher than in the phaenograins (16.7 %). However, porosity in the individual grains varies strongly between 2 and 59 % (unpubl data).

The volume fractions of the phaenograins and the groundmass are compared with those calculated for cement and interstitial space in the inital paste mixture (w/c-ratio 0.35, 3.33 [kg/l], -2‰ shrinkage, fig. 9). Initially the cement occupied a volume of 51.4%. The difference to the volume of the phaenograins (43.6%) reflects the amount of cement grains that could not be identified as such (7.8%). These fully hydrated grains are thus included in the undifferentiated groundmass. They represent the volume of IP that can not be distinguished from OP by means of FIB and SEM imaging.

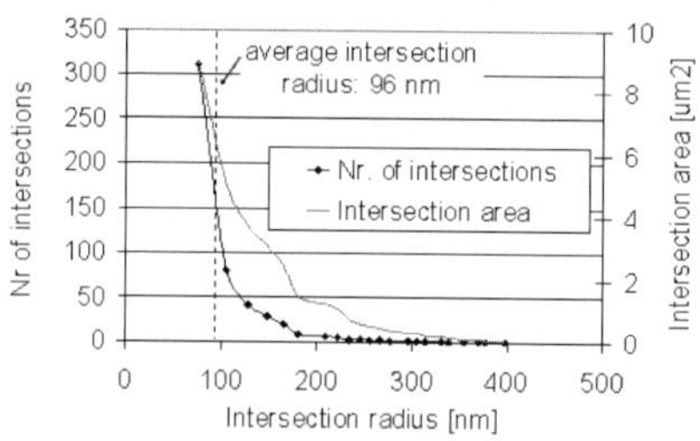

Figure 11. Cumulative size distribution of pore intersections through the dense hydration layer (average from 40 grains)

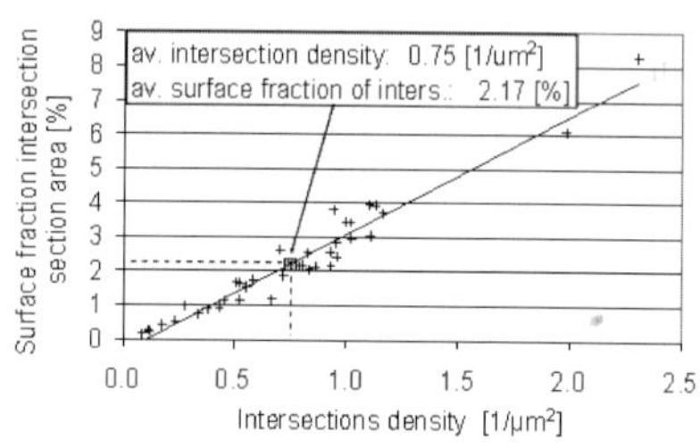

Figure 12. Intersection density (nr of inters. per surface area) vs. surface fraction of the intersections

4.3. Connectivity between porosity in hollow phaenograins and capillary pores in the groundmass

According to Figure 9, 1/3 of the total porosity is hosted in the hollow phaenograins. For consideration of permeability and transport properties it is thus crucial to know to what extent this phaenograin porosity is communicating with the capillary pore network in the groundmass. From a qualitative inspection of SEM-images the phaenograin porosity appears to be separated from the groundmass by a dense hydration layer which covers the phaenograins (For a qualitative discussion see Hadley et al., 2000; Kjellsen & Atlassi, 1999; Kjellsen et al., 1997; Scrivener et al., 2004). The connection between the two types of porosity can be quantified using the intersection of the pore skeleton through the surface layer of the phaenograins (Fig. 10). Thereby the number and the size of intersections represent the basic statistical parameters. The cumulative histogram of the intersection size distribution (Fig. 11) is clearly dominated by the smallest size class. For 3D-data with a better voxel resolution of 20 nm (unpubl. data) the average intersection radius is even smaller (i.e. 47 nm).

The surface fraction of the intersections (i.e. the area fraction of pores that intersect the dense phaenograin surface layer) is generally below 4 %, with an average of 2.17 % (fig. 12). This surface fraction is independent from the amount and from the size of pores within the grains. However, there is a linear correlation between the surface fraction of intersections with the intersection density (i.e. the nr of intersections per surface area). This correlation is due to the observed uniform size distribution of the intersections. Hence, connectivity between porosity in phaenograins and capillary pore network in the groundmass is dominated by small pathways below 100 nm radius. The dense hydration layer between the two types of porosity represents a membrane with approximately 2 % of these fine pore intersections.

5. CONCLUSIONS

Based on high resolution 3D-data from FIB-nt, in conjunction with extended image analysis techniques, different types of porosity with dimensions in the sub-μm range can be quantified. Also topological aspects of connectivity related to the permeability of the interface between the different types of porosity and the corresponding microstructural reservoirs can be characterized. This opens new possibilities for microstructure analysis in order to establish the relevant relationships between engineering parameters (mix design) and the final materials properties.

6. REFERENCES

Barberon, F., Korb, J.-P., Petit, D., Morin, V. & Bermejo, E., 2003. Probing the surface area of a cement-based material by nuclear magnetic relaxation dispersion. *Phys. Rev. Letters*, 90, 116103-4.

Crumbie, A. K., 2001. SEM microstructural studies of cementitious materials: Sample preparation of polished sections and microstructural observations with backscattered images - artefacts and practical considerations. In: *Proceedings of the 23rd international conference on cement microscopy* (eds Jany, L. A. & Nisperos, A. G.), pp. 320-341, ICMA, Albuquerque, USA.

Diamond, S., 2000. Mercury porosimetry - An inappropriate method for the measurement of pore size distributions in cement-based materials. *Cement and Concrete Research*, 30, 1517-1525.

Diamond, S. & Bonen, D., 1993. Microstructure of Hardened Cement Paste - a New Interpretation. *Journal of the American Ceramic Society*, 76(12), 2993-2999.

Hadley, D. W., Dolch, W. L. & Diamond, S., 2000. On the occurrence of hollow-shell hydration grains in hydrated cement paste. *Cement and Concrete Research*, 30, 1-6.

Holzer, L., Indutnyi, F., Gasser, P., Münch, B. & Wegmann, M., 2004. 3D analysis of porous $BaTiO_3$ ceramics using FIB nanotomography. *Journal of Microscopy*, 216(1), 84-95.

Kjellsen, K. O. & Atlassi, E. H., 1999. Pore structure of cement silica fume systems - Presence of hollow-shell pores. *Cement and Concrete Research*, 29(1), 133-142.

Kjellsen, K. O., Lagerblad, B. & Jennings, H. M., 1997. Hollow-shell formation - an important mode in the hydration of Portland cement. *Journal of Materials Science*, 32, 2921-2927.

Kjellsen, K. O., Monsoy, A., Isachsen, K. & Detwiler, R. J., 2003. Preparation of flat-polished specimens for SEM-backscattered electron imaging and X-ray microanalysis - importance of epoxy impregnation. *Cement and Concrete Research*, 33, 611-616.

Plasssais, A., Pomies, M. P., Lequeux, N., Korb, J. P., Petit, D., Barberon, F. & Bresson, B., 2005. Microstructure evolution of hydrated cement pastes. *Physical Review E*, 72(4), Art. No. 041401, Part 1.

Scrivener, K., Galluci, E., Füllmann, T., Pignat, C. & Navi, P., 2004. The challenge of quantification of cementitious systems. In: *Concrete Science and Engineering - RILEM PRO 36* (eds Kovler, K., Marchand, J., Mindess, S. & Weiss, J.), RILEM.

Taplin, J. H., 1959. A method for following the hydration reaction in Portland cement paste. *Australian Journal of Applied Science*, 10, 329-345.

THREE DIMENSIONAL ANALYSIS OF AIR VOID SYSTEMS IN CONCRETE

E.N. Landis and D.J. Corr
Department of Civil & Environmental Engineering, University of Maine, Orono, Maine 04469 USA; Center for Advanced Cement-Based Materials, Northwestern University, Evanston, IL 60208 USA

Abstract: The addition of an entrained air void system in concrete to reduce damage from freeze-thaw cycles may be considered one of the great technological advances in building materials in the last 50 years. This is despite the fact that we are not completely clear how and why it is so effective. To help shed light on these questions, we have applied a high-resolution 3D imaging technique to analyze an entrained air void system in concrete. X-ray microtomography allows us to image the internal structure of materials at spatial resolutions approaching 1 micron. 3D image analysis techniques can be used to extract quantitative measurements from the images. Using these techniques, the entrained air system in four different concretes was measured. Sample measurements include void-size distributions, air bubble spacing factors, and connectivity of the void system. These measurements allow us to challenge conventional assumptions on the internal void system in concrete.

Key words: image processing; tomography; entrained air

1. INTRODUCTION

The volume of H_2O increases by 9% during the phase change of liquid water to solid ice[1]. Early theories on the damage of concrete due to freezing and thawing were based on this volumetric change and a closed-vessel model for the pores in concrete: for saturated hydrated cement pastes, solidification resulted in an expansion that greatly exceeded the tensile strain capacity of the cement paste and cracking resulted.

The closed vessel model for the porosity of concrete is an inaccurate one, as first pointed out by Powers[2,3]. Although the hydraulic conductivity of the hardened cement paste is high, water flow is possible, governed by Darcy flow. Powers argues that as ice begins to form in a pore, an increase in pressure develops which simultaneously retards further ice formation (according to the Gibbs-Duhem equation), and forces water out of

M.S. Konsta-Gdoutos, (ed.), Measuring, Monitoring and Modeling Concrete Properties, 517–524.

the pore according to Darcy's law. If an escape boundary is present, in the form of an external surface or large pore, the escaping water has a location where it can solidify without generating pressure. If the distance from the freezing pore to the escape boundary is not great, the pressures exerted will not be large enough to cause damage, and if they are, cracking of the matrix results. This description of damage is known as the hydraulic pressure theory. Powers also developed the notion of a critical spacing factor which describes the maximum distance water can travel from a freezing pore to an escape boundary without causing damage to the paste. This is the core effectiveness of the entrained air voids – they provide escape boundaries for every location within the hydrated cement paste, as long as distances are below the critical spacing factor.

The hydraulic pressure theory, however, fails to explain an important observation made by Powers and Helmuth: the contraction of a concrete specimen at a constant freezing temperature[4]. This observation is not consistent with water flowing away from the site of freezing, an effect that would cause expansion in the bulk material, and led to the development of the osmotic pressure theory[5]. As ice forms in a solution, such as concrete pore fluid, the solid ice is pure water, and solute ions are ejected into the remaining solution as solidification occurs, which concentrates the remaining pore fluid. Osmotic theory states that the pore fluid in the surrounding smaller, unfrozen pores will drain towards this concentrated solution to equilibrate the solution. The driving force here is difference in chemical potential, which also dictates that undercooled water in small pores will tend to drain towards the sites of freezing in large pores to release latent heat of fusion and attain lower energy states. This is the dominant mechanism for ice lens formation[6], which has been observed in young cement pastes, and a related micro ice lens formation process has been proposed as a possible damage mechanism[7]. All of these theories indicate that pore water flows towards the site of freezing in the concrete matrix, in apparent disagreement with the hydraulic pressure theory.

Recent efforts by a number of research groups have resulted in further explanation of frost damage, due to crystallization pressure[8,9]. In order for an ice crystal to exist in a pore, it must adopt a curvature dictated by the pore size and the contact angle between the pore wall and crystal[10]. If this curvature is positive (the crystal is concave), the pressure inside the crystal increases and the equilibrium temperature between the crystal and its melt decreases. As the temperature drops, the crystal penetrates narrower pores and adopts larger curvatures, which increase the pressure on the leading edge of the crystal. Away from the leading edges of the crystal, the bulk crystal has a smaller curvature, but must remain at the same pressure for equilibrium. This additional pressure is provided by resistance from the pore wall and surrounding matrix in contact with the crystal, and is known as crystallization pressure.

There is clear evidence of the beneficial role of entrained air voids in the protection of concrete from internal freezing and thawing damage. Air voids are incorporated into the concrete composite through the use of an air entraining admixture (AEA). The active components of commercial AEA produced from vinsol resin are typically anionic surfactants, bipolar molecules with the anionic end being hydrophilic and the polymer end of the being hydrophobic[11,12]. These molecules tend to align themselves along air-water interfaces within the concrete mixture, where they reduce the surface tension of the air-water interface. This stabilizes the bubbles that naturally form during mixing. A typical entrained air void system has well-dispersed voids 20-200 mm in size. The critical parameter of the air void system is not the total volume of air, but the spacing factor of the air void system, as small, well-dispersed voids provide more comprehensive protection than large isolated air voids, even if the total air volume is the same. Critical spacing factors are usually specified at 200 – 250 μm, and are measured using ASTM C457, a two-dimensional microscopy technique.

The spatial distribution of air voids in the entrained system is instrumental to the effectiveness of the protection. Whatever the presumed mechanism of damage, the proximity of an area of cement paste to an escape boundary in the form of an air void plays a significant role in determining the probability of damage. The standard method for evaluating this proximity effect is the critical spacing factor, defined as the mean half distance between two adjacent air voids. The spacing factor is usually found using manual 2D image processing such as the point count or linear traverse methods of ASTM C457. However, the reduction of the three-dimensional air void system to a two dimensional image, and then a one-dimensional linear traverse or non-dimensional point count can result in only rough estimates of the actual distance water must flow during freezing[13].

This paper addresses the issue of 2D imaging of 3D air void systems through the use of microstructural data obtained using x-ray microtomography, a high resolution 3D imaging technique.

2. X-RAY MICROTOMOGRAPHY

X-ray microtomography is a three-dimensional imaging technique that uses a series of radiographic images to reconstruct a map of an object's x-ray absorption[14]. The technique is identical in practice to medical computed axial tomography (CAT) scans, except that microtomography achieves much higher spatial resolution by combining extremely bright, monochromated synchrotron radiation with high quality optics and x-ray detection. Microtomographic scans result in volumetric images with possible spatial resolutions approaching one micron. A schematic illustration of the microtomography system used in this work is shown in Figure 1. For the system used in this work, the field of view had a fixed width of 1024 pixels. Variations in spatial magnification were possible through the use of different microscope objective lenses that focused the x-ray image on to the CCD detector. As detailed below, two different magnifications were used in this work to obtain voxel sizes of 4 and 1 microns.

Synchrotron-based microtomography has been used extensively over the past 15 years to study pore structure in a variety of heterogeneous materials[15,16,17]. The significance of the technique is in its ability to capture true three-dimensional internal structure at relatively high resolution.

The images produced by tomographic scans and subsequent reconstruction are grayscale volumes where the pixel intensity is roughly proportional to the object's density. An example cross sectional "slice" image of a small shard of concrete is shown in Figure 2. In the figure one can see many of the important features of the microstructure: air voids (dark), sand particles (gray), and unreacted cement grains (white specks). .It should be emphasized that this is one slice out of the hundreds that are produced in a single scan, as the data is truly three dimensional.

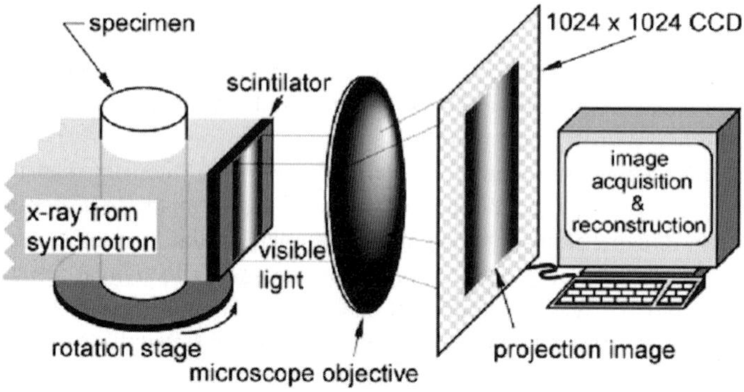

Figure 1. Schematic illustration of microtomography imaging system

3. IMAGE PROCESSING

The digital data produced by the tomographic scans allows us to employ an extensive library of image analysis and image processing techniques. In this work, several basic image analysis steps were utilized. First, the images were segmented to distinguish void space from solids. Second, the resulting void space is analyzed for connectivity to identify individual void objects, as well as the connectivity between the void objects.

For image segmentation, we exploit the fact that microtomography produces images where the voxel intensity is proportional to the density of the material at that point in space. (A voxel is a "volume element", or a 3D pixel.) For 8-bit images such as those considered here, each voxel takes on a value between 0 and 255. In this case zero is black, corresponding to minimum density, and 255 is white, corresponding to maximum density. Thus for the slice image of Figure 2, the air surrounding the specimen, as well as the pore space inside the specimen is dark. Higher density portions of the specimen, such as the unhydrated cement particles, are lighter.

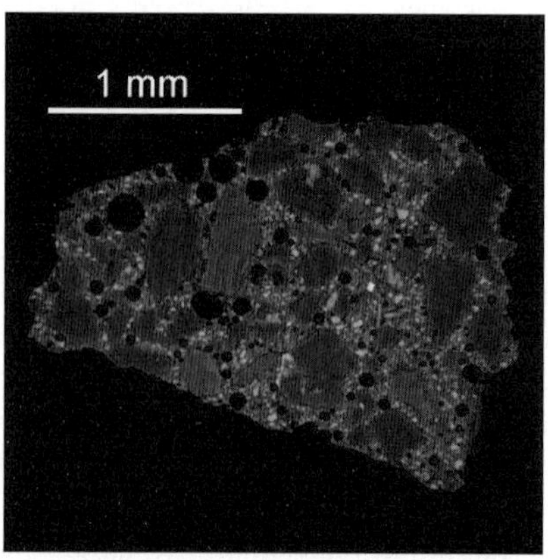

Figure 2. Tomographic slice image of small concrete shard

From an image such as this, a pixel intensity histogram, as shown in Figure 3 can be constructed showing the distribution of pixel intensities in the image. The two peaks of the images correspond to the air space (left peak) and solid material (right peak). Because of this clear distinction between intensity distributions, the image can be segmented into void and solid by setting a threshold intensity at the minimum point between the peaks. Any pixel with an intensity below the threshold is considered to be void space, while all others are considered to be solid. Through this segmentation procedure, the binary image of Figure 4 is produced. In this image, the black void space is clearly distinguished from the white solid. Again, we should emphasize that although this example is presented in two dimensions, our analysis ultimately is three dimensional.

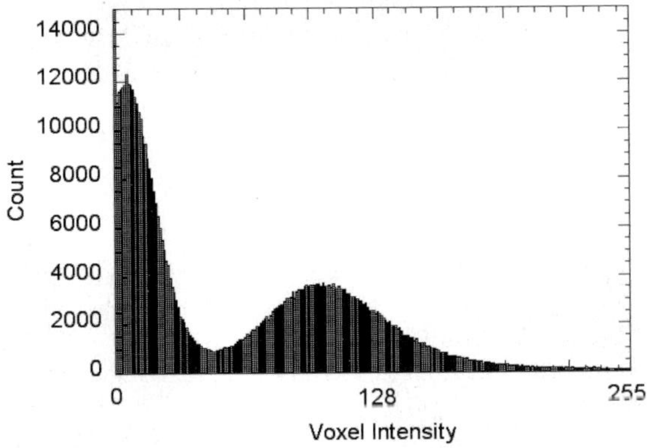

Figure 3. Pixel intensity histogram for image of Figure 2

4. AIR VOID ANALYSIS

A representative digital section was taken from a tomographic image of a conventional concrete mix. The mix had a w/c ratio of 0.42, with a measured air content of 6.2%. The representative section is rendered in Figure 5. Figure 6 shows and isolated air void system established using the segmentation methods described above. These figures illustrate the three dimensional nature of the data.

In this work we seek to measure the spacing factor in this volume using a series of additional image processing routines. In order to measure spacing factors, we employ a two-part algorithm that is illustrated in two dimensions in Figure 7. First we take the original grayscale data and segment it such that black is void and white is solid, as shown in Figures 7(a) and 7(b). Tho the binary image we then apply a distance transform18. A distance transform is an image in which the intensity of a pixel is proportional to the distance from the nearest void object. This is illustrated in Figure 7(c). In this image, brighter pixels represent those that are farther away from a void space. Thus, the distance transform actually tells us the distance from any point in the vol-ume to the closest air pocket. This distance information can be collected as shown in the histogram of Figure 8, which shows the distribution of distances in the concrete sample.

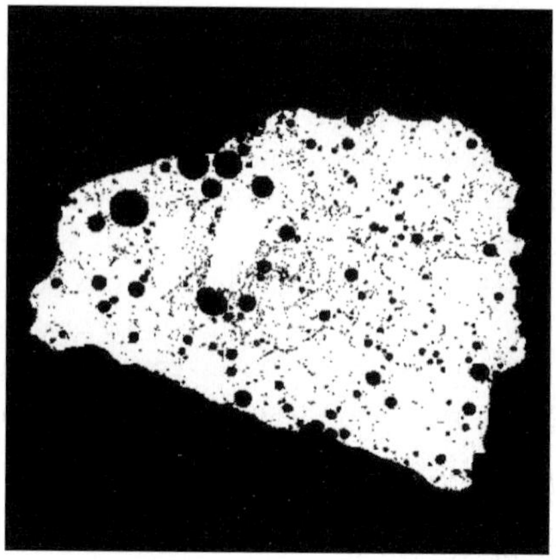

Figure 4. Binary representation of the image of Figure 2

Figure 5. 3D rendering of data

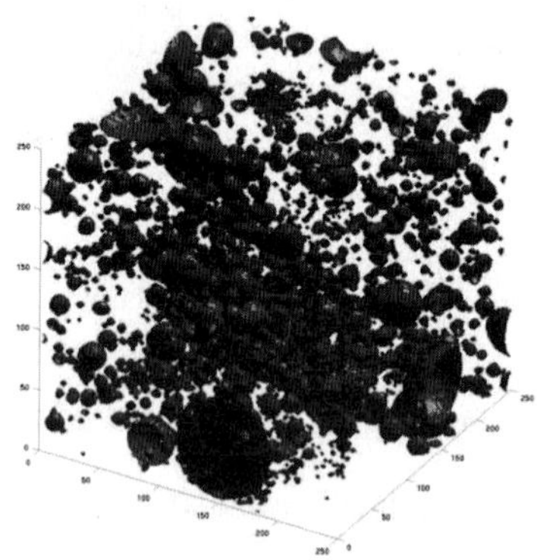

Figure 6. Isolated air void system

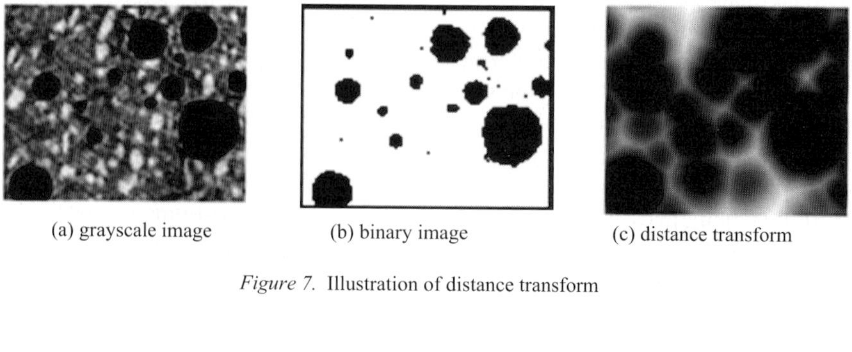

(a) grayscale image (b) binary image (c) distance transform

Figure 7. Illustration of distance transform

It should be clear that this kind of information could be used to quantify the air void system in concrete. Unlike surface methods, the information presented here is three dimensional, and therefore can provide a much more complete picture of the air voids. In future work we will look at ASTM standard as a basis for comparison.

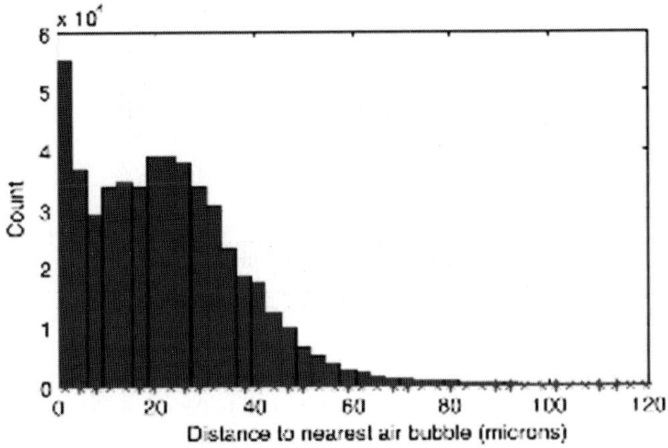

Figure 8. Histogram of distances to nearest air bubble for 400 x 320 x 320 μm volume of cement paste

Acknowledgments

Parts of this research were conducted at the National Synchrotron Light Source, Brookhaven National Laboratory, which is supported by the U.S. Department of Energy, Division of Materials Sciences and Division of Chemical Sciences (DOE contract number DE-AC02-76CH00016.

6. REFERENCES

1. Lock G.S.H., The Growth and Decay of Ice, Cambridge University Press. Cambridge, 1990.
2. Powers T.C., A Working Hypothesis for Further Studies of Frost Resistance of Concrete. ACI Journal, Proceedings. **41**(3) 1945, pp. 245-272.
3. Powers T.C., The Air Requirement of Frost Resistant Concrete, Proceedings, Highway Research Board **29**, 1949, pp.184-211.
4. Powers T.C. and Helmuth R.A., Theory of Volume Changes in Hardened Portland Cement Paste During Freezing, Proceedings, Highway Research Board. **32**, 1953, pp. 285-297.
5. Powers T.C., Resistance of Concrete to Frost at Early Ages, Research Bulletin, Portland Cement Association, No. 41, 1956.
6. Corr D.J., Monteiro P.J.M., and Bastacky J., Observations of Ice Lens Formation and Frost Heave in Young Portland Cement Paste, Cement and Concrete Research, **33**(10) 2003, pp. 1531-1537.
7. Setzer M.J., Micro Ice Lens Formation, 3rd International Bolomey Workshop: Pore Solution in Hardened Cement Paste, 2000, Aedificatio Publishers, pp. 89-112.
8. Scherer G.W., Crystallization in Pores, Cement and Concrete Research **29**, 1999, 1347-1358.

9. Chatterji S. and Thaulow N., Unambiguous Demonstration of Destructive Crystal Growth Pressure, Cement and Concrete Research, **27** (6), 1997, pp. 811-816.
10. Scherer G.W., Freezing Gels, Journal of Non-Crystalline Solids, Vol. 155, 1993, pp. 1-25.
11. Hewlett P.C., editor, Lea's Chemistry of Cement and Concrete. Arnold Publishing, London. 1988, pp. 839-844.
12. Ramachandran V.S., editor, Concrete Admixtures Handbook: Properties, Science, and Technology. Noyes Publications, New Jersey. 1984, pp. 269-302.
13. Pleau R. and Pigeon M., The Use of the Flow Length Concept to Assess the Efficiency of Air Entrainment with Regards to Frost Durability: Part I – Description of the Test Method, Cement Concrete and Aggregates, **18**(1), 1996, pp. 19-29.
14. Flannery, B.P., Deckman, H.W., Roberge, W.G. & D'Amico, K.L., 'Three-dimensional x-ray microtomography', Science, **237** (1987) 1439-1444.
15. Auzerais, F.M., Dunsmuir, J., Ferreol, B.B., Martys, N., Olson, J., Ramakrishnan, T.S., Rothman, D.H. & Schwartz, L.M., 'Transport in sandstone: A study based on three dimensional microtomography', Geophysical Research Letters, **23**(7), (1996) 705-708.
16. Rintoul, M.D., Torquato, S., Yeong, C., Keane, D.T., Erramilli, S., Jun, Y.N., Dabbs, D.M. & Aksay, I.A., 'Structure and transport properties of a porous magnetic gel via x-ray microtomography', Physical Review E, **54**(3) (1996) 2663-2669.
17. Bentz, D.P., Quenard, D.A., Kunzel, H.M., Baruchel, J., Peyrin, F., Martys, N.S. & Garboczi, E.J., 'Microstructure and transport properties of porous building materials: Three-dimensional x-ray tomographic studies', Materials and Structures, **33** (2000) 147-153.
18. Gonzalez, R.C. & Woods, R.E., 'Digital image processing'. (Prentice Hall, Upper Saddle River, NJ, 2002).

Concrete Deterioration, Repair and Rehabilitation

CALCULATION OF STRUCTURAL DEGRADATION DUE TO CORROSION OF REINFORCEMENTS

J. Rodríguez,[1] L. Ortega,[1] D. Izquierdo[2] and C. Andrade[2]
[1]*GEOCISA, 10 &12 St.Los Llanos de Jerez, 28820 Coslada, Madrid, Spain;* [2]*IETCC, 4 St.Serrano Galvache, 28033 Madrid, Spain*

Abstract: Reinforced concrete structures can deteriorate due to corrosion of reinforcements. The main structural consequences due to corrosion are: a) loss of steel cross section which can be generalized or localized; b) loss of steel ductility, a phenomenon attributed to hydrogen generation due to acid production during the corrosion process; c) loss of steel/concrete bond due to the gap and cracks created by the steel section reduction and d) the cracking of concrete cover due the pressure generated by the oxide production, which aims in a loss of load-bearing cross section of concrete. All these effects influence the progressive loss in structural capacity. The loss rate is directly linked to the corrosion rate which in turn depends on the climate. Assuming that the corrosion rate is averaged at a yearly basis, the rate of structural deterioration can be predicted by considering residual load-bearing areas. Examples will be presented on corroded beams at different degrees, in order to illustrate de principle of recalculation based in the reduced cross section and the consideration of the bond loss.

Key words: structural degradation, corrosion of reinforcement

1. INTRODUCTION

In order to make an adequate management of concrete structures and, therefore, an efficient conservation policy, it is necessary to know during its life whether their security is enough or not. The evolution of the security will acts as a base for the need of strengthening or repair. Thus, the needing of adequate tools for assessing structures has become a subject of crucial economical and social interest.

The aging of concrete structures is primarily produced by a continuous impact in the material of the surrounding ambient, thus it is essential the appraisal of the deterioration mechanisms, and how they influence the safety of the whole structure.

M.S. Konsta-Gdoutos, (ed.), Measuring, Monitoring and Modeling Concrete Properties, 527–536.

Among the different deterioration mechanisms occurring in concrete structures, the corrosion of reinforcement is the most detrimental one. Their consequences, whether due to concrete carbonation or to an excessive chloride content in the concrete, can be classified into three main groups (Figure 1):

- Those, which affect the reinforcement section, reducing the effective area and ductility.
- Those, which are related to concrete integrity.
- Those, which affect the interaction concrete – reinforcement due to the bond reduction.

The knowledge of the actual state and future evolution of these consequences is essential for the assessment of a structure suffering reinforcement corrosion.

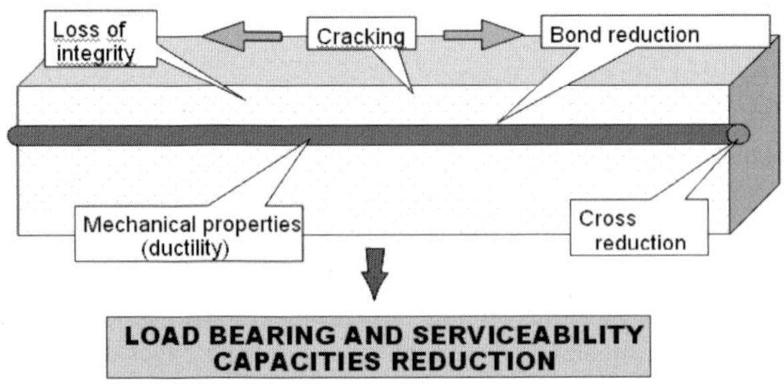

Figure 1. Reinforcement corrosion effects on concrete structures

Several efforts have been made during last years to quantify these effects. Examples are the FIB bulletin no. 243 (1) or the Swedish Road Administration Manual (2). The simplified procedure here proposed, has been developed in the BRITE – EURAM project BE – 4062. Other parteners in this project have been: BCA (UK), Lund University (Sweden) Cementa (Sweden) and CBI (Sweden).

2. SIMPLIFIED ASSESSMENT

The needing of adequate tools for a fast, economic and safe enough assessment of a structure is shown by an increase in the structural management programs developed last years. As it is shown in Figure 2, the simplified assessment developed is an empirical procedure based on the application of several indexes, experimentally developed, that can reflect the principal aspects involved in the residual life calculation. These indexes are:

- Corrosion Index (deterioration mechanism evolution): It tries to represent the damage level caused by corrosion on reinforcement. This index is related to the measurement of corrosion rate or corrosion current in the structure and the type of corrosion (generalised or localised).

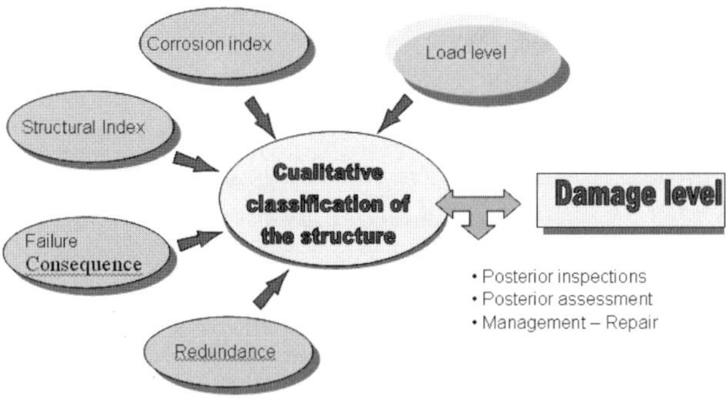

Figure 2. Simplified assessment general schema

In addition, visual inspection can provide an external damage classification in different levels, as is shown in Table 1. These indexes can vary from *Negligible* (when no external damage are detected or small rust spots are presented) to *High* (when a generalised *spalling* and cracking are detected).

Table 1. Visual damage rating

External damage	Description
Negligible	No external damage, few external rust spots.
Low	Small damages (crack width < 0.3 mm)
Moderate	Crack width > 0.3 mm following reinforcing map
High	Generalised cracking and spallig.

Source: BE-4062 (1995)

- Structural index: Their function is to take into account the influence of corrosion in the structural typology studied. This parameter depends on the type of the element (beam or column) and the geometrical and mechanical characteristics of the element (longitudinal reinforcement, transversal reinforcement, size, general load level). For bending moments Figure 3 shows the source of the Structural index.

In column elements, the proposed procedure is similar to that given for beams. A transversal reinforcement index is obtained, function of the rebar diameter and the spacing. This value tries to characterise the possibility of buckling in the longitudinal bars. A second factor, function of the cover/column size ratio represents the loss of bearing capacity due to *spalling*. In addition, it has to be taken into account:

- Failure consequences, determined through the importance of the structure or their risk of victims.
- In the case of assessment of a part of a whole structure, the hyperstaticity is considered in the final value of the Damage level.
- Also, it is necessary to take into account the actual load level of the element, due to possible existence of oversize elements.

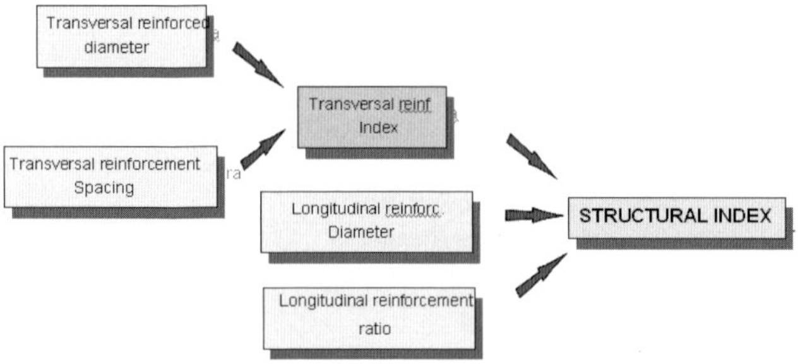

Figure 3. Structural Index procedure for beams

These indexes (corrosion and structural) are weighed adequately in order to obtain a Damage Level of the whole structure, which is ranked into *Negligible*, *Medium*, *Severe*, *Very Severe,* leading into Table 2 which in addition contains the "urgency of intervention" in years.

Table 2. Damage classification and urgency of intervention in years

External Damage	**Damage Level**			
	Negligible	Medium	Severe	Very Severe
Negligible	> 10	6-10	4-6	
Low	6-10	4-6	2-4	1-2
Moderate	4-6	2-4	1-2	1-2
High		1-2	0-1	0-1

Although the indexes above have been obtained trough an extensive experimental procedure, a calibration phase is needed. Thus, an Innovation project funded by the DGXIII (CONTECVET) is being developed in order to calibrate these indexes in an extensive amount of real structures. The objective of CONTECVET is mainly the validation for each deterioration mechanism and the provision of a new and calibrated Manual for each structural typology (bridges, buildings, nuclear plants, etc.)

3. DETAILED ASSESSMENT

The main objective is residual safety level determination, in order to establish and adequate intervention program with a high degree of information available. On the other hand, it can be also used for the calibration procedure of the indexes above presented.

The essential difference between assessment of an existing structure and design phase is the amount of available information. In existing structures it can be possible, although quite expensive in many cases, to increase the amount of data by means of *in situ* test and measurements in order to reduce the uncertainties. This higher degree of knowledge on the structure can be used through a reliability analysis for the possibility of reduction of the partial safety factors (3)(4). These reduced partial safety factors are now implemented in the present methodology of assessment:

Three are the main aspects to be analysed in a structural assessment:

- Deterioration process evaluation.
- Action, or better action effect, evaluated on the structure. For instance, it is advisable a better knowledge on the dead load present on the structure in order to reduce significantly their partial load factor.
- Safety or serviceability limit state verification.

The last aspects will be applied bellow for the case of reinforcement corrosion.

3.1 Deterioration process evaluation

Corrosion rate: The penetration of attack by corrosion P_X in the steel is the main parameter necessary to allow a correlation of the damage level with the general effects on the composite section concrete – steel (5). The P_X can be measured in residual diameter or estimated by means the Polarisation Resistance Rp, method (6)(7), (Figure 4). The Rp gives the instantaneous corrosion current, Icorr, which integrated over time gives the total decrease in bar diameter. Thus:

$$P_X = I_{corr} \cdot t \tag{1}$$

The determination of $\mathbf{I_{corr}}$ depends on the environment evolution. Thus, several strategies may be used for the determination of $\mathbf{I_{corr}}$ and the loss of section with time P_X (8).

- Several measurements in time with different environmental conditions, obtaining an averaged $I_{corr,\ ave}$ value to be used in the calculation.
- A single value in any environmental condition and calibration of the maximum expected $I_{corr,\ max}$, by means of the electrical resistivity, r-I_{corr} graph.
- A single value in worst environmental conditions (wet and rainfall). By means of the wetness time parameter $\mathbf{w_t}$, the value can be averaged during the time life (table 3).

The value experimentally obtained can be compared with Table 3 provided (source BRITE-4062), where averaged values of Icorr are given for exposure classes of EN206.

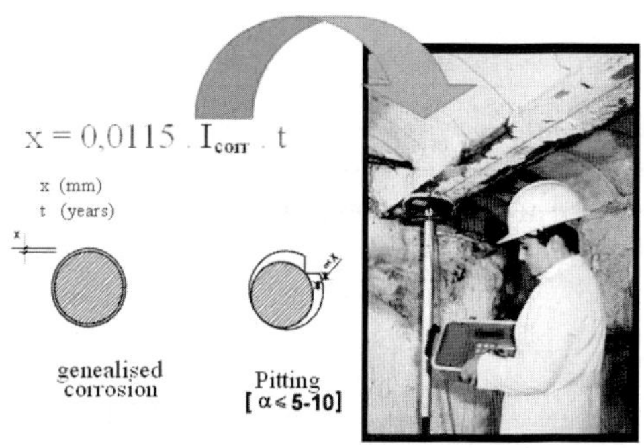

Figure 4. $\mathbf{I_{corr}}$ measurement and relationship between $\mathbf{I_{corr}}$ and loss of section **x**

Table 3. Averaged corrosion currents and wetness periods w_t for the exposure classes of EN206

EXPOSURE CLASS		I_{corr} (μA/cm²) (mean value)	w_t [-]
0	No risk of corrosion, very dry	0	0
CARBONATION			
XC1	Dry or permanent wet	0	0
XC2	Wet rarely dry	0.4	1
XC3	Moderate humidity	0.2	0.5
XC4	Cyclic wet dry	0.5	0.75
CHLORIDE AMBIENTS			
XD1	Moderate humidity	0.4	0.5
XD2	Wet, rarely dry	3.0	1
XD3	Cyclic, wet and dry	3.0	0.75
XS1	Airborne salt conditions	3.0	0.5
XS2	Submerged	Not present	
XS3	Tidal, splash and spray zones	7.0	1

Effective steel section reduction. The reduction in steel cross section is calculated through expression (2) where a is the "pitting factor". The a values are different if the corrosion is homogeneous (a = 2) than for pitting corrosion (5< α < 10) (8).

$$\Phi = \Phi_0 - \alpha P_x \qquad (2)$$

Cover cracking (9). The oxides generated in the corrosion process provoke an tensional state in the concrete cover that will produce final cracks, reducing consequently the cross section of the concrete element and therefore their load bearing capacity. Several empirical expressions have been developed, that can evaluate the crack width of the cover, as a direct function of the corrosion attack **x** and several geometric and mechanical parameters (5)(2).

$$w = 0.05 + \beta \left\lfloor P_x - P_{x_0} \right\rfloor \qquad [w \leq 1.0 \text{ mm}] \qquad (3)$$

Where:

- **w** is the crack width in mm,
- $\mathbf{P_x}$ is the attack penetration in microns.
- $\mathbf{P_{X_0}}$ is an attack corresponding to the crack initiation and,
- β is a factor depending on the bar position. (Table 4)

Table 4. β Values for crack width calculations

	Mean Values		Characteristic Values	
	Upper Bars	Lower Bars	Upper Bars	Lower Bars
β	0.0086	0.0104	0.01	0.0125

The value of $\mathbf{Px_0}$ can be estimated trough expression (4) on Table 5.

Table 5. **Px_0** Expressions for crack initiation

	Mean Values	Characteristic Values
	$Px_0 = a + b_1\ c/f + b_2\ f_{c,sp}$ (3)	
a	74.5	83.8
b_1	7.3	7.4
b_2	-17.4	-22.6

Where:

- **Px_0** is the attack in mm,
- **c/ϕ** is the cover diameter ratio.

Loss of bond. (10) The concrete – steel bond is the responsible of the bar anchorage in the element ends and the composite behaviour of both elements. However, corrosion provokes a reduction in bond due to the cover cracking and stirrups corrosion. Finally a limit state of bond can be achieved. Three main aspects should be considered:

- *Residual bond assessment.* Table 6 shows empirical expressions obtained by (7)(10) based on test type developed by (11) that allow to obtain realistic residual bond values. All of them are expressed depending on the attack penetration **P_X**.

Table 6. Relationship between bond and P_X in mm

	Bond strength (MPa)	
	With stirrups	**No stirrups**
Mean Values	$5.25 - 2.72\ P_X$	$3.00 - 4.76\ P_X$
Characteristic Values	$4.75 - 4.64\ P_X$	$2.50 - 6.62\ P_X$

For intermediate cases where the amount of stirrups is low, bellow the actual minimum, or the stirrups capacity can be strongly reduced by corrosion effect, expressions of Table 7 may be applied.

Table 7. Bond Values for cases of intermediate amount of stirrups (expression (4))

	Mean Values	**Characteristic Values**
f_b	$8.25 + m(1.10+P_x)$	$10.04 + m(1.14+P_x)$
m	$-4.76 + 2.04(\rho/0.25)$	$-6.62 + 1.98(\rho/0.25)$
ρ	$n\ [(\phi_w - \alpha x)/\phi]^2$	$n\ [(\phi_w - \alpha x)/\phi]^2$

Where:

- ϕ is the initial longitudinal diameter in mm.
- ϕ_w is the transversal diameter in mm.
- n is the number of transversal reinforcements.
- α depends on the type of corrosion.
- f_b bond strength

These expressions are of application with P_X values between 0,05 and 1 mm with $\rho \leq 0.25$.

- *Influence of external pressures* that can be present due to external supports. In this case, expressions similar to that presented in Eurocode 2 have been developed by (5). These are:

$$f_b = (4.75 - 4.64\ P_x)/(1 - 0.098\ p) \quad (5)$$

where f_b is the bond in MPa, P_x is the corrosion attack in mm y **p** the external pressure in the bond zone (MPa). This expression can be used for the bond evaluation of the rebar at element ends.

Figure 5 shows an application of expressions (4) and (5) in a reinforcing bar diameter of 20mm without stirrups (curve 3) or with 4f8 stirrups. Curves 1 and 2 correspond to the bond with a reduction at the end of the element without pressure (1 with homogeneous corrosion and 2 with pitting corrosion), curve 4 corresponds to an external force of 5 MPa.

- *Relationship between bond and crack width.* Several expressions have been developed for relating the residual bond with the crack width (Table 8).

Table 8. Relationship bond f_b (MPa) and crack width w (mm)

	Stirrups	No stirrups
Mean values	f_b =18 - 0.52 w	f_b =3.19 - 1.06 w
Characteristic values	f_b =4.66 - 0.95 w	f_b =2.47 - 1.58 w

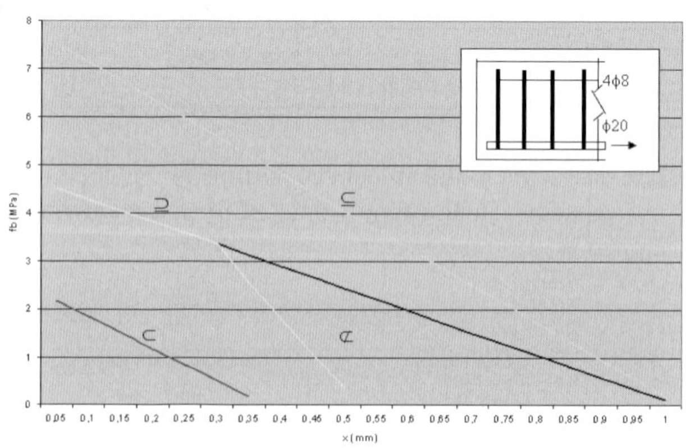

Figure 5. Residual bond as a function of $\mathbf{P_x}$

3.2 Limit state verification

3.2.1 Ultimate limit state

For slab and beams, a conservative value of the ultimate bending moment can be achieved by using the classical models but reducing the steel section and the concrete section spalled or cracked. A possible reduction due to bond deterioration must be considered, specially if the corrosion attack is on the tensile zone of the beams.

Although, shear and bending moment are supposed to have the same safety in the design design phase, for beams without corrosion the shear formulation is considered to be conservative, whereas for corroded ones, several factors may induce a premature fail of shear, such as:

- Small diameters on stirrups.
- Lower cover for stirrups.
- Spalling of cover.

In order to check the ultimate axial effort of a column element, (12) the reduction should be applied on the reduced concrete section in the case of spalling and if there are not stirrups, a reduction in the longitudinal bars subjected to compression due to risk of buckling should be also taken into account.

3.2.2 Serviceability limit state

The serviceability limit states to be checked should be:

- Exterior aspect of the structures (rust, spalling).
- Cracking of cover due to corrosion or excessive loading.
- Excessive deflections.

For the deflection and crack checking due to loading, the same expressions provided by Eurocode 2 can be used, but reducing the steel section and that of the concrete due to spalling.

4. CONCLUSIONS AND FINAL REMARKS

At present there are not adequate recalculation tools for the economic and safe management of concrete structures. Such a tool is still in its beginnings and only empirical or qualitative appraisal methodologies can be found in the literature.

Through the EU projects "The residual service life of concrete structures" (BRITE 4062) and the Contecvet, an important advancement from present situation has been made. Thus, methodologies for simplified and refined assessment have been developed and recalculation tools have been derived based in an extensive research on the structural behaviour of corroding elements.

The methodology starts by a preliminary inspection, that tries to identify the damage level of the structure and the environmental characteristics that surrounds the structure. Then, it is needed a structural assessment on the element in order to evaluate the intervention urgency. This structural assessment can be performed at two different levels:

- **Simplified assessment**: An empirical procedure using several indexes from which, a general damage level of the structure is obtained. As a direct result of their application the specialist should be able to decide if more studies are needed.
- **Detailed assessment** That allows a complete verification of the element safety in a similar manner as that proposed by the Limit State theory. The aspects covered are:

1. Action effect assessment.
2. Material properties and their damage level.
3. Load bearing capacity and serviceability verification

This detailed assessment can be performed using classical design concrete models but, reducing adequately the final characteristics of the composite section reinforcement–concrete.

Acknowledgement

The BRITE – EURAM project BE4062 *"The service life of reinforced concrete structures"* funded by the EU, was performed in collaboration with BCA (UK), GEOCISA, Lund and Cementa (Sweden)

The innovation project IN30901I CONTECVET, will be finished on 2001 in collaboration with BCA, NCP, TRL from UK, CBI, Lund, Skanska, VUAB, BV y SNRA from Sweden and Iberdrola, Enresa, Dirección General de Arquitectura y Vivienda de la Generalitat Valenciana, Instituto Eduardo Torroja and Geocisa from Spain.

5. REFERENCES

1. FIB (CEB-FIP): *'Strategies for Testing and Assessment of Concrete Structures'.* Work group 5.4 'Assessment, Maintenance and Repair', (1998)
2. Swedish National Road Administration. *'Bridge measurement and condition assessment'.*(1996)
3. Tanner, P.; Ortega, L.: *Evaluación de la seguridad estructural futura de puentes del pasado.* Jornadas sobre Nuevas Técnicas de Evaluación Estructural, Rehabilitación y Refuerzo de Estructuras, IABSE Spanish group (1999)
4. Izquierdo, D., Andrade C.: *Probabilistic assessment concrete structures affected by rebar corrosion: Case Study.* Int. Conference on In situ behaviour of structures Lassy (Rumanie) (2000)
5. Rodríguez, J., Ortega, L.M., Casal, J. y Díez, J.M : *Corrosion of reinforcement and service life of concrete structures*, Proceedings of Durability of Building Materials and Components Stockholm (1996) 117-126, Vol.I, Edited by C. Sjöstrom, E&FN Spon Publishers.
6. Feliú S., González J.A., Feliú S. Jr. and Andrade C : *Confinement of the electrical signal for in situ measurement of polarisation resistance in reinforced concrete.* ACI Materials Journal, Sept.-Oct. pp.457-460 (1990)
7. Rodríguez, J., Ortega, L.M., García M., Johansson, L., K. Petterson: *On-site corrosion rate measurements in concrete structures using a device developed under the Eureka project EU-401,* Proceedings "Concrete Across borders". Odense (Denmark) June (1994) 215-226.
8. González, J.A., Andrade C., Alonso C. y Feliú S. : *Comparison of rates of general corrosion and maximum pitting penetration on concrete embedded steel reinforcement.* Cement and Concrete Research. Vol. 25, nº 2, pp. 257-264 (1995).
9. Alonso C., Andrade C., Rodríguez J. y Díez J.M. : *Factors controlling cracking of concrete affected by reinforcement corrosion.* Materials and Structures, Vol. 31 pp 435-441 (1998)
10. Rodríguez, J.; Ortega, L.M.; García, A.M.*; Assessment of structural elements with corroded reinforcement.* Proceedings Corrosion and Corrosion Protection of steel in concrete, Sheffield (U.K) (1994) 171-185 - R.N. Swamy Ed. Sheffield Ac. Press.
11. Chana P.S. : *A test method to establish realistic bond stresses.* Magazine of Concrete Research, Vol. 42, Nº 151, pp. 83-90 (1990).
12. Rodríguez, J.; Ortega, L.M.; Casal, J.*: Load carrying capacity of concrete columns with corroded reinforcement*-Proceedings of Corrosion of Reinforcement in Concrete Construction Conference, Cambridge (U.K) July)Ç(1996) 220-230. Edited by C.L. Page, P.B. Bamforth, J.W. Figg, SCI Publisher.

ARCHAEOLOGICAL MUSEUMS OF RETHYMNON AND HERAKLEION

Pilot Diagnostic Studies of Corrosion of Steel Reinforcement in Concrete

G. Batis, A. Moropoulou, M. Chronopoulos, Ch. Mavronikolas, A. Athanasiadou, A. Bakolas, P. Moundoulas and E. Aggelakopoulou

School of Chemical Engineering, Materials Science and Engineering Section, National Technical University of Athens, 9 Iroon Polytechniou St., Zografou Campous 15780, Athens

Abstract: In this work the cases of Archaeological Museums of Rethimnon and Herakleion are examined. Corrosion examination has been succeeded with non destructive techniques such as Fiber Optics Microscopy (FOM), Infra Red Thermography (IR – Th), Ultrasound Velocity (US), Determination of Reinforce Corrosion Potential, Concrete Specific Electrical Resistance and Concrete Carbonation Depth. Destructive techniques were also applied in lab such as X- Ray diffraction, Thermal Analysis, Total Chloride Ions Measurement and Mercury Intrusion Porosimetry. The results obtained reveal that concrete reinforcement is in both cases in the energetic state and corrodes. In Herakleion's Museum the oxidation of the reinforcement is mainly due to concrete carbonation. On the other hand, in the case of Rethimnon Museum the presence of the sandstone in contact with the reinforcement creates a specific situation. The presence of chloride ions is the main reason of corrosion in this case. Chloride ions have been transferred in the concrete mass, mainly due to the marine environment and the close relation to the sandstone, which is affected by salt solutions through out the years. In both cases interventions are proposed in order to inhibit the reinforcement corrosion, in addition with a strength-increase repair solution. In both cases the interventions proposals are decided with care to the sensitive building materials of the museums, such as the sandstone in Rethimnon.

1. INTRODUCTION

The modern built environment of the early 20th century, constructed mainly by reinforced concrete present, nowadays, serious damage problems. The durability of the concretes structures is, essentially, depended to internal factors (concrete characteristics, microstructure, macroscopical defects, presence of cracks and fissures) and to external

M.S. Konsta-Gdoutos, (ed.), Measuring, Monitoring and Modeling Concrete Properties, 537–544.

factors (environmental factors - aggressive urban or marine atmosphere). The degradation of concrete can be observed both in the concrete mass (existence of microcracks and fissures, loss of binding material) and the reinforcement bars (corrosion products, expansion, decrease of mechanical strength, e.t.c.).

More particular, the concrete structures degrade through the following degradation actions:

The formation of the calcium hydroxide that takes place during the hydration of cement generates an alkaline environment in the concrete that render the steel reinforcement passive. [1,2] In the case that diffusion of CO_2 takes place in a humid environment, formation of $CaCO_3$ occurs. This formation decreases the pH value, favoring the corrosion of the reinforcement and increases the concrete porosity that raises the diffusion of pollutants (CO_2, Cl^-) in the concrete [3].

The formation of expansive products such as ettringite and thaumasite by the reaction of the gypsum with the calcium aluminosilicates hydrates (CSH). These high molecular weight substances produce defects and fissures in the concrete mass, due to the generation of high expansive stresses [4].

The alkali silica reaction within specific siliceous constituents, which sometimes are contained in the aggregate material and the alkali hydroxides released during the hydration of cement. The reaction product is an alkali – silicate gel that displays a variable capacity of swelling caused by the absorption of further moisture. Such swelling within hardened concrete can cause cracking and overall expansion [5].

The existence of the pre-mentioned fissures and cracks favor the diffusion of the chloride ions in the concrete. When the ratio of Cl^-/OH^- is higher than a specific value [Cl^-/OH^-=0,63] [6], a galvanic element is formed in the reinforcement and the corrosion initiates [3]. The products of the reactions that are taking place provoke corrosion in some areas with consequent volume expansion that generates tensile stresses. The result is the creation of fissures and spalling along these specific areas that favor further pollutants diffusion in the concrete.

In order to study the degradation process of concrete structures, both non-destructive techniques, in situ, and destructive techniques, in lab, should be performed [7], [8].

2. DESCRIPTION OF BUILDINGS – MACROSCOPICAL OBSERVATIONS

The archaeological Museum of Herakleion, general aspect of which is presented in Figure 1, is built in many different periods. The reinforced concrete present extended decay, due to weathering. There is the need of emerge interventions in order the Museum to take the shape of the original concept and to reverse the incompatible interventions of the last 15 years.

On the other hand the Archaeological Museum of Rethimnon is now standing where the old city jails where situated until 1964. The city jails where demolished and only the outer walls where kept. The building presents immediate conservations and partial reconstruction not only due to the weathering of the materials, but also due to the false use of rooms such as storage space in the first floor and exhibition room to the basement. In Figure 2 a general aspect of the Museum is presented.

Figure 1. Aspect of the entrance (1950)

Figure 2. General aspect of Rethimnon Archaeological Museum

3. EXPERIMENTAL AND RESULTS

The buildings were examined using Non-destructive techniques that were applied in situ, Fiber Optics Microscopy (Pico Scopeman Moritex), Infrared Thermography, Ultrasound technique (Pundit 6, CNS Farnell), reinforcement corrosion potential, concrete specific electrical resistance, concrete carbonation depth.

Destructive Techniques that were applied in laboratory include Granulometric Analysis, Mercury Intrusion Porosimetry (Posimeter 2000, Fisons Instruments), X-Ray Diffraction (Diffraktometer D5000, Siemens), Simoultaneous Thermal Analyses (DTA/TG Netsch 409), determination of total soluble salts. In Table 1, the sampling of the materials used for the performing of the analytical methods, is presented.

3.1 In situ measurements

With the aid of fiber optics microscopy a direct visual evidence of deterioration state is accomplished. Figures 3, 4 present the materials surface observed by the Fiber Optics Microscopy. Regarding the concrete surface at Heraklion Museum, it can be observed increase of the total porosity, loss of binding material, as well as, corrosion of the steel reinforcements (Fig. 5).

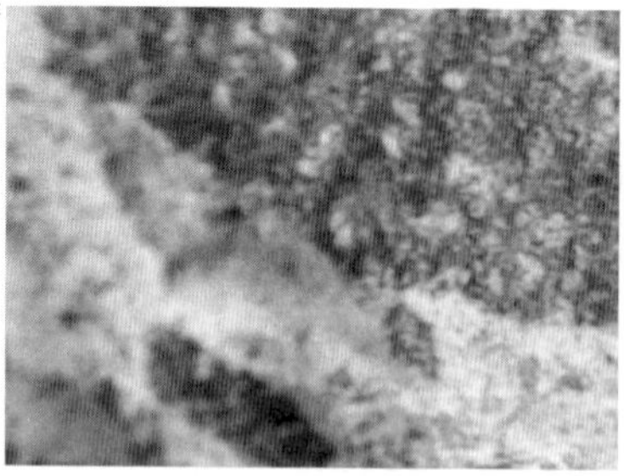

Figure 3. Concrete Surface

Figure 4. Steel Corrosion

The same results are observed also for the concrete in Rethymnon Museum, increase of total porosity and loss of binding material, while in some cases there is evidence of bio-corrosion (Figure 5).

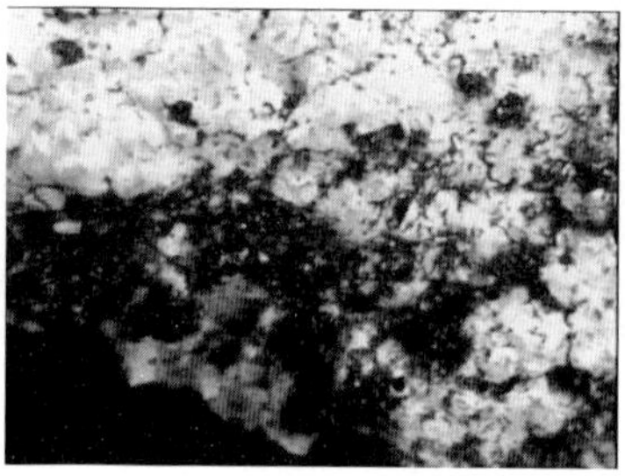

Figure 5. Bio-corrosion

The infrared thermography provides information about the humidity of the surfaces, as well as the differences due to decay and incompatible materials. In the figure presented for the Heraklion Museum (Fig. 6) the mortar/ plaster interface is shown. The compatibility between them is rather satisfactory. On the other hand on Rethimnon Museum there are thermographs showing the salts decay on the masonry and the retension of humidity on the masonry. The ultrasound velocity transmission provides information about the quality and the homogeneity of the examined materials. Measuring the ultrasound velocity transmission, the detection of materials discontinuities and estimation of their conservation state could be accomplished. The measurements were done using the indirect method and the obtained velocities for the materials are reported in Tables 1 and 2.

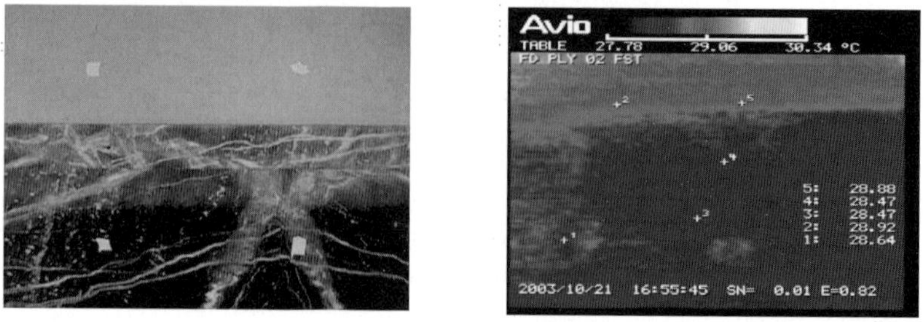

Figure 6. Marble / Plaster interface thermograph

Figure 7. North out aspect (salts) thermograph

Table 1. Ultrasound velocity for Heraklion Museum

	Material	Position	Velocity (m/sec)
1	Surface Plaster	Main entrance 22 cm above marble	715
2	Marble	Main entrance,	4798
3	Concrete	Eastern side	1132

The concrete examined in both cases, presents low values of ultrasound velocity transmission in comparison with sound concrete (Velocity=3800-4900m/sec). This reveals that the concrete present a lot of voids and discontinuities and it could be characterized as extremely decayed. Extremely low is also the value for the surface plaster.

Table 2. Ultrasound velocity for Rethimnon Museum

	Material	Position	Velocity (m/sec)
1	Concrete	Inside the building near the stairs	745
2	Sandstone	North West Side Height :0,30 cm	1638

Determination of reinforcement corrosion potential

The measurements were accomplished using a Cu/CuSO4 electrode. According to ASTM C 876-87, values greater than –200mV, exhibit that the reinforcement is in passivation state. When the potential fluctuates between -200mV and -350mV the corrosion probability is at 50%. Values greater than -350mV correspond to energetic state, leading to the corrosion of steel. Almost all the potential measured are greater than -350mv and therefore the steel is in energetic state where corrosion is occurring.

Measurement of concrete specific electrical resistance

The measurement of concrete specific electrical resistance took place at the same position where reinforcement potentials were measured, using the four point method. The values of specific electrical resistance are low, indicating that the concrete is full of moisture and chloride anions. As a result, the high corrosion rate is inevitable.

Measurement of concrete carbonation depth

According to the results reported in the Tables, the carbonation front reaches the reinforcement for all the structural units. This indicates that the concrete is in advanced state of decay. In Table 3 the results for Heraklion Museum are reported, while in Table 4 the results for Rethimnon Museum are shown.

Table 3. Results for Heraklion Museum

Position	**Corrosion Potential (mV)**	**Specific Electrical Resistance (Ohms.cm)**	**Carbonation Depth**
(Stairs), East Side	-363		
Basement, East Side	-393	2,12	
Prop, North Side	-318	2,09	Beyond overlay
Prop, South Side	-304	2,15	Beyond overlay

Table 4. Results for Rethymnon Museum

Position	Corrosion Potential (mV)	Specific Electrical Resistance (Ohms.cm)	Carbonation Depth
Basement	-375	2,21	Beyond overlay
Basement	-316	2,13	Beyond overlay
Beam in the loft	-364	2,09	Beyond overlay

3.2 Laboratory Measurements

Grain size distribution

For the realisation of grain size distribution, concrete samples undergone cycles with the use of liquid azote (6). This consists of immersion of samples in water for 1 hour, afterwards immersion in liquid azote, and then heating at 105×C. The whole procedure continued until the total disintegration of samples. After the grain size distribution analysis, it was calculated the binder/total ratio. The ratio values were at the 1/10 range, testifying to the loss of binding material. Representative diagrams of two samples (one for both building) are presenting in Figures 10 and 11.

X-Ray Diffraction Analysis

XRD and DTA/TG determine the composition of reaction products formed, or to find out the carbonation state. The X-Ray Diffraction results for the samples examined are showing that in Heraklion Museum the concrete consists of calcite, quartz and dolomite. For the Rethymnon Museum calcite, halite, and moschobite is detected in almost all samples of the concrete.

Thermal Analyses DTA/TG

DTA analysis reveals the presence of calcite in two samples examined from the Heraklion Museum. The aggregates for these two samples are of limestone sources. On the other hand in all samples for stones for the Rethymnon Museum it is proved that these are limestones. For the plasters as well as the concrete samples the results show that they are fully carbonated, and especially for the plasters samples considerable amount of structurally bound water, which can classify them as hydraulic plasters.

Determination of total soluble salts

The determination of total soluble salts was accomplished according to NORMAL 13/83, while the identification of Cl^- and $SO4^{--}$ was performed by spot tests. The results show that in Heraklion Museum the values for Cl^- are relatively low 0,4 – 0,6%, while for all the measurements in Rethymnon Museum considerable amount Cl^- is detected from 2 – 5%.

Microstructural results

The data from porosimetry of the examined materials shows that in all samples tested for both Museums the concrete present higher values of total porosity than those of a healthy concrete, which implies the decay. The decay of the concrete can be also observed by the higher values of Average Pore Radious, while in tha case of Rethimnon Museum some differentiation is observed due to the presence of salts.

4. CONCLUSIONS - RECOMMENDATIONS

From the results of the Non Destructive techniques, as well as from the laboratory tests the following are concluded for the Museum of Heraklion:

- The concrete is of poor quality and with high total porosity. In addition the carbonation rate has already overlapped the steel reinforcements. This fact is mainly due to the decay of the plasters from the surfaces of the walls especially in the East and North side of the building. The decay of the plasters leaves the steel exposured to the carbon dioxide of the atmosphere. The corrosion potential measurements indicate that the steel is already in active area for corrosion.

On the other hand from all the experimental results in situ and in lab, for the Museum of Rethymnon the following could be obtained:

- The steel of the reinforcements in Rethymnon Museum is in rather bad situation than those of Heraklion. This is occurring because in addition with the bad quality of the concrete and the natural weathering of the materials the concentrate of soluble salts, and especially the Cl^- is making the steel more vulnerable to corrosion.

In both cases though the recommendations are the same. Two methods are proposed to the authorities of the Ministry of Culture. In all cases of conservation interventions or reconstruction the quality of the concrete must be the number one priority in order to provide the maximum protection to the steel. But in all cases also the corrosion of steel must be stopped. There are two methods with different results in terms of time protection and also in terms of cost. The first is to provide corrosion inhibitors to the reinforced concrete after the cleaning of the decayed parts. This could provide the building another 20–30 years of life, and it is less expensive. The other one which confront the problem for more than 100 years is the cathode protection. This method is much more expensive, but it is recommended since the Museums is not just buildings but living monuments of our times.

5. REFERENCES

1. Moropoulou, A., Batis, G., Chronopoulos, M., Spanos, Ch., "Investigation of the marine atmosphere - Concrete structures interaction and its impact to the reinforcements", Scienza e Beni Culturali IX, ed. G. Biscontin, D. Mietto, Publ. Libreria Progetto Editore Padova (1993) pp.389-402.
2. Collepardi M., Fratesi R., Moriconi G., Pauri M., "The influence of carbonation on the corrosion of steel in reinforced concrete", 2nd International Symposium on Cement and concrete BISCC Beijing, China, 1989, Vol.2, pp. 185-197.
3. Fratesi R., Moriconi G., "La corrosione delle armature metalliche nelle strutture in calcestruzzo armato", Scienza e Beni Culturali IX, ed. G. Biscontin, D. Mietto, Publ. Libreria Progetto Editore Padova (1993) pp.203-210.
4. Lea F.M. "Lea' s chemistry of cement and concrete", 3rd Edition, Arnold Ltd, London, 1988.
5. Francis Young J., "Cement Based Materials", Current Opinion in Solid State and Materials Science, Volume 3, Issue 5, October 1998, Pages 505-509.
6. Hussain S. E., Rasheeduzzafar A., Al-Mussallam A., and Al-Gahtani A.S., "Factors Affecting Threshold Chloride for Reinforcement Corrosion in Concrete." Cement and Concrete Research, Vol.25, No.7, 1995, pp.1543-1555.
7. Moropoulou, A., Koui, M., Avdelidis, N.P., Achilleopoulos, N., "NDT for materials quality control, environmental impact assessment and management of cultural heritage", INSIGHT, J. of the British Institute of non-destructive testing, 41, No 6 (1999), pp. 362-368.
8. A. Moropoulou, B. Christaras, M. Koui, N.P. Avdelidis, Th. Tsiourva, Ch. Kourteli, "Integrated Non-destructive evaluation for the protection of Cultural Heritage", of the 2nd Int. Conf. on Emerging Technologies in Non Destructive Testing, Athens, Greece, (2000), pp. 323-333.

EFFICIENCY OF TRADITIONAL AND INNOVATIVE PROTECTION METHODS AGAINST CORROSION

F. Tittarelli and G. Moriconi
Department of Materials and Environment Engineering and Physics, Technical University of Marche, Via Brecce Bianche, 60131 Ancona, Italy

Abstract: The corrosion resistance of cracked concrete specimens reinforced with bare, stainless, or galvanized steel plates are compared with the corrosion behavior of bare steel reinforcement embedded in concrete specimens coated with a flexible polymer-cement based mortar either before or after specimen cracking and with those related to bare and galvanized reinforcement embedded in hydrophobic concrete. The specimens were exposed to increasingly aggressive environments: forty days of full immersion in a 3.5% sodium chloride aqueous solution were followed by five months of wet-dry cycles using a 10% sodium chloride aqueous solution. The results for the full immersion condition show that negligible corrosion rates were detected in all the cracked specimens, except those treated with the flexible polymer-cement mortar before specimen cracking and the hydrophobic concrete specimens especially when bare steel reinforcement are adopted. On the other hand, high corrosion rates were measured in all cracked specimen exposed to wet-dry cycles, except for those with stainless steel reinforcement and those coated with the flexible polymer-cement mortar after specimen cracking, as was expected, and, surprisingly, for the galvanized steel reinforcement embedded in hydrophobic concrete.

Key words: cracked concrete; steel reinforcement corrosion; stainless steel; galvanized steel; polymer-cement coating; silane-based hydrophobic agents; hydrophobic concrete.

1. INTRODUCTION

Static, dynamic, and cyclic loading as well as shrinkage, creep, and thermal stress can cause cracking of reinforced concrete structures. Moreover, in precast concrete cracks can also be produced by mechanical shock or flexural stress induced during transportation, lifting, and mounting. Cracks greatly increase the concrete surface permeability[1-2], since they represent preferential paths for penetration of aggressive

M.S. Konsta-Gdoutos, (ed.), Measuring, Monitoring and Modeling Concrete Properties, 545–555.

agents such as chloride ions and atmospheric carbon dioxide[3], which promote corrosion of steel. After the initial induction period, the degradation process accelerates rapidly[4] (propagation stage), since the deterioration mechanisms have a destructive expansive nature[5-7].

It is well known that the cost of repairing reinforced concrete structures during the induction period of the corrosion process is generally much lower than the rehabilitation cost during the propagation period[8]. Therefore, concrete technology is continuously developing new methods to prevent the onset of deterioration in reinforced concrete structures. Obviously, the costs needed to obtain durable concrete rise with the required durability level, but it is important to provide an adequate level of protection in relation to the structure service life, avoiding unnecessary expenses[9-10].

Traditional protection methods against corrosion, such as reducing the water-cement ratio to reduce concrete porosity and permeability, the use of pozzolanic additions to reduce chloride diffusion, the use of corrosion inhibitor admixtures and the surface coating of steel bars (galvanizing or epoxy coating), often fail when cracks reach the reinforcing bars[11-12]. So, other methods have been proposed to increase the corrosion resistance of cracked reinforced concrete, such as the use of stainless steel bars[13] and, more recently, concrete protection through flexible polymer-cement coatings[14] or surface hydrophobic coatings[15-16] due to their ability to make concrete less susceptible to water saturation, since water is the main agent for environmental attack.

The most limiting factor of stainless steel reinforcement is its high cost, even if cheaper materials have been recently developed which show good behavior[17] and even if there is a threshold chloride concentration up to 20 times higher than that of bare steel. On the other hand, the good results obtained by polymer-cement coatings are mainly based on their impermeability, bond strength, and their flexibility, which allows the coating to bridge the cracks of the concrete substrate. However, mechanical stress due to accidental loads and weathering can cause a flexibility loss, leading to microcracks in the coating.

On the other hand, the effectiveness of hydrophobic surface treatment in time depends on the alkali resistance of the used compounds, their penetration depth, their resistance to atmospheric agents, and the integrity of the structure[18-22]. Therefore, to optimize the utilization of hydrophobic agents, they have been recently introduced in the concrete mixture directly in order to make both the surface and the whole concrete bulk hydrophobic[23-25]. The use of these admixtures appears, indeed, a promising and relatively cheap prevention method against the environmental attack of reinforced concrete structures, but, due to their relative "early age", further investigation has to be carried out in order to ensure their real effectiveness against corrosion of reinforced concrete.

The aim of this work is to compare the efficiency of traditional methods used to mitigate corrosion of cracked reinforced concrete, such as the use of galvanized or stainless steel, with more innovative ones, such as coating the concrete surface with a polymer-cement based mortar, used either as a preventive or as a restorative method, or the introduction in the concrete mixture of a hydrophobic admixture.

2. EXPERIMENTAL

Forty two prismatic specimens of highly porous concrete (w/c = 0.80) were manufactured in order to highlight the different types of corrosion behavior:

- six specimens were reinforced with bare steel plates, as reference;
- six specimens were reinforced with galvanized steel plates;
- six specimens were reinforced with stainless steel plates;
- twelve specimens were protected by a polymer-cement coating and reinforced with bare steel plates;
- six hydrophobic concrete specimens were reinforced with bare steel plates.
- six hydrophobic concrete specimens were reinforced with galvanized steel plates

The coating, if any, was applied after 1 month of air drying of the concrete specimens. Half of the specimens for each group were cracked by flexural stress (crack width of about 1 mm) after an additional week. However, in order to estimate the efficiency of the polymer-cement coating as a restorative method against corrosion, three specimens were cracked before coating, while the other three were coated before cracking in order to estimate the efficiency of the surface coating as a preventive method.

The concrete specimens were then exposed to increasingly aggressive environments: initial full immersion in a 3.5% NaCl aqueous solution simulating a marine environment was followed by wet-dry cycles in a 10% NaCl aqueous solution simulating a bridge deck treated with deicing salts. The corrosion resistance of the specimens was evaluated by corrosion electrochemical potential and short-circuit current measurements.

2.1 Materials

A commercial Portland cement type CEM II/A-M 32.5 was used. Crushed aggregate (15 mm maximum size) and natural sand (2 mm maximum size) were used. The hydrophobic admixture was a 30% aqueous emulsion of butyl-ethoxy-silane.

The mixture proportions for the polymer-cement coating were 1 part of 2-ethylhexyl acrylate polymer latex (50% water), 1 part of Portland cement type CEM II/A-L 42.5 and 2 parts of fine sand (0-0.2 mm). Then the w/c of this coating as well as the polymer/cement was 0.50.

Different steel plates were used to reinforce the concrete specimens:

- bare steel plates;
- stainless steel plates (AISI 304);
- hot dip galvanized steel plates.

2.2 Concrete mixtures proportions

Concrete with w/c = 0.80 was used for all the specimens. The following amounts of concrete ingredients were mixed:

- cement 288 kg/m^3
- water 230 kg/m^3
- sand 600 kg/m^3
- coarse aggregate 1167 kg/m^3.

Hydrophobic concrete was manufactured by adding 2% of the active ingredient in the hydrophobic admixture by mass of cement in the concrete mixture. In order to obtain similar microstructure exposed to the aggressive environment, the strength loss due to the hydrophobic admixture[20,22] was compensated for by reducing the w/c to 0.75.

2.3 Reinforced concrete specimens

All specimens were prismatic (100×100×400 mm). Half of the specimens, kept uncracked to act as cathodes for the short-circuit current measurements, were reinforced with single steel plates (70×1×360 mm) embedded at mid depth (Figure 1). The other half of the specimens, which were cracked and acted as evaluation test specimens, were reinforced with two steel plates not in contact with each other. The two steel plates (70×1×360 mm and 70×1×120 mm, Figure 1) were placed at 70 mm and 30 mm, respectively, from the specimen side containing a preformed notch, whose function was to initiate a crack reaching the smallest plate under flexural loading[20]. This plate acted as the anode during the experiment, while the longer steel plate served to control the crack width. The electrical connections required for corrosion monitoring through electrochemical measurements were carried out as described in previous works[24,26]. All the specimens were kept for 48 hours at 100% R.H. and, after demoulding, they were air dried for 1 month at room temperature. The coating, if any, was then applied on the specimens. After an additional week of air-curing, half of all the specimens were stressed by bending, by loading the specimen surface opposite the notch (Figure 1) to initiate the development of a crack. Crack width of 1 mm was obtained with sufficient accuracy by slowly varying the applied load.

Some observations can be made about the specimens protected with the polymer-cement coating before cracking (as a preventive method against corrosion). First, no failure of the polymer-cement coating was visible after the concrete substrate was cracked, thus apparently confirming the good flexibility properties of the coating. Second, after the flexural loading, the produced crack width could not be measured, but its size could be reasonably assured by the load reached to crack the concrete specimen.

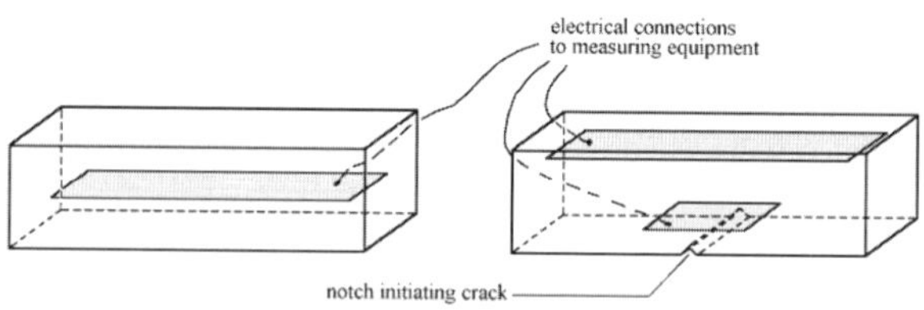

Figure 1. Prismatic reinforced concrete specimens

2.4 Testing of specimens

After the drying period, all the specimens, sound or cracked, were completely submerged in a 3.5% NaCl aqueous solution, taking care to maintain constant atmospheric oxygen saturation through adequate recycling. After forty days of immersion, the specimens were exposed to weekly wet-dry cycles, characterized by two days of full immersion in a 10% NaCl aqueous solution followed by 5 days of air drying, up to about 6 months.

The corrosion resistance of the different steel plates was monitored by measuring their corrosion potential with respect to a reference saturated calomel electrode (SCE). Moreover, during the immersion period, the short-circuit current was measured between the smallest plate (anode), embedded in the cracked specimen and reached by the crack tip, and the same type of steel plate (cathode) placed in the corresponding sound specimen. The reported values are the averages of the measurements carried out on three specimens of each type.

3. RESULTS AND DISCUSSION

Due to a large scale difference in the potential measurements between steel, either bare or stainless, and galvanized steel reinforcement, the results are separately discussed.

3.1 Steel plates

Figure 2 shows the free corrosion potential of the anodic steel plates embedded in the cracked specimens as a function of the test time. By assuming that potential values lower than -450 mV/SCE indicate a relatively high corrosion risk of the steel reinforcement, stainless steel and the polymer-cement coating applied after cracking guarantee adequate protection whatever the aggressive exposure condition. This is not true, as expected, with the reference bare steel or with bare steel in hydrophobic concrete. On the other hand, when the concrete specimen is preventively protected by a polymer-cement coating applied before cracking, the corrosion risk does not seem to be reduced, demonstrating that the corrosion behavior is not consistent with the apparent coating integrity observed.

The short-circuit currents (Figure 3) measured in the full immersion condition are, in any case, very low as a consequence of the low oxygen availability, which slows down the kinetics of the corrosion process with the exception of those related to reinforcing steel bars in hydrophobic concrete. This different behavior is explained by admitting that the gaseous oxygen diffuses better through the open pores of the hydrophobic concrete with respect to the water saturated pores of the reference mixture, feeding in this way the cathodic reaction of the corrosion process[25]. They remain negligible in the wet-dry cycle condition only for stainless steel, thus assuring prevention of corrosion, and for the concrete protected by polymer-cement coating applied after cracking, hence confirming the efficiency of this application as a successful repair method.

On the other hand, the short-circuit current values measured in this condition obviously become unacceptably high for bare steel, but they also appear abnormally high for the bare steel anodic plate embedded in the concrete specimen protected with the polymer-cement coating applied before cracking. In fact, some corrosion products already appeared at the end of the full immersion period of these specimens, thus indicating that coating stretching induced by flexural loading can compromise the efficiency of this preventive method against corrosion. Steel plates in hydrophobic concrete do not modify significantly the short circuit currents values when the exposure condition changes from full immersion to wet-dry cycle in the chloride solution.

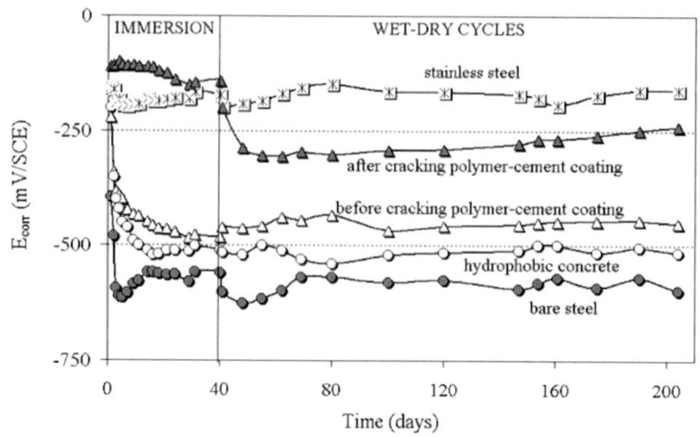

Figure 2. Free corrosion potential of the anodic steel plate embedded in cracked concrete specimens as a function of the test time

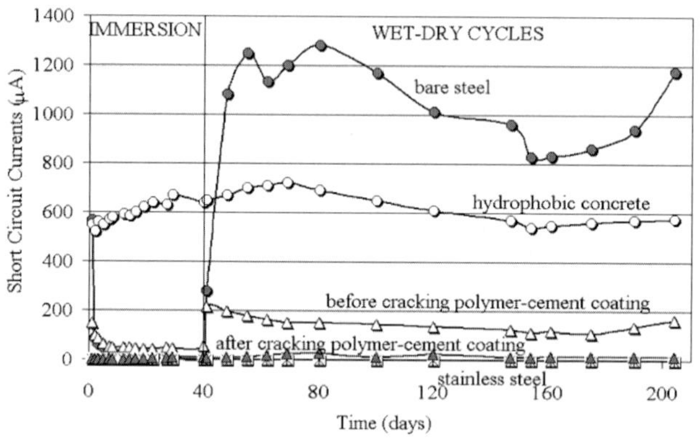

Figure 3. Short circuit currents of steel plate embedded in cracked concrete specimens as a function of the test time

3.2 Galvanized steel plates

Figure 4 shows that the active corrosion potential assumed by the galvanized steel plate during the full immersion period moves towards passivation values when embedded in hydrophobic concrete. When the cracked concrete specimens reinforced with galvanized steel are exposed to wet-dry cycles the anodic potential rises in any case even if the potential values become rapidly typical of the passive state only in the case of hydrophobic concrete (Figure 4).

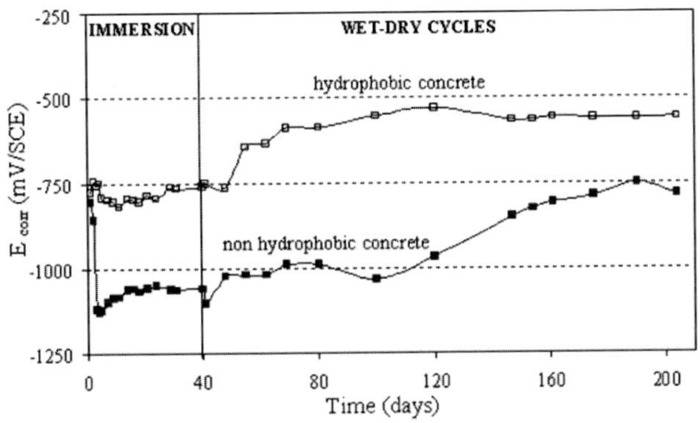

Figure 4. Free corrosion potential of the anodic galvanized steel plate embedded in cracked concrete specimens as a function of the test time

The short-circuit currents measured in the full immersion condition for galvanized steel embedded in hydrophobic concrete (Figure 5), rapidly decrease and remain at negligible values after the cracked concrete specimens are transferred to the wet-dry cycles environment. On the other hand, the negligible currents measured in the full immersion condition for galvanized steel in ordinary concrete, due to poor oxygen availability in water saturated concrete, suddenly rise to very high values before gradually decreasing as long as the anodic galvanized steel plate reaches less negative potential values when exposed to wet-dry cycles (Figures 4 and 5).

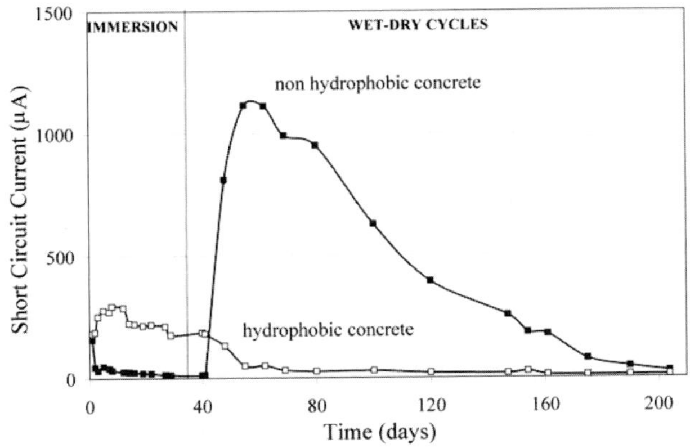

Figure 5. Short circuit currents of galvanized steel plate embedded in cracked concrete specimens as a function of the test time

3.3 General comparison

The comparison is drawn on the basis of the short-circuit current evaluation as a function of the potential difference between the steel plates respectively acting as anode (cracked specimen) and cathode (uncracked specimen).

In the full immersion condition (Figure 6), it can be observed that, at equal electromotive force of the corrosion process, the highest corrosion currents were monitored for bare and galvanized steel embedded in hydrophobic concrete. Lower currents were found in all other concrete specimens, except for those protected with the polymer-cement coating applied before the specimen cracking.

In the wet-dry cycles condition (Figure 7), a very high current was recorded for bare steel; it is high for galvanized steel, which up to its corrosion potential reaches more positive values, and also for bare steel in hydrophobic concrete; it is unacceptable for bare steel embedded in concrete protected by the polymer-cement coating applied before concrete cracking; it is definitely low when this protection is applied after the concrete cracking and in the case of galvanized steel embedded in hydrophobic concrete; it is practically insignificant for stainless steel.

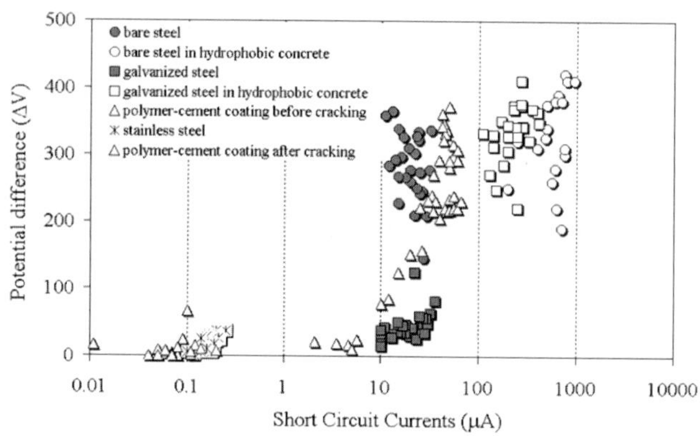

Figure 6. Potential difference versus short-circuit current for the full immersion condition

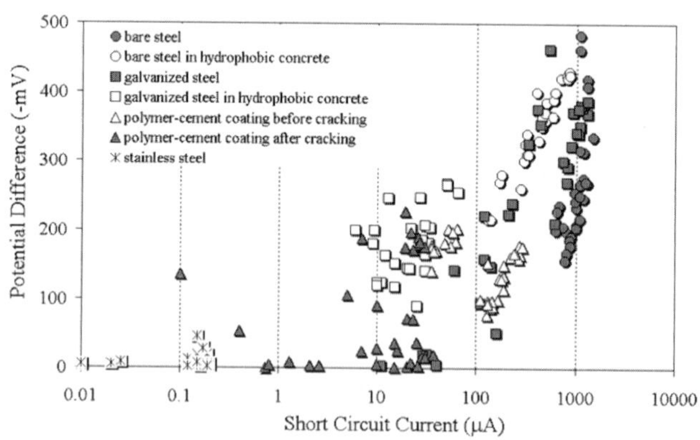

Figure 7. Potential difference versus short-circuit current for the wet-dry cycles condition

4. CONCLUSIONS

In this paper the corrosion resistance of traditional methods used to mitigate corrosion of steel (galvanizing or using stainless steel) embedded in cracked concrete is compared with more innovative techniques (protection of concrete by a polymer-cement coating and bare or galvanized steel in a hydrophobic concrete) when cracked reinforced concrete is exposed to increasingly aggressive environments: namely, full immersion in a chloride aqueous solution (representing a marine environment) followed by wet-dry cycles in a more chloride concentrated aqueous solution (representing deicing salts treatment).

The results obtained show that the efficiency of the tested innovative methods to mitigate corrosion of steel in cracked concrete is lower than the use of stainless steel bars. However, these techniques show good prospects especially when technical-economical considerations are taken into account, despite the poor efficiency observed in certain situations.

In particular, the good behavior of cracked concrete protected by a polymer-cement coating is not fully confirmed when this technique is applied as a preventive method to reinforced concrete before cracks occur due to coating stretching which compromises its impermeability through the occurrence of visually undetectable microcracking.

On the other hand, the hydrophobic admixture makes significantly worse the corrosion behavior of both bare and galvanized steel in the full immersion condition, but it increases surprisingly the corrosion resistance of galvanized steel reinforcement in cracked concrete when exposed to the very aggressive, and more common, wet-dry cycles condition. Therefore, the use of galvanized steel reinforcement in hydrophobic concrete shows good prospects especially when technical-economical considerations are taken into account.

5. REFERENCES

1. P. K. Mehta, Concrete Technology at the Crossroads - Problems and Opportunities, *Proceedings of the V. M. Malhotra Symposium on Concrete Technology: Past, Present, Future*, S.Francisco, U.S.A., 1994, edited by P. K. Mehta (ACI Publication SP-144, Detroit, Michigan, U.S.A., 1994), pp. 1-30.
2. R. Francois, G. Arliguie, and A. Castel, Influence of Service Cracking on Service Life of Reinforced Concret, *Proceedings of the Second International Conference on Concrete under Severe Conditions, CONSEC '98*, Tromsø, Norway, 1998, edited by O. E. Gjørv, K. Sakai and N. Banthia (E & FN Spon, London, U.K., 1998), pp. 143-152.
3. C. L. Page, and K. W. J. Treadaway, Aspects of the Electrochemistry of Steel in Concrete, *Nature*, **297**, 109-115 (1982).
4. K. Tuutti, Corrosion of Steel in Concrete, Swedish Cement and Concrete Research Institute, Stockolm, Report 4/82, 1982, p. 468.
5. M. L. Allan, and B. W. Cherry, Factors Controlling the Amount of Corrosion for Cracking in Reinforced Concrete, *Corrosion*, **48**(5), 426-430 (1992).
6. M. Middel, Microcracking of Concrete Due to Chemical Attack, *Proceedings of the International Conference on Corrosion and Corrosion Protection of Steel in Concrete*, Sheffield, U.K., 1994, edited by R. N. Swamy (Sheffield Academic Press, U.K., 1994), pp. 548-560.
7. T. Ueda, Y. Sato, Y. Kakuta, and H. Kameya, Analytical Study on Concrete Cover Cracking Due to Reinforcement Corrosion, *Proceedings of the Second International Conference on*

Concrete under Severe Conditions, CONSEC '98, Tromsø, Norway, 1998, edited by O. E. Gjørv, K. Sakai and N. Banthia (E & FN Spon, London, U.K., 1998), pp. 678-687.
8. W. R. De Sitter Jr., Costs for Service Life Optimization, the Law of Five, *CEB-RILEM International Workshop on Durability of Concrete Structures*, Copenhagen, Denmark, 1983, *CEB Bulletin d'Information*, No.152, pp. 131-134 (1984).
9. W. E. Ellis Jr., and G. D. Shelton, How Much Durability Do You Need (and Do You Have to Pay For)?, *Proceedings of the Fifth CANMET/ACI International Conference on Fly Ash, Silica Fume, Slag and Natural Pozzolans in Concrete*, Milwaukee, U.S.A., 1995, Supplementary Papers, pp. 513-529.
10. P. R. W. Vassie, A Discussion of Methods for Preventing Reinforcement Corrosion in Bridges, in: *Corrosion of Reinforcement in Concrete Construction*, edited by C. L. Page, P. B. Bamforth and J. W. Figg (SCI Special Publication No.183, 1996), pp. 654-661.
11. M. Collepardi, R. Fratesi, G. Moriconi, L. Coppola, and C. Corradetti, Use of Nitrite Salt as Corrosion Inhibitor Admixture in Reinforced Concrete Structures Immersed in Sea-water, *Proceedings of the International RILEM Symposium on Admixtures for Concrete – Improvement of Properties*, Barcelona, Spain, 1990, edited by E. Vazquez (Chapaman & Hall, London, U.K., 1990), pp. 279-288.
12. K. Sakai, and S. Sasaki, Ten Years Exposure Test of Precracked Reinforced Concrete in a Marine Environment, *Proceedings of the Third CANMET/ACI International Conference on Durability of Concrete*, Nice, France, 1994, edited by V. M. Malhotra (ACI Publication SP-145, Detroit, Michigan, U.S.A., 1994), pp. 353-369.
13. R. N. Cox, and J. W. Oldfield, The Long Term Performance of Austenitic Stainless Steel in Chloride Contaminated Concrete, in: *Corrosion of Reinforcement in Concrete Construction*, edited by C. L. Page, P. B. Bamforth and J. W. Figg (SCI Special Publication No.183, 1996), pp. 662-669.
14. L. Coppola, C. Pistolesi, P. Zaffaroni, and M. Collepardi, Properties of Polymer-Cement Coatings for Concrete Protection, *Proceedings of the Fifth CANMET/ACI International Conference on Superplasticizers and Other Chemical Admixtures in Concrete*, Roma, Italy, 1997, edited by V. M. Malhotra (ACI Publication SP-173, Farmington Hills, Michigan, U.S.A., 1997), pp. 267-285.
15. K. H. Wong, E. Weyers, and P. D. Cady, The Retardation of Reinforcing Steel Corrosion by Alkyl-Alkoxy Silane, *Cement and Concrete Research*, **13**(6), 778-788 (1983).
16. E. McGettigan, Application Mechanism of Silane Weatherproofers, *Concrete International*, Vol. 12, No. 10, 66-68 (1990).
17. L. Bertolini, P. Pedeferri, and T. Pastore, Stainless Steel in Reinforced Concrete Structures, *Proceedings of the Second International Conference on Concrete under Severe Conditions, CONSEC '98*, Tromsø, Norway, 1998, edited by O. E. Gjørv, K. Sakai and N. Banthia (E & FN Spon, London, U.K., 1998), pp. 911-918.
18. P. A. M. Basheer, Surface Treatments for Concrete, *Proceedings of COST 509 Workshop on Corrosion and Protection of Metals in Concrete*, edited by C. Andrade, R. Cox and B. Pukl (Sevilla, Spain, 1995).
19. A. J. J. Calder, and Z. S. Chowdhury, A Performance Specification for Hydrophobic Materials for Use on Concrete Bridges, in: *Corrosion of Reinforcement in Concrete Construction*, edited by C. L. Page, P. B. Bamforth and J. W. Figg (SCI Special Publication No.183, 1996), pp. 556-566.
20. D. J. Cleland, and P. A. M. Basheer, Management and Maintenance of Surface Treatments for Concrete, *Proceedings of COST 521 Workshop on Corrosion of Steel in Reinforced Concrete Structures*, Utrecht, The Netherlands, 1998, edited by R. B. Polder (1998).
21. H. deVries, R. B. Polder, and H. Borsje, Durability of Hydrophobic Treatment of Concrete, *Proceedings of the Second International Conference on Concrete under Severe Conditions,*

CONSEC '98, Tromsø, Norway, 1998, edited by O. E. Gjørv, K. Sakai and N. Banthia (E & FN Spon, London, U.K., 1998), pp. 1341-1350.

22. M. R. Jones, R. K. Dhir, and J. P. Gill, Concrete Surface Treatment: Effect of Exposure Temperature on Chloride Diffusion Resistance, *Cement and Concrete Research,* **25**(1), 197-208 (1995).
23. N. P. Mailvaganam, Miscellaneous Admixtures, in: *Concrete Admixtures Handbook*, edited by V.S. Ramachandran (Noyes Publications, Park Ridge, New Jersey, U.S.A., 1984), pp. 518-524.
24. R. Fratesi, G. Moriconi, F. Tittarelli, and M. Collepardi, The Influence of Hydrophobic Concrete on the Corrosion of Rebars, *Proceedings of the Fifth CANMET/ACI International Conference on Superplasticizers and Other Chemical Admixtures in Concrete*, Roma, Italy, 1997, edited by V. M. Malhotra (ACI Publication SP-173, Farmington Hills, Michigan, U.S.A., 1997), pp. 105-122.
25. F. Tittarelli, G. Moriconi, and R. Fratesi, Influence of Silane-Based Hydrophobic Admixture on Oxygen Diffusion Through Concrete Cement Matrix, *Proceedings of the Sixth CANMET/ACI International Conference on Superplasticizers and Other Chemical Admixtures in Concrete*, Nice, France, October 10-13, 2000, edited by V. M. Malhotra (Publication SP-195, American Concrete Institute, Farmington Hills, Michigan, U. S. A., 2000), pp. 431-445.
26. R. Fratesi, G. Moriconi, and L. Coppola, The Influence of Steel Galvanization on Rebars Behaviour in Concrete, in: *Corrosion of Reinforcement in Concrete Construction*, edited by C. L. Page, P. B. Bamforth and J. W. Figg (SCI Special Publication No.183, 1996), pp. 630-641.

CORROSION OF STEEL IN CRACKED CONCRETE: EXPERIMENTAL INVESTIGATION

M. Bi and K. Subramaniam
Department of Civil Engineering, Steinman Hall, City College of the City University of New York, Convent Avenue at 140th Street, New York, NY 10031, USA

Abstract: The polarization response of a steel bar with an established macrocell is presented. A simple circuit-based model, which allows for predicting the polarization response of the macrocell is described. The results indicate that there is a spatial variation in the potential relative to the crack. It is shown that the applied current is primarily confined to the active steel located near the crack.

Key words: Corrosion; Polarization; Macrocell, Circuit-model

1. INTRODUCTION

Corrosion of steel reinforcement is one of the main causes of damage in concrete structures. Thus, the detection and prevention of corrosion of steel in concrete is important for the condition assessment and rehabilitation of reinforced concrete structures. Often the corrosion process of steel embedded in concrete is influenced by factors such as cracks in the concrete. Cracks in the concrete are often produced due to action of loads, restrained shrinkage or thermal gradients. Once a crack is formed, it provides an easy access for ingress of water, chloride ions and oxygen to the steel surface. Furthermore, a crack introduces a physical discontinuity in the material medium and also produces spatial variation in the ionic concentration of ions, oxygen and moisture throughout the length of steel bar. Relatively little work has been done to understand the mechanism of corrosion of steel embedded in cracked concrete[1-5].

This paper presents the results of an experimental program which aims to investigate the influence of a crack on the corrosion mechanism and the polarization response of steel in reinforced concrete. It is shown that steel embedded in cracked concrete forms a macro-cell, which results in a spatial variation in the potential of the steel bar. The results from external polarization of steel embedded inside cracked concrete are also presented. Finally, it is shown that due to the low polarization resistance of the active area close to the crack, the applied current is confined to the active area.

M.S. Konsta-Gdoutos, (ed.), Measuring, Monitoring and Modeling Concrete Properties, 557–562.

2. OBJECTIVES

The objectives of the work presented in this paper are: (a) To study the influence of a crack on the spatial variation of potential for steel embedded in concrete; and (b) To study the influence of external polarization (potential) on the response of the cracked specimen.

3. MATERIALS AND METHODS

Three concrete specimens were cast to study the response of steel embedded in cracked concrete. The specimen geometry and dimensions are shown in Figure 1. A segmental steel bar with three segments was used. The central segment 5.1cm in length, labeled A, was positioned close to the mid-span. The other two segments, labeled C_1 and C_2, had lengths equal to 53.3cm and were positioned on either side of the central segment. The segments were electrically isolated along their length. The segments were connected to external switches such that the electrical connections between the segments could be controlled from the outside (as shown in Figure 1).

Type 1 Portland cement (ASTM C 150) was used for all specimens. The cement: sand: aggregate ratio of concrete by weight was 1:1.70:2.42 and the water/cement ratio was 0.45. An air entraining agent, MB-VR by Master Builders (ASTM C 494), was also used. The concrete cover depth for all specimens was equal to 25.4 mm. Plain carbon steel bars with diameter equal to 12.7 mm were used for all specimens. The steel bars were cleaned with acetone and polished using 600-grit Sic paper. Electrical connections were made with each steel bar by soldering a copper wire close to one of its ends. Embedded Ag/AgCl reference electrode were installed close to each steel segment prior to placing the steel in the form. The embedded reference electrode is referred to "ERE" in this paper. A crack was introduced at the time of casting using a thin plastic sheet placed in the middle of the specimen.

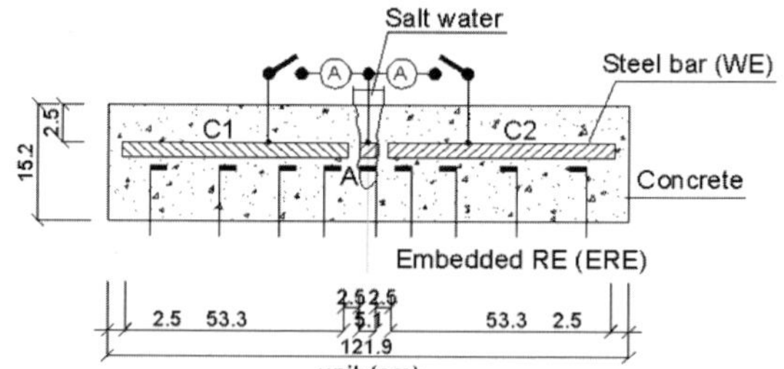

Figure 1. The specimen to study corrosion in cracked concrete

All specimens were demolded 24 hours after casting and cured in a 100% relative humidity (RH) chamber for 90 days following which they were subjected to periodic wetting and drying cycles. Each wetting-drying cycle involved three days of wetting followed by a four day drying period. During wetting, the specimens were subjected to 100%RH at 23°C. The drying comprised of exposing the specimens to the laboratory environment, which was maintained at 50% RH and 23°C. At 90 days age, specimens were subjected to 3-point flexural loading to initiate a crack along the plastic sheet. During the loading procedure, as the crack faces moved apart, the plastic sheet was pulled out from the inside of the concrete. After the crack was created, the specimens the periodic wetting-drying cycles were continued for the duration of testing.

4. MEASUREMENTS

Initial measurements were performed with the external switches in the open position. This configuration is referred to as "open" mode in this paper. Half-cell potential and linear polarization measurements were performed on every steel segment. Before each electrochemical measurement, the concrete specimens were covered with a thin sponge soaked in 3% NaCl solution. When the potential values recorded from the different segments (A, C_1 and C_2) were found to be stable over a period of one month, the external switches were moved to the closed position. The configuration with all electrical connections between the segments established is referred to as the "closed" mode. After connecting the steel segments, the macrocell current and potential were measured and recorded continuously. When the macrocell values were found to be relatively constant with time, polarization measurements were preformed in the closed mode.

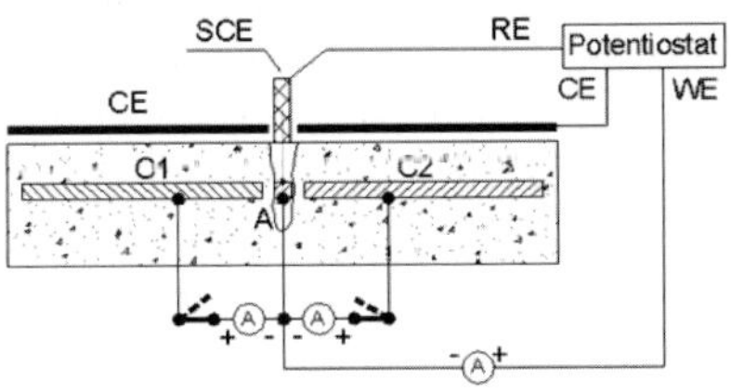

Figure 2. Test Configuration for polarization measurements

In both the open and closed modes, half cell potential and polarization measurements were performed at the end of each wetting period within one wetting-drying cycle. Linear polarization of the specimen in the closed mode was performed using a setup shown in Figure 2. The length of CE was equal to the length of specimen. During the polarization, the current between the segments was continuously recorded, in addition to the applied potential and current recorded by the Potentiostat.

5. EXPERIMENTAL RESULTS

The potential profiles measured using a SCE on the concrete surface and the ERE along the length of the specimen in the open and the closed modes are shown in Figure 3. In the open mode, the potential difference between the segment at crack and segments away from crack were 250 mV ~300 mV. The measured potentials from the open mode suggest that the steel segment at the crack is undergoing active corrosion while the steel segments away from the crack are in passive state.

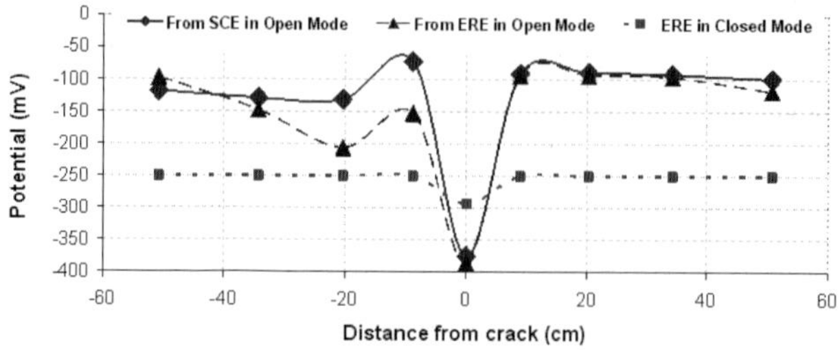

Figure 3. Potential measurements in the Open and Closed modes

The potential of the segments in the closed mode recorded using the ERE are also shown in Figure3. It can be seen that after connection the potential of segment A increases to a more positive value, almost 150 mV more positive than its potential prior to connection. With time there appears to be a permanent shift away from the potentials in open configuration. A steady macrocell current approximately equal to 20 μA was recorded between the segment A and the segments C_1 and C_2. The potential shift from the open to the closed modes therefore corresponds with the polarization produced by the macrocell current. The segment A is the macrocell anode while C_1 and C_2 form the macro-cell cathode. Finally, there appears to be a constant potential difference of 50mV between the macrocell anode and cathode for the steady state.

The effective medium resistance and the polarization resistance were determined following the procedure recommended by Gonzalez et al.[6] The polarization resistances, R_p, for the steel segments A and Cs (C_1 and C_2) obtained in the open mode were equal to 29480 and 1288120 Ohm-cm^2, respectively. It can be seen that the magnitude of R_p of steel away from crack is in the order of 10^6, which is consistent with the findings of Feliu et al.[7] for steel in a passive state. The value of R_p of steel at crack provides a value which is in the order of 10^4. This value is consistent with those reported by Feliu[7] for steel undergoing active corrosion. The total effective medium resistance for the steel segments A and Cs were each found to be approximately equal in magnitude at 300 Ohms.

Results of linear polarization of the specimen in the closed state are shown in Figure 4. It can be seen that following initial nonlinearity, the change in current is directly proportional to the change in the applied potential. In the linear range, the applied potential varied from 7mV below to 8mV above the open circuit potential of the macro-cell in the steady state. The corresponding change in applied current for this time period is equal to 8.9 μA. The changes in the current in this time period between the segments A-C_1 and A-C_2 were equal to 3.5 μA and 4.7 μA, respectively.

6. DISCUSSION

A simple circuit, shown in Figure 5, is proposed to explain the polarization response of the steel in cracked concrete. In this model, R_1 and R_{pA} correspond to the resistance of the medium and the total polarization resistance of the anode, respectively.

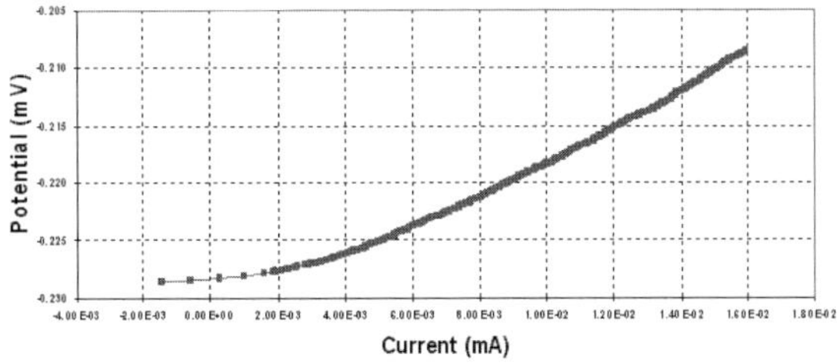

Figure 4. Linear polarization scan in the closed mode

R_2 and R_{pC} correspond to the concrete media resistance and the total polarization resistance at the cathode, respectively.

The total polarization resistance is obtained by dividing the R_p with the surface area of the polarized steel bar. The values of R_1 and R_{PA} are obtained from the LPR scan of Segment A in the open mode. The value of R_2, R_{pC} are obtained from the LPR scan preformed on the segments C_1 and C_2 in the open mode. In the circuit, R_3 represents the effective medium resistance between the macrocell anode and the cathode and is obtained by dividing the steady macrocell potential difference between the anode and the cathode by the total current between the segments.

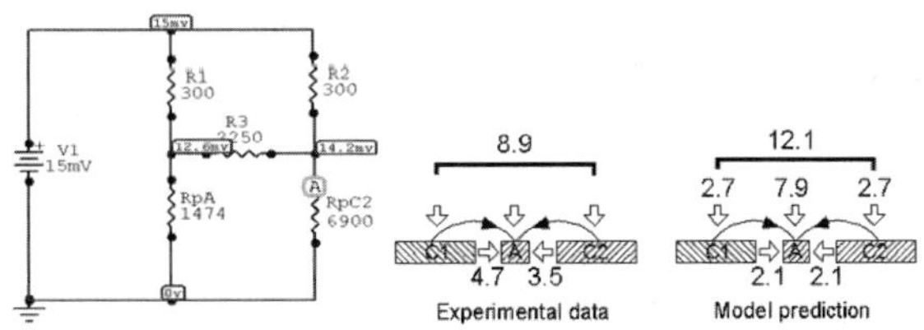

Figure 5. (a) Circuit model; (b) current change due to polarization

In this circuit model the external polarization is represented as a potential applied between the steel bar and the concrete surface. The concrete surface in contact with the CE is treated as an equi-potential surface. Experimental evidence suggests a linear response between the incremental applied potential and the incremental currents in the system for the applied potential sweep between -7 and 8 mV. The external polarization is shown as a constant potential difference of 15mV between the concrete surface and the steel bar. It can be seen that the change in the polarization current predicted by the circuit is compares favorably with that measured experimentally (8.9μA). Further, the change in the current between the anode and cathode is 2.1μA, which agrees well with the measured value of 3.5μA. Therefore, it can be concluded that the linear polarization response of the macrocell can be predicted adequately using the equivalent circuit. The equivalent circuit also allows us to determine the currents through the macrocell anode and the cathode as a result of the external current due to the polarization. The model predicts that for a total polarization current of 12.1 μA, 7.9μA flow through the anode. The results of the analysis indicate that there is a variation in the spatial distribution of the applied current with respect to the crack. The current imposed by the application of external polarization is primarily confined to the active area located at the crack. Similar observations have been reported previously by the authors using other procedures[8]. Therefore, the use of linear polarization to estimate the corrosion of steel in cracked concrete requires a careful consideration of the active steel at the crack. The polarization resistance estimated using a large counter electrode under the assumptions of uniform corrosion would severely under-estimate the local corrosion rate.

7. CONCLUSIONS

Based on the results obtained, the following conclusions can be drawn: (a) The corrosion of steel in cracked concrete is spatially inhomogeneous. The steel in the vicinity of the crack undergoes active corrosion, while the steel away from the crack is in a passive state; and (b) There is a spatial variation in the applied current with respect to the crack. The current imposed by the application of external polarization is confined to the active area close to the crack.

8. REFERENCES

1. Schiessl, P., Raupach, M, ACI Mat'ls J., v. 94, n. 1, pp. 56-62, 1997
2. Otsuki, N., et al., ACI Mat'ls J., v. 97, n. 4, pp. 454-464, 2000
3. Elsener, B., Cement and Concrete Composites, v. 24, pp. 65-72, 2002
4. Berke, N.S., et al., Corrosion 93, No. 322, 1993
5. Arya. C, Cement and Concrete Research, v.26, n. 3, pp. 345-353, 1996
6. Gonzalez, J.A., et al., Corrosion Science, v25, n10, pp. 917-930, 1985
7. Feliu, S., Corrosion Science, v29, n1, pp. 105-113, 1989
8. Bi, M., and Subramaniam, K. V., in the proceedings of the 3rd International Conference on Construction Materials: Performance, Innovations and Structural Implications, Vancouver, Canada, August 22-24, 2005.

CRITERIA AND METHODOLOGY FOR DIAGNOSIS OF CORROSION OF STEEL REINFORCEMENTS IN RESTORED MONUMENTS

A. Moropoulou, G. Batis, M. Chronopoulos, A. Bakolas, P. Moundoulas, E. Aggelakopoulou, E. Rakanta, K. Lambropoulos and E. Daflou

School of Chemical Engineering, Materials Science and Engineering Section, National Technical University of Athens, 9 Iroon Polytechniou St., Zografou Campous 15780, Athens

Abstract: In this work criteria and methodology are proposed for the diagnosis of the durability of steel reinforcement's in restored monuments. An integrated methodology is applied for the assessment, of traditional buildings from several areas of Greece, such as Rhodes (Kallithea Spa), Chios (Nea Moni Monastery) and Symi (The Bell Tower of St. John Prodromos). Non–destructive techniques were applied in situ (Fiber Optics Microscopy, Infrared Thermography, ultrasound technique, determination of reinforce corrosion potential, concrete specific electrical resistance and concrete carbonation depth) along with Destructive Techniques in laboratory (Mercury Intrusion Porosimetry, X–Ray Diffraction, Thermal Analysis, Determination of soluble salts. The applied methodology allows the assessment of the conservation interventions, and concludes for their effectiveness. Proper actions are proposed either for preservation of the restored monuments or for the restoration where it is needed. The marine atmosphere in all three cases has caused corrosion to the reinforcements. The results show that the conservation interventions in all cases are improper and actions must be taken in order to reverse the corrosion phenomena and to preserve the historic character of the monuments. Compatible restoration innervations are proposed in order to diminish the corrosion and keep the durability of the buildings examined, as well as to avoid the decay of the original materials.

1. INTRODUCTION

Reinforced concrete has been one of the most widely used building materials in the 20th century architecture, but now, at the beginning of the 21th century, the durability and the service life of concrete has become a worldwide concern.

Durability is the ability of building, its parts, components and materials to resist the action of degrading agents over a period of time [1]. Service life refers to the period of

M.S. Konsta-Gdoutos, (ed.), Measuring, Monitoring and Modeling Concrete Properties, 563–573.

time in which all-essential performance characteristics of a properly maintained service item exceeds the minimum acceptable values [2]. Lack of concrete's durability manifests in the form of cracking, spalling, loss of strength, or loss of mass. Most important causes of deterioration are the following [3]:

- corrosion of reinforcing steel
- frost action in cold climates
- chemical effects on hydrated cement paste from external agents (i.e. water containing carbon dioxide, sulphates or chlorides)
- physical-chemical effects from internal phenomenon, such as alkali-agreggate reaction, and salt weathering.

Many of the structures constructed at the previous century present extended deterioration, in many cases without reaching their intended service lifetime and the cost for their maintenance and repair is extremely high. It is estimated that approximately 50% of the expenditure of the construction industry in Europe is spent in repairs. A large percentage is due to the deterioration of concrete structures. In the United States conservative estimates show that the cost to rehabilitate deteriorating concrete structures is in the 100 billion dollars range [4].

Maintenance and renovation of concrete structures are key issues for a sustainable built environment. Nowadays there is a growing need for knowledge of durability and service life data of new or existing building. The durability of concrete is an important input for the design, operation and maintenance of such structures. In particular an integrated methodology for the study of these structures, but also a plan for the evaluation and the management of such data is of great importance.

In this study, three buildings exposed to aggressive marine environment are examined: the spa at Kallithea in Rhodes, constructed mainly with reinforced concrete, the Bell tower of the Church of Saint John Prodromos in Simi Island, where reinforced concrete was used for repair works and Nea Moni Monastery in Chios island. The construction and repair works for the two first buildings were completed in the 1st half of the 20th century. The cement of that period was not as fine, and did not contain high amount of C3S, as today's cement. Moreover its compressive strength continued to increase after 28 days, while modern cement achieve most of their expected strength within this period [5].

The examination of reinforced concrete, included both non-destructive techniques applied in situ, along with Destructive Techniques in laboratory. This is an integrated methodology for the characterization of materials and structures and the environmental impact evaluation on them, serving to determine the cause of deterioration leading to poor durability. The knowledge of the state of deterioration and its causes contributes to the:

- understanding of the action of decay agents as well as the resistance of materials to those factors. This knowledge is essential for the selection of the most appropriate maintenance and repair interventions on the damaged structures. In this way their service life will extend, avoiding possible failure problems of the structure in the future. Concrete structures as part of modern architectural heritage should be preserved with the aid of appropriate materials and methods.
- collection of data on the durability parameters and service life of reinforced concrete structures of the early 20th century placed in an aggressive marine environ-

ment, such as that of Mediterranean islands. Additionally, this study provides information about the construction technology and workmanship, as well as the quality of concrete structures at Greek Islands.

2. DESCRIPTION OF BUILDINGS – MACROSCOPICAL OBSERVATIONS

The Rotonda and the Dome of the Spa Complex in Kallithea Rhodes are presented in Figures 1-3. Figures 1 and 2 exhibit the waiting lodge and the dome of Rotonda building while Figure 3 presents the Dome building, that is currently repaired using gunite. Rotonda building was constructed using, mainly, reinforced concrete and sandstone. The reinforced concrete present extended decay, and almost all the columns exhibit extended areas of cracks, in parallel direction to the bars, due to their corrosion that has provoked the generation of tensile stresses to the concrete Figure 4.

Figure 1. The waiting lodge of Rotonda

Figure 2. The Dome of Rotonda

The concrete presents loss of binding material and the aggregates can be shown, macroscopically (Figure 5). Furthermore, several layers of plasters were used in time, of different colored and nature, as presented in Figure 6. The plasters exhibit an intensive decay and in several areas they have detached from the wall.

Figures 7 and 8 present the Bell Tower of the Church of Saint John Prodromos in Simi, which was constructed in the early twenties, with stones, marble and historic mortars. During the 2nd world war it was severely damaged. In 1946 there were restoration works carried out using reinforced concrete as columns as well as for the floor. The reinforced concrete has deteriorated, presenting cracks due to the reinforcement corrosion as shown in Figure 9. The repair of these parts is crucial for the structural integrity of the building.

Figure 3. The Dome of the Spa Complex

Figure 4. Tensile Stresses due to the corrosion of the steel

Figure 5. Loss of binding material and aggregates

Figure 6. Several layers of plasters

Figure 7. The Bell tower of the Church of Saint John Prodromod in Simi

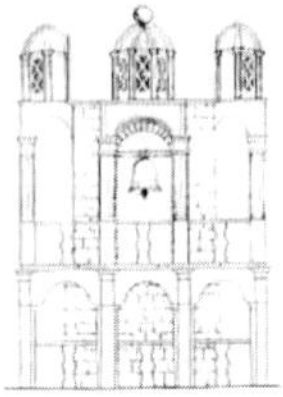

Figure 8. East side of the The Bell tower

Figure 9. Cracks due to reinforcement corrosion

3. EXPERIMENTAL AND RESULTS

The buildings were examined using: Non-destructive techniques that were applied in situ, Fiber Optics Microscopy (Pico Scopeman Moritex), Ultrasound technique (Pundit 6, CNS Farnell), reinforcement corrosion potential, concrete specific electrical resistance, concrete carbonation depth.

Destructive Techniques that were applied in laboratory: Granulometric Analysis, Mercury Intrusion Porosimetry (Posimeter 2000, Fisons Instruments), X-Ray Diffraction (Diffraktometer D5000, Siemens), Simoultaneous Thermal Analyses (DTA/TG Netsch 409), determination of total soluble salts. In Table 1, the sampling of the materials used for the performing of the analytical methods, is presented.

Table 1. Materials Sampling

Type	Sample point	Co.	Remarks
Rotonda Building			
Concrete	8th column	5	Intensive decay – Loss of binding material. The reinforcement was visible
	6th column	2a	
	7th column	4b	
Reinforcement	6th column	2b	Intensive corrosion
	9th column	10	
Bell tower			
Concrete	1st floor column	1c	Intensive decay
	1st floor column base	2c	
	1st floor column	3c	
Reinforcement	2nd floor column	B4	Intensive corrosion

3.1 In situ measurements

With the aid of fiber optics microscopy a direct visual evidence of deterioration state is accomplished. Regarding the concrete surface at Rotonda building, it can be observed loss of binding material, creation of fissures and cavities, as wall as, deposits of salts.

Regarding the concrete of the bell tower, they present extensive cracking, salts deposit, and biodeterioration products.

The ultrasound velocity transmission provides information about the quality and the homogeneity of the examined materials. Measuring the ultrasound velocity transmission, the detection of materials discontinuities and estimation of their conservation state could be accomplished. The measurements were done using the indirect method and the obtained velocities for the materials are reported in Table 2.

Table 2. Ultrasound velocity transmission measurements

Material	Sample Position	Velocity (m/sec)
Rhodes		
Concrete	6th Column	2564
	3rd Column	1176
	Arc	2174
	Left hall	1949
Reinforcment	6th column	4651
	Entrance	2941
Simi		
Concrete	2st floor column	1407
	1st floor column	1355
	1st floor column	1488
	2nd floor Flooring	2209
Reinforcement	1st floor column	3346
	1st floor column	2511

The concrete examined, presents low values of ultrasound velocity transmission in comparison with sound concrete (Velocity=3800-4900 m/sec). This reveals that the concrete present a lot of voids and discontinuities and it could be characterized as extremely decayed. In addition, the reinforcement bars present in general low velocity values, because of their extended corrosion, although that the 6th column of Rotonda, exhibits higher value that is attributed to its better conservation state.

Determination of reinforcement corrosion potential

Table 3 reports the corrosion potential values. The measurements were accomplished using a Cu/CuSO4 electrode. According to ASTM C 876-87, values greater than -200 mV, exhibit that the reinforcement is in passivation state. When the potential fluctuates between -200 mV and -350 mV the corrosion probability is at 50%. Values greater than -350 mV correspond to energetic state, leading to the corrosion of steel. Almost all the potential measured are greater than -350mV and therefore the steel is in energetic state where corrosion is occurring.

Measurement of concrete specific electrical resistance

The measurement of concrete specific electrical resistance took place at the same position where reinforcement potentials were measured, using the four point method. The values of specific electrical resistance are low (Table 3), indicating that the concrete is full of moisture and chloride anions. As a result, the high corrosion rate is inevitable.

Measurement of concrete carbonation depth

According to the results reported in Table 3, the carbonation front reaches the reinforcement for all the structural units. This indicates that the concrete is in advanced state of decay.

Table 3. Reinforcement corrosion potential, concrete specific electrical resistance and concrete carbonation depth of Rotonda columns and Nea Moni Monastery

Column	Corrosion Potential (mV)	Specific electrical resistance (ohm.m)	Carbonation depth (mm)
Rhodes			
8 inner	-324	51	99
9 inner	-300	63	105
7 inner	-350	45	119
2 outer	-389	49	120
4 outer	-361	66	130
5 outer	-370	37	135
Simi			
2nd floor (right)	-360	54	Has gone beyond the protective layer
2nd floor (left)	-356	61	Has gone beyond the protective layer
1st floor(right)	-350	45	Has gone beyond the protective layer
1st floor(left)	-376	41	Has gone beyond the protective layer
1st floor(center)	-370	55	Has gone beyond the protective layer
Chios			
Outside of drum – level B			
Horizontal bracket 1	-245	68	--
Parametric junction	-300	41	--
Horizontal bracket 2	-250	49	--
Outside of drum – level A			
Parametric junction	-150	67	--
Corona	-145	9.8	--

3.2 Laboratory Measurements

Grain size distribution

For the realisation of grain size distribution, concrete samples undergone cycles with the use of liquid azote (6). This consists of immersion of samples in water for 1 hour, afterwards immersion in liquid azote, and then heating at 105×C. The whole procedure continued until the total disintegration of samples. After the grain size distribution analysis, it was calculated the binder/total ratio. The ratio values were at the 1/10 range, testifying to the loss of binding material.

X-Ray Diffraction Analysis

XRD and DTA/TG determine the composition of reaction products formed, or to find out the carbonation state (7). Table 4 present the X-Ray Diffraction results for the samples examined. In concrete samples examined, calcite and quartz were detected, as main components and dolomite as an accessory mineral. In some samples, portlandite, halite, hillebrandite, tobermorite, albite, and biotite were, also, identified. In addition, the patinas of the reinforcement bars consisted of iron oxides (magnetite), which are products of the corrosion process.

Table 4. X-Ray Diffraction Analysis

Sample	Co.		Composition
Rhodes			
Concrete	2a	bin	Calcite, Quartz, Dolomite
	5	bin	Calcite, Quartz, Dolomite, Clorite, Portlandite
	4b	Tot	Calcite, Quartz, Dolomite, Calcium Chloride Hydrate, Clorite, Hillebrandite
Reinforcement	13		Magnetite
	14		Magnetite
Simi			
Concrete	2c	Tot	Calcite, Quartz, Halite, Tobermorite
	3c	Tot	Calcite, Quartz, Halite, Albite
	1c	bin	Calcite, Quartz, Dolomite, Biotite
Reinforcement	B4		Magnetite

Thermal Analyses DTA/TG

DTA analysis reveals the presence of calcite and dolomite in all samples. The calcite could be attributed to the aggregates, to the carbonation of the free $Ca(OH)_2$ (compound present in cement) and to the decomposition process of the hydraulic cement phases while the dolomite could be a compound present in the aggregates used. Furthermore, quartz was detected.

TG results, expressed as percentage weight loss for the investigated materials, are reported in Table 5 (mean value of three measurements). Measurements were performed at the total sample of concrete as far as at their fraction lower than 63m (binder). The percentage weight loss corresponding to temperatures lower than 120°C is attributed to the loss of physical absorbed water in samples, the 120-200°C range corresponds to the bound water of some components of the sound concrete but in our case where the concrete is carbonated it corresponds to the loss of crystallized salts bound water (gypsum). Furthermore the measured weight loss in the temperature range of 200-600°C is due to the loss of chemical bound water of cement hydrated phases or hydrate minerals. Lastly, the loss that occurred in temperatures higher that 600°C is attributed to loss of CO_2 of calcite and dolomite thermal decomposition.

It could be observed that all the concrete samples present low percentage of weight loss in the area of 120-200°C except in sample 2a the high percentage could be attributed to the presence of gypsum. This hypothesis is in accordance with the spot results test. In the range of 200-600°C, higher percentage loss is observed in the binding materials samples (especially for 2a sample), probably because of the presence of hydraulic phases. In all other cases, the percentage weight loss values were much lower. For temperatures greater than 600°C, concrete binding materials present a high percentage loss weight, due to the decomposition of the hydraulic phases and the carbonation of the $Ca(OH)_2$. In order to find the percentage of calcium carbonate, which derives from the binder, at the total, this percentage loss of 600-1000°C, is multiplied with the percentage of binder (derived from the grain size distribution) and the ratio 100/44 (44: molecular weight of CO_2).

Table 5. TG results expressed as percentage weight loss

Sample	Code		Weight Percentage Mass Loss (%)per Temperature Range(oC)				
			25-120	120-200	200-600	<600-1000	% CaCO3 (total) due to the binder
Rhodes							
Concrete	2a	tot	0,98	0,56	2,96	17,34	
		bin	2,41	2,19	7,51	20,86	4,34
	4b	bin	1,45	1,15	5,05	29,85	7,095
	5	bin	1,70	1,44	6,32	22,63	3,487
Simi							
Concrete	3c	tot	0,49	0,38	2,08	23,68	
	2c	tot	0,17	0,18	1,67	30,13	
	1c	bin	0,49	0,95	5,16	24,74	6,545

Determination of total soluble salts

The determination of total soluble salts was accomplished according to NORMAL 13/83, while the identification of Cl^- and $SO4^{--}$ was performed by spot tests. The results are reported in Table 6. The measured conductivity is rather low for all samples. Spot tests reveal the presence of chlorides in all samples while some of them present also sulphates.

Microstructural results

Table 7, reports the data from porosimetry of the examined materials. The changes of microstructure are due to the carbonation process. The concrete samples present different microstructure. Sample 5 seems to be in good conservation state while the sample 4b present extended decay, with high values of total cumulative volume, porosity, specific surface area and pore radius average.

Table 6. Total Soluble Salts and Spot Tests results

Sample	Co.	SST %	Cl-	SO4--
Rhodes				
Concrete	2a	3,54	+	++
	5	3,61	++	+
Simi				
Concrete	2c	2,86	+	-
	3c	2,32	+	-
Chios				
mortar	--	--	+++	--

-: traces, +: low, ++: medium, +++: high

Table 7. Porosimetric results

Sample	Co.	CV (mm3/g)	As (m2/g)	rav. (μm)	(g/cm3)	P (%)
Rhodes						
Concrete	4b	104,96	4,58	0,374	2,17	22,77
	5	48,83	2,86	0,015	2,32	11,32
Simi						
Concrete	2c	67,65	2,94	3,845	2,21	14,95
	3c	83,96	6,53	0,015	2,15	18,05

CV: Total Cumulative Volume (mm3/g), : Bulk Density (g/cm3), P%: Total Porosity (%), As: Specific Surface Area (m2/g), rav: Pore Radius Average (μm)

4. RECOMMENDATIONS

The main degradation problems that these concrete structures present at severe marine environment are the carbonation of concrete, in synergy with the presence of chloride anions, and the corrosion of reinforcement steel.

In the phase of conserving and protecting these structures most appropriate methods to confront this problem of the reinforced concrete are the realkalisation of the carbonated concrete (accompanied probably with chloride anions extraction), the passivation of steel, using corrosion inhibitors, and the cathodic corrosion protection of steel (8,9). The realkalisation technique and the chloride extraction present the disadvantages of high cost and long time of application. On the other hand the re-passivation of steel, using corrosion inhibitors can be applied easily. The concrete surface is spayed with the appropriate solution containing inhibitor (1 lt/m^2), applied at three layers. The most appropriate period of such application is summer time, because at that time concrete is dry. The finishing plaster of the surface should contain inhibitors too. The cathodic protection is applied with the use of a titanium anode mesh and the application of direct current, considering the reinforcement steel as negative pole and the mesh as positive pole.

This method is more expensive comparing to the use of inhibitors, but is the only method that can protect the structure for longer time.

In the phase of designing of a new structure of reinforced concrete that will be exposed to that severe environment the following recommendations regarding the concrete composition should be consider. Firstly the concrete paste should have the lowest permitted water/cement ratio, in order to obtain lower values of the final porosity (10) and the thickness of steel should have greater values than the ones calculated. In addition the mixture should contain inhibitors. Finally if the building it is decided to have lifetime period beyond a century then cathodic protection should be applied.

5. CONCLUSIONS

From the diagnosis of the deterioration state of the concrete it is concluded that the carbonation front has reached the reinforcement of these structures with changes to the porosity. This state of deterioration is confirmed from the results of porosimetry, ultrasound test, measurement of carbonation depth, as well as from the XRD and DTA/TG results. This fact contributes to the corrosion of reinforcements, provoking the spalling and the fissures in the concrete structures.

The concrete shows variable states of decay, due to the specific orientation of the surfaces, and also because of the technology and application techniques of concrete.

The concrete structures of the 1st half of the 20th century belong to an important modern architectural heritage, present severe decay problems and must be urgently repaired with appropriate materials and methods.

6. REFERENCES

1. ISO 15686, "Buildings service life planning, Part1, general principles",Draft for Public Comment, BSI UK, 1998, pp. 62
2. Nireki, T., "Service life design", Construction and building Materials, Vol. 10, No. 5, pp. 403-406, 1996
3. Mehta, P.K.,"Durability of concrete fifty years of progress?", 2nd Int. Conference of Durability of Concrete", Vol.I, Montreal, Canada 1991, pp.1-32 (Ed. V.M.Malhotra).
4. Long A. E., Henderson G. D., Montgomery F. R., "Why assess the properties of near-surface concrete?", Construction and Building Materials, Vol. 15, Issues 2-3, March-April 2001, pp. 65-79
5. Aïtcin P.C., "Cements of yesterday and today; Concrete of tomorrow", Cement and Concrete Research, Vol. 30, Issue 9, September 2000, pp.1349-1359
6. Rossi P.P., Cimitan L., "Studio delle techniche di disgregazione per le indagini diagnistiche delle malte", I quaderni dell'Ismes, No 279, 1991
7. Grattan-Bellew P. E., "Microstructural investigation of deteriorated Portland cement concretes", Construction and Building Materials, Vol. 10, Issue 1, February 1996, pp. 3-16

USING THE CHLORIDE MIGRATION RATE TO PREDICT THE CHLORIDE PENETRATION RESISTANCE OF CONCRETE

S.W. Cho and S.C. Chiang

Department of Building Engineering & Architecture at China Institute of Technology, Taiwan, ROC; Institute of Materials Engineering, National Taiwan Ocean University, Keelung, Taiwan, ROC.

Abstract: An experimental investigation is conducted to study the relationship between the chloride diffusion coefficient and charge passed. In this study, the concrete specimens made with different w/c (ranging from 0.35 to 0.65) and different slag contents (ranging from 0% to 70%) were used in the RCPT, ACMT, and ponding tests. From the test results, the charge passed shows no direct correlation with the diffusion coefficient obtained from the ponding test when concrete containing slag, but it indicates fair correlation when concrete without mineral admixtures. However, whether or not containing slag, it shows a linear correlation between the steady-state diffusion coefficients from ACMT and the non-steady-state diffusion coefficient from the ponding test. It appears that the ACMT may be more useful method than RCPT, when corresponding to long-term chloride diffusion test.

Keywords: Chloride diffusion coefficient; Pomding test; ACMT; RCPT.

1. INTRODUCTION

Both the RCPT and ponding test are widely used to predict the ability of chloride penetrate concrete. Ponding test is standardized by AASHTO and ASTM as AASHTO T259 and ASTM C1543. This test must consume 90 days and the procedures are too complicated, because of sample must be obtained in each depth and analyzed the chloride content. In 1981, Whiting developed RCPT method to specify the chloride permeability of concrete and standardized by AASHTO and ASTM as AASHTO T277 and ASTM C1202. This test used electrical conductive to classify the chloride permeability without analysis and only 6 hours test period. From early researches [1], they found the total charge passed of RCPT is correlated close with the integral chloride content of ponding test. But recent years, the mineral admixtures such as fly ash and slag are usually added into concrete instead of cement. Some researchers [2,3] pointed out that both

M.S. Konsta-Gdoutos, (ed.), Measuring, Monitoring and Modeling Concrete Properties, 575–581.

tests were poorly correlated and the RCPT might produce misleading results when concrete with mineral admixture. Therefore ASTM C1202 recommended when test method was applied types of concrete, the correlation between the RCPT and ponding test have been established. However the ponding test is not a simple and short term test. Several modified methods have been created to predict chloride penetration [4,5]. Apply an external electrical field to accelerate chloride migration is most used for modified methods. The objective of this study is to determine the chloride permeability of the different concrete mixes with various test methods. In this study the modified method of accelerated chloride migration test (ACMT) is used to determinate the steady-state diffusion coefficient for concrete. The total charge passed is obtained from RCPT. And the non-steady state diffusion coefficient is calculated from ponding test. The relationship between the diffusion coefficients from different coefficient and the total charge passed are compared and correlated.

2. EXPERIMENTAL PROGRAM

2.1 Mix proportions and specimen preparation

In this study, concrete is made with Type 1 cement, water, slag, river sand, and crushed stone (maximum size: 10-mn). The concrete mix proportions are summarized in Table 1. Notation for the mixes is that the first letter A, B, C, and D is the water-to-cementitious material ratios (W/CM) of 0.35, 0.45, 0.55, and 0.65, and the number is percentages of slag (0%, 20%, 40%, 50%, 70% of cement was replaced by slag). All cylindrical specimens (ϕ100x200 mm) were cast and cured in water for 90 days. Three discs of 50 mm thick for the ponding test and RCPT and 30 mm thick for the ACMT were obtained by cutting from central portion of the cylinder.

2.2 Ponding test

The ponding test is similar to the test described in AASHTO T259. Each 50 mm thick discs are air dried then the lateral surface of discs was coated with epoxy. After epoxy hardening, the specimens were sealed outside edge using an acrylic ring to create the dam for salt solution, and the dam were continuous ponded to a depth 1.5-cm with 3% sodium chloride solution for 90 days. The solution was removed after 90 days exposure. And all salt crystal build-up on the surface was completely removed by wire brushing. After brushing, three cylindrical specimens (20 mm x 50 mm) were obtained by coring from the ponding specimen. Each cylindrical specimen was cutting from the surface into ten 5-mm thick slices at dry condition. And then the slices were dried to constant mass and grounded to pass a 300m sieve for chloride analysis. The results of chloride analysis could plot chloride content profile. Non-steady-state chloride diffusion coefficient D_p could be calculated according to the second Fick's law with fitting the data of profile.

Table 1. Mixture proportions

Mix	W/CM	Slag (%)	Unit content: kg/m^3				
			Cement	Slag	Water	Coarse Aggregate	Fine Aggregate
A0	0.35	0	695	0	243	733	593
A20	0.35	20	550	137	241	733	593
A40	0.35	40	408	272	238	733	593
A50	0.35	50	338	338	237	733	593
A70	0.35	70	201	468	234	733	593
B0	0.45	0	540	0	243	743	719
B20	0.45	20	428	107	241	743	719
B40	0.45	40	318	212	239	743	719
B50	0.45	50	264	264	237	743	719
B70	0.45	70	157	366	235	743	719
C0	0.55	0	442	0	243	743	801
C20	0.55	20	351	88	241	743	801
C40	0.55	40	261	174	239	743	801
C50	0.55	50	216	216	238	743	801
C70	0.55	70	129	300	236	743	801
D0	0.65	0	374	0	243	743	858
D20	0.65	20	297	74	241	743	858
D40	0.65	40	221	147	239	743	858
D50	0.65	50	184	184	239	743	858
D70	0.65	70	109	255	237	743	858

2.3 Accelerated chloride migration test (ACMT)

The accelerated chloride migration test is a modified version of the ASTM C1202 method. The most different between RCPT and ACMT is the migration cell. The cell of the ACMT was a 4.5 liter container, but the solution volume of cells used in RCPT is about 250 ml. The increasing volume may avoid the Joule effect during the test period, and the accumulative chloride ion concentration can be measured periodically form the sodium hydroxide solution by the potentiometric titration method. The test period is depended on the quality of concrete (usually 7~10 days). And the chloride ion diffusion coefficient D_s is calculated by the Nernst-Plank equation.

3. RESULTS AND DISCUSSION

3.1 Total charge passed

The total charge passed is obtained from RCPT for all mixes are listed in Table 2. Figure 1(a) shows the total charge passed and the W/CM ratio with different slag contents. For a given W/CM ratio, the total charge passed is decreased with increasing the slag content. For concrete containing more than 40% slag, the total charge passed significant decreases. Figure 1(a) also shows the previous study [6]. In the previous study, the total charge passed of four concretes containing 25% fly ash of the same W/CM ratio and aggregate volume are determined using the RCPT. For concrete containing 25% fly ash in each W/CM ratio, the total charge passed is lower than concrete containing 50% slag. It appears that the charge passed decreases with concrete containing mineral admixtures.

Table 2. Total charge passed , and diffusion coefficient from ACMT and ponding test

Mix	RCPT, Total charge passed (coulomb)	Diffusion coefficient (Þ 10-8 m2/hour)	
		Ponding, non-steady-state, D_p	ACMT, steady-state, D_s
A0	5311	2.44	17.51
A20	4685	--	13.92
A40	2208	--	7.06
A50	1543	1.36	6.19
A70	1034	--	5.2
B0	7568	5.51	29.14
B20	5630	--	20.29
B40	2387	--	12.43
B50	1825	1.47	7.29
B70	1122	--	6.98
C0	9639	8.37	39.48
C20	6355	--	20.47
C40	2709	--	13.37
C50	2148	1.69	10.66
C70	1350	--	7.73
D0	10442	12.74	56.02
D20	7114	--	31.41
D40	2992	--	15.83
D50	2565	2.34	15.61
D70	1686	--	13.87

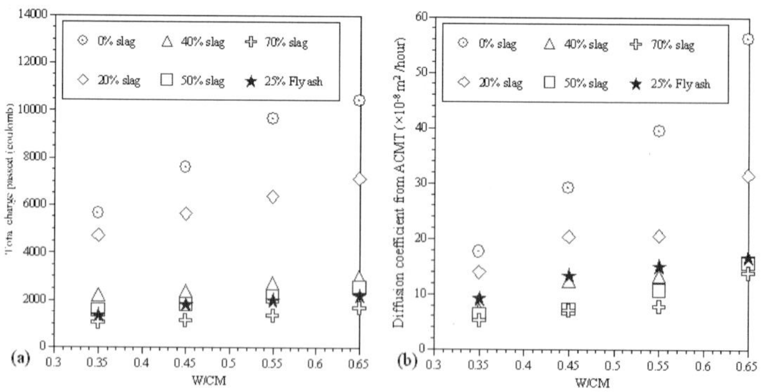

Figure 1. (a) The relationships between total charge passed and W/CM; (b) The relationships between the steady state chloride diffusion coefficient from ACMT and W/CM

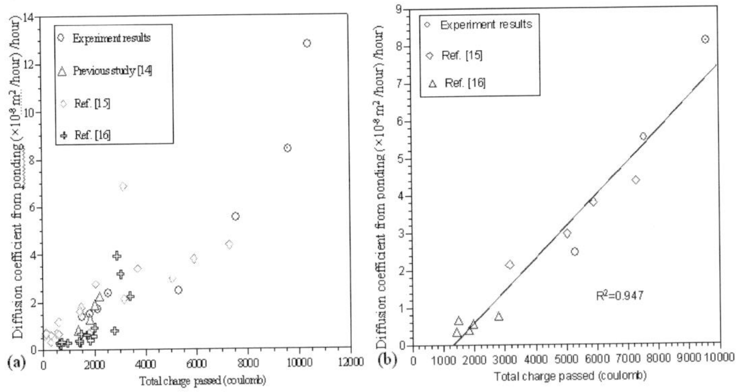

Figure 2. (a) Correlation between the chloride diffusion coefficient from ponding test and total charge passed; (b) Considering concrete mixes without mineral admixtures

3.2 Non-steady-state diffusion coefficients from ponding test

From the ponding test results, the non-steady-state diffusion coefficient is decreased with concrete containing slag. A comparison between the results of ponding test and the RCPT is shown in Figure 2(a). It shows that the results can be separated into two groups. One is concrete containing slag and the other is concrete without mineral admixtures. In Figure 2(a), some data of concrete containing different mineral admixtures were collected from McGrath and Hooton's research [7], and Sherman et al. [8] are also plotting in this figure. It shows no obvious correlation between the non-steady-state diffusion coefficient and the total charge passed. When only considering concrete mixes without mineral admixtures, Figure 2(b) shows a linear relationship between the non-steady-state diffusion coefficient and the total charge passed. Several researches pointed out [7,9,10] that for concrete containing mineral admixtures may changed the chemical composition of the pore solution and microstructure of concrete and affect the measuring current of RCPT.

3.3 Steady-state diffusion coefficient from ACMT

The diffusion coefficient obtained from ACMT for all mixes are listed in Table 2. Figure 1(b) is the relationship between the steady-state diffusion coefficient and W/CM with different slag contents, and the previous results [14] are also shown in the figure. It can be seen in Figure 1(b), the steady-state diffusion coefficient decreases with increasing the slag content. For concrete containing 25% fly ash in each W/CM ratio, the steady-state diffusion coefficient is lower than concrete containing 20% slag.

In Figure 3(a), the steady-state diffusion coefficients from ACMT are compared with the total charge passed from RCPT for all mixes, and the previous results of fly ash are also shown in the figure. It shows that no obvious correlation between the steady-state diffusion coefficient and the total charge passed, such as Figure 2(a).

Figure 3(b) shows the relationship between the steady-state diffusion coefficients from ACMT (D_s) and the non-steady-state diffusion coefficient from the ponding test

(D_p), and the previous results are also shown in the figure. By linear regression, the empirical relationship between D_s and D_p is statistically derived as:

$$f(x) = 0.241x - 1.235 \times 10^{-8} \quad (1)$$

The coefficient R^2 for the Equation (1) is 0.972. It appears that D_s correlates linearly with D_p regardless of concrete with or without containing mineral admixtures. When comparing Figure 3(b) and Figure 2(a), it appears that the ACMT may be more useful method than RCPT, when corresponding to long-term chloride diffusion test.

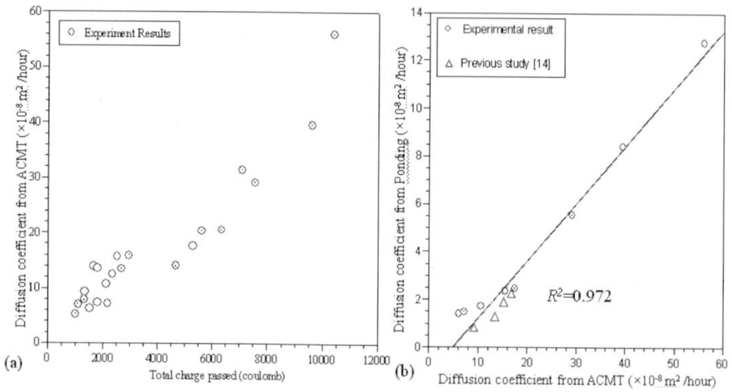

Figure 3. (a) Correlation between the steady state chloride diffusion coefficient from ACMT and total charge passed; (b) Correlation between the steady state chloride diffusion coefficient from ACMT and the non-steady state chloride diffusion coefficient from ponding test

4. CONCLUSIONS

The conclusions derived from the experimental investigation are presented below.

1 The charge passed obtained from RCPT and the steady-state diffusion coefficients obtained from ACMT are decreasing with increasing the slag content.

2 When concrete containing mineral admixtures, there is no obvious correlation between the non-steady-state diffusion coefficient obtained from ponding test and the total charge passed obtained from the RCPT. But a good correlation was observed in plain cement concrete mixes.

3 In ACMT, there is no the steady-state diffusion coefficients obvious correlation between the steady-state diffusion coefficient and the total charge passed obtained from the RCPT.

4 The plain cement concrete and concrete containing mineral admixtures, a good correlation is observed between the steady-state diffusion coefficients from ACMT and the non-steady-state diffusion coefficient from the ponding test. It appears that the ACMT may be more useful method than RCPT, when corresponding to long-term chloride diffusion test.

Acknowledgment

The financial support of National Science Council, ROC, under the grants NSC 94-2211-E-157-004 and NSC 93-2211-E-157-004 is gratefully appreciated.

5. REFERENCES

1. D. Whiting, "Rapid Measurement of the chloride permeability of concrete", Public Roads 45(3) (1981) 101-112.
2. R. Feldman, L. R. Prudencio, and G. Chan, "Rapid chloride permeability test on blend cement and other concretes: correlations between charge, initial current and conductivity", Con. Bulid. Mater. 13 (1999) 149-154.
3. T. H. Wee, A. K. Suryavanshi, and S. S. Tin, "Evaluation of rapid chloride permeability test (RCPT) results for concrete containing mineral admixture", ACI Mater. J. 97(2) (2000) 221-232.
4. O. Truc, J. P. Oilvier, and M. Carcasses, "A New Way for Determining the Chloride Diffusion Coefficient in Concrete from Steady State Migration Test", Cem. Con. Res., 30(2) (2000) 217-226.
5. C. C. Yang, and S. W. Cho, "An electrochemical method for accelerated chloride migration test of diffusion coefficient in cement-based materials", Mater. Chem. Phy. 81(2) (2003) 116-125.
6. C. C. Yang, and L. C. Wang, "The diffusion characteristic of concrete with mineral admixture between salt ponding test and accelerated chloride migration test", Mater. Chem. Phy., 85(2) (2004) 266-272.
7. P. McGrath, and R. D. Hooton, "Re-evaluation of the AASHTO T259 90-day salt ponding test", Cem. Con. Res., 29 (1999) 1239-1248.
8. M. R. Sherman, D. B. McDonald, and D. W. Pfeifer, "Durability aspects of precast prestressed concrete Part 2: Chloride permeability study", PCI J., (1996) 76-95.

PORE-SIZE DISTRIBUTION IN BLENDED CEMENT PASTES USING NMR TECHNIQUES

M. Katsioti,[1] M.S. Katsiotis[2], M. Fardis[2], G. Papavassiliou[2] and J. Marinos[3]
[1] *School of Chemical Engineering, NTUA, 9 Heroon Polytechniou St., 157 73,Greece;* [2] *Institute of Materials Science, NCSR 'Demokritos', 153 10 Ag. Paraskevi, Attikis, Greece;* [3] *Hercules G.C.Co, Greece*

Abstract: The microstructure of a porous system is an important characteristic of the material that largely determines not only its mechanical properties but also its transport properties and durability performance. In particular, for cement the porosity is strongly correlated with strength, and the permeability and the rate at which ions and gases diffuse through the material are of major importance for durability. In this work, Nuclear Magnetic Resonance (NMR) has been used in order to determine the pore size distribution in blended cements. The pore-size distribution was measured by the NMR cryoporometry method which is based on the well-known freezing point depression of water when confined inside the pore matrix of a material. It is demonstrated that this technique can be applied in cementitious materials to probe the microstructure of the main hydration product: the cement gel. The influence of various mineral additives to the pore-structure and the transport of water have been assessed for a series of blended cements pastes. The pastes were left to cure at curing conditions of 20±2^{0}C for 28 days. In addition, the pore size distribution of the pastes was also determined by means of mercury porosimetry in order to compare the two methods used.

Keywords: Blended cements, mineral additives, pore size distribution, NMR cryoporometry, freezing point depression, mercury porosimetry.

1. INTRODUCTION

Porosity and pore size distribution are determining factors of the mechanical and physical properties of cement. In particular, for cement paste the porosity is strongly correlated with strength, and the permeability and the rate at which ions and gases diffuse through the material are of major importance for durability.

M.S. Konsta-Gdoutos, (ed.), Measuring, Monitoring and Modeling Concrete Properties, 583–590.

The traditional methods for measuring pore size distributions in cement pastes, such as sorption isotherms and mercury intrusion porosimetry, suffer from major disadvantages mainly associated with the preliminary drying of the sample.[1]

Nuclear Magnetic Resonance (NMR) techniques have become a potential tool for the study of the physicochemical properties of cement. In particularly an *NMR-cryoporometry* technique that probes non-destructively the pore size distribution of porous solids has been well established.[2-8] This method is based on the well-known freezing point depression of water and other liquids when confined inside the pore matrix e.g. the capillary pores of a material.[9,10] Low temperature calorimetry on materials with confined liquids is based on the same property and has been used in order to access the pore structure of saturated porous building materials.[11]

2. FREEZING POINT OF LIQUIDS CONFINED IN POROUS MEDIA

The lowering of the freezing point of an adsorbed liquid has been described by many formulas in the literature[10], most of which are based on the Kelvin equation. Batchelor and Foster[9] proposed the following equation:

$$\Delta T = \frac{T_{observed}}{L}\frac{2M}{r}\left\{\frac{\gamma_l}{\rho_l} - \frac{\gamma_s}{\rho_s}\right\}T_{normal} \qquad (1)$$

where L is the latent heat of fusion, M is the molecular weight, r is the radius of the capillaries, and are the surface energy and density measured at the normal melting point T_0, and $T = T_{\text{normal}} - T_{\text{observed}}$ is the depression of the freezing point.

Based on Eq. 1, a calibration curve has been constructed for the freezing point depression of pure water confined inside the pores as a function of the pore radius. This is shown in Figure 1.

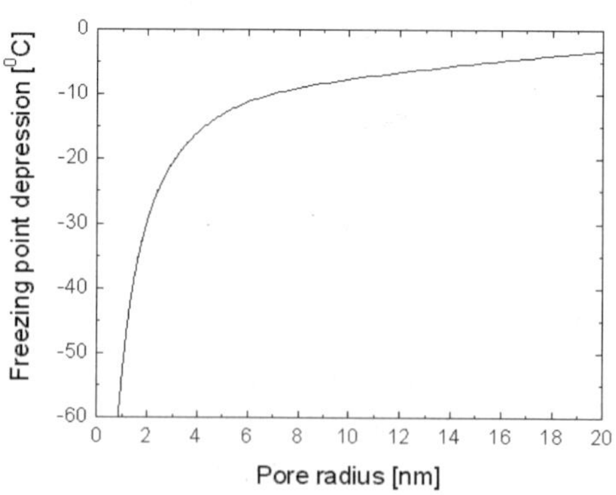

Figure 1. Freezing point depression of pore water as a function of pore radius, based on Eq. 1

Applying the NMR method to evaluate the pore size distribution relies on the different intensities of the ^{1}H NMR signal arising from non-frozen and frozen liquid. In an NMR experiment it is possible with the proper setting of the NMR spectrometer to allow the observation of the signal coming from the liquid only. Therefore in the NMR cryoporometry method, the NMR signal attributed to the liquid part of the water is recorded as a function of temperature. Then using the calibration curve of Figure 1, the observed temperature is converted into pore radius, and the derivative of the liquid signal gives the pore size distribution dV/dr. This evidently reflects the pore size distribution of the cement pore matrix.

3. EXPERIMENTAL

3.1 Sample preparation

In this work, the NMR-cryoporometry method has been used in order to determine the pore size distributions of a series of blended cements (A, B, C) each containing 25% of various mineral additives [fly ash (A), pozzolan (B), alunite (C)] and 75% CEM I-42.5. Three different samples were each prepared by mixing 100 parts of blended cement with ~30 parts of water (31 for pozzolan, 28 for alunite, 32 for fly ash). A fourth sample (control - D) consisting only of CEM I-42.5 was also prepared by mixing 100 parts of cement with 28 parts of water. All the samples were mixed and placed separately to harden in sealed glass tubes. The curing duration was 28 days at curing conditions of of 20±2C, the tubes were filled with water under a small pressure to assure "fast" diffusion of the water into the cement paste. In addition, the pore size distribution of the samples was also determined by means of mercury porosimetry in order to compare the two methods used.

3.2 NMR measurement procedures

The ^{1}H NMR experiments were performed on a conventional pulsed spectrometer at 2.35 T equipped with an Oxford 1200 continuous flow cryostat for measurements in the range 100-300 K. A two pulse spin-echo sequence was used with a delay of =30 sec between the pulses. The spin-echo amplitude was monitored as the sample was gradually cooled from room temperature with a rate of about 4 K/hr. Since the NMR signal from the ice part of the water has been decayed after the τ =30 μsec delay, the observed spin-echo signal originates from the total non-frozen liquid part of the water. It should be stretched that this method is a non-destructive technique which does not require previous drying of the samples.

3.3 Mercury Porosimetry

The pore size distribution analysis were carried out by means of Mercury Porosimetry. The test was concerned with the distribution of the pore volume in relation to their radii. Assuming that all the pores are cylindrical, pore radii were determined on the basis of the well known Washburn equation:

$$P = -2K\frac{\cos\varphi}{r} \tag{2}$$

where:

P: absolute pressure exerted (kg/cm^2)

K: surface tension of mercury (=0,48N/m 25°C)
ϕ: mercury contact angle (=141°)
r: pore radius (Å)

4. RESULTS AND DISCUSSION

All the samples were chemically and mineralogically (by XRD) analyzed and the results are shown in Table 1 and in Figures 2 – 4, respectively.

Table 1. Chemical Analysis of mineral additives and control sample

Component	Content (weight %)			
	CEM I-42.5	Fly ash	Pozzolan	Alunite
SiO_2	19.74	30.27	67.10	0.03
Al_2O_3	5.20	14.05	13.74	41.74
Fe_2O_3	3.45	6.26	3.36	0.13
CaO	63.05	30.11	2.75	-
MgO	2.38	4.14	0.92	0.45
K_2O	0.53	1.19	2.70	11.00
Na_2O	0.25	0.53	2.50	-
SO_3	3.10	6.52	-	36.00
CaO_f	-	6.27	-	-
LOI	2.88	1.55	7.00	8.55

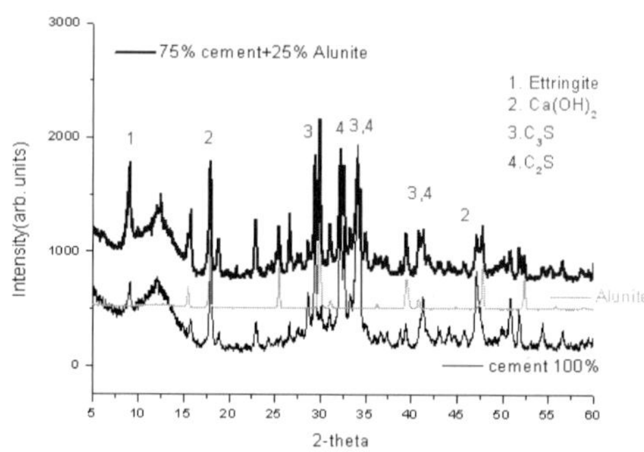

Figure 2. X-Ray Diffraction of blended cement with 25% alunite (A) at 28 days

The 28 days pastes' XRD analysis has shown that the characteristic peaks of $Ca(OH)_2$ and ettringite exist and also that the blended cement with alunite contained more ettringite and less portlandite [$Ca(OH)_2$] than the blended cement which contained pozzolan or fly ash. The presence of alunite favours the formation of ettringite probably because of the presence of Al^{3+} in alunite.

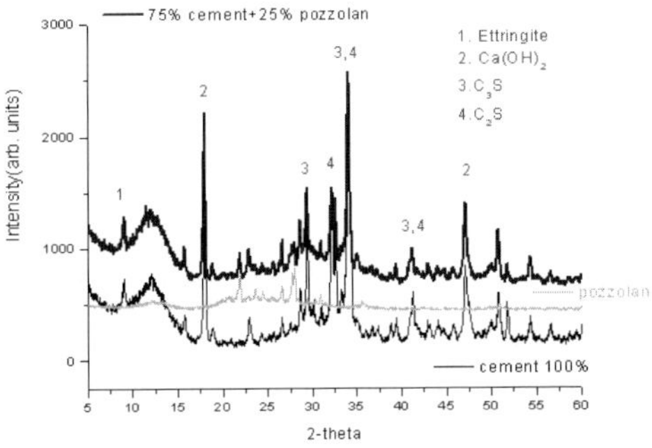

Figure 3. X-Ray Diffraction of blended cement with 25% pozzolan (B) at 28 days

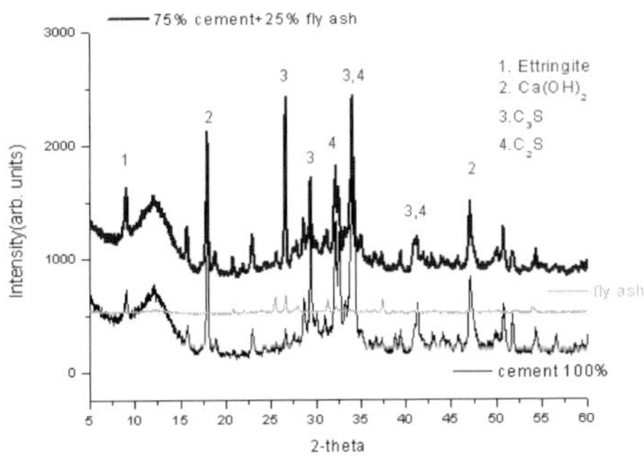

Figure 4. X-Ray Diffraction of blended cement with 25% fly ash (C) at 28 days

The NMR experimental results for the four cement samples are shown in Fig. 5. The normalized NMR signal attributed to the liquid part of the water is shown as a function of temperature. It can be seen that the water inside the pore structure of the samples remain liquid well below the normal freezing point of ordinary bulk water.

Measurement of the fraction of liquid as a function of temperature yields the pore size distribution. The pore volume v(r) is a function of the pore diameter r. If the pores are filled with liquid, the melting temperature of the liquid $T_m(r)$ is related to the pore size distribution by:[2]

$$\frac{dv}{dr} = \frac{dv}{dT_m(r)} \frac{dT_m(r)}{dr} \tag{3}$$

The measurement of $dv/dT_m(r)$, provided $dT_m(r)/dr$ is known for the liquid used, will allow the pore size distribution to be determined. $dT_m(r)/dr$ is known from Eq. 1. The pore size distributions for the blended cements are shown in Figure 6.

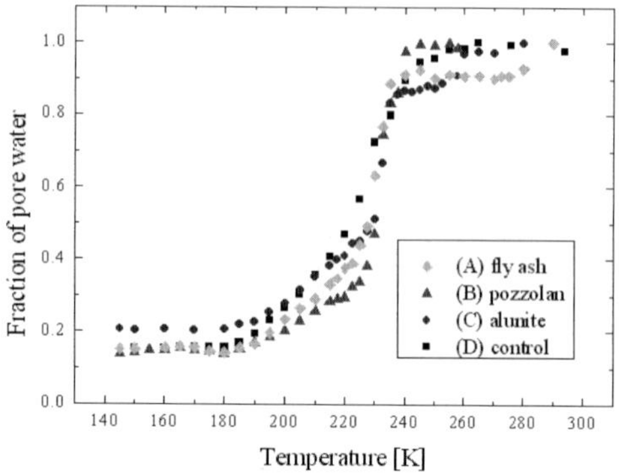

Figure 5. Volume of liquid (pore volume) as a function of temperature for the cement samples, as obtained from the NMR experiments. The vertical scale is arbitrary

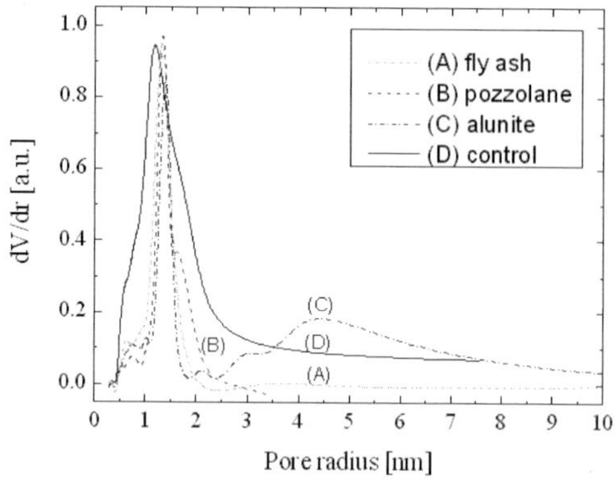

Figure 6. Pore size distributions for blended cements as obtained from the NMR measurements (at 28 days). The vertical scale is arbitrary

The pore size distributions are characterized by a maximum in the region of 1 -2 nm. This size corresponds to the so-called gel porosity, which is substantially smaller than the capillary porosity measured by the mercury porosimetry. It is clearly seen that the NMR cryoporometry method is sensitive to the porosity of the gel cement product. Our results are in agreement with water sorption experiments in hardened cement pastes.[11] The pore-size distribution in the microcapillaries of hardened cement paste shows a pronounced maximum at ~ 2 nm.[11] Regarding the influence of the mineral additions to the gel porosity of the cement paste, it is observed from Figure 6 that in all blended cements the pore size distribution in the 1 nm range is narrower than the control OPC. Moreover, the alunite has considerable porosity at the 4 nm range. The results from mercury porosimetry are shown in graph form below. In Figure 7 the pore size distribution is presented while the cumulative pore volume is presented in Figure 8.

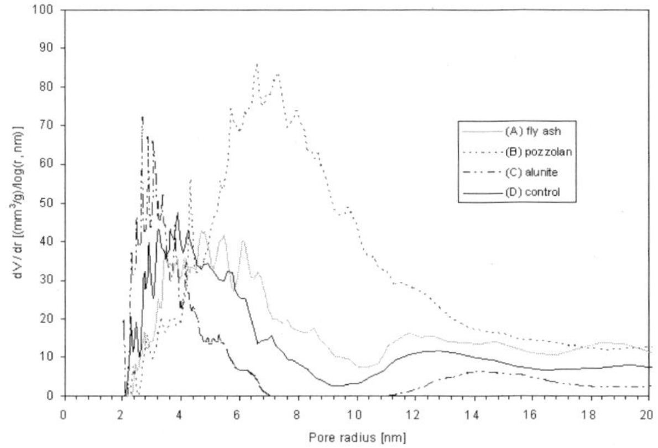

Figure 7. Pore size distribution for blended cements as obtained from mercury porosimetry (at 28 days)

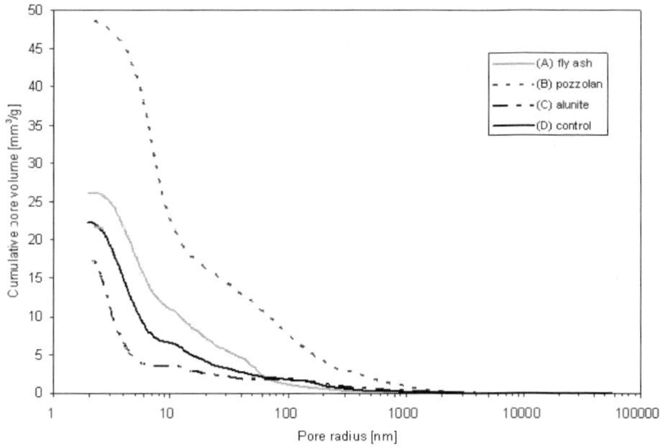

Figure 8. Cumulative pore volume for blended cements as obtained from mercury porosimetry (at 28 days)

4. CONCLUSIONS

From the results of NMR-cryoporometry, the control sample shows a wider distribution of all the pores. Alunite shows a more narrow distribution compared to the fly ash and pozzolan as far as the small pores are concerned (1-2nm) and a wider distribution regarding the bigger pores (4nm).

As far as the mercury porosimetry is concerned, the results showed the pore radius of the samples ranged from 2 to 56000 nm. The blended cement with alunite presents the lowest pore volume (17.9 mm^3/g) and also the most narrow maximum frequency radius of pores (2.4 nm).

Acknowledgements

The authors are grateful to the Hellenic Cement Research Center (HCRC) LTD for their assistance in the experiments and in specific, they would like to thank Z. Tsibouki, manager of HCRC, V. Kaloidas and J. Karagiannis. They also express their warmest thanks to Dr V. Psycharis from the Institute of Material Science, NCSR Demokritos.

5. REFERENCES

1. H.F.W. Taylor, *Cement Chemistry*, (Academic Press, London, 1990).
2. J. H. Strange, M. Rahman, E. G. Smith, Characterization of porous solids by NMR, *Phys. Rev. Lett.* **71**(21), 3589-3591 (1993).
3. F. Milia, M. Fardis, G. Papavassiliou, and A. Leventis, NMR in porous materials, *Magn. Reson. Imaging.* **16**(5/6), 677-678 (1998).
4. J.-Y. Jehng, D. T. Sprague, and W. P. Halperin, pore structure of hydrating cement paste by magnetic resonance relaxation analysis and freezing, *Magn. Reson. Imaging.* **14**(7/8), 785-791 (1996).
5. R. Schmidt, E. W. Hansen, M. Stocker, D. Akporiaye, and O. H. Ellestad, Pore size determination of MCM-41 mesoporous materials by means of ^1H NMR spectroscopy, N_2 adsorption, and HREM. A preliminary study, *J. Am. Chem. Soc,* **117**(14), 4049-4056 (1995).
6. D. Akporiaye, E. W. Hansen, R. Schmidt, and M. Stocker, Water-saturated mesoporous MCM-41 systems characterized by ^1H NMR, *J. Phys. Chem.* **98**(7), 1926-1928 (1994).
7. R. Holly, J. Tritt-Goc, N. Pislewski, C. M. Hansson, and H. Peemoeller, magnetic resonance microimaging of pore freezing in cement: Effect of corrosion inhibitor, *J. Appl. Phys.* **88**(12), 7339-7345 (2000).
8. R. Valiullin and I. Furo, Phase separation of a binary liquid mixture in porous media studied by nuclear magnetic resonance cryoporometry, *J. Chem. Phys.* **116**(3), 1072-1076 (2002).
9. R. W. Batchelor, A. G. Foster, The freezing point of adsorbed liquids, *Trans. Faraday Soc.* **40**, 300-305 (1944).
10. J. R. Blachere, J. E. Young, The freezing point of water in porous glass, *J. Am. Ceram. Soc.* **55**, 306-308 (1972).
11. D. H. Bager, and E. J. Sellevold, Ice formation in hardened cement paste. I. Room temperature cured pastes with variable moisture contents, *Cem. Concr. Res* **16**, 709-720 (1986).

INVESTIGATION OF CKD – BFS IN REINFORCEMENT CORROSION PROTECTION

A. Routoulas, S. Kalogeropoulou, P. Pantazopoulou and P. Koulouris
Technological Educational Institute of Piraeus, Physics, Chemistry & Materials Technology Department, P. Ralli and Thivon 250, 122 44, Egaleo, Athens, Greece

Abstract: The aim of this study is the examination of concrete performance when cement kiln dust is used as cement replacement in mortars with blast furnace slag and corrosion inhibitors and the determination of their optimum proportions. The test results indicated that specimens containing CKD up to 10% with the incorporation of BFS exhibited better or equal reinforcement corrosion protection than the reference ones. Moreover, the addition of the corrosion inhibitor in all cases further improved the anticorrosion performance of the composite cements.

Key words: cement kiln dust; blast furnace slag; alkanolamines; reinforcement corrosion; strain gauge.

1. INTRODUCTION

The use of suitable Type-II additives such as natural or artificial pozzolans is a well established practice for the improvement of concrete properties. Blast-Furnace Slag (BFS), a waste product in the manufacture of cast iron, is produced by rapid cooling of molten slag at the exit of the furnace. It is a mixture of lime, silica and alumina, that is, the same oxides that make up Portland cement, but not in the same proportions. It has also nearly the same hydraulic ability as Portland cement (Neville, 1975). Cement Kiln Dust (CKD), which is a by-product of low-alkali cement manufacturing process, mainly contains calcium carbonate and can be classified as Type-I addition according to the new European codes, increases the pH value of concrete as a result of its high alkalinity. Due to its fineness, in many cases it causes environmental pollution in the form of dust. Therefore, the exploitation of this material will provide valuable environmental benefits. In a previous study (Batis et al., 1996) the incorporation of CKD as a filler addition in cement improved the durability of the material under aggressive corrosion conditions. The presence of pozzolans in cementitious materials reduces the level of free chlorides and thus provides protection to the reinforcement against chloride-induced corrosion. However, the pozzolans also react with the OH^- causing a reduction in the pH value of the concrete,

M.S. Konsta-Gdoutos, (ed.), Measuring, Monitoring and Modeling Concrete Properties, 591–596.

which in its turn leads to an increase in carbonation and a relative reduction in the durability of the hardened material. The introduction of CKD, which is an inert material, does not produce the above reaction. On the contrary it causes a relative increase of alkalinity (Batis et al., 1996; Konsta and Shah, 2003), which is expected to counteract the above-mentioned phenomenon encountered with the sole use of pozzolans (Buchwald and Schulz, 2005).

Alkanolamine-based corrosion inhibitors can make use of the porosity of the concrete, by moving through the pore structure of the concrete to reach the surface of reinforcing steel, where they form a protective film. They also reduce chloride ion ingress into concrete. They are classified as mixed inhibitors, because they influence both the cathodic and the anodic process of corrosion (Broomfield, 1999; Bjegovic et al., 1999). Consequently, the performance and durability of reinforced concrete incorporating CKD in combination with BFS, whether an alkanolamine-based inhibitor is present or not, is evaluated.

2. EXPERIMENTAL

A reference Portland cement (PC) and composite cements containing BFS and / or CKD were used to produce mortar test specimens at a standard water/cement ratio of 0.50. The chemical composition of the cementitious materials is given in Table 1. The reinforcing steel used in casting the mortar test specimens had the following chemical composition of steel (% wt): C:0.16, Mn:0.77, S:0.019, P:0.019, Si:0.19, Ni:0.14, Cr:0.22, Cu:0.26 and Mo:0.054. The aggregate used was Greek sand of diameter 250 m < d < 4 mm. Drinking water from Athens water supply network and an alkanolamine-based corrosion inhibitor were also used for the specimens' preparation. Twelve categories of specimens were cast for each type of measurement. The proportion of the materials used and their category codes are given in Table 2.

The corrosion testing specimens (Batis et al., 2005) used were cylindrical, 40 mm in diameter and 100 mm in height, with one axially embedded 10 mm diameter steel rebar and prepared according to ISO/DIS 8407.3. The mortar test specimens for the Strain Gauge (Routoulas and Batis, 1999) measurements were in the form of 80 mm x 80 mm x 100 mm prisms. The embedded steel bars (diameter 12 mm) were also prepared according to ISO/DIS 8407.3. Specimens were cured (20°C, 100 % humidity) for 7 days and were afterwards immersed in the corrosive environment of 3.5 % wt NaCl solution.

The sensors used for measurement of corrosion rate of reinforcing steel are strain gauge (SG) type KM-30-120 of KYOWA. In each specimen two SG sensors were embedded. The first one, horizontally mounted near the rebar, measured the swelling of the specimen due to cumulative of corrosion and other parameters (creep, wetting), which change the specimen's volume. The second SG, vertically mounted far from the rebar, compensated the parameters of specimen volume variation except corrosion (Routoulas and Batis, 1999; Colombo, 1986). Impressed anodic potential of 1.0, 1.5, and 2.0 V was applied for several days for the acceleration of the corrosion process of the reinforced specimens immersed in 3.5 % NaCl solution. The test set–up included a potentiostat for applying the anodic potential, the SG bridge amplifier circuit and the multimeter for SG resistance measurements ((Routoulas and Batis, 1999).

Table 1. Chemical composition of materials (%)

	SiO_2	Al_2O_3	Fe_2O_3	CaO	MgO	K_2O	Na_2O	SO_3	$CaO_{(f)}$	LOI
PC	20.67	4.99	3.18	63.60	2.73	0.37	0.29	2.44	2.41	1.52
BFS	33.34	11.37	2.44	41.30	6.90	0.32	0.38	0.12	-	0.09
CKD	13.68	4.36	2.30	42.59	1.23	0.79	0.28	0.10	-	

Table 2. Categories of specimens - composition proportions (wt)

CODE	PC	Sand	Water	BFS	CKD	Inh-M
RE	1.00	3.00	0.50	-	-	-
REM	1.00	3.00	0.50	-	-	0.01
SD0-5	0.95	3.00	0.50	-	0.05	-
SDM0-5	0.95	3.00	0.50	-	0.05	0.01
SD65-0	0.35	3.00	0.50	0.65	-	-
SDM65-0	0.35	3.00	0.50	0.65	-	0.01
SD62-5	0.33	3.00	0.50	0.62	0.05	-
SDM62-5	0.33	3.00	0.50	0.62	0.05	0.01
SD60-10	0.30	3.00	0.50	0.60	0.10	-
SDM60-10	0.30	3.00	0.50	0.60	0.10	0.01
SD55-15	0.60	3.00	0.50	0.55	0.15	-
SDM55-15	0.60	3.00	0.50	0.55	0.15	0.01

3. RESULTS AND DISCUSSION

The comparative diagram obtained by the SG technique for the reference specimens as well as for all composite specimens is illustrated in Figure 1 as a function of time. Specimens tested under similar conditions of fixed anodic potential are classified according to their corrosion behavior. The initial time is the moment of application of anodic potential to the specimen. In all cases there is an increase in SG values, related to corrosion. During the first 18 days of the anodic potential application, a low rate of expansion of the specimens RE and SD0-5 was observed. The time of corrosion initiation as indicated by an increase in the SG value for the specimens SD65-0, SD55-15, SD62-5 and SD60-10 was the 4th day, where cracks were noted from the 5th day, with one-day delays, respectively. The better corrosion resistance of SD0-5 is evident compared to the other composite specimens.

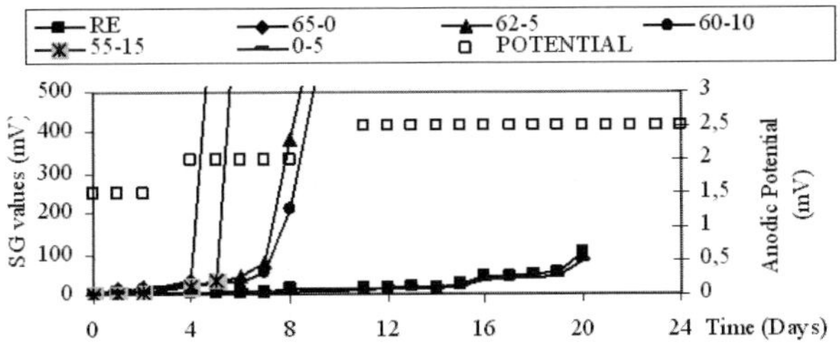

Figure 1. SG-values versus time

Measurements obtained by the SG technique as a function of time for the reference specimens and composite specimens with the corrosion inhibitor are illustrated in Figure 2. The incorporation of the alkanolamine-based inhibitor in the composite specimens retarded their cracking for about 4-7 days, without changing their ranking.

The results of the gravimetric determination of mass loss of rebars for all SG-specimens after exposure in the corrosive environment of 3.5% wt. NaCl solution are given in Figure 3. The use of CKD alone provided protection of reinforcement against induced corrosion as this is expressed by the lowest mass loss of steel. The use of BFS alone, as well as the combined use of BFS and CKD in the percentage of 15%, resulted in higher mass losses, whereas the combined use of BFS and CKD in the percentages of 5% and 10% provoked mass losses almost equal to that of the reference specimen. It is obvious that the addition of BFS does not offer protection to the reinforcing steel, but again the lowering of BFS and the increase in the CKD percentage results in an improvement of the abovementioned situation. This could partly be attributed to the fact that the incorporation of the finely ground CKD in the mix results in a relative reduction of the pore size and the effective porosity as it was shown in previous studies (Konsta and Shah, 2003). This, in turn, leads to an improvement of the durability characteristics of the external layers of the hardened material. However, it should be noted that mortars containing slag are known to be sufficiently durable to corrosive conditions such as immersion to NaCl solution, but with an at least three-month delay, because they retard to obtain their maximum strength as compared to pure PC mortars. Therefore, and given the short-term duration of the SG measurements, it was considered necessary to cross-examine the results of this method with others with a longer duration of time. The corrosion tendency of cylindrical mortar specimens partially immersed in 3.5% w/w NaCl solution was estimated by monitoring the half-cell potentials according to ASTM C876 – 87, using a saturated calomel electrode as reference electrode (Figure 4).

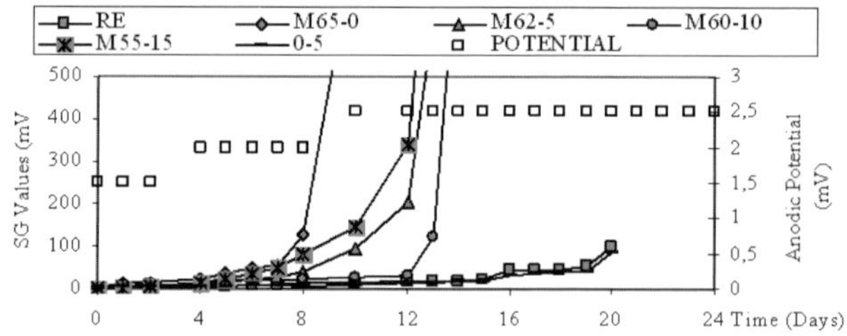

Figure 2. SG-values for specimens with the alkanolamine-based inhibitor

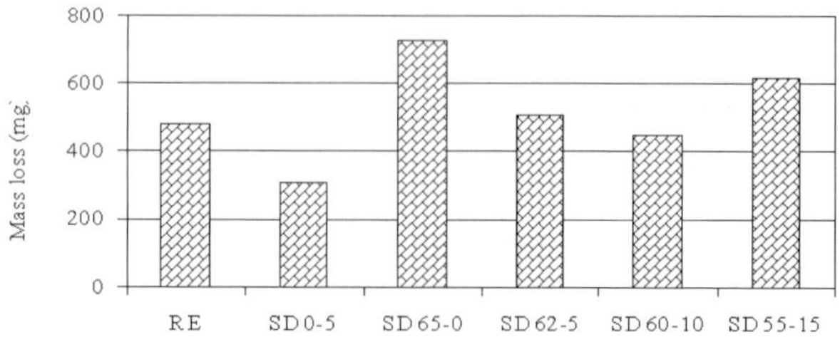

Figure 3. Mass loss of steel reinforcement in SG specimens

SD62-5 and SD60-10 specimens present the lowest electronegative values, whereas SD55-15 specimens present the greatest, which along with the most electronegative final potential values denotes an increase in the electrochemical activity of the system.

The other categories of specimens range between the two extremes, with small differences in potential values. The incorporation of the alkanolamine-based inhibitor again reduced in all cases the values of the corrosion potential denoting amelioration in concrete performance (Figure 5).

Potentiodynamic polarization measurements (Batis et al., 2005) were carried out for specimens partially immersed in 3.5% wt NaCl solution for six months. From Figure 6 it is obvious that SD62-5 and SD60-10 specimens present the lower corrosion current densities, even lower than the reference ones, whereas specimens with BFS and 15% CKD present increased values. Specimens with BFS alone present also increased corrosion rates. Finally, the addition of the alkanolamine-based inhibitor in mortar specimens resulted in reduced corrosion rates, the reduction being more intense in pure slag specimens. It seems that the behaviour of specimens containing BFS is comparatively better than that from the SG measurements, which can be attributed to the fact that the necessary time for the evolution of its beneficial properties has elapsed.

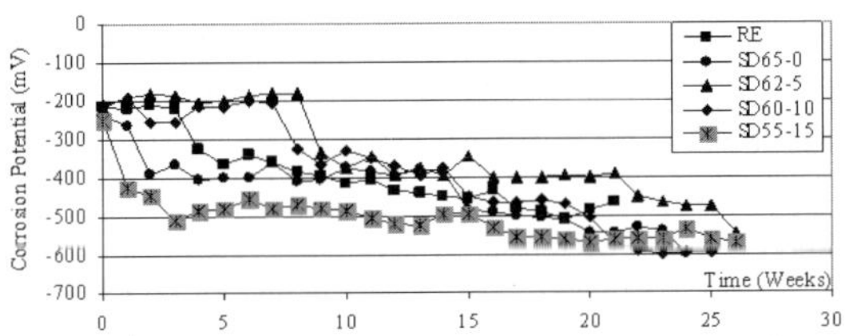

Figure 4. Half-cell potentials on steel reinforcement

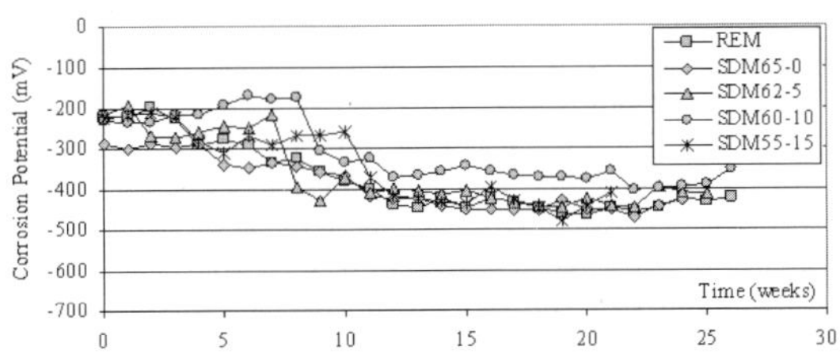

Figure 5. Half-cell potentials on steel reinforcement for specimens with inhibitor

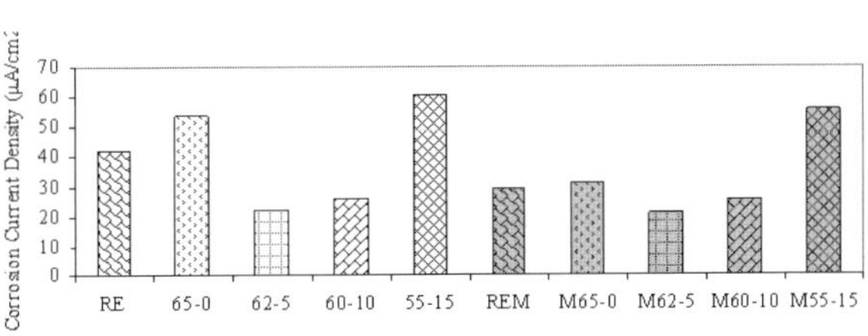

Figure 6. Corrosion current density as a function of time

4. CONCLUSIONS

Based on the so far results, it seems that the use of BFS alone, as well as the combined use of BFS and 15% CKD, provoked higher corrosion rates than the use of BFS together with CKD in percentages of 5% and 10%, which resulted in a protection level equal or even better than that of the reference specimen.

The use of CKD alone provided the lowest corrosion rate. It is obvious that the addition of BFS at early ages does not offer protection to the reinforcing steel, but the lowering of BFS percentage and the CKD addition in percentages of 5% to 10% results in an improvement of their anticorrosive behavior. The addition of CKD up to 10% compensates the low protection offered by BFS by a relative reduction of the pore size leading to an improvement of the durability characteristics of the external layers of the hardened material. The incorporation of the alkanolamine-based inhibitor in the composite speci mens provoked a further enhancement of the protection offered.

Acknowledgements

The present work forms part of the Archimedes research program, funded by the European Community Fund and the Greek Ministry of Education.

5. REFERENCES

1. Batis, G. Katsiamboulas, A. Meletiou, C. A., and Chaniotakis, E., 1996, Durability of reinforced concrete made with composite cement containing kiln dust. Proc. Int. Conf. *Concrete for Environment Enhancement and Protection,* Ed. R.K. Dhir and T.D. Dyer, E & FN Spon, London, pp. 67-72.
2. Batis, G., Pantazopoulou, P., Tsivilis, S., and Badogiannis, E., 2005, The effect of metakaolin on the corrosion behavior of cement mortars, *Cement Concrete Comp.* **27**(1):125-130.
3. Bjegovic, D., Ukrainczyk, V., and Miksic, B., 1999, Corrosion protection of existing concrete structures, Proc. Int. Conf. *Infrastructure Regeneration and Rehabilitation Improving the Quality of Life through Better Construction*, Ed. Swamy RN, Sheffield, pp.725-733.
4. Broomfield, J. P., 1999, Corrosion inhibitors for steel in concrete, *Concrete.* **33**(6):44-47.
5. Buchwald, A., and Schulz, M., 2005, Alkali-activated binders by use of industrial by-products, Cement Concrete Res. **35**:968-973.
6. Colombo, G., 1986, *Automazione Industriale*, Vol. 4. Torino: Dott. Giorgio.
7. Konsta-Gdoutos, M. S., and Shah, S. P., 2003, Hydration and properties of novel blended cements based on cement kiln dust and blast furnace slag, *Cement Concrete Res.* **33**:1269-1276.
8. Neville, A. M., 1975, *Properties of Concrete*, Pitman, Surrey.
9. Routoulas, A., and Batis, G., 1999, Performance evaluation of steel rebars corrosion inhibitors with Strain Gauges, *Anti-Corros Method M.*, **46**(4):276 -283.

STRAIN MONITORING METHOD USING IN FREEZE-THAW TEST OF RC

H. Pengfei
Department of Building Engineering, Tongji University, 1 Floor, No. 52, Long 1220, Beijing Road (West), Jingan District, Shanghai, 200040, P.R.China

Abstract: The damage experiments of reinforced concrete (RC) samples under cyclic freeze-thaw were investigated in this study. Influence of cyclic freeze-thaw on the damage evolvement of RC was studied by measuring the change of rebar strain and concrete strain. Experimental results showed similar rule in damage evolvement but different damage rate between the normal-strength concrete (C45, the 28-day compressive strength of 52 MPa) and the high-strength concrete (C70, the 28-day compressive strength of 77 MPa). Meanwhile, an interesting phenomenon is revealed that the amplitude of cyclic strains increase with the numbers of freeze-thaw cycles within a fixed temperature difference.

Key words: strain monitoring; freeze-thaw; RC

1. INTRODUCTION

Reinforced concrete (RC) elements always are subjected to synergistic effects of the environmental and mechanical loads. The study on synergistic effects is undoubtedly meaning and vital in terms of the consummation of the RC materials and structure theory and for purposes of the insurance of the safety, durability as well as life prediction of RC structures [1-2]. As a result, it has drawn increasing attention from material scientists and structural engineers [3-13] all over the world. However, most of the contemporary research activities mainly focus on the durability study of concrete under synergistic effects of two or three factors. References [3-13] investigated and revealed that rebar concrete was caused by Cl^- ion penetration in concrete under load, and the synergistic effect mechanism of rebar corrosion and freeze-thaw cycles on concrete, and developed an analysis method and a series of measures to prevent the reliability of RC structures from the reduction caused by the synergistic effects mentioned above. Sun et al. [6] studied the damage mechanism of concrete under synergistic effects of freeze-thaw cycles, external bending stress and salt solution. Zhou et al. [7] studied the influence of an external flex-

M.S. Konsta-Gdoutos, (ed.), Measuring, Monitoring and Modeling Concrete Properties, 597–604.

ural load on the frost resistance of air-entrained and non-air-entrained Mortar beam specimens with and without silica fume, which were placed in a specially designed apparatus and subjected to a flexural stress ranging from 0 to 50 percent of the modulus of rupture. Zhou et al. [7] found that the samples were immersed along with the apparatus in the chambers inside the freezing-thawing machine and the deterioration processes that may occur, including uniform and homogeneous microcracking, growth of a critical crack and surface scaling at high water-to-cementitious materials ratios. Literature [4-6] found that concrete surface would severely scale under the synergistic effects of freeze-thaw cycles and being dipped in sodium chloride solution with mass percent of 3.5%. Schneider et al. [8] investigated the mechanism of the synergistic effects of two factors, i.e. dead load and ammonium nitrate corrosive solution attack. The performance variation of cementitious composite material subjected to the synergistic effects of compressive load and sulphate attack was investigated by Zivia et al. [9] who presented the mechanism of concrete under high stress ratio state and subjected to high content sulphate attack. Mohamed and Hansen [10] presented a damage model to simulate the process of concrete subjected to load and freeze-thaw cycles.

Experimental data indicated that under the synergistic effects of freeze-thaw cycles and deicing salt corrosion, the rebar surface in RC structure would easily corrode, and microcracks in concrete would initiated and developed, and as a result, the whole strength and stiffness of RC would decline. The damage was a kind of failure process caused by the interaction of thermal stress and corrosion penetration. Therefore, the thermal stress distribution under freeze-thaw cycles, internal stress distribution under bending stress of RC, and the redistribution of these stresses with the development of damage were of importance to the investigation of the damage mechanism of RC elements, determination of failure source and lifetime prediction. However, strain monitoring method of reinforced concrete under synergistic effects of cyclic freeze-thaw, deicing-salt attack, rebar corrosion and bending stress was rarely reported in published literatures.

In order to study this problem, RC beam samples under the synergistic effects of cyclic freeze-thaw, deicing-salt attack, rebar corrosion and bending stress were investigated using a comprehensive experimental method [14]. It was found that C45 concrete (the compressive strength of 28 days was 52 MPa) severely cracked under the synergistic effects, especially on the rebar-concrete interface, which probably caused by the stress distribution and damage evolution of the interface. As shown in Figure 1, rebar corrosion was grave which was caused by the Cl^- ion penetration of deicing-salt solution. The residual flexural strength of reinforced concrete decreased with the damage factors' addition, in another word, the synergistic effects of freeze-thaw, deicing salt attack, rebar corrosion and four point bending load was remarkably harmful than single effect in the deterioration process.

2. EXPERIMENTAL

2.1 Materials used

Locally available Type I Portland cement and fly ash collected from Beijing were used as cementitious materials. Strength properties, including measured flexural and compressive strength of cement paste varying with curing age were shown in Figure 2. Chemical and physical properties of Type I Portland cement and fly ash used were reported in Tables 1-2, respectively

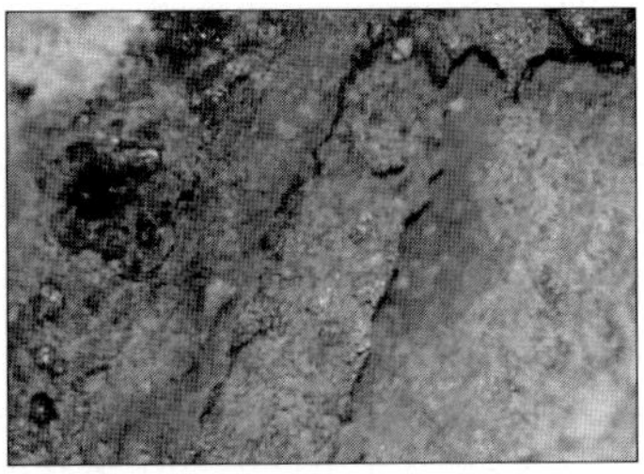

Figure 1. Optical micrograph of the profile at the surface of rebar in C45 RC after 50 freeze-thaw cycles in 3% NaCl solution and constant bending stress of 100 hours

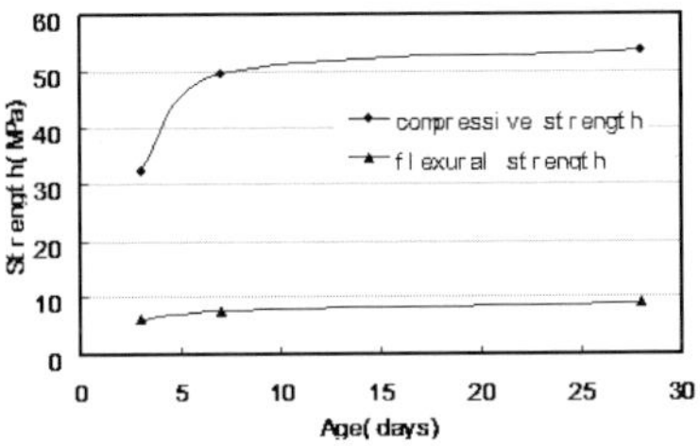

Figure 2. Measured flexural and compressive strength of cement paste varying with curing age

Washed river sand with a specific gravity of 2.62 and water absorption of 1.46% and crushed granite with a nominal maximum size of 25 mm and a specific gravity of 2.61 were used as the fine and coarse aggregates, respectively. The specific gravity and specific surface area of fly ash used were 1.82 g/cm^3 and 126 m^2/g, respectively.

Table 1. Composition of Portland cement used in this study

Chemical composition	SiO_2	Al_2O_3	Fe_2O_3	CaO	MgO	SO_3	K_2O	Na_2O	TiO_2	Loss	SUM
Weight ratio to total Portland cement (%)	22.15	7.93	2.12	61.45	3.78	0.25	0.79	0.13	0.38	0.98	99.96

2.2 Specimen design and preparation

Specimens with sizes of 100 mm×100 mm×400 mm were designed and prepared to investigate the synergistic effects of cyclic freeze-thaw, deicing-salt attack and bending stress on behavior of RC (curing ages=28 days, curing in tap water refreshed every week). Table 3 reports the mixture proportions of two types of concrete used in this study. During moulding, the cubes of concrete (C45 and C70, respectively, as shown in Table 3) were mechanically vibrated. After 24 hours, the specimens were removed from the mould and subjected to water curing for 7, 28 and 180 days. After curing, the specimens were tested for compressive strength using a compression testing machine with 2000 kN capacity. The tests were carried out on six specimens and average compressive strength values were obtained. The ratios of water to cementitious materials, i.e. w/cm, were 0.50, and 0.31, respectively. A naphthalene sulfonate-based high-range water reducer (HRWR) was utilized to improve the workability of fresh concrete.

Table 2. Composition of fly ash used in this study

Chemical compositions	Weight ratio to total Portland cement (%)	Physical properties	Ratio (%)
SiO_2	53.80	Fineness (residual of 45m Mesh)	2.8
Al_2O_3	30.75		
Fe_2O_3	5.12		
CaO	3.25	Loss on ignition (LOI)	1.87
MgO	0.99		
SO_3	1.22		
K_2O	0.73	Moisture Content	2.58
Na_2O	0.51		
C	2.15		
TiO_2	0.52	Water Requirement	91
Loss	0.91		
SUM	99.95		

2.3 Device and procedures undertaken

A forced pan mixer was used throughout the experiments. Mixing procedures were as follows: (1) powders and sand were mixed for 30s; (2) premixed water with various chemical admixtures were then introduced and homogenized for 90 s; (3) the said materials were then thrown in a gravel and were mixed for 60 s; and (4) a rest was held for 60 s to allow the chemical admixtures to initiate. Test apparatus was shown in detail in Figure 3. Figure 3(a) includes every parts of the test apparatus. Figure 3(b) is the section reinforcement, Figure 3(c) is four-point bending of RC specimen, and Figure 3(d) is the distribution of strain gauges for measuring the deformed content of concrete and rebar under such severe test condition. In detail, the No.1, 2 and 3 is for the measurement of rebar deformation, and the No.4 and 5 strain gauges is for the measurement of concrete deformation.

2.4 Compressive strength and elastic modulus of concrete

The compressive strength and elastic modulus of concrete were indicated in Table 4. With the gradual progress of the second hydration reaction of fly ash with $Ca(OH)_2$ increasing with age, it was noted that the compressive strength of C45 (the compressive strength of 28 days was 52 MPa) increased 10 MPa from 28 days to 180 days. However, the elastic modulus of C45 was lower than that of C70 (the compressive strength of 28 days was 77 MPa) due to the higher strength.

Table 3. Mixture composition of concrete

Concrete type	C45	C70
Cement (kg/m^3)	296	495
Fly ash (kg/m^3)	74	55
Fine aggregate (kg/m^3)	775	607
Coarse aggregate (kg/m^3)	1070	1078
Water (kg/m^3)	185	171
HRWR* (kg/m^3)	2.22	6.60
Water/binder ratio W/(C+F)	0.50	0.31
S/A (%)	42	36

* HRWR: High-range water-reducing admixture; S/A: Weight ratio of fine aggregate to total aggregate.

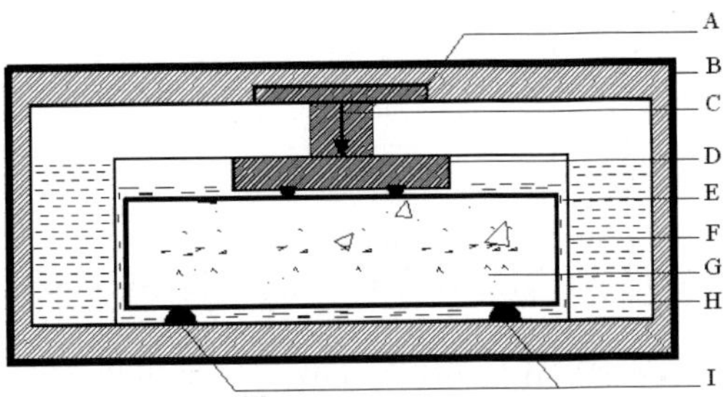

Figure 3(a). Special test apparatus

A. load sensor; B. Freezing-thawing machine; C. loading; D. load balance block, E. NaCl solution (3% in mass); F. stainless steel trough; G. reinforced concrete specimen; H. antifreeze solution; I. Support

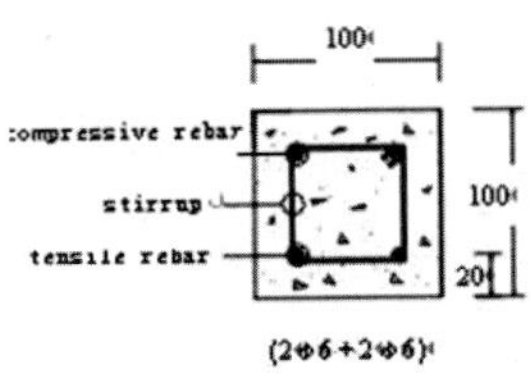

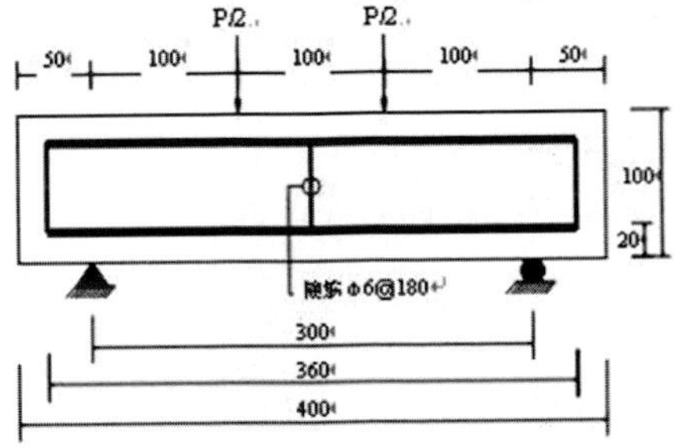

Figure 3(b). Section reinforcement *Figure 3(c).* Four-point bending of RC specimen (mm)

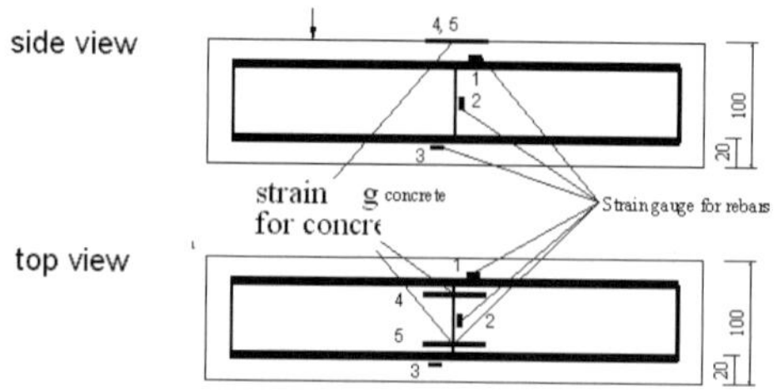

Figure 3(d). Strain gauge distribution

Table 4. Compressive strength and elastic modulus of concrete

Concrete type	Compressive strength (MPa)			Elastic modulus(28d) (GPa)	Diffusion Coefficients of Chloride [10^{-8} cm^2/s]
	7d	28d	180d		
C45	35	52	62	34.1	4.15
C70	58	77	96	37.2	2.02

3. EXPERIMENTAL RESULTS AND DISCUSSIONS

As shown in Figure 4, the experimental setup was used to do four-point bending test of concrete or reinforced concrete with sizes of 100 mm×100 mm×400 mm as described in Figure 3. Two kinds of reinforced concrete (C70 and C45, two kinds according to Chinese code [12]: 2Φ10 + 2Φ6, Φ6@180 and 2Φ6 + 2Φ6, Φ6@180) suffered various synergistic effects of freeze-thaw and deicing salt attack and constant four point bending load (two kinds of bending stress ratios: 0.35 and 0.6). As shown in Figure 5, there is an interesting phenomenon found by strain measurement.

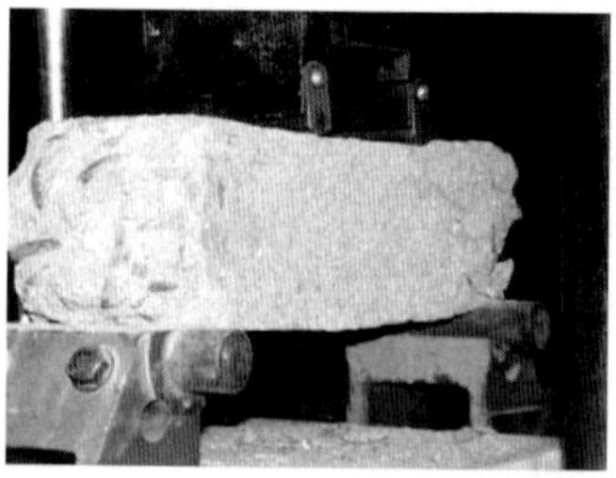

Figure 4. Experimental setup

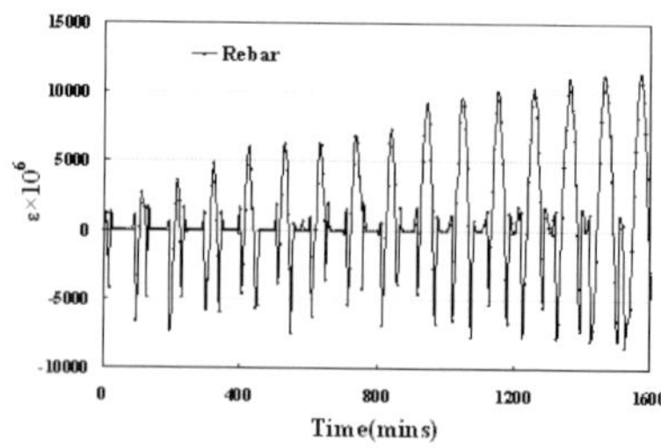

Figure 5. The strain of tensile rebar in C70 RC increasing with the freeze-thaw cycles under the synergistic effects of freeze-thaw and deicing salt attack and four-point bending load

In detail, under the fixed temperature variation, the absolute strain value of tensile rebar in C70 RC increases step by step with the freeze-thaw cycles, after subjected to the synergistic effects of freeze-thaw and deicing salt attack and four point bending load. A probable reason for the phenomenon mentioned above is cyclic strain softening. In detail, microcracks of concrete and rebar germinate and develop under the action of cyclic strain controlled by temperature fatigue stress and cyclic stress relaxation. Then the dynamic elastic modulus decreases gradually and at the same time the cyclic absolute strain value increases. As the above only a coarse explanation, the mechanism of this phenomenon need further study.

In this study, the failure mechanism of RC specimen was caused by crack propagation interactions of thermal stress, surface corrosion, rebar corrosion and mechanical load in nature, which were the synergistic effects of cyclic freeze-thaw, deicing-salt attack and bending stress. Under the synergistic action of cyclic freeze-thaw and deicing-salt attack, especially after corrosion resolution gradually penetrated into inner of concrete. In the process of every freeze-thaw cycle, the volume variation of corrosion resolution during freezing period made the inner structure of concrete gradually released, which resulted in the deduction of penetration resistance, i.e. the reduction of concrete's ability to protect rebar. After the sum of thermal stress and mechanical stress exceeded the gradually decreasing tensile strength, concrete cracking would occur. The crack form and propagation accelerated rebar corrosion, which caused the gradually decrease of RC beam. Therefore, with the increase of freeze-thaw cycles, concrete cracking caused by the action of cyclic freeze-thaw and deicing-salt attack, and rebar corrosion become two interacting factors, which increased the failure rate of RC. Finally, due to the reduction of the RC's capacity, it could not meet the strength need of normal use state.

4. CONCLUSIONS

The deterioration mechanism of reinforced concrete under synergistic effects of cyclic freeze-thaw, deicing-salt attack, rebar corrosion and constant bending stress were investigated using a comprehensive experimental method.

Under the fixed temperature variation, the absolute strain value of tensile rebar in C70 RC increases step by step with the freeze-thaw cycles, after subjected to the synergistic effects of freeze-thaw and deicing salt attack and four point bending load.

5. REFERENCES

1. P. K. Mehta. Durability - Critical Issues for the future. Concrete International, 1997; 19(7):27-31.
2. Adam Neville. Consideration of durability of concrete structures: Past, present, and future. Materials and structures, 2001; 34(3): 114-118.
3. Cerny, R.; Drchalova, J.; Rovnanikova, P. The effects of thermal load and frost cycles on the water transport in two high-performance concretes. CCR Journal, 2001, 31(8): 1129-1140.
4. Yoon, Sanchun;, Wang, Kejin; Weiss, W. Jason; Shah, Surendra P. Interaction between loading, corrosion, and serviceability of reinforced concrete. ACI Materials Journal, 2000, 97(6): 637-644.

5. Mu, R.; Miao, C.; Luo, X.;Sun, W. Interaction between loading, freeze–thaw cycles, and chloride salt attack of concrete with and without steel fiber reinforcement. CCR Journal, 2002, 32(7): 1061-1066.
6. W. Sun, Y. M. Zhang, H. D. Yan. Damage and damage resistance of high strength concrete under the action of load and freezing-thawing cycles. Cement and Concrete Research, 1999,29(6):1519-1523.
7. Zhou Yixia; Cohen, Menashi D.; Dolch, William L. Effect of external loads on the frost resistant properties of mortar with and without silica fume. ACI Material Journal, 1994,91(6): 595-601.
8. Schneider U., Chen S. W. Behavior of high performance concrete under ammonium nitrate solution and sustained load. ACI Materials Journal, 1999,96(1): 47-51.
9. Zivia V., et al. The behavior of cement composite under compression load at sulphate attack. Cement and Concrete Research, 1994,24(8): 1475-1484.
10. Ashraf Ragab Mohamed, Will Hansen, Micromechanical modeling of concrete response under static loading—Part II: Model prediction for shear and compressive loading. ACI M Journal,1999,96(3): 354-358.
11. Yiwang Bao; Shengbiao Su; Jow-Lay Huang. An uneven strain model for analysis of residual stress and interface stress in laminate composites. Journal of composite materials, 2002, 36(14): 1769-1778.
12. Chinese Standard GB50010—2002, Code for Design of Concrete Structure, China Architecture and Building Press, Beijing, China, 2002.
13. Popov EP., Nagarajan S., Lu Z.A., Mechanics of Materials(2nd Edition). Prentice-Hall Press,London, 1978.
14. Huang Pengfei, Yao Yan, Bao Yiwang,etc. The deterioration mechanism study of reinforced concrete under the synergistic effects of freeze-thaw, deicing salt attack, rebar corrosion and four point bending load. The sixth national conference of concrete durability of China, May, 2004, Shenzhen, China

EVALUATION OF ORGANIC CORROSION INHIBITOR EFFECTIVENESS INTO THE CONCRETE

E. Rakanta[1], E. Daflou[1] and G. Batis[2]
[1]*Chemical Engineer, National Technical University of Athens;* [2]*Professor of Chemical Engineering School, Section of Material Science and Engineering, NTUA, 9, Iroon Polytechniou Str. Zografou Campus, 157 80 Athens, Greece*

Abstract: The aim of this study is to examine the protective effect of N, N' Dimethylaminoethanol (DMEA - organic corrosion inhibitor), against rebar corrosion in the presence of chloride ions. The inhibiting properties of DMEA based corrosion inhibitor were evaluated in saturated $Ca(OH)_2$ solution and in cement mortar specimens. The electrochemical measurements were performed using saturated calcium hydroxide $Ca(OH)_2$ solution, contaminated with chloride ions, simulating the concrete interstitial electrolyte. Corrosion parameters, such as corrosion rate, corrosion current density I_{corr}, polarization resistance R_p and pitting potential E_{pitt} of reinforcing steel have been evaluated by electrochemical measurements and compared with that obtained from metal loss determination. Results indicated that the addition of N, N' dimethylaminoethanol to the saturated $Ca(OH)_2$ solution as well as into the concrete contaminated with chloride ions, decreases the corrosion rate of steel reinforcement.

Keywords: Organic corrosion inhibitor; DMEA; reinforcement corrosion; electrochemical evaluation.

1. INTRODUCTION

Corrosion of reinforcing steel embedded in concrete is becoming a significant structural and financial problem. As it is known, in Greece, many historical buildings and structures are located in coastal regions where the weather is characterized by pollutants such as Cl^- and carbon dioxide, CO_2. This leads to an increased incidence of spalling, delimination and as a consequence the deterioration of concrete in reinforced structures. The use of chemical admixtures, which acts as corrosion inhibitors, is a method for preventing and delaying the onset of rebar corrosion[1,2].

M.S. Konsta-Gdoutos, (ed.), Measuring, Monitoring and Modeling Concrete Properties, 605–611.

An ideal corrosion inhibitor has been defined as "a chemical compound, which, when added in adequate amounts to concrete, can prevent corrosion of embedded steel and has no adverse effect on the properties of concrete". Nowadays chemical corrosion inhibitors present an easily implemented solution to the growing problem of corrosion of reinforcing steel in concrete. However, to be considered viable, these additives should not only prevent or delay the onset of corrosion, they must not have any detrimental effect on the properties of concrete itself, such as strength, setting time, workability and durability.[3] Drawbacks of corrosion inhibiting admixtures are that they may not remain in the repair area, potentially reducing the concentration of the inhibitor below necessary values and secondly, when used in a limited area long a continuous reinforcing bar, there is the potential for micro cell corrosion development[4,5]. The effect of N, N' dimethylaminoethanol (DMEA) on the corrosion of steel due to chloride ingress was experimentally investigated in this paper. The aim of this study was to examine the protective effect of DMEA (organic corrosion inhibitor), against rebar corrosion in cement mortar specimens and in saturated $Ca(OH)_2$ solution containing corrosion inhibitor in the presence of chloride ions.

2. MATERIALS AND EVALUATION METHODS

2.1 Mortar specimens

The test specimens were prepared with cement, sand, and water in ratio 1:3:0.6. The mean value of the sand grains diameter was 250 $\mu m < d < 4mm$. Greek Portland cement was used in all the specimens. Cylindrical steel rebars of type S500s, Vanadus/ELOT 971 with dimensions of 12mm in diameter and 10 mm high were used for all test specimens. Drinking water from Athens water supply network and DMEA corrosion inhibitor, were used in the preparation of the specimens. DMEA based corrosion inhibitor was in liquid form and the solid content was 29% by weight and it was mixed in the mixing water at two dosages of 1% and 2% by weight of cement. Chloride contaminated cement mortar was made by addition of sodium chloride to the mixing water, thus producing cement mortar specimens with 1%, 2%, 3%, 4% and 5% chloride by weight of cement. Cement mortar specimens treated with corrosion inhibitor in concentrations of 1% and 2% by weight of cement and there were prepared with 1.5%, 2%, 2.5% and 2%, 2.5%, 3%w.t chlorides by weight of cement, respectively.

2.2 Steel reinforcements in saturated $Ca(OH)_2$ solution

For the electrochemical measurements, disk–shaped steel specimens were prepared from deformed reinforcing steel, nominally 12mm in diameter and a thickness of 10mm. The steel disks were cleaned according to the ISO/DIS 8407.3 standard. Prior to the electrochemical tests, the steel specimens were exposed to saturated $Ca(OH)_2$ solution containing salt and inhibitor preparation for 24 hour. The code names and the composition for the different sets of saturated $Ca(OH)_2$ solutions used in this study as a corrosion medium for the electrochemical measurements are shown in Table 1.

Table 1. Results of electrochemical measurements of steel reinforcements in artificial concrete pore solution

Code name	Chemical composition of corrosion medium		Electrochemical Parameters			
			LPR Technique			Cyclic Polarization
	Corrosion inhibitor (%w.t)	NaCl concentration (%w.t)	Icorr (μA/cm^2)	Rp (kOhms)	Corrosion Rate (mpy)	Pitting Potential (mV)
SCH 0_0	0%	0%	0.4870	44,57	0.4461	535
SCH 0_1		1%	1.732	12.54	1.586	419
SCH 0_2		2%	5.730	3.790	5.246	185
SCH 0_3		3%	6.042	3.594	5.532	119
SCH 0_4		4%	10.74	2.021	9.836	117
SCH 0_5		5%	13.24	1.640	12.12	85
SCH 1_0	1%	0%	0.2494	87.08	0.2283	660
SCH 1_1		1%	0.763	28.44	0.6992	570
SCH 1_2		2%	4.056	5.353	3.714	496
SCH 1_3		3%	5.235	4.148	4.793	233
SCH 1_4		4%	5.453	3.982	4.993	131
SCH 1_5		5%	5.640	3.850	5.164	114
SCH 2_0	2%	0%	0.172	126.244	0.1574	757
SCH 2_1		1%	0.717	30.284	0.6564	658
SCH 2_2		2%	2.896	7.497	2.6515	458
SCH 2_3		3%	3.202	6.781	2.9317	263
SCH 2_4		4%	4.298	5.0521	3.9352	220
SCH 2_5		5%	5.195	4.1797	4.7565	134

The electrochemical measurements of steel rebars in artificial concrete pore solution were contacted according to Batis et al.[5] Methods used to assess cement mortar specimens' performance against corrosion included the measurement of corrosion potential, corrosion rate, and mass loss.

3. RESULTS AND DISCUSSION

3.1 Steel reinforcement embedded in contaminated cement mortar

The corrosion potential of rebars in concrete is a good indication of either the depassivation or activation of corrosion. The corrosion potential measurements given in Figures 1 and 2 are for specimens without and with inhibitor respectively. According to the guidelines involved in ASTM C 876, the probability of corrosion initiation is greater than 90% when corrosion potentials are more negative than -350mV relative to the copper – copper sulfate (CSE) and -270mV relative to the saturated calomel electrode (SCE).

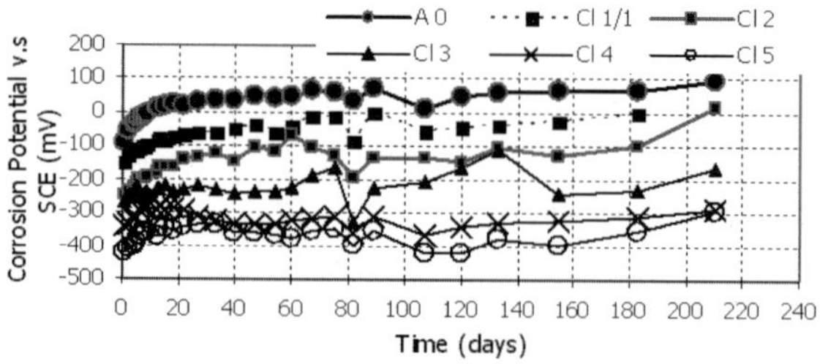

Figure 1. Half cell Potential measurements versus time in chloride ions contaminated specimens

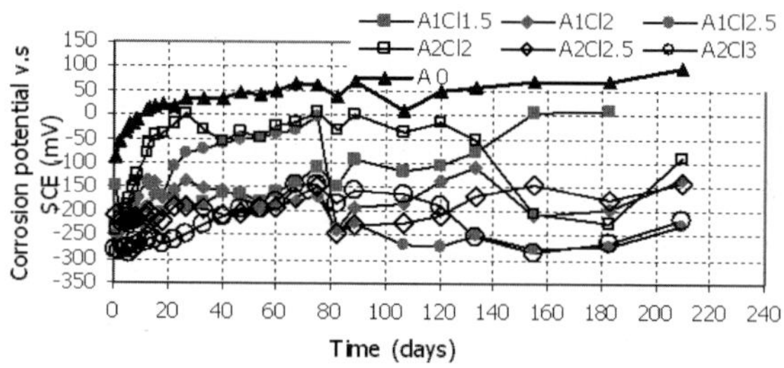

Figure 2. Half cell Potential measurements versus time in inhibitor treated mortar specimens

Regarding the untreated cement mortar specimens contaminated with chloride ions, it is observed that as the concentration of the chloride ions increases, the corrosion potential values shift towards to more electronegative values.

On the contrary, for the inhibitor treated cement mortar specimens contaminated with chloride ions, it is evident that all these values of corrosion potential measurements are in the range of -300 mV to -100mV vs. SCE. These values suggest a high probability of 90% of a passive stage. The results of mass loss measurements of reinforcing steel, after seven months of exposure to chloride solution are given in Figures 3 and 4.

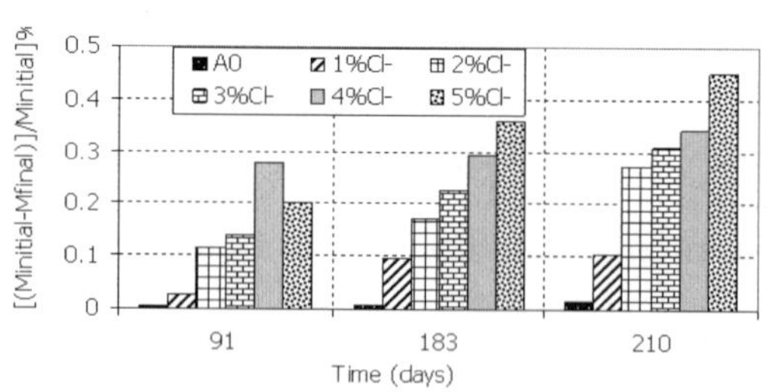

Figure 3. Mass loss measurements versus time, in chloride ions contaminated specimens

From these results, the improvement of the corrosion performance of steel rebars when the amino alcohols corrosion inhibitor was added is evident. The DMEA decreases the steel rebar mass loss after 91, 183 and 210days of exposure by about 17.96%, 62.29% and 48.90%, respectively, in inhibitor treated cement mortar specimens (2% by weight of cement).

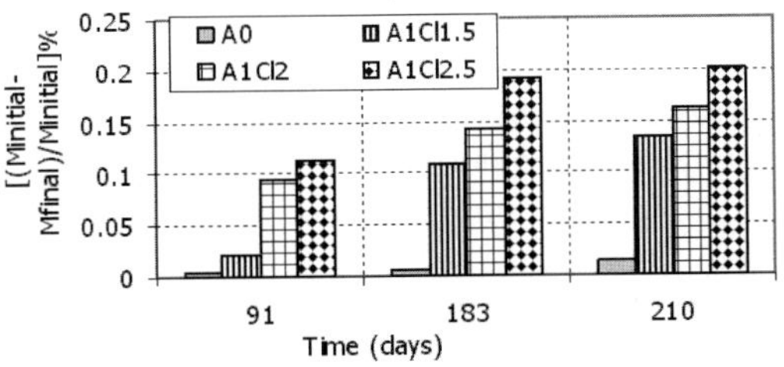

Figure 4. Mass loss measurements versus time, in inhibitor treated mortar specimens (1% per weight of cement)

It has been reported that DMEA provides protection by forming an adsorption layer on the steel surface, hindering steel dissolution[7]. Moreover DMEA was found to partially displace chloride from the iron oxide surface. These observations suggest that inhibition of corrosion occurs through mechanism whereby DMEA displaces chloride ions and forms a durable passivating film.

In this view, although the aminoalcohols adsorb on non–corroding sites which may seem more cathodic than anodic, they can just as easily be said to adsorb on potentially anodic sites[7]. Overall, of weight loss and corrosion potential measurements performed on mortar specimens indicated that the addition of DMEA - based corrosion inhibitor reduces the rate of corrosion in chloride - contaminated cement mortars (2% DMEA by weight of cement).

3.2 Steel reinforcement in saturated calcium hydroxide solution

The electrochemical measurements of steel rebars in artificial concrete pore solution allowed to evaluate the behavior and the protective efficiency of the DMEA based corrosion inhibitor. This behavior was assessed through linear polarization and cyclic polarization techniques. Table 1 shows the results of electrochemical parameters which were measured by LPR and cyclic polarization techniques, for all the types of artificial pore solution with varying concentration of corrosion inhibitor and varying amounts of Cl^- ions. The test results indicate that DMEA retards reinforcement corrosion in the presence of chloride ions. After 24 hours of immersion, the corrosion rates were 12.12, 5.164, 4.756, 4.565 mpy in solutions containing 0%, 1%, 2% and 3% corrosion inhibitor, respectively, and 5% NaCl. It is also observed that the polarization resistance, R_p, of steel reinforcements decreases as the chloride content increases. This is due to the presence of chloride ions in the working solution, which are typically responsible for the local breakdown of the passive layer on steel surface. The corrosion current density of steel reinforcement in saturated $Ca(OH)_2$ solution containing various amounts of corrosion inhibitors and chlorides is presented in Figure 5. It is shown that corrosion current density on steel in solutions without inhibitor is higher than that in the solutions with the inhibitor. This indicates that N, N' dimethylaminoethanol inhibitor retards reinforcement corrosion.

The critical potential for pit initiation is a function of the nature and concentration of aggressive and inhibiting molecules. Furthermore, the pitting potential is more positive the lower the aggressive anions (catalyzing the anodic metal dissolution) concentration and higher the inhibiting molecules (inhibiting the anodic metal dissolution) concentration are. Corresponding to Table 1, in saturated $Ca(OH)_2$ solution with 2%wt DMEA, a straight increase of current intensity at approximately 754mV is observed, whereas the curves with Chloride ions at varying amounts of 1%, 2%, 3%, 4% and 5%w.t show a significant increase of corrosion current density at approximately 658, 458, 263, 220 and 134mV, respectively. As expected, chloride ions are typically responsible for the breakdown of passive film on the rebar surface. Chloride–induced reinforcement corrosion tends to be localized corrosion process, with the original passive surface being destroyed locally under the influence of chloride ions. In cyclic polarization curves the probability of localized corrosion is related to the difference between the repassivation, E_{rep}, and corrosion potentials, E_{corr}, thus the higher the difference the lower the probability of pitting corrosion.[8,9,10] Overall, the experimental results obtained on the steel rebars immersed in saturated calcium hydroxide solution with varying amounts of chloride ions show that, in absence of DMEA inhibitor the corrosion activity on the steel surface strongly increases with the increase of chloride ion content, leading to high corrosion rates and low pitting potentials as observed in Table 1.

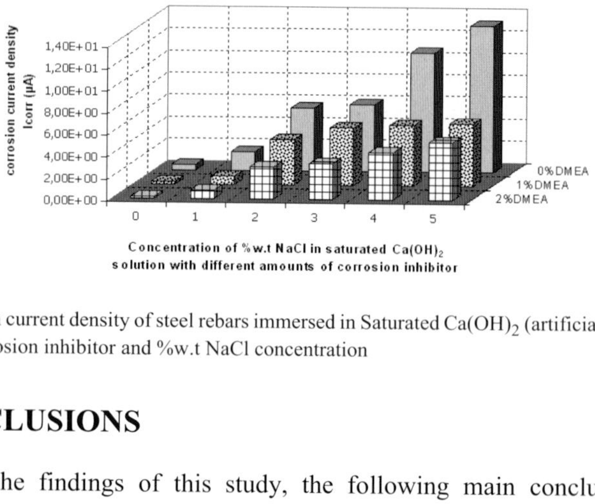

Figure 5. Corrosion current density of steel rebars immersed in Saturated $Ca(OH)_2$ (artificial concrete pore solution) vs. %w.t corrosion inhibitor and %w.t NaCl concentration

4. CONCLUSIONS

Based on the findings of this study, the following main conclusions could be obtained:

1. The corrosion potentials, in cement mortar specimens contaminated with chloride ions, shift to more negative values as the chloride concentration increases. On the other hand, when corrosion inhibitor is added, the corrosion potentials shift towards positive values.
2. The corrosion current density on steel in solutions containing Cl^- ions is higher than that in solutions with the same Cl^- ion and no inhibitor is higher than that on steel in the saturated $Ca(OH)_2$ and inhibitor. This indicates that DMEA retards reinforcement corrosion.
3. The corrosion rate of steel reinforcement increases with an increase in the chloride concentration. However it decreases with increasing the concentration of corrosion inhibitor.

4. Overall, the addition of N, N' dimethylaminoethanol into the artificial concrete pore solution contaminated with NaCl, decreases the corrosion rates of steel reinforcements due to the fact that this type of organic inhibitor forms a stable interfacial layer on steel surface, which is able to keep interface under passive state. However, when artificial concrete pore solution is contaminating with high concentrations of chloride ions (4% and 5%w.t NaCl), layer becomes unstable and consequently the efficiency of corrosion inhibitor decreases.

5. REFERENCES

1. C.K. Nmai, Multi – functional organic corrosion inhibitor. *Cem. & Con. Comp.*, **26**, 199-207 (2004).
2. F. Wombacher, U. Maeder, B. Marazzani, Aminoalchohols based mixed corrosion inhibitors, *Cem. & Con. Comp.* **26**, 209-216 (2004).
3. S.M. Trepanier, B.B. Hope, C.M. Hansson. Corrosion inhibitors in Concrete. Part III: Effect on time to chloride – induced corrosion initiation subsequence corrosion rates of steel in mortar, *Cem. & Con. Res.*, **31**, 713-718 (2001).
4. L. Mammoliti, C.M. Hansson, B.B. Hope. Corrosion inhibitors in concrete, PART II: Effect on chloride threshold values for corrosion of steel in synthetic pore solutions. *Cem. & Con. Res*, **29**, 1583–1589 (1999).
5. G. Batis, E. Rakanta, B. Theodoridis, K.K. Sideris, K. Psomas and X. Barbari. Influence of N-N Dimethyloaminoethanol Corrosion Inhibitor on Carbonation and Chlorid – Induced Corrosion of Steel. In: Malhotra, editor. Proceedings Seventh CANMET/ACI, Berlin: ACI SP-217, 469-478, (2003).
6. J.P. Broomfield. *Corrosion of steel in concrete understanding, investigation and repair* (E & FN SPON, 1997).
7. J. M. Gaidis, Chemistry of corrosion inhibitors, *Cem. & Con. Comp.*, **25**, 181-189 (2003).
8. M Saremi, E. Mahallati, A study on chloride – induced depassivation of mild steel in simulated concrete pore solution, *Cem. & Con. Res.*, **32**, 1915-1921 (2002).
9. L L. Dhouibi, E. Triki, A. Raharinaivo, The application of electrochemical impedance spectroscopy to determine the long – term effectiveness of corrosion inhibitors for steel in concrete, *Cem. &Con. Comp.*, **24**, 35-43 (2002).
10. S. Qian, D. Cusson, Electrochemical evaluation of the performance of corrosion – inhibiting systems in concrete bridges, *Cem. & Con. Comp.*, **26**, 217-233 (2004).
11. O. Troconis de Rincon, O. Perez, E. Paredes, Y. Caldera, C. Urdaneta, I. Sandoval, Long – term performance of ZnO as a rebar corrosion inhibitor, *Cem. & Con. Comp.*, **24**: 79-87 (2002).

USING CHLORIDE CONCENTRATION AND ELECTRICAL CURRENT TO DETERMINE THE NON-STEADY-STATE CHLORIDE DIFFUSIVITY FROM MIGRATION TEST

C.C. Yang and S.C. Chiang
Institute of Materials Engineering, National Taiwan Ocean University, Keelung, Taiwan, ROC.

Abstract: The electrochemical technique is applied to accelerate the chloride migration in concrete to determine the chloride migration coefficient by measuring the chloride concentration and the electrical current. Sixteen concrete mixtures were used in this investigation. The non-steady-state migration coefficient was determined based on the measurements of the breakthrough time for chloride ion penetrating over through the specimen. A comparison of the non-steady-state migration coefficient obtained from the chloride concentration and the electrical current using the assumption of chloride concentration (C/C_0). From the experimental results, a linear correlation between the non-steady-state migration coefficient obtained from the chloride concentration and the electrical current was obtained.

Keywords: Durability; ACMT; Transport properties; Current.

1. INTRODUCTION

The conventional diffusion experiment requires considerable time to obtain the steady-state chloride flow across the specimen. An effective method is to apply an external electrical field for accelerating chloride penetration.

The rapid chloride permeability test, designed as ASTM C1202-97[1], was the proposed for rapid qualitative assessment of chloride permeability of concrete. Tong and Gjϕrv[2] found that a good correlation between the electrical conductivity and the diffusivity obtained from both the steady-state and the non-steady-state migration method. Tang and Nilsson[3] introduced an accelerated non-steady-state migration test method, the penetration depth of chloride ions can be used to determine the chloride diffusivity. Castellote et al.[4] have established a linear relationship between the chloride concentration and conductivity in the anodic compartment from the migration test. Yang and Cho[5] have used the

M.S. Konsta-Gdoutos, (ed.), Measuring, Monitoring and Modeling Concrete Properties, 613–618.

ACMT method to measure the chloride migration coefficient in steady state. However, the steady-state migration coefficient obtained in ACMT needs to use liquid samples to analyze the chloride concentration and the test duration may cost two weeks or one month for get the constant chloride flux.

The main objective is to evaluate the correlation between the non-steady-state migration coefficients from the chloride concentration and the electrical current for different type of mineral admixtures (fly ash and slag) with w/b ratios.

Table 1. Composition of C series mixtures

Mix	w/b	Unit content (kg m^{-3})			
		Water	Cement	Fine aggregate	Coarse aggregate
C35	0.35	178	509	946	806
C45	0.45	199	443	946	806
C55	0.55	216	392	946	806
C65	0.65	228	351	946	806

2. EXPERIMENTAL PROGRAM

2.1 Materials and specimen preparation

Sixteen concrete mixtures were used, and the control mixture (C series) proportions are summarized in Table 1. Twenty percent of cement was replaced by fly ash in F series; forty percent of cement was replaced by slag in S series. In SF series, thirty percent of cement was replaced by fly ash (9%) and slag (21%). Four w/b ratios (0.35, 0.45, 0.55 and 0.65) were used in each series. The same amount of coarse aggregate 806 kg/m^3 and fine aggregate 946 kg/m^3 were used for all mixes.

For each mix, a number of cylindrical (100×200 mm) were cast. After demolding, the specimens were cured in water for 28 days. The ACMT specimen (30-mm thick slice) was obtained by sawing the mid-portion of the cylindrical specimen. In order to avoid heterogeneity, the specimens were vacuum-saturated following the specification in ASTM C1202 prior to test.

2.2 Experimental procedure

The accelerate chloride migration test (ACMT) setup is illustrated in Figure 1(a). The specimen was placed between two acrylic cells, with one of the cells filled with 0.30N NaOH solution (4.5 liter) and the other was with 3.0% NaCl solution (4.5 liter). Brass meshes are placed on two sides of the specimen as the electrode. The electrodes were connected to a 24-V DC power source and the current circulating between anode and cathode was recorded by a data logger. The chloride ion concentration in anode cell was determined by potentiometric titration.

3. RESULTS AND DISCUSSION

3.1 Non-steady-state migration coefficient from ACMT

A typical result of chloride concentration gain in anode cell from ACMT is plotted in Figure 1(b) as a function of time. It shows three distinct stages exist, non-steady state, transition period and steady state with respect to the change of the chloride concentration.

In non-steady-state condition, the chloride ions have not yet reached the anode cell. The non-steady-state migration coefficient is determined based on measurements of the time-span for chloride ion penetration through specimen when the chloride ion reaches the edge of the anode cell. In order to determine the breakthrough time of the chloride ions to penetrate over through the specimen, a regression analysis is used as a basis for the determination of the duration of the chloride ion to penetrate over through the specimen as:

$$C = a \times t^{b} \qquad \text{for transition period} \tag{1}$$

where C is the chloride concentration gain in anode cell, t is the elapsed time, a and b are experimental constants. The duration of chloride ion penetrating through specimen can be determined from Eq. (1). A typical result of chloride concentration and current is plotted in Figure 2 as a function of time. In Figure 2, the breakthrough time for the chloride front t_p corresponding the $C/C_0 = 0.5\%$ (C_0 is the source of chloride concentration), was calculated and listed in Table 2. The non-steady-state migration coefficient (M_n) can be calculated from the modified Fick's second law and expressed by[2,4] :

$$M_n = \frac{1}{\beta}\left[\frac{x - \alpha\sqrt{x}}{t}\right] \tag{2}$$

where $\beta = \frac{|z|FE}{RT}$, $\alpha = 2\sqrt{\frac{1}{\beta}}erf^{-1}\left(1 - \frac{2C}{C_0}\right)$, z is the electrical charge of chloride, F is the Faraday constant, E is the strength of the electric field between the anode and cathode, R is the universal gas constant, T is absolute temperature, and erf^{-1} is the inverse of error function. Based on the modified Fick's second law by using the breakthrough time from the chloride front t_p with the $C/C_0 = 0.5\%$[6], the non-steady-state migration coefficient M_{nc} obtained from ACMT are calculated from Eq. (2), and the results are listed in Table 2.

3.2 Electrical current in ACMT

In Figure 2, the chloride ions penetrate the specimen from upstream to downstream whereas hydroxide ions leave the sample under the electrical field effect. The electrical current was reduced when the chloride ions were in the process of migration through the saturated pores in specimen and had not reached the anode cell. In this stage, the ions (OH^- and SO_4^{2-} et.) in pore solution would migration toward the anode cell when the electrical field applied.

At the upstream interface, the chloride concentration in the sample is increasing, but the hydroxyl concentration is decreasing faster because the hydroxyl ions are more mobile. At the downstream interface, the sodium ions are replacing the potassium ions.

Moreover, because the potassium ion has greater mobility than sodium, therefore, the ionic strength at both ends of the sample decrease, decreasing the overall conductivity, leading to a reduction in the current.[7] After non-steady state, the chloride ions start to reach the downstream and causing the electrical current increase. While comparing the trend of electrical current and chloride concentration before steady state (Figure 2), it can be observed that the turning point (P_t) of electrical current-time curve in early transition period. The time corresponding to the turning point are marked as the breakthrough time for the chloride front, and the breakthrough time (t_i) basis on electrical current for all mixes are listed in Table 2.

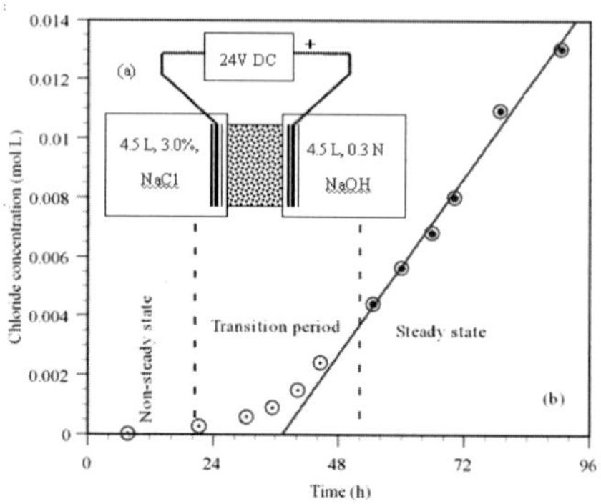

Figure 1. (a) ACMT setup, and (b) chloride concentration vs time

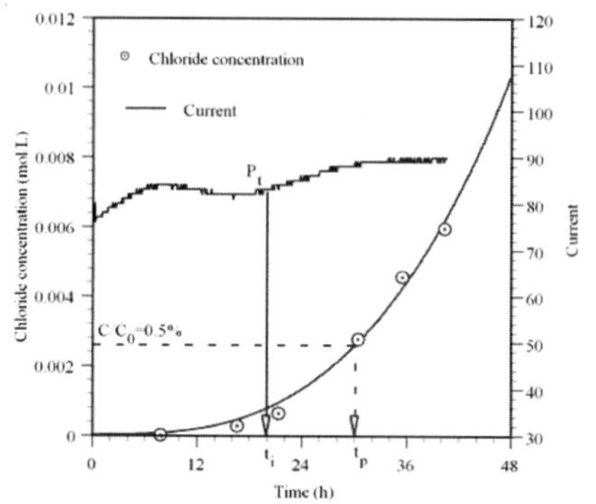

Figure 2. The evolution of current and chloride concentration

Table 2. The breakthrough times and non-steady-state migration coefficients

Mix	Breakthrough time (h)			Migration coefficient ($\times10^{-12}m^2/s$)	
	t_p	t_i		M_{nc}	M_{ni}
C35	45.9	40.4		5.03	5.72
C45	31.6	30.5		7.16	7.42
C55	27.7	19.0		9.93	14.46
C65	15.8	10.0		14.12	22.34
F35	74.0	65.9		3.32	3.72
F45	47.4	35.6		5.28	7.03
F55	30.1	19.0		8.85	14.02
F65	20.6	12.1		12.13	20.69
S35	104.2	93.9		2.25	2.49
S45	85.5	93.9		3.02	2.75
S55	60.2	50.0		3.81	4.59
S65	49.9	44.7		4.95	5.53
SF35	126.3	104.4		2.14	2.58
SF45	79.1	65.9		3.56	4.27
SF55	39.6	30.5		5.92	7.68
SF65	28.8	19.0		6.93	10.51

3.4 Comparison of diffusivity from the different measurement techniques

By using the breakthrough times t_i in Table 2 and $C/C_0 = 0.5\%$, the non-steady-state migration coefficient, M_{ni} was calculated from Eq. (2) and listed in Table 2. A comparison of the non-steady-state migration coefficient (M_{ni}) obtained from ACMT basis on electrical current and the non-steady-state migration coefficient (M_{nc}) basis on chloride concentration is presented in Figure 3. By linear regression analysis, the empirical relationship between M_{ni} and M_{nc} is statistically as:

$$M_{nc} = 0.561 M_{ni} + 1.374 \times 10^{-12}, \tag{3}$$

where M_{nc} and M_{ni} are in 1×10^{-12} m^2/s. From linear regression, the correlation coefficient (R^2) for the model (Eq. (3)) is obtained as 0.976. It can be found that a very good correlation between the M_{ni} and the M_{nc}. However, the magnitude of the non-steady-state migration coefficient from electrical current (M_{ni}) is about twice times higher than from chloride concentration (M_{nc}). In ACMT, the non-steady-state or steady-state migration coefficient obtained from chloride concentration basis on the breakthrough time or constant chloride flux have to sampling for analyzed the chloride concentration, this kind of test analysis is laborious and expensive, even time-consuming. In order to avoid the sampling and analyzing chloride concentration during the ACMT, a data logger can be used to record the electrical current and plot the current-time curve. When the turning point (P_t) is observed in current-time curve as show as in Figure 2, the time (t_i) corresponding to the turning point can be obtained. The non-steady-state migration coefficient on the basis of t_i and $C/C_0 = 0.5\%$ can be calculated from Eq. (2).

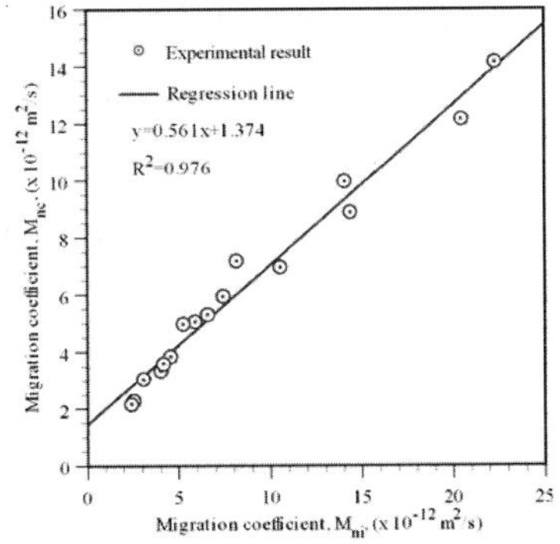

Figure 3. A comparison of the migration coefficient from current and chloride concentration

4. CONCLUSIONS

Based on the results obtained from the present experimental investigation, the following conclusions can be drawn:

1. A good correlation was observed between the non-steady-state migration coefficient M_{ni} obtained from electrical current and M_{nc} from the chloride concentra-

tion. The value of M_{ni} gave a chloride migration coefficient that was about twice higher than M_{nc}. The value of M_{ni} may vary due to different measurements and assumptions.

2. The breakthrough time (t_i) on the basis of electrical current can be used to determine the non-steady-state migration coefficient in ACMT. For a given contained mineral admixtures of concrete, measurement on basis of electrical current appear to provide a good basis both for estimation of chloride migration coefficient and for routine control of concrete quality.

Acknowledgment

The financial support of the National Science Council, ROC, under grant NSC 93-2211-E-019-017 is gratefully appreciated.

5. REFERENCES

1. Standard test method for electrical indication of concrete's ability to resist chloride ion penetration, ASTM C 1202-97, American Society for Testing and Materials, (1994).
2. L. Tong, O. E. GjØrv, Chloride diffusivity based on migration testing, Cem. Con. Res. 31(7) (2001) 973-982.
3. L. Tang, L. Nilsson, Rapid determination of the chloride diffusivity in concrete by applying an electrical field, ACI mater. J. 89(1) (1992) 49-53.
4. C. Andrade, Calculation of chloride diffusion coefficients in concrete from ionic migration measurements, Cem. Concr. Res. 23 (1993) 724-742.
5. C. C. Yang, S. W. Cho, The relationship between chloride migration rate for concrete and electrical current in steady state using the accelerated chloride migration test, Mater. Stru. 37(271) (2004) 456-463.
6. P. Halamickova, R. J. Detwiler, D. P. Bentz, E. J. Garboczi, Water permeability and chloride ion diffusion in portland cement mortars: relationship to sand content and critical pore diameter, Cem. Con. Res. 25(4) (1995) 790-802.
7. E. Samson, J. Marchand, K. A. Snyder, Calculation on ionic diffusion coefficients on the basis of migration test results, Mater. Stru. 36 (2003) 156-165.

CONCRETE REPAIR ACCORDING TO THE NEW EUROPEAN STANDARD

F. Dehn
Institute for Materials Research and Testing (MFPA Leipzig), Leipzig, Germany

Abstract: At present, the existing national standards in Europe are being converted to European regulations. This also affects repairing reinforced and prestressed concrete structures. This article will specifically concern the new repair concepts according to the European standard EN 1504 while also delving into the protection and repair of reinforced and prestressed concrete structures in general.

Key words: Concrete repair, EN 1504, reinforced concrete structures, prestressed concrete structures

1. INTRODUCTION

At present, the existing national standards in Europe are being converted to European regulations. This also affects repairing steel and prestressed concrete structures. This article will specifically concern the new repair concepts for according to the European standard EN 1504 while also delving into the protection and repair of reinforced and prestressed concrete structures in general. Due to its limited volume, this article will not go into design work such as strengthening.

This article will start off with a description of the current progress of the development of the European sets of rules and then explain the general classification of the repair principles set forth in laws and rules. It will also differentiate between traditional and more recent principles such as electrochemical principles.

2. THE CURRENT PROGRESS OF DEVELOPING THE SETS OF RULES FOR PROTECTION AND REPAIR

At present, the European EN 1504 series of standards for protecting and repairing concrete components are on the verge of completion and being subsequently introduced on the national level. The EN 1504 series of standards consist of the 10 main standards below and 60 test standards.

M.S. Konsta-Gdoutos, (ed.), Measuring, Monitoring and Modeling Concrete Properties, 619–624.

Table 1. The structure of the EN 1504 series of standards

EN 1504-1	General Scope and Definitions
prEN 1504-2	Surface Protection Systems (Concrete Coatings)
prEN 1504-3	Structural and Non-Structural Repair (Mortars)
prEN 1504-4	Structural Bonding (Steel Plates and Fibre Reinforced Polymers)
prEN 1504-5	Concrete Injection
prEN 1504-6	Grouting to Anchor Reinforcement or to Fill External Voids
prEN 1504-7	Reinforcement Corrosion Protection
prEN 1504-8	Quality Control and Evaluation of Conformity
ENV 1504-9	General Principles for the Use of Products and Systems
prEN 1504-10	Site Application of Products and Systems and Quality Control of the Works

Parts 2 through 7 are what are known as "Harmonised Product Standards" including the rules for CE labelling protection and repair products that may be brought into circulation in Europe. They are binding on all European countries.

The exact schedule for introducing them is still under discussion since Germany is targeting a package solution for introducing parts 2 through 7. That would mean that all parts of the series of standards would be introduced at the same time as soon as the last part is published. Since not all parts of the substandards are published simultaneously, it would also be conceivable to introduce them one after another over a period of time of 1 to 2 years. The latter scenario would have the drawback that the groups of substances of surface protection, mortar, concrete injections and corrosion protection of reinforcement could not be introduced simultaneously.

The single states are responsible for using the products labelled with CE. The latest from the discussion is that the single parts of EN 1504 should be handled as follows:

It is presently being discussed to apply the existing sets of rules “Protecting and Repairing Concrete Components” of the DAfStb (German Committee for Reinforced Concrete) in the version revised in October 2001 or the harmonised ZTV-ING of the BMVBW (Federal Ministry for Transportation, Building and Housing), particularly for bridge structures so that the existing repair concepts would essentially be retained after introducing the European EN 1504 series of standards while the primary changes would be in the product designation and their quality assurance systems.

3. PRINCIPLES OF REPAIR ACCORDING TO THE NATIONAL AND EUROPEAN SETS OF RULES

The electrochemical principles of how steel corrodes in concrete form the basis for defining the repair principles of DafStb's presently applicable directive of “Protecting and Repairing Concrete Components”. As is generally known, this consists of three subprocesses, namely anodic iron dissolution, cathodic oxygen reduction and the electrolytic subprocess. If only one of these subprocesses is prevented, corrosion comes to a halt. This can be done in the following fashion:

Preventing the anodic subreaction

This can be done in a variety of different ways. One possibility is restoring the alkaline milieu in the environment of the reinforcement. A second possibility is forcing the reinforcement to function cathodically in a closed control circuit (cathodic corrosion protection). A third possibility is separating the electrolyte from the steel with an effective steel coating, thus preventing the anodic subprocess.

Preventing the cathodic subreaction

Under real-life situations, the cathodic subprocesses can only be prevented in rare cases so that the DAfStb guideline does not envision this option as a principle of repair.

Preventing the electrolytic subprocesses

The corrosion rate can be reduced to virtually negligent values by lowering the water content in the concrete since all transport processes are inhibited in the concrete.

This gives rise to the following basic principles of corrosion protection:

- **R** Restoring active corrosion protection by repassivating the reinforcement or permanently realkalinising the concrete in the environment of the reinforcement.
- **W** Lowering the water content to values which ensure that the electrolytic subprocess is prevented to the extent that the further corrosion rate is reduced to a non-damaging extent.
- **C** Coating steel surfaces to prevent the anodic (and cathodic) subprocess in the area of the repaired steel surfaces.
- **K** Cathodic corrosion protection to force the reinforcement to act exclusively or mostly cathodically.

Table 2. The present progress of the envisioned classification of the parts of the standard for applications subject to building supervision in Germany

Part of EN 1504	Abbreviations	How planned for introduction in Gemany
prEN 1504-1	Definitions	-
prEN 1504-2	Surface Protection	+ German application standard
prEN 1504-3	Mortar/Concretes	+ German application standard
prEN 1504-4	Adhesives	Technical Approval
prEN 1504-5	Injection	+ German application standard
prEN 1504-6	Anchoring Mortar	Technical Approvals
prEN 1504-7	Steel Coating	+ German application standard
prEN 1504-8	Quality Assurance	-
ENV 1504-9	Planning Principles	+ German application standard
prEN 1504-10	Execution	+ German application standard

The EN 1504 series of standards add more principles for corrosion protection of reinforcement and the range of principles to ensure concrete durability to the range of repair principles according to the German guideline. Part 9 of EN 1504 gives a total of 11 repair principles that are classified according to concrete protection (principles 1-6, refer to Table 3) and reinforcement protection (principles 7-11, refer to Table 4):

There are a variety of methods for the repair principles specified in Tables 3 and 4 that are compiled in Tables 5-12 as per EN 1504-9. For the sake of completeness and clarity, we also specified the methods not governed by the EN 1504 series of standards as well as methods for which there are not any tested methods at present such as Method 9.1 to cover all potential methods in future.

Table 3. The repair principles for concrete as per EN 1504, part 9

Number of principle	Abbreviation
Principle 1 [IP]	Protection against ingress of substances
Principle 2 [MC]	Moisture Control (regulation of concrete's water balance)
Principle 3 [CR]	Concrete Restoration
Principle 4 [SS]	Structural Strengthening
Principle 5 [PR]	Physical Resistance
Principle 6 [RC]	Resistance to Chemical

Table 4. The repair principles for reinforcement as per EN 1504, part 9

Number of principle	abbreviation
Principle 7 [RP]	Preserving or restoring passivity
Principle 8 [IR]	Increasing Resistivity
Principle 9 [CC]	Cathodic Control
Principle 10 [CP]	Cathodic Protection
Principle 11 [CA]	Control of Anodic Areas

Tables 5-12 show the diversity of potential repair concepts very strikingly. EN 12696 governs cathodic corrosion protection of steel in concrete while there is a draft of EN 14038 available for electrochemical realkalinisation and chloride extraction handling for reinforced concrete anticipated for publication in two parts.

4. SUMMARY

No restrictions are likely in the potential repair conceptions when changing over the sets of rules for protecting and repairing structural concrete components. On the contrary, this will even make an extensive range of new conceptions possible.

There are interesting and innovative conceptions for repairing reinforced and prestressed concrete structures such as the new electrochemical methods. But there are also a series of other innovations such as in the strengthening of existing concrete structures.

Table 5. Methods for the IP principle of repair

Principle			Methods for enacting the principle
Number	Abbr.	Definition	
1	IP	**"Ingress Protection"**	1.1 Impregnating (sealing): capillary suctioning of liquid products that harden in the pore system and block it
			1.2 Film-forming coating with or without crack bridging effect
			1.3 Local crack bridging
			1.4 Filling cracks
			1.5 Transforming cracks in joints plate lining membrane sealing

Table 6. Methods for the MC principle of repair

Principle			Methods for enacting the principle
Number	Abbr.	Definition	
2	MC	**"Moisture Control"**	2.1 Hydrophobising 2.2 Film-forming coating 2.3 External lining 2.4 Electrochemical treatment (applying a potential difference in certain structural component zones to reinforce or weaken water diffusion)

Table 7. Methods for the CR principle of repair

Principle			Methods for enacting the principle
Number	Abbr.	Definition	
3	CR	**"Concrete Restoration"**	3.1 Applying mortar by hand 3.2 Placing of concrete as per EN 206 3.3 Applying shotcrete 3.4 Replacing structural components

Table 8. Methods for the SS principle of repair

Principle			Methods for enacting the principle
Number	Abbr.	Definition	
4	SS	**"Structural Strengthening"**	4.1 Replacing or supplementing built-in or external reinforcement 4.2 Inserting reinforcement into drilled holes or milled slits 4.3 Adhering brackets made of steel or fibre-reinforced plastic 4.4 Increasing the component thickness by concreting 4.5 Injecting cracks and defects 4.6 Filling cracks and defects 4.7 Arranging external tendons

Table 9. Methods for the PR and RC principle of repair

Principle			Methods for enacting the principle
Number	Abbr.	Definition	
5	PR	**"Physical Resistance"**	5.1 Plastering finishes or coatings 5.2 Strengthening impregnation
6	RC	**"Resistance to Chemicals"**	6.1 Plastering finishes or coatings 6.2 Pore-filling impregnation

Table 10. Methods for the PR principle of repair

Principle			Methods for enacting the principle	
Number	Abbr.	Definition		
7	RP	**"Preserving or Restoring Passivity"**	7.1	Increasing concrete cover
			7.2	Replacing carbonated concrete or pollutant-containing concrete
			7.3	Electrochemical realkalinisation of carbonated concrete
			7.4	Realkalinisation of carbonated concrete by diffusion from alkaline zones
			7.5	Electrochemical chloride extraction

Table 11. Methods for the IR and CC principle of repair

Principle			Methods for enacting the principle	
Number	Abbr.	Definition		
8	IR	**"Increasing Resistivity"**	8.1	Lowering the concrete's moisture content of concrete (by coating or lining)

Table 11. Methods for the IR and CC principle of repair (continued)

Principle			Methods for enacting the principle	
Number	Abbr.	Definition		
9	CC	**"Cathodic Control"**	9.1	Limiting the oxygen content in the potential cathodic zones to a non-damaging extent (by water saturation or coating)

Table 12. Methods for the CP and CA principle of maintenance

Principle			Methods for enacting the principle	
Number	Abbr.	Definition		
10	CP	**"Cathodic Protection"**	10.1	Generating suitable electrical potential difference in the structural component lining)

RETROFIT OF CONCRETE MEMBERS WITH EXTERNALLY BONDED PREFABRICATED SFRCC JACKETS

A. Ilki, D. Akgun, O. Goray, C. Demir and N. Kumbasar
Istanbul Technical University, Civil Engineering Faculty, 34469, Istanbul, Turkey

Abstract: Many existing reinforced concrete structures, built before recent seismic design codes, can not exhibit a ductile behavior during earthquakes, and may experience severe structural damage. The main reason of nonductile behavior is insufficient and inadequately detailed transverse reinforcement. In developing countries, concrete quality of structural members is not sufficient as well. External confinement of structural members can overcome this deficiency. In this study, a retrofit technique in terms of external confinement of concrete members by using prefabricated panels of steel fiber reinforced cementitious composites (SFRCC) is investigated. For this purpose, concrete members with rectangular cross-section were externally confined by using prefabricated SFRCC panels and tested under axial loads. The concrete compressive strength of original specimens was around 10 MPa, while the compressive and tensile strengths of the prefabricated SFRCC panels were around 100 and 14 MPa, respectively. Test results showed that external confinement with prefabricated panels provided significant enhancement in ductility, as well as compressive strength enhancement. The investigated technique seems to be a promising method due to ease in application, relatively less hindrance to the building, high efficiency in the case of non-circular members, better quality control due to prefabrication and relatively less cost with respect to its alternatives.

Key words: compression; concrete; ductility; fiber; prefabrication; retrofit; seismic.

1. INTRODUCTION

Insufficient ductility and low quality concrete are among the most common deficiencies of relatively older existing reinforced concrete structures, particularly in the developing countries in seismic areas. There are also many structures, which were designed only considering vertical loads. Thus, many of these structures are weak in terms of lateral strength, stiffness and ductility requirements of the recent seismic design codes. Therefore, retrofit of these types of structures may be necessary for reduction of losses

M.S. Konsta-Gdoutos, (ed.), Measuring, Monitoring and Modeling Concrete Properties, 625–632.

after earthquakes. To overcome these problems, steel or concrete jacketing and external confinement with fiber reinforced polymer (FRP) composite sheets are the most preferred methods. Concrete or steel jacketing can hinder the functions of the structures during construction, as well as causing losses in space. FRP composites may not be feasible for structural members with rectangular cross-sections of relatively higher cross-sectional aspect ratios due to insufficient flexural stiffness, as well as being relatively more expensive. Consequently, there is a need for research on economical and occupant friendly member retrofitting techniques. SFRCCs can be used when high tensile strength and reduced cracking are required. The fracture energy of these materials is much higher than ordinary concrete[1]. Three decades of research has enabled the production of SFRCCs with good strength, stiffness, crack control, toughness and energy absorption capacity[2]. Besides these advantages, low workability of SFRCC is an obstacle for many applications on site, including utilization of SRFCCs in structural retrofitting. To overcome the difficulty of low workability, the idea of prefabrication of SFRCC was first utilized by Alaee[3] and Alaee and Karihaloo[4] in terms of beam retrofit. For column seismic retrofitting, the prefabricated SFRCC panels were first used by Ilki et al.[5] in terms of external confinement for improving ductility, to the best knowledge of the authors. Details of other studies related with retrofitting using cast-in-place steel fiber reinforced cementitious composites can be found elsewhere[6-10].

In this study, twelve specimens were tested under concentric compression to investigate the effect of external confinement on the axial behavior of low strength concrete members. External confinement was applied in terms of prefabricated SFRCC panels. The main test parameters were the cross-section shape of the original specimens and the thickness of the prefabricated SFRCC jackets. Test results showed that both ductility and strength of the specimens increased remarkably due to introduction of SFRCC jackets.

2. EXPERIMENTAL DETAILS

2.1 Outline of the specimens and testing setup

Twelve specimens with rectangular cross-section were tested under compression using a 5000 kN capacity Amsler machine. The deformations of the specimens were measured by displacement transducers and strain gages in both vertical and horizontal directions. The details of the specimens are given in Figure 1 and Table 1. The production phases of the specimens are given in Figure 2. In Table 1, b and h are width and depth of the cross-section, t is the thickness of the SFRCC jacket, j is the age of the original specimens at the day of testing, $f'_{c,j}$ is the standard cylinder compressive strength at the day of testing, $f'_{co,j}$ and $f'_{cc,j}$ are the member compressive strengths at the testing day for original and retrofitted specimens, respectively. $f_{SFRCC,k}$ and $f_{t,SFRC,k}$ represent compressive and tensile strengths of the SFRCC panels at the day of testing and k is the age of panels. The height of the SFRCC panels were intentionally designed less than the height of the specimens to prevent direct axial loading of these panels. The main purpose of the retrofitting is not increasing the axial capacity by enlarging the cross-section, but increasing ductility by providing confinement.

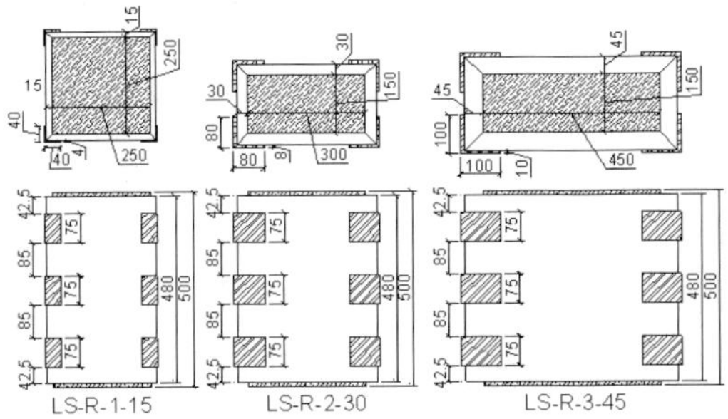

Figure 1. General appearance of the specimens

Table 1. Details of the specimens

Specimen	b	h	t	j	$f_{c,j}$	$f_{co,j}$	f_{cc}	k	$f_{SFRCC,k}$	$f_{t,SFRCC,k}$	Steel
	cm	cm	mm	day	MPa	MPa	MPa	day	MPa	MPa	corner
LS-R-1-0	25	25	0	106	10.3	8.7	-	-	–	–	-
LS-R-1-15	25	25	15	319	10.9	-	11.6	56	111.3	13.7	40.40.4
LS-R-1-30	25	25	30	315	10.9	-	11.9	58	111.4	13.8	80.80.8
LS-R-1-15-NSC	25	25	15	321	10.9	-	11.0	52	110.9	13.7	40.40.4
LS-R-2-0	15	30	0	106	10.3	9.3	-	-	–	–	-
LS-R-2-15	15	30	15	315	10.9	-	12.9	58	111.4	13.8	40.40.4
LS-R-2-30	15	30	30	321	10.9	-	13.7	52	110.9	13.7	80.80.8
LS-R-2-30-C	15	30	30	325	10.9	-	13.7	49	110.7	13.6	80.80.8
LS-R-3-0	15	45	0	180	10.5	10.0	-	-	–	–	-
LS-R-3-15	15	45	15	327	10.9	-	12.4	47	110.5	13.6	40.40.4
LS-R-3-30	15	45	30	321	10.9	-	12.3	57	111.3	13.7	80.80.8
LS-R-3-45	15	45	45	165	10.5	-	11.1	83	93.8	8.7	100.100.10

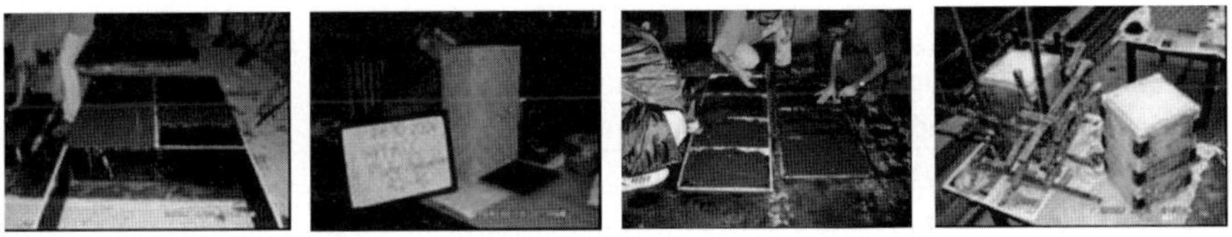

Figure 2. Production phases of the specimens

2.2 Material characteristics

The specimens were cast using the mix-proportion given in Table 2, intentionally for obtaining low strength concrete. The standard cylinder (150×300 mm) compression tests were carried out at the ages of 28, 90, 180 and 360 days, Figure 3. Note that elasticity modulus of concrete was between 14000 and 14700 MPa between the ages of 28 days to 360 days, and member tests were carried out between the ages of 106 and 327 days.

Table 2. Mix-proportion of low strength (LS) concrete (kg/m^3)

Mix	Cement	Sand	Water	Crushed stone
LS	150	923	210	1074

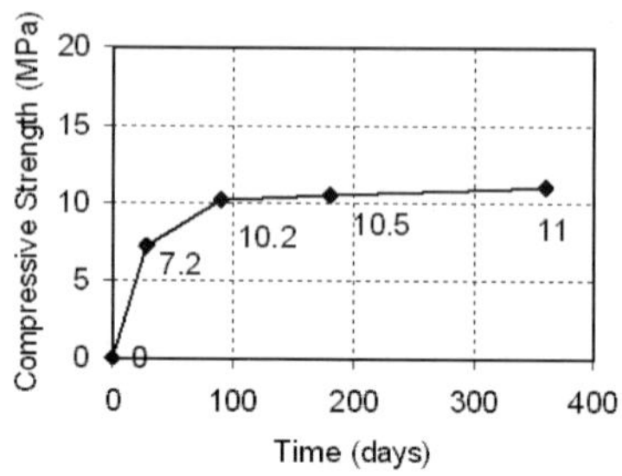

Figure 3. Compressive strength-time relationship for low strength concrete

The mix-proportion of SFRCC is given in Table 3, for Mix I and Mix II. In mixtures, ordinary Portland cement (Lafarge) with 28 days compressive strength of 42.5 MPa was used. Microsilica was produced by Elkem Materials with a mean particle size smaller than 500 μm, and the specific gravity was 2.3 kg/dm^3. The silica sand was produced by Sisecam in Turkey, and the particle size was between 62.5 and 250 μm. As admixture, Glenium 51 hyperplasticizer was used. Fibers were Dramix ZP 305, with diameter of 0.55 mm and length of 30 mm, (aspect ratio:55). The density and tensile strength of fibers were 7.85 kg/dm^3 and 1100 MPa, respectively. Note that all the specimens were cast using Mix II, except LS-R-3-45, which was cast using Mix I. The average standard cylinder compressive strengths of Mix I and II are given for different ages in Figure 4. The tensile strengths of SFRCC determined by splitting tests are given in Table 4. All material characteristics were determined as averages of at least three specimens.

Table 3. Mix-proportions of SFRCC mixes I and II (kg/m^3)

Mix	Cement	Silica sand I	Silica sand II	Sand (≤1mm)	Silica fume	Water	Admixture	Steel fiber
Mix I	981	392	392	-	196	245	29	157
Mix II	928	278	278	278	186	204	33	314

3. TEST RESULTS

Test results are mainly presented with axial stress-axial strain relationships, (Figure 5). It should be noted that the vertical axis is given as dimensionless axial stress, determined as the ratio of axial stress to the axial strength of original (unretrofitted) specimen of the corresponding comparison. As seen in these relationships, both ductility and axial strength increased remarkably, due to the presence of the SFRCC jackets.

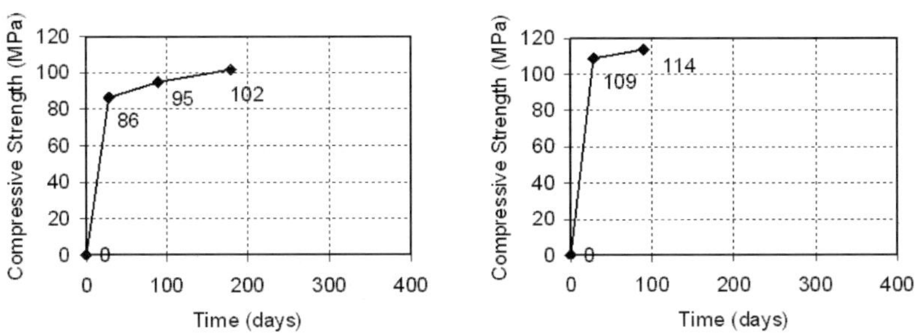

Figure 4. Compressive strength-time relationship for SFRCC mixes

Table 4. Splitting tensile strength of SFRCC mixes at 28 and 90 days of age (MPa)

SFRCC	28 days	90 days
Mix I	8.6	8.8
Mix II	13.3	14.2

The passive lateral pressure exerted by the SFRCC jackets after the specimens deformations in axial and transverse directions, changes the uniaxial stress state to a triaxial stress state, which in turn provides enhancement in ductility and axial strength. The specimen, which was jacketed with only SFRCC panels (LS-R-1-15-NSC), without any additional steel corners around the corners of the section, did not perform well due to premature separation of the SRFCC panels around one of the corners. Both ductility and strength enhancements are more pronounced in the case of square specimens and rectangular specimens with the cross-sectional aspect ratio of 2. As expected, when the cross-sectional aspect ratio is larger, the enhancement in ductility and strength is less due to reduced flexural stiffness of the SFRCC panels as a result of higher cross-sectional dimensions. As seen in Figure 6, repeated loading did not negatively influenced the behavior; the axial stress-axial strain relationships of specimens LS-R-2-30, which was tested under monotonic loads, and LS-R-2-30-C, which was tested under repeated axial loads, are almost identical. For observing the effect of the external confinement with SFRCC panels on toughness, the areas under stress-strain relationships were calculated for square specimens. As seen in Figure 7, the areas under the curves increase significantly due to presence of SRFCC panels, as well as achieving relatively higher axial strains without a drastic loss of strength.

Damages of some specimens are presented in Figure 8. As seen, with the exception of the specimen LS-R-1-15-NSC, which failed prematurely due to separation of panels around the corners, all specimens failed with a major vertical crack right after the steel corners. This also shows that steel corners are effective to resist concentrated tensile stresses around the corners.

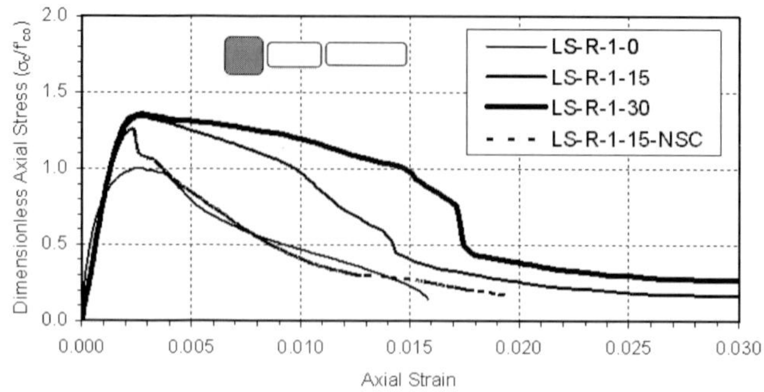

(a)

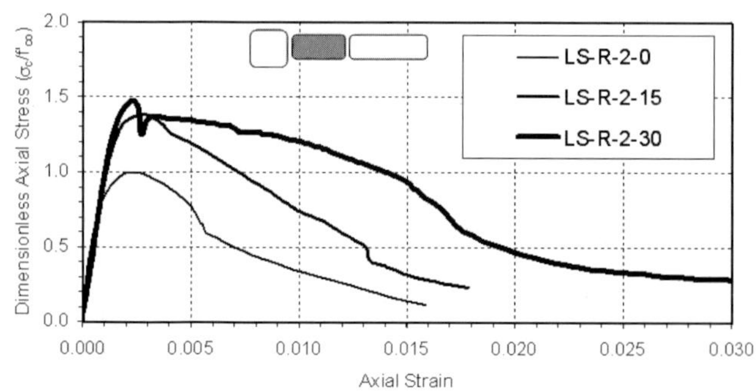

(b)

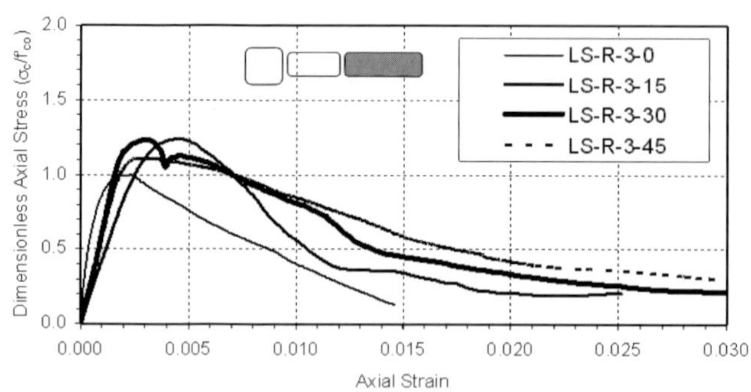

(c)

Figure 5. Stress-strain relationships for LS-R-1, LS-R-2 and LS-R-3 type specimens

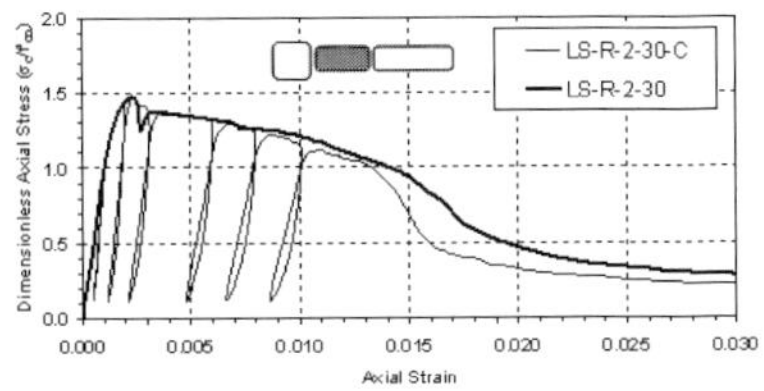

Figure 6. Stress-strain relationships of LS-R-2-30-C and LS-R-2-30

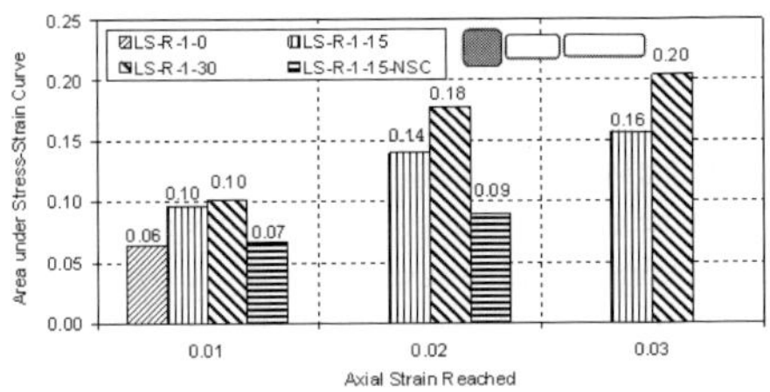

Figure 7. Areas under stress-strain relationships for square specimens

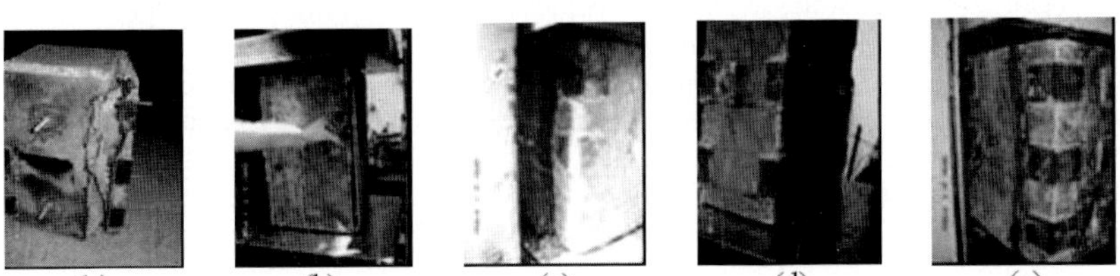

Figure 8. (a) LS-R-1-15, (b) LS-R-1-15-NSC, (c) LS-R-1-30, (d) LS-R-2-15, (e) LS-R-2-30

4. CONCLUSIONS

There are many existing reinforced concrete structures, which suffer from insufficient ductility. The utilization of conventional techniques like concrete or steel jacketing are not occupant friendly, while fiber reinforced polymer composite wrapping is relatively expensive and ineffective in the cases of rectangular cross-section. In this study, prefabricated SFRCC panels are used for external jacketing of concrete members. Test results showed that significantly enhanced ductility is possible by the introduction of SFRCC panels, as well as increased axial strength. The method is both occupant friendly and economical with respect to available techniques, which are used for enhancing the ductility of existing structural members.

Acknowledgments

The financial support of Istanbul Technical University, Beksa, Yapkim Degussa, Sise Cam and Akcansa Companies, and the valuable comments of Prof. M.A. Tasdemir, C. Sengul, and B. Akcay are gratefully acknowledged.

5. REFERENCES

1. F. Bayramov, A. Ilki, C. Tasdemir, M.A. Tasdemir, and M. Yerlikaya, An optimum design for steel fiber reinforced concretes under cyclic loading, in: *Proc. of 5th Int. Conf. on Fracture Mech. of Concrete and Concrete Structures*, (Colorado, 2004), pp. 1121-1128.
2. M.K. Lee, R.J. Lark, and B.I.G. Barr, A State of the Art Review on HPFRCC, Report for Subtask 15, Sustainable Advanced Materials for Road Infrastructure (SAMARIS), 2003.
3. F.J. Alaee, *Retrofitting of Concrete Structures Using High Performance Fibre Reinforced Cementitious Composite (HPFRCC)*, (PhD Thesis, Cardiff University, Cardiff, 2002).
4. F.J. Alaee, and B.L. Karihaloo, Retrofitting of reinforced concrete beams with CARDIFRC, *Journal of Composites for Construction* **7**(3), 174-186 (2003).
5. A. Ilki, E. Yilmaz, C. Demir, and N. Kumbasar, Prefabricated SFRC jackets for seismic retrofit of non-ductile reinforced concrete columns, in: *Proc. of 13th World Conf. on Earthquake Engineering*, (Vancouver, 2004), on CD-ROM.
6. N. Krstulovic-Opara, and M.J. Shannag, Slurry infiltrated mat concrete (SIMCON)-based shear retrofit of reinforced concrete members, *ACI Struct. Journal* **96**(1), 105-114 (1999).
7. M.J. Shannag, S. Barakat, F. Jaber, Structural repair of shear-deficient reinforced concrete beams using SIFCON, *Magazine of Concrete Research* **53**(6), 391-403 (2001).
8. M.J. Shannag, S. Barakat, M. Abdul-Kareem, Cyclic behavior of HPFRC-repaired reinforced concrete interior beam-column joints, *Mat. and Structures* **35**, 348-356 (2002).
9. E. Dogan, and N. Krstulovic-Opara, Seismic retrofit with continuous slurry-infiltrated mat concrete jackets, *ACI Structural Journal* **100**(6), 713-722 (2003).
10. M.H. Harajli, and A.A. Rteil, Effect of confinement using fiber-reinforced polymer or fiber-reinforced concrete on seismic performance of gravity load-designed columns, *ACI Structural Journal* **101**(1), 47-56 (2004).

INTERNAL STRESS AND CRACKING IN STONE AND MASONRY

G.W. Scherer
Princeton University, Civil & Env. Eng./PRISM, Eng. Quad. E-319, Princeton, NJ 08544 USA

Abstract: Internal stresses arise during weathering of stone and masonry as a result of crystallization of salts and ice, swelling of clay inclusions, and thermal expansion, among other causes. In this paper, we review the origin of the stresses and examine the unresolved questions regarding several of these mechanisms.

Key words: crystallization pressure, frost, salt, ice, clay, swelling pressure, stress, fracture

1. INTRODUCTION

Several phenomena are known to contribute to stress and damage during weathering of stone and masonry, but the details of the mechanisms by which they act are not clear. In this paper, we examine several important weathering processes (thermal expansion of calcite, freeze/thaw cycles, salt crystallization, swelling of clay inclusions), reviewing what is understood and identifying aspects that remain to be resolved. Only by understanding these mechanisms in detail can we hope to develop methods for preventing, rather than repairing, damage from weathering.

2. CALCITE THERMAL EXPANSION

The low porosity of marble makes it relatively resistant to weathering mechanisms other than acid rain.[1,2,3] However, the porosity of marble can increase after heating/cooling cycles, owing to the peculiar thermal expansion behavior of the constituent crystals of calcite (calcium carbonate): they expand along two crystallographic directions, but contract along the third as the temperature increases.[4] As a result, each of the tightly packed grains in a sample of marble changes its *shape*, not only its size, during thermal cycles. Thermal expansion stresses tend to open cracks along grain boundaries, leading to an increase in porosity, as shown in Figure 1. Opening of grain boundaries not only makes marble susceptible to invasion of moisture, it also causes expansion that can lead to warping of a thin plate of stone, if the grain boundary damage only occurs on one face.

M.S. Konsta-Gdoutos, (ed.), Measuring, Monitoring and Modeling Concrete Properties, 633–641.

Warping of marble cladding has occurred on several notable buildings in recent years, requiring costly replacement. The cause of this phenomenon is controversial. Winkler[2] claims that it results from relaxation of residual geological stresses that are released when the stone is quarried, leading to slow creep. If this were so, it would be surprising to see all of the slabs on a building warp in the same direction, since there would be no control of their orientation when they are mounted on the building. It could be caused by weathering of one side of a thin slab that causes opening of grain boundaries. If expansion from weathering were the cause, then the slabs would be expected to warp in the same direction, as is generally observed; however, they should become convex as cracking of grain boundaries expands the outer surface, and that is not always so. In the case of the Finlandia Hall,[4] the original slabs became concave outward, as if the protected side had expanded. This could indicate that the warping results from mechanical constraint: if a plate expands uniformly, but if constrained near an edge, it will bow toward the opposite edge, as indicated in Figure 2.

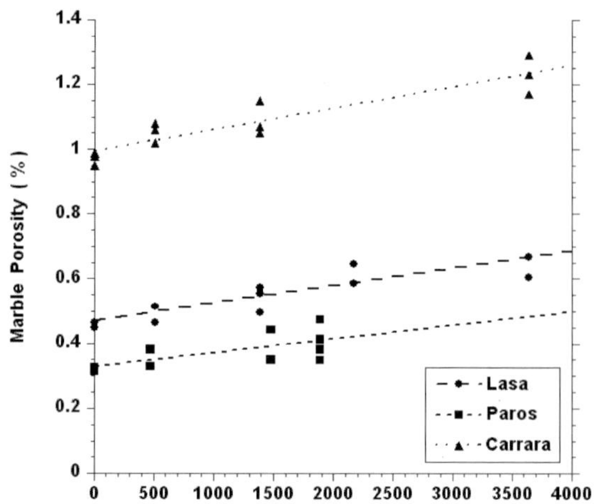

Figure 1. Change in porosity versus number of cycles between +35C and -15C for three types of marble. Data from Table. 3, p. 13 of ref. 5

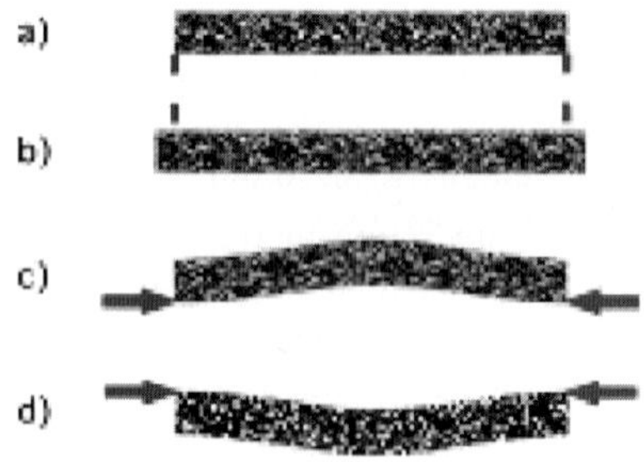

Figure 2. Warping of marble by constraint: a) original size; b) after uniform expansion; c) expansion constrained by clamping at lower edge; d) expansion constrained by clamping at upper edge

3. FREEZE/THAW DAMAGE

Damage from freezing is commonly thought to result from the 9% increase in volume as water transforms into ice. The volume change can be highly destructive, if the liquid is entirely confined (as when a bottle of milk is forgotten in the freezer). A small object, such as a sculpture, might be chilled uniformly over its exterior surface, so that ice entraps the interior liquid. However, even in this extreme case, damage will result only if the saturation of the pores exceeds 91%; otherwise, the volume expansion can be accommodated by the vacant pores. A larger object, such as the wall of a monument, usually freezes from one direction (*viz*., the exterior), so the ice pushes the water ahead. On the pore scale, it is possible that ice propagating through the pore network of a stone will surround small pockets whose entries are very small; in that case, subsequent freezing would generate local stresses and damage.[6] It is not clear how important this mechanism is, but it could be quite significant when the freezing solution contains salt: since the salt cannot be incorporated into the ice crystal, the unfrozen brine becomes increasingly concentrated, and droplets of brine might become surrounded by ice. As the temperature drops, ice formation within the brine droplet would result in volume expansion that would exert stress on the surrounding stone.

Another mechanism for generation of stress during freezing is hydraulic pressure, as the volume change of the ice forces displacement of the adjacent liquid. Pressure builds up as the liquid is forced through the pores, and the pressure can be destructive if the growth is fast and the permeability of the body is low.[7] The pores in stone are generally too large to permit large stresses to arise in this way. Moreover, even when the pores are small (as in concrete), the rate of growth of ice rapidly decreases owing to the release of heat of fusion, which raises the temperature of the water/ice interface. The rate of growth is eventually controlled by the diffusion of heat, so it decreases in proportion to $t^{-1/2}$, with the result that the rate of displacement of the water is not sufficient to cause high stresses.

Frost damage can result from crystallization pressure, which is the stress exerted on the pore walls by the growing crystal.[8,9,10] Although damage from growing crystals is widely recognized, some authors express puzzlement as to how a crystal can exert stress on a pore wall once it has made contact.[11] The answer, probably first recognized by Taber,[12] is that there is a thin film of liquid that surrounds the crystal, so that material is always available to permit growth (see Figure 3). The reason that growth is inhibited is that there is a repulsive pressure (or, disjoining force) between the crystal and the pore wall that is exerted across the liquid film.[13] The disjoining pressure between ice and minerals has at least two sources. First, the van der Waals forces are repulsive in this case, because the Hamaker constants decrease in order from ice to water to mineral,[14] so that a pressure of tens of MPa would be required to push ice into direct contact with most mineral.[9] Second, water molecules pack with a different structure in the vicinity of an interface,[14,15] and this structure is probably not the same adjacent to ice and mineral surfaces; therefore, the liquid structure is disrupted when ice approaches a mineral and the energy of the system consequently rises. Analyses based on equilibrium thermodynamics indicate that crystallization stresses in the megapascal range (comparable to the tensile strength of stone) are expected when crystals grow in pores with diameters below 100 nm. Such pores are rare in stone, so this mechanism does not seem to account for the observed damage.

To understand frost damage in stone, it might be necessary to assume non-equilibrium conditions. For example, consider the reasonable case in which an ice crystal extends through the pores of a stone over a volume of tens or hundreds of microns; a layer (perhaps a nanometer or so) of liquid surrounds the ice, owing to disjoining forces. When the temperature drops, the crystal tends to grow, so it pushes toward the wall against the disjoining pressure. Under equilibrium conditions, the chemical potential in the liquid film will be uniform, so the molecules of ice near the pore wall (W in Figure 3) and near the pore entry (E), which are in contact with the liquid, must also have the same potential. If the pore entry diameter is assumed to be in the micron range (so that its curvature has a negligible effect on the melting point of the crystal), then the pressure on the ice at E is negligible; ice near W will melt, and the molecules will be transferred to E as the temperature drops, so that the internal pressure of the crystal remains constant. However, to establish this equilibrium requires transport of molecules through a nanometric film over a distance equal to the size of the ice crystal. If this process is slow compared to the rate of cooling, then high transient stresses could develop. The characteristic time, t_e, for equilibration over a distance x by diffusion is approximately $t_e \approx x^2/D$, where D is the diffusion coefficient. For bulk water, $D \approx 2 \times 10^{-9}$ m^2/s, but the diffusivity is reported to be reduced by a factor of $\sim 10^3$ in films between mineral grains and $\sim 10^2$ in a thin film on a silica surface so we estimate $D \approx 10^{-11}$ - 10^{-12} m^2/s in the water film around the ice. For diffusion through the film to equilibrate the pressure in an ice crystal with $x = 10$ μm would only take $t_e \approx 10$-100 s, so it should easily remain in equilibrium with natural decreases in temperature (~5°C/h). However, if the isolated region of ice were 1 mm in diameter, the equilibration time would rise to the order of a day, so substantial stresses could be exerted on the pore wall during the period of equilibration of the ice. This phenomenon might cause damage on a very large scale, when ice is trapped in fissures in natural rocks, as well as creating cracks on the millimeter or centimeter scale in monuments.

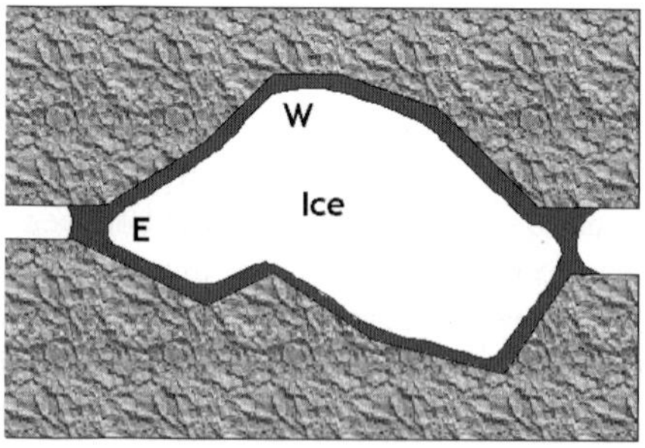

Figure 3. Isolated macroscopic crystal of ice, only in contact with surrounding film of water. As the temperature drops, it exerts high stress on the surrounding stone, limited only by the disjoining pressure in the film. The chemical potentials of ice and water must be equal near the pore entry (E) and near the pore wall (W)

4. SALT CRYSTALLIZATION

Salt crystals are also separated from the pore wall by a film of solution,[12] but the van der Waals forces are attractive in this case, so the repulsion must arise from electrostatic foreces or ordering of the water and ions against the solid surfaces. As in the case of ice, theories based on equilibrium thermodynamics[9] predict megapascal stresses only in pores smaller than ~100 nm. However, under nonequilibrium conditions, it is possible for an isolated macroscopic crystal of salt to exert high stresses.[18,19]

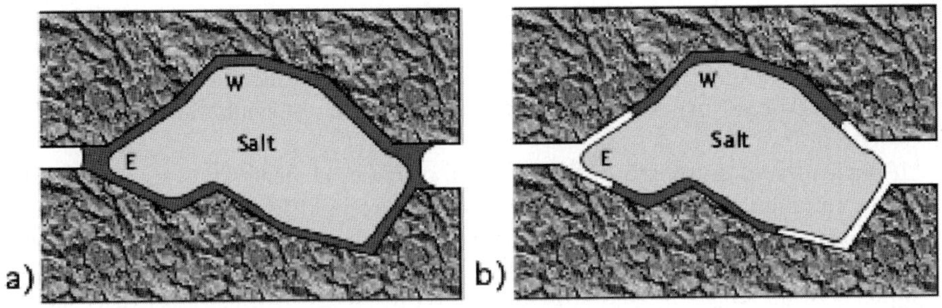

Figure 4. Macroscopic crystal of salt: a) surrounded by solution or b) partially dried

Figure 4a shows a macroscopic salt crystal surrounded by a solution. If the supersaturation were to increase owing to a drop in temperature, the situation would be similar to that in Figure 3. Under equilibrium conditions, growth would occur near E to consume the supersaturation, and the ions in the film surrounding the crystal would diffuse to the site of growth; no significant stress would exist in this case. However, if the crystal were too large to permit the diffusion process to keep up with the rate of increase of supersaturation (corresponding to the rate of decrease in temperature), then high crystallization pressure could be exerted on the pore wall during the equilibration period.

The distribution of salt in the pores of a stone is controlled by several independent kinetic processes: supply of water (say, by rising damp or a leaky roof), evaporation, diffusion of dissolved ions away from the site of evaporation (*i.e.*, toward the source of invading water), and nucleation and growth of crystals. As water evaporates from a body with a pore size distribution, the larger pores empty first, while the smaller ones remain full by using their greater capillary suction to drain liquid from their larger neighbors. At equilibrium, the order of drainage during drying is identical to the order of invasion of the pores in a mercury intrusion experiment, since both processes involve the intrusion of a non-wetting fluid (*viz.,* air during drying and Hg during porosimetry). Therefore, the increasingly concentrated salt solution retreats into smaller and smaller pores in a predictable way. If the rate of evaporation is fast compared to the rate of redistribution of water among neighboring pores, then the large and small pores may empty almost simultaneously, but that is not expected for stones, owing to their high permeability. When the concentration of the remaining fluid (which, in the assumed state of equilibrium, has a uniform composition) reaches the solubility limit, macroscopic crystals become stable. However, for crystals to form in a pore with radius r, the solubility product, Q, must satisfy the Ostwald-Freundlich equation,[21]

$$\frac{2\gamma_{CL}}{r-\delta} = \frac{RT}{V_C} \ln\left(\frac{Q}{K}\right) \tag{1}$$

where γ_{CL} is the crystal/liquid interfacial energy, δ is the thickness of the film of solution between the crystal and the pore wall, R is the ideal gas constant, T is absolute temperature, V_C is the molar volume of the crystal, and K is the equilibrium solubility product. If the solution initially filling the pores is dilute, then a great deal of evaporation must occur to raise the concentration to the solubility limit; at that point, the solution will have retreated into relatively small pores, so a high Q may be needed to initiate crystallization. On the other hand, if the initial solution is nearly saturated, then precipitation may occur in the largest pores after very little evaporation has occurred. The stress generated during this process of evaporation at equilibrium has recently been analyzed by Coussy.[22]

If nucleation of crystals is difficult, then a different sequence of events is expected. The concentration will continue to rise until the threshold for nucleation is exceeded, whereupon crystals will form in all the liquid-filled pores. Crystals in the smallest pores will eventually have their growth arrested by the pore walls, while those in larger pores continue to grow. Since the solubility of a crystal increases as its size decreases, crystals that form in the smaller pores tend to dissolve as the solute diffuses to the larger crystals; this process of Ostwald ripening continues until the salt is transferred to the largest accessible (*i.e.*, solution-filled) pores[19]. The equilibrium state is the same as that discussed previously, but higher transient stresses are exerted in the smaller pores before those crystals dissolve. The kinetics of the transient may be further complicated by diffusion of solute away from the drying surface. Buenfeld et al.[23] succeeded in modeling the chloride distribution subject to simultaneous wicking, continuous evaporation, and solute diffusion. However, Mayer and Wittmann[24] concluded that the rate of diffusion would usually overwhelm the natural (cyclical) rate of evaporation, so that it would be difficult for the pore solution to reach supersaturation. In such cases, crystallization pressure would only occur where isolated pockets of solution are left behind after the evaporation front passes, leading to the situation depicted in Figure 4b. Trapping of liquid during drying requires that a cluster of pores is surrounded by small pore entries; this can happen in any body with a distribution of pore sizes, but is particularly likely within clay inclusions which have much smaller pores than the surrounding stone.

5. SWELLING CLAY INCLUSIONS

Many sedimentary rocks contain clays, as isolated inclusions, as films around grains, or both. If the clays swell on exposure to moisture, damaging stresses can result. Some stones show expansions of several percent, but the failure strain of stone is typically ~0.1%, so much smaller expansion can be destructive.[27] Clays expand when water molecules enter between the sheets to hydrate the charge-compensating alkali ions.[28] Since the initial spacing of the sheets is ~1 nm, insertion of a single layer of water molecules causes an expansion of ~30%. Suppose that the clays form a film of thickness d around grains with diameter D, as in Figure 5a; if the strain in the clay is ε_C, then the strain in the stone is

$$\varepsilon_S = \frac{d\,\varepsilon_C}{D+d} \tag{2}$$

Therefore, when $d/D << 1$, the expansion of the stone can be orders of magnitude smaller than that in the clay. If the clay constitutes the cementing phase in the stone, the softening of the clay is directly translated into softening of the stone. For example, Portland Brownstone only expands by $\varepsilon_C \approx 5 \times 10^{-4}$, but the modulus decreases by more than a factor of 2 when it is wet.[27]

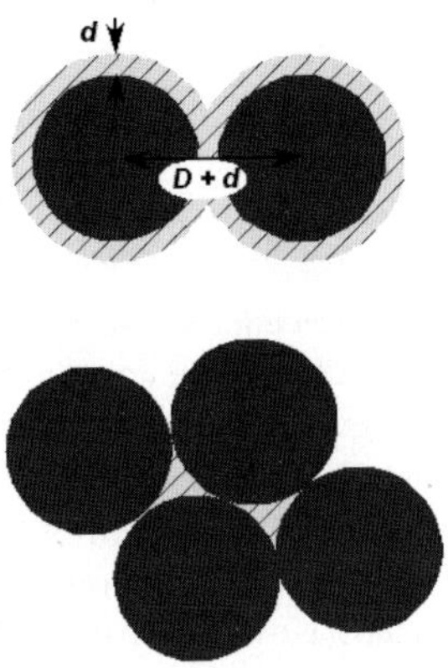

Figure 5. Clay may form a layer around grains of stone (a) or be isolated as inclusions between grains (b)

If the clay is rigidly trapped as inclusions between grains, as in Figure 5b, then it may not be able to expand at all. If it is loosely packed in the dry state, then it may swell and exert stress on the surrounding stone. In that case, the strain in the stone can be calculated in the same way as one would calculate the thermal expansion of a composite containing isolated inclusions with an expansion coefficient larger than that of the matrix. For example, using the composite sphere model[29] (where the clay is the inner sphere and the stone is the outer sphere), the strain of the stone is estimated to be

$$\varepsilon_S \approx \frac{2\,v_C\,\varepsilon_C}{1+K_S/K_C} \tag{3}$$

where v_C is the volume fraction of clay inclusions (assumed to be $<< 1$) and K_S and K_C are the bulk moduli of the stone and clay, respectively. If the clay is very compressible ($K_S >> K_C$), then the expansion of the clay causes little deformation of the stone.

If a clay-bearing stone is wetted by rain, its surface swells; when the water has only penetrated a small fraction of the depth of the body, the expansion of the surface is entirely suppressed by the dry interior, so high compressive stresses develop in the surface. If the body is thin, so that the water can penetrate to its center, then the dry interior is forced to stretch as the surrounding volume expands, so high tensile stresses develop in the interior. During drying, the stresses would reverse: contraction of the surface would create high tensile stresses near the outside of the body, while the wet interior would experience increasing compressive stresses. In this case, the most likely damage pattern would be "mud-cracking" (i.e., randomly oriented fissures) on the surface. This situation might occur for a tombstone or small sculpture, but not for the wall of a building. The compressive gravitational load on a building would suppress the expansion of the stone entirely (except for corners and protruding decorations), so the expansion during wetting would increase the compressive stress, but the interior would never experience tension; during drying, the compressive stresses would diminish to zero, but no tension would develop at the surface. If swelling of clays causes damage to a wall, it must result from buckling under compressive stress, which is consistent with the pattern of damage seen in structures made of swelling stone. Buckling will only occur if there are buried flaws that are several times wider than the thickness of the buckling layer. Such flaws could result from crystallization of salt or ice.

6. CONCLUSIONS

Several mechanisms are known to cause stresses, and many of their essential features are know, but some details are not clear. The cause of warping of thin marble plates is not understood. Freeze/thaw damage may have several causes, but most are not expected to cause high stresses in stone; it may be that trapping of liquid filled pores is important, particularly when dissolved salts lead to formation of brine pockets in the ice. Salt damage is one of the most serious, but least understood, causes of stone deterioration. How are large supersaturations produced? How easily do crystals redistribute from smaller to larger pores? What is the origin of the disjoining force? In the case of clay expansion, it is not clear whether damage requires clays to be in the cement between grains, or whether isolated inclusions can contribute. Once these details are understood, it may be possible to develop treatments that attack the cause, rather than the symptoms, of these damage mechanisms.

7. REFERENCES

1. E.M. Winkler, Weathering of crystalline marble, pp. 717-721 in *The engineering geology of ancient works, monuments and historical sites: preservation and protection*, Editor: Marinos, Paul G.; Koukis, George C. (A.A. Balkema, Rotterdam, 1988)
2. E.M. Winkler, *Stone in Architecture*, 3rd. ed. (Springer, Berlin, 1997)
3. K. Lal Gauri and J.K. Bandyopadhyay, *Carbonate Stone, Chemical behavior, durability and conservation* (Wiley, New York, 1999)
4. S. Siegesmund, K. Ullemeyer, T. Weiss, and E.K. Tshegg, Physical weathering of marbles caused by anisotropic thermal expansion, *Int. J. Earth Sci.* **89**, 170-182 (2000)
5. G.G. Amoroso and V. Fassina, *Stone Decay and Conservation* (Elsevier, Amsterdam, 1983)
6. S. Chatterji, Aspects of the freezing process in a porous material-water system. Part 1. Freezing and the properties of water and ice, *Cem. Concr. Res.* **29**, 627-630 (1999)
7. T.C. Powers, The air requirement of frost-resistant concrete, *Proc. Highway Res. Board*, **29**, 184-211 (1949)
8. D.H. Everett, The thermodynamics of frost damage to porous solids, *Trans. Faraday Soc.*, **57**, 1541-1551 (1961)

9. G.W. Scherer, Crystallization in pores, *Cement Concr. Res.* **29** (8) 1347-1358 (1999); Reply to discussion of *Crystallization in pores*, G.W. Scherer, *Cement and Concr. Res.* **30** (4) 673-675 (2000)
10. G.W. Scherer and J.J. Valenza II, Mechanisms of Frost Damage, in *Materials Science of Concrete, Vol. VII*, eds. J. Skalny and F. Young (American Ceramic Society, 2005) 209-246
11. S.Z. Lewin, The mechanism of masonry decay through crystallization, pp. 120-144 in *Conservation of Historic Stone Buildings and Monuments* (National Acad. Press, Washington, DC, 1982)
12. S. Taber, The growth of crystals under external pressure, *Am. J. Sci.* **41**, 532-556 (1916)
13. C.W. Correns, Growth and dissolution of crystals under linear pressure, *Disc. Faraday Soc.* **5**, 267-271 (1949)
14. J. Israelachvili, *Intermolecular & Surface Forces*, 2nd ed. (Academic, London, 1992)
15. P. Gallo, M.A. Ricci, and M. Rovere, Layer analysis of the structure of water confined in vycor glass, *J. Chem. Phys.* **116** (1) 342-346 (2002)
16. P.K. Weyl, Pressure solution and the force of crystallization - A phenomenological theory, *J. Geophys. Res.* **64** (11) 2001-2025 (1959)
17. W.P. Halperin, S. Bhattacharija, and F. D'Orazio, Relaxation and dynamical properties of water in partially filled porous materials using NMR techniques, *Magnetic Res. Imaging* **9**, 733-737 (1991)
18. G.W. Scherer, Stress from crystallization of salt in pores, in *Proc. 9th Int. Cong. Deterioration and Conservation of Stone, Vol. 1*, ed. V. Fassina (Elsevier, Amsterdam, 2000) 187-194
19. G.W. Scherer, Stress from crystallization of salt, *Cement Concr. Res.* **34**, 1613-1624 (2004)
20. G.W. Scherer, Fundamentals of drying and shrinkage, in *Science of Whitewares*, eds. V.E. Henkes, G.Y. Onoda, and W.M. Carty (Am. Ceram. Soc., Westerville, OH, 1996) 199-211
21. J. Freundlich, *Colloid & Capillary Chemistry* (Methuen, London, 1926) 154-157
22. O. Coussy, Deformation and brittle fracture from drying-induced crystallization of salts, submitted to the *Journal of the Mechanics and Physics of Solids*
23. N.R. Buenfeld, M.T. Shurafa-Daoudi, and I.M. McLoughlin, Chloride transport due to wick action in concrete, *Chloride Penetration into Concrete,* ed. L.O. Nilsson and M.P. Olliver (RILEM, Paris, 1997) 315-324
24. G. Mayer and F.H. Wittmann, Ein Modell zur Beschreibung des Wasser- und Salztransports in Mauerwerk, *Int. Zeitschrift für Bauinstandsetzen* **2** (1) 67-82 (1996)
25. J.R. Dunn and P.P. Hudec, Water, clay, and rock soundness, *Ohio J. Science* **66** (2) 153-168 (1966)
26. C. Rodriguez-Navarro, E. Sebastian, E. Doehne, and W.S. Ginell, The role of sepiolite-palygorskite in the decay of ancient Egyptian limestone sculptures, *Clays Clay Minerals* **46** (4) 414-422 (1998)
27. G.W. Scherer and I. Jimenez Gonzalez, Characterization of Swelling in Clay-Bearing Stone, in *Stone decay and conservation*, SP-390, ed. A.V. Turkington (Geological Soc. Am., 2005) 51-61
28. H. van Olphen, *An Introduction to Clay Colloid Chemistry*, 2d ed. (Wiley, NY, 1977)
29. G.W. Scherer, *Relaxation in Glass and Composites* (Wiley, New York, 1986; reprinted by Krieger, Malabar, FL, 1992)
30. J.W. Hutchinson, M.Y. He, A.G. Evans, The influence of imperfections on the nucleation and propagation of buckling driven delaminations, J. Mech. Physics Solids 48, 709-734 (2000)

THE CONTRIBUTION OF HISTORIC MORTARS ON THE EARTHQUAKE RESISTANCE OF BYZANTINE MONUMENTS

A. Moropoulou, K. Labropoulos, P. Moundoulas and A. Bakolas
National Technical University of Athens, School of Chemical Engineering, Section of Materials Science and Engineering, 9 Iroon Polytechniou, 157 80, Zografou Campus, Athens, Greece

Abstract: Structural studies to assess the earthquake resistance of Byzantine monuments have proved that their static and dynamic behaviour depend on the mechanical, chemical and microstructural properties of the masonry mortars and bricks. The crushed brick/lime concrete is classified as an advanced cement-based composite, explaining the longevity of Byzantine monuments. More specifically, the Byzantine monument of the Church of St. Michaels in Vydubytskyi Monastery (11th c.), in Kiev has been studied to provide insights on its effective dynamic properties, during major earthquake in the area. The bricks and mortars are investigated taking under consideration the construction technique known as "concealed course", by optical microscopy (OM), scanning electron microscopy (SEM), transmission electron microscopy (TEM), X-ray diffraction (XRD) and thermal analysis (TG-DTA). They were found to allow for continuous stresses and strains due to the presence of the amorphous hydraulic formations (CSH) at the crushed-brick powder/binder interfaces and in the binding matrix, and be able to absorb greater energy without initiations of fractures. The interpretation of the amorphous nature of the hydraulic formations is compared with the use of fine siliceous sources to accelerate the formation of hydraulic compounds. The determination of the mortar properties indicated that they are of considerable mechanical strength and durability. The results show major similarities with these of the Byzantine monuments in Istanbul (Theodosian Walls and Hagia Sophia – 6th - 11th c.), giving evidence to earthquake resistant construction techniques and materials. Hence, a reverse engineering approach, complying with the deduced criteria is proposed in order to reproduce compatible restoration concrete for earthquake protection of monuments.

Key words: earthquake resistance; Byzantine monuments; crushed brick mortars

M.S. Konsta-Gdoutos, (ed.), Measuring, Monitoring and Modeling Concrete Properties, 643–652.

1. INTRODUCTION

Earthquake engineering has benefited significantly from structural studies of historic buildings that have withstood large earthquakes in the past. The structural preservation of such historic buildings involves a wide array of tasks. Initially, the existing structural system and the materials it is composed of have to be surveyed. Specifically for historic monuments the materials' properties and microstructure are characterized with non-destructive techniques. The survey and identification of existing materials, together with historical documentation, will reveal the original materials[1], any zones of previous repairs[2], weaknesses in the structure, cracking and discontinuities. Once the evaluation of the mechanical properties of these diverse materials is accomplished, those data are then fed into finite element models of the structure which are subjected to linear and non-linear dynamic analyses to evaluate the resistance of the structure to seismic loads.[3] These analytical investigations are supported by data obtained from ambient vibration testing, soil and foundation investigations and measurements by strong motion instrumentation. Based on the results from these experimental and analytical investigations, the most appropriate and an effective structural intervention on the historic building is designed with the aim to improve its structural worthiness.

Structural studies to assess the earthquake resistance of Byzantine monuments have proved that their static and dynamic behaviour depend on the mechanical, chemical and microstructural properties of the mortars and bricks used for the masonry.[4] The crushed brick/lime concrete is classified as an advanced cement-based composite, explaining the longevity of Byzantine monuments, and is characterized by a very long curing period. Previous structural studies to determine the earthquake worthiness of Hagia Sophia in Istanbul, an outstanding example of Byzantine architecture and a testimony to the longevity of such historic structures, show a decrease of 5–10% in the building's natural frequencies, as the amplitude of accelerations increase and return to their initial values, a behaviour imposed by the non-linear nature of the masonry. The characterization of the historic mortars has revealed that they allow for continuous stresses and strains due to the presence of the amorphous hydraulic formations (CSH) at the crushed-brick powder/binder interfaces and in the binding matrix, and be able to absorb greater energy without initiations of fractures.[5] Similar earthquake resistant behaviour has been observed in other Byzantine monuments, such as those of the 11–13th century in Kiev. In particular, the Church of St. Michael in the Vydubytskyi Monastery (11th c.) has been studied to provide insights into the building's efficient dynamic behaviour which allowed it to withstand successfully an earthquake of magnitude 6 on the Richter scale which occurred in the area in 1230 AD.

2. BUILDING MATERIALS

The church of St. Michael was built in the 1070s in the Monastery in Vydubytskyi, at the vicinity of ancient Kiev. The original eastern part of the original church was destroyed, however, the western part is still preserved. In the 18th c. a new eastern part of the building was constructed and a Baroque-type dome was built above it. The width of the masonries is approximately 1.1m. The construction technique used for the 10th - 12th c. period was that of the "concealed course" in which every second course of bricks was slightly recessed and covered with mortar (Figure 1).[6]

The dimensions of the bricks were approximately 3.6–4.2cm × 25–29cm × 33–37 cm. The bricks were manufactured from local clays and loam and were baked between 1000-1200°C. The lime mortar contained fill material made of clay mass which was baked at low temperatures (<750°C) and subsequently crushed and sieved to obtain the finer fractions, called locally "tsemianka" (Figure 2). The thickness of the mortar layer between two successive bricks was approximately 5cm, with the result that the thickness of the mortar layer visible from the front, taking into account the concealed course construction technique, was approximately 14 cm. The thickness of the brick and mortar layers and the brick thickness to mortar thickness ratio are typical of the Byzantine period.[7] The outer surface of the church was originally covered with plaster consisting of lime mortar with carefully sieved crushed ceramic mass, whereas the inner surface was covered with frescoes.

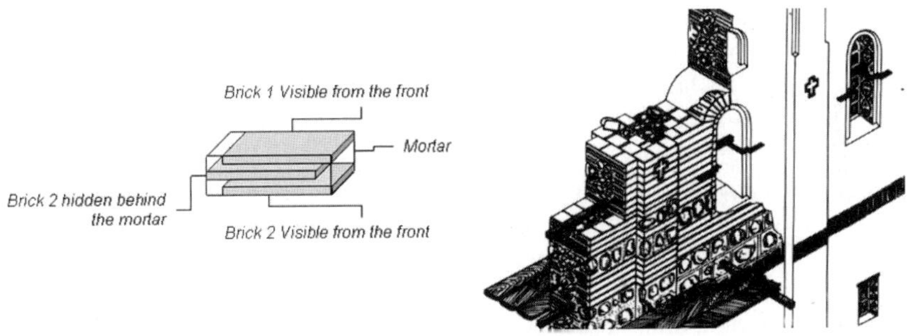

Figure 1. Schematic of the "concealed course" construction technique in which every second course of bricks was slightly deepened and covered with mortar of more than 5-cm thickness

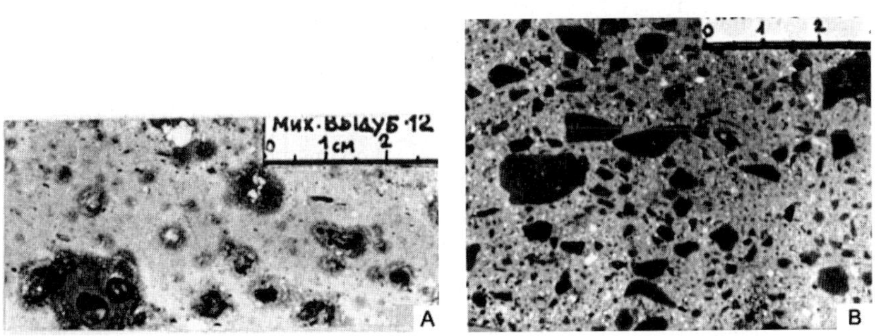

Figure 2. (A) Typical section of the brick used in the Church of St. Michael in the Monastery in Vydubytskyi. (B) Typical section of the lime mortar with ceramic fill material used in the same building

Longitudinal wooden beams were placed in the central part of the walls in order to prevent deformations in brickwork and to strengthen the structure (Figure 1). Moreover, they improved the dynamic behaviour of the building and were. The wooden beams were installed at various levels in the masonries and in particular at the walls' bedding, in the abutment of arches and arched openings of doors and windows, and in the bases of drums and domes. Similar wooden strengthening ties have also been discovered in historic buildings in Greece.[8]

3. EXPERIMENTAL

A petrographical characterization of the mortar components was accomplished by optical microscopy. Thin sections of mortar samples were observed and photographed at various magnifications using a polarizing Zeiss microscope. The microphotographs taken allowed for the petrographical–mineralogical characterization of the mortar constituents, microscopic observations of the different mineral phases in the matrix as well as the qualitative distribution between binder and aggregate. The mortars were also subjected to an XRD analysis, in order to identify the mineral constituents of the mortars. Finely pulverized mortar samples were examined by a Siemens D-500 X-ray diffractometer, with an automatic adjustment and analysis system and a Diffract-EVA quality analysis software. To facilitate the direct observation of various spectra, a diffraction interval between 2θ-5 and 2θ-60, with a step of 0.02 was used.

Thermo gravimetric analyses (TG/DTG) were performed by a Mettler TG 50 thermo balance - thermal analyser system to reveal thermal transformations such as dehydration, dehydroxidation, oxidation, and decomposition. The TG-DTA analyses were compared with similar differential thermal analyses (DTA) performed with a Perking Elmer Thermo analyser DTA 1700, to aid in the identification of the thermal transformations. The temperature ranges and the relative weight losses, under heating, were found to be a reliable tool for the characterization of these materials.[9] In particular, the temperature ranges are relative to the weight loss due to adsorbed water (120°C), when there are no particular hydrated salts, to the loss of chemically bound water (200–600°C), when there are no other compounds that undergo weight loss in this temperature range and the loss of CO_2 (>600°C) due to the decomposition of carbonates. Weight losses at reaction temperatures transferred more or less near to 750°C, render the loss of CO_2, not from primary calcite, but from re-carbonated lime.[10]

The thermal analyses were used to calculate the structurally bound water percentage vs. the CO_2/H_2O ratio (weight loss % >600°C / weight loss % 200-600°C) and compare the processing technologies of the traditional mortars. The dehydroxylated clays act as a 'pozzolan' which imparts early strength to the mortar.[11] The phenomenon becomes more complex and occurs when hydraulic reactions take place at the aggregate/matrix interface. The DTA and TG-DTG analyses can identify the dehydration of calcium silico-aluminate phases, giving clear evidence of the hydraulic phases at the mortar matrix, rather than that of pure lime.

Microscopic observations of mortar samples were performed, under a scanning electron microscope (SEM) Philips 515, with secondary electron emission, in order to examine the microstructure and the texture of the mortars. Energy Dispersive X-ray microanalysis (EDX), provided elementary semi-quantitative analysis. Transmission electron microscopy (TEM) was performed to identify amorphous phases, using a Philips CM20/ST TEM. The TEM samples were prepared using the dipping method from colloidal suspension of finely ground particles of mortar samples. The particles of the sample powder were suspended in methanol and ultrasonicated for 30 min. A carbon-Formvar-coated Cu grid was dipped into the suspension, air dried and stored for further use.

Due to the fragile nature of the materials and their relatively small size of the specimens, dynamic elastic moduli of the materials were evaluated by using a Portable Ultrasonic Non-Destructive Indicating Tester (PUNDIT) with transducers of 54 kHz. The tensile strength (fmt,*k*) measurements were performed according to the fragments test method.[12,13] In particular, small historical mortar fragments (gravel size) were taken from the joints and arranged in a special mould within a strong matrix (an epoxy resin or a much stronger mortar). A standard steel pin is then dragged along the specimen under a standard load. The width of the scratch is measured with optical microscopes and the tensile strength of the specimens is evaluated using calibrated tensile strength vs. scratch width curves.

4. RESULTS

The petrographical study for the St. Michael samples revealed that light coloured ceramic fragments and powder, as well as small rounded quartz grains are embedded in a finely crystallized matrix, and are well adhered to it (Figure 3) . The calcitic matrix is well and homogeneously processed without any fossils and shows the coherence of hydraulic mortars. Similar observations are made for samples from Hagia Sophia. It is interesting to note that in both cases any cracks present are filled with secondary crystallized calcitic material indicating a self-healing effect to earthquake damage.

Thc X-ray Diffraction spectra of the matrix from crushed brick samples from St. Michael in Kiev and the Theodosian walls in Istanbul show significant similarities (Figure 4). In particular, they reveal abundance in calcite, sufficient amounts of quartz, some dolomite and feldspars and presence of calcium silicate hydrate (CSH) and hematite (Hm) traces. The mortar sample from the Byzantine Walls in Istanbul differs only in the presence of Aragonite, due most probably to different climatic conditions governing the crystallization of $CaCO_3$. The spectra reveal the presence of either hydraulic lime or pozzolana admixed with aerial lime, rich in CSH. The diversification of the spectra regarding CSH and Hm indicates the different provenance of the pozzolana or the clay used.

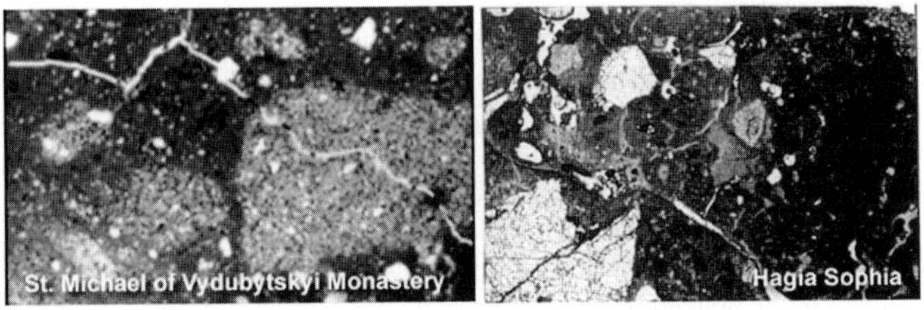

Figure 3. Optical microscopy of samples from St. Michael of Vydubytskyi Monastery and Hagia Sophia, Istanbul, were it is evident that the ceramic fragments and powder are embedded I a finely crystallized matrix, and any cracks have been filled with secondary crystallized calcitic material

Figure 5 shows the results of the percentage of structurally bound water vs. the CO_2/H_2O ratio, for binder samples from Hagia Sophia in Istanbul, the Theodosian Walls in Istanbul, and the Church of St. Michael in Kiev. The results refer to the binder fraction (<63 μm), and reveal an inverse linear trend. Taking into account that a small percentage of the aggregate is included in the binder fraction (<63 μm) as observed by optical microscopy and from the percentages of CO_2 and H_2O binder, the binder / aggregate ratio for the mortars of Byzantine monuments in Istanbul take values between 1:2 to 1:4.[5] Correspondingly, the binder/aggregate ratio for the St. Michael Church is approximately 1:3, a value similar to the binder/aggregate ratio for the Byzantine monuments in Istanbul and typical of the Byzantine period artificial pozzolanic mortars.[7]

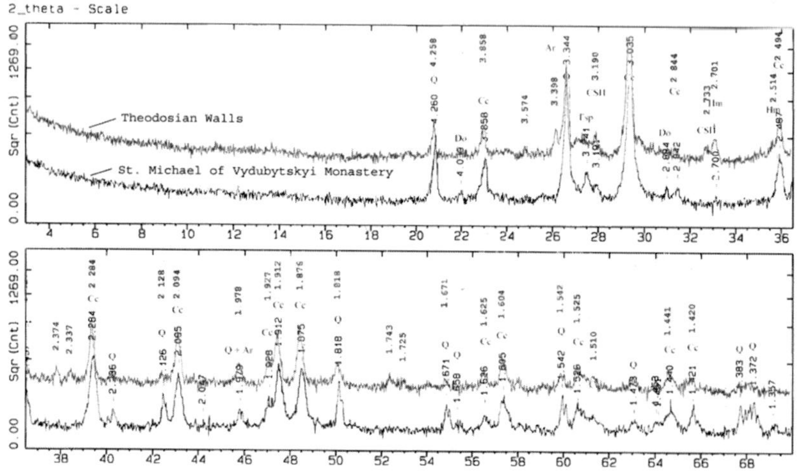

Figure 4. X-ray diffraction spectra of the matrix of the crushed brick samples from Byzantine monuments in Kiev (St. Michael of Vydubytskyi Monastery) and Istanbul (Theodosian Walls). Legend: *Q: quartz; Cc: calcite; Do: dolomite (St. Michael); Ar: aragonite, $CaCO_3$, 3.40, 1.98, 3.27 (5-453) (Theodosian Walls); Fsp: feldspars (Sanidin), $NaKAl(Si_3O_8)$, 3.24, 3.21 (10-357) (St. Michael); CSH: calcium silicate hydrate, $Ca_3(SiO_3OH)2{\cdot}2H_2O$ (St. Michael) 3.19, 2.84, 2.23 (9-454), (Theodosian Walls) 3.19, 2.84, 2.74, 2.23 (9-454); Hm: hematite: Fe_2O_3 (St. Michael) (13-534); (Theodosian Walls) 2.69, 1.69, 2.51 (13-534)*

The considerable durability of this type of historic mortars and their efficient dynamic behaviour are related to the physico-chemical adhesion and cohesion bonds that develop at the matrix and at the binder /aggregate interface, respectively.[14] These bonds have been studied for the Hagia Sophia crushed brick and lime mortar, using EDX microanalysis at the brick / mortar interface. Figure 6 shows the elemental distribution of Ca, Si and Fe at the above interface. In this figure, it is revealed that carbonates (Figure 6a) are substituted for calcium silicates (Figure 6b) with simultaneous compaction of the calcite content in the boundary. The boundary transition area appears to have a thickness of 30-50 μm, and it can be observed that in this transition zone the Ca element decreases, whereas Si species increase as we move from the binder matrix to the main body of the brick. The boundary surface is evidenced, where the physicochemical penetration of $Ca(OH)_2$ to the zones immediately adjacent to the brick determines the alkaline environment.[15] In such an environment, amorphous silicates develop as the reaction product

(Figure 6e). In Figure 6c the presence of Fe is characteristic of the ceramic fragments which usually present the texture of an initial vitrification stage attributed to low firing temperatures.

Transmission electron microscopy was employed to provide information regarding the development of the amorphous calcium silicate and silicate hydrate gel between the crystalline phases of the calcite and the dispersed ceramic fragments. Figure 7a reveals that an amorphous gel-like phase is developed which has a foil-like morphology similar to that of the outer product of the hydration of tricalcium silicate (Ordinary Portland C-S-H) in the presence of silica fume.[16] In the case of artificial pozzolanic mortars, this phase is produced under the pozzolanic reaction of slaked lime with fine reactive siliceous sources at temperatures <100°C.

Figure 7b demonstrates, that the reaction limiting factor for the pozzolanic reaction is the dissolution of the siliceous sources.[17] Specifically, when the traditional siliceous sources were replaced by Tetra-Ethyl-Ortho-Silicate (TEOS), a more reactive fine siliceous source, the microstructural analysis revealed that an enhanced reaction rate was indeed present, and offered similar morphologies of the resultant C-S-H phases as compared to the traditional pozzolanic reaction with the crushed brick fragments.[17]

The dynamic elastic moduli estimated for various brick, mortar and composite samples from the Church of St. Michael of the Vydubytskyi Monastery, Kiev, and Hagia Sophia, Istanbul are shown in Table 1. The results show significant similarities. The small difference in the elastic modulus of the composite may be attributed to the different brick thickness/mortar thickness ratios. In the case of the Church of St. Michael the ratio had a value of approximately 1.2, whereas in the case of Hagia Sophia the ratio had a higher value between 1.5 - 2.0. However, both cases of Byzantine composites can be considered to be as early reinforced concrete.[7] The tensile strength of mortar samples from both cases obtained values between 1.25 - 2 MPa , and 0.7 - 1.5 MPa for the Church of St. Michael and Hagia Sophia respectively. Both cases show tensile strength values typical for artificial pozzolanic mortars.[7]

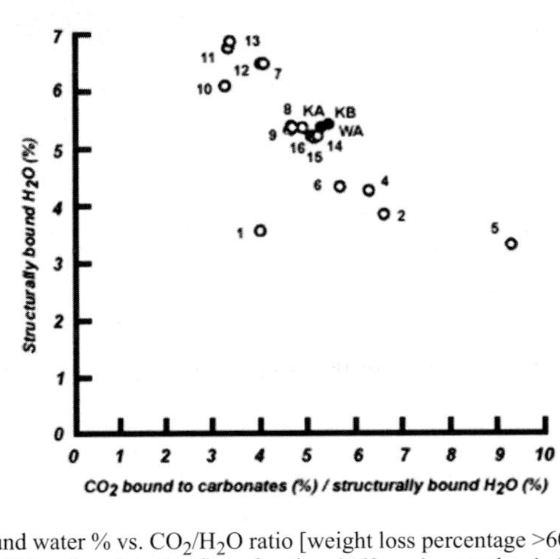

Figure 5. Structurally bound water % vs. CO_2/H_2O ratio [weight loss percentage >600°C / weight loss percentage between 200 and 600°C] referred to the finer fraction (<63 μm): samples 1–16, Hagia Sophia; samples WA, WB, Theodosian Walls; samples KA and KB, St. Michael Church, Kiev

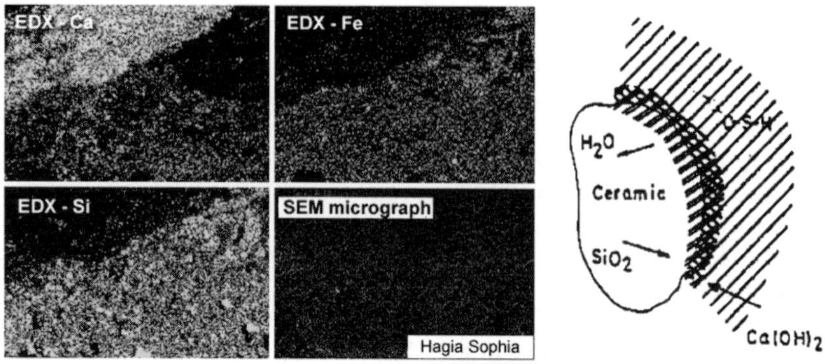

Figure 6. (Left) EDX microanalysis of a crushed brick and lime mortar sample from Hagia Sophia. Distribution of elements (a-c) Ca, Si, and Fe respectively. (d) SEM micrograph of the brick / mortar interface. (Right) A schematic of the boundary zone

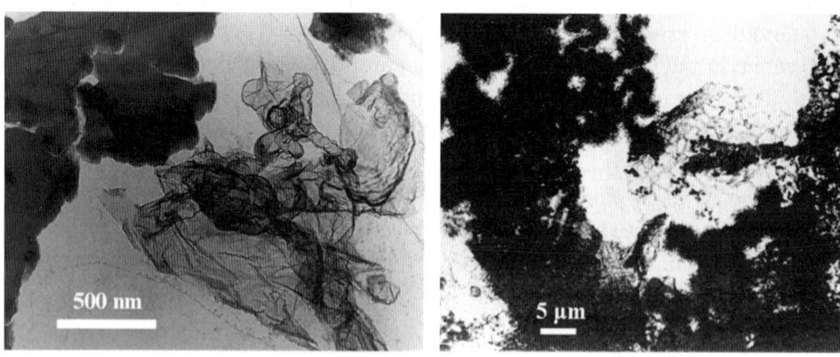

Figure 7. TEM micrographs: (a) Traditional mortar of Hagia Sophia revealing the morphology of the binding C-S-H phase developed by pozzolanic reaction. (b) Use of TEOS paste as fine siliceous source - the formation of a C-S-H phase is evident even after 10 min of hydration

Table 1. Dynamic elastic moduli of samples from the Church of St. Michael of the Vydubytskyi Monastery, Kiev, and Hagia Sophia, Istanbul

Dynamic Elastic Moduli (GPa)		Church of St. Michael of the Vydubytskyi Monastery, Kiev	Hagia Sophia, Istanbul
Brick:	E_B	1 – 3	3.1
Mortar:	E_M	0.6 – 0.7	0.66
Composite:	E_C	2.2 – 2.5	1.83

5. CONCLUSIONS

The mechanical, chemical and microstructural properties of the masonry mortars and bricks influence significantly the static and dynamic behaviour of Byzantine monuments and are responsible for the durability and longevity of these structures. The study of the processing technology, properties, composition, and microstructure of such materials can be useful in the design of restoration materials for monuments, and provide valuable knowledge in the improvement of modern construction materials. The crushed brick/ lime concrete is classified as an advanced cement-based composite. The amorphous calcium – silicate – hydrate phase that develops during the pozzolanic has a similar morphology to that of the outer product of the hydration of tricalcium silicate of Ordinary Portland Cement. The reaction limiting factor for the pozzolanic reaction is the dissolution of the siliceous sources. The considerable durability of this type of historic mortars and their efficient dynamic behaviour are related to the physico-chemical adhesion and cohesion bonds that develop at the matrix and at the binder /aggregate interface, respectively. These mortars were found to allow for continuous stresses and strains due to the presence of such an amorphous hydraulic formations (CSH) at the crushed-brick powder/binder interfaces and in the binding matrix, and be able to absorb greater energy without initiations of fractures. The determination of the mortar mechanical properties indicated that they are of considerable mechanical strength and durability. The results show major similarities among the Byzantine monuments studied (Istanbul - Theodosian Walls and Hagia Sophia, and Kiev – St. Michael Church). Thus, a reverse engineering approach, complying with the deduced criteria can reproduce compatible restoration concrete for earthquake protection of monuments.

6. REFERENCES

1. A. S. Cakmak, M. Erdik, and A. Moropoulou, A joint program for the protection of the Justinian Hagia Sophia, *Proc. Fourth Int. Symp. on the Conservation of Monuments in the Mediterranean Basin*, Rhodes, **4**, (1997).
2. A. Bakolas, G. Biscontin, and E. Zendri, Characterization of mortars from traditional buildings in seismic areas, *PACT Revue du groupe europeen d'etudes pour les techniques physiques, chimiques, biologiques et mathematiques appliquees a l'archcologie* **56** 17–29 (1998).
3. E. Durukal, A. S. Cakmak, O. Yuzugullu, and M. Erdik, Assessment of the earthquake performance of Hagia Sophia, *PACT Revue du groupe europeen d'etudes pour les techniques physiques, chimiques, biologiques et mathematiques appliquees a l'archeologie* **56** 49–57 (1998).
4. A. S. Cakmak, A. Moropoulou, and C. A. Mullen, Interdisciplinary study of dynamic behaviour and earthquake response of Hagia Sophia, *Soil Dyn. Earthquake Engng* **14** (9) 125–133 (1995).
5. A. Moropoulou, A. S. Cakmak, and G. Biscontin, Crushed brick lime mortars of Justinian's Hagia Sophia, in: *Materials issues in art and archaeology V*, edited by Vandiver PB, Druzik JR, Merkel JF, Stewart J, **462** (MRS, 1997).
6. A. Moropoulou, A. S. Cakmak, and N. Lohvyn, Earthquake resistant construction techniques and materials on Byzantine monuments in Kiev, *Soil Dynamics and Earthquake Engineering* **19**, 603-615 (2000).

7. A. Moropoulou, A. Bakolas, and S. Anagnostopoulou, Composite materials in ancient structures, *Cement and Concrete Composites* **27,** 295-300 (2005).
8. P. Touliatos, The historic construction as a complete structural system, its decay in time and compatible restoration interventions, *Proc. of the INCOMARECH Conf., Athens*, (1998).
9. A. Moropoulou, A. Bakolas, and K. Bisbikou, Characterization of ancient, Byzantine and later historic mortars by thermal analysis and X-ray diffraction techniques, *Thermochim Acta* **269/270,** 779–795 (1995).
10. J. E. Adams, and A. W. Kneller, Thermal analysis of medieval mortars from gothic cathedrals in France, *Engineering geology of ancient works Engineering Geology of Ancient Works*, Balkema, Rotterdam 1019–1026 (1988).
11. A. R. Livingston, E. P. Stuzman, R. Mark, and M. Erdik, Preliminary analysis of the masonry of Hagia Sophia Basilica, *Materials issues in art and archaeology III.* Mat Res Soc, Pittsburgh, 721–736 (1992).
12. E. S. Katsaragakis, A new tensile test for concrete *Mater Struct* **20,** 120-125 (1987).
13. Th. Tassios, C. Vachliotis, and C. Spanos, In situ strength measurements of masonry mortars, *Proc Int. Conf. on Repair and Strengthening of Stone Masonries*, ICCROM, Athens, 53–61 (1989).
14. A. Moropoulou, A. Bakolas, and K. Bisbikou, Physico-chemical adhesion and cohesion bonds in joint mortars imparting durability to the historic structures, *Construction and Building Materials* **14,** 35-46 (2000).
15. A. Moropoulou, A. S. Cakmak, G. Biscontin, A. Bakolas, and E. Zendri, Advanced Byzantine cement based composies resisting earthquake stresses: the crushed brick / lime mortars of Justinian's Hagia Sophia, *Construction and Building Materials* **20** (8), 543-552 (2002).
16. I. G. Richardson, The nature of the hydration products in hardened cement pastes, *Cem Concr Comp* **22,** 97-113 (2000).
17. A. Moropoulou, A. S. Cakmak, K. C. Labropoulos, R. Van Grieken, and K. Torfs, Accelerated microstructural evolution of a calcium-silicate-hydrate (C-S-H) phase in pozzolanic pastes using fine siliceous sources: Comparison with historic pozzolanic mortars, *Cement and Concrete Research* **34,** 1-6 (2004).

MOISTURE AND ION TRANSPORT IN LAYERED PLASTER/SUBSTRATE COMBINATIONS: AN NMR STUDY

L. Pel, J. Petkovi and H. Huinink
Eindhoven University of Technology, Department of Applied Physics, P.O. Box 513, 5600 MB Eindhoven, The Netherlands

Abstract: It has been investigated how transport and accumulation of salt in a plaster depends on the underlying substrate material. Therefore the moisture and sodium profiles have been measured non-destructively with a Nuclear Magnetic Resonance (NMR) technique during drying of plaster/ substrate systems. The same plaster is applied on two substrates of which the pores are either an order of magnitude larger or smaller than those of the plaster. The moisture and salt transport and the salt accumulation differed significantly for these two systems. In a plaster/ Bentheimer sandstone system (the pores of the plaster are smaller than those of the substrate) all salt is removed from the substrate and accumulates in the plaster. In a plaster/calcium-silicate brick system (the substrate has a considerable amount of pores that are smaller than those of the plaster) some salt crystallizes in the plaster layer, but a significant amount of salt crystallizes within the substrate itself.

Keywords: Moisture and ion transport, plasters, NMR

1. INTRODUCTION

The choice of restoration plasters, suitable for long-time protection and performance, is an important conservation problem. The performance of specially developed plasters is not always satisfactory.[1,2] The durability of a plaster and its ability to protect the underlying masonry strongly depend on its transport properties with respect to salt and moisture. Although salt damage has been investigated intensively for several decades, the mechanisms that control salt crystallization in porous media are poorly understood. A better understanding of the transport of water and dissolved ions during drying and salt crystallization in plasters and the underlying substrate is necessary for understanding salt damage and for developing plasters that meet the requirements with respect to durability and protection.

M.S. Konsta-Gdoutos, (ed.), Measuring, Monitoring and Modeling Concrete Properties, 653–659.

In the present recommendations for the application of plasters on salt-loaded substrates[3] the influence of the substrate materials is not taken into account. The aim of this study is to investigate the influence of a difference in pore-size distribution between the plaster and substrate layer on the water and salt transport and the salt accumulation during drying. Until now, the details of the drying process of two-layer materials has not been investigated very extensively. Using a Nuclear Magnetic Resonance (NMR) technique we are able to measure non-destructively the time evolution of water and dissolved ions in these layered materials during drying.

In section 2 the moisture and salt transport during drying will be discussed. In section 3 the Nuclear Magnetic Resonance (NMR) technique will be explained. Finally in section 4 we will discuss the measured combined salt and moisture in plaster/substrate systems.

2. THEORY

During the drying process moisture will be transported to the drying surface. In general during drying two stages can be distinguished. During the first period, moisture transport is fast and occurs through the water network. During the second period characterized by the receding drying front, water near this drying front is present in the form of isolated clusters, and transport occurs through the vapor phase. The water clusters evaporate because of the large difference in relative humidity between the vapor near the clusters and the air at the drying surface of the material.

In a two layer material the situation is more complicated. Over the interface of two materials the capillary pressure is constant. Hence there will be jump in moisture content over the interface[4]. The material with the largest pores is emptied first. In the case of plaster/substrate systems, the plaster will dry first if it has larger pores than the substrate. On the other hand, if the plaster has smaller pores than the substrate, the substrate will dry first.

Initially the salt will be transported by advection to the drying surface. As a consequence a salt peak will built up and a salt concentration profile will develop. But as soon a concentration gradient develops diffusion will start to level off the accumulation. Hence during a drying experiment there will be a competition between advection, which transports salt to the top of the sample and thereby causes accumulation, and diffusion, which levels off any accumulation. The competition between advection and diffusion can be characterized by the Péclet number Pe, given by[4]:

$$Pe = \frac{|U|L}{D} \qquad (1)$$

where $|U|$, L and D are the water velocity, the length scale of interest, and the ion diffusivity, respectively. For $Pe \gg 1$ advection dominates the salt transport, which happens at sufficiently high drying rates, whereas for $Pe \ll 1$ diffusion dominates the salt transport.

3. NUCLEAR MAGNETIC RESONANCE

In a nuclear magnetic resonance (NMR) experiment the magnetic moments of the nuclei are manipulated by suitably chosen radio frequency fields, resulting in a so-called spin-echo signal. The amplitude of this signal is proportional to the number of nuclei excited by the radio frequency field. NMR is a magnetic resonance technique, where the resonance condition for the nuclei is given by:

$$f = \gamma B_o \tag{2}$$

In this equation f is the frequency of the radio frequency field, γ is the gyromagnetic ratio (γ= 42.6 MHz /T for ^1H and 11.3 MHz /T for ^{23}Na) and B_o is the externally applied static magnetic field. Because of this condition the method can be made sensitive to one type of nuclei and, therefore to hydrogen (and thus to water) or sodium.

For the experiments described here a home-built NMR scanner[5,6] is used, which incorporates an iron-cored electromagnet operating at a field of 0.78 T. In order to perform quantitative measurements a Faraday shield is placed between the tuned circuit of the probe head and the sample[5]. In addition the quality factor of the LC circuit is chosen rather low ($Q \approx 40$), to suppress the effects of the (electrically conducting) NaCl solution. The sample, a cylinder with a length of 50 mm and a diameter of 20 mm, was placed inside a closed Teflon holder. A constant magnetic field gradient of up to 0.3 T/m was applied using Anderson coils, giving a one-dimensional resolution of the order of 1 mm for water and 3.5 mm for Na. The spin-echo experiments were performed at a fixed frequency, corresponding to the centre of the RF coil. The sample is moved vertically through the magnet with the help of a step motor. The sample is sealed at all sides, except for the top over which air with a relative humidity of 5% is blown. In this way a one-dimensional drying process is created. While the sample is drying, first the moisture content in the small region of the sample near the centre of the RF coil is measured. Next, the frequency is changed from 33 MHz (^1H) to 9 MHz (^{23}Na) and the Na concentration in that region is measured. After these two measurements the sample is moved in the vertical direction by the step motor and the moisture and Na concentration are measured again. The measurement time for the moisture content was 1 min whereas it took 4 min to measure the Na content with a similar signal to noise ratio. This procedure is repeated until a complete moisture and Na profile has been measured. A time stamp is given to each measurement point. Measuring an entire Na concentration profile takes approximately 3 h. Since the typical time of a drying experiment is several days, the variation of the moisture and ion profiles during a single scan can be neglected. With NMR settings used in these experiments only the Na nuclei in the solution are measured, i.e., no signal is obtained from NaCl crystals.

4. COMBINED MOISTURE AND SALT TRANSPORT

Experiments were done on two different plaster-substrate systems. Bentheimer and calcium-silicate brick were used as substrates. The plaster had the same composition in both systems: lime:cement:sand = 4:1:10 (v/v). The two substrates were selected because of the significant differences in their pore-size distributions, i.e., Bentheimer has a mean pore size of 30 μm, whereas the calcium silicate brick has predominantly pores at 12 nm. The systems were dried during seven days in a plastic box and during 21 days at 40° C and 65 % relative humidity. After that the systems were carbonated during a few weeks, until full carbonation was reached, in a box with 3 % CO_2 (v/v) at 50 % relative humidity. The pore sizes and open porosities of the materials have been measured by means of mercury intrusion porosimetry[7]. The pores in the plaster were found to be in the order of 0.5 μm. Hence the pores of the plaster are an order of magnitude smaller than those of Bentheimer sandstone and an order of magnitude larger than the nanometer pores of the calcium-silicate brick.

To study the salt transport in the plaster/Bentheimer sandstone system, the sample was initially saturated with a NaCl solution, $c = 4$ mol l^{-1}. In Figure 1 we have plotted water profiles for several times during the drying process. Two drying stages are observed. The drying of the salt loaded sample is much slower compared to that of a pure water saturated one which can be partly explained by the dependence of the drying rate on the relative humidity. The presence of salt decreases the relative humidity near the liquid-air interface and, consequently, decreases the drying rate. Apart from this, in the presence of salt no receding drying front is observed (Fig. 1a). This might be caused by a change in the wetting properties in the presence of NaCl.

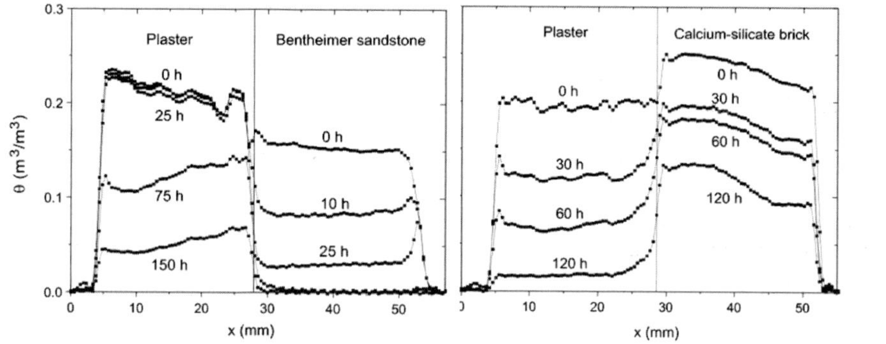

Figure 1. Moisture profiles in the plaster/Bentheimer sandstone system and calcium-silicate brick plaster system during drying. The samples were initially saturated with a NaCl solution ($c = 4$ mol l^{-1}). Dry air is blown over the top of the sample ($x = 4$ mm)

In Figure 2a the Na profiles are presented for several times during the drying process. The total amounts of dissolved Na in the plaster and the Bentheimer sandstone as a function of the drying time are shown in Figure 2b. Inspection of Figure 2 shows that during the first drying stage ($t < 25$ h) the Na in the Bentheimer sandstone remains uniformly distributed, while the total amount of dissolved Na in the system decreases. The calculated Na concentration in the Bentheimer sandstone does not exceed the initial concentration of 4 mol l^{-1}. The amount of dissolved Na in the plaster increases during the first 25 hours of drying. The resulting salt distribution in the plaster is not uniform, but a Na peak develops at the drying surface, see figure 2a. The estimated Na concentration at the top of the plaster after 25 hours of drying is 6 mol l^{-1}, which is around the solubility limit of NaCl. At this stage, crystallization at the air/plaster interface is visually observed.

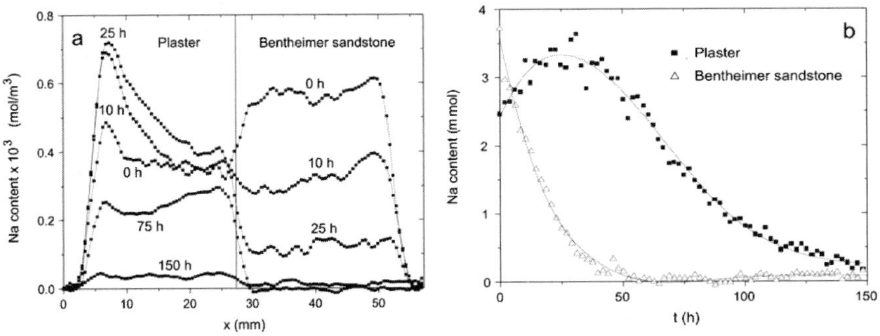

Figure 2. a) Profiles of dissolved Na in the plaster/Bentheimer sandstone system. b) Total amounts of dissolved Na ions in the plaster and Bentheimer sandstone as a function of the drying time

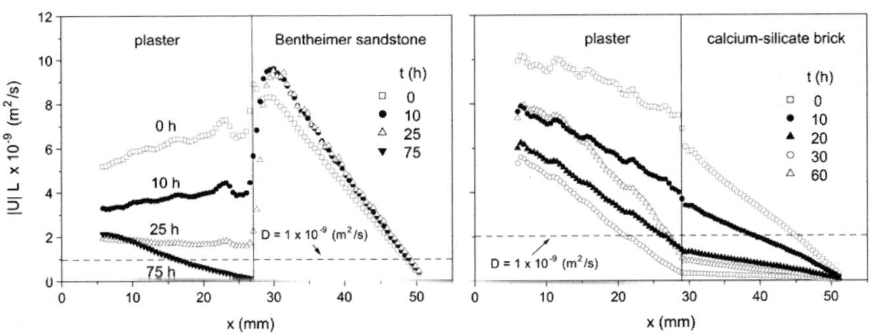

Figure 3. Water velocity multiplied by the thickness of the plaster layer, |U|·L, as a function of the position in the sample for several times during the drying process of the plaster/Bentheimer sandstone system and the plaster/calcium-silicate brick system. The horizontal line indicates the situation where |U|·L equals a diffusion constant of 10^{-9} m^2 s^{-1}, i.e., Pe = 1

These results indicate that diffusion plays a minor role in this stage of the drying process. This is confirmed by the velocity profiles that are calculated from the experimentally observed moisture profiles. In figure 3 the absolute value of the water velocity multiplied by the thickness of the plaster layer |U|·L is shown as a function of the position for several drying times. The horizontal line indicates the situation where |U|·L equals a diffusion constant of 10^{-9} m^2 s^{-1}. This value is roughly equal to the bulk diffusion constant of dissolved Na. It follows from Eq. 1 that above this line |U|·L > D ($Pe > 1$), in which case advection dominates and ions are transported to the drying surface. Below the line |U|·L < D ($Pe < 1$): diffusion transport dominates and ions tend to be uniformly distributed within the various layers of the sample.

During the first drying stage ($t < 25$ h) |U|·L in the plaster decreases but does not vary much with x, since the water is distributed rather uniformly and is constant in time. During this stage the plaster acts only as a transport medium for moisture and ions from the Bentheimer sandstone to the drying surface. During this stage, |U|·L always exceeds D, which indicates that advection dominates. Ions that are present in or transported to the plaster layer will never be able to diffuse back to the Bentheimer sandstone. Therefore all salt will crystallize within the plaster. During the second drying stage ($t > 25$ h) the Bentheimer sandstone is rather dry and the plaster starts to dry. The salt that is present in the plaster will crystallize and, therefore, the amount of dissolved Na decreases. Hence these measurements indicate that almost all salt will be transported from the Bentheimer sandstone into the plaster, which was confirmed by independent chemical analysis[7].

The second system measured was plaster/calcium-silicate brick. In this case the pores of the plaster are an order of magnitude larger than pores of the calcium-silicate brick. The sample was again initially saturated with a NaCl solution, $c = 4$ mol l^{-1}. The water profiles are presented in Figure 4a for several times during the drying process. The total amounts of water in the plaster and the calcium-silicate brick during drying are shown in Figure 4b. Three drying stages can be distinguished, similar to the case of the pure water[7]. The main influence of the salt is that the drying is much slower than in the case of pure water and that again no receding drying front is present. Possible causes of these characteristics have already been discussed above.

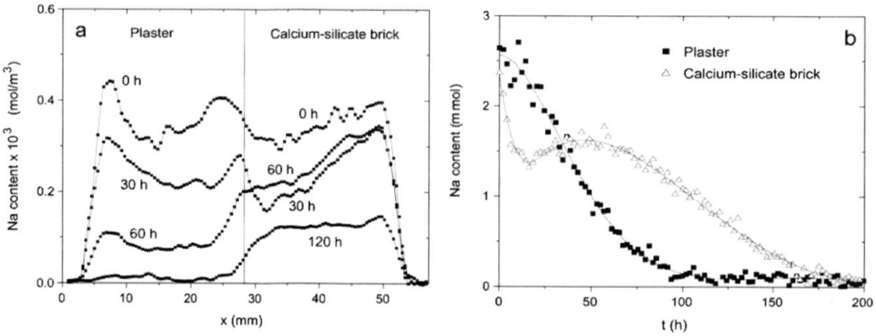

Figure 4. a) Profiles of dissolved Na in the plaster/ calcium-silicate brick system. The data were recorded during the same drying experiment. b) Total amounts of dissolved sodium ions in the plaster and the calcium-silicate brick as a function of the drying time. The meaning of the time t' is explained in the text

During the first drying stage ($t < 12$ h) the amount of Na in the plaster stays constant within experimental accuracy, whereas the amount of Na in the calcium-silicate brick decreases. The velocity profiles are plotted in figure 3b. This figure shows that, advection dominates during the first drying stage; in both materials $|U|\cdot L$ is larger than D ($Pe > 1$), with the obvious exception of a small region near the sealed end of the substrate, where $U=0$. As a result, salt is transported from the calcium-silicate brick to plaster, where it accumulates near the drying surface.

During the second drying stage (12 – 100 h), the amount of dissolved Na in the plaster decreases. If we estimate the NaCl concentration in the plaster from the amounts of water and dissolved Na it has reached the solubility limit ($c = 6$ mol l^{-1}), which implies that crystallization will occur in the plaster. At the same time the amount of dissolved Na in the calcium-silicate brick increases.

Because of the slower drying of the calcium-silicate brick the diffusive transport becomes more important. This is confirmed by the velocity profiles, plotted in figure 3b. During this stage, $|U|\cdot L$ in calcium-silicate brick is lower than D ($Pe < 1$) and diffusion dominates advection. Back diffusion and redistribution of the salt occurs and salt accumulates in the calcium-silicate brick. The amount of dissolved Na continues to increase until the moment that the solubility limit is reached and crystallization occurs. During the third drying stage ($t > 100$ h) no dissolved Na is present in the plaster, because the plaster is essentially dry. The quantity of dissolved Na in calcium-silicate brick decreases due to crystallization. Hence the NMR results on this brick system suggest that salt crystallizes everywhere in the sample. This was confirmed by independent chemical analysis[7].

Acknowledgements

Part of this research was supported by the Dutch Technology Foundation (STW), the Priority Program Materials Research (PPM) and the Center for Building and Systems TNO-TUE.

5. REFERENCES

1. T. Wijffels, C. Groot, and R. V. Hees, "Performance of restoration plasters," in *11th Int. Brick/Block Masonry Conf., Shanghai, China*, pp. 1050–1062, 1997.
2. B. Lubelli, R. van Hees, and C. Groot, "The performance of a restoration plaster in the field: Investigation and monitoring of two case studies," in *International Workshop: 'Repair Mortars for historic masonry', Delft, The Netherlands*, 2005.
3. W*TA Merkblatt 2-2-91 - Sanierputzsysteme, Wissenschaftlich technischen Arbeits-gemeinschaft für Bauwerkserhaltung und Denkmalpflege*. 1992.
4. J. Bear and Y. Bachmat, *Introduction to Modeling of Transport Phenomena in Porous Media*, vol. 4. Dordrecht, The Netherlands: Kluwer, 1990.
5. K. Kopinga and L. Pel, "One-dimensional scanning of moisture in porous materials with NMR," *Rev. Sci. Instrum.*, vol. 65, pp. 3673–3681, 1994.
6. L. Pel, H. Huinink, and K. Kopinga, "Ion transport and crystallization in inorganic building materials as studied by nuclear magnetic resonance," *Appl. Phys. Lett*, vol. 81, pp. 2893–2895, 2002.
7. J. Petkovi, Moisture and ion transport in layered porous building materials: a Nuclear Magnetic Resonance study, Ph.D. thesis, Eindhoven University of Technology, the Netherlands (2005).

FREEZING OF SALT SOLUTIONS IN SMALL PORES

M. Steiger
University of Hamburg, FB Chemie, Martin-Luther-King-Platz 6, 20146 Hamburg, Germany

Abstract: The thermodynamics of liquid water and ice in small pores in the presence of dissolved salts are reviewed. The salt influence is dominant at pore sizes above 10–30 nm while interfacial effects control the behavior in smaller pores. The paper also provides a brief discussion of growth pressures in small pores.

Key words: freezing temperature; ice; crystallization pressure; sodium chloride

1. INTRODUCTION

The interaction of salts and salt mixtures with water in the liquid, gaseous and solid state causes a number of different phase transitions including crystallization or hydration of salts and the crystallization of ice. Crystal growth associated with these processes is generally considered as an important damage mechanism in porous building materials such as stone, brick or concrete. The crystallization of ice in a porous material is not only strongly influenced by the presence of dissolved electrolytes but is also affected by pore size. This paper reviews the thermodynamics of liquid water and ice in small pores in the presence of dissolved salts. The paper also provides a discussion of the crystallization pressure generated by growing ice crystals.

2. VAPOR–LIQUID EQUILIBRIUM

Equilibrium between water vapor and liquid water in an aqueous solution requires equality of the chemical potentials in the liquid phase (μ_l) and in the gas phase (μ_g). The chemical potentials of liquid water and ice are given by

$$\mu_l = \mu_l^\circ + RT\ln a_w = \mu_g = \mu_g^\circ + RT\ln(p_w/p_w^\circ) \tag{1}$$

where a_w is the water activity of the solution, p_w is the water vapor partial pressure over the solution and is the saturation vapor pressure. For pure water at equilibrium we

M.S. Konsta-Gdoutos, (ed.), Measuring, Monitoring and Modeling Concrete Properties, 661–668.

obtain $a_w = 1$ and $p_w = p_w^\circ$. In the case of a salt solution at equilibrium, $a_w = p_w / p_w^\circ$, i.e. the water activity equals the ratio p_w / p_w° which is by definition the relative humidity, *RH*. For water in a porous material the degree of saturation is defined as the fraction of the pore space filled with liquid water. In a water saturated material the liquid–vapor interface is flat and $a_w = RH$. However, in an unsaturated porous material, a curved liquid–vapor interface is formed resulting in a pressure decrease in the liquid phase which is given by Laplace's law,

$$\Delta p = p_l - p_a = 2\gamma_{lv}/r_{lv} = 2\gamma_{lv}\cos\theta/r_P \tag{3}$$

where p_l is the pressure in the liquid phase, p_a is the ambient pressure, γ_{lv} and r_{lv} are the interfacial energy and the radius of curvature of the liquid–vapor interface, θ is the contact angle and r_p is the radius of a cylindrical pore. Note that by definition the radius of curvature of a concave surface is negative. For the chemical potential of liquid water under the reduced pressure p_l we obtain

$$\mu_l = \mu_l^\circ + RT\ln a_w + 2\gamma_{lv}\bar{V}_w/r_{lv} \tag{4}$$

where $\bar{V}_W$ is the partial molar volume of water in the solution. Hence, at equilibrium, the water vapor partial pressure over a solution in an unsaturated porous material is given by:

$$\ln\left(p_w/p_w^\circ\right) = \ln a_w + 2\gamma_{lv}\bar{V}_w/\left(RTr_{lv}\right) \tag{5}$$

Equation (5) yields the depression of the water vapor partial pressure due to the influence of both dissolved salts and the curvature of the liquid–vapor interface in small unsaturated pores. In the derivation of Eq. (5) it is assumed that a_w is not a function of pressure. Strictly, however,

$$d\ln a_w/dp = \left(\bar{V}_w - \bar{V}_w^\circ\right)\Delta p/RT \tag{6}$$

where $\bar{V}_w^\circ$ is the molar volume of pure water. It has been shown[1] that the quantity ($\bar{V}_w - \bar{V}_w^\circ$) is very small even at high solute concentrations. Therefore, the pressure dependence of a_w can be neglected. Also, $\bar{V}_w$ in Eq. (5) may be replaced by $\bar{V}_w^\circ$. Figure 1 depicts the equilibrium relative humidities of aqueous NaCl as a function of pore size for $\theta = 0°$, i.e. $|r_{lv}| = r_p$.

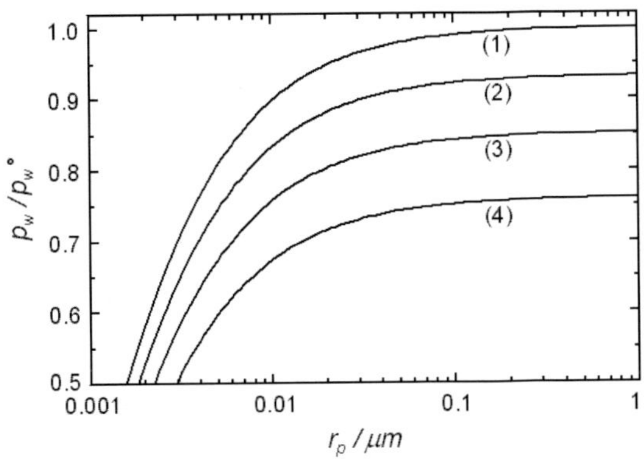

Figure 1. Equilibrium water vapor pressures at 20°C as a function of pore size of (1) pure water and sodium chloride solutions with molalities: (2) 2 mol kg^{-1}, (3) 4 mol kg^{-1} and (4) 6 mol kg^{-1}

The molar volume of pure water was taken from Kell.[2] Water activities were calculated using the ion interaction equations of Pitzer[3] The surface tensions of the solutions were calculated using the following equation

$$\gamma_{lv} = \gamma_{lv}^{\circ} + gm \tag{7}$$

where γ_{lv}° is the interfacial tension of pure water, m is the molality of the solution and g − 1.66 J kg m^{-2} mol^{-1} is an empirical constant which was determined using tabulated surface tensions of sodium chloride solutions.[4] Values of γ_{lv}° were taken from Cini et al.[5] The curves in Figure 1 show that there is a contribution of both influences, pore size and dissolved salts, to the decrease in the water vapor pressure. However, while the interfacial effects are only relevant at pore sizes < 10–30 nm, the reduction in water activity is independent of pore size.

3. FREEZING TEMPERATURE

Considering the equilibrium between ice and liquid water the equilibrium constant K in a bulk solution is given by:

$$\ln K = -(\mu_l - \mu_s)/RT = \ln a_w \tag{9}$$

where μ_l and μ_s are the chemical potentials of liquid water and ice. The temperature dependence of the equilibrium constant may be evaluated from thermochemical data of ice and liquid water at subzero temperatures.[6,7] In our previous work,[8] using the Pitzer equations for the calculation of water activities and activity coefficients in bulk solutions, we calculated freezing temperatures and solubilities in solutions of a number of salts and salt mixtures commonly present in building materials. As an example, Figure 2 depicts the phase diagram of the NaCl–H_2O system at low temperature. Curve (a) represents values of the equilibrium constant K for the liquid water–ice equilibrium in bulk solutions (i.e. $a_w = K$). This curve limits the range of stable existence of a solution at high water activity, i.e. high humidity or low solution concentration. The upper concentration limit is given by the solubility of the crystalline salts. In the *RH*–temperature diagram depicted in Figure 2 this solubility limit is represented by the equilibrium humidities of the saturated solutions of NaCl and NaCl · $2H_2O$, respectively. At relative humidities below the saturation value, a solution cannot exist. Therefore, at low temperatures a solution can only exist in a range of relative humidity and crystallization damage in building materials may be caused by either an increase or a decrease in the ambient *RH*.

The freezing temperature of salt solutions in very small pores is affected by a number of additional influences. The growth of an ice crystal in a cylindrical pore with radius r_p is illustrated in Figure 3. The radius of the crystal is $r_c = r_p - d$ where d is the thickness of an unfrozen liquid film separating the ice crystal and the pore wall.[9]

The liquid phase pressure p_l is determined by the ambient *RH* according to Eqs. (3)–(5). With decreasing *RH* the pressure in the liquid phase decreases reaching the minimum value at $r_{lv} = r_p$. For example, in an unsaturated pore with radius $r_p = 5$ nm the minimum pressure reached at 81% *RH* amounts to about –30 MPa. As freezing is associated with a considerable change in total volume, the reduced pressure in small pores will also affect the liquid water–ice equilibrium. In addition, the melting temperature is also influenced by the crystal–liquid interface which is also subject to strong curvature in small pores. Finally, a crystal growing in a pore might generate stress if confined. Then, the crystal is under anisotropic, non-hydrostatic stress which affects the melting temperature as well.

For a small crystal of irregular shape and under anisotropic stress, there is no uniform value of the chemical potential. It is convenient, however, to define the chemical potential of the solid in the solution as a surface property.[10] In doing so, each face of the crystal is assigned an individual value of the chemical potential depending on both the enhanced pressure due to growth against the pore wall and its curvature.[11,12] In effect, this results in different melting points of individual crystal faces. We obtain for the chemical potential of an ice a crystal in an aqueous pore solution:

$$\mu_s = \mu_{s,0} + (2\gamma_{lv}/r_{lv})V_s + \gamma_{lc}(dA/dV)V_s + \Delta p_c V_s \tag{10}$$

In Eq. (10) $\mu_{s,0}$ is the chemical potential of a large ice crystal at the reference pressure and γ_{cl} is the crystal–liquid interfacial free energy. A and V are the surface and the volume of the growing crystal and $dA/dV = 2/r_c$ in the case of a spherical crystal. The second term on the right-hand side accounts for the reduced pressure in the liquid phase. The third term describes the increase in the chemical potential with decreasing crystal size. Finally, the last term reflects the influence of anisotropic stress, where Δp_c is the crystallization pressure. In Eq. (10) both phases are considered as incompressible. Combining Eqs. (4), (9) and (10), one obtains the following expression for the equilibrium constant of the liquid water–ice equilibrium in small unsaturated pores:

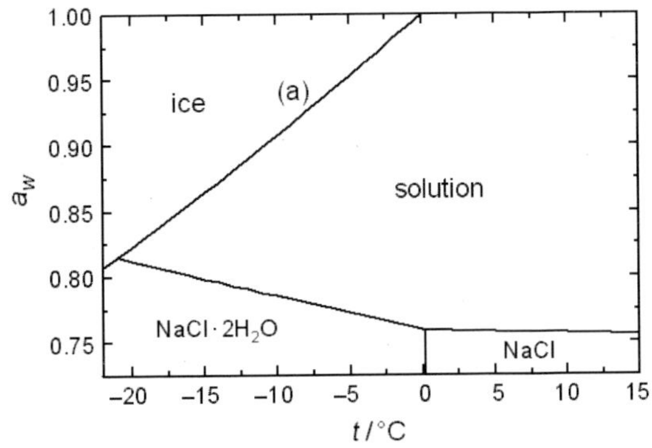

Figure 2. Freezing point depression of NaCl(aq) and saturation relative humidities of NaCl · $2H_2O$ and NaCl

$$RT\ln K = RT\ln a_w + 2\gamma_{lv}/r_{lv}\left(\bar{V}_l - V_s\right) - \gamma_{lc}\left(dA/dV\right)V_s - \Delta p_c V_s \qquad (11)$$

As the values of K continuously decrease with decreasing temperature (c.f. Figure 2), all negative terms on the right-hand side of Eq. (11) cause a depression of the freezing temperature. As discussed before, ln a_w is always negative and dissolved salts depress the freezing temperature. Similarly, the third term reflecting the influence of crystal size is always negative for $r_c > 0$ (convex curvature) resulting in an additional freezing temperature depression. In contrast, the second term is always positive due to the fact that both r_{lv} and $\left(\bar{V}_l - V_s\right)$ are negative. Therefore, under conditions of negative pressure, there is an increase in freezing temperature. The last term in Eq. (11) accounts for the influence of anisotropic stress on the freezing or melting temperature at a loaded crystal face. In the case of a crystal face under enhanced pressure, there is always a decrease of the freezing temperature.

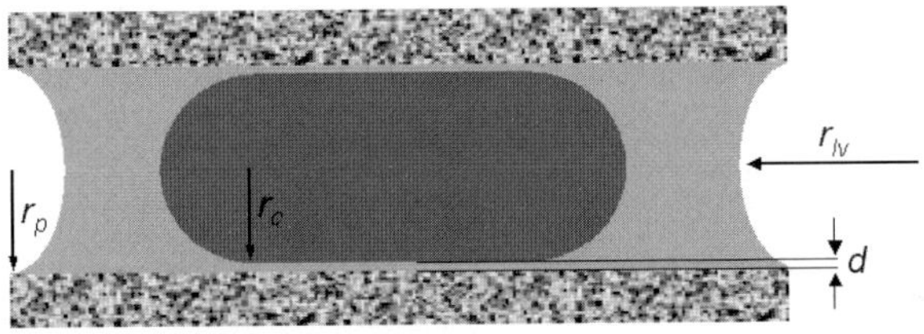

Figure 3. Confined crystal with radius r_c in unsaturated cylindrical pore with radius $r_p = r_c + d$. The curvature of the liquid–vapor interface is $|_{lv}$ with $|r_{lv}| > r_p$.

Figure 4 depicts the freezing temperatures in saturated and unsaturated pores as a function of pore size calculated from Eq. (11) for pure water and NaCl solutions. In the calculations spherical geometry of the ice crystal was assumed, i.e. $dA/dV = 2/r_c$. The thickness of the unfrozen film was assumed as $d = 1$ nm. Molar volumes of supercooled water were taken from Kell[2] while the molar volume of ice ($V_s = 19.65$ cm^3 mol^{-1}) was assumed to be independent of temperature. Values of γ_{cl} were taken from Brun et al[13] and interfacial tensions and water activities of NaCl solutions were calculated as described before.

The curves in Figure 4 indicate that in pores larger than about 10 nm the freezing temperature depression is largely controlled by the presence of dissolved NaCl. In even smaller pores the freezing temperature rapidly decreases mainly as a result of crystal size, i.e. the influence of the interfacial energy of the crystal–liquid interface. In contrast, the pressure decrease in the liquid phase of unsaturated pores causes a minor increase only in the freezing temperature (see dotted curves in Figure 4). This is due to the fact that the difference in the molar volumes of liquid water and ice entering the second term on right-hand side of Eq. (11) is small.

4. CRYSTALLIZATION PRESSURE

According to Eq. (11) anisotropic stress affects the equilibrium freezing temperature of the loaded face of a confined crystal. Comparing the equilibrium constants for the loaded crystal face and the same face without additional load, we obtain the following expression for the crystallization pressure:

$$\Delta p_c = RT/V_s \ln\left(a_w/a_{w,0}\right) \tag{12}$$

where $a_{w,0}$ is the equilibrium water activity of the crystal face of the given curvature under the pressure of the liquid phase.

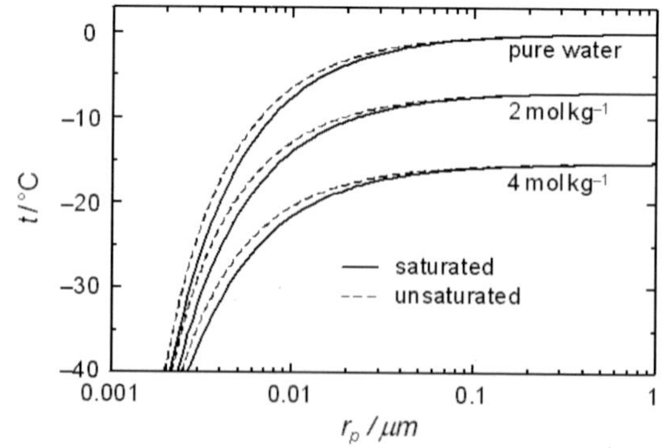

Figure 4. Freezing temperatures in water and NaCl(aq) as a function of pore size in saturated and unsaturated ($|r_{lv}| = r_p$) porous materials

It follows from Eq. (12) that a confined ice crystal can only generate stress if in contact with a solution of greater water activity a_w than its equilibrium water activity under the liquid phase pressure. Hence, the crystallization pressure is directly related to the degree of supersaturation of the pore solution with respect to ice crystallization, where $a_{w,0}$ is the water activity of a solution just saturated with ice. Note that Eq. (12) is entirely consistent with an analogous equation for the crystallization pressure of growing salt crystals.[11]

As long as the confined crystal is under anisotropic stress, the pore solution, which is just in equilibrium with the loaded face of the crystal, is supersaturated in respect to the unloaded faces. In this case, therefore, the crystallization pressure does not represent an equilibrium situation. The solution is metastable and the degree of supersaturation cannot be maintained due to growth on the unloaded faces of the crystal. However, an equilibrium situation can evolve in a very small pore. Consider the crystal depicted in Figure 3. Assuming that the unloaded hemispherical tips of the crystal are just saturated with the pore solution, the water activity a_w of the solution is fixed and may be calculated by using Eq. (11) with $dA/dV = 2/r_c$. However, the same solution is supersaturated with respect to the cylindrical side of the crystal where $dA/dV = 1/r_c$. Using Eq. (12), the degree of supersaturation and, therefore, the crystallization pressure can be calculated by replacing a_w and $a_{w,0}$ with the equilibrium water activities at the tip and at the side, respectively, of the cylindrical crystal. This yields the very simple expression for the equilibrium crystallization pressure in a cylindrical pore:[11]

$$\Delta p_c = \gamma_{cl} / r_c \tag{13}$$

This equation is in entire agreement with the equation of Scherer.[9] The same approach can be used to treat the situation in small pores of different geometry where a permanent equilibrium crystallization pressure can evolve.[11] Nonetheless, also in such small pores the driving force for the generation of stress is always supersaturation. In the case of a large pore, however, equilibrium is not established between the loaded and the unloaded faces of the crystal and the evolution of crystallization pressure is the result of a non-equilibrium situation. The generation of stress is then a dynamic process largely controlled by kinetic influences such as the rates of cooling, diffusion and crystal growth.

5. REFERENCES

1. C. Monnin, The influence of pressure on the activity coefficients of the solutes and the solubility of minerals in the system Na–Ca–Cl–SO_4–H_2O to 200°C and 1 kbar, and to high NaCl concentration, *Geochim. Cosmochim. Acta* **54**, 3265–3282 (1990).
2. G. S. Kell, Effects of isotopic composition, temperature, pressure, and dissolved gases on the density of liquid water, *J. Phys. Chem. Ref. Data* **6**, 1109–1131.
3. K. S. Pitzer, in: *Activity Coefficients in Electrolyte Solutions*, edited by K. S. Pitzer (CRC Press, Boca Raton, 1991), pp. 75–153.
4. A. A. Abramzon and R. D. Gaukhberg, Surface tension of salt solutions, *Russ. J. Appl. Chem.* **66**, 1473–1480.
5. R. Cini, G. Loglio and A. Ficalbi, Temperature dependence of the surface tension of water by the equilibrium ring method, *J. Coll. Interface Sci.* **41**, 287–297 (1972).
6. T. F. Young, The complete calculation of activity coefficients from freezing point data, *Chem. Rev.* **13**, 103–110 (1933).
7. I. M. Klotz and R. M. Rosenberg, *Chemical thermodynamics, basic theory and methods* (Benjamin/Cunnings, Menlo Park CA, 3rd ed., 1972), pp. 374–378.
8. M. Steiger, in: *Proceedings of the 10th International Congress on Deterioration and Conservation of Stone*, edited by D. Kwiatkowski, R. Löfvendahl (ICOMOS, Stockholm, 2004), pp. 179–186.

9. G. W. Scherer, Crystallization in pores, *Cem. Concr. Res.* **29**, 1347–1358 (1999).
10. J. C. M. Li, R. A. Oriani and L. S. Darken, The thermodynamics of stressed solids, *Z. Phys. Chem. N.F.* **49**, 271–290 (1966).
11. M. Steiger, Crystal growth in porous materials: I. The crystallization pressure of large crystals, *J. Crystal Growth* **282**, 455–469 (2005).
12. M. Steiger, Crystal growth in porous materials: II. Influence of crystal size on the crystallization pressure, *J. Crystal Growth*, **282**, 470–482 (2005).
13. M. Brun, A. Lallemand, J.-F. Quinson and C. Eyraud, A new method for the simultaneous determination of the size and the shape of pores, *Thermochim. Acta* **21**, 59–88 (1977).

EFFECT OF THE PORE SIZE DISTRIBUTION ON CRYSTALLIZATION PRESSURE

G. Chanvillard and G.W. Scherer
Lafarge Laboratoire Central De Recherche, 95 Rue Du Montmurier B.P. 15, St Quentin Fallavier 38291 FRANCE; Princeton University, Civil & Env. Eng./PRISM, Eng. Quad. E-319, Princeton, NJ 08544 USA

Abstract: A novel graphical method is introduced for evaluation of the state of equilibrium of a solution crystallizing within a porous medium. This tool makes it easy to anticipate the effects of changes in pore size distribution, initial concentration of salt, or type of salt, on the crystallization pressure at equilibrium.

Key words: salt crystallization, stress, supersaturation, pore size distribution

1. INTRODUCTION

Salt crystallization is widely recognized as a cause of weathering of porous building materials, including stone, mortar and concrete.[1,2,3] A century of research has revealed the key mechanisms involved in such phenomena,[4,5,6] and the thermodynamic aspects of the problem are well understood.[7,8] However, predicting damage to porous media by salt crystallization requires consideration of crystal growth kinetics and fracture mechanics, and no detailed theory is yet available, although progress is being made.[9] In this paper we consider the influence of a pore size distribution on crystallization pressure, using a novel graphical approach that will be described in detail in a future publication.[10]

2. SUPERSATURATION AND PRESSURE

The solubility product for a crystal whose chemical formula is of the form $A_aB_bC_c$ is

$$K = \{a_A\}^a \{a_B\}^b \{a_C\}^c \tag{1}$$

where $\{a_i\}$ is the activity of ion type i. The equilibrium value of the solubility product, K_0, is the value of K when a macroscopic crystal is in equilibrium with a solution.

M.S. Konsta-Gdoutos, (ed.), Measuring, Monitoring and Modeling Concrete Properties, 669–673.

If $K > K_0$, the solution is said to be supersaturated, and the crystal will grow until the supersaturation is consumed (i.e., until K is reduced to K_0). The supersaturation is written as $\beta = K/K_0$.

Chemical equilibrium is reached when $\beta = 1$, leading to a macroscopic crystal with a radius of curvature equal to zero. As the curvature, κ_{CL}, of the crystal/liquid interface increases (i.e., as the radius decreases), equilibrium requires a higher salt concentration in the solution, according to the Ostwald-Freundlich equation[11]:

$$\gamma_{CL}\,\kappa_{CL} = \frac{RT}{v_C}\ln(\beta) \tag{2}$$

where γ_{CL} is the crystal/solution interfacial energy, R is the gas constant, T is the absolute temperature, and v_c is the molar volume of the crystal. Consequently, the smaller the pore, the higher β must be to allow a crystal to penetrate[8].

As crystals grow they consume ions from the solution and consequently its supersaturation decreases. At a given supersaturation, equilibrium requires that all the crystals present have the same radius of curvature. If two crystals (one large and one small) are in the same solution, the smaller crystal will be more soluble, so it will dissolve as the larger one grows. Of course, crystal growth and dissolution depend on solute transport by diffusion. This can be the critical mechanism leading to a transient state of equilibrium.

The pore size distribution can be represented as a cumulative curve of the pore volumes, beginning with the larger pores. If $V_p(r)\,dr$ is the volume of pores with radii between r and $r+dr$, then the cumulative volume fraction of pores larger than r is

$$V_p^c(r_p) = \frac{\int_{r_p}^{\infty} V_p(r_p)\,dr_p}{V_p^T} \tag{3}$$

The volume is normalized by the total porous volume, V_p^T. During crystallization, the pore volume will be filled beginning with the larger pores. The normalized pore size distribution is therefore equivalent to a degree of salt saturation in the pores.

Crystals growing in pores are separated from the pore walls by a film of solution, owing to disjoining forces that oppose contact of the dissimilar crystals. The portion of the crystal in contact with the pore liquid, but not against the pore wall, has a curvature dictated by Eq. (2). Elsewhere, the curvature of the crystal is dictated by the shape of the pore wall, even though it is in contact with the same solution. To preserve equilibrium, a mechanical pressure must be applied on the crystal by the pore wall. In the simple case of a cylindrical pore (and neglecting the thickness of the solution film), the pressure on the pore wall is expressed by[3,12]

$$P_w = \gamma_{CL}\left(\frac{2}{r_p} - \frac{1}{r_p}\right) = \frac{\gamma_{CL}}{r_p} \tag{4}$$

where $2/r_p$ is the curvature of the hemispheric ends and $1/r_p$ is the curvature of the pore wall. For a crystal of arbitrary shape, the pressure varies from point to point on the pore wall, according to the difference in curvature at that point and at the pore entry where the free crystal is in contact with the pore liquid.

Now consider the case of a crystal that has filled a large pore and continues to grow into smaller pores (as a result of an increase in β brought about by evaporation or a change in temperature). The smaller pore radius controls the curvature of the advancing (hemispherical) end. The pressure on the pore wall increases in the larger pores and becomes proportional to the difference between the local curvature and that of the hemispherical end in the pore being invaded. The average pressure exerted on the pore walls by salt crystallization in the pores can be evaluated[13,14] as a pore volume-weighted average of the local pressure over all pore radii containing crystals. If β is in equilibrium with radius r_p, then the average pressure in the body is

$$\bar{P}(r_p) = \frac{\int_{r_p}^{r_{max}} V_p(r) P_W(r) dr}{V_p^T} \tag{5}$$

The smaller r_p is, the higher the average pressure exerted on the pore walls. It is clear that when the average pressure increases, the risk of damage to the porous medium also increases.

3. EQUILIBRIUM DIAGRAM

Equilibrium of a crystal in a porous medium can be represented on a unique graph, illustrated in Figure 1 . The four axes correspond to the relative volume (porosity or salt saturation), molality of the solution, supersaturation, and the pore radius. The average pressure versus pore radius is also included on this graph (dashed curve) introducing an additional axis called the pressure axis.

The two curves on the left side of Figure 1 are only dependent on the chemistry of the crystal, whereas the two curves on the right are specific to the porous medium. The top-right curve (Quandrant I) represents the solution's molality in the pores and its evolution when the salts precipitate. The bottom-right curve (Quadrant IV) describes the pore size distribution of the porous medium. Any change in the pore size distribution will affect this curve and, consequently, the average pressure curve. The unique state of equilibrium is indicated by the rectangular dashed frame.

Beginning in Quadrant I and turning counter clockwise, the functions are successively $m(V_p^c)$, β(m), $r_p(\beta)$, and $V_p^c(r_p)$. Each of these four functions is monotonic, meaning that the signs of their derivatives are constant. Suppose that we pick a particular value of the relative volume, say $\overline{V}_p^c$. The molality of the solution at the point where the salt fills that volume fraction is $m(\overline{V}_p^c)$, the supersaturation of the solution at that point is $\beta(m(\overline{V}_p^c))$, the size of the crystal that would be in equilibrium with that supersaturation is $r_p(\beta(m(\overline{V}_p^c)))$, and the cumulative pore volume up to that pore size is $V_p^c(r_p(\beta(m(\overline{V}_p^c))))$. If the system is in equilibrium, then the latter quantity must be equal to $\overline{V}_p^c$. It can be shown that the root $\overline{V}_p^c$ exists and is unique. Convergence to the solution can be graphically illustrated: starting at any point on one of the curves and circulating clockwise, as in Figure 2, the loop inevitably con-verges on the equilibrium rectangle.

This diagram makes is easy to anticipate the consequences of changes in the parameters. For example, increasing the initial molality of the solution in the pores (i.e., shifting the curve $m(V_p^c)$ upward) raises the equilibrium crystallization pressure by a predictable amount. Similarly, the effects of changes in the type of solute, which affects $\beta(m)$, or in the shape of the pore size distribution can be quantified.

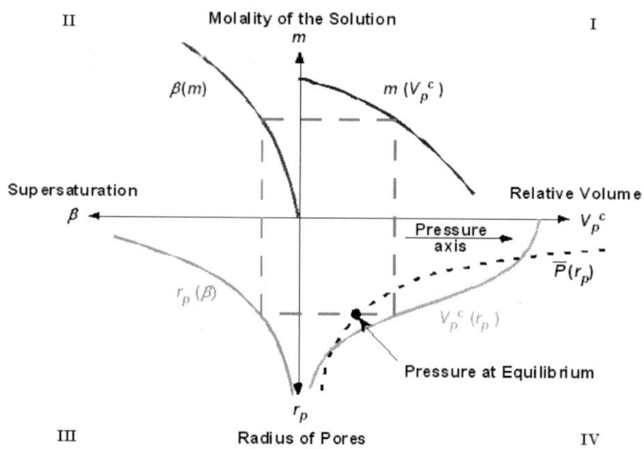

Figure 1. Four quadrant diagram showing the parameters controlling equilibrium. Quadrant I contains molality (m) as a function of fraction of pore volume crystallized (V_p^c), Quadrant II shows supersaturation (β) as a function of molality, Quadrant III shows the pore size (r_p) in equilibrium at supersaturation β, and Quadrant IV shows the cumulative pore size distribution. Also shown in Quadrant IV is the crystallization pressure from eq. (5) versus r_p. The vertices of the rectangle identify the values of each of the quantities at the state of equilibrium

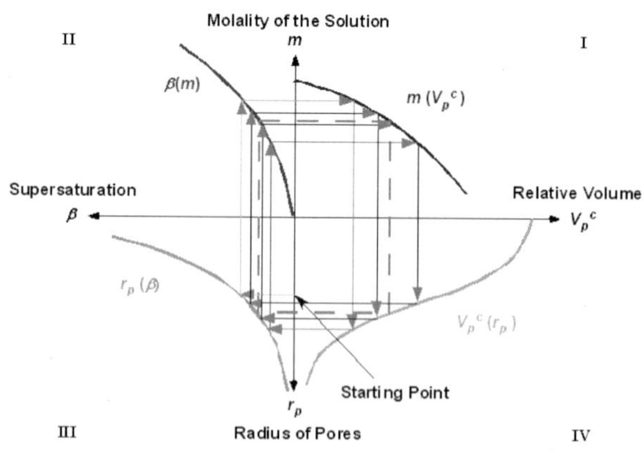

Figure 2. Demonstration of convergence on equilibrium from an arbitrary starting point

4. CONCLUSIONS

A novel graphical approach facilitates the prediction of crystallization pressure, taking account of the effects of supersaturation and the shape of the pore size distribution. Application of this tool to the stress caused by sodium sulfate in limestone will be presented in a forthcoming paper.

Acknowledgments

The authors are grateful to Lafarge for supporting G.C. as a visiting scientist at Princeton University, where this project was undertaken.

5. REFERENCES

1. K. Zehnder and A. Arnold, "Crystal growth in salt efflorescence", J. Cryst. Growth 97 (1989) 513-521
2. A.E. Charola, G.E. Wheeler and R.J. Koestler, "Treatment of the Abydos reliefs: Prelimi-nary investigations", pp. 77-88 in Fourth Int. Cong. Deterioration and Preservation of Stone Objects, ed. K.L. Gauri and J.A. Gwinn (University of Louisville, Louisville, KY, 1982)
3. G.W. Scherer, "Crystallization in pores", Cement Concr. Res. 29 [8] (1999) 1347-1358
4. J. Lavalle, "Recherches sur la formation lente des cristaux ? la temp?rature ordinaire", Compte Rend. Acad. Sci. (Paris) 36 (1853) 493-495
5. S. Taber, "The growth of crystals under external pressure", Am. J. Sci. 41 (1916) 532-556
6. I.S. Evans, "Salt crystallization and rock weathering", Rev. G?omorphologie dynamique XIX [4] (1969-70) 153-177
7. R.J. Flatt, "Salt damage in porous materials: how high supersaturations are generated", J. Cryst. Growth 242 (2002) 435-454
8. G.W. Scherer, "Stress from crystallization of salt", Cement Concr. Res. 34 (2004) 1613-1624
9. O. Coussy, "Deformation and brittle fracture from drying-induced crystallization of salts", Journal of the Mechanics and Physics of Solids
10. G. Chanvillard and G.W. Scherer, "Crystallization of salts in a porous medium: first-order factors for internally generated pressure", to be published in J. Crystal Growth
11. J. Freundlich, Colloid & Capillary Chemistry (Methuen, London, 1926) pp. 154-157
12. D.H. Everett, "The thermodynamics of frost damage to porous solids", Trans. Faraday Soc., 57 (1961) 1541-1551
13. B. Zuber, J. Marchand, A. Delagrave,and J.P. Bournazel, "Ice formation mechanisms in normal and high-performance concrete", J. Mater. Civil Eng. (Feb. 2000) 16-23
14. R. Rossi-Manaresi and A. Tucci, "Pore structure and the disruptive or cementing effect of salt crystallization in various types of stone", Studies in Conservation 36 (1991) 53-58
15. G. Chanvillard and G.W. Scherer, "Quantification of crystallization pressure in lime-stone", to be published in J. Crystal Growth

OPTIMIZATION ASSESSMENT OF COMPATIBLE REPAIR BYZANTINE CONCRETE FOR THE HISTORIC STRUCTURES' RESTORATION INTERVENTION

E. Aggelakopoulou, A. Moropoulou and A. Bakolas
Section of Materials Science and Engineering, School of Chemical Engineering, National Technical University of Athens, 9 Iroon Politechniou Str., 15782, Zografou, Greece

Abstract: The present work is dealing with the optimization assessment of repair Byzantine concrete addressed to restoration interventions on thick joints brickwork masonries of Byzantine era. The design of these materials was based on the data obtained by the characterization of historic concrete of typical Byzantine era monuments that presented a great durability in time. Traditional materials (hydrated lime, hydraulic lime, pozzolanic additions, sand and brick fragments) are used for the concrete preparation. The mechanical characteristics of concrete syntheses are evaluated by using mechanical tests (compressive strength, flexural strength, static modulus of elasticity) and ultrasonic technique (ultrasonic velocity, dynamic modulus of elasticity) at the time of 1, 3, 6 and 12 months. The obtained results indicate that the concrete prepared by a high mixing ratio of lime/metakaolin present sufficient values of compressive strength, flexural strength and static modulus of elasticity.

1. INTRODUCTION

Characteristic monuments of Byzantine era that presented a great durability in time and an excellent behavior under earthquake stresses are the Hagia Sophia in Istanbul, and the church of St. Michael in Kiev [1,2,3]. A thorough study in the structural materials revealed that they presented a similar nature and production technology. The study of these buildings could become a valuable tool for the decoding of construction techniques and materials and therefore the reproduction of materials with analogous behavior that could be used for the restoration interventions on structures of Byzantine era.

Regarding the masonry structure type of these two monuments, the structural materials were bricks and the mortar joint was about 1-1.5 times the brick thickness, up to 4-5 cm. Furthermore the mortars used presented a binder of hydraulic nature. The aggregates were coarse and composed by a mixture of ceramic fragments and sand. Especially, in the

M.S. Konsta-Gdoutos, (ed.), Measuring, Monitoring and Modeling Concrete Properties, 675–684.

case of Hagia Sophia, the aggregates dimensions are in the range of 0-16mm and this is the reason that this kind of mortar is considered as concrete. Regarding their physico-chemical and mechanical characteristics they both exhibited a high tensile strength, a low value of elasticity modulus, low values of density, high value of hydraulicity [1,2,3,4].

The main goal of this research is the design, production and evaluation of concrete for restoration interventions of monuments thick joints brickwork masonries of Byzantine era that simulate the historic concrete, regarding the physico-chemical and mechanical data. In addition, as secondary goals, the effect of aggregates and binder nature to the mechanical characteristics of concrete synthesis is, also, studied.

2. EXPERIMENTAL

Traditional types of materials were used for the concrete production in order to assure the physico-chemical compatibility to authentic materials. Lime powder ($Ca(OH)_2$: 89%, $CaCO_3$: 5%, CaO Hellas) and natural hydraulic lime (NHL3.5-Z according to CEN EN 459-1) were used as binding materials along with cement (I/45, TITAN Cement Industry) for comparative reasons.

The pozzolanic additions used were either earth of Milos (EM) – a natural pozzolan derived by the island of Milos in Greece, or metakaolin, an artificial high reactive pozzolanic addition [5] (Metastar 501 of IMERYS Minerals L.t.d). Table 1 reports the chemical composition, the physical properties of materials used for concrete preparation. In addition, the percentage of total and reactive silica (EN 197-1, EN 196-2) for the two pozzolanic additions is presented along with the values of specific surface area, measured by the adsorption of nitrogen method, according to Brunauer-Emmett-Teller (BET) method.

The grain size distribution of the pozzolanic additions is determined by laser CILAS 715 method. Metakaolin is the finest pozzolanic addition with cumulative passing percentage at 64 μm up to 100% and at 16 μm up to 95.6%. On the other hand, Earth of Milos presents more coarse grain size distribution with a cumulative passing percentage at 64 μm up to 88.1%.

The aggregates used consist of a mixture of sand (calcitic and quartz origin) and ceramic fragments. The former has already been detected in historic mortars samples resulted to the production of lightweight, low-modulus of elasticity materials, due to its lower bulk density in respect to the sand aggregates [6].

Table 1. Physico-chemical characteristics of materials used for the concrete preparation

MT	SiO_2 (%)	Al_2O_3 (%)	Fe_2O_3 (%)	CaO (%)	MgO (%)	K_2O (%)	Na_2O (%)	SO_3 (%)	LOI (%)	Tot. SiO_2 (%)	React. SiO_2 (%)	S.S.A. (m^2/g)
MK	51.70	40.60	0.64	0.71	0.96	2.00	0.31	0.11	1.19	54.19	44.59	13.83
EM	69.66	12.21	2.34	2.01	0.70	3.28	0.62	-	7.35	70.15	52.89	4.34
L	0.17	0.18	0.07	70.06	2.35	-	-	0.77	25.60	-	-	-

MT: Material, MK: Metakaolin, EM: Earth of Milos, L: Hydrated Lime, LOI: Loss of ignition, Tot. SiO_2: total silica percentage, React. SiO_2: reactive silica percentage, S.S.A.: Specific Surface Area

In order to achieve an analogous grain size distribution to Hagia Sophia historic one 4 types of aggregates were mixed:

- Sand of quartz origin with the following fractions of grain size (0.063-0.5 mm, 0.5-1 mm, 1-2 mm, 3-6 mm)
- Sand of calcitic origin with a grain size fraction of 2-4 mm.
- Coarse calcitic gravel with a grain size fraction of 2-16 mm.
- Ceramic Fragments disposed in two grain size fractions (0-8, 2-16 mm)

Figure 1 presents the grain size distribution of aggregates mixes compared to the Hagia Sofia historic concrete.

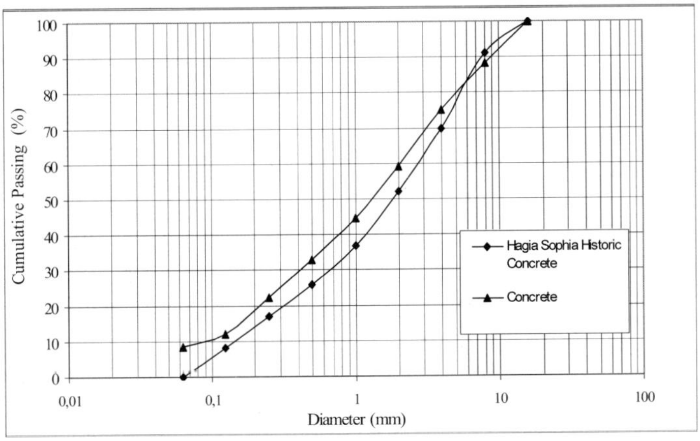

Figure 1. Grain size distribution of aggregates compared to the Hagia Sofia Historic Concrete

Table 2 states the materials mixing proportions as percentage per weight (%, p.w.) for the concrete preparation. The amount of water that was added in the syntheses was determined through the criteria of slump cone test. The acceptable value of slump was determined as lower than 40 mm, according to EN 12350-2 (Testing Fresh Mortar – Part 2: Slump Test). In that way, the amount of water was the minimum that could be added and the syntheses presented almost the same consistency.

Once the concrete was prepared, it was molded in moulds of 10x10x50cm, using a vibrator table with the intention of accomplishing a sufficient compaction. Then, they were stored in a moist curing chamber of relative humidity RH>95% and temperature T=20±2°C for 7 days in the case of concrete 3-9 and 14 days in the case of concrete 1-2. Afterwards, they were demoulded and stored in a chamber of standard conditions (RH=50±1%, T=20±2°C) till the testing day.

The concrete mechanical and chemical characteristics were evaluated using the following techniques:

- Fiber Optics Microscopy for the evaluation of the adhesion of aggregates to the binder matrix and the homogeneity of the binder at 12 months of curing time

- Flexural strength, using prisms of 10x10x50cm according to ASTM C 78-00 (Triscan 100, max load: 100 KN, load rate: 0.05 mm/min) at the time of 1, 3, 6 and 12 months
- Compressive strength tests using portions of beams broken in flexure according to ASTM C 116 (Form – Test Type: 110/300, max load: 300 KN load rate: 0.2 MPa/s) at the time of 1, 3, 6 and 12 months
- Compressive strength tests using cylinders (height: 30 cm, diameter: 15 cm) according to ASTM C116-99 (Impact CTE 19, maximum load: 2000 KN, load rate: 1 KN/s) at the time of 12 months
- Ultrasonic technique (CNS Farnell-Pundit 6, transducers frequency: 54 KHz) for the estimation of ultrasonic velocity propagation through concrete and the dynamic modulus of elasticity (Ed) at the time of 1, 3, 6 and 12 months

Table 2. Mixing Design of Concrete - Materials mixing proportions as percentage per weight (%, p.w.)

	Code	L	EM	MK	C	NHL	Aggregates (%.p.w.)		
							Gr	Snd	CF
1	EM2.Gr.S.	10	20				7	63	
2	EM2.CF.S.	10	20					35	35
3	MK1.Gr.S.	15		15			7	63	
4	MK1.CF.S.	15		15				35	35
5	MK1.CF.	15		15				35	35
6	MK05.CF.S.	20		10				35	35
7	LCem.CF.S.	15			15			35	35
8	NHL.CF.S.					30		35	35
9	MK2.5.CF.S.	27,5		2,5				35	35

L: Hydrated Lime, EM: Earth of Milos, MK: Metakaolin, C: Cement, NHL: Natural Hydraulic Lime, Gr: Gravel, Snd: Sand, CF: Ceramic Fragment

The concrete mechanical and chemical characteristics were evaluated using the following techniques:

- Fiber Optics Microscopy for the evaluation of the adhesion of aggregates to the binder matrix and the homogeneity of the binder at 12 months of curing time
- Flexural strength, using prisms of 10x10x50cm according to ASTM C 78-00 (Triscan 100, max load: 100 KN, load rate: 0.05 mm/min) at the time of 1, 3, 6 and 12 months
- Compressive strength tests using portions of beams broken in flexure according to ASTM C 116 (Form – Test Type: 110/300, max load: 300 KN load rate: 0.2 MPa/s) at the time of 1, 3, 6 and 12 months
- Compressive strength tests using cylinders (height: 30 cm, diameter: 15 cm) according to ASTM C116-99 (Impact CTE 19, maximum load: 2000 KN, load rate: 1 KN/s) at the time of 12 months

- Ultrasonic technique (CNS Farnell-Pundit 6, transducers frequency: 54 KHz) for the estimation of ultrasonic velocity propagation through concrete and the dynamic modulus of elasticity (Ed) at the time of 1, 3, 6 and 12 months

3. RESULTS

In general, all the concrete syntheses present a uniform distribution of the aggregates in the whole concrete mass and a good homogeneity in binder matrix (Photograph 1: concrete 5). Photographs 2 and 3 present the beam core of the concrete 5 and 3, obtained by FOM, where one could notice that a better adhesion of the binder matrix to the aggregates is occurred in the case of ceramic aggregates than to the gravels/sand aggregates.

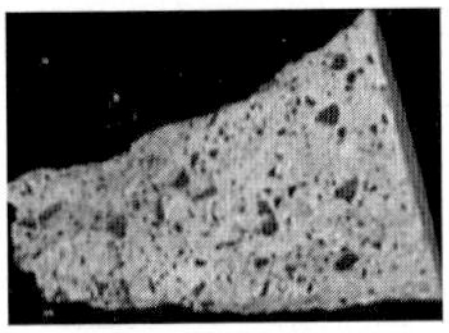
Photo 1. Concrete 5

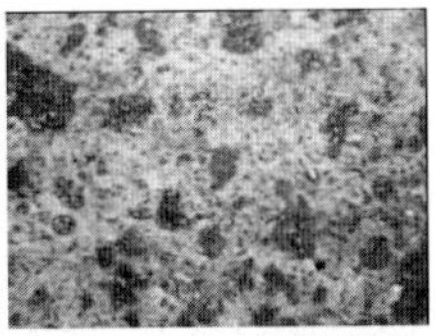
Photo 2. Concrete 5 (FOM)

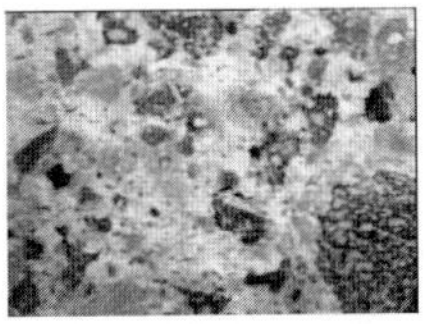
Photo 3. Concrete 3 (FOM)

Figures 3 and 4 report the flexural and compressive strength data for concrete at the time of 1,3, 6 and 12 months of curing. Regarding the effect of binder nature to concrete mechanical strength it could be noticed that in the case of using EM as a pozzolanic addition, the concrete presents low values of compressive strength (6.2 & 5.8 MPa for the syntheses 1 & 2, respectively at 12 months of curing time) and flexural strength (0.44-0.54 MPa). This fact could be attributed to the low reactivity of this pozzolanic addition regarding the $Ca(OH)_2$ consumption. Moreover, in the case of EM concrete the maximum value of compressive strength is gained by the time of 6 months whereas the flexural strength increases till 12 months of curing time.

On the other hand, MK concrete exhibit a wide range of compressive strength (8.4-26.5 MPa) and flexural strength (1.42-2.82 MPa) values at 12 months of curing time, fact that could become a valuable tool for the design of restoration concrete, taking into account each time the historic structure' s specific characteristics. MK concrete syntheses (3-6) present high values of mechanical strength. By the time of 3 to 6 months MK concrete present the maximum value of mechanical strength while beyond this period it is decreased. By the time of 1 month these mortars gain the 90-100% of the final compressive strength and the 79-100% of the final flexural strength and therefore it could be used as a pozzolanic addition for restoration mortars/concrete production in order to ameliorate the early strength of hydrated lime mortars.

Comparing the concrete syntheses 3, 6 and 9 as far as the 3, 4 and 5, it could be figured that by increasing the lime percentage and the ceramic fragments percentage, the mechanical strength is reduced.

Almost all MK concrete syntheses (concrete 3, 4, 5, 6) present too high values of compressive strength compared to the one of traditional handmade bricks of brickwork masonry. Regarding the mechanical compatibility, only concrete 9 presents a sufficient value of compressive and flexural strength at 12 months of curing.

Furthermore, some results could be drawn regarding the two different pozzolanic additions that are tested. From the obtained mechanical strength results and DTA/TG results, it could be said that metakaolin presents a much higher reactivity in reacting with lime compared to the natural pozzolanic addition. This fact resulted to a total increase on the compressive the flexural strength up to 400% and it could be attributed to the metakaolin fine grain size distribution, chemical composition and high specific surface area.

Concrete produced by mixing hydrated lime and cement (Concrete 7) exhibit an increase in compressive strength values till the time of 12 months whereas a decrease in flexural strength is observed from 6 to 12 months. Regarding the mechanical behavior of hydraulic lime concrete, it is observed that the strength is increased till the 12 months of curing time. Though, both syntheses present too high values of mechanical strength regarding the strength of traditional structural materials.

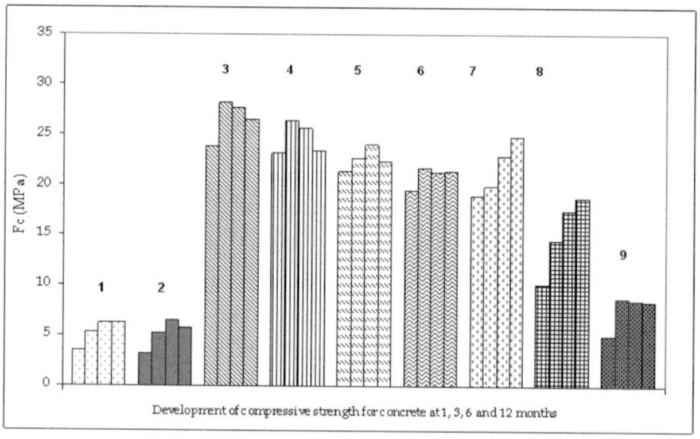

Figure 3. Compressive strength for concrete at the time of 1,3, 6 and 12 months of curing

Figure 5 presents the data of static modulus of elasticity for concrete. In general, it could be said that the *Est.* data are in accordance with the mechanical strength ones, meaning that concrete with high mechanical strength present, also, high values of modulus of elasticity. Furthermore, it could be observed that concrete produced by earth of milos present low values of static modulus of elasticity ($E_{st.}$: 337-570 MPa) while the metakaolin concrete exhibit much higher values (1950-5650 MPa).

In addition, it could be discerned that by using ceramic fragments as aggregates instead of conventional aggregates, the value of modulus of elasticity is decreased. Though, synthesis 5 where the aggregates were all ceramic fragments exhibit a decrease in $E_{st.}$ value up to 50% compared with synthesis 4 where the ceramic fragments comprise the 50% of the total percentage of aggregates used.

Furthermore, by using hydrated lime in a higher proportion in concrete the static modulus of elasticity is decreased. Synthesis 4 exhibits value of static modulus of elasticity up to 4190 MPa while syntheses 6 equal to 2419 MPa. Finally, concrete prepared by natural hydraulic lime or by mixing hydrated lime with a low percentage of metakaolin presents the lowest value of static modulus of elasticity (E_{st}. ~824 MPa).

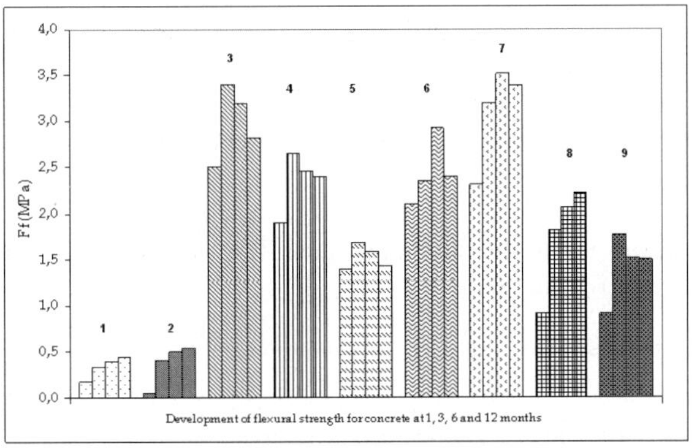

Figure 4. Flexural strength for concrete at the time of 1,3, 6 and 12 months of curing

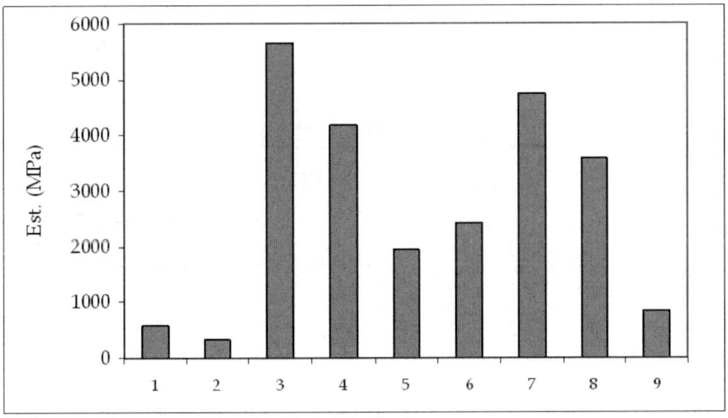

Figure 5. Static modulus of elasticity for concrete at 12 months of curing time

Table 3 present the data of concrete's physical and mechanical characteristics. Comparing the data of apparent density, the minimum value is reported for concrete synthesis 5 (~1.56 g/cm^3) where only ceramic fragments are used. On the other hand the maximum value is stated for concrete syntheses 1 (~1.95 g/cm^3) where the aggregates of sand and gravel are used as aggregates. Furthermore, concrete syntheses 9 exhibits low value of apparent density (~1.63 g/cm^3), fact that could be attributed to the use of a high percentage of hydrated lime and the use of ceramic fragments up to 50% in the total fraction of aggregates.

Concerning the ultrasonic velocity, it could be observed that it exhibits similar trend with the mechanical strength. More specifically, the obtained data drawn that in general, concrete that present high values of ultrasonic velocity, present also high values of mechanical strength.

Regarding the dynamic modulus of elasticity, the maximum value is reported for natural hydraulic lime concrete while high values are reported for syntheses 3 and 7. On the contrary, the lowest values are reported for concrete of earth of milos and concrete syntheses 9 (7539 MPa).

Comparing, the data of static modulus of elasticity with the dynamic one, it could be observed that the data differ very much, though, they present similar trend. The ratio of Ed./Est. varies in the range 3-17, while these data seem to be close as the ultrasonic propagation velocity increases.

Finally, it could be noticed that there is a correlation between the compressive strength and the ultrasonic pulse velocity (Figure 7) as far as the ultrasonic velocity with compressive strength (Figure 8).

Table 3. Physico-mechanical characteristics of concrete at 12 months of curing time

	Code	dapp. (g/ cm^3)	SD	Vu.s. (m/s)	SD	Ed. (MPa)	Est. (MPa)
1	EM2.Gr.S.	1,95	0,04	1970	126	6811	570
2	EM2.CF1.S.	1,74	0,02	1909	128	5720	337
3	MK1.Gr.S.	1,87	0,02	3328	137	18676	5650
4	MK1.CF1.S.	1,77	0,01	2753	78	12080	4190
5	MK1.CF1.	1,56	0,01	2618	50	9604	1950
6	MK05.CF2.S.	1,69	0,03	3077	57	14401	2419
7	LCem.CF2.S.	1,83	0,04	3214	225	17001	4744
8	NHL.CF2.S.	1,85	0,03	3469	139	20031	3574
9	MK2.5.CF2.S.	1,62	0,01	2274	104	7539	824

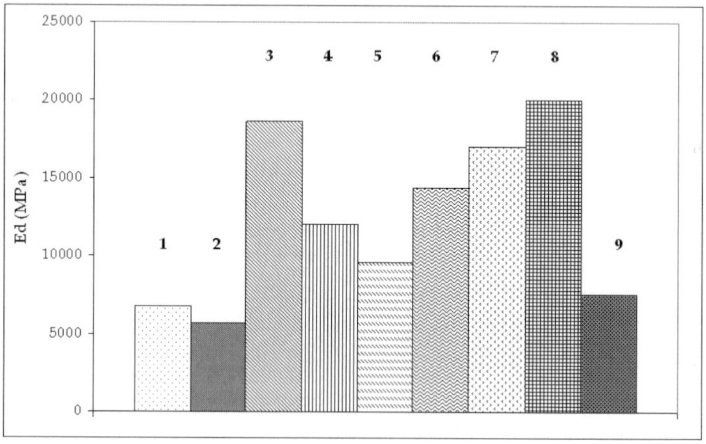

Figure 6. Concrete dynamic modulus of elasticity (Ed) in 3 months of curing

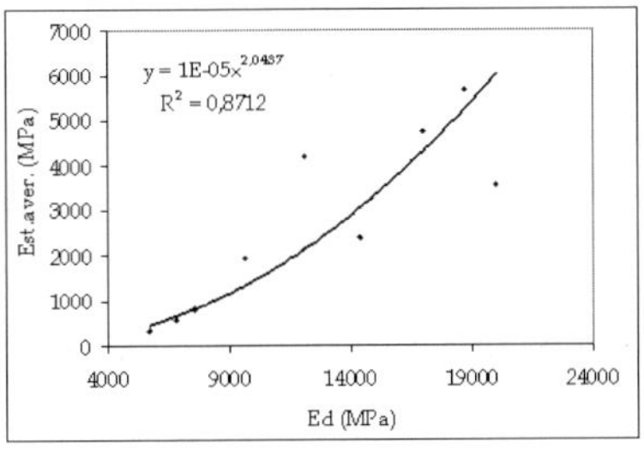

Figure 7. Dynamic modulus of elasticity versus static modulus of elasticity

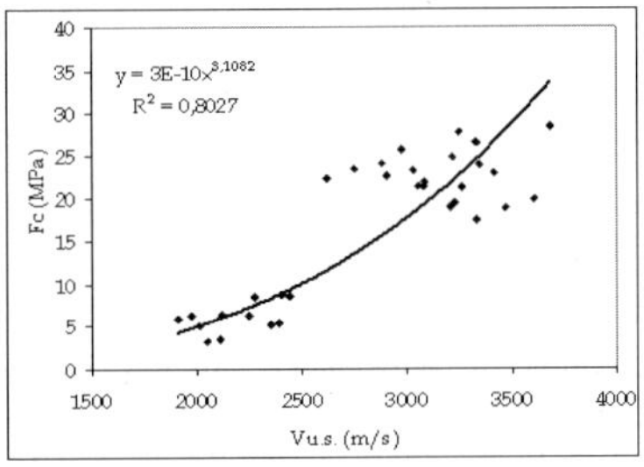

Figure 8. Ultrasonic velocity versus compressive strength

4. CONCLUSIONS

From the obtained results the following conclusive remarks can be point out:

- The natural pozzolanic addition-earth of milos presents low reactivity regarding the $Ca(OH)_2$ consumption resulted to low values of final mechanical strength and mechanical strength acquisition.
- The artificial pozzolanic addition-metakaolin presents high reactivity regarding the $Ca(OH)_2$ consumption due to its physicochemical and mineralogical characteristics. Therefore, it could be used as a pozzolanic addition (in small percentages) in hydrated lime restoration mortars in order to ameliorate the early strength of lime mortars.
- Concrete prepared by mixing lime/metakaolin/ceramic fragment/sand:27.5/2.5/35/35 (p.w.%), present sufficient mechanical strength (F_c: 8.4 MPa, F_f: 1.50 MPa and Est.: 824 MPa), in 12 months of curing time, assuring in that way the mechanical compatibility with the traditional structural materials of Byzantine structures.
- Concrete produced by natural hydraulic lime or lime/cement in mixing ratio (p.w.):1/1 or by mixing lime and metakaolin in ratio 1/1 or 2/1 (p.w.) present too high values of mechanical strength, fact that could provoke a mechanical incompatibility problem in historic brickwork masonries.
- By increasing the hydrated lime percentage in the mixture and the percentage of ceramic fragments in the total fraction of aggregates, a decrease in compressive and flexural strength, dynamic and static modulus of elasticity and apparent density occurs.
- There is a correlation between the compressive strength and the ultrasonic velocity as far as the ultrasonic velocity with compressive strength.
- The ratio of Ed./Est. varies in the range 3-17, while these data seem to be close as the ultrasonic propagation velocity and the compressive strength increases.

Acknowledgments

The Authors would like to thank the TITAN S.A. Cement Industry - Research and Development Department, for the technical support for the experiments accomplishment and their valuable scientific contribution. They, also, would like to thank the CENTER OF PUBLIC CORPORATION - Testing, Research and Standards Center for the mechanical tests accomplishment as far as the EARTHQUAKE PLANNING AND PROTECTION ORGANIZATION of GREECE and the GENERAL SECRETARY FOR RESEARCH AND DEVELOPMENT OF GREECE for their financial support of this research.

5. REFERENCES

1. Cakmak, A.S., Moropoulou, A., Mullen, C.A., Interdisciplinary study of dynamic behavior and earthquake response of Hagia Sophia, J. Soil Dynamics and Earthquake Engineering, Vol 14, No 9, (1995), pp. 125-133.
2. Moropoulou, A., Cakmak, A.S., Biscontin, G., Crushed brick lime mortars of Justinian's Hagia Sophia, Materials Issues in Art and Archaeology V, Materials Research Society, Vol. 462 (1997), pp.307-316.
3. Moropoulou, A., Cakmak, A.S., Lohvyn, N., "Earthquake resistant construction techniques and materials of byzantine monuments in Kiev", Soil Dynamics and Earthquake Engineering, 19 (2000) pp. 603-315.
4. Moropoulou, A., Lohvyn, N., "Earthquake resistant byzantine (11th c.) Church of St. Michael in Kiev", PACT, J. European Study Group on Physical, Chemical, Biological and Mathematical Techniques Applied to Archaeology, 58 (2000) pp. 53-69.
5. Moropoulou, A., Bakolas, A., Aggelakopoulou ., "Evaluation of pozzolanic additionic activity of natural and artificial pozzolanic additions by thermal analysis", Thermochimica Acta, 422/1-2, (2004), In Press.
6. Livingston A.R., Stuzman E.P., Mark R., Erdik M. Preliminary analysis of the masonry of Hagia Sophia Basilica, In: Materials Issues in Art and Archaeology III, Mat. Res. Soc., Pittsburgh, 1992, p. 721-736

EVALUATING THE POTENTIAL DAMAGE TO STONES FROM WETTING AND DRYING CYCLES

I. Jiménez González and G.W. Scherer

University of Granada, Spain; Princeton University, Dept of Civil & Env. Eng/ Princeton Materials Institute, Eng. Quad. E-319, Princeton, NJ 08544 USA

Abstract: The literature on stone conservation often mentions that clay-containing stones can be damaged over time through cycles of wetting and drying (Félix 1988). Several studies demonstrate the deleterious action of these cycles on stones consolidated with ethyl silicates [Félix and Furlan (1994), Félix (1995)]. However, to our knowledge, only one study (Wendler et al. 1996) demonstrates that these cycles can damage unconsolidated stone. The procedure is rather long and probably this is the reason for which so little work has been done to examine the importance of this damage mechanism.In this paper, we present a testing machine that has been developed to automate and accelerate the rate at which stone samples may be submitted to these cycles. Direct measurement of swelling indicates that swelling increases with the number of cycles, indicating progressive damage. However, the swelling can be durably reduced, although not completely eliminated by swelling inhibitors.We use a novel technique to examine the behavior of swelling stones, which consists in measuring the warping of a thin stone plate placed on two supports and which is wetted from above. Deflection and relaxation of the plate can be analyzed to extract free swelling, the ratio of wet to dry modulus and the sorptivity of the stone. However, agreement with separate measurements requires introducing a separate kinetic expression for the rate of swelling.

Key words: sandstone; swelling clays; consolidation; wetting and drying; swelling; sorptivity; warping; swelling inhibitors, fatigue, elastic modulus

1. INTRODUCTION

Alteration of clay-bearing stones is often attributed to stresses arising from cycles of swelling and shrinking of the clays. However, to the best of the authors' knowledge, apart from stones consolidated with ethyl silicates [Félix and Furlan (1994), Félix (1994, 1995)], only one study successfully demonstrates this [Wendler et al. (1996)].

M.S. Konsta-Gdoutos, (ed.), Measuring, Monitoring and Modeling Concrete Properties, 685–693.

The experiments of Wendler et al. clearly establish that wetting and drying cycles can damage clay-bearing stones. Furthermore, they demonstrate that swelling inhibitors may actively reduce this damage, although they do not eliminate it completely. In previous work we have presented analysis of the stress development during wetting and drying cycles, showing to a first approximation that stresses should increase with free swelling strain, but that those could be significantly reduced by stress relaxation, the values of which were characterized by bending techniques (Jiménez González and Scherer 2004). Based on these results, in this paper we examine the main effect of this damage mechanism by determining the evolution of free swelling during cycles of wetting and drying, as well as the effect of swelling inhibitors during such cycles.

In addition, we use the warping technique (Scherer and Jiménez González 2005) to examine the variation of sorptivity and the ratio of the wet to dry modulus during these tests.

2. MATERIALS AND METHODS

Portland Brownstone, a coarse ferruginous sandstone quarried in the Connecticut River valley and widely used in the North East of America, was obtained from Pasvalco Co. (Closter, NJ, USA). This stone has been reported to suffer extensive degradation due to swelling clays. It is mostly composed of quartz grains coated by iron oxide films with a variable amount of feldspar and mica (flakes of muscovite), with a cementing phase mostly made of silica and clays. It shows evident bedding planes and samples discussed in this paper were cut so that water ingress would take place in the direction parallel to the bedding planes

By swelling inhibitors, we refer to products that limit or eliminate the swelling that clay-bearing stones undergo when exposed to water or humidity. We have shown that best results are obtained when such products are formulated as a mixture of various small organic compounds (Jiménez González & Scherer 2004). The mixture used in this paper involves a 1,3 Diaminopropane dihydrochloride ($H_2N(CH_2)3NH_2.2HCl$) the use of which was first suggested by Snethlage and Wendler (1991) and a corrosion inhibitor for concrete based on aminoalcohols. The stones were treated by partially submerging them in a solution after having been oven dried. Further details are available elsewhere (Jiménez González & Scherer 2004). After this treatment the samples remain hydrophilic.

A home-made dilatometer was used to measure the linear expansion of the samples (Jiménez González & Scherer 2004). Samples were plates of about 100×22×3.6 mm with the bedding perpendicular to their longest dimension. The oven-dried samples (60°C) were placed on end in a stainless steel sample holder. They were held in place by four plastic screws, which were found to stabilize the samples without preventing their swelling. The sample and sample holder were placed in a glass container and the pushrod of an LVDT was lowered on top of the sample to allow displacement measurement. After data acquisition was started, deionized water was poured around the sample until it reached the upper surface of the sample. As soon as the surface gets wet, the sample starts swelling.

Sorptivity measurements were performed with an electronic balance with a 0.001 g resolution and connected to a computer for a data acquisition (Scherer and Jiménez González 2005). A dish of water was raised into contact with the bottom of the sample and the weight change was continuously monitored.

To test fatigue resistance to wetting and drying cycles, as well as the duration of the swelling inhibition treatments, we built a special machine that can submit stone samples to a large number of cycles. This was necessary, because the initial manual testing with up to about 20 cycles did not show any significant change in mechanical properties of the stones.

The machine is illustrated in Figure 1. It consists of two parallel belts on which rings of stainless steel springs allow to fix up to 70 thin stone plates. Rotation of the wheels on which the belts are fixed brings the samples successively into a water bath and into a zone in which fans dry the sample. The duration of the cycles can be easily adjusted. However, in our experiments, the machine was set up so that the impregnation lasted 30 minutes and drying lasted 60 minutes, for the case of Portland Brownstone. The water bath has a large volume and a constant flow of water to avoid any contamination of the untreated samples by possible washout of the swelling inhibitors from the treated samples.

Figure 1. Wetting and drying testing machine. General view (a), detail of the bath and sample holder (b)

Details for the warping technique, which we have introduce for characterizing stones, can be found elsewhere (Scherer and Jiménez González 2005). In short, it consists in measuring the deflection of a thin plate of stone placed horizontally on two supports. The plate warps upwards as a result of adding water on its upper surface, and this is measured by a LVDT. A mathematical analysis described later gives the swelling strain, sorptivity and ratio of wet to dry modulus from the time dependent deflection of the sample.

3. RESULTS

3.1 Swelling

Results of the evolution of the free swelling strain obtained by direct measurement are reported in Figure 2. Evolution of the free swelling strain (direct measure) during the cycling. The samples Pb-22, 23 and 20 all received a treatment with the swelling inhibition mixture. The samples Pb-7 and 43 were not treated. The data indicate that this treatment reduces the initial swelling of this stone by about 42%. The data collected after 100, 200, and 700 cycles of wetting and drying show overall a gradual increase of swelling with the number of cycles regardless of whether they have been treated or not.

This increased swelling can be attributed to damage, in that a decrease of the material's stiffness will lower its ability to resist the swelling pressure caused by the wetting of the clays. From the perspective of conservation practice however, the important information of these results is that the effect of the treatment maintains the swelling of these samples well below their initial value, even after 700 cycles. Consequently, the treatment we propose reduces durably, but not indefinitely, the expansion and the associated damage. The aging of the treated samples may be due to the residual swelling strains or to a partial washout of the applied products.

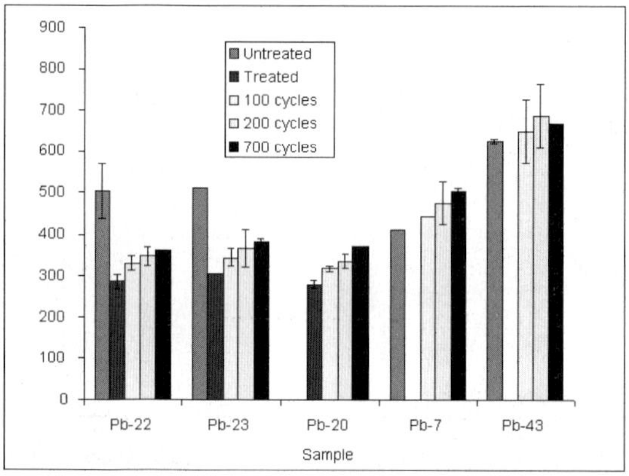

Figure 2. Evolution of the free swelling strain (direct measure) during the cycling

3.2 Warping

3.2.1 Swelling strain

In addition to doing direct measurements of swelling during these tests, we have also performed warping measurements on the same samples. The analysis of the warping experiment leads to the following expression for the height of the deflection, Δ, as a function of the depth to which the water has penetrated:

$$\Delta = \left(\frac{3w^2\varepsilon_{fw}}{4h}\right)\left(\frac{r(1-d)d}{d^4(1-r)^2 - 4d^3(1-r) + 6d^2(1-r) - 4d(1-r) + 1}\right) \qquad (1)$$

where w is the span of the plate, ε_{fw} is the free swelling strain, h is the plate thickness, d is the depth of penetration normalized by h, and $r = E_{wet}/E_{dry}$ is the ratio of the wet to the dry modulus.

The maximum of this curve is given by:

$$\Delta_{\max} = \left(\frac{3w^2 \varepsilon_{fw}}{16h} \right) \tag{2}$$

Thus, in principle the height of the maximum deflection should be a direct measure of the free swelling strain. In fact, we find that although this measurement of the free swelling is correlated to the direct measurements discussed earlier, there is not a one-to-one relation, as can be seen in Figure 3.

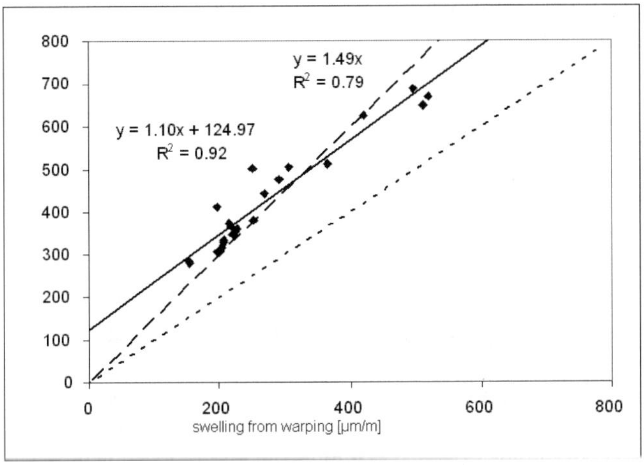

Figure 3. Relation between the free swelling strain obtained by the direct measurement and the warping measurement

Indeed, we find that direct measurements of swelling strains are on average about 50% higher than those obtained from the warping measurement (dashed line). In fact, there is a better correlation if we admit the existence of an offset in the warping measurement. In that case the linear relation between both measurements is close to unity (solid line). Possible causes for this offset are discussed later in the paper.

3.2.2 Sorptivity and modulus ratio

Equation (1) can also be used to estimate the sorptivity and the ratio of wet to dry modulus. For this we write the rate of water ingress as:

$$d = \frac{S}{h}\sqrt{t} \tag{3}$$

where S is the sorptivity and t is the time.

It can be shown that the initial slope, a_0, of deflection versus $\sqrt{t}$ is equal to

$$a_0 \equiv \frac{d^2}{d\sqrt{t}} = \left(\frac{3w^2 \varepsilon_{fw}}{4h} \right) \frac{rS}{h} = \frac{4\Delta_0 rS}{h} \tag{4}$$

The time needed to reach the maximum deflection is:

$$t_{\Delta\max} = \left(\frac{h}{S\sqrt{1+r}} \right)^2 \tag{5}$$

Thus equations (4) and (5) can be used with the measured values of the initial slope and time to maximum deflection to obtain the ratio between wet and dry modulus, r, and the sorptivity S.

It turns out here again that there is a discrepancy between the values estimated by the warping measurement and direct measurements; however, in this case the direct measurements were performed on different samples than the warping tests, owing to experimental requirements. The values of r are in the range of 0.35 for direct measurement, while they are estimated at about 0.6 by warping. Direct measurement of sorptivity is in the range of 0.01 cm/s$^{1/2}$, while values from warping are about 0.025 cm/s$^{1/2}$. Possible reasons for this are discussed below.

4. DISCUSSION

4.1 Warping Measurements

We have found, contrary to our preliminary experiments with this technique (Scherer and Jiménez González 2005), that there is a discrepancy between the three parameters estimated from warping measurements with respect to independent measurements of the same parameters. In the particular case of the swelling strain the situation cannot be attributed to sample-to-sample difference, because the same samples were used for both measurements. These discrepancies reveal the complexity of the kinetics of expansion, so they deserve careful study. To analyze the possible origin of these discrepancies, we now focus our attention only on the samples measured before treatment or cycling. One way of doing this is to see whether there are dependencies among the different parameters.

From the data in Figure 4, we can see that sorptivity clearly decreases when the swelling strain increases, while the modulus ratio is relatively unaffected by it.

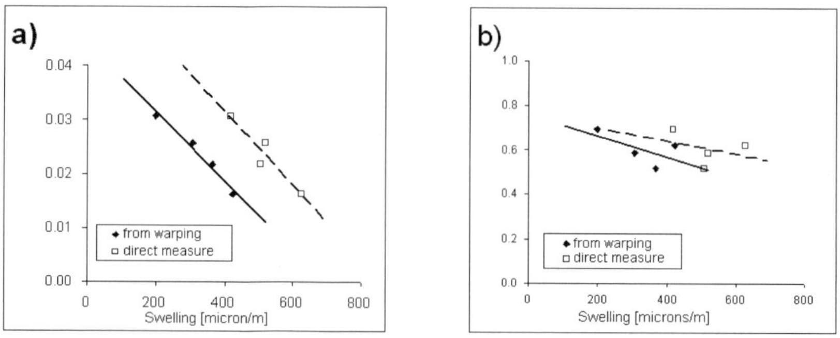

Figure 4. Dependence of sorptiviy (a) and modulus ratio, r (b) on the swelling strain. Values are plotted with respect to both the direct swelling strain measurement and the one inferred from warping

Sorptivity experiments show that there is an initial stage of faster water ingress. This complicates the analysis of the water experiments by requiring the introduction of a kinetic expression for the changes in sorptivity and for the rate of expansion, which are not necessarily linked in a trivial way one to the other. Finally, there is also the possibility that the softening of the wet stone is time dependent which would also account for the discrepancy in ratios of wet and dry modulus.

As the simplest approach, we will assume first that the kinetics of sorptivity change are the same as those of swelling. We have found that sorptivity curves could be well fitted in the following way:

$$d = \left(\frac{\Delta_S \alpha \sqrt{t}}{1 + \alpha \sqrt{t}} + S\sqrt{t} \right) \frac{1}{h} \tag{6}$$

where Δ_S is the intercept of the linear regression to the sorptivity curve, when plotted as height of rise versus square root of time, and $\Delta_S \alpha$ is the difference between the sorptivity at time zero and the one at steady state.

Using, an average value of 0.13 cm/s$^{1/2}$ for the sorptivity at steady state, values from independent measurements of swelling strain and modulus ratio, we fit the initial part of the curve by adjusting the values of Δ_S and α. The agreement is good, but clearly insufficient at longer times (Figure 5).

At this stage our treatment for the delayed expansion just multiplies the swelling strain of the wet part by the hyperbolic part of the equation (6): $\alpha \sqrt{t} \,/\, \left(1 + \alpha \sqrt{t}\right)$. Any further adjustment that improves the fit to the deflection after the maximum spoils the fit at short times.

From the fitting parameters, we predict that the sample will be completely saturated at ~280 s. From the slope of deflection versus time (secondary axis in Figure 5), we determine a minimum at 230 s, which can reasonably be attributed to a change in deflection mechanism when the water reaches the other side of the sample. The similarity of these estimates supports the validity of the fitting parameters we obtained.

Recent results by Wangler (2005) suggest that other factors, including evaporative cooling of the liquid pool on the sample, and capillary pressure from the pore liquid, may contribute to the post-peak deflection. Studies are underway to quantify those effects.

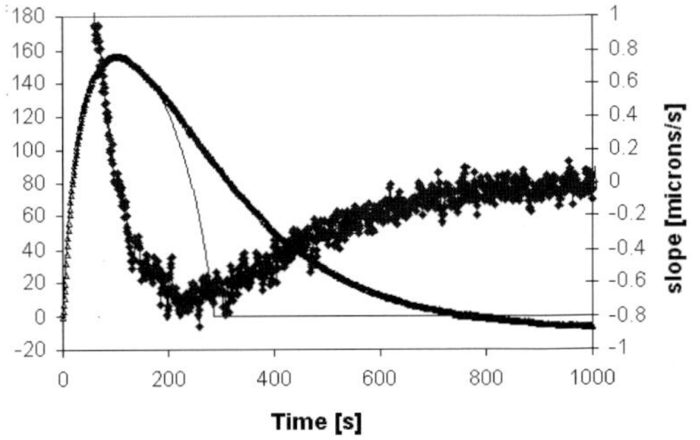

Figure 5. Comparison of a warping curve with the fitted function using the same time dependent function for sorptivity and swelling

5. CONCLUSIONS

We have examined the durability of a clay bearing stone exposed to accelerated cycles of wetting and drying. From direct measurements of swelling it appears that samples either treated or not with a swelling inhibitor, show an increased expansion over time. However, even after 700 cycles the swelling of the treated samples remains significantly under values before the application of the product, which indicates that this treatment has a durable effect in reducing damage from wetting and drying cycles which increases with the extent of swelling.

In addition we have examined in more detail the warping test by performing swelling and warping tests on the same samples. Both tests show the same trends but differ quantitatively as do values for sorptivity and modulus ratio. Agreement can be improved if the warping test is analyzed by introducing a time dependent change of sorptivity and swelling based on similar kinetics. The pertinence of the fitting parameters is strengthened by the fact that they provide a satisfactory estimate of the time the water takes to cross the sample in this test. Additional work is needed to describe the rest of the curve, taking account of other phenomena that contribute to deformation of the sample.

Acknowledgements

The authors would like to thank Joe Vocaturo (Princeton University) for his excellent work in designing and building the wetting/drying machine. We thank Dr. Andreas Queisser, Jérôme Constantin and Dr. Bénédicte Rousset at the Expert Center pour la Conservation du Patrimoine Bâti, in Lausanne, Switzerland for providing IJG with space to perform some of the experiments. We thank Dr. Robert J. Flatt for valuable discussions regarding interpretation of the data. Financial support for Inmaculada Jiménez González was generously provided by the Samuel Kress Foundation and VIP Restoration, Inc.

6. REFERENCES

1. Félix, C., 1988, Comportement des grès en construction sur le plateau suisse (Performance of Sandstones in Construction on the Swiss Plateau). In *LCP Publications 1975-1995*, Montreux, R. Pancella Ed., EPFL, 833-841
2. Félix, C.; Furlan, V., 1994, Variations dimensionnelles des gres et calcaires liees a leur consolidation avec un silicate d'ethyle (Dimensional changes of sandstones and limestones related to their consolidation with an ethyl silicate). In *3rd international Symposium on the conservation of Monuments in the Mediterranean Basin.* Edited by V. Fassina, F. Zezza. Venice, 22-25-June
3. Félix C., 1995, Choix de gres tenders du Plateau Suisse pour les travaux de conservation (Choice of soft sandstones from the Swiss plateau for conservation work). In *Conservation et restauration des biens culturels, Actes du Congres LCP*, Montreux, Septembre 1995, R. Pancella Ed., EPFL, 45-71.
4. Jiménez González, I.; Higgins, M. and Scherer, G.W., 2002, Hygric swelling of Portland Brownstone. In *Materials Issues in Art & Archaeology VI, MRS Symposium Proc.*, eds P.B. Vandiver, M. Goodway and J.L. Mass (Material Res. Soc.), Warrendale, PA, Vol 712: 21-27.

5. Jiménez González, I. and Scherer, G.W., 2004, Effect of swelling inhibitors on the swelling and stress relaxation of clay bearing stones. In *Environmental Geology*, 46: 364-377
6. Scherer, G.W. and Jiménez González, I., 2005, Characterization of swelling in clay-bearing stone. In Turkington A.V., ed., *Stone decay in the architectural environment:* Geological Society of America Special Paper 390: 51-61.
7. Wangler, T., 2005, Princeton University, private communication
8. Wendler, E., Charola, A.E., and Fitzner, B., 1996, Easter Island tuff: Laboratory studies for its consolidation. In *Proceedings of the 8th International Congress on Deterioration and Conservation of Stone*, ed J. Riederer. Berlin, Germany, 2, 1159-1170.

ASSESSMENT OF ATMOSPHERIC POLLUTION IMPACT ON THE MICROSTRUCTURE OF MARBLE SURFACES

A. Moropoulou, E.T. Delegou, E. Karaviti and V. Vlahakis

National Technical University of Athens, School of Chemical Engineering, Lab of Materials Science and Engineering,Iroon Polytechniou 9, 15780 Zografou, Athens, Greece.

Abstract: In this work, digital processing of SEM images is utilized in order to assess the impact of atmospheric pollution on the microstructure of pentelic marble surfaces. The investigation samples were collected by the pentelic marble surfaces of the historic buildings of National Archaeological Museum and National Library in Athens, Greece. The investigated surfaces disclosed the representative decay patterns of black–grey crusts, washed out surfaces, etc, that usually develop in polluted urban atmosphere like the Athens' centre. Beside the mineralogical/ morphological and chemical investigation that took place by the means of SEM-EDX, SEM-image-analysis program EDGE was applied for the estimation of three evaluation parameters of marble microstructure. EDGE program, which was developed by the US Geological Survey, is a computer program which analyzes scanning-electron-microscopy images for measuring the fractal dimension of the exposed surfaces of stone specimens cut in cross section. The near-surface fracture density of the stone can also be computed, while the shape factor, a surface roughness factor, results from the traced fractal dimension. Moreover, the parameter of friability index is introduced, representing the physicochemical and physicomechanical stability of the stone surface. The combined use of SEM-EDX and EDGE program provided results which can classify the presented decay patterns not only according to their chemical composition and morphological characteristics, but regarding the micro-structural parameters of surface roughness, near-surface fracture density and surface friability as well. Therefore, a better and more thorough description of the impact of atmospheric pollution on marble microstructure is accomplished, leading to better decision making on conservation planning of decayed stones.

Key words: marble; pollution impact; microstructure; SEM; digital processing.

M.S. Konsta-Gdoutos, (ed.), Measuring, Monitoring and Modeling Concrete Properties, 695–701.

1. INTRODUCTION

The last decades, deterioration of stone monuments and historic buildings in urban environments has been accelerated due to the aggressive role of air pollutants, leading to the formation of different decay patterns in terms of compositional and aesthetical alterations. Weathering of calcareous stones, especially marbles, is usually caused by the impact of SO_2 that produces gypsum on the surface of the stone through the mechanism of sulfation of calcite. Black-grey crusts are a commonly observed type of such a surface deformation, constituting of gypsum matrix that encloses black depositions (soot, dust, aluminosilicates e.t.c.)[1-6]. Black-grey crusts are located in sheltered areas, while washed-out surfaces are formed on areas exposed to rain. In this case, the gypsum formed is dissolved and removed by water, causing the disintegration of the surface. A third case of marble deterioration is when thermal cycles cause decohesion of calcitic crystals and formation of intergranular microcracks. Normal temperature changes generate a physical mechanism of successive compression and expansion, affecting profoundly the cohesion of the structure. Fissures presented around the mineral contacts may easily be penetrated by agents as sulphuric acid solutions which, subsequently, cause further disintegration of the calcitic fabric[7-9].

In order to assess the impact of environmental effects on stone and to propose appropriate cleaning interventions, it is important to determine an objective characterization of the microstructure[10]. In this work, the representative forms of marble deterioration mentioned above were examined so as to perform a decay diagnosis. The surfaces were located on two historic buildings in the center of Athens, the National Library of Greece and the National Archaeological Museum[11-12]. Mineralogical/chemical analysis by means of SEM/EDX was performed. In addition, morphological parameters of the surfaces were investigated by the means of image processing.

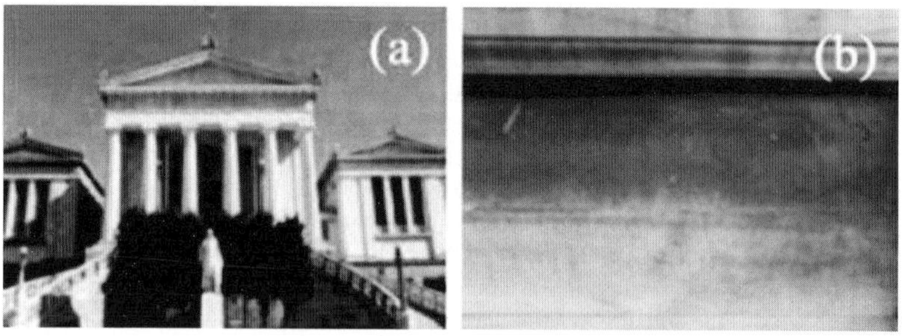

Figure 1. (a) National Library of Greece and (b) the investigation areas

2. EXPERIMENTAL

2.1 Sampling

Samples displaying black-grey crust as well as those presenting intergranular fissures were collected from the façade of the National Library (Fig.1). In Figure 1, the dark area sheltered by a cornice depicts the presence of the crust.

Figure 2. (a) National Archaeological Museum and (b) the investigation area

At the semi sheltered surface located below, marble maintains a brighter color, while physicomechanical processes provoked superficial crumbling. Washed-out surface sample was collected by the north side of a column of the National Archaeological Museum (Fig. 2), and more specifically from the left side of the flutting, an area exposed to rainfall[9-11]. Specimens were cut in cross-sections in order to evaluate the morphological characteristics of the surface in depth.

Table 1. Surface codes

Surface code	Decay pattern	Building
Pm1	Black-grey crust	National Library of Greece
Pma	Intergranular fissures with dust fall	National Library of Greece
Kn3cd	Washed-out surface	National Archaeological Museum

2.2 Scanning electron microscopy analysis

Analyses were carried out using a Scanning Electron Microscope (SEM) with the following characteristics: Type of instrument; JEOL JSM-5600, Energy Dispersive X-ray Microanalysis system (EDX); OXFORD LINK^{TM} ISIS^{TM} 300, Conditions of microanalysis; Accelerating Voltage 20 KV, Beam current: 0.5nA, Lifetime: 50 sec, Beam diameter < 2μm.

2.3 Digital image processing

Image processing was accomplished by the software package MORPH-II, which provides fractal analysis of the specimen exposed surface based on the cross-sectional profile of the exposed surface. MORPH-II was developed for the analysis of back-scattered electron-micrograph images stored in a binary file format that represents images with lateral resolution 2μm/pixels. Measurements were made by computer analysis of 100x SEM images of core cross sections. The program used for the image analysis was EDGE.EXE that provides integrated image calibration, image editing, fracture density measurement, and the measurement of linear distances between any two planes through the image. In this algorithm, the fate of a given cell is determined by the relationship between the signal intensity of the cell and that of its first- and second-nearest neighbors. Two numbers between 0 and 255 delimit the range of gray values defining the states PORE and MASS; pixels with gray values less than or equal to the lower threshold (T_L) are defined as PORE. Similarly, pixels with gray values equal to or greater than the upper threshold (T_U) are defined as MASS. The range of gray values falling between the thresholds parameters belong to the state EDGE. During the calibration procedure, the user can actively adjust the upper and lower thresholds to alter the population of pixels identified as EDGE based on his experience with electron micrographs and knowledge of the physical structure of the material under examination[13-14]. The method used for the determination of the fractal dimension is based on a Richardson structured walk at a fixed contour level along the profile of the exposed surface of the specimen. The Richardson effect asserts that an irregular contour line can be approximated with a broken line made up of N intervals of length r, where[15]:

$$N = \mu D\, r^{-D} \tag{1}$$

If the value of the exponent D is constant within a range of values for the yardstick r, the line is said to be fractal and the exponent D is called the fractal dimension that will range from 1 to 2. The algorithm uses the slope of the log (N) vs. log (r) curve to estimate the fractal dimension for a given contour. If the interfacial surface between the mass and pore space is truly fractal, the boundary trace will be self-similar at all scales. The shape factor for measuring the surface roughness (Γ) ranges from 0 to about 10, where 0 represents a perfectly smooth plane and 10 corresponds to a surface exhibiting an irregular morphology over the range 10 to 10^5 μm. The shape factor is based on the fractal nature of the exposed surface, although the fractional part of the Richardson dimension is used as the scale-independent parameter. If we measure the length, L, along the trace of the surface cross section (number of pixels in the trace of the exposed surface) relative to a reference state L^* (the Euclidean length of the trace of the exposed surface, converted to pixel units by use of the factor 512/22.5 pixels/cm), we can define the shape factor, Γ(μm) by the ratio $\mu_D/\mu_D{}^*$ and use the following algorithm for its computation:

$$\Gamma \approx (L/L^*)\, \varepsilon^{(DD^*)} \tag{2}$$

The fracture density (F.D.) is a measure of the fraction of the stone volume filled by fractures, crevices, and pore space. The results of program EDGE computations are displayed on a PC monitor, reported as the percentage of pixels identified as components of the fractures in the window calculated until 100 μm under the surface area. The third variable related to surface microstructure is the friability index (F.I.), which represents a correlation coefficient reflecting the relation between the fracture density and the surface roughness. It also refers to the friability of the stone on the microscale. The friability index is given by[13-14]:

$$FI = \sqrt{(FD)^2 + (\Gamma_R)^2} \tag{3}$$

3. RESULTS AND DISCUSSION

3.1 SEM/EDX Analysis

The encrustrated sample (code name Pm1) presented a relatively intense relief, disclosing a cohesive gypsum layer (20 to 80μm width). This gypsum crust consisted of microcrystals (diameter <10μm) mixed with dust fall (alluminosilicates) and metal oxides. The calcitic matrix was separated at length from the crust by a discontinuity of approximately 3 μm width (Fig. 3). From EDX analysis, S was not traced within the matrix, meaning that the sulphation of calcite did not proceed beyond the fracture. On the contrary, at the sample originated from the zone that was partially sheltered (code name Pma); a distinct layer of gypsum was not presented. Cyclic compression and expansion enforced on the surface provoked the formation of fissures and microcrackings at the boundaries of crystals. SO_2 attack caused the formation of microcrystalline and cryptocystalline gypsum accumulated within the cavities, that resulted in a further expansion of their size and finally, led to the structural decohesion and the detachment of marble grains (Fig. 4). At the washed-out surface the gypsum formed was dissolved and washed by the rainfall. This procedure favored superficial exfoliation and loss of authentic material. The calcitic matrix appeared severally decayed as fractures were found even at the depth of 500 μm (Fig. 5).

3.2 SEM-image Processing

Results of the digital image processing are reported in Table 2.

Table 2. Results of digital image processing using EDGE.EXE program

Surface/ Sample code	Fractal Dimension	Fracture density F.D. (%)	Shape Factor (Γ)	Friability Index F.I. (%)
Pm1	1.114	23.2	3.48	41.8
Pma	1.226	46.0	5.64	72.8
Kn3cd	1.132	20.0	4.30	47.0

The surface indexes of black-grey crust (Pm1) retained lower values of fractal dimension and shape factor in comparison to the other decay patterns, describing a relatively rough relief of the gypsum crust. In parallel, the relatively high value of fracture density can be attributed to the porosity of the layer of the microcrystalline gypsum. The high value of friability index though indicated the physicomechanical and physicochemical instability of the examined surface revealing that the surface friability refers to possible loss of gypsum grains and dust fall. The surface presenting intergranular fissures (Pma) displayed the highest values of all the examined micro structural indexes. The surface held the highest value of shape factor ascribing the extremely high surface roughness. Furthermore, the value of the fracture density was extremely high quantifying and demonstrating the extreme fatigue that took place due to the combined action of physicomechanical and physicochemical decay mechanisms. Additionally, the extremely high value of the friability index pointed out the great physicomechanical and physicochemical instability of the examined surface revealing that the surface friability refers to possible loss of both calcite and gypsum grains. The washed-out surface (kn3cd) presented intermediary values compared to the previous ones. Acid rain fall created a rather rough surface displaying a relatively low fracturing value, which however is a significant one if we consider that it describes exclusively the calcitic matrix. Moreover, the high value of friability index indicates the physicomechanical and physicochemical instability of the examined surface revealing that additional loss of calcitic grains is possible.

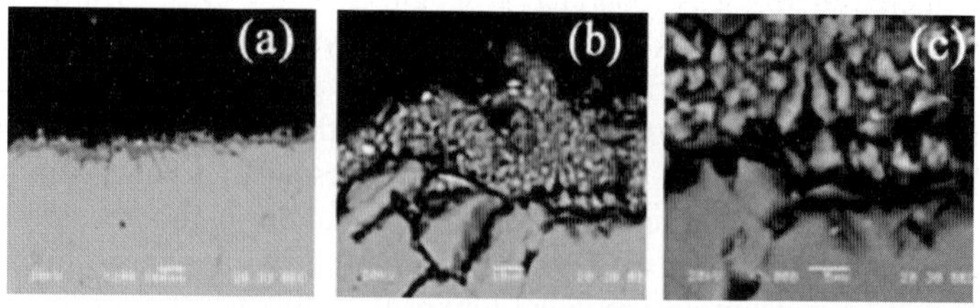

Figure 4. SEM images of Pm1 sample

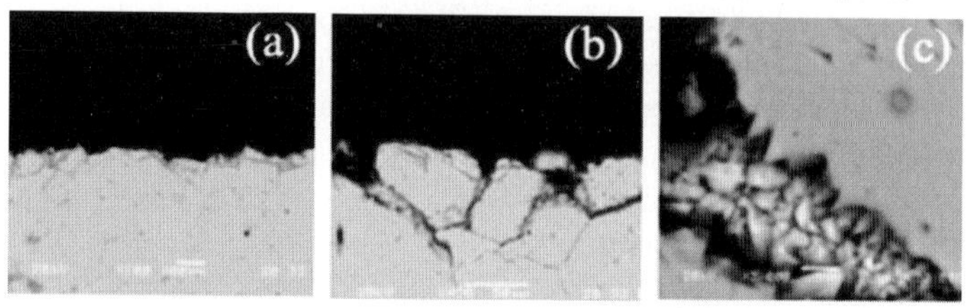

Figure 5. SEM images of Pma sample

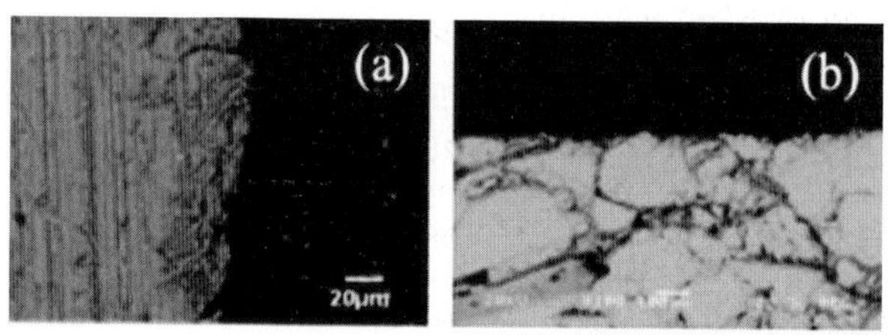

Figure 6. SEM images of Kn3cd sample

4. CONCLUSIONS

SEM image processing, combined with the results derived from chemical analysis by means of EDX, characterized the different forms of marble decay in depth. Although the EDGE program was primarily created for the assessment of cleaning interventions on stone surfaces, in this work classification of distinct micro structural characteristics of different decay patterns was possible. The contribution of a morphological evaluation that includes quantified parameters which describe roughness, fracturing and friability can be proved essential in order to accomplish an integrated decay diagnosis. Therefore, a better and more thorough description of the impact of atmospheric pollution on marble microstructure is accomplished, leading to better decision making on conservation planning of decayed stones.

5. REFERENCES

1. Moropoulou A., Bisbikou K., Torfs K., Van Grieken R., Zezza F. and Macri F., Origin and growth of weathering crusts on ancient marbles in industrial atmosphere, Atmospheric Environment 32 (6), pp. 967-982, (1998).

2. Bugini R., Laurenzi Tabasso M. and Realini M., Rate of formation of black crusts on marble. A case study, Journal of Cultural Heritage 1, pp. 111-116 (2000).
3. Elfving P., Panas I., Lindqvist O.: Model study of the first steps in the deterioration of calcareous stone: I. Initial surface sul-phite formation on calcite, Applied Surface Science 74, pp. 91-98, (1994).
4. Skoulikidis Th. and Charalambous D.: Mechanism of sulphation by atmospheric SO2 of the limestone and marbles in the ancient monuments and statues, II. Hypothesis concerning the rate determining step in the process of sulphation, and its experimental confirmation. British Corrosion Journal, 16 pp. 70-76 (1981).
5. Maravelaki P., Zafiropoulos V., Kalaitzaki M., Kilikoglou V., Fotakis C., in J. Riederer Ed. Deterioration and Conservation of Stone, Proceedings of 8th International Congress, Berlin, p. 1395 (1996).
6. Maravelaki P., G. Biscontin G., Origin, characteristics and morphology of weathering crusts on Istria stone in Venice, Atmospheric Environment 33, pp. 1699-1709 (1999).
7. Amorosso G., Fassina V.: Stone decay and conservation, Amsterdam, Elsevier Science Publishers B.V. (1983).
8. Winkler E.M., Stone in architecture; Properties and durability, 2nd Edition, Berlin-Heidelberg, Springer –Werlag (1997).
9. Moropoulou A., Delegou E.T., Avdelidis N.P., Koui M., Non-destructive investigation of architectural surfaces in polluted urban atmosphere, the 39th conference of BINDT, Buxton, UK, pp. 143-148 (2000).
10. Moropoulou A., Delegou E. T., Karaviti E., Vlahakis V., Digital processing of SEM images for the assessment of evaluation indexes of cleaning interventions on pentelic marble surfaces, 10th Euroseminar on Microscopy Applied to Building Materials (2005).
11. Moropoulou A., Koui M., Delegou E.T., Avdelidis N.P., Bakolas, A., Giabanis, D., Kouris S., Stefanou, J., Fotoniata E., Tzamalis A., Decay Diagnosis and Conservation Interventions Planning for the facades of the National Library of Greece Historic Building, Lab of Materials Science and Engineering, Final Research Working Paper, National Technical University of Athens (2000).
12. Moropoulou A., Koui M., Delegou E.T., Bakolas A., Karoglou M., Rapti E., Kouris S., Mauridis Th., Ypsilanti E., Avdelidis N.P., Ntae, D., Conservation Interventions Planning for the main façade of the National Archaeological Museum, Lab of Materials Science and Engineering, Final Research Working Paper, National Technical University of Athens (2000).
13. Mossotti V.G. and Eldeeb A.R., MORPH-2, A software package for the analysis of scanning electron micrograph (binary formatted) images for the assessment of the fractal dimension of exposed stone surfaces, U.S. Geological Survey, Open file report: 00-013 (2000).
14. Mossotti V.G. et al.,The effect of selected cleaning techniques on Berkshire Lee marble; A scientific study at Philadelphia City Hall, U.S. Geological Survey, Professional Paper 1635, Virginia, U.S.A (2002).
15. Russ J.C., Fractal Surfaces, 1st Edition, New York and London, Plenum Press (1994).

CONTROLLING STRESS FROM SWELLING CLAY

T.P. Wangler, A.K. Wylykanowitz and G.W. Scherer
Princeton University, Civil & Env. Eng./PRISM, Eng. Quad. E-319, Princeton, NJ 08544 USA

Abstract: Many sedimentary rocks contain clays that swell on exposure to moisture, producing stresses from differential strain. Wendler and Snethlage showed that the swelling can be reduced by treatment with α–ω diamino alkanes. In this paper, we present results showing that mixtures of such molecules are more effective than any single molecule, and that better results are obtained by applying smaller molecules before the larger ones.

Key words: clay, swelling pressure, stress, fracture, surfactant, intercalation

1. INTRODUCTION

Many sedimentary rocks contain clays that cause swelling when the stone is exposed to moisture[1], resulting in damage to the civil infrastructure[2] as well as to monuments and works of art.[3,4] Expansion of clays results from the presence of alkali ions between the charged layers in the crystal structure of the clay.[5] Water molecules are electrostatically attracted to those ions, so they penetrate between the layers and surround the alkali ions, resulting in expansion. The pressure needed to prevent invasion of a monolayer of water molecules can be hundreds of MPa in bentonite, but this drops drastically as additional layers of water molecules accumulate.[6] The clays can form a layer surrounding grains in the stone,[7] so their swelling is directly reflected in expansion of the stone; moreover, the softening of the wet clay reduces the stiffness of the stone.[8] The damage is generally attributed to differential strain from expansion of the wet region of the stone, but the clay may also contribute to damage from salt[9] by creating small pores that are particularly susceptible to high crystallization pressure.[10]

Wendler and Snethlage[11,12,13] demonstrated that the expansion of clays could be substantially reduced by treatment with α–ω diamino alkanes (hereafter called DAA), which are molecules having amine groups at each end of an alkane chain. Whereas similar molecules with a single amine group cause increased swelling, the α–ω structure allows the molecule to bond to adjacent sheets in the clay structure, binding them together. The charged sites on the clay layers are randomly distributed,[14] so Gonzalez and Scherer[15] reasoned that no single molecular size would allow binding to all of those

M.S. Konsta-Gdoutos, (ed.), Measuring, Monitoring and Modeling Concrete Properties, 703–708.

sites. They demonstrated that the swelling of Portland Brownstone was reduced more by a treatment with a mixture of surfactants than with any one alone. This observation was confirmed in a study of sandstones by Velo-Simpson;[16] in addition, she found that the order of treatment with the surfactants had some influence on the swelling. In this paper, we present results of a study using sequential or simultaneous treatment with DAA having alkane chain lengths ranging from 2 to 10 carbons. This provides a direct test of the importance of the size and dispersity in size, as well as the order of application, of the surfactant on the amount of reduction in swelling.

2. EXPERIMENTAL PROCEDURE

The surfactant treatments employ diaminoalkane molecules (Acros Organics) dissolved in water for easy application. The DAA has a linear carbon chain linking two amine end groups, with the general formula $H_2N\text{-}(CH_2)_n\text{-}NH_2$. The shorter chains (n=2,3,4) are shipped as salts with hydrochloric acid forming the amine salts, or $HCl\text{-}H_2N\text{-}(CH_2)_n\text{-}NH_2\text{-}HCl$; the diaminohexane (n=6) was not neutralized. The solubility of these molecules decreases as the carbon chain length increases, but is adequate to achieve concentrations of about 0.5 M for $n \leq 10$. To indicate the number (n) of carbons in the alkane chain, we will identify them as DAAn; thus, diaminopropane will be written as DAA3.

The stone used in this work (supplied by Dr. George Wheeler, Metropolitan Museum of Art, New York City) is a sandstone from Aztec Ruins National Monument in New Mexico (http://www.nps.gov/azru/).

Samples were cut on a diamond saw to approximate dimensions 3x3x12 mm. The expansion was measured using a differential mechanical analyzer (Perkin-Elmer DMA7). The DMA has a temperature control unit that is raised up around the sample; within that unit, the sample is separated from the heating coil by a nickel cup. The cup was filled with water, so that the sample was immersed when the unit was raised; temperature was maintained at 30°C.

Surfactant treatments were applied by placing the stone sample on a layer of glass beads in a petri dish, then adding the solution to the dish until it just touched the bottom of the sample. The dish was hermetically sealed and left for two hours, during which the stone was saturated by capillary rise of the solution. The sample was then left in the hood for an hour and then placed in a convection oven at 60°C overnight.

The elastic modulus of the stones was measured in three-point bending using apparatus described in detail elsewhere.[15,17] To obtain the static Young's modulus, the sample was driven up and down at a rate such that relaxation would be negligible; samples were run dry and while immersed in a water bath. To measure the kinetics of stress relaxation, a displacement was imposed suddenly (stabilized in ~0.5 s) and held constant for several hours while the force on the beam was monitored continuously. Samples used for these tests were in the form of rectangular plates with dimensions of 2-3 x 5-10 x 50 mm, and were immersed in water during the test.

3. RESULTS

The untreated stone showed an expansion of 1.33 x 10^{-3} in water, which is about three times as large as the swelling of Portland Brownstone[15]. The effect of treatment with DAA is summarized in Table 1. All of the surfactant treatments cause impressive changes in swelling, but some patterns emerge. A second treatment always helps, presumably because the stone is thereby exposed to more amine groups. For example, a single treatment with a mixture of DAA2 and DAA3 reduces the expansion by 73%, and a second treatment with the same solution provides a total reduction (compared to the untreated stone) of 83%. Multiple treatments with mixtures provide no advantage over sequential treatments with individual surfactants. For example, treating twice with a mixture of DAA2 and DAA6 gives about the same improvement (85%) as sequential treatments with DAA2 followed by DAA6 (84%). Again, these sequences expose the stone to the same total molarity of amine. This is not the whole story, though, because a 5% solution of DAA2 + DAA3 contains less amine than DAA2 + DAA6, but it causes a slightly greater reduction in swelling; therefore, the size of the molecule is significant. The best result was obtained with sequential treatment with DAA2, 3, 4, and 6 (91%). Of course, this quadruple treatment exposes the stone to more surfactant. What is most striking is that reversing the order of treatment substantially reduces the effectiveness of the treatment, even though the amount of exposure to amine is the same. This suggests that the order of addition can be important, perhaps because the smaller molecules open the interlayer space, so that it is more easily entered by the larger molecules. This may also explain why a second treatment with DAA2 + DAA6 makes such a great improvement: the first exposure to DAA2 may open the interlayer space so that both amines penetrate during the second treatment.

Table 1. Effect of treatment with diaminoalkanes

Treatment[a]	**$M(NH_2)$**[b]	**Strain (%)**[c]	**% Reduction**
None	0	0.1330	-
2+3	0.72	0.0356	73
2+3 / 2+3	0.72 / 0.72	0.0222	83
2 / 3	0.76 / 0.68	0.0198	85
3 / 2	0.68 / 0.76	0.0171	87
2+6	0.80	0.0490	63
2+6 / 2+6	0.80 / 0.80	0.0197	85
2 / 6	0.76 / 0.80	0.0209	84
4 / 6	0.62 / 0.80	0.0350	74
2+3+4+6	0.72	0.0252	81
2+3+4+6 / 2+3+4+6	0.72 / 0.72	0.0228	83
2 / 3 / 4 / 6	0.76 / 0.68 / 0.86 / 0.80	0.0119	91
6 / 4 / 3 / 2	0.80 / 0.86 / 0.68 / 0.76	0.0360	73

[a] Integer represents number (n) of carbons in alkane; plus (+) indicates components of mixture; slash (/) separates sequential treatments. [b] Molarity of amine groups in solution. [c] Strain measured when oven-dried sample immersed in deionized water.

The static elastic modulus of the stone dry or saturated with water is shown in Table 2. Young's modulus drops by about a factor of 4 when the stone is wet, owing to the softening of the clay. Treatment with the DAA mixture (2+3+4+6) raises the modulus by ~30% in the dry state and 100% in the wet state.

Table 2. Static Young's modulus, *E* (GPa)

Condition	Untreated	Treated
Dry	7.55	10.3
Wet	1.87	3.91

Damage to monuments from swelling of stone occurs during wetting, owing to buckling of the wetted surface[18]. The maximum stress, which occurs when the wetted layer is thin[15], is given by

$$\sigma_S = -\frac{E_{wet}\,\varepsilon_S}{1-\nu_{wet}} \tag{1}$$

where ν_{wet} is Poisson's ratio for the wet stone, and can be estimated to be ~0.25. As shown in Table 3, treatment with DAA reduces the compressive stress by as much as 70-80%, even though the static modulus is increased. Note that these stresses are all calculated using the modulus of the stone treated sequentially with DAA 2/3/4/6, which probably has a higher modulus than the other treated stones; consequently, the actual stress reduction may be greater than shown in the table.

Table 3. Effect of treatment on stress during swelling

Treatment[a]	Stress (MPa)[b]	% Reduction
None	-3.3	-
2+3	-1.9	44
2+3 / 2+3	-1.2	65
2 / 3	-1.0	69
3 / 2	-0.89	73
2+6	-2.6	23
2+6 / 2+6	-1.0	69
2 / 6	-1.1	67
4 / 6	-1.8	45
2+3+4+6	-1.3	60
2+3+4+6 / 2+3+4+6	-1.2	64
2 / 3 / 4 / 6	-0.62	81
6 / 4 / 3 / 2	-1.9	43

[a] Integer represents number (n) of carbons in alkane; plus (+) indicates components of mixture; slash (/) separates sequential treatments. [b] Calculated from eq. 1 using the measured strain for each treatment and the elastic modulus (3.9 GPa) measured on the sample treated with once with 2/3/4/6.

The stress relaxation behavior of the wet stone, before and after treatment with the DAA mixture (2+3+4+6), is shown in Figure 1. As was observed previously[15], the treatment raises the elastic modulus, but accelerates the rate of relaxation, probably by a stick-slip mechanism. This effect further reduces the stresses resulting from swelling in treated stone.

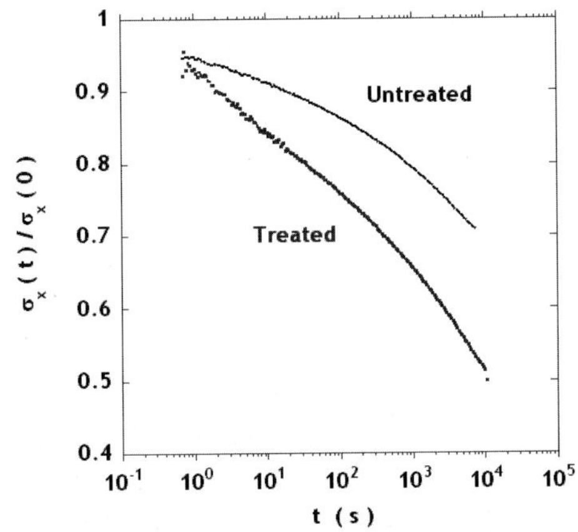

Figure 1. Stress relaxation of wet stones in three-point bending, before and after sequential treatment with DAA 2/3/4/6

4. CONCLUSIONS

The use of a mixture of α–ω diamino alkanes with different molecular sizes offers significantly better performance than any single molecule. The treatment also changes the viscoelastic properties of the stone, increasing the stiffness. To reduce the stress caused by hygric swelling, the treatment must reduce the expansion more than it raises the modulus. This has proved to be the case in our experiments, so the treatments are expected to reduce damage caused by swelling of the stone.

Acknowledgments

The authors are indebted to the Seaver Foundation for their support of T. Wangler.

5. REFERENCES

1. J.R. Dunn and P.P. Hudec, Water, clay, and rock soundness, Ohio J. Science 66 (2) 153-168 (1966)
2. B. Brattli and E. Broch, Stability problems in water tunnels caused by expandable minerals. Swelling pressure measurements and mineralogical analysis, Eng. Geology 39, 151-169 (1995)
3. C. Rodriguez-Navarro, E. Hansen, E. Sebastian, and W.S. Ginell, The role of clays in the decay of ancient Egyptian limestone sculptures, J. Am. Inst. Conservation 36 (2) 151-163 (1997)

4. F. Veniale, M. Setti, C. Rodriguez-Navarro, S. Lodola, Procesos de alteraci?n asociados al contenido de minerales arcillosos en materiales petreos (Role of clay constituents in stone decay processes), Materiales de Construcci?n, 51 (263-264) 163-182 (2001)
5. H. van Olphen, An Introduction to Clay Colloid Chemistry, 2d ed. (Wiley, NY, 1977)
6. F.T. Madsen and M. M?ller -Vonmoos, The swelling behaviour of clays, Appl. Clay Sci. 4, 143-156 (1989)
7. E.N. Caner and N.J. Seeley, The clay minerals and the decay of limestone, pp. 1-34 in Deterioration and Protection of Stone Monuments (RILEM, 19)
8. G.W. Scherer and I. Jimenez Gonzalez, "Characterization of Swelling in Clay-Bearing Stone", in Stone decay and conservation, SP-390, ed. A.V. Turkington (Geological Soc. Am., 2005) 51-61
9. J.P. McGreevy and B.J. Smith, The possible role of clay minerals in salt weathering, CATENA Braunschweig 11, 169-175 (1984)
10. G.W. Scherer, Crystallization in pores, Cement Concr. Res. 29 (8) 1347-1358 (1999); "Reply to discussion of Crystallization in pores", G.W. Scherer, Cement and Concr. Res. 30 (4) 673-675 (2000)
11. E. Wendler, D.D. Klemm, and R. Snethlage, Consolidation and hydrohobic treatment of natural stone, in Proc. 5th Int. Conf. on Durability of Building Materials and Components, eds. J.M. Baker, P.J. Nixon, A.J. Majumdar, and H. Davies (Chapman & Hall, London, 1991) 203-212
12. R. Snethlage and E. Wendler, Surfactants and akherent silicon resins - New protective agents for natural stone, in Mat. Res. Soc. Symp. Proc. Vol. 185 (Mater. Res. Soc., Pittsburgh, PA, 1991) 193-200
13. E. Wendler et al., Saving our Architectural Heritage, eds. Baer and Snethlage (Wiley, 1997) Ch. 10
14. D.M.C. MacEwan and M.J. Wilson, Interlayer and intercalation complexes of clay minerals, in Crystal Structures of Clay Minerals and their X-ray Identification, Mineralogical Society Monograph No. 5, eds. G.W. Brindley and G. Brown (Mineralogical Soc, London, 1980) Ch.3
15. I. Jimenez Gonzalez and G.W. Scherer, Effect of swelling inhibitors on the swelling and stress relaxation of clay-bearing stones, Environmental Geology 46, 364-377 (2004)
16. M.L. Velo-Simpson, Falling Apart: Understanding the damage mechanism of Bollingen and Krauchthal stone, Senior Thesis, Dept. Chem. Eng., Princeton University, 2004
17. W. Vichit-Vadakan and G.W. Scherer, Measuring Permeability of Rigid Materials by a Beam-Bending Method: II. Porous Vycor, J. Am. Ceram. Soc. 83 (9) 2240-2245 (2000); Erratum, J. Am. Ceram. Soc. 87 (8) 1614 (2004)
18. G.W. Scherer, Internal stress and cracking in stone and masonry, these proceedings

FRPs and Textiles in Cement Composites

TENSION STIFFENING IN GFRP REINFORCED CONCRETE BEAMS

R. Al-Sunna,[1,2] K. Pilakoutas,[1] P. Waldron[1] and T. Al-Hadeed[2]

[1] *University of Sheffield, Department of Civil and Structural Engineering, Sheffield, S1 3JD, United Kingdom;* [2] *Royal Scientific Society, Building Research Center, Amman, Jordan*

Abstract: The relatively low modulus of elasticity and the different surface treatment of fibre reinforced polymer (FRP) rebars have a direct impact on the spacing of cracks and the profile of bond stress between cracks. Hence, a tension stiffening model for FRP rebars is likely to be very different from that of steel rebars. In this paper, tension stiffening behaviour is investigated through the test results of six reinforced concrete (RC) beams with glass FRP (GFRP) reinforcement covering a wide range of reinforcement ratios. It is found that the stabilized cracking phase is reasonably characterized by an average effective modulus of elasticity. Tension stiffening for GFRP is much lower than for steel rebars. A tension stiffening model is proposed for an RC tie with the particular GFRP rebars used.

Keywords: beam; concrete; fibre reinforced polymer; tension stiffening.

1. INTRODUCTION

FRP reinforcement for concrete has been developed to replace steel in special applications, particularly in corrosive environments. Compared to steel, FRP generally has a lower modulus of elasticity, which leads to higher reinforcement strains, wider cracks and larger deflections. Therefore, the design of FRP RC is often governed by the serviceability limit state.

The prediction of deflection and crack width for RC flexural members requires proper evaluation of tension stiffening. Tension stiffening is a term that accounts for the beneficial effect of bonded concrete acting in tension between cracks on the reinforcement. At a crack, all the tensile force is carried by the reinforcement, whereas, between cracks, some of the tensile force is transferred, through bond, to the surrounding concrete. This results in a local reduction of the reinforcement stresses and strains. Hence, between cracks, the concrete stiffens the reinforcement, and causes the reinforcement to

M.S. Konsta-Gdoutos, (ed.), Measuring, Monitoring and Modeling Concrete Properties, 711–718.

have an average strain that is less than the strain at a crack or of the bare bar. Therefore, the reinforcement appears to have a higher effective modulus of elasticity [1-3].

For steel RC, CEB [1] and fib [2] provide a basically identical simplified tension stiffening model for a tie under pure tension in the form of a stress-average strain relationship. The model with fib [2] terminology, which uses force instead of stress, is shown in figure 1. In a flexural member, the tension steel and the surrounding concrete are considered as a tension tie. However, when dealing with deflections, it is more common to consider the effects of tension stiffening at both the section and member levels, by providing a transition between the un-cracked and cracked moments of inertia or curvatures at any load level, as discussed by Bischoff [4].

The relatively low modulus of elasticity and the different surface treatment of FRP rebars have a direct impact on the spacing and the profile of bond stress between cracks. Hence, the tension stiffening model for FRP rebars is likely to be very different from that of steel rebars. Furthermore, due to the numerous possible combinations of resin matrix, fibre and surface treatments a different tension stiffening model may be required for every particular rebar, or some form of categorization may be needed.

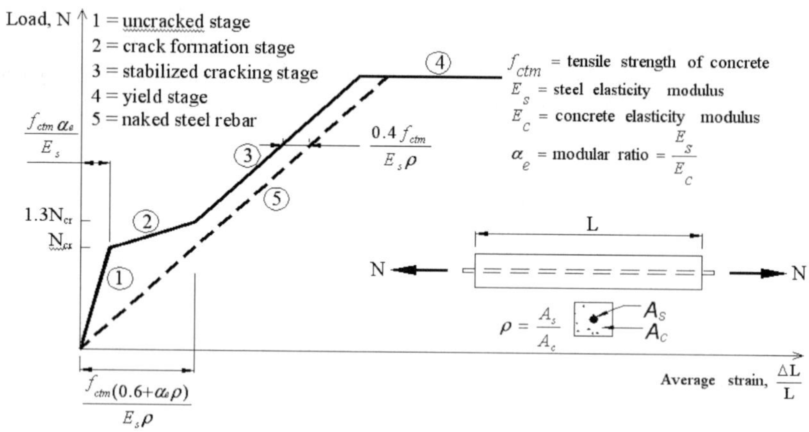

Figure 1. Load-strain relationship of a tension tie with steel reinforcement [2]

An extensive research program on the flexural behaviour of FRP RC members was carried out at the Royal Scientific Society (Jordan) in collaboration with the University of Sheffield (UK). In this paper, tension stiffening behaviour is investigated through test results of six RC beams with GFRP reinforcement covering a wide range of reinforcement ratios. A tension stiffening model is then proposed for an RC tie with the particular GFRP rebars used.

2. TEST SPECIMENS AND PROCEDURE

Aslan 100 GFRP rebars were used for the main flexural reinforcement in the beams. The surface treatment of these rebars is characterized by helically over-wound fibres and sand coating (Figure 2). The tensile properties were obtained by testing representative samples under uni-axial tension and are shown in Table 1. The 28-day cube compressive strength of the concrete was 38 MPa.

Figure 2. Aslan 100 GFRP rebar

Table 1. Tensile properties of GFRP rebars

Rebar type	Rebar size	Nominal diameter, (mm)	Modulus of elasticity, (MPa)	Tensile strength, (MPa)
	#3	9.53	42750	665
GFRP	#4	12.70	41600	620
	#6	19.05	41970	670

The test specimens consisted of three series of GFRP RC beams. To ensure repeatability, each series comprised two identical beams. The beam series were designated as BG#, where; B stands for beam; G refers to GFRP; while # is the series number. The two identical beams, within each series, were identified by adding a or b to the series name. Each beam series had a different reinforcement ratio. The beam sections were under-reinforced, close to balanced or over-reinforced, with failure occurring by rupture of bars or crushing of concrete. The geometric and reinforcement details of the test beams are shown in Figure 3 and Table 2.

All beams were tested under four-point loading. The load was applied centrally by a 600 kN hydraulic jack, and a distribution beam was used to distribute the load to the two third-span points. To allow for the evaluation of tension stiffening, a crack inducer, (20 x 0.6) mm, was used to ensure that a crack would occur at midspan. Then, a total of ten closely-spaced strain gauges were used to measure rebar strains on either side (Figure 3). Thus, it was possible to obtain strain profiles between one forced crack at midspan and two contiguous naturally occurring cracks. Four additional strain gauges were used to measure rebar strain under one of the loading points, at the third-points of one shear span, and at one support.

The testing was carried out using load control. The load was increased at a rate of 1 kN/min and was paused at about 5 kN intervals to mark and measure the cracks and to take notes. Two load cycles were performed. In the first cycle, the load was increased to a nominal service load level, which corresponded to a stress of about 45% the concrete compressive strength in the top concrete fibre at midspan. In the second cycle, the load was increased until failure. Failure occurred either by rupture of bars or by crushing of concrete. All data (force, strains and deflections) were collected by a data acquisition system and downloaded to a PC every second.

Table 2. Reinforcement details of the test beams

Rebar type	Series designation	Beam designation	Rebar details	Reinforcement ratio
GFRP	BG1	BG1a	2&9.53	0.00432
		BG1b		
	BG2	BG2a	2&12.7	0.00772
		BG2b		
	BG3	BG3a	4&19.05	0.0391
		BG3b		

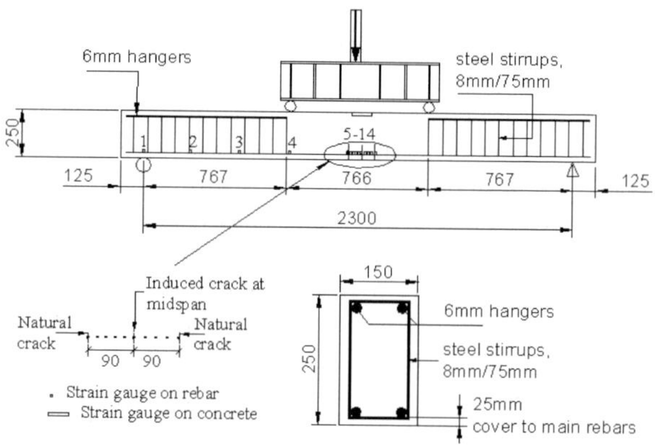

Figure 3. Test setup and geometric and reinforcement details of the beams

3. TEST RESULTS AND DISCUSSION

Figure 4 shows a typical distribution of strains along the rebar and around a crack inducer. From the strain profiles at every load level, the average strain between the cracks was determined by numerical integration. Figure 5 shows the experimentally derived stress at a crack versus the average strain for one beam from every beam series.

In the pre-cracking and crack formation phases, the stress-average strain relationship for an FRP RC tension tie should not be different from that of a steel RC tie. However, in a flexural member, a tension tie develops progressively with cracking, until it becomes a fully defined tie at the end of the crack formation phase. Hence, the first part of the stress-average strain relationship does not have any resemblance with a tension tie; it is a curve that shows reduction of stiffness as additional cracks form. However, the end of the curve clearly identifies the beginning of the stabilized cracking phase.

The stabilized cracking phase is clearly identifiable in Figure 5, with the stress-average strain relationship being essentially linear up to failure. Also, all the beam series follow essentially the same stabilized cracking line.

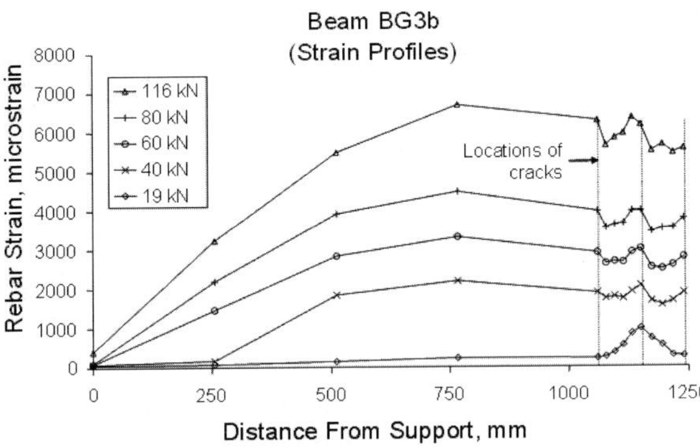

Figure 4. Typical strain profile along the rebar and around the crack inducer

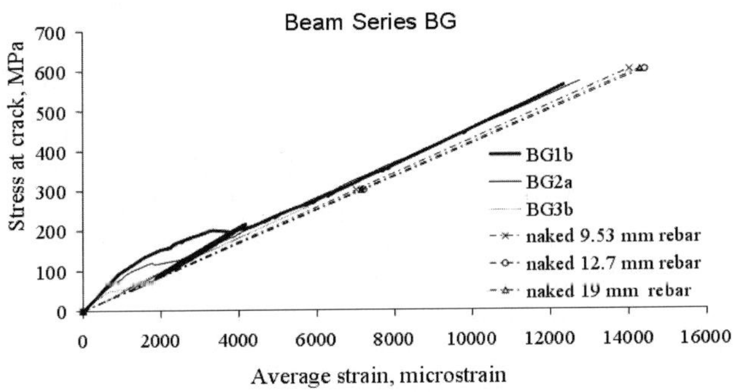

Figure 5. Experimentally derived stress-average strain relationship for the GFRP RC beams

Therefore, contrary to the case of steel reinforcement, the amount of GFRP reinforcement seems to have negligible effect on tension stiffening in the stabilized cracking phase; at least as in the tested beams where the reinforcement ratio is greater than about 0.4%, which corresponds to a ratio of about 1.25% for the RC tie in the tension zone. A lower reinforcement ratio would be lower than the minimum ratio stipulated in ACI 440 [5].

At any load level, it is possible to define an effective modulus of elasticity (E_{eff}) as the ratio between the stress at the crack and the average strain between cracks. Table 3 shows the ratio of E_{eff} to the naked rebar modulus (E_f) at the beginning and end of the stabilized cracking phase. An average E_{eff} of $1.10E_f$ offers a reasonable approximation for the whole stabilized cracking phase.

Table 3. Effective modulus in the stabilized cracking phase

Beam ID	Naked rebar modulus, E_f, "MPa"	Beginning of stabilized cracking phase		End of stabilized cracking phase		Stabilized cracking phase
		Effective Modulus, E_{eff}, "MPa"	E_{eff}/E_f	Effective Modulus, E_{eff}, "MPa"	E_{eff}/E_f	Average E_{eff}/E_f
BG3b	41970	47715	1.14	44548	1.06	1.1
BG2a	41600	49889	1.20	44636	1.07	1.14
BG1b	42750	492280	1.15	45370	1.06	1.11

In comparison, E_{eff} of a steel RC tie would depend on the reinforcement ratio (ρ). Based on Figure 1, the following relations can be defined.

At the beginning of a stabilized cracking phase: $E_{eff} = \dfrac{1+\alpha_e\rho}{0.6+\alpha_e\rho}E_s$

At the end of a stabilized-cracking phase: $E_{eff} = \dfrac{\varepsilon_y}{\varepsilon_y-(0.4f_{ctm}/E_s\rho)}E_s$

In which, ε_y is the strain at yield and all the other terms are as defined in Figure 1. So, for instance, for f_{ctm}=3.5 MPa, α_e=7 and ε_y=2500 $\mu\varepsilon$: E_{eff} would reduce from 1.60 E_s to 1.39 E_s, for ρ = 1%; and E_{eff} would reduce from 1.31 E_s to 1.03 E_s, for ρ= 10%. Therefore, for steel rebars, concrete provides considerable tension stiffening that gradually reduces with increasing stress in the rebar, whereas, for GFRP rebars, tension stiffening is much lower, which is due to the relatively high levels of rebar strain developed.

In view of the above, a tension stiffening relationship is proposed for an RC tie with the GFRP reinforcement used, as shown in Figure 6. More work is required to refine this relationship to account for the proportions of the flexural element considered, the concrete strength and for reinforcement ratios lower than those used herein, though any such lower ratios may not be viable in practice.

The proposed tension stiffening relationship was used to predict the behaviour of two RC tension ties with similar GFRP reinforcement, which were tested by Sooriyaararachchi [6]. The two ties were 1500 mm long, with a square cross section of 100 mm and 150 mm. The reinforcement ratio was 1.27 %, and the concrete strength was around 50 MPa, for both ties. The proposed tension stiffening relationship was also used to construct a moment-curvature relationship for one of the tested beams (beam BG3b), which was then used to predict the deflection behaviour of the beam. Figures 7 and 8 show that the prediction compares reasonably well with the experimental results in both cases.

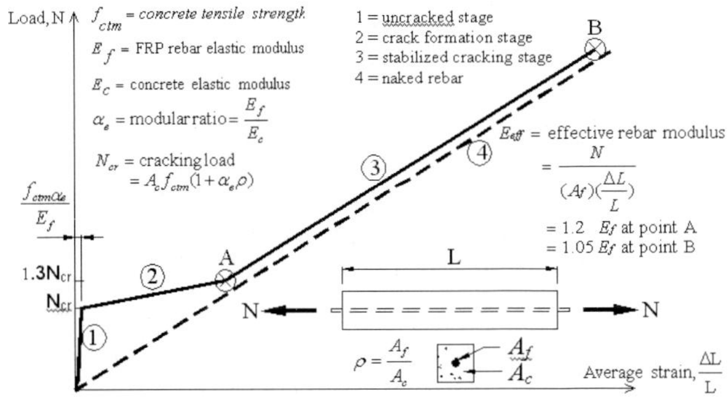

Figure 6. Proposed load-strain relationship for a tension tie with GFRP reinforcement

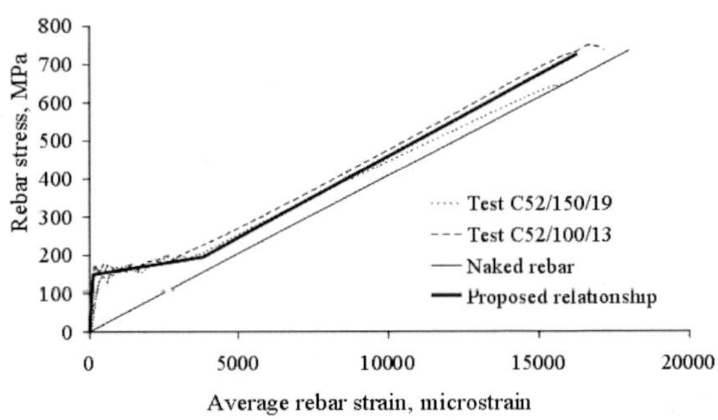

Figure 7. Tests of GFRP RC tension ties by Sooriyaararachchi [7]

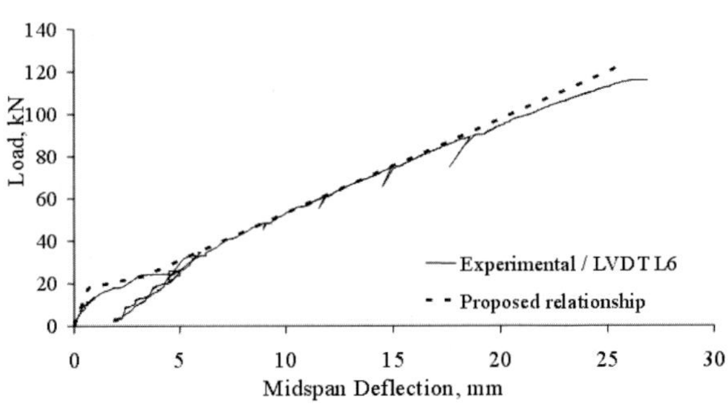

Figure 8. Midspan deflection of beam BG3b

4. CONCLUSIONS

- In a constant flexure zone, measuring rebar strains on either side of a preformed crack can be used to investigate tension stiffening during flexural tests.
- For GFRP rebars, and for a practical range of reinforcement ratios, the stabilized cracking phase is reasonably characterized by an average effective modulus of elasticity (E_{eff}), which is about 10% higher than the modulus of the bare rebar.
- Due to the relatively high levels of rebar strain involved, tension stiffening is much lower for GFRP than for steel rebars.
- A tension stiffening relationship is proposed for an RC tie with the GFRP reinforcement used. More work is required to refine this relationship to account for the proportions of the flexural element considered, the concrete strength and very low reinforcement ratios.

Aknowledgments

The authors wish to acknowledge the financial support provided by The Higher Council for Science and Technology (Jordan), the Royal Scientific Society (Jordan), the Centre for Cement and Concrete at the University of Sheffield (UK) and the Karim Rida Said Foundation (UK).

5. REFERENCES

1. (CEB) Comite Euro-International du Beton, CEB-FIP Model Code 1990, Thomas Telford Services Ltd., 437 pp. (1993).
2. (fib) Federation internationale du beton, Structural Concrete, Textbook on Behaviour, Design and Performance (Bulletin 1), Federation Internationale du Beton, Lausanne, Switzerland, 224pp (1999).
3. ACI Committee 224, Cracking of Concrete Members in Direct Tension (ACI 224.2R-92, Reapproved 1997), American Concrete Institute, Farmington Hills, Mich, 12 pp (1992).
4. Bischoff, P. H., Reevaluation of Deflection Prediction for Concrete Beams Reinforced with Steel and Fibre Reinforced Polymer Bars, Journal of Structural Engineering, ASCE, 131(5), pp. 752-767 (2005).
5. ACI Committee 440, Guide for the Design and Construction of Concrete Reinforced with FRP Bars (ACI 440.1R-03), American Concrete Institute, Farmington Hills, Mich, 41 pp (2003).
6. Sooriyaararachchi, H., Pilakoutas, K. and Byars, E., Tension Stiffening Behaviour of GFRP-Reinforced Concrete, The Fourth Middle East Symposium on Structural Composites for Infrastructure Applications (MESC-4), Alexandria, Egypt, 20-23 May (2005).

CURVED NON FERROUS REINFORCEMENT FOR CONCRETE STRUCTURES

M. Guadagnini, T. Imjai and K. Pilakoutas

The University of Sheffield, Department of Civil and Structural Engineering, Sir Frederick Mappin Building, Mappin Street, S1 3JD, Sheffield, UK

Abstract: The mechanical properties of composite reinforcing elements differ from those of conventional steel rebars in a number of ways. As a result, several issues need to be taken into account when designing concrete elements reinforced with composite reinforcement. For example, it has been proven experimentally that the capacity of curved composite elements can be considerably lower than that of their straight counterparts. This paper reports on an experimental study that was carried out at the University of Sheffield to investigate the capacity of curved composite reinforcing elements. Direct pull-out tests were performed on thermoplastic composite strips embedded in concrete cubes. A total of 47 specimens and 19 different configurations were tested. The parameters that were investigated included geometry of the bend, surface treatment, embedment length and concrete strength. This paper presents the preliminary findings of this experimental investigation as concerns the problems related to the use and performance of curved FRP bars in concrete.

1. INTRODUCTION

Steel bars have been used historically as reinforcement for concrete structures to compensate for the low tensile strength of the concrete. Owing to its ductile behaviour, steel reinforcement can be used effectively both in straight elements or cold formed into various shapes for detailing of particular structural elements and critical structural connections. Nevertheless, when concrete structures come into contact with carbonic acid deriving from the carbon dioxide present in the atmosphere or are exposed to chloride rich environments, such as those created by the presence of seawater or the frequent use of de-icing salts, steel reinforcement is susceptible to corrosion. Corrosion of the reinforcement can lead ultimately to premature deterioration of the mechanical performance of the structure and subsequent failure.

At present, corrosion of steel reinforced concrete structures is considered to be the most significant factor in limiting the life expectancy of RC structures in North America, Europe, the Middle-East and other parts of the world. In Europe alone, the annual cost of

M.S. Konsta-Gdoutos, (ed.), Measuring, Monitoring and Modeling Concrete Properties, 719–728.

repair and maintenance of these infrastructures is estimated to be over €30 billions and, in the United States, the overall costs associated with damages due to reinforcing steel corrosion have been estimated at more than $150 billions per year [11].

In the past thirty years, a range of different measures have been taken to reduce the corrosion related problems. Practical solutions include: the use of an increased concrete cover; the use of additives and inhibitors to make the concrete more impermeable and non conductive; and the use of different types of reinforcing materials, most commonly stainless steel and epoxy coated bars. The most recent attempt to increase the durability of structures has come with the introduction of composites in the construction industry as reinforcing material for concrete.

Manufactured with a combination of mineral fibres immersed in a polymeric matrix, composite materials, or fiber reinforced polymer (FRP) reinforcements as they are better known in the construction industry, display excellent resistance to environmental factors such as freeze-thaw cycles, chemical attack and temperature variations. Above all, composites can be engineered to be highly corrosion resistant. Although FRPs are already used extensively in various sectors of the construction industry (e.g. for the strengthening and repair of existing structures), their use as internal reinforcement for concrete is limited only to specific structural elements and does not extend to the whole structure. The reason for the limited use of FRPs as internal reinforcement can be partly related to the commercial availability of curved or shaped reinforcing elements suitable for the detailing of structural connections or to efficiently resist internal forces such as shear and torsion. The high production costs that are associated with the manufacture of FRP curved elements have generally reduced the interest in using FRPs for these types of applications. In addition, as it will be discussed below, various studies have shown that the mechanical performance of bent portions of composite bar is reduced dramatically as the maximum tensile strength that can be carried along the bend can be as low as 40% of the maximum tensile strength that can be developed in straight elements [2,3].

This paper presents and discusses the potential issues related to the use of curved FRP bars in concrete and reports the results of an experimental study that aims at investigating the performance of such reinforcing elements.

2. MECHANICAL PERFORMANCE OF BENT BARS

The large majority of bent reinforcing bars used in concrete structures are provided by the manufacturer in form of bent pieces that have been pre-cut at the factory according to design specification. Bending is rarely done on site and generally is only carried out when made necessary by construction conditions. Whether bending occurs on site or at the factory, conventional steel reinforcing bars offer major practical advantages. Due to their elastoplastic behaviour, steel bars can be easily formed by cold bending, and hence, most detailing needs can be easily met at a very low cost.

The two basic factors that limit the values of bending radii prescribed in construction practice guidelines are failure of the bar and crushing of the concrete inside the bend. To avoid inducing excessive strain levels in the reinforcing bar upon bending, minimum bending radii, r, are specified in the various codes of practice. Figure 1a, for example, summarizes the provisions recommended by the British Standard Institute [4] and the

American Concrete Institute [5] for steel bars. When such values of bending radius to bar diameter ratios are used in standard construction applications, the bent portion of the steel bar can be considered to be fully plastic and strain values ranging from 10% to 20% are induced in the outermost fibres (Figure 1b). Due to the mechanical characteristics of the material, however, steel reinforcement is capable of sustaining even higher levels of induced strain and carrying high levels of stress along the bend.

Although the values of bending radius generally recommended by existing design guidelines induce acceptable levels of strains in steel bars without affecting their mechanical performance, the stress that can be carried by curved FRP reinforcement has been proven to be affected by its geometry and the way in which stresses flow within the material. As reported by various researchers [6,7,8,9], the tensile strength of FRP rods is largely reduced when subjected to a multiaxial state of stress. This phenomenon can often become an issue whenever non-straight unidirectional composite elements are used as structural materials, and especially when the fibres are designed to carry high tensile stresses, since premature failure can occur at the corner portion of the composite. This reduction in the strength of the composite, therefore, needs to be carefully taken into account, since it may impose a further limitation to the maximum value of strain that can be safely sustained by the reinforcement. For example, Morphy *et al.* [2] observed that the total tensile strength of composite stirrups is reduced to as much as 40% of the maximum uniaxial tensile strength of the composite in the direction of the fibres. In a subsequent study [3], the authors also found that in order to achieve a capacity of at least 50% of the maximum strength of the composite, a minimum bending radius of not less than four times the effective diameter of the bar should be used.

The reduction in strength that occurs at the corners of a FRP bar has been quantified using empirical models such as that proposed by the Japanese Concrete Institute, described by Eq. (1) [10]. In this equation, the strength of the bent portion, ffb, is expressed solely as a function of the uniaxial tensile strength of the composite, ff, and the bar geometry (i.e. bar diameter, d, and bend radius, r). The parameter represents a probability factor that can range from 0.05 to 0.092, corresponding to a confidence interval of 95% and 50%, respectively. A value of $\alpha = 0.05$ is suggested in the Japanese recommendations.

$$f_{fb} = \left(\alpha \frac{r}{d} + 0.3 \right) f_f \leq f_f \tag{1}$$

Based on the above equation, a limit to the maximum strength that can be developed in bent bars, such as shear links, is hence imposed. Eq. 1 and the relevant strength limit are also adopted in the different design recommendations that have been proposed for FRP RC structures [11,12,13,14].

3. EXPERIMENTAL PROGRAMME

With the above issues in mind, pull-out tests were conducted on curved, thermoplastic composite strips embedded in concrete cubes to investigate the maximum strength that can be developed in their bent portion. A total of 47 specimens and 19 different configurations were tested as part of this experimental programme.

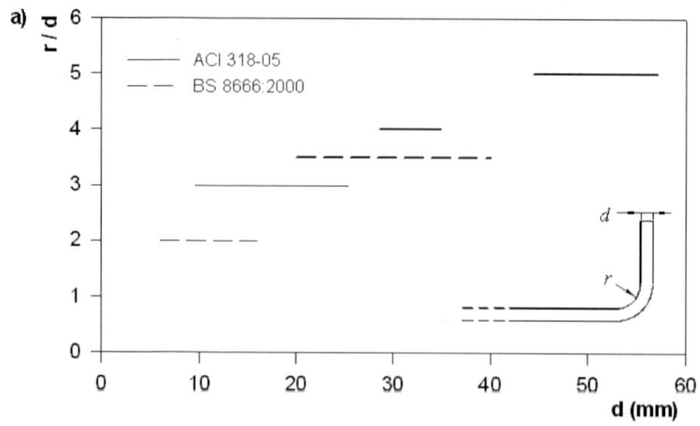

Figure 1a. Typical values of bending radii for steel reinforcing bars according to British Standard and American Code

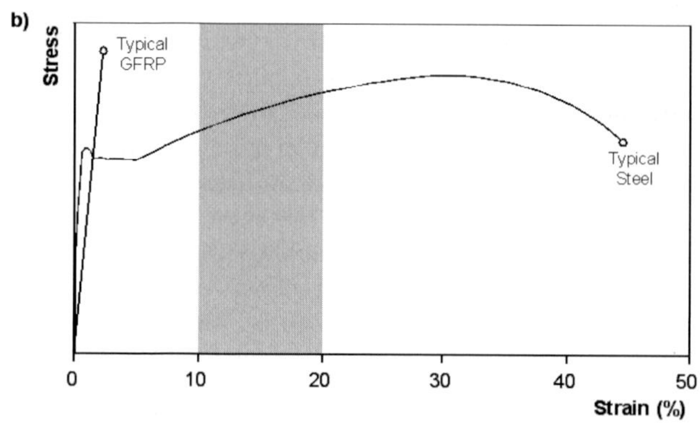

Figure 1b. Plastic strain induced in the region of the bend.

Different bending radius to strip thickness ratios, r/t, were investigated, as specified in Table 2. Different concrete strengths, f_{cu}, embedment lengths, tail lengths, t_l, and surface treatments were also examined. Type 2 and Type 3 specimens, which differ in terms of the length of the bonded portion of the vertical length provided (see Figures 2a and 2b), can be considered to be representative of a bent bar embedded in a section of an element where cracks occur at different locations.

Specimen preparation and test set-up

The specimens were manufactured as shown in Figure 2. An un-bonded length of 60 mm was adopted for Type 2 specimens, whilst Type 3 specimens were left un-bonded along the entire length of their vertical leg. The un-bonded region was created by wrapping several layers of cling film, which were treated with demoulding agent, around the composite strip. A minimum un-bonded length of 60 mm was provided to minimize the effect of concrete surface cracking on the development of bond stresses during pull-out. Foil-type electrical strain gauges were positioned at various locations along the bent composite strips (see Figure 2) to monitor the variation of strains along the un-bonded and bonded portion of the strip. The slip of the loaded end of the composite strip was measured during the tests using two Linear Variable Differential Transformer transducers (LVDTs), positioned at diametrically opposed locations on the strip, to account for possible eccentricity of the applied load and the resulting bending of the strip during pull-out. The distance between the LVDTs and the point at which the bar was fully embedded in the concrete was measured prior to each of the tests in order to determine the elastic extension of the bar that needed to be subtracted from the measurement of the loaded end slip. Two types of concrete were used to manufacture the test specimens: a normal strength concrete (N) with an average cube strength of 45 MPa; and a high strength concrete (H) with an average cube strength of 95 MPa. An aggregate with a maximum size of 10mm was used in the design of both concrete types.

The 10 mm wide reinforcing strips utilized in this testing programme were manufactured from unidirectional thermoplastic GFRP plates with a nominal thickness, *t*, of 3 mm. The mechanical and physical properties of the composite are listed in Table 1. The strips were bent to the desired shape by applying heat and moulding them around a specially designed device equipped with interchangeable corner inserts to allow for the fabrication of the required bend radius to thickness ratios. Nine of the specimens were coated with silica sand to investigate the effect of surface treatment on the bond properties between the composite and the concrete, as well as the overall capacity of the bent specimens.

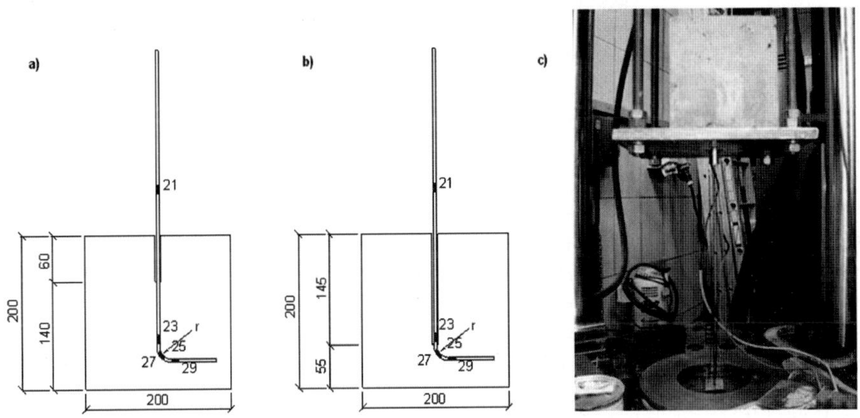

Figure 2. Location of strain-gauges for Type 2 (a) and Type 3 (b) and experimental set-up (c)

4. EXPERIMENTAL RESULTS

The experimental results obtained for all of the test specimens are summarized in Table 2 in terms of: 1) maximum load, F_{max}; 2) maximum stress developed in the vertical leg of the composite strip, f_{max}; 3) maximum average stress calculated for each set of specimens, $f_{max,avg}$; 4) the respective ratio between the average stress and ultimate tensile strength, f_f, of the composite, $f_{max,avg}/f_f$; and 5) the observed mode of failure.

Table 1. GFRP thermoplastic strips: mechanical/physical properties as given by the manufacturer

Property	
Tensile strength (MPa)	720
Tensile modulus (GPa)	28
Ultimate strain (%)	1.9
Glass content (%v/v)	35
Density (g/cm^3)	1.48

The majority of the test specimens (85%) failed due to rupture of the composite strips in the proximity of the bend (BR) whilst the remainder of the specimens failed prematurely due to crushing of the composite strip within the grips of the testing machine (GR). The maximum stress values measured for those specimens that did not fail in the region of the bend were not considered when computing the average stress values shown in Table 2.

Figures 3a and 3b show typical strain measurements along the composite, for Type 2 and Type 3 specimens, as a function of the stress developed in the vertical leg of the strip. The composite strips used in the two specimens reported in figures 3a and 3b have the same geometry but differ in terms of the length of composite embedded in the concrete cube. As can be observed, for both specimen types, the strains developed in the bent region (strain gauges 25 and 27) increase more rapidly than those induced directly in the vertical leg (strain gauges 21 and 23). The level of pull-out load at which strains start to develop in the region of the bend, however, is a function of the embedment length and the quality of the bond between the composite and the concrete. For example, in the Type 2 specimen (see Figure 3a), strains start to develop in the region of the bend only after a certain level of stress is reached in the vertical leg. In addition, although they increase at a faster rate, the strains developed in the bend are generally lower than those developed in the vertical leg. In the Type 3 specimen (see Figure 3b), however, the strains in the bent region start to develop as soon as the pull-out load is applied to the specimen and rapidly surpass those measured in the vertical leg of the composite, thus affecting, more severely, the maximum sustainable load. Figures 4a and 4b present the average values of maximum stress that were developed in the vertical leg of the composite strip for different specimen configurations as a percentage of the ultimate strength of the composite and as a function of the ratio between the bending radius and the thickness of the composite. As it will be discussed below, the maximum stress level that can be developed in the vertical leg of both Type 2 and Type 3 specimens increases as the bending radius to strip thickness ratio (r/t) increases.

5. DISCUSSION OF THE RESULTS

The capacity of bent composite strips can be expressed in terms of the maximum stress that can be developed in the vertical leg of the strip when failure at the bend occurs. Figure 5 shows the variation in the capacity of the bent specimens that was observed for the entire set of the tested specimens as a function of the bending radius to strip thickness ratio.

Table 2. Summary of experimental results

Specimen ID	r/t	t_d/t	Surface treatment	F_{max}	f_{max}	$f_{max,ave}$	$f_{max,ave}/f_f$	f_{cu}	Failure Mode
SM-2-10-N-2	2	10	SM	7.1	233	236	0.33	45	BR
	2	10	SM	6.8	239			45	BR
SM-3-10-N-2	3	10	SM	8.2	300	309	0.43	45	BR
	3	10	SM	8.7	318			45	BR
SM-4-10-N-2	4	10	SM	10.5	327	324	0.45	45	BR
	4	10	SM	9.3	314			45	BR
	4	10	SM	9.2	332			45	BR
SM-5-10-N-2	5	10	SM	11.3	383	370	0.51	45	BR
	5	10	SM	11.1	373			45	BR
	5	10	SM	9.9	354			45	BR
SC-3-10-N-2	3	10	SC	9.1	307	316	0.44	45	BR
	3	10	SC	10.5	326			45	BR
SC-5-10-N-2	5	10	SC	10.9	391			45	GR
	5	10	SC	13.0	442			45	GR
	5	10	SC	12.2	411	411	0.57	45	BR
SM-3-10-H-2	3	10	SM	9.9	334	340	0.47	95	BR
	3	10	SM	9.4	346			95	BR
SM-5-10-H-2	5	10	SM	10.9	402	402	0.56	95	BR
	5	10	SM	11.6	395			95	GR
SC-3-10-H-2	3	10	SC	10.3	332	367	0.51	95	BR
	3	10	SC	12.6	403			95	BR

Table 2. Summary of experimental results (continued)

Specimen ID	r/t	TL/t	Surface treatment	F_{max}	f_{max}	$f_{max,ave}$	$f_{max,ave}/f_f$	f_{cu}	Failure Mode
SC-5-10-H-2	5	10	SC	8.1	258			95	GR
	5	10	SC	13.1	464	464	0.64	95	BR
SM-3-5-N-2	3	5	SM	8.3	292	299	0.42	45	BR
	3	5	SM	9.0	306			45	BR
SM-5-5-N-2	5	5	SM	8.9	300			45	GR
	5	5	SM	8.8	307			45	GR
	5	5	SM	10.9	334	334	0.46	45	BR
SM-3-12-N-2	3	12	SM	9.1	323	324	0.45	45	BR
	3	12	SM	10.4	325			45	BR
SM-3-15-N-2	3	15	SM	10.4	349	352	0.49	45	BR
	3	15	SM	9.0	331			45	GR
	3	15	SM	11.3	355			45	BR
SM-2-10-N-3	2	10	SM	5.53	194	183	0.25	45	BR
	2	10	SM	5.75	196			45	BR
	2	10	SM	4.86	158			45	BR
SM-3-10-N-3	3	10	SM	7.84	283	280	0.39	45	BR
	3	10	SM	8.48	285			45	BR
	3	10	SM	8.01	271			45	BR
SM-4-10-N-3	4	10	SM	8.87	310	301	0.42	45	BR
	4	10	SM	9.02	290			45	BR
	4	10	SM	9.07	303			45	BR
SM-5-10-N-3	5	10	SM	10.04	321	316	0.44	45	BR
	5	10	SM	9.33	317			45	BR
	5	10	SM	9.54	311			45	BR
SM-3-10-H-3	3	10	SM	8.17	268	281	0.39	95	BR
	3	10	SM	8.99	294			95	BR

In general, the tests results indicated that the capacity of bent specimens increases as the bend radius increases. However, the capacity of the specimens was found to vary considerably for a given geometry and appears to be affected by the nature of the bond developed between the composite and the concrete. Overall, Type 2 specimens were able to sustain higher values of pull-out load than Type 3 specimens, and specimens embedded in normal strength concrete generally failed at lower load levels than those embedded in high strength concrete.

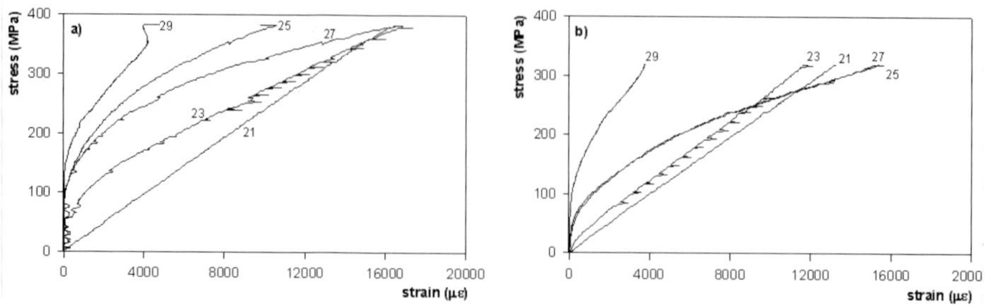

Figure 3. Stress-strain relationships along the curved composite strips of specimens a) SM-5-10-N-2 (Type 2) and b) SM-5-10-N-3 (Type 3)

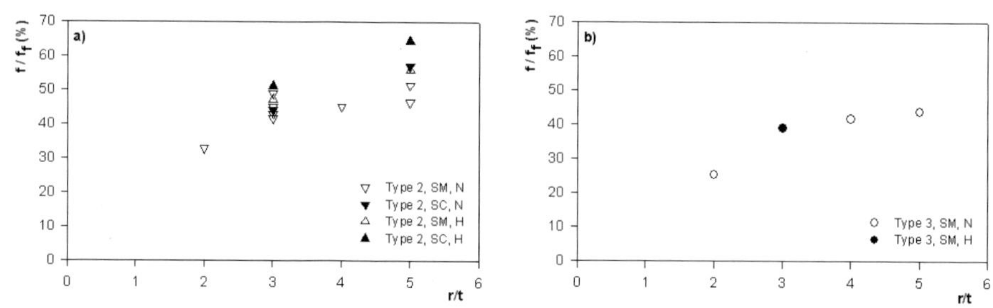

Figure 4. Variation of the average values of maximum stress developed in the vertical leg of a) Type 2 specimens and b) Type 3 specimens

Stress values developed in the vertical leg of the strips as low as 25% of the ultimate strength of the composite were observed at failure for specimens with an *r/t* value of 2. In some instances, however, the capacity of the bend in specimens with an *r/t* value of 5 was more than 60% of that of a straight element.

Figure 5 also shows the variation in the strength of the bent specimens according to the equation proposed in the JSCE design recommendations (see Eq. 1), which currently is used in all of the main design recommendations for the design of FRP RC elements (the shaded area demarked by dashed lines). As can be observed, the current design equation does not adequately describe the variation in bend capacity that was observed experimentally. Moreover, it would appear that Eq. 1 could overestimate the bend capacity of the composite strip that was used in this study. Sheata et al. [3] have also reported a similar tendency for a commercial type of CFRP reinforcement. In the present study, however, acceptable predictions were obtained for those specimens for which a better bond between the composite and the concrete was ensured either by sand coating the strips or through the use of a high strength concrete.

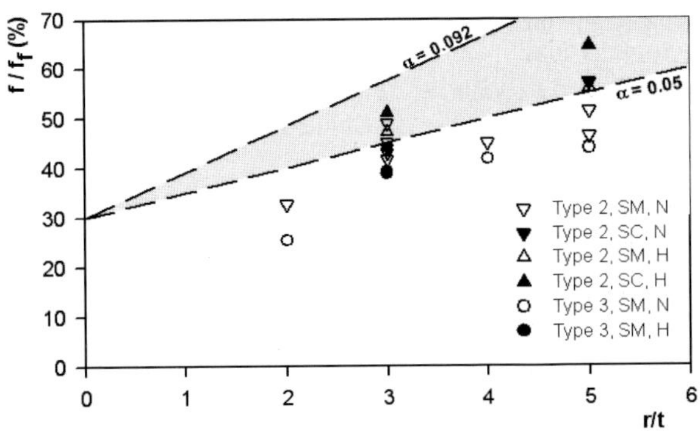

Figure 5. Comparison of test results with predictions according to current design recommendations (Eq. 1).

6. CONCLUDING REMARKS

Based on the preliminary results of the pull-out tests on curved thermoplastic composite strips presented in this paper, the following observations can be made:

1 The capacity of the bent portion of the composite appeared to be mainly a function of the geometry of the test specimens, namely the bending radius.

2 The bend capacity of the test specimens varied between 25% and 64% of the ultimate strength of the composite.

3 Values of r/t greater than 4 are required to guarantee a minimum bend capacity of 40% of the ultimate strength of the composite.

4 In a significant number of cases, the equation included in the current design recommendations for concrete structures reinforced with FRP was found to overestimate the bend capacity of the composite strip used in this study. Moreover, the capacity of the bent specimens does not seem to vary linearly with the r/t ratio, as defined in Eq. 1, and does not appear to be solely a function of the bend geometry. Rather, bond characteristics appeared to be important in controlling the development of stresses along the embedded portion of the composite and in dictating its ultimate behaviour.

Acknowledgments

The author wishes to acknowledge the financial assistance of the European Union for the Marie Curie Research Training Network En-Core, and the CRAFT RTD project CurvedNFR.

7. REFERENCES

1. Federal Highway Administration, "The Status of the Nation's Highway Bridges: Highway Bridge Replacement and Rehabilitation Program and National Bridge Inventory," Thirteenth Report to the United States Congress, Washington, D. C (1997).

2. Morphy, R., Sheata, E., and Rizkalla, S., "Bent Effect on Strength of CFRP Stirrups", Third International Symposium on Non-Metallic (FRP) Reinforcement for Concrete Structures, Sapporo, Japan, 19-26 (1997).
3. Sheata, E., Morphy, R. and Rizkalla, S., "Fibre Reinforced Polymer Shear Reinforcement for Concrete Members: Behaviour and Design Guidelines", Can. J. Civ. Eng./Rev. can. génie civ. **27(5),** 859-872 (2000)
4. British Standards Institution, Specification for Scheduling, Dimensioning, Bending and Cutting of Steel Reinforcement for Concrete, BS8666:2000, BSI, London (2000).
5. American Concrete Institute (ACI), "Building Code Requirements for Reinforced Concrete and Commentary ACI 318-05/R-05", ACI Committee 318, Farmington Hills, MI, USA. (2005).
6. Ehsani, M. R., Saadatmanesh, H., and Tao, S., "Bond of Hooked Glass Fiber Reinorced Plastic (GFRP) Reinforcing Bars to Concrete" Materials Journal, 122(3), 247-257 (1995).
7. Maruyama, T., Honma, M., and Okamura, H., "Experimental Study on the Diagonal Tensile Characteristics of Various Fiber Reinforced Plastic Rods." Transactions of the Japan Concrete Institute, 11, 193-198 (1989).
8. Mochizuki, S., Matsuzaki, Y., and Sugita, M., "Evaluation Items and Methods of FRP Reinforcement as Structural Elements." Transactions of the Japan Concrete Institute, 11, 117-131 (1989).
9. Nagasaka, T., Fukuyama, H., and Tanigaki, M., "Shear Performance of Concrete Beams Reinforced with FRP Stirrups." Transactions of the Japan Concrete Institute, 11 (1989).
10. Japan Society of Civil Engineering, "Recommendation for Design and Construction of Concrete Structures using Continuous Fiber Reinforcing Materials", JSCE, Tokyo, Japan (1996).
11. American Concrete Institute (ACI), "Guide for the Design and Construction of Concrete Reinforced with FRP Bars ACI 440.1R-03", ACI Committee 440, Farmington Hills, MI, USA (2003).
12. ISIS Canada (ISIS), "Manual No. 3 - Reinforcing Concrete Structures with Fibre Reinforced Polymers (FRPs)", ISIS Canada, Winnipeg, Manitoba, Canada (2001).
13. Institution of Structural Engineers (ISE), "Interim Guidance on the Design of Reinforced Concrete Structures Using Fibre Composite Reinforcement", IStructE, SETO Ltd., London (1999).
14. Canadian Standards Association (CSA), "Canadian Highways Bridge Design Code, Section 16 - Fibre Reinforced Structures" (1996).

FAILURE AND INSTABILITY ANALYSIS OF FRP-CONCRETE SHEAR DEBONDING USING STOCHASTIC APPROACH

K. Subramaniam, M. Ali-Ahmad and M. Ghosn
Department of Civil Engineering, Steinman Hall, City College of the City University of New York, Convent Avenue at 140th Street, New York, NY 10031, ph. (212) 650-6569, fax: (212) 650-6965, email:ksubram@ce.ccny.cuny.edu.

Abstract: The debonding of FRP from concrete substrate is studied using a stochastic finite element simulation of the direct shear test. The instability at final failure in the debonding of FRP from concrete is shown to be the result of snapback. The local variations in the interface fracture properties are shown not to significantly influence the severity of snapback.

Keywords: snapback, instability, debonding, interface fracture, FRP

1. INTRODUCTION

Both flexure and shear strengthening applications using externally attached FRP rely on shear stress transfer across the FRP-concrete interface. It has been shown that when FRP bridges a pre-existing flexure or shear-flexure crack in concrete, very high interfacial shear stresses are generated close to the edge of the crack (Leung 2004). This stress concentration results in the initiation of an interfacial crack between FRP and concrete. The opening of the pre-existing crack faces in concrete results in the propagation of the interfacial crack, leading to the debonding of a portion of the FRP from the substrate. This shear debonding response of the FRP has previously been studied using direct-shear tests, which allowed for studying the debonding phenomenon and also to extract meaningful material parameters for shear response of the interfacial bond (Ueda et al. 1999, Taljsten 1997, Ali-Ahmad et al. 2004 and 2005).

An experimental investigation of the debonding failure between the concrete and the FRP in a direct-shear test was conducted using a high-resolution, full-field optical technique (Ali-Ahmad et al. 2004 and 2005). The direct-shear test geometry was chosen because it provides a realistic representation of the shear debonding in a beam and allows for a fundamental evaluation of the debonding mechanism free from other geometric

M.S. Konsta-Gdoutos, (ed.), Measuring, Monitoring and Modeling Concrete Properties, 729–734.

effects. In the test an axial force is applied to the FRP sheet bonded to a concrete block which is restrained from movement (shown in the inset of Figure 1). A typical load response associated with debonding of the FRP sheet from the concrete substrate is shown in Figure 1a. The displacement of the load point at the edge of the bonded area is referred to as global slip, d. It was noticed that the debonding is initiated when the load response becomes nonlinear. The load then levels off and remains approximately constant at P_{crit} with increasing global slip up to failure. The crack reaches a critical length at P_{crit} following which it propagates in a self-similar manner at a constant applied load. When the load levels off at P_{crit}, there is a constant stress transfer length between the concrete and the FRP sheets associated with the crack. The stress transfer zone advances with the bonded length of the FRP as the crack propagates. The nonlinear interfacial material law that exhibits softening behavior was then established as shown in Figure 1b (Ali-Ahmad et al. 2004 and 2005). The final failure of the specimen was produced by the complete separation of the FRP composite sheet from the concrete substrate and it occurs at a high load in a sudden and uncontrollable manner.

Any design guidelines that are established for the application of FRP repairs of concrete structures that rely on the transfer of the shear stresses across the interface require an understanding of the instability at failure. Further, a large scatter were observed in the parameters of the material law for readings taken within the same specimen or for readings taken from different specimens prepared using the same concrete mix and adhesives (Ali-Ahmad et al. 2004 and 2005). The role of the material property variations on the debonding response of FRP must also be studied. These issues are addressed in this paper through a finite element simulation of the debonding response of FRP from concrete substrate.

Figure 1. (a) Direct shear test load response; (b) interface cohesive law

2. OBJECTIVES

The objectives of this paper are to develop a numerical model for studying the shear debonding at the concrete-FRP interface. The specific objectives are: (a) to trace the equilibrium path of the entire load response during debonding of FRP from concrete; (b) to study the effect of the variation of the fracture parameters on the load response of FRP during debonding from the concrete substrate.

3. NUMERICAL ANALYSIS

The finite element discetization of the specimen geometry is shown in Figure 2b. In the simplistic model, the FRP composite sheet is modeled using one-dimensional axial elements. The bond at the FRP-concrete interface is modeled using spring elements with a simplified bilinear response, which was obtained from the experimentally determined material law for the interface (Figure 2c). In the bilinear representation, a linear ascending shear stress versus local slip is obtained up to the peak shear stress, τ_{max}, following which a linear softening corresponding with the initial tangent of the actual softening response is assumed. The average values of the parameters of the bilinear response were obtained experimentally and identified as: τ_{max}=5.03 MPa, s_o=0.048 mm and s_f=0.23mm (Ali-Ahmad et al. 2005). In the numerical representation, the concrete substrate is assumed to be rigid, based on the observed experimental results (Ali-Ahmad et al 2005). The material response of the CFRP was considered to be linear elastic up to rupture.

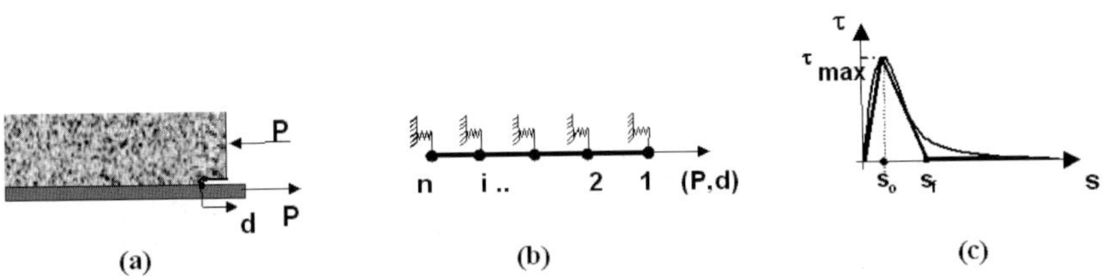

Figure 2. (a) Direct shear test geometry; (b) finite element model; and (c) bi-linear material law of FRP/concrete interface

Numerical analyses were performed for a bonded length and width of the CFRP equal to 150mm and 46 mm, respectively. The FRP bonded length was discretized into one-dimensional elements, each of length 1mm. Analysis was performed considering a variation in the fracture parameters. The analysis was started using the classical load-control procedure with Newton-Raphson method up to an initial point within the linear elastic range. The arc length method was then applied for all subsequent points along the loading curve.

4. VARIATION OF FRACTURE PARAMETERS

The fracture parameters of the material law were treated as random variables to study the effect of the uncertainties in determining these parameters on the ultimate capacity and the failure mechanism of the FRP-concrete blocks. Previous test showed that Young's modulus of the FRP composite laminate is approximately constant and therefore can be assumed to be deterministic. The parameters that define the material law are maximum shear stress, τ_{max}, the corresponding local relative slip, s_o, and the local slip at failure, s_f. All these parameters are considered as random and their means, standard deviations and correlation coefficients are provided in Table 1. It is clearly seen that the correlation between the fracture parameters is significant. To determine the probability distribution type, the data points are plotted on normal probability papers. The plots showed that all three random parameters could be adequately modeled using normal probability distribution. For instance, Figure 3 plots the fracture parameter, s_f, on normal probability paper.

Table 1. Statistical data of the fracture parameters

Random variables	$x=\tau_{max}$	$y=s_o$	$z=s_f$
Mean,	5.85	0.0384	0.18
Std. Dev.,	1.23	0.0107	0.05
Corr. Coeff., ρ	ρ_{xy}=-0.87	ρ_{yz}=0.981	ρ_{xz}=-0.897

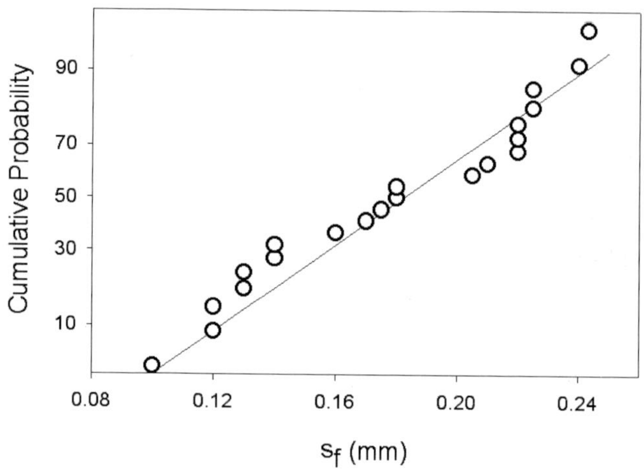

Figure 3. Relative slip sf plotted on normal probability paper

5. RESULTS OF ANALYSIS

To study the effect of the randomness in the three parameters a Monte Carlo simulation was performed taking into consideration the correlation between the parameters. Pseudo-random data was generated for the fracture parameters ($x=\tau_{max}$, $y=s_o$) by correlating them to the third parameter ($z=s_f$) and a random variable r, using the relationships:

$$\begin{aligned} x &= \sigma_x(a_1 Z + R_1) + \mu_x \\ y &= \sigma_y(a_2 Z + R_2) + \mu_y \\ z &= \sigma_z Z + \mu_z \end{aligned} \quad (1)$$

where μ_x, σ_x, μ_y, and σ_y are the means and standard deviations of x and y respectively; Z, R_1 and R_2 are standard normal random variables with means equal to zero and standard deviations equal to 1; a_1 and a_2 are given by the following expressions:

$$a_1 = \frac{\rho_{xz}}{\sqrt{1-\rho_{xz}\rho_{xz}}}$$
$$a_2 = \frac{\rho_{yz}}{\sqrt{1-\rho_{yz}\rho_{yz}}} \tag{2}$$

The analysis procedure described previously was applied for each set of parameters to determine the load versus global slip response of FRP bonded to concrete. Every spring of Figure 2b was assigned one set of fracture parameters generated randomly, which differ from those of the other springs. This would simulate the effect of the random nature of the fracture parameters at the local level. The load versus global slip response for the specimen analyzed herein is shown in Figure 4. The fluctuations in the load response are due to the effect of the randomness of the fracture parameters. The results of the numerical analysis suggest that the uncontrolled and sudden response obtained experimentally at ultimate failure in the shear debonding of FRP from concrete is associated with snap-back instability. Snap-back results in a simultaneous decrease in load and displacement and is associated with a sudden release of elastic energy, which is stored in the fully debonded FRP. This release of energy makes the final failure very catastrophic. It is noticed that the overall load response is not affected by local variations of the interface fracture properties. While the variation in the local material properties produces a variation in the global debonding response, the severity of snapback is not affected by these variations.

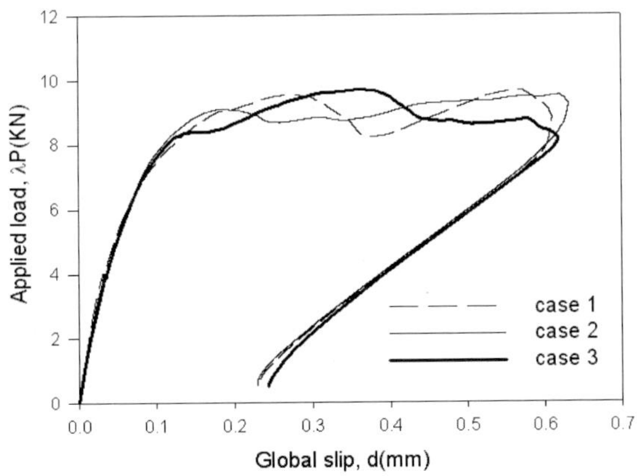

Figure 4. Load versus global slip response of FRP bonded to concrete generated for three different sets of fracture parameters

6. CONCLUSIONS

Based on the results obtained, the following conclusions can be drawn: (a) The results of the numerical analysis demonstrate that the shear failure of the FRP-concrete bond from concrete may lead to snap-back instability; (b) The randomness in the fracture parameters does not significantly affect the severity of snapback at ultimate failure.

7. REFERENCES

1. Ali-Ahmad, M., Subramaniam, K., and Ghosn, M. (2004). "Fracture Analysis of the Debonding between FRP and Concrete using Digital Image Correlation." *Proceedings of FRAMCOS-5 International Conference on Fracture of Concrete and Concrete Structures/ Vail*, Colorado, USA, 787-793.
2. Ali-Ahmad, M., Subramaniam, K., and Ghosn, M. (2005). "Experimental Investigation and Fracture Analysis of Debonding between Concrete and FRP." *ASCE, Journal of Engineering Mechanics*, accepted for publication.
3. Leung, C.K.Y. (2004). "Fracture Mechanics of Debonding Failure in FRP-Strengthened Concrete Beams," *FraMCoS-5 conference*, pp. 12-16.
4. Taljsten, B. (1997b). "Defining Anchor Lengths of Steel and CFRP Plates Bonded to Concrete," *International Journal of Adhesion and Adhesives*, 17(4), pp. 319-327.
5. Ueda, T., Sato, Y., and Asano, Y. (1999). "Experimental Study on Bond Strength of Continuous Carbon Fiber Sheets," *Proceedings of the FRPRCS-4*, pp. 407-416.

BOND CHARACTERISTICS AND STRUCTURAL BEHAVIOR OF INORGANIC POLYMER FRP

Ch. Papakonstantinou and P. Balaguru
University of Massachusetts Dartmouth, Dept of Civil and Environmental Engineering, North Dartmouth, MA 02747, USA; Rutgers University, Dept of Civil and Environmental Engineering, Piscataway, NJ 088854, USA

Abstract: Results reported in this paper deal with the bond properties between concrete and an inorganic polymer Fiber Reinforced Polymer (FRP). Tests were conducted using reinforced and plain concrete beams strengthened with carbon fibers and inorganic matrix. Beams were tested under flexural monotonic loading. None of the beams failed by FRP delamination, which is the most common mode of failure for organic polymer FRP strengthened beams. Test results and analysis with particular focus on failure mechanism are presented.

Keywords: Inorganic polymer, bond, concrete, flexure, FRP, carbon fibers

1. INTRODUCTION

Acceptance of high strength composites in the construction industry has grown at a rapid pace during the last decade. Fiber reinforced polymers (FRPs) are very popular for repair and rehabilitation because they are lightweight, corrosion resistant, and they also exhibit high strength. Their low weight reduces the duration and cost of construction. The composites can be applied layer by layer or as a thin plate. In all commercially available systems, the adhesives are organic polymers (Polyesters, Vinyl-esters, Epoxies, Phenolics), which exhibit high strength and adhesion [1]. However, they are easily susceptible to fire and they also emit toxic gases (CO or CO_2) when subjected to fire [1,2]. Given these disadvantages it is potentially hazardous to use any type of these adhesives as a strengthening system in buildings. Some compositions are also vulnerable to degradation under ultraviolet light, leading to long-term durability problems.

Since carbon and glass fabrics can withstand normal fire exposure and are durable under ultraviolet light, the organic polymers used to attach these fabrics to concrete become the weak link. The inorganic matrix discussed in this paper is not combustible and does not degrade under ultraviolet light. The inorganic adhesive is being evaluated

M.S. Konsta-Gdoutos, (ed.), Measuring, Monitoring and Modeling Concrete Properties, 735–741.

for use in aircrafts and the infrastructure [3-8]. Temperature exposure tests conducted with carbon composites made using the inorganic matrix have shown that only 37% of the composite's initial flexural strength is lost after 1 hour of exposure at 800^{0}C [3]. The inorganic matrix is a low-viscosity resin, suitable for penetrating carbon or glass fiber sheets and fabrics. It is prepared by blending an aluminosilicate powder with a water-based activator. At room temperature, it has a pot life of about 3 h.

The inorganic matrix was used with carbon sheets to strengthen non reinforced and reinforced concrete beams. The behavior of these beams is presented in the following paragraphs. A special focus is given on the failure mechanism and the bond between the concrete substrate and the inorganic matrix.

1.1 Properties of the Inorganic Matrix

The matrix to be evaluated was developed for use in aircraft structures and modified for use as a coating material and adhesive for brick, concrete, wood, and steel. The cementitious part is a potassium aluminosilicate. The resin hardens to an amorphous (glassy) structure at moderate temperatures of 800 to 150^{0}C. Hardeners have been developed to obtain a room temperature cure in about a day. The research conducted so far has focused on the mechanical, thermal, and durability properties of composites made with carbon, glass, and steel [3-8]. The unique features of the matrix are as follows:

- The resin is prepared by mixing a liquid component with silica powder. Fillers and hardening agents can be added to the powder component. The two components can be mixed to a paint consistency. Since the matrix is water based, tools and spills can be cleaned with water. All of the components are nontoxic and no fumes are emitted during mixing or curing. The excess material or material removed from the old application can be discarded as general waste.
- The pot life varies from 30 minutes to 3 hours for compositions that cure at room temperature. Compositions that require heat for curing at 800 to 1500^0 C can be stored for weeks.
- Common application procedures such as brushing and spraying can be used for the application. The product was successfully used to coat bridge substructures by brush and a sprayer in Rhode Island, USA.
- The matrix bonds well with carbon and glass fibers and these fibers can be used as reinforcements. Tows and fabrics made out of carbon and glass can also be attached to existing structures.
- The matrix can withstand temperatures up to 1000^0 C, and is not affected by UV radiation. Fire tests show that the flame-spread index is zero. Since the air permeability is low, the matrix also protects the material it is covering by reducing the amount of oxygen for combustion.

2. REINFORCED CONCRETE BEAMS

A total of eight reinforced concrete beams were constructed and tested in four-point flexure over a span of 3,000 mm with loads at one-third points. The primary variables were carbon length, end anchorage, and the amount of strengthening reinforcement. The beams constructed for the current study were of identical span, width and depth (3,000x200x300 mm) and effective cover, Figure 4. Strengthening was done using 2, 3, and 5 layers of unidirectional carbon sheets. The concrete was made using a laboratory mixer. The compressive strength was 47.3 MPa and was obtained using 150-mm diameter cylinders. The flexural reinforcement used in the current program consisted of two No. 4 bars. No. 3 stirrups at 95-mm spacing were used in order to avoid shear failure. The experimental program was designed to simulate the research conducted at the Universite de Sherbrooke, Québec, Canada [9], thus reducing the number of beams to be tested for comparing the organic and inorganic matrices.

The organic matrix is more than 1,000 times tougher than the inorganic matrix. Its strain capacity is more than 65 times greater than that of the inorganic matrix. Despite differences in mechanical properties, the performance of the inorganic strengthening system was comparable to that of the organic system. While the performance was comparable, the failure mode was different. All beams strengthened with inorganic matrix failed by rupture of the carbon fiber sheet. On the contrary all beams strengthened with organic FRPs failed by delamination of the carbon reinforcement. This surprising result suggests that there are important differences in the load transfer mechanism.

During the experiments no small interfacial cracks were present when the inorganic system was used. The following points explain this behavior: (i) The matrix toughness is comparable to the toughness of the cement paste. It is not sufficiently tough to maintain the bond with aggregate that is necessary to produce closely spaced interfacial cracks. At high strain, the matrix fails, rather than the cement paste. (ii) Beginning at the matrix cracking strain of 0.0007 mm/mm, cracks in concrete propagate through the matrix of the composite, resulting in increased carbon stress and reduced composite stiffness.

The principal differences in load transfer mechanism are that a tough matrix disintegrates the bond between cement and aggregates, while a brittle matrix results in intermittent bond loss and local regions of stress concentration in the carbon. The behavior of the brittle matrix is similar to the bond behavior of steel reinforcement in concrete. The bond fails adjacent to cracks, facilitating local bond slip in the carbon while bond is maintained intermittently. This mechanism reduces the tensile strain on the concrete at the interface. Complete delamination does not occur because the interface is strong in shear when it is not highly strained. This concept deserves further investigation because of its importance for the further study of bond failure in externally strengthened beams.

3. PLAIN CONCRETE BEAMS

Concrete prisms with dimensions of 50 X 50 X 330mm were cast using steel molds and table vibrators and cured at 100% relative humidity for at least 28 days. The cured samples were prepared for strengthening using the standard procedures. This involves sand blasting and then cleaning with a wire brush. Once the surface was cleaned, a thin layer of the inorganic matrix was applied to fill the small air voids and to create a smooth surface. Pre-cut carbon tows and sheets were impregnated with the matrix and placed on the prepared concrete surface and bonded using grooved rollers. A second layer of the matrix was applied as a protective coating. The samples were cured for at 24 hours at room temperature followed by 24 hours at 80^0C. The elevated temperature was used to ensure adequate curing in a two-day period. One day curing at 80^0C is equivalent to about one week curing at room temperature.

For the organic matrix, the manufacturer's recommendations were followed. The cleaned surface was coated with primer and cured for three days. The matrix and the carbon were placed on the primed surface and bonded using rollers. After applying the top coat, the samples were left at room temperature for at least seven days before testing.

The prisms were tested using a center point load over a span of 300mm. An MTS testing machine with computer controls was used to measure load and midspan deflections. The tests were run under displacement control at the rate of 0.1 mm/mm.

As expected, the control beams behaved linearly elastic up to peak load, followed by brittle failure. Each of the strengthened beams carried considerable load after the development of the crack at the mid span. Only one crack developed which widened until failure of the specimen. The crack width at failure increased with an increase in reinforcement area. An increase in deflection at failure was also experienced as the crack width increased.

In the case of the organic matrix, the crack at midspan extended at the interface towards the supports. In each case, failure occurred because of debonding of the reinforcement. After the tests were completed, the debonded tows and fabrics could easily be peeled away from the sample. It can be hypothesized that a thin layer of concrete near the interface was weakened by the shear produced by the carbon fibers.

In the case of the inorganic matrix, the failure was always caused by fracture of the carbon fibers. It should be noted that the inorganic matrix is much more brittle than the organic matrix and hence microcracks occur within the matrix. Therefore, the fibers act individually rather than as a single plate reducing the possibility of delamination. The reinforcement was not easily peeled away from the prisms after failure as in the case of the organic matrix.

The absence of effective load transfer between the fibers within the composite reduces the shear stress at the interface. The positive effect is the absence of debonding failure. The negative effect is the inefficient use of carbon fibers. The carbon fibers at the microcracks experience higher stresses, resulting in fiber failure at lower average strains as compared to organic matrices. This aspect of the failure pattern is further discussed in a later section.

The load-deflection behavior of beams strengthened with inorganic and organic matrices is presented in Figures 1 and 2, respectively. In Figure 1, the behavior of beams strengthened with 1, 2, and 3 tows, 1 and 2 layers of fabric, and control beams are presented. In Figure 2, an additional curve for the unidirectional carbon sheet is also presented. A careful review of these curves leads to the following observations.

The load-deflection behavior of these plain concrete beams is different from the curves corresponding to the strengthened reinforced concrete beams [8]. The major difference could be due to the single crack that occurs in plain concrete beams. This difference in behavior should be considered when high strength composites are used to strengthen non-reinforced concrete and masonry. The test results also provide an opportunity to understand the mechanism of load transfer near the crack.

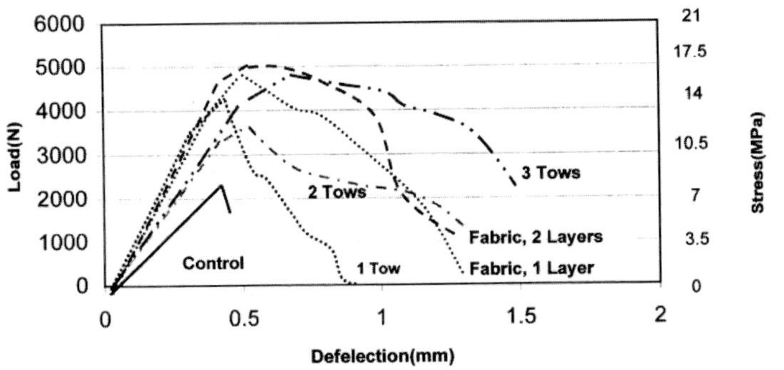

Figure 1. Load-deflection behavior, inorganic matrix

As expected, the beams strengthened with the organic matrix had larger deflections before failure. As mentioned earlier, the inorganic matrix cracks at strains of about 0.007mm/mm. In addition, the silicate matrix bonds with the concrete chemically by transfer of CaOH and KOH between parent concrete and the adhesive, resulting in the absence of a well defined interlaminar layer. Therefore, when the concrete cracks, the cracks will go through the repair layer, transferring the forces to the carbon fibers. As a result, the debonding of the repair plate at the crack is much less in the inorganic matrix and the average strain in the plate at failure is also lower.

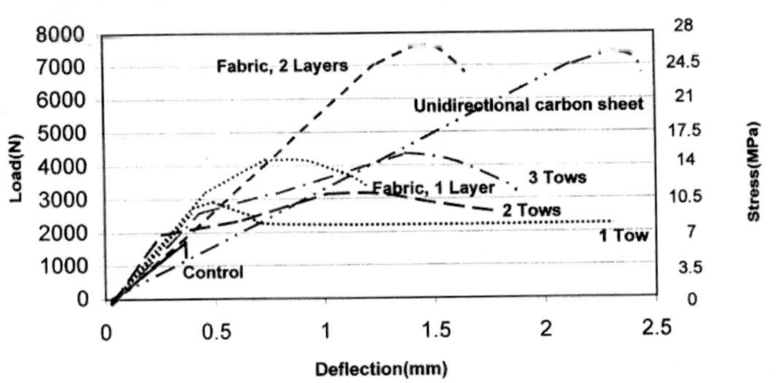

Figure 2. Load-deflection behavior, organic matrix

Increase in the failure loads is compared in Figure 3. The increases are presented as a factor of control strength for easy comparison. In beams reinforced with carbon tows, the strength increase for the inorganic and organic matrices is not significantly different. In the case of the inorganic matrix, the beam reinforced with only one tow had a greater strength increase than the beam reinforced with two tows. Initially, this was assumed to be an experimental error. Additional tests confirmed the accuracy of this result. Strength analysis presented later and close observation of the failed specimens lead to the following hypothesis.

When inorganic matrix is used for repair, the amount of fibers and their distribution plays an important role near the crack and load transfer. In the case of one tow, the fibers are well spread out and are very close to the bonded surface. As soon as the crack forms in the parent concrete, it penetrates the repair matrix. The authors believe that this penetration occurs when the crack width is only a fraction of a millimeter. This process enables the contribution of the carbon fibers at very low crack width and hence the tension force contribution of the concrete is not lost. Strength calculations confirm that, on the onset of cracking, both carbon and tension zone concrete contributes to strength capacity. The disadvantage in this load transfer mechanism is the reduction in ductility of the specimen. When more fibers are added, the load transfer becomes much less localized and an increased crack width opening occurs before failure. This results in an increased ductility but less utilization of the carbon fibers. The average stress and strain of carbon at failure reduces as the amount of fibers is increased. Ductility can also be improved by adding short fibers to the matrix [5]. The stress and strain failures also decrease with an increase in fiber content in organic matrices. Inorganic matrices provide strength increases up to 216 percent where as organic matrices provide as much as 400 percent increase.

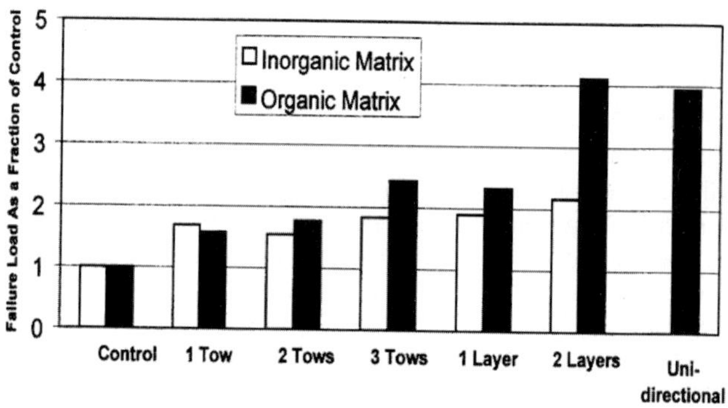

Figure 3. Increase in failure loads

4. CONCLUSIONS

Based on the experimental and analytical results presented in this paper and observations made during the testing, the following conclusions can be drawn:

- Both inorganic and organic matrices can be effectively used for strengthening non-reinforced and reinforced concrete.
- The load transfer mechanisms are different for inorganic and organic matrices: The inorganic matrix composite develops micro cracks and failure occurs by fracture of the carbon. The load transfer from fiber to fiber within the composite plate is not very efficient, especially when fiber areas are large. The organic matrix plates fail by delamination.
- For both organic and inorganic matrices, the average fiber strain at failure reduces with an increase in fiber area.
- When carbon composites are used to strengthen non-reinforced systems, load redistribution through a crack plays a major role in the failure mechanism. This aspect should be considered in repair design.

- The inorganic and organic systems provide comparable performance with respect to the increase in strength of reinforced concrete beams.
- For inorganic systems, the strength increase for unit carbon area seems to increase with the number of carbon layers.
- The inorganic system and the organic system improved post-crack stiffness equally well. It was found that each strengthening system provided approximately the same post-crack stiffness increase for unit carbon area.
- For reinforced concrete beams, the inorganic system increased the post yield stiffness more than the organic system. It was found that the inorganic matrix system provided higher post yield stiffness increase for unit carbon area.
- All of the beams strengthened with the inorganic matrix system failed by carbon rupture. In contrast, the beam strengthened with the organic matrix system failed by delamination. This difference can be explained by differences in the load transfer mechanism between composite and parent concrete.

5. REFERENCES

1. Edwards, K. L., "An overview of the technology of fibre-reinforced plastics for design purposes," Materials & Design, Volume 19, Issues 1-2, 1 February 1998, pp. 1-10
2. Autian, J. "Toxicologic Aspects of Flammability and Combustion of Polymeric Materials",. Journal of Fire and Flammability, Vol. 1, No.239, July 1970.
3. Foden, A., Lyon, R., and Balaguru, P., "A High Temperature Inorganic Resin for Use in Fiber Reinforced Composite," Proceedings, First International Conference on Composites in Infrastructure, Tucson, Az, January 1996, pp.166-177.
4. Davidovits, J., "Geopolymers: Inorganic Polymeric New Materials," Journal of Thermal Analysis, V.37, 1991, pp.1633-1756.
5. Foden, A., Balagurn, P., and Lyon, R., "Mechanical Properties of Carbon Composites Made Using an Inorganic Polymer," ANTEC, 1996, pp.3013-3018.
6. Lyon, R. E., Balaguru, P., et al., "Fire Resistant Alumino Silicate Composites," Fire and Materials, V.21, 1997, pp.67-73.
7. Hammell, J.A., Balaguru, P., Lyon, R.E., "Strength Retention of Fire Resistant Aluminosilicate-Carbon Composites under Wet-Dry Conditions," Composites Part B: Engineering, v.31, Issue 2, March 2000, Pages 107-111
8. Toutanji, H., Gomez, W., "Durability Characteristics of Concrete Beams Externally Bonded with FRP Composite Sheets," Cement and Concrete Composites, V.19, 1997, pp.351-358.
9. M'Bazaa, I. M., Missihoun, M., and Labossiere, P. (1996). ''Strengthening of reinforced concrete beams with CFRP sheets.'' Proc., 1st Int.Conf. on Compos. in Infrastruct., ICCI'96, Dept. of Civ. Engrg. and Engrg. Mech., Univ. of Arizona, Tucson, Ariz., 746–759.

EFFECT OF CONCRETE COMPOSITION ON FRP/ CONCRETE BOND CAPACITY

J. Pan and C.K.Y. Leung
Department of Civil Engineering, HKUST, Clear Water Bay, HK, Clear Water Bay, Kowloon, HK

Abstract: The bond capacity is considered to be most related to shear softening relation along concrete/FRP interface. In existing models, it is strongly dependent on the compressive (or tensile strength) of the concrete itself. However, theoretical analysis indicates that the shear stress distribution along the FRP/concrete interface at ultimate debonding failure is usually dominated by the frictional part, and aggregate interlocking is leading to residual stress. Size and content of aggregates are therefore suspected to affect the interlocking effect. To investigate the effect of aggregates on the bond capacity, ten different compositions of concrete have been used to prepare specimens for the direct shear test. The bond capacity is found to be related to concrete surface tensile strength, and the aggregate content. Based on the results, an empirical expression using neural network is derived to calculate the interfacial fracture energy in the shear test. The bond capacity can then be calculated according fracture mechanics based model. Good agreement has been obtained between the simulation and experimental results.

Keywords: FRP plates; direct shear test; concrete composition; debonding

1 INTRODUCTION

For a flexural strengthened RC beam, it may fail by FRP debonding from the bottom of a major flexural crack in the span, as shown in Figure 1. It is usually observed that the failure occurs in the concrete and a thin layer of concrete is attached on the surface of the debonded FRP plate. Hence, for concrete members strengthened with externally bonded FRP, the concrete/FRP interfacial behavior should be an important focus of investigation.

This kind of crack-induced debonding failure is often studied with the direct shear test, which involves a FRP plate bonded on a concrete prism. Recently, many studies have been undertaken to study the bond behavior between concrete and FRP sheets.

M.S. Konsta-Gdoutos, (ed.), Measuring, Monitoring and Modeling Concrete Properties, 743–748.

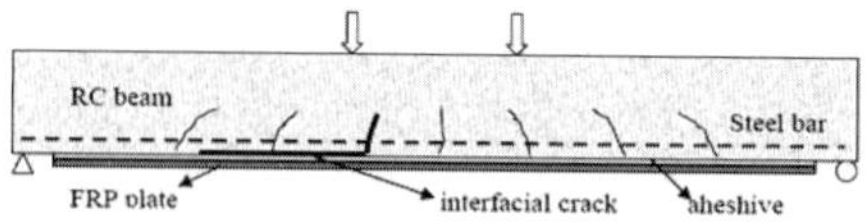

Figure 1. FRP debonding from the bottom of a major flexural crack in the RC beam

From the literature, plenty of experiments have been conducted by using single shear tests (e.g. Taljsten 1997, Bizindavyi and Neale 1999), double shear tests (e.g. Neubauer and Rostásy 1997) and modified beam (e.g. Laura D.L. et al. 2001) to simulate the bond behavior of the FRP plates in the real RC beams. Theoretical work on the bond behavior has also been carried out and a number of empirical models (Tanaka 1996, Maeda et al. 1997) and analytical models (e.g. Chen and Teng 2001) have been developed, as well as some practical design proposals. These works have led to improved understanding of the failure characteristics of the FRP-to-concrete joints. Based on most existing models, interfacial fracture energy is an important parameter related to the effectiveness of strengthening. The interfacial fracture energy is often taken to be a function of the compressive or surface tensile strength of concrete. However, this has never been proved and is also questionable from the physical point of view because fracture energy is dependent on the shear slip relation at concrete/FRP interface.

The objectives of this study are: (1) to investigate the effect of material properties, such as concrete compressive strength, splitting tensile strength, and concrete surface strength on the bond capacity; (2) to investigate the effect of aggregate content on the bond capacity. To achieve these objectives, ten different compositions of concrete with crushed aggregate have been used to prepare specimens for the direct shear tests. Based on the experimental results, the correlations between the material mechanical parameters and the ultimate bond capacity are evaluated, as well as the contents of aggregates. Then, neural network is employed to develop an empirical approach to predict the interfacial fracture energy at debonding based on the test results.

2 EXPERIMENTAL PROGRAM

2.1 Specimen preparation and material properties

To investigate the concrete composition on the bond capacity in direct shear test, ten batches of concrete with different mixing proportions were cast to prepare the concrete prisms. The mixing proportions for each batch of specimens are listed in Table 1. The size of the concrete prisms and FRP plate was shown in Figure 2. The initial 50mm was left unbonded to avoid wedge failure of concrete due to shear stress. In order to record the strain variation along the FRP plate during the loading process, nine strain gauges were placed on the FRP plate with a center-to-center space of 30mm, as shown in Figure 2.

To study the correlations between the concrete material properties and the bond capacity, the mechanical properties of concrete, such as compressive strength, tensile strength, surface tensile strength and mode I fracture energy, were measured together with the ultimate bond capacity. Figure 3 shows the test setups for surface tensile test and three-point bending test, which used to measure the fracture energy in Mode I. The mechanical properties of the concrete are summarized in Table 1. The FRP used in the tests is the Reno composite material system, whose tensile strength is up to 4200MPa in fiber direction. The Young's modulus of the FRP is about 235GPa. The FRP thickness is about 0.11mm per ply according to the production specifications.

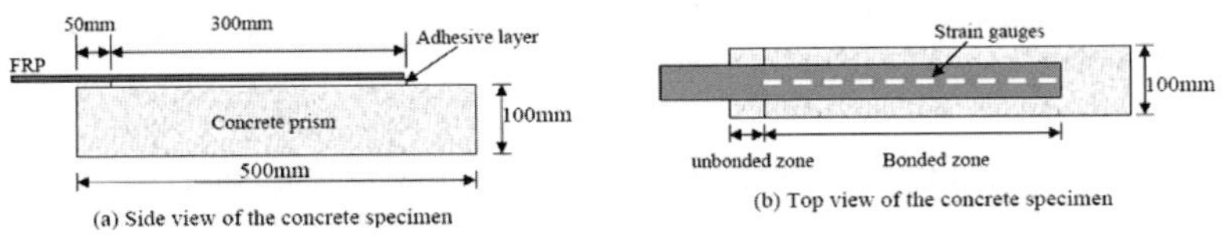

Figure 2. Dimensional information about the tested specimens

Table 1. Experimental results of material properties of the concrete

Concrete composition	Mixing proportion	Aggregate content (%)	Compr. Strength (MPa)	Tensile Strength (MPa)	Surface tensile strength (MPa)	Fracture Energy in Mode I (MPamm)
M1	1:0.5:1.5:2.6	46.4	43.1	4.18	2.85	0.129
M2	1:0.6:1.5:2.6	45.6	35.2	3.17	2.15	0.115
M3	1:0.5:1.5:1.5	33.3	57.5	3.69	3.04	0.134
M4	1:0.6:1.5:1.5	32.6	38.6	3.38	1.63	0.157
M5	1:0.4:1.5:2.6	47.3	61.5	4.48	2.78	0.174
M6	1:0.5:1.5:0.5	14.3	47.4	3.76	2.26	0.119
M7	1:0.54:2.55:1	19.6	47.1	3.68	2.04	0.077
M8	1:0.5:2.25:1.25	25.0	44.7	3.26	2.23	0.117
M9	1:0.5:2:1.5	30.0	52.4	3.99	2.73	0.121
M10	1:0.5:1.5:2	40.0	57.9	4.49	2.47	0.156

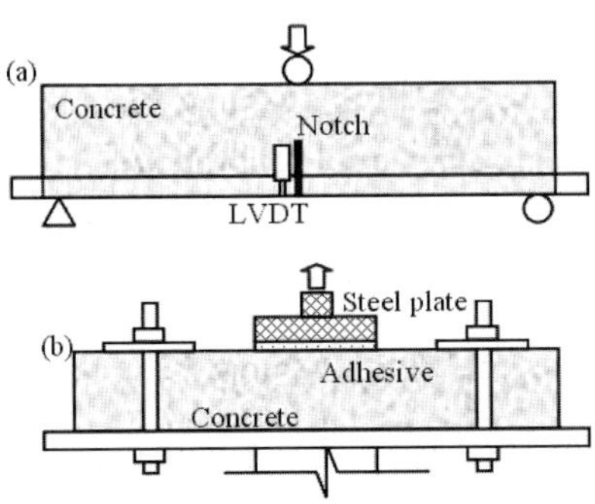

Figure 3. Test setups for (a) three-point bending test (b) and pull-out test

Figure 4. Test setup for the direct shear test

2.2 Test setup and test procedure

As for the setup of the direct shear test, a steel frame which could hold the concrete specimen tightly in the vertical direction was designed, as shown in Figure 4. During the testing process, a LVDT was used to measure the global displacement of the FRP plate bonded on concrete. The test was conducted under displacement control with loading rate of 0.1mm/min. An automatic data logger was employed to collect the strains along the FRP plate and the displacement from the LVDT, as well as load and stroke data from the MTS machine.

3 EXPERIMENTAL RESULTS AND DISCUSSIONS

3.1 Effects of concrete properties on the bond capacity

In the experimental program, a total of 30 specimens were tested to investigate the effect of concrete composition on the bond behavior between the FRP plate and concrete. Correlations between the mechanical properties of concrete and the bond capacity are then investigated.

From the test results, the compressive strength for various batches of concrete varies from 35.2MPa to 61.5MPa. The tensile strength of each batch of concrete has been obtained through the splitting tension test. The tensile strength values of the concrete are between 3.17MPa and 4.49MPa. It is found that there seems to be no sound correlations between concrete compressive strength or splitting tensile strength and the bond capacity. Mode I fracture failure was calculated for all specimens based on three-point bending test. Since debonding failure can be considered as failure due to the Mode II propagation of an interfacial crack, it is of interest to see if it is affected by the Mode I fracture energy as well. However, the test results show there is no obvious correlation between the fracture energy and the ultimate bond capacity.

In the present experiments, the surface tensile strength of each concrete composition is measured by the pull-out test. The surface tensile strength values for all specimens range from 1.625MPa to 3.04MPa. Figure 5 shows the correlation between the surface tensile strength and the bond capacity for all concrete compositions. Phenomenologically, the bond capacity of the specimens increases with the concrete surface tensile strength. For the concrete prism bonded with the FRP plate under direct pulling, the concrete at the concrete/adhesive interface is under shear. The locally damage of concrete is most related to the surface tensile strength.

3.2 Effects of aggregate content

Figure 6 shows that the ultimate bond capacity increases with the aggregate content of the concrete compositions. In the concrete, the amount of aggregate on the failed surface increase with the aggregate content, and bond between the aggregate and the FRP plate is stronger than that between the paste and the FRP plate. Also, the interfacial friction in the debonded region depends on interlocking and abrasion effects along the damaged interface. Increasing the aggregate content can significantly slow down the softening behavior in the debonded zone, which leads to improvement of the ultimate bond capacity. Thus, the proportion of aggregate in specimens is an important parameter affecting the bond capacity.

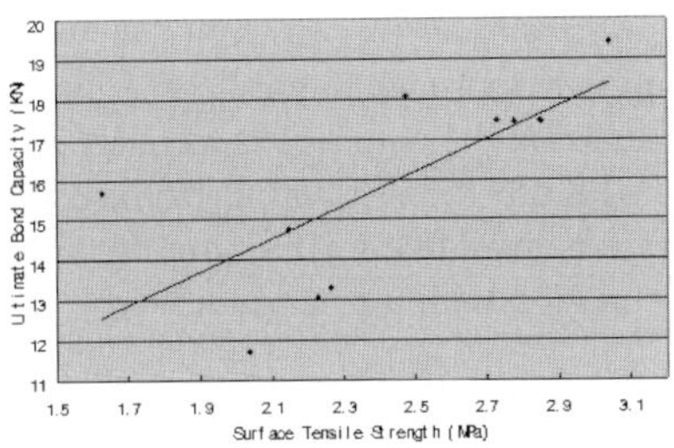

Figure 5. Correlation between concrete surface tensile strength and the bond capacity

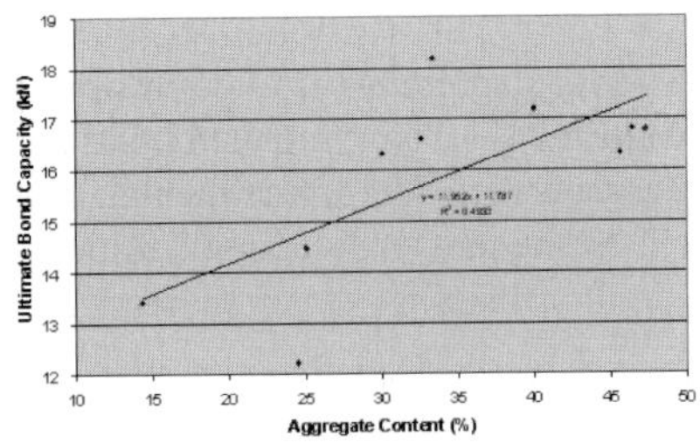

Figure 6. The correlation between aggregate content and the bond capacity

4. AN EMPIRICAL APPROACH USING NEURAL NETWORK

According to the test results, the bond capacity is mostly related to the concrete surface tensile strength and aggregate content, indicating that the interfacial fracture energy is significantly affected by the two parameters. The interfacial fracture energy for each specimen can be calculated according to the following bond strength model (Taljsten 1996). To obtain an empirical relation between the interfacial fracture energy and the two parameters, an approach using neural network is proposed in this section.

$$P_u = b_p\sqrt{2E_p t_p G_f/(1+\alpha)}\text{, where }\ \alpha = E_p t_p b_p/(E_c t_c b_c) \tag{1}$$

In this study, a program using Matlab 6.1's built-in neural network Toolbox was developed to conduct the analysis. Figure 7 shows the comparisons between the simulated bond capacity from the neural network and the experimental results. It is found that the simulated P_{ulti} are very close to the experimental results since all points in Figure 7 lie close to the 45-degreee line. This indicates that the network has generalized underlying information well.

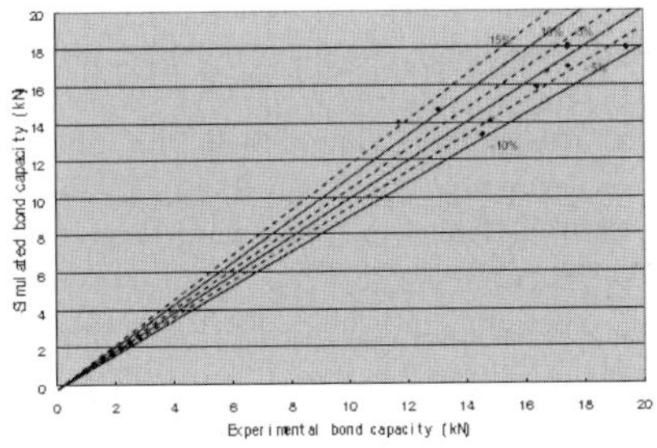

Figure 7. Comparison between the network simulated P_{ulti} and experimental results

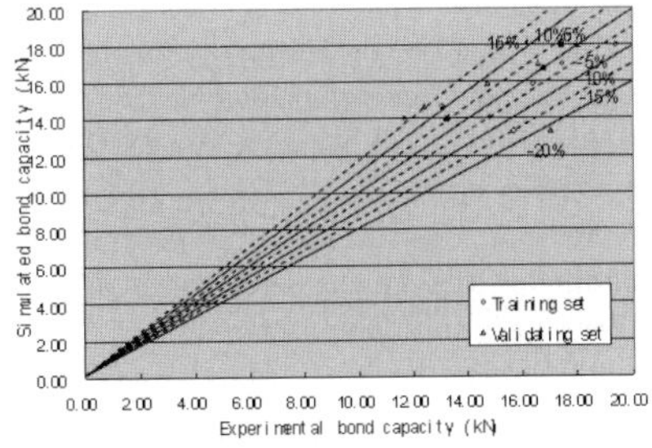

Figure 8. Comparison between the bond capacities from empirical and experimental results

With neural network analysis, an empirical equation for the interfacial fracture energy is given by:

$$G_f = (-1.0053a + 1.3432)(0.4245 f_{ctm} - 0.0163) \tag{2}$$

The above expression provides a simple empirical approach to predict the interfacial fracture energy G_f. The calculated bond capacities from the empirical approach and Equ (1) are compared to the experimental results in Figure 8 and good agreement can be obtained. At the part, only a small amount of data is available for training the network. However, the feasibility of generating a design equation with the neural network is illustrated.

5 CONCLUSIONS

In this paper, experimental investigations have been conducted to study the effect of the concrete composition, and properties on FRP debonding behavior in the direct shear test. It is found that the bond capacities of the specimens are mostly dependent on the concrete surface tensile strength, aggregate content. Although other material properties (such as compressive strength, splitting tensile strength, and fracture energy) may have some effect on the debonding behavior, the test results do not reveal clear correlations between the bond capacity and these mentioned parameters. The neural network is employed to derive an empirical expression for obtaining the interfacial fracture energy G_f in terms of surface tensile strength and aggregate content. The bond capacity can then be calculated according to the fracture mechanics based model. Good agreement has been obtained between the simulated results using the empirical expression and the experimental results. With more data available in the future, the empirical expression can be further refined and improved through additional training with the neural network approach.

6. REFERENCES

1. Taljsten, B. (1997) "Strengthening of beams by plate bonding." *J. Mat. in Civ. Engrg.*, ASCE, 9(4), 206-212
2. Taljsten B. (1996) "Strengthening of concrete prisms using the plate bonding technique" *International journal of Fracture*, 82: 253-266
3. Bizindavyi L. and Neale K.W., (1999) "Transfer length and Bond strengths for composites bonded to concrete." *Journal of composites for construction*, Vol. 3, No. 4. pp.153-160
4. Laura D.L., Brian M., and Antonio Nanni (2001) "Bond of Fiber-reinforced Polymer Laminates to Concrete" *ACI material Journal*, V. 98, No.3, pp. 256-264
5. Chen, J.F., and Teng, J.G. (2001) "Anchorage strength models for FRP and steel plates bonded to concrete." *J. Struct.Engrg.*, 127(7), 784-791

MECHANICAL PROPERTIES OF HYBRID FABRICS IN PULTRUDED COMPOSITES

A. Peled,[1] B. Mobasher[2] and S. Sueki[2]
[1]*Structural Engineering Department, Ben Gurion University, Beer Sheva, Israel;*
[2]*Department of Civil and Environmental Engineering, Arizona State University, Tempe, AZ, USA*

Abstract: Cement based composites were developed with a combination of fabric types using cast (hand lay up) process and pultrusion (impregnated) method. The influences involved with the impregnation process using the pultrusion technique was studied by pullout and tensile tests. Hybrid composite laminates with combination of low modulus PE or PP fabrics with high modulus AR glass fabric were developed and tested in tension. The effect of various fabric types in suppressing the localization and crack bridging mechanisms were studied as well as the microstructure of these composites. It was found that the processing methods and fabric type significantly affect the bond as well as the tensile performance of the composite. Hybrid com posites combining low modulus and high modulus fabrics found to influence the mechanical performance of the composite. Combinations of PE and Glass fabrics gave composite with better ductility than a single glass and with greater strength than the PE single composite.

Key words: cement composites; fabric; processing; pultrusion; hybrid; tensile; pullout

1. INTRODUCTION

There is a growing interest in the use of fabrics as reinforcement for thin sheet cement composites. In addition to ease of manufacturing, non-linear geometry of individual yarns within the fabric results in excellent bond development and mechanical anchorage. These characteristics result in improved strength due to strain-hardening behavior even though the reinforcing yarns have a low modulus of elasticity[1,2]. The wide variety of fabric production methods such as: weaving, knitting, breading and non-woven allows great flexibility in fabric design. This flexibility enables controlling of fabric geometry, yarn geometry and orientation of yarns in the fabric in various directions as well as yarns material combinations (hybrid).

Hybrid systems with two or more fiber materials are being used to combine the benefits of each fiber into a single composite product. Optimized performance of hybrid thin

M.S. Konsta-Gdoutos, (ed.), Measuring, Monitoring and Modeling Concrete Properties, 749–761.

sheet composite in the hardened state, with respect to strength and toughness, has been studied by several investigators using combination of different fiber types with low and high modulus of elasticity[3-6]. The use of hybrid fiber reinforcement is particularly promising in fabric-cement composites. Fabrics allow several ways to combine different yarn types in a composite, i.e. different yarns in x direction, y direction or in both directions. Various contents of each yarn type at each direction are also possible. Moreover, one can combine several fabrics layers in one composite where each layer is made from a different single material.

Pultrusion process is an efficient way to produce fabric-cement composites laminates. This method is based on a relatively simple set up using low cost equipment and results in uniform production[7]. In this study cement based composites were developed with a combination of different layers of single fabric types. The composites were prepared by the cast (hand lay up) process and pultrusion (impregnated) method. The influences involved with the impregnation process using the pultrusion technique was studied by pullout and tensile tests. Composite laminates with combination of low modulus fabrics PE or PP with high modulus AR glass fabric were developed and tested in tension. The effect of various fabric types in suppressing the localization and crack bridging mechanisms were studied as well as the microstructure of these composites.

2. EXPERIMENTAL PROGRAM

2.1 Fabric types

Three different fabrics were used for this study made from high modulus AR Glass, low modulus polypropylene (PP) and low modulus polyethylene (PE). The AR Glass was a bonded fabric in which perpendicular set of yarns (warp and weft) was glued together at the junction points, having 4 yarns per cm in both directions of the fabric. The entire fabric was coated with sizing. The PP was a weft insertion warp knitted fabric in which the yarns in the warp direction were knitted into stitches to assemble together straight yarns in the weft direction which were the reinforcing yarns in the composite. This fabric was made from multifilament yarns with 8 yarns per cm in the reinforcimg direction (weft yarns) and 0.8 stitches per cm in the perpendicuar direction (warp yarns). The PE was a woven fabric, where the warp and the fill (weft) yarns pass over and under each other, made from monofilament with 22 yarns per cm in the reinforcing direction (warps) and 6 yarns per cm in the perpendicular direction (fills). The properties and geometry of the yarns made up the fabrics are presented in Table 1.

Table 1. Properties and geometry of yarns made up the fabrics

Yarn type	Yarn nature	Tensile strength, MPa	Modulus of elasticity, MPa	Filament size, mm	Bundle diameter, mm
PE	Monofilament	260	1760	0.25	0.25
PP	Bundle	500	6900	0.04	0.40
AR-Glass	Bundle (coated)	1360	78000	0.0135	0.80

2.2 Composite preparation

2.2.1 Pullout Tests

The specimens for pullout tests were produced by hand lay up of a single layer of fabric in the center of the cement matrix, along the specimen length. The pulled out yarns in the fabrics were the same as the reinforcing yarns in the tension specimens. The length of the specimen was equal to the embedded length of the fabric in the cement matrix: 12.7 mm long (0.5 inch) for all fabrics PE, AR Glass, and PP. Two sets of specimens were prepared for the PP and glass systems. In the first set, "clean" fabrics without cement were embedded in the cement matrix. In the second set, the fabrics were first impregnated in the cement bath using the pultrusion process; then the impregnated fabrics were embedded in the cement matrix. Table 2 presents the different tested systems.

2.2.2 Tensile Tests

Composite laminates were prepared by the cast and pultrusion process. In the pultrusion method the fabrics were passed through a slurry infiltration chamber, and then pulled through a set of rollers to squeeze the paste between the fabric openings while removing excessive paste. The fabric-cement composite laminate sheets were then formed on a plate shaped mandrel resulting in samples with width of 20 cm, length of 33 cm (continuous length) and thickness of about 1 cm. Each cement board was made with 8 layers of fabrics. Two sets of cement boards were prepared: (i) of single fabric boards where all the 8 layers were made from the same fabric; and (ii) hybrid boards where 4 layers of fabrics were made from the high modulus AR Glass fabric and the other 4 layers were from low modulus fabric either PE or PP. The four layers of the low modulus fabrics were located at the center of the composite where two layers of Glass fabrics were located at the outer sides of the composite (bottom and top). Another set of specimens were prepared by the cast process, for comparison, where the fabrics were located in the cement matrix by hands. In the case of the cast specimens only single fabric composites were prepared.

After forming the pultruded samples, a pressure was applied on top of the fabric cement laminates to improve penetration of the matrix in between the opening of the fabrics. A constant pressure of 15 KPa was used for all specimens. Most of this pressure was reduced within 1 hour after the pultrusion process to a level corresponding to 100 N load (1.7 KPa). This pressure was maintained up to 24 hours from the pultrusion process. In all cases the matrix was made from 42% cement, 5% silica fume and 0.1% superplasticizer by volume, and 50% water by volume as the water/cement ratio by weight was 0.37. All specimens were demolded at 24 hours after casting and were cut to at least 5 specimens from each system. The specimens were cured for 3 days in 80°C, 100% RH and then stored in room environment until testing in tension at 7 days after the pultrusion process.

3. TESTING

3.1 Pullout Tests

Pullout tests were carried out in an Instron testing machine at a crosshead rate of 0.25 mm/sec. The test was continued until the embedded fabric was completely pulled out.

Load-slip curves were recorded. In all of the fabric systems, 8 yarns were pulled out together. For more details on pullout tests see Ref. 8. In order to achieve ease of handling as well as uniformity in load application, a 25.4 mm free length (1 inch) of fabric was used between the pulled yarns and the grips. The compliance calculation was based on measuring the slope of the ascending portion of the experimental pullout curve as reported in[8].

Replicate experimental load-slip curves were used to obtain a mean representative average pullout load-slip curve for each test series. The average and standard deviations of the pullout load at each slip level were calculated and used to construct a mean representative curve. In Figure 1 four experimental curves (measured responses) and the average curve of these specimens are shown for PP fabric. This figure also shows the load-slip curves after reducing the deformation due to the free fabric located between the grip and the cement matrix during pullout tests (corrected response). These average load slip responses of each system were used to calculate the nominal shear bond strength, τ_{nom} assuming that there is very limited penetration of the cement matrix in between the filaments of the bundle. This represents a single reinforcing yarn with a contact perimeter corresponding to the whole bundle contact area, (n=1, d = deq = $\sqrt{4A/n\pi}$). By assuming constant shear strength along the embedded length mainly due to the friction, the nominal parameter nom was calculated using the following equation:

$$\tau_{nom} = \frac{P_{max}}{n\pi dl} \tag{1}$$

Where: P_{max} = maximum pullout load of the bundle, n = number of bundles pulled out, d = equivalent bundle diameter, and l = bundle embedded length. This approach offers a single average parameter for the strength of the interface. A three parameter model for the interface of fabrics has also been presented which defines the interface properties by the stiffness (shear stiffness), an adhesional shear strength, τ_{max}, and a frictional shear strength, τ_{frc}. The area under the curve was also calculated, referred here as the toughness. The toughness values provide a measure of total energy consumption during pullout.

3.2 Tensile Tests

Tensile tests were performed using a closed loop control direct tensile tests on a MTS testing machine with a capacity of 89KN. The rate of cross head displacement was set at 0.008 mm/sec. Metal plates with dimension of 25x30 mm and 1 mm thick were glued on the gripping edges of the specimen to minimize localized damage and allow better load transfer from the hydraulic grips. At least five replicate samples of each category were tested and results reported reflect the average and standard deviation values. Typical stress-strain curves representing the tensile behavior of individual composites were chosen for comparison. Note that with the PP, PE and hybrid composites no failure was occurred and the tests were ended after reaching a strain level of 0.07.

Several parameters were calculated: (1) the initial modulus of the composite (at the elastic region); (2) the post cracking tensile strength; and (3) the toughness (area under the load-displacement curve, up to 0.07 strain when applicable).

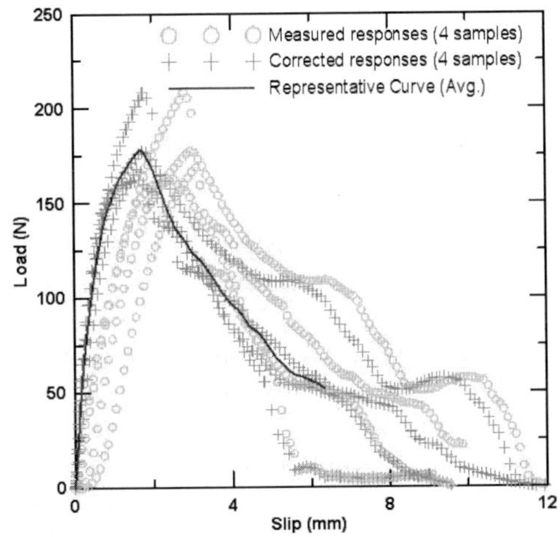

Figure 1. Pullout behavior of PP impregnated fabric

3.3 Crack Pattern Observations

The crack pattern after tensile tests was studied for the different composites by optical microscope. The cracks were observed for the single and for the hybrid systems. Observations were made at the side of the laminated composites and at the cross sections of the composites.

3.4 Microstructure Characteristics

Microstructure characteristics of the different composites were observed, using a Scanning Electron Microscopy (SEM). For these observations, fragments of specimens obtained after tensile tests were dried at 60°C and gold-coated. Microstructural features such as matrix penetration in between the opening of the fabrics and in between the filaments of the bundle were evaluated.

4. RESULTS

4.1 Pullout Characteristics

Impregnated fabrics, using the pultrusion process (pultruded), are compared with the non impregnated fabrics (cast). The nominal (τ_{nom}), adhesional (τ_{max}), and frictional (τ_{frc}) shear parameters for the different tested specimens, cast and pultruded (impregnated and non-impregnated) fabrics, are presented in Table 2 and Figure 2a. The frictional and adhesional parameters show approximately the upper and lower bounds of the shear strength for each interface. It is clearly shown that the nominal shear strength correlates approximately with the average of the two shear parameters for majority of the systems, and the larger the nominal shear strength, the larger the difference between the adhesional and frictional components.

Table 2. Bond strengths of the different fabric and processing systems

Specimen type	Fabric type	Processing	Bond Strength, MPa		
			τ_{nom}	τ_{max}	τ_{frc}
G105	AR-Glass	Cast	1.97	3.41	1.24
GP105	AR-Glass	Pultrusion	1.78	2.96	1.14
PP105	Polypropylene	Cast	1.39	2.76	1.05
PPP105	Polypropylene	Pultrusion	2.38	4.24	1.50
PEP105	Polyethylene	Pultrusion	0.90	1.10	0.83

Figure 2a and Table 2 clearly show that the pultruded composite with the PP fabrics (Sample PPP105) shows substantially higher bond strengths in comparison to the other samples. Such improved bonding is observed for all shear parameters, nominal, adhesional and frictional. However, the adhesional parameter is far above all the others. Moreover, the difference between the adhesional and frictional shear values is significant in this PP pultruded system. These results suggest that the pultrusion process highly improves the bond between the fabric and the cement matrix and this bond is not necessarily controlled by constant friction. The cast PP system (PP105) shows much smaller bond strength values as compared with the pultruded PP system. However, in the case of the Glass system the shear values of the pultruded and cast are similar (GP105 and G105, respectively) with small advantage for the cast system. The PE fabric exhibits the lowest bond values and hardly no difference between τ_{nom}, τ_{max}, and τ_{frc} (PEP105). This may suggest that the bond between the PE fabric and the cement matrix is governed mainly by friction and constant along the entire interface. Note that the PE system was prepared by the pultrusion process only.

The toughness values, calculated as the area under the load-slip curve, are shown in Figure 2b. It is observed that the PP pultruded system exhibits the best performance as compared with the other fabrics and processing systems. This increased toughness implies improved energy absorption for the pultruded PP systems. Note that according to Figure 2b, a majority of the energy absorption takes place during the frictional sliding region of the response which correlates with the pullout of the fabric. Such behavior is most significant with the Pultruded PP composite as the toughness due to the frictional sliding is about 4 times than the toughness at the maximum pullout load.

The initial bond stiffness values are presented in Figure 2c vs. the ultimate pullout loads, for the different systems. These results show that both the strength and initial stiffness for various systems correlate for majority of the systems indicating that samples with a higher bond stiffness have a higher ultimate strength. In general, it can be concluded that the pultrusion process significantly influences the bond properties of the PP system, including shear stress parameters and energy absorption during pullout. No significant influence of the production process on pullout properties and bond strength is observed for the glass composites. Moreover, the PP pultruded system shows improved bond values as compared with the Glass systems.

4.2 SINGLE FABRIC LAMINATED COMPOSITES

4.2.1 Tensile behavior

Summary of average tensile results of the different composites are shown in Table 3. The tensile behavior of the different fabrics and processing is presented in Figure 3. The tensile results clearly show the benefit of the pultruded PP composite, similar to the pull-out results. Greater tensile performance of the pultruded PP composite than that of the other fabrics is observed; in both strength and toughness, mainly at large strains above a value of 0.04. The pultruded Glass composite shows relatively brittle behavior as compared with a more ductile behavior of the Pultruded PE and PP composites. The ductile composites, PP and PE were loaded to strain levels as high as 0.06 with no apparent failure. The improved behavior of the pultruded PP composite (Figure 3a) as compared with the PP cast specimen (Figure 3b) is obvious. No such significant differences are observed between the pultruded and cast Glass systems, similar to the pullout results (Figure 2).

The toughness of the pultruded PP composite is the greatest among the other systems (Table 3), more than 3 times greater than the pultruded Glass and PE. This implies better energy absorption of the pultruded PP system. It should be noted that the PP fabric has much lower modulus of elasticity and strength than that of the Glass but greater performance than the PE fabric (Table 1), however the tensile behavior of the pultruded PP composite, at least at high strains, is much greater even than that of the Glass composite. Such improved performance of the PP composite can be explained based on the improved bonding of this system indicated by pullout (Figure 2).

4.2.2 Microstructure

Figure 4 presents SEM micrographs of the composites with the different fabrics, PE, PP and Glass produced with the pultrusion process. It is clear in this figure that the geometry of the different fabrics is not the same. These differences in fabric geometry may explain the differences in tensile behavior and bonding of the various systems as follows describe.

The Glass is relatively open fabric and its yarns are glued at the junction points providing strong connections between its two sets of yarns (Figure 4a). The yarns of the glass are coated with sizing preventing from the cement matrix to penetrate in between the filaments of the bundle. The stiff and strong junctions of the Glass fabric can benefit the mechanical performance of the composite; however, the coating of the bundles may be detrimental from a bonding point of view (Figure 2), as only the coating at the perimeter of the bundle is in contact with the cement matrix.

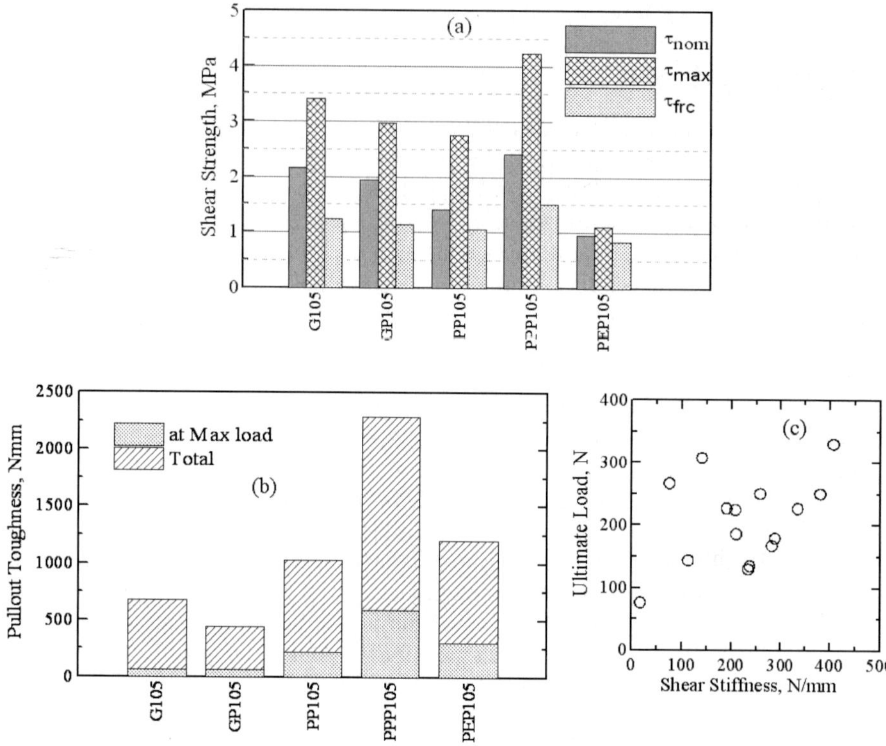

Figure 2. Pullout properties of pultruded and cast systems for the different fabrics: (a) shear stresses, (b) pullout toughness, and (c) correlation between pullout load and pullout stiffness

Table 3. Tensile properties of the different composites

Fabric Type	Processing	Tensile Strength, MPa	Ultimate strain, mm/mm	Toughness, MPa
PE	Cast	5.0	0.066	0.284
PE	Pultrusion	7.0	0.075	0.408
Glass	Cast	11.5	0.036	0.324
Glass	Pultrusion	18.5	0.030	0.359
PP	Cast	11.2	0.068	0.517
PP	Pultrusion	24.0	0.075	1.424
Hybrid PE-Glass	Pultrusion	9.3	0.024	0.397
Hybrid PP-Glass	Pultrusion	14.4	0.050	0.609

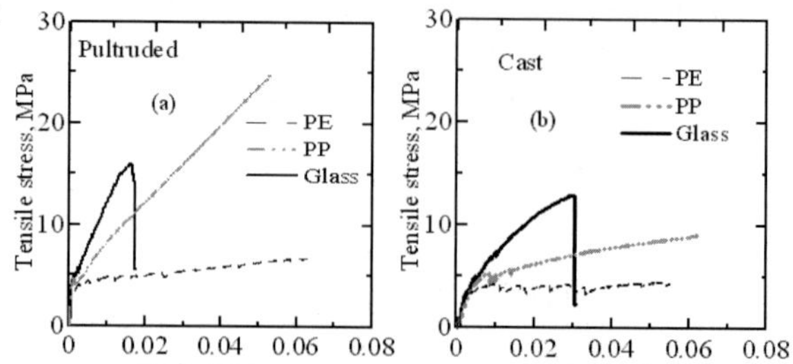

Figure 3. Tensile behavior of the composites with the different fabrics produced with the (a) pultrusion, and (b) cast process

(a) (b) (c)

Figure 4. SEM observations of the composites with the different fabrics: (a) Glass, (b) PP, and (c) PE

The PP is a knitted fabric in which the reinforcing yarns are connected by stitches (loops) at the junction points. The reinforcing yarns are having multifilament nature (Figure 4b). Due to the bundle nature of the yarns the cement matrix can penetrate in between the filaments of the bundle providing large surface area with the matrix, leading to improved bonding. However, such improved bonding can take place only if the bundles within the fabric are open enough to allow matrix penetration; as the stitches in the knit fabric may tight the bundles and reduce cement penetrability. Such low cement penetrability is clearly observed in Figure 5a which shows cross section of a bundle within PP knit fabric placed in a cement matrix by hands, using the cast process. This low penetrability reduces bond properties (Figure 2) and tensile behavior (Figure 3b). However, in the case where the composite was prepared with the pultrusion process a better penetrability of the cement is observed (Figure 5b), due to the intensive impregnation process of the fabric in the cement matrix. Such improved penetration can provide strong bonding with the cement matrix (Figure 2), leading to the improved tensile performance of the pultruded PP fabric composite (Figure 3a).

The PE has the highest density among the other two fabrics but its yarns are made from monofilament giving relatively small contact surface area with the cement matrix (Figure 4c). In this fabric the two sets of yarns are connected by friction while they pass under and above each other. The resulted crimp shape of the yarns in this fabric may provide quit strong mechanical anchoring with the cement matrix[1] but the monofilament nature of the reinforcing yarns gives relatively small interface area with the cement matrix, leading to low bonding as compared with the multifilament PP fabric. This is in addition to the low performance of the yarns made this fabric (Table 1).

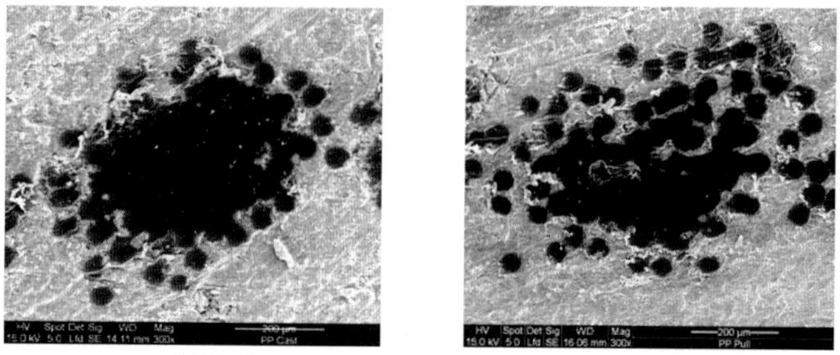

(a) 200μm (b)

Figure 5. Cross sections of PP fabrics in cement matrix produced by different methods: (a) hand lay up of fabric, (b) impregnated fabric (pultrusion)

4.2.3 Crack pattern

The diversity in tensile behavior of the various fabric composites can also observed when comparing the crack pattern of the different systems occurred during tensile loading, as presented in Figure 6. This figure presents the pultruded systems. The crack of the PE composite is developed along the weak areas of the composite, i.e., at the curvature of the reinforcing crimped yarn where the transverse yarns are located (Figure 6a). During tensile tests the crimped reinforcing yarns are whiling to straighten causing stresses concentration at the matrix near the vicinity of the yarn curvature; at these areas also the perpendicular yarns are located which can cause farther weakening of the matrix. This can explain some of the low tensile performance of the PE composite (Figure 3). A relatively large and wide cracking through the specimen is observed with the Glass composite as shown in Figure 6b, which exhibits side view of this composite (along the reinforcing yarns). Also here the cracks are developed through the weak points of the composite, where the perpendicular yarns are located. Delamination between the reinforcing yarns of the fabric and the cement matrix is clearly observed in this figure, suggesting relatively poor bond strength of the Glass fabric with the cement matrix. It should be noted that the glass bundles are coated with sizing prior to composite production, and this coating is located at the fabric-cement interface.

No delamination is observed with the pultruded PP composite, giving a close contact between the reinforcing yarns and the cement matrix after tensile loading (Figure 6c, exhibiting side view of the composite). With the PP, relatively thin and dispersed cracks are observed. Crack arresting by the reinforcing yarns is also clear in these figures. These observations are correlated with the improved bonding (Figure 2) and improved tensile behavior of the pultruded PP composite (Figure 3a).

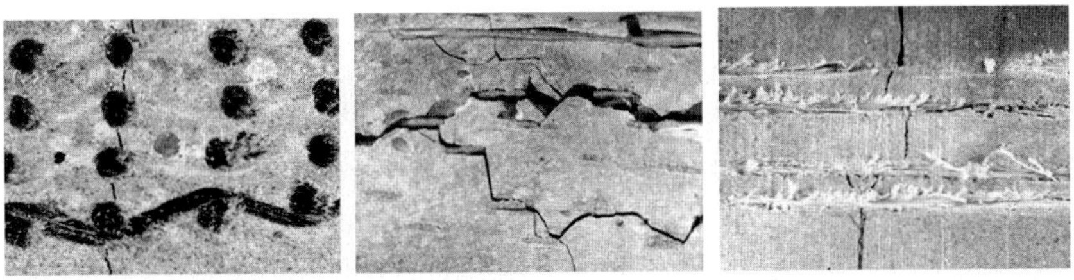

(a) (b) (c)

Figure 6. Crack pattern of the different composites (side view): (a) Glass (b) PE, and (c) PP

4.3 Hybrid Laminated Composites

Combining the brittle and strong Glass system with the ductile PE system can lead to a composite stronger than the PE and ductile than the Glass. The improved strength of such combination is evident in Figure 7a, which shows the tensile behavior of hybrid composite made with layers of PE and Glass fabrics as compared with the single PE fabric system. The strength of the hybrid system is twice as that of the PE system, however it is lower than that of the Glass (Figure 3a). The ductile behavior of the hybrid composite is also clear in Figure 7a, as compared with the Glass composite (Figure 3a).

The pattern of the developed cracks in the Glass-PE hybrid composites is presented in Figure 7b. Note that the Glass fabrics are located at the surfaces of the laminated composite, 2 layers at each side (bottom and top), as 4 layers of PE fabrics are located at the center of the composite. A difference in crack pattern at the glass fabric zone than at the PE fabric zones is observed (Figure 7a). Multiple cracking is observed at the PE zone (middle of the composite), where only few cracks are observed at the Glass zones, bottom and top of the composite. A clear delamination between the glass zone and the PE zone is observed (at the center of Figure7a). Such delamination can occur due to the differences in the mechanical properties of the PE and Glass fabrics, as the PE can sustain much larger strains than the Glass. During tensile loading the PE is highly lengthen while the change in the length of the glass fabric is much smaller, leading to separation and sliding between the two fabrics regions.

Hybrid performances of composites contain Glass and PP fabrics are presented in Figure 8. The PP-Glass hybrid composite is more ductile than the Glass composite but does not show any significant advantage when comparing with the PP single material composite. When observing the cracks of this hybrid composite no significant difference in crack pattern is seen for the two fabrics regions (Figure 8a). Also delamination is not observed between the two materials zones. This suggests that the strains of the composite at the two material regions during loading are similar implying good contact between the Glass and the PP regions.

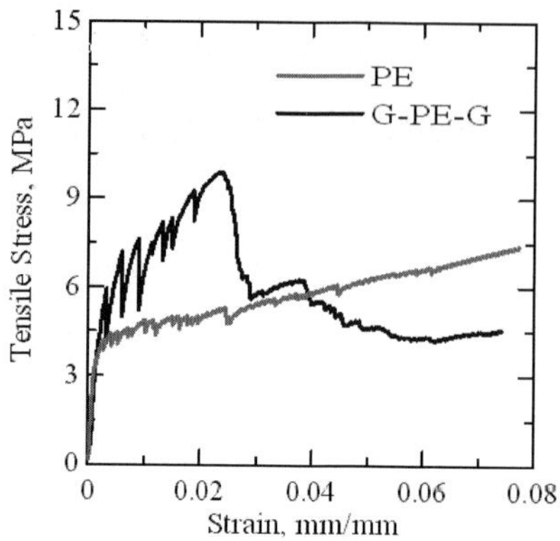

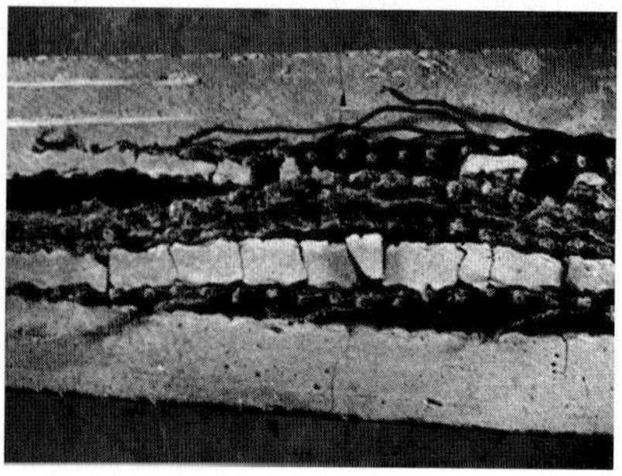

Figure 7. Hybrid PE-Glass composites: (a) tensile behavior, and (b) crack pattern

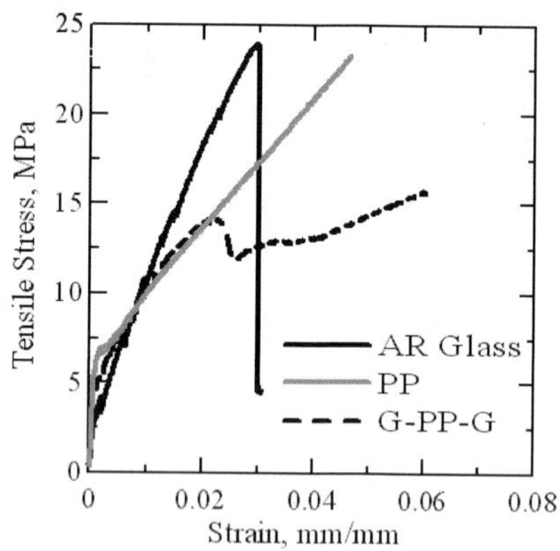

Figure 8. Hybrid PP-Glass composites: (a) tensile behavior, and (b) crack pattern

5. SUMMARY AND CONCLUSION

Hybrid composites made from brittle Glass fabrics and ductile PE fabrics performed better than Glass in their ability to sustain strains and stronger than a single PE fabric composite. Pultruded PP fabric composites showed the best performance at large strains of about 0.04. Pultruded Glass fabric composites performed the best at relative small strains. Combinations of PP and Glass fabrics gives composite with better ductility than a single Glass composite, but does not show any benefit in performance compared with the PP single composite.

The improvement in tension of the pultruded PP composite is due to improve bonding induced by the intensive fabric impregnation during the pultrusion process. The processing methods can highly affect the bond between fabrics and cement matrix depending on the nature and geometry of the fabric and its yarns. The pultrusion technique increases bond strength where non-coated multifilament is used to produce the fabric, as obtained in this work for the PP fabric. Coated or monofilament yarns do not highly influences by the processing methods (Glass and PE fabrics). The dominant shear stresses at the interface of the PP pultruded system is govern mainly by the adhesional parameter suggesting that the shear stresses along the pultruded fabric-cement interface are not necessarily controlled by constant friction.

Acknowledgement

The authors would like to thank Nippon Electric Glass Co., Ltd., and Karl Mayer Ltd. for their cooperation for providing the fabrics used in this study. The National Science Foundation, program 0324669-03, and the BSF (United States Israel Binational Science Foundation) are acknowledged for the financial support in this research.

6. REFERENCES

1. A. Peled, A. Bentur, D. Yankelevsky, Effect of woven fabrics geometry on the bonding performance of cementitious composites: mechanical performance, *Advanced Cement Based Materials J.*, **7**(1), 20-27 (1998)
2. A. Peled. A. Bentur, Fabric structure and its reinforcing efficiency in textile reinforced cement composites, *Composites, Part A* **34**, 107-118 (2003).
3. G. Xu, S. Magnani, D.J. Hannant, Tensile behavior of fiber-cement hybrid composites containing Polyvinyl Alcohol fiber yarns," *ACI Materials J.* **95**(6), 667-674 (1998).
4. M. Kakemi, D.J. Hannant, and M. Mulheron, Filament fracture within glass fiber strands in hybrid fiber cement composites, *J. of Materials Science*, **33**, 5375-5382 (1998).
5. B. Mobasher and C.Y. Li, Mechanical properties of hybrid cement-based composites" ACI Materials J., **93**(3), 284-293 (1996)
6. A. Peled, M. Cyr, and S.P. Shah, Hybrid fibers in high performances extruded cement composites, in: Fibre Reinforced Concrete – BEFIB 2004, edited by M. Di Prisco, R. Felicetti, and G.A. Plizzari, G.A., pp. 139-148 (PRO 39, RILEM, 2004).
7. A. Peled, and B. Mobasher, Pultruded Fabric-Cement Composites, *ACI Materials J.,* **102**(1), 15-23 (2005)
8. S. Sueki, C. Soranakom, B. Mobasher, A. Peled, Pullout-slip response of fabrics embedded in a cement paste matrix, *ASCE Journal of Materials in Civil Engineering*, In press.

IMPROVING THE BOND CHARACTERISTICS OF A STRAND EMBEDDED IN A CEMENTITIOUS MATRIX

B.-G. Kang,[1] B. Banholzer[2] and W. Brameshuber[3]

[1]*Dipl.-Ing., Research assistant at the Institute of Building Materials Research, RWTH Aachen University, Germany;* [2]*Dr.-Ing., Research assistant at the Institute of Building Materials Research, RWTH Aachen University, Germany;* [3]*Prof. Dr.-Ing., Professor at the Department of Engineering and Head of the Institute of Building Materials Research, RWTH Aachen University, Germany*

Abstract: This paper shows a preparation and manufacturing technique which improves the overall load carrying capacity of a strand/fine grained concrete matrix system by improving the bond and thus the load transfer between the inner filaments of the strand. A cement paste based on ultra fine grained cement is used before the actual placing of the strand in the surrounding fine grained concrete to achieve a more homogeneous penetration of the strand by cement particles. This in turn enhances the inner bonding characteristics important to the overall load carrying capacity of the system.

Key words: bond characteristics; impregnation, pull-out test

1. INTRODUCTION

Concrete is relatively brittle, and its tensile strength is typically only about one tenths of its compressive strength. Regular concrete is therefore normally reinforced with steel reinforcing bars. Such reinforced concrete structures are used in a wide field in almost every area of structural engineering. However, a replacement of the steel reinforcement in concrete structures with for example technical textiles is rapidly becoming an interesting option for construction facilities. Placing multidimensional fabrics made of alkali resistant glass (AR-glass) instead of the usual steel bars in the main load directions of the concrete, might enable the engineer in the future to design and build light and slender structures with a high load carrying capacity if the basic load carrying mechanisms of this composite are known, e.g. the bond mechanisms between the strands a fabric consists of and the surrounding matrix.

M.S. Konsta-Gdoutos, (ed.), Measuring, Monitoring and Modeling Concrete Properties, 763–768.

The mechanisms predominating the failure of the system during for example a pull-out test - which is a typical test to determine the bond characteristics of a composite - are quite complex because an AR-glass strand represents an element of reinforcement quite different from other fibrous reinforcements used in cement based composites, especially steel fibers. AR-glass strands consist of several hundred individual filaments approximately 10 to 30 μm in diameter which are immediately sized after the drawing process and loosely assembled in form of a flattened bundle. This non-uniform nature of the strands makes the bond characteristics inherently variable [1]. See also Figure 1.

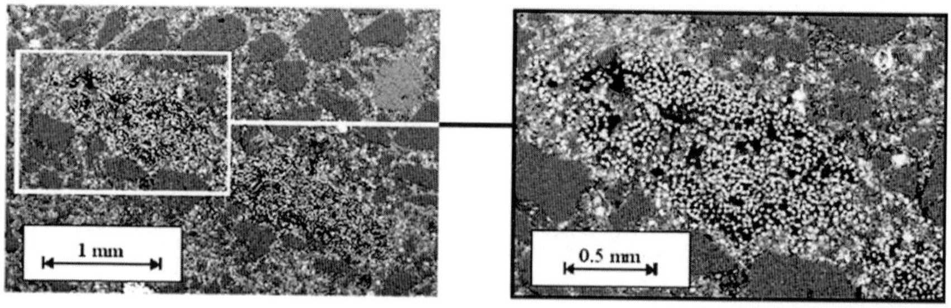

Figure 1. SEM micrograph of an untreated 2400 tex AR-glass strand in a fine grained concrete

In many cases, however, the original compact flattened bundle is loosened during the placing and manufacturing of the composite, and hence matrix may penetrate up to a certain degree into the core of the strand [2,3]. Nevertheless, the formation of hydration products within the strand is initially limited. The uncontrolled penetration leads to a different formation of the inner and outer bond characteristics, and hence the failure mechanism after exceeding the maximum pull-out load is described as a so-called "telescopic failure", i.e. a successive break down layer by layer from the sleeve to the core filaments [1]. As a result the real, higher tensile strength of the strand is not activated during the pull-out test and regardless of the embedded length only a small amount of the original tensile capacity of the strand is used. For example, a filament of a 2400 tex Vetrotex AR-glass strand has a tensile strength of 1,473 N/mm^2 [1] corresponding to a force of about 1,320 N which the strand is supposed to carry, in contrast to the 260 N it actually takes in a pull-out test.

Because the main parameter influencing the failure process and hence the load carrying behavior of a strand in fine grained concrete is the amount of matrix penetrating its core, which affects the extent to which the full tensile strength of a strand is reached, this study concentrates on preparation and manufacturing techniques which improve the inner bond and thus the overall load carrying capacity of the composite. To achieve a more homogeneous penetration of the strand, ultra fine grained cement pastes are used in this study to saturate the strand before the actual placing in the original fine grained concrete matrix. During this procedure ultra fine cement particles can penetrate the strand, suppress the air enclosed within the strand, hydrate at a similar rate than the surrounding concrete and that way also lead to an improved outer bond. In pull-out tests the effectiveness of this method is demonstrated.

2. SPECIMEN PREPARATION AND EXPERIMENTAL METHODS

For specimen preparation one series of 2400 tex Vetrotex AR-glass strands is cast in a fine-grained concrete matrix. The dimensions of the mould are 50 x 50 x L mm^2 with L being the embedded length of the strand. L is chosen for the presented tests to be 30 mm. The strand is situated horizontally in the mould. A slight pre-stress is applied to align the strand. In a second test series a commercial ultra fine grained cement paste (95 % of the cement particles feature a particle size less than 13.1 μm) which is commonly applied as injection material is used to saturate the strands before the actual placing into the fine grained concrete matrix. Due to the extreme fineness of the material a special mixer with a "dissolver-disk" is used to assure a sufficiently good dispersion. The water/cement ratio of the cement paste amounts to 0.55. A superplasticizer (0.6 % of the cement mass) is added to assure a good workability and to guarantee high flowability. The cement paste is then mixed for ten minutes with 3000 rotations per minute. Subsequently the strand is kept for 2 minutes in this paste under vibration using a compaction table with an amplitude of 0.7 mm and a frequency of 50 Hz. It is placed finally - still in a wet stage - within the mould. The fine grained concrete used (PZ-0899-01) features a maximum grain size of 0.6 mm, a water / binder ratio of 0.4, and a binder content of 700 kg/m^3. For more information on the mixture design of the used fine grained concrete, please refer to [4]. After casting, the specimens are compacted and cured in the mould for 24 hours. The specimens are finally stored at 20°C and 95% RH until testing at a total age of 28 days.

The pull-out tests (pull-push test) are carried out using a universal testing machine (Instron 5566) at a displacement rate of 0.1 mm/min until a maximum displacement of about 1.7 mm is reached. For illustration see Figure 2. The pull-out load and the cross head displacement are recorded every 2 N force increment resulting in a load versus displacement relationship P(Ω).

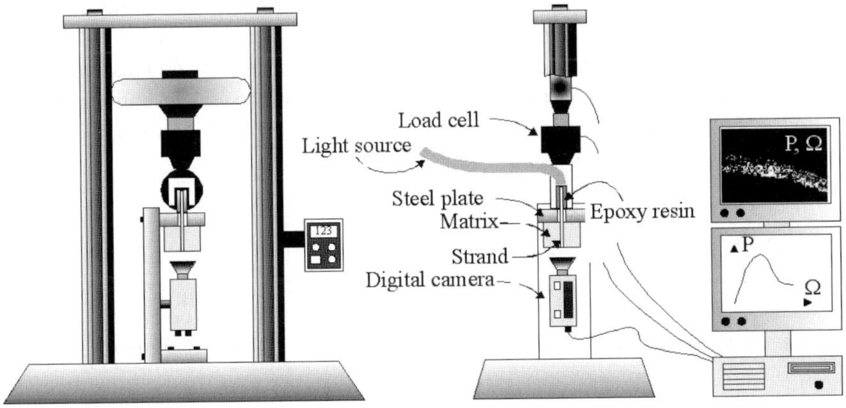

Figure 2. Test set-up for a pull-out test on a strand

During the actual pull-out test the specimen is additionally exposed by an artificial light source from the front via an epoxy resin block used as load introduction (see Figure

2). By using a small charge-coupled device camera (CCD-camera) and a zoom lens, the strand can be distinguished on the rear of the specimen from the surrounding matrix due to its exposure and therefore bright appearance (Figure 2). After a tensile failure a filament is no longer capable of transferring light, and therefore the bright appearance vanishes in the image of the next load step. Thus this test is called FILT test (Failure Investigation by Light Transmission). A numerical image analyzing routine developed by "ECM Datensysteme Ltd" according to the needs is applied during this study, the optical image recorded by the CCD-camera can be analyzed and converted in a binarized image (black and white pixels). A numerical procedure has been developed, which allows the evaluation of these binarized images with regard to the number of white pixels the image contains. By using this computer routine it is possible to plot the number of pixels recorded versus the corresponding loads, and displacements from the pull-out test respectively. Since only an optical image of the CCD-camera visualising the filaments is analysed and converted in a binarized image outlining the filaments as white pixels, the number of these identified pixels can be directly related to the cross-sectional area of the strand and thus to the number of filaments in a strand. For more information see [1].

3. TEST RESULTS

In Figure 3 the pull-out test results are presented for the non-saturated and saturated strands embedded in the fine grained concrete matrix.

The presented results clearly show the effects of the saturation of the glass strands by cement paste on basis of ultra fine grained cement before the actual placing in the fine grained concrete. A maximum load of around 260 N is reached at a displacement of 0.2 mm if the untreated strand is pulled out of the fine grained concrete; see the diagram on the left of Figure 3. Thus the effectiveness of this material combination can be calculated as around 20 % (compare "Introduction"). This effectiveness can be drastically improved up to 72 % if a saturated strand is used (maximum pull-out load of around 950 N at a displacement of 0.6 mm). A comparison of both diagrams also indicates this significant rising in the load carrying capacity of the system. In addition the increase of the initial stiffness of the load versus displacement relations of the treated in comparison to the untreated systems proofs that more filaments are activated and participate in the load transfer from the beginning of the pull-out test.

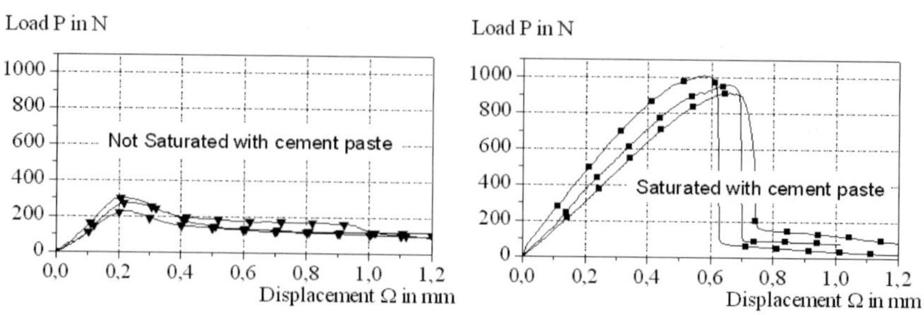

Figure 3. Pull-out test results of non-saturated and saturated strands in fine grained concrete

This is due to the more homogeneous penetration of cement paste into the strand which leads to an improved bond quality of the inner filaments. Investigations carried out by means of scanning electron microscopy (SEM) confirm this statement. Figure 4 shows that the ultra fine grained cement paste penetrates the complete strand and encapsulates almost all inner filaments. Micro-cracks within the penetrated cement paste (black areas) are possibly due to different shrinkage and hardening characteristics of the cement paste and the fine grained concrete.

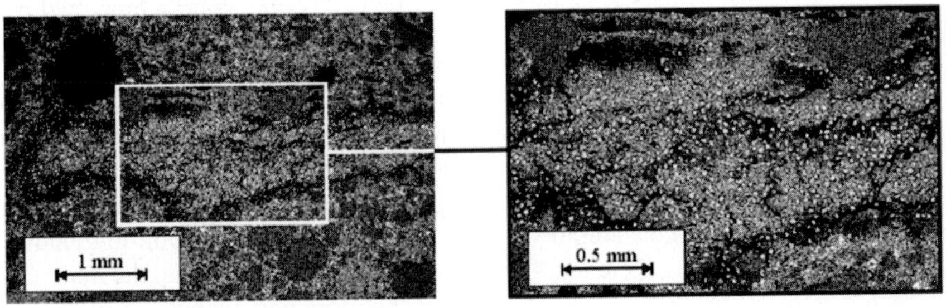

Figure 4. SEM micrograph of an 2400 tex AR-glass strand saturated with an ultra fine grained cement paste and placed subsequently in a fine grained concrete

However, as the pull-out tests shows, these do not affect the overall mechanical performance at all. This microstructure is in contrast to the microstructure of the composite made with the not saturated strand, compare Figure 1. In that case almost no fine particles of the fine grained concrete penetrate the strand and many of the inner filaments are not or only loosely in contact with the surrounding system which gives reason for their minor contribution to the load transfer during the pull-out. A similar conclusion can be drawn from the results of the FILT test as shown in Figure 5.

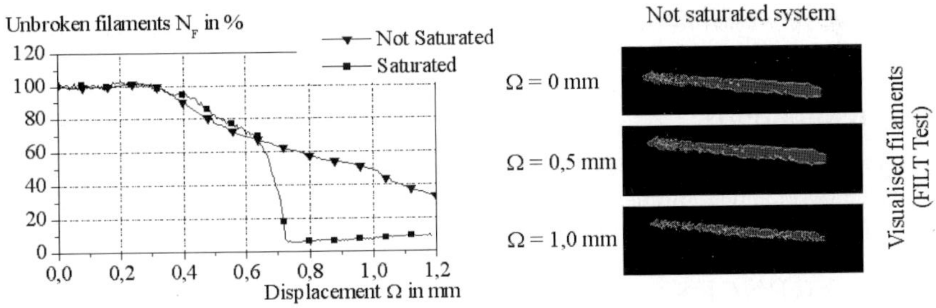

Figure 5. Results of the FILT test: Unbroken filaments versus displacement diagram NF() and binearized image series (FILT test) of the not saturated system

These results reveal that almost no filaments of both kinds of composite fail in tension up to an introduced displacement of approximately 0.25 mm which corresponds well to the displacement where the maximum pull-out load is reached in the tests on the not saturated strand/matrix systems. At that displacement the telescopic failure of the strands, which is visualized by a series of binearized images on the right of Figure 5 for the not saturated system, is also initiated. However, this failure process develops in a remarkably different way for the two composites investigated. Whereas the filament failure equally progresses for the not saturated system with the ongoing pull-out test, more than 50 percent of the filaments experience a sudden rupture at a displacement of approximately 0.7 mm for the saturated system. This result also clearly confirms the afore made statement that more filaments are activated and participate in the load transfer from the beginning of the pull-out test and as a consequence experience a more sudden failure.

4. SUMMARY AND CONCLUSIONS

The presented results clearly outline that the load carrying capacity of a strand/fine grained concrete matrix systems can be significantly improved if the strand is saturated with a cement paste based on ultra fine grained cement before the actual placing in the fine grained concrete. The principal reason for this increase in efficiency is the improved penetration capacity of cement particles into the core of the strand and the so enhanced bonding characteristics of the inner filaments. Pull-out test results show that the generally insufficient efficiency of the untreated strand/matrix system (20 % of the strand's tensile strength) can be improved considerably by this treatment and lead to pull-out loads corresponding to around 72 % efficiency. Investigations by SEM verify that the main mechanism which leads to this increase in efficiency is the more homogeneous penetration of cement paste into the strand.

From this results some conclusion may be drawn which can be used to identify future fields of research. First of all, to use the overall capacity of textile reinforced materials, the bond of the inner filaments with the surrounding system has to be improved drastically. Either the strands of which a fabric consists are saturated by cement paste based on ultra fine grained cement as done in this study or by other methods to achieve a participation of the inner filaments in the load transfer process. Therefore, within the Collaborative Research Center 532 "Textile reinforced concrete – Basics for the development of a new technology" gel particles on polymer basis will be investigated in future research which are applied on the strand before casting, swell in contact with the mixing water, and thus open up the closely assembled filaments to allow a significantly improved penetration by the original fine grained concrete.

Acknowledgements

This project is part of the Collaborative Research Center 532 "Textile reinforced concrete – Basics for the development of a new technology" and sponsored by the Deutsche Forschungsgemeinschaft (DFG). The support is gratefully acknowledged.

5. REFERENCES

1. Banholzer, B.:"Bond behaviour of a multi-filament yarn embedded in a cementitious matrix" In: Schriftenreihe Aachener Beiträge zur Bauforschung, Institut für Bauforschung der RWTH Aachen, Nr. 12, (pdf file at: http://www.bth.rwth-aachen.de), 2004.
2. Majumdar, A.J.: "The Role of the Interface in Glass Fibre Reinforced Cement" In: Cement and Concrete Research 4, No. 2, pp. 247-266, 1974.
3. Bartos, P.: "Brittle-Matrix Composites Reinforced with Bundles of Fibres. London" In: Proceedings of the First International RILEM Congress from Materials Science to Construction Materials Engineering. Vol. 2, pp. 539-554,1987.
4. Brockmann, T.: "Anforderungen und Eigenschaften zementgebundener Feinbetone. Aachen : Lehrstuhl und Institut für Massivbau" In: Textilbeton. 1. Fachkolloquium der Sonderforschungsbereiche 528 und 532, pp. 82-98, 2001.

ASPECTS OF MODELING TEXTILE REINFORCED CONCRETE (TRC) IN 2D

J. Hegger and O. Bruckermann

Institute of Structural Concrete, RWTH Aachen University, Mies-van-der-Rohe-Str. 1, D-52064 Aachen, Germany

Abstract: The paper discusses the fundamental effects occurring in Textile Reinforced Concrete (TRC) at a crack bridged by a fabric at an angle. A Finite Element model comprising these effects is presented and their influence on the load bearing behavior of TRC is analyzed.

Key words: Textile Reinforced Concrete, Crack Bridging, Finite Elements.

1. INTRODUCTION

Although, in principle, one of the basic advantages of Textile Reinforced Concrete (TRC) is the possible alignment of the fibers with the direction of the tensile stresses, this cannot always be realized in construction elements. This is either due to practical reasons or because the direction of the principle stresses leading to the first cracks is not known a priori, for instance if there are different loading cases. Therefore, among the quite well established models for the behavior of TRC with aligned fibers, additional models are required describing the behavior of the textiles at an angle with respect to the load.

It is well known that an inclination of the textile leads to a reduced tensile strength of the composite[1-6]. With an increasing angle α the strength decreases as the comparison of experimental results found in the literature reveals (Figure 1). The loss of strength is caused by bending moments and local lateral pressure at the crack edge, both leading to an earlier failure of the filaments. However, this evident insight is not sufficient for the description of the total stress-strain curve of the composite. In addition to the increasing damage of the textile, the behavior of the remaining unbroken filaments has to be considered as well. In the following, the basic effects occuring due to loading a textile at an angle are discussed and an appropriate Finite Element model is presented. All considerations are based on a two-layered idealization of rovings, i.e. core and sleeve filaments differing in the quality of bond to the matrix[7-9].

M.S. Konsta-Gdoutos, (ed.), Measuring, Monitoring and Modeling Concrete Properties, 769–776.

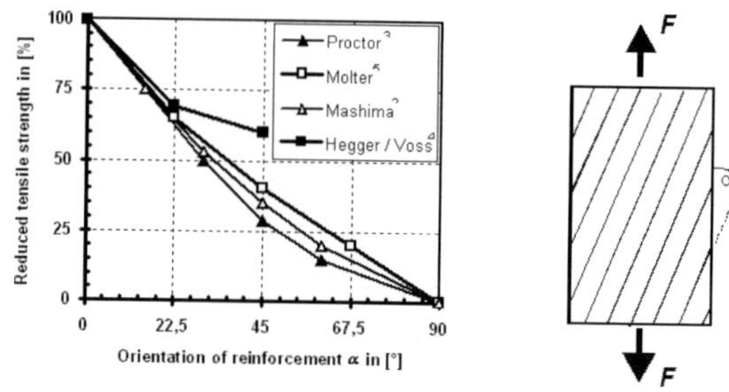

Figure 1. Reduced tensile strength due to inclination of fiber

2. BASIC EFFECTS

2.1 Alignment of textile with the direction of the load

The angular flexibility of the filaments leads to their partial alignment with the direction of the load, see Figure 2. If a filament is perfectly embedded in the matrix, the angle φ between the load direction and the textile becomes zero. In the present study this is assumed for all filaments belonging to the sleeve. However, depending on the textile binding, there might be only very few filaments with a perfect embedment and the outmost filaments may even debond from the matrix[2].

Because of voidage in the roving and the imperfect embedment in the matrix, the core filaments do not align as much as the sleeve filaments. Their bending point is not exactly located at the crack edge, but at a certain distance from the crack inside the matrix[6,10]. Thus, the angle φ_c is larger than zero, but limited by the textile slope α.

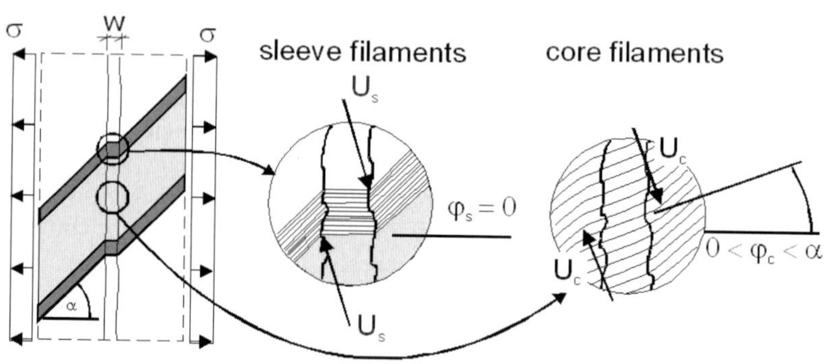

Figure 2. Alignment of the textile with the direction of the load

2.2 Increase of bond capacity due to lateral pressure at the crack edge

If the filaments are bent into the direction of the load, the lateral forces U_s and U_c according to Figure 2 are acting on the sleeve and the core, respectively. The equilibrium yields

$$U = F \cdot \sqrt{2 \cdot (1 - \cos(\alpha - \varphi))} \tag{1}$$

where F is the normal force of the sleeve and the core filaments, respectively, at the crack. The stresses equivalent to the lateral forces cannot be evaluated, since the areas U_s and U_c are acting on are unknown. However, if we define the fictitious lateral pressure $p_\perp$ acting on a certain length k as

$$p_\perp = \frac{\sigma_0}{k} \cdot \sqrt{(1 - \cos(\alpha - \varphi))} \tag{2}$$

where σ_0 is the normal stress at the crack, we can qualitatively take into account that the lateral pressure is inversely proportional to the length k and is dependent on the angles α and φ by the factor $\sqrt{(1 - \cos(\alpha - \varphi))}$. It is likely that the lateral pressure locally enhances the bond performance of the filaments leading to higher textile stresses at the crack if the same average strain is applied to the composite. If a coulomb-friction model is assumed, the increased bond performance can be expressed as

$$\tau^*_{\max} = \tau_{\max} \cdot (1 + \beta \cdot \frac{\sigma_0}{k} \cdot \sqrt{(1 - \cos(\alpha - \varphi))}) = \tau_{\max} \cdot (1 + \beta \cdot p_\perp) \tag{3}$$

which is illustrated in Figure 3. The parameter β has been introduced in order to control the linear increase of the maximum bond stress with increasing fictitious lateral pressure.

2.3 Damage of the filaments

Besides the higher stresses of the textile at the crack due to increased bond, the filaments are subjected to bending moments resulting from the lateral forces. This leads to a damaging of the filaments at early stages right after the matrix has cracked. Since the stress distribution within the cross-section of the roving is not uniform, first the sleeve filaments break whereas the damage of the core filaments starts later and progresses more slowly. In the model presented below, the damage is considered as a function of the angle α and the actual crack width w_{cr}. Thereby, separate damage laws for the sleeve and the core can be defined. This approach is of course phenomenological, i.e. the damage is not calculated by the mechanical force, but by geometrical values. The advantage is that the effect of damage on the load bearing behavior can be analyzed independently of the other effects mentioned.

3. FINITE ELEMENT MODEL

In order to study the influence of the effects described in section 2, the two dimensional plane stress element *fe_q4_tex2*, see Figure 4, has been developed. This element is a macro-element consisting of several sub-elements for the components matrix, textile and bond.

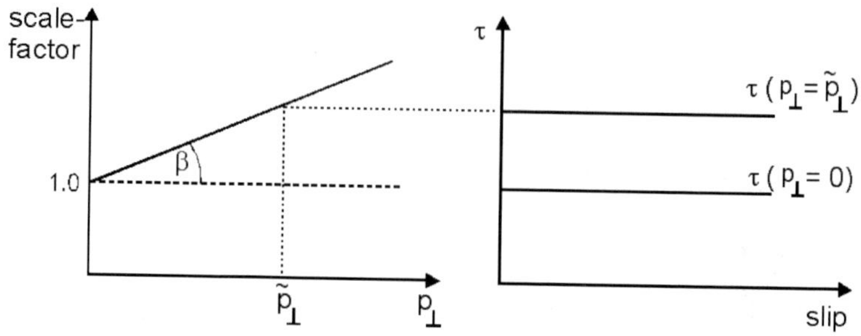

Figure 3. Model for locally enhanced bond

Even though it is possible to generate such a structure with standard elements in an ordinary FE-program, this element has several advantages. In the first place, only a simple 2D-mesh instead of a mesh with coupled rebar, interface and quadrilateral disk elements is required, considerably simplifying the set-up of the model. Secondly, from the programming point of view, the element is only one object giving all subelements direct access to the status of the other subelements. This easily allows defining interactions between the components, of which there are the following four:

- The textile damage depends on the crack width of the matrix.
- If the strain of the matrix at an edge of the element is larger than the crack strain, the lateral forces U_s and U_c acting between textile and matrix are calculated depending on the inclination of the textile and the actual textile stress. These forces turn the action line of the textile into the angle φ according to Figure 2.
- The stiffness of the bond springs changes dynamically, depending on the lateral forces. Thereby, the length k in Eqs. (2)-(3) is equal to half the element length.
- The stiffness of the bond springs decreases with increasing damage of the textile, assuming that broken filaments do not transfer bond stresses any longer.

If the additional 2D-effects are "switched off", the behavior of the textile is exactly the same as for the *Two-Subroving-Model* [9,10].

4. PARAMETRIC STUDY

As a first test, the element *fe_q4_tex2* was used to model a crack bridge with a textile inclination of 45° and a crack spacing s_{rm} of 7.2 mm as it was observed in tensile tests on 8 mm thick disks reinforced with two layers of the biaxial AR-glass fabric *MAG-07-03* [11]. In the following parametric study, the influence of the additional 2D-effects on the crack bridging behavior shall be demonstrated. Since this influence is qualitatively the same for many different fabrics, this is done here without giving all used parameters in detail. Figure 5 shows a sketch of the finite element mesh.

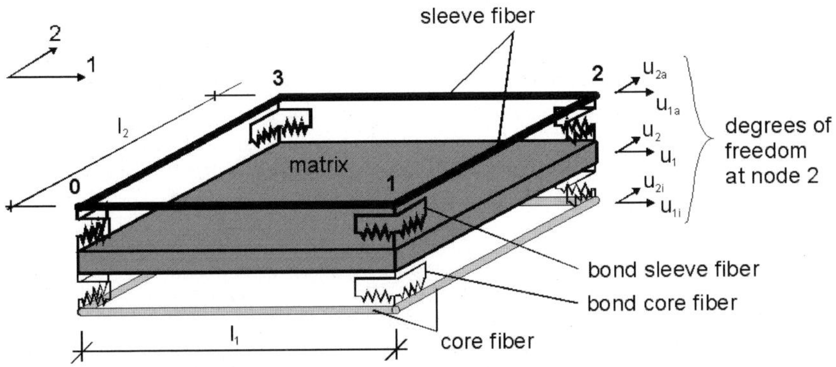

Figure 4. 2D Finite Element *fe_q4_tex2* for concrete with biaxial textile reinforcement

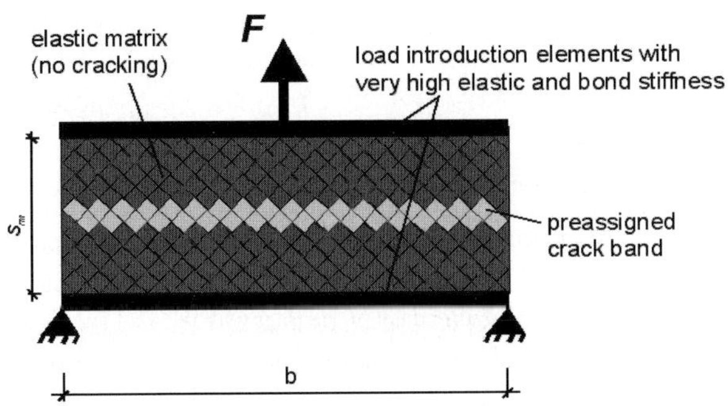

Figure 5. Finite Element model of crack bridge

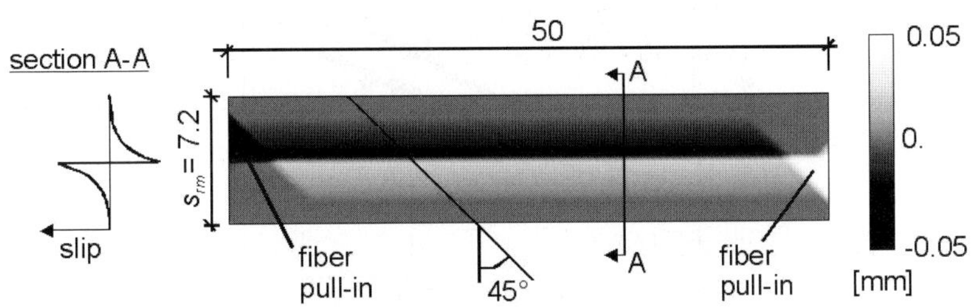

Figure 6. Slip of the core-fiber of the mean direction of the fabric

The stresses calculated by the model are always smaller than the theoretical crack bridging stresses of a specimen with infinite width, because at the edges the fibers are pulled into the specimen. As an example, Figure 6 shows the slip of the core-fiber of the mean direction of the fabric reaching its maximum value at the left and right side of the specimen. This side-effect has to be kept in mind when comparing the results of this model to other models such as those based on Representative Volume Elements where the same displacements of all components at the boundaries are presumed.

4.1 Influence of alignment with load direction

The alignment of the fiber with the direction of the load is controlled by the angles φ_S and φ_C according to Figure 2. The decrease of these angles leads to a stiffer behavior of the crack bridge, see Figure 7. The diagram shows the load over the strain for a full alignment of the sleeve-fiber (φ_S=0°) and varying angles of the core-fiber. If all filaments do fully align, the maximum load becomes larger than in the initial case (no alignment at all) by the factor 1/(cos 45°) = 1.41.

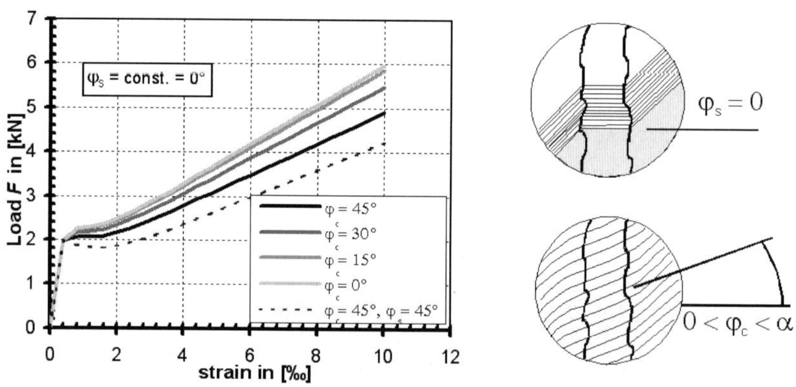

Figure 7. Influence of alignment of the fiber with the load direction

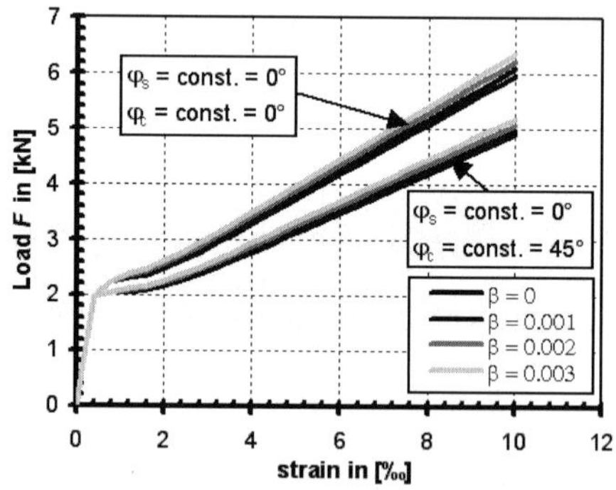

Figure 8. Influence of parameter β

4.2 Influence Of Enhanced Bond Due To Lateral Pressure

Figure 8 shows that, as expected, the parameter β according to Eq. (3) has a stiffening effect on the crack bridge. Its influence grows, if the lateral pressure due to alignment of the fiber increases as the comparison of the cases φ_C=0° and φ_C=45° reveals.

4.3 Influence of damage function

In Figure 9 the effects of three linear damage laws for the sleeve-fiber on the load bearing behavior are compared. Contrary to the other effects, the damage leads to a softening of the crack bridge and is the main influence on the tensile strength.

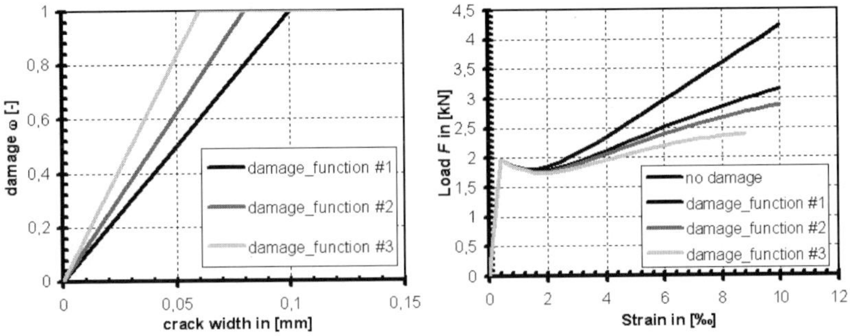

Figure 9. Influence of linear damage of the sleeve-fiber

Acknowledgement

The German Research Foundation (DFG) is gratefully acknowledged for the financial support of this project within the scope of the collaborative research center SFB 532 "Textile Reinforced Concrete - Foundation of a New Technology".

5. REFERENCES

1. P. Bartos, Bond in Glass Reinforced Cements, *Bond in Concrete*, Elsevier Applied Science, pp. 60-72, (London, 1982).
2. M. Mashima, D.J. Hannant and J.G. Keer, Tensile Properties of Polypropylene Reinforced Cement with Different Fiber Orientations, *ACI Materials Journal* **87**, pp. 172-178 (1990).
3. B.A. Proctor, A Review of the Theory of GRC, *Cement & Concrete Composites* **12**, pp. 53-61 (1990).
4. J. Hegger and S. Voss, Tragverhalten von Textilbeton unter zweiachsialer Beanspruchung, Proceedings of the 2nd Colloquium on Textile Reinforced Structures (CTRS2), pp. 313-324 (Dresden, Germany, 2003).
5. M. Molter, Zum Tragverhalten von textilbewehrtem Beton, PhD-thesis, Institute of Structural Concrete, RWTH Aachen University, Germany (2005).
6. F. Jesse, Tragverhalten von unidirektionalen und textilen Bewehrungen aus Multifilamentgarnen in einer zementgebundenen Matrix, PhD-thesis, TU Dresden, Germany (2004).
7. S. Ohno and D.J. Hannant, Modelling the Stress-Strain Response of Continuous Fibre Reinforced Cement Composites, *ACI Materials Journal* **91**, pp. 306-312 (1994).
8. U. Häußler-Combe and F. Jesse, Rechnerische Untersuchungen zum einaxialen Tragverhalten von Textilbeton, *Bauingenieur* **3**, pp. 131-141 (2005).
9. J. Hegger, N. Will, O. Bruckermann and S. Voss, Load-Bearing Behaviour and Simulation of Textile Reinforced Concrete, *Materials and Structures*, (accepted for publication 2005).

10. J. Hegger, O. Bruckermann and S. Voss, AR-Glass and Carbon Fibers in Textile Reinforced Concrete – Simulation and Design, Special Publication on thin fiber reinforced concretes, ACI Spring Convention New York, April 2005, to be published.
11. J. Hegger and S. Voss, Textile Reinforced Concrete - Bearing Behavior, Design, Applications, 3rd Int. Conference CCC2005, Lyon, France, pp. 1139 – 1146 (2005).

TRC-SPECIMENS MODELED AS A CHAIN OF CRACKS BRIDGED BY BUNDLES

Study of impact of local scatter on global tensile strength

R. Chudoba, M. Vorechovsky, J. Jerabek and M. Konrad

Chair of Structural Statics and Dynamics, Aachen University of Technology, Mies-van-der-Rohe-Str. 1, 52056 Aachen, Germany

Abstract: The present paper shows the correspondence between the short-range size effect occurring in crack bridge and the long-range size effect observed on a textile reinforced concrete (TRC) tensile specimens. In the analysis of the effects we exploit the fact that the specimen acts as a chain of crack bridges in its failure state. The cracks are bridged by bundles consisting of several thousands of filaments that are embedded in a cementitious matrix. We include the effect of scatter of filament properties over the bundle cross section in a crack bridge and study its influence on the ultimate failure of the reinforced specimen.

Key words: fiber bundle model, statistical size effect, chain of bundles

1. INTRODUCTION

The crucial aspect in determining the ultimate load carrying capacity of textile reinforced concrete (TRC) specimens is a correct description of the hot-spots of strain and damage, e.g. of the crack bridges. In the final stages of the tensile loading with a finished crack pattern the specimen may be viewed as a chain of crack bridges with its strength governed by the "weakest-link" concept. The experimental results on tensile TRC specimens show a significant loss of bundle efficiency that are usually ascribed to damage or low penetration of the bundle by the matrix combined with insufficient anchorage in the boundary layers (Hegger et al., 2005). This paper contributes to the discussions about the reasons for the strength reduction by studying the weakest link effect in a chain of crack bridges with scatter of strength.

The chained crack bridges are realized by multi-filament bundles exhibiting a considerable scatter of strength due to a highly heterogeneous nature of the material structure in the yarn, in the bond layer and in the cementitious matrix. As documented in Chudoba et al. (2005) on experimental and numerical studies, these sources of random-

M.S. Konsta-Gdoutos, (ed.), Measuring, Monitoring and Modeling Concrete Properties, 777–783.

ness can lead to a substantial reduction of the bundle strength especially for extremely short nominal lengths occurring in a case of a crack bridge. Therefore, the evaluation of the impact of varying material properties of bridges in a serial ordering on the global strength is inevitable.

In order to demonstrate the correspondence between the statistics of the global response and the local scatter in the crack bridge we first review the weakest link concept in Sec. 2. The simple modeling framework allows us to study the change of the ultimate strength with an increasing number of cracks N for different levels of scatter of local filament properties in the crack bridges. In Sec. 3 we use an analytical strain based fiber bundle model to evaluate the probability density function of a single crack bridge strength. Another example in this section uses a more sophisticated model including the debonding effect. Both examples document the general applicability of the procedure for evaluating the chain effect in a TRC loaded in tension.

2. GENERAL DETERMINATION OF TRC STRENGTH STATISTICS

As documented in Fig. 1 the tensile specimen exhibits very fine crack pattern. Obviously, the ultimate failure is governed by the weakest-link statistics. Therefore, the survival probability of the chain with N cracks may be obtained as a product of survival probabilities of the individual cracks:

$$1 - P_{f,N} = \left(1 - P_{f,1}\right)^{N} \tag{1}$$

where $P_{f,1} \equiv F_1(\sigma)$ represents the failure probability of a single crack bridge (cumulative strength distribution). The load level for a given number of crack bridges N and probability of failure $P_{f,N}$ can be computed with the inverse cumulative strength distribution function of a crack bridge:

$$\sigma = F_1^{-1}\left(1 - \sqrt[N]{1 - P_{f,N}}\right) \tag{2}$$

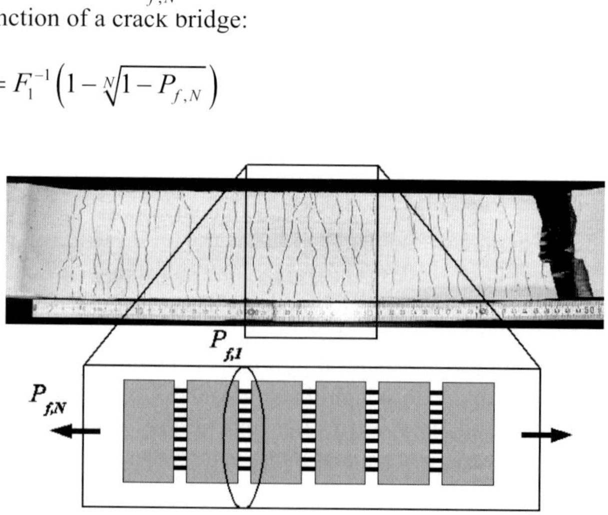

Figure 1. Crack pattern of a failed tensile specimen reductions by studying the weakest link effect in a chain of crack bridges with scatter of strength

In case of normally distributed strength of a crack bridge (short filament bundle) the load level of a chain of bridges with failure probability $P_{f,N}$ reads simply:

$$\sigma = \Phi^{-1}\left(1 - \sqrt[N]{1 - P_{f,N}}\right) \tag{3}$$

This formula provides a general procedure for estimating the strength statistics of TRC with N crack bridges, each with the failure probability distribution $P_{f,1} \equiv F_1(\sigma)$. This distribution is obtained using the statistical strain-based bundle model (Phoenix and Taylor 1973) as follows

1. Given a single-filament response function (constitutive law) $q(e;\boldsymbol{\theta})$ as a function of strain e and a vector of random (or deterministic) quantities $\boldsymbol{\theta}$ with their corresponding distribution functions $G_{\theta,i}(\theta_i)$ compute the mean response of a filament within the bundle (normalized bundle force) as a k-fold integral over k-number of nondeterministic variables of the model $q(e;\boldsymbol{\theta})$:

$$\mu_{\boldsymbol{\theta}}(e;\boldsymbol{\theta}) = \int_{\boldsymbol{\theta}} q(e;\boldsymbol{\theta})\,\mathrm{d}G_{\boldsymbol{\theta}}(\boldsymbol{\theta}) \tag{4}$$

2. Find the local maximum of the mean response (bundle strain $e = e^*$ at which the maximum force is attained). This can be done either by seeking the stationary point of $\mu_{\boldsymbol{0}}(e;\boldsymbol{\theta})$ in case of analytical expression, see examples in Chudoba et al. (2005) and Voechovský & Chudoba (2005) or numerically by seeking the peak force value.

3. Evaluate the mean bundle response function at bundle strain $e = e^*$ to get the mean bundle strength: $\mu_\theta^* = \mu_\theta(e^*)$ and compute the bundle strength variance as:

$$\Gamma_\theta^* = \Gamma_\theta(e^*, e^*) = \int_{\boldsymbol{\theta}} \left[q(e^*;\boldsymbol{\theta}) - \mu_{\boldsymbol{\theta}}^*\right]^2 \mathrm{d}G_{\boldsymbol{\theta}}(\boldsymbol{\theta}) \tag{5}$$

4. Estimate the whole cumulative density function $P_{f,1}$. In most cases it suffices to consider Gaussian distribution for the "middle" part (core) (0.1 to 0.9 percentiles) so that the probability of failure at a given load level reads:

$$P_{f,1}(\sigma) \equiv F_1(\sigma) = \Phi\left(\frac{\sigma - \mu_\theta^*}{\sqrt{\Gamma_\theta^*}}\right) \tag{6}$$

where $\Phi(\bullet)$ = standard Gaussian cumulative distribution function.

Note that the distribution (of bundle strength $P_{f,1}$) can be estimated numerically by means of Monte Carlo simulation. One can evaluate the bundle response N_{sim} times and save the peak forces (of all simulations). Then the CDF of bundle strength can be estimated by an empirical cumulative histogram of the peak sample.

3. EXAMPLES FOR SELECTED CRACK BRIDGE MODELS IN A CHAIN

In order to demonstrate the procedure and the significance of the size effect in a TRC specimen we now provide two examples of a crack bridge model with the response function represented both analytically and numerically.

Example 1. Bundle model with perfect clamping: The scatter of filament stiffness parameters leads to a reduced strength of a bundle with a very short length. In a crack bridge the strength reduction is in particular caused by the scatter of filament lengths and their delayed activation (slack). Besides that, the scatter of filament strength across the bundle leads to a further reduction of the bundle strength (Smith and Phoenix 1981). In order to demonstrate the effect on an example we consider a crack bridge with a scatter of filament lengths. In particular, we introduce the relative difference of the filament length with respect to the nominal length as $\lambda = (l - l_i)/l$ with a uniform distribution, i.e. $G_\lambda(\lambda) = \lambda/\lambda_{\max}$ where $0 \le \lambda \le \lambda_{\max}$. The other parameters of the filament, i.e. Young's modulus E, area A and breaking strain ξ are considered constant. For the chosen distribution, it is possible to derive analytical formulas for the bundle mean strength and its variance (see Chudoba et al. 2005) at a given control strain e as

$$\begin{aligned} \mu_\lambda(e;\lambda) &= \int_\lambda q(e;\lambda)\,\mathrm{d}G_\lambda(\lambda) \\ &= \begin{cases} EAe\ln(1+\lambda_{\max})/\lambda_{\max} & 0 \le e \le \xi \text{ (linear)} \\ EAe\dfrac{\ln(1+\lambda_{\max}) - \ln(e/\xi)}{\lambda_{\max}} & e > \xi \text{ (nonlinear)} \end{cases} \end{aligned} \tag{7}$$

The calculated normalized mean load strain diagrams are exemplified in Fig. 2 for several levels of scatter represented by $\lambda_{\max}$. Obviously, the crack bridge strength/efficiency rapidly decreases with an increasing $\lambda_{\max}$.

In the derivation of the crack bridge strength distribution we shall exploit the fact that the maximum mean bundle strength is attained at the global strain $e^* = \xi$ for the uniform distributions with $\lambda_{\max} \le 1.71$ and has the simple form:

$$\begin{aligned} \mu_\lambda(e^*) &= \int_\lambda q(e^*;\lambda)\,\mathrm{d}G_\lambda(\lambda) \\ &= EA\xi\ln(1+\lambda_{\max})/\lambda_{\max} \end{aligned} \tag{8}$$

The corresponding variance (see Phoenix and Taylor 1973) is obtained as

$$\begin{aligned}\Gamma_{\lambda}^{*} &= \int_{\lambda}\left[q\left(e^{*};\lambda\right)-\mu_{\lambda}\left(e^{*}\right)\right]^{2}\,\mathrm{d}G_{\lambda}\left(\lambda\right)\\ &=\left(EA\xi\right)^{2}\left[\frac{1}{1+\lambda_{\max}}-\frac{\ln^{2}\left(1+\lambda_{\max}\right)}{\lambda_{\max}^{2}}\right]\end{aligned} \tag{9}$$

With reference to the central limit theorem we can expect the convergence of the crack bridge strength distribution to the normal distribution, see Eq. 6. The verification of the convergence to the Gaussian distribution has been done by Monte Carlo simulation in connection with the deterministic bundle model described in Chudoba et al. (2005). With the crack strength distribution at hand we can approach to the quantification of the chain statistics. Using Eq. (3) we quantify the interaction of the short-range size effect due to a local scatter $(\lambda_{\max})$ with the chaining of crack bridges in a tensile specimen. In Fig. 4 we plot the crack bridge efficiency (reduction of strength with respect to a perfect crack bridge) for the failure probability $P_{f,N}$ = 0.5 (median) and the levels of the scatter parameter $\lambda_{\max}$ studied previously in Fig. 2.

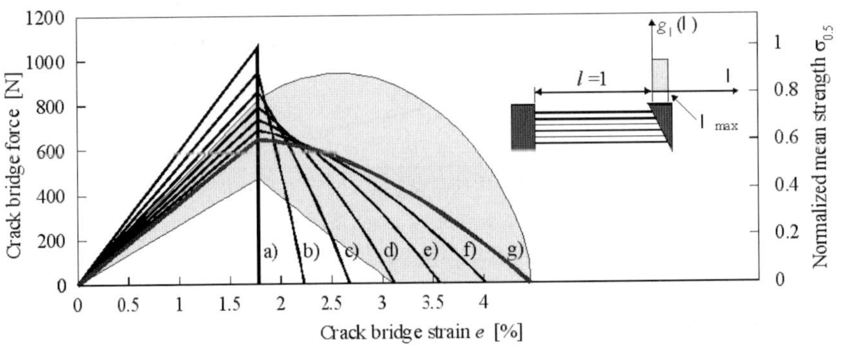

Figure 2. Mean force-strain diagrams of one crack bridge with uniform distribution of additional fiber length λ ∈ (0, $\lambda_{\max}$). $\lambda_{\max}$= a) 0.0, b) 0.25, c) 0.5, d) 0.75, e) 1.0, f) 1.25 and g) 1.5 plotted with a scatterband (mean ± standard deviation)

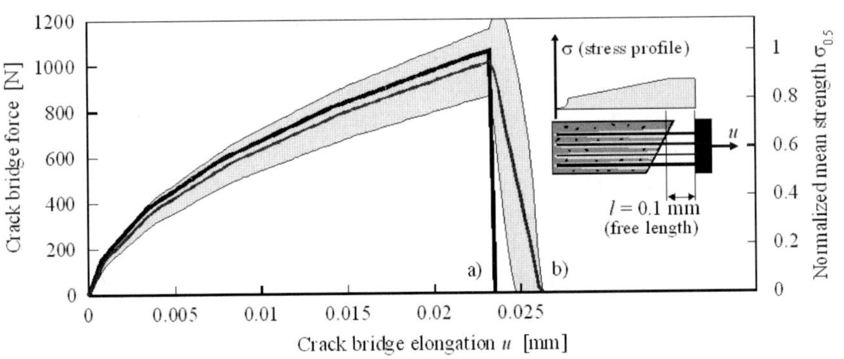

Figure 3. Mean force-strain diagrams ± one standard deviation of one crack bridge with debonding model and with uniform distribution of additional fiber length λ∈(0, $\lambda_{\max}$).$\lambda_{\max}$ = a) 0.0, b) 1.5

Example 2. Bundle model with debonging: The procedure for evaluating the total strength described in Sec. 2 can be used with more complex idealizations of a crack bridge taking into account further failure and damage mechanisms. In addition to filament rupture considered in the previous model we now include the influence of debonding between filament and matrix. The response function $q(e;\boldsymbol{\theta})$ is represented by a finite element idealization of the shear lag with a cohesive interface between the filament and matrix. In order to evaluate the integrals (Eqs. 7 and 8) a general numerical integration tool has been implemented to obtain the statistical characteristics of a crack bridge strength (Konrad et al., 2006). We remark that the crack bridge model provides the possibility to study the impact of the variability in any of the model parameter(s) on the statistics of the overall bundle response (load displacement diagram), not only on the length that is used in this paper.

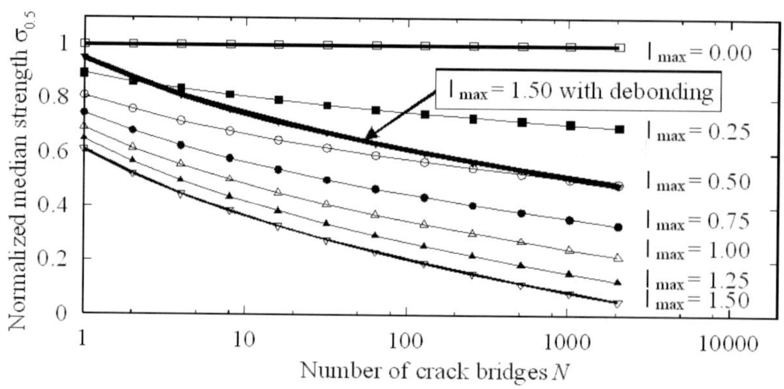

Figure 4. Median chain strength for varying number of cracks and scatter $\lambda_{\max}$ ($P_{f,N}$ = 0.5)

With $G_{\lambda}(\lambda)$ defined as in the example 1, we now include the debonding of a filament from the matrix as an additional effect. Again we keep all other parameters including the interface characteristics, constant. The resulting load-displacement diagram displayed in Fig. 3b for $\lambda_{\max} = 1.5$ shows a significantly higher mean crack bridge strength (1008 N, see Fig. 3b) than in the case of a perfect bond (648 N, see Fig. 2g). The reason for such an increased strength is the homogenizing effect of the debonding causing a more uniform stress distribution across the bundle (more filaments can act simultaneously before they break). We also note that there is no significant reduction in the scatter of strength: 173.99 N for perfect bond and 145.08 N with included debonding.

The performance of a chain of crack bridges with and without debonding is compared for $\lambda_{\max} = 1.5$ in the semi-logarithmic plot in Fig. 4. Due to a similar amount of scatter, the slope of the two size-effect curves is almost the same. In other words, while the local debonding improves the mean strength by introducing stress redistribution during the failure process, the decay of strength with the increasing number of cracks remains almost the same. The size effect curve is simply shifted upwards.

4. CONCLUSIONS

The paper shows the correspondence between local scatter in a crack bridge of a textile-reinforced tensile specimen and the resulting reduction of the specimen tensile strength. The presented approach provides a rough estimation and explanation of the strength reduction of tensile specimens with dry yarn reinforcement with a high amount of imperfections in the bundle structure and in its bond to the cementitious matrix. If we consider a usual range of lengths of tensile structural elements of order of magnitude 1-6 m and average crack distance of 0.01-0.02 m, the realistic range of N is 50-600. As demonstrated by the two examples, the local scatter in a crack bridge significantly affects the load bearing capacity of textile reinforced specimens and, thus, the statistical size-effect resulting from the weakest link failure must be an inherent part of dimensioning rules for the discussed type of composite. While the size effect was demonstrated with a single source of randomness, in reality the crack bridge exhibits several mutually interacting sources of randomness that have been deliberately disregarded here.

Acknowledgments

The present work has been carried out in the framework of the project Simulation of Bond and Crack Behavior of TRC at the Meso Level included in the SFB 532: Textile reinforced concrete: foundation of a new technology sponsored by the German research foundation.

5. REFERENCES

1. Chudoba, R., Voechovský, M. and Konrad, M., 2005, Stochastic modeling of multi-filament yarns I: Random properties within the cross section and size effect. *International Journal of Solids and Structures*, in print.
2. Hegger, J., Voss, S., Bruckermann, O., 2005, Load-bearing behaviour and simulation of textile reinforced concrete. *Materials and Structures*, in print.
3. Konrad, M., Jeábek, J., Voechovský, M. and Chudoba, R., March 2006, Evaluation of mean performance of cracks bridged by multi-filament yarns. In: *EURO-C, Computational Modeling of Concrete Structures*. Mayrhofen, Austria, under preparation.
4. Phoenix, S. L., Taylor, H. M., 1973, The asymptotic strength distribution of a general fiber bundle. *Advances in Applied Probability* **5**: 200–216.
5. Smith, R. L., Phoenix, S. L., 3 1981, Asymptotic distributions for the failure of fibrous materials under series-parallel structure and equal load-sharing. *J. Appl. Mech.* **48**: 75–82.
6. Voechovský, M., Chudoba, R., 2005, Stochastic modeling of multi-filament yarns II: Random properties over the length and size effect. *Int. J. Solids and Structures*, in print.

Author Index